Biometeorology

Volume 6

This Series is now being indexed by SCOPUS

Biometeorology has its linguistic origins in Greek; bio refers to life while meteoros makes reference to the study of phenomena near and above the earth's surface. These are the essential elements of biometeorology, such that it is concerned with the interaction between living organisms and variations in atmospheric processes at a range of time and space scales.

Since its emergence as a discipline in the mid 1950s, Biometeorology has grown to have great societal relevance. Much of its subject matter focuses on how biophysical, human, animal and plant systems are influenced by change and variability in the atmospheric environment. Understanding such relationships provides insights into how the provisioning, supporting and regulating services provided by the biosphere may be affected by atmospheric variations and ultimately the societal consequences of atmosphere related changes in biophysical and ecosystem processes.

The purpose of this series in biometeorology is to:

- complement the material published in the International Society of Biometeorology's International Journal of Biometeorology,
- to communicate the interdisciplinary philosophy and science of biometeorology to as wide an audience as possible,
- introduce scientists and policy makers to the societal relevance of and recent developments in its subfields.
- demonstrate how a biometeorological approach can provide entry points to the understanding and possible solution of cross-cutting environmental issues such as climate change, desertification, deforestation, biodiversity loss, land degradation, emerging diseases and the availability of clean water and air.

The series in biometeorology comprises theme-based volumes organised around subject matter within the subfields of biometeorology, which broadly include human, animal, plant, ecosystems, tourism, artificial or indoor climates, air quality and urban. In addition to volumes that synthesise the state of knowledge in the subfields, a primary focus will be on important, new and emerging topics in biometeorology that have a bearing on the health of the biosphere and thus human society. An editorial board, the broad role of which will be to elicit and review suggestions for volume themes, will oversee the series.

Glenn McGregor

Heatwaves

Causes, Consequences and Responses

Springer

Glenn McGregor
Department of Geography
Durham University
Durham, UK

ISSN 1877-5284 ISSN 2452-1558 (electronic)
Biometeorology
ISBN 978-3-031-69905-4 ISBN 978-3-031-69906-1 (eBook)
https://doi.org/10.1007/978-3-031-69906-1

Cover illustration: Orange Futures. Artist: Stephen Irving, Eaglescliffe, County Durham, England, 2024

This Springer imprint is published by the registered company Springer Nature Switzerland AG
The registered company address is: Gewerbestrasse 11, 6330 Cham, Switzerland

Foreword by Gerald Mills

As I write this foreword in the summer of 2024, there is ample evidence of heatwaves and their impacts around the world. During the Hajj pilgrimage in mid-June, over 1000 people died from heat stroke as air temperature exceeded 50°C, while in early July a historic heatwave affected 28% of the US population (over 90 million). This event took place as a hurricane made landfall on the Gulf Coast; Beryl is the earliest recorded category 5 hurricane (windspeeds of over 250 km/hr) in 100 years, due in part to unseasonably warm Atlantic Ocean temperatures. These events are a harbinger of global climate change, which projects global warming and changes in the frequency, intensity, and duration of heatwaves; in other words, not only will the average global temperature increase, but so too will the hot extremes. A science of heatwaves is needed to understand these processes and to cope with the impacts. *Heatwaves: Causes, Consequences and Responses* is the first comprehensive examination of the state of this science and how it may be improved.

Land, water and air temperatures are outcomes of complex processes associated with the absorption and transfer of energy; shifts and periodic changes in temperature fields therefore represent fundamental alterations to energy exchanges within and among components of the Earth-Atmosphere System and must be understood within this context. However, unlike other 'natural' hazards, such as landslides, storms and sea-level rise that have defined spatial expressions, the causes and impacts of heatwaves occur across all scales from global to local. Moreover, it is impossible to confine the discussion of the heatwave hazard to just its expression (heat), as it is often correlated with a range of other hazards, such as droughts, forest fires, and air and water quality. In fact, heatwaves can initiate a chain of disruptions to complex systems such as energy generation and transmission and transport networks, which compound the impacts. Adapting to heatwaves is also a challenge given that the associated heat stress represents the ambient environment for all lifeforms. For humans especially, the urban environments we have enhance heat stress (the urban heat island effect) during larger-scale heatwave events; the public health impact within cities varies considerably based on the exposure and vulnerability of the population, which is strongly linked to the condition of housing and urban greening. To cope with these events, various warning systems have been developed that use meteorological models to inform the population about the imminent threat and appropriate responses. Ironically, a conventional way of responding in wealthy economies is simply to refrigerate indoor space and create a climate 'refuge'. However, this technology is not a panacea as it contributes to global warming and is simply unaffordable for the great majority of the planet's population who live in hot climates where heat exposure is greatest. Managing heatwaves meets the definition of a 'wicked problem' for policymakers owing to the difficulty in disentangling causal relationships and the challenge of developing appropriate responses.

Heatwaves: Causes, Consequences and Responses covers the state of knowledge on the physical processes, the methods of analysis and the impacts for natural and human ecosystems; critically, it includes heatwaves on land, in water and in the atmosphere, which are coupled at different space and time scales. It examines the responses to heatwave events, which not surprisingly are human-centric and have a strong emphasis on adaptation and mitigation in urban landscapes, where the risk to health and infrastructure is greatest. The material in this book is of value to a range of physical and social science disciplines with interests in climate and weather, public health, natural ecosystems and risk management (Table 1.3 is a comprehensive list of these disciplines). Its 15 chapters set out the current science including our understanding of the causes of heatwaves, our knowledge of the consequences and the management systems to cope.

Importantly, the scope of the book permits an assessment of the weaknesses in the current multi-disciplinary approach to heatwave study. While each field of study has acquired insights into specific aspects of heatwaves, other productive lines of inquiry have been neglected. Moreover, as each discipline has developed its own lexicon and methods, the transfer of knowledge has been impeded—for example, the literature uses various definitions of what constitutes a 'heatwave'. More broadly our science has sparse information on historic heatwaves and their impacts, so that establishing climate norms against which to examine change is difficult; similarly, we have a limited understanding of the economic costs associated with heatwave events and little information on the benefits of early warning systems and adaptation and mitigation measures. This book goes well beyond a typical subject text by making the case for a new interdisciplinary science that can set an integrated research agenda, foster communication (among the sciences and with the public) and develop evidence-based responses. Given the immense challenges that global climate change presents, which cannot be addressed through narrow fields of study, this book provides a map for a new type of discipline to address one of the major environmental risks that the planet faces.

Associate Professor, School of Geography
University College Dublin
Dublin, Ireland

Gerald Mills

Foreword by Kristie L. Ebi

Heatwaves: Causes, Consequences and Responses is a tour-de-force that effectively integrates multidisciplinary knowledge to understand and manage the population health and health system risks posed by heatwaves in a changing climate, or the heatwave imperative. Tens of thousands of people are estimated to succumb to heat annually; the actual number is likely closer to 100,000 or more. Nearly all those deaths are preventable. How high ambient temperatures affect us physiologically is understood, as are exacerbating factors such as age, level of fitness, presence of chronic diseases, pregnancy, use of certain prescription drugs, urban heat islands, living in certain regions or in marginalized communities, and employment outdoors. This understanding is being incorporated into heatwave early warning and response systems, although there are questions about the extent to which current systems explicitly consider the consequences of climate change. The planet has entered an era where virtually all heatwaves are increasing in frequency, intensity, and length because of climate change. *Heatwaves: Causes, Consequences and Responses* arrives at a critical moment with communities and researchers seeking answers for how to effectively prepare for and respond to the challenges of the heat imperative in the context of a changing climate. Prof. McGregor covers the breadth and depth of topics that communities and researchers need to understand, in text that is engaging, informative, and widely accessible. He also provides an impressive list of references for anyone curious about more details.

Prof. McGregor makes a compelling case for focusing primarily on population health and well-being, while acknowledging the importance of heatwaves for nature, infrastructure, and livelihoods. The 15 chapters cover the weather and climate phenomena that lead to heatwaves; the causes and consequences of heatwaves; and short- and longer-term approaches to effectively manage the risks from a whole of society perspective. Prof. McGregor even wades into the debate of whether the term heatwaves or heat waves more effectively communicates the potential challenges, advocating effectively for the former.

Prof. McGregor begins with an in-depth discussion of the physical properties of heatwaves and their atmospheric and environmental drivers to provide a basis for understanding the consequences for human health and for formulating effective responses to managing the risks. He explains to non-meteorologists the different processes over multiple temporal and spatial scales that can result in a heatwave, with the special challenges faced in urban areas highlighted because of population density and because urban infrastructure interacts with heatwaves to increase local temperatures.

While heatwaves are frequently defined operationally by ambient temperature, humidity and other environmental variables affect an individual's experience of the weather conditions, such as whether they are in the sun or shade, or windy or calm conditions, as do the activities individuals undertake and the clothing they wear. The differences between heat strain and heat stress are clearly described. Prof. McGregor discusses the differences among the large numbers of thermal comfort indices, highlighting those with most relevance for addressing the heat imperative. This is most helpful for the research community and to those with responsibility for designing and implementing heatwave early warning and response systems. Increases in the geographic extent of heatwaves mean that coordination across communities is becoming increasingly important so that, for example, a heatwave warning in one community is comparable to a warning in an adjacent community. In some larger countries, there is national coordination to facilitate increasing awareness of the health risks of heat and uptake of warning when they are issued. International coordination is beneficial across borders.

Effectively managing the risks of heatwaves requires more than just understanding the physical processes. Prof. McGregor explains the importance of understanding the consequences for individuals and communities. That understanding arises from conducting a heatwave impact assessment that includes stakeholders from the start. By their nature, these impact assessments are multidisciplinary undertakings, requiring input from physical and social scientists to provide useful and useable insights into the nature of the risks and effective options for protecting health and well-being. Conducting these assessments is made more challenging by the fact that the health and well-being impacts of heatwaves can be direct or indirect and can propagate over space and time. Directly, exposure to high ambient temperatures can stress the heart, lungs, kidneys, and other organ systems, resulting in increased hospitalizations and deaths. About half of all deaths from a heatwave are due to cardiovascular causes because of the high prevalence of heart disease and the sensitivity of the heart to internal heating. Indirectly, heatwaves can reduce crop yields that result in a higher prevalence of food insecurity. Because undernourished children are at higher risk of a range of infectious diseases, there can be consequences across a wider region than the heatwave occurred. Prof. McGregor provides practical advice on how to achieve each step of the heatwave impact assessment.

Prof. McGregor illustrates how high ambient temperatures are an all-of-society issue, affecting human and animal health, worker productivity, infrastructure (e.g., buildings, electricity generation, road and rail transport, etc.), economic activities, and ecosystems and ecosystem services.

Despite the understanding of widespread impacts, a challenge when conducting heatwave impact assessments is the significant underreporting of heatwaves in many vulnerable regions, particularly Africa. If heatwaves are reported, then access to health data may be limited; in many low- and middle-income countries, health data are reported monthly, which is too course a scale to accurately quantify the human toll. A further challenge is that vulnerability and the capacity to manage heatwaves varies within and across communities, requiring local knowledge when quantifying impacts and designing measures to protect those who are more vulnerable.

Prof. McGregor provides a critical understanding of the breadth and depth of the evidence base for interventions to protect human health and well-being, facilitating transmission of lessons learned and best practices so that communities can rapidly assess the modifications needed for interventions implemented elsewhere. Effective interventions are produced in collaboration with vulnerable populations, ensuring their perspectives are incorporated from the beginning. This means that the general shape of an intervention will be similar from location to location, although the specifics will depend on the context. Governance is an important determinant of effective interventions.

Proactive interventions are required for all communities to prepare for the heatwaves of the future, within the context of multi-hazard planning. Practical advice is provided for the challenge of identifying metrics for evaluation, particularly in the context of a changing climate. Determining whether interventions increase resilience to climate change will take years, yet evidence is needed today. Investments in heatwave early warning and response systems are urgently needed to address the increasing challenges of extreme heat. Prof. McGregor also discusses the imperative of reducing greenhouse gas emissions to decrease the challenges with the heat imperative later this century.

Prof. McGregor provides timely and valuable insights for practitioners to researchers on the chain of understanding from the physical phenomena of heatwaves through to the actions needed to protect and promote human health and well-being, highlighting key data and knowledge gaps. *Heatwaves: Causes, Consequences and Responses* provides the information needed to support increasing resilience to the health risks of heatwaves in a changing climate.

Professor, Center for Health and the Global Environment (CHanGE)
University of Washington
Seattle, WA, USA
krisebi@uw.edu

Kristie L. Ebi

Preface

The idea for this book entitled *Heatwaves: Causes, Consequences and Responses* arose from a lived experience, that being the monumental August 2003 European heatwave event. At the time, I was in the south of England. While enduring the ferocity of the record-breaking heat, it occurred to me that this was exceptional not only in climatological terms but also in a societal sense as it became obvious via various news outlets of the consequences of the extreme heat across the UK and Europe. Alarmingly, as I watched the news, the human toll began to reveal itself in stark numbers. Hospitals across Europe became overwhelmed with cases of heat-related illnesses. Tragically, many lives were lost, particularly among the elderly and vulnerable populations. The images of heatwave fuelled wildfires were equally disturbing. Notwithstanding the importance of heatwave events that preceded it, there is no doubt the 2003 European heatwave was an unambiguous wake-up call to the profound impacts of heatwaves on people's daily lives, society, the economy, and the environment. The salience of heatwaves as a serious threat continues to this day as evidenced by reports of the impacts of severe heatwaves across India, the United States and Saudi Arabia in 2024. Despite their growing threat, heatwaves remain an underappreciated and poorly understood phenomenon. That heatwaves continue to be largely overlooked in comparison to their more spectacular counterparts such as tropical cyclones and floods serves as the motivation for this book which aims to provide an overview of our current scientific understanding of heatwaves, in particular their physical causes, societal and environmental consequences, and our responses to them.

Although this book is manly about atmospheric heatwaves, marine heatwaves are touched upon throughout with an occasional excursion into the topic of heatwaves in terrestrial water bodies. The material presented is a mixture of an assessment of the literature on heatwaves as of early 2024 and original research. With regard to the former, by necessity the amount of literature cited is large and the reference lists are long but unlikely exhaustive. The book begins by examining the nature of heatwaves as meteorological and climatological phenomena followed by material on the physical processes that give rise to heatwaves, with modes of climate variability, synoptic scale atmospheric processes and land-atmosphere interactions covered. Attention is then turned to the consequences of heatwaves and specifically how a heatwave impact assessment might be conducted along with material on the way exposure and vulnerability may moderate the impacts of heatwaves as well as highlighting the myriad ways that heatwaves impact human systems and the environment. Under the chapeau of heatwave responses, a case is made for a clearer distinction between heatwave adaptation and mitigation options along with the presentation of some ideas about what might precipitate heatwave-related actions at the individual to national government level. Given the view that early warning systems may play a critical role in reducing heatwave-related risk, a dedicated chapter is given over to this topic. The book concludes with a synthesis of the foregoing material and suggests some directions for future enquiry.

My hope in writing this book, which has had a much longer gestation period than I would have preferred, is it will elevate the urgency of the heatwave imperative and serve as a worthwhile resource for researchers, practitioners, policymakers and students working to understand and tackle the complex challenges associated with heatwaves. Additionally, I hope *Heatwaves: Causes, Consequences and Responses* raises interest amongst those curious about heatwaves as meteorological and climatological phenomena and weather and climate extremes in general.

Durham, UK Glenn McGregor
June 30, 2014

Acknowledgements

I would like to express my thanks to the Japan Society for the Promotion of Science (JSPS) and the United Kingdom Daiwa Foundation for separately funding periods of book-related research and writing in Japan. Their commitment to advancing knowledge and promoting UK-Japan relations and fostering academic and cultural exchange has not only significantly enriched my work, but has been instrumental in writing this book. Related to this, I am deeply grateful to Associate Professor Asuka Suzuki-Parker for hosting me while working in Japan at Rissho University and for processing the data on which some of the material on Japanese heatwaves presented in this book is based.

I would also like to extend my sincere gratitude to Durham University for granting me research leave, which was unquestionably crucial for the completion of this book. The time provided during this period was invaluable, allowing me to focus on my research and writing without interruption. The supreme efforts of Mr Chris Orton of the Durham University's Department of Geography's Cartographic Unit in drafting many of the figures for this book are also acknowledged.

Importantly, I wish to acknowledge all the 'heatwave conversations' I have had with colleagues from across the globe over the years. Such conversations have provided invaluable insights into the 'heatwave problem' and have been vital to the development of the ideas presented in this book.

Lastly a huge thanks to my wife Miyoko for her unwavering care and understanding during the time it has taken to complete this book. Your encouragement over the years and support of 'all things climatological' that I take on has been a constant source of motivation.

Contents

Part I Heatwaves as Weather and Climate Phenomenon

List of Figures

List of Tables

Part I
Heatwaves as Weather and Climate Phenomenon

Chapter 1
Heatwaves

Abstract Emphasising the often-overlooked deadliness of heatwaves compared to more dramatic extreme events like cyclones and floods, this chapter introduces the concept of the 'heatwave imperative' and sets out the reasons why there is an urgent need to address the research and management challenges posed by heatwaves. It also posits that a new type of science, 'heatwave science' is required to understand the complex interactions between climatic processes, societal dynamics, and natural systems. The growing trend and impact of heatwave events is highlighted and set in the context of the shifting statistical distribution of global air temperature anomalies, with the increase in extreme heatwave events largely attributed to human induced climate change. The chapter also presents the book's conceptual framework and describes why the focus is on the here and now of heatwaves as opposed to heatwave futures. The penultimate section of this Chapter outlines the book's structure which is organised into four parts covering heatwaves as weather and climate phenomenon, the physical causes of heatwaves, the consequences of heatwaves in the form of impacts on human and natural systems and the responses to heatwaves via the development and implementation of adaptation and mitigation options. The Chapter closes by providing reasons why the compound noun 'heatwave' is preferred over 'heat wave' and used throughout this book.

Keywords Heatwaves · The heatwave imperative · Heatwave science · Heatwave attribution and climate change · Climate system processes

1.1 Introduction

Heatwaves are a natural hazard, natural because they originate from natural processes that occur in the Earth's climate system manifesting broadly as atmospheric or marine heatwaves. From a human perspective, heatwaves are considered a hazard because if they interact with human related activity systems, they may engender a range of impacts related to health, loss of livelihood, social and economic disruption including

G. McGregor, *Heatwaves*, Biometeorology 6,
https://doi.org/10.1007/978-3-031-69906-1_1

critical infrastructure, or environmental damage. Heatwaves are also extreme events because the values attained by the physical variables that describe them are unusually high, and in statistical terms, are associated with a low probability of occurrence. In this sense, atmospheric heatwaves are generally characterised as several days of abnormally high air temperatures and radiative heat load often accompanied by low or elevated levels of humidity (dry and wet/humid heatwaves respectively) and little ventilation. In the case of marine heatwaves, the extremes relate to sea surface temperatures. Heatwaves can have profound impacts on human and natural systems. Yet in the context of more spectacular, violent, and visibly impactful extreme meteorological events such as tropical cyclones, tornadoes, flash floods or storm surges, heatwaves are often overlooked as mere infrequent occurrences of unusual summer heat with minimal significance deserving scant attention. This existential dilemma may no longer be warranted as it becomes increasingly obvious that heatwaves are one of the most 'silent' and lethal of all natural hazards rendering harm to both animate and inanimate objects, especially those that are exposed, vulnerable and unable to respond. What's more, heatwaves have become more frequent over recent decades and are likely to continue to do so due to anthropogenic climate change posing unprecedented challenges for society and ecosystems. Worryingly, despite the threats posed by heatwaves and notwithstanding research progress over the past 10–15 years, our understanding of the physical drivers and dynamics of these events as well as their impacts and how society might respond in terms of managing heatwave related risk remains underdeveloped. These knowledge gaps, or in other words the 'heatwave imperative', serve as the motivation for this book which represents an attempt to summarise current knowledge related to the causes and consequences of heatwaves and responses to them. Before providing an overview of the framing and how this book is structured the following sections provide some background material considered relevant to the wider topic of heatwaves.

1.2 The Heatwave Imperative

The heatwave imperative refers to the pressing need to address the research and management challenges posed by heatwaves. The imperative to do so is a compelling one. This is because heatwaves as a serious threat to society, economy and the environment are not likely to go away. Indeed, they form an important part of society's relationship with climate and have always done so as the historical record attests. Take for example the major heatwaves with significant consequences across Australia and the eastern seaboard states of America in 1896 (Fig. 1.1). Other noteworthy historical heatwaves include the Indian and Russian heatwaves of 1878 and 1911 respectively for which high death tolls were recorded as well as those across Europe in 1473 and 1540 (Camenisch et al., 2020; Lipfert et al., 2024; Wetter et al., 2014).

Since the turn of the century heatwaves have become an ever-increasing and significant feature of the summer or warm season climate for some regions. For Europe which has experienced a warming trend surpassing twice the global average,

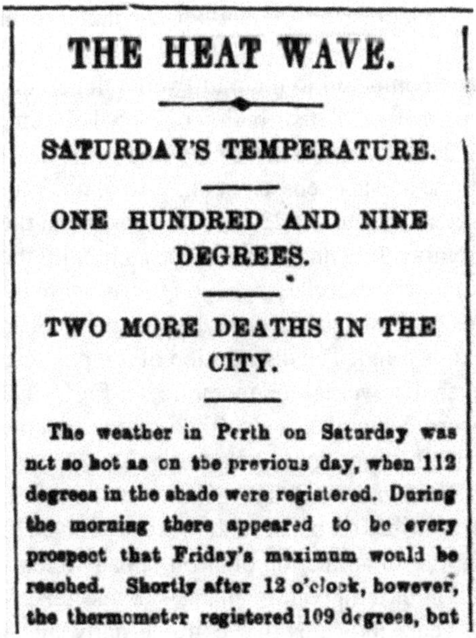

THE HEAT WAVE.

SATURDAY'S TEMPERATURE.

ONE HUNDRED AND NINE DEGREES.

TWO MORE DEATHS IN THE CITY.

The weather in Perth on Saturday was not so hot as on the previous day, when 112 degrees in the shade were registered. During the morning there appeared to be every prospect that Friday's maximum would be reached. Shortly after 12 o'clock, however, the thermometer registered 109 degrees, but

Fig. 1.1 West Australian newspaper heading (January 6, page 4) reporting the major January 1–4, 1896 heatwave in Perth (left) and an artist's impression of people sleeping outside on roofs during the New York heatwave of the same year (right). Source: Trove, National Library of Australia provided via Climate History Australia (https://climatehistory.com.au/) (left) and US Library of Congress (https://www.loc.gov/rr/print/) (right)

with temperatures rising at approximately 0.5 °C per decade, 23 out of the 30 most intense heatwaves since 1950 have occurred post-2000, with five occurring in the period 2021–2023 alone. In concert, extreme heat has emerged as the primary driver of weather- and climate related fatalities across Europe, with a significant upturn since 2000. Indeed, over the past two decades, heat-related deaths have soared by approximately 30% with 94% of nearly 1000 European regions demonstrating an alarming uptick in heat-related fatalities (C3S, 2024). These are worrying facts. Further, in the absence of turning the present global warming situation around the planet's average temperature is expected to exceed the all-important threshold of 1.5 °C above pre-industrial levels in the early to mid-2030s (Diffenbaugh & Barnes, 2023; Giménez et al., 2024; Lee et al., 2021), a threshold that bears significance for the future occurrence of heatwaves and their changing nature (Allen et al., 2018; Thompson et al., 2023). In essence, society may have to learn to live with heatwaves—the heatwave quotidian. To do so will require a better understanding of the mechanisms of heatwave formation so we can be smarter about predicting and preparing for heatwaves over a range of time scales. Additionally, it is imperative to deepen our understanding of heatwave impacts and evaluate how various adaptation and mitigation options can reduce the associated risks (e.g., health, economic, environmental). This is particularly urgent given the worrying possibility that existing measures may soon become inadequate in addressing the anticipated increase in heatwave-related impacts under climate change.

That we are living in a different climate compared to the mid-twentieth century is clear from a consideration of the shifting statistical distribution of global air temperature anomalies as portrayed in Fig. 1.2. Based on NASA's Goddard Institute of Space Studies (GISS) surface temperature analysis (Lenssen et al., 2019), this plot plainly shows how air temperatures between 1962 and 2022 have departed from the average for 1951–1980. The shift in the temperature anomalies to the right with the transition from the twentieth to twenty-first century reflects the long-term increase in global temperatures—human activities and associated increases in greenhouse gas emissions are responsible for driving this trend. The distribution of temperature anomalies for 2022 indicates for that year the average temperature on Earth had increased to 1.2 °C above the reference period average. It also shows how extreme temperatures, especially extreme heat, have become more common in the twenty-first century compared to the situation in 1962 and 1982. This is reflected in a flattening and broadening of the temperature distribution curve with the curve for 2022 stretching out and 'bulking up' to the right in the direction of strong positive temperature anomalies. Moreover, an evaluation of temperature changes across terrestrial surfaces worldwide for the past 70 years indicates that extreme daily mean temperatures are increasing at a swifter pace compared to the average rate of temperature rise of the 6 months of highest background warmth in areas spanning Europe, western regions of North America, sections of southeast Asia, and much of

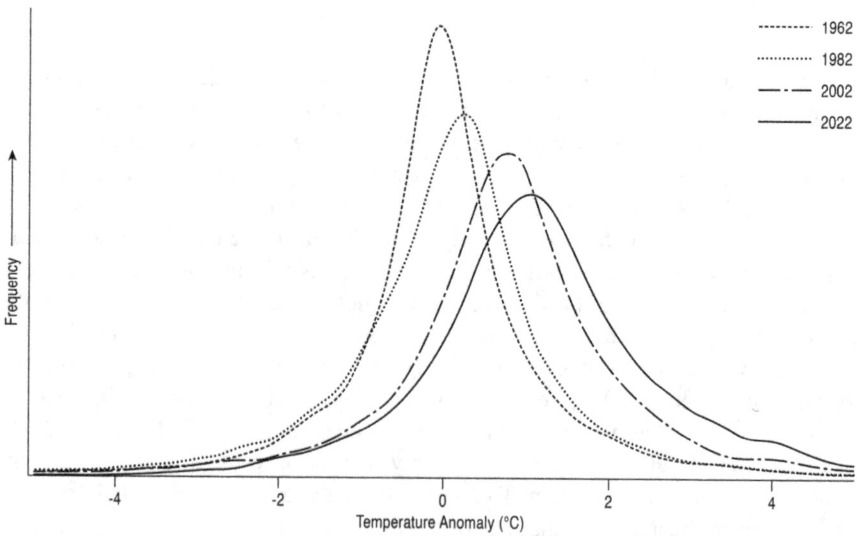

Fig. 1.2 Distribution of air temperature anomalies (°C) for 1962, 1982, 2002 and 2022. The reference period is 1951–1980 with the associated mean represented by an anomaly value of zero. The scaleless vertical axis represents increasing frequency in a relative sense. Adapted from NASA's Scientific Visualization Studio. https://svs.gsfc.nasa.gov/5065/#section_credits (accessed 08/04/2024)

South America (Huntingford et al., 2024). Notably, these regions have witnessed significant heatwaves in the last decade.

Corroborating the assertion of living in a changed climate, is confirmation that 2023 was the warmest year on record with the 2023 global temperature recorded as 1.45 °C higher than the average of the 1850–1900 pre-industrial period making 2023 significantly warmer than the 1.29 °C increase recorded for 2016. Even more worrying is the immediate past decade of 2014–2023 has been the warmest ever (WMO 2024). Perhaps unsurprisingly, 2023 was marked by the occurrence of numerous unrelenting heatwaves—a significant portion of the globe witnessed at least 20 additional heatwave days compared to the 1991–2020 average (Perkins-Kirkpatrick et al., 2024). Prominent and record-breaking warm season heatwaves in 2023 included extraordinary spring heatwaves in South America, and extensive heatwaves across Europe, Africa, and Asia, as well as an extended event spanning the southeastern USA and Central America (Perkins-Kirkpatrick et al., 2024) (Fig. 1.3).

Notably 2023 followed on from the years 2020–2022 also noteworthy for heatwaves across several regions of the globe with perhaps one of the most exceptional being the June 2021 heatwave covering the Pacific Northwest of the United States and British Columbia in Canada. Amongst the 25 heatwaves with the greatest magnitude over the period 1950–2022 (Table 1.1) this event (Western Canada in Table 1.1) is ranked as the second 'hottest'. In fact, five of the top 10 heatwaves in terms of magnitude with estimated return periods greater than 600 years, have occurred since the turn of the century (ranks 2, 3, 6, 8 and 9 in Table 1.1). While the more visible impacts for some of the twenty-first century events are relatively well known, especially in terms of the human death toll, very limited if any impact data are available for other major events occurring in the top echelon of heatwaves. This does not mean the absence of heatwave related impacts during these events. Rather it is an indication of insufficient knowledge about heatwave related risks and the associated lack of the systematic chronicling of impact data. This is especially so in low- and middle-income countries where attention to matters other than heatwaves and associated risks often takes precedence. Of course, it would be misleading to suggest that the situation in high income countries is markedly better as many of these countries lag some distance behind the ideal of a fully heatwave literate population and indeed government. The imperative to improve the situation related to heatwave monitoring and literacy and associated actions to understand the nature of heatwave impacts and how to minimise heatwave related risk is an obvious one for all countries.

The twenty-first century heatwaves noted in Table 1.1 form part of a continuing upward trend in atmospheric heatwave occurrence. For example, Perkins-Kirkpatrick and Lewis (2020) found a significant increase in the frequency of heatwaves across nearly all regions of the globe, with no region experiencing a notable decrease. While their analysis reveals little alteration in the average intensity for most regions, apart from the Amazon, Mediterranean, Northeast Brazil, Southeast South America and West Asia regions, the seasonal cumulative intensity of heatwaves has increased at a rate comparable to the frequency for most regions. Furthermore, the most severe

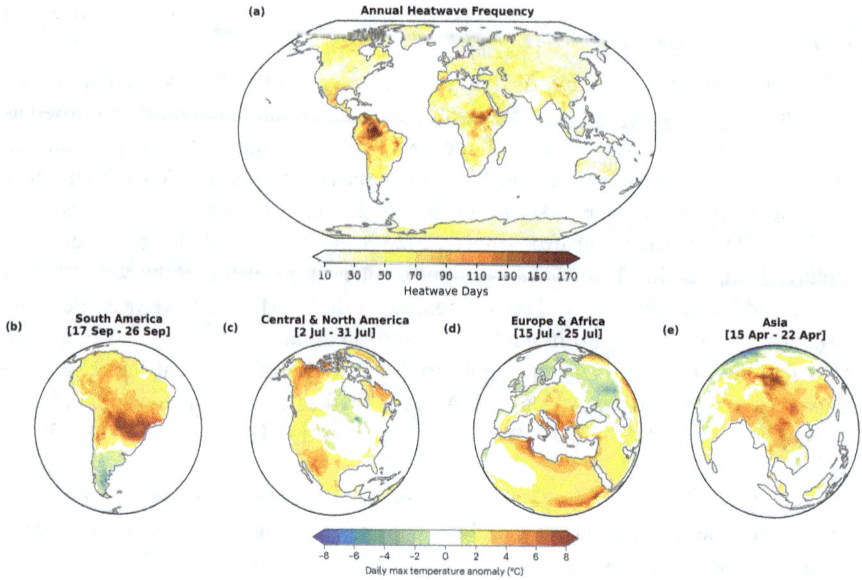

Fig. 1.3 (**a**) Number of heatwave days (a day is considered part of a heatwave when it is one of at least three consecutive days where the daily maximum temperature is above the local 90th percentile computed from a 15-day moving window relative to 1991–2020), calculated using the European Reanalysis 5 (ERA5) data set. (**b**) Average daily maximum temperature anomalies for South America during 17th–26th September, 2023. (**c**) As for (**b**), but for Central and North America during 2nd–31st July, 2023. (**d**). As for (**b**), but for Europe and North Africa during 15th–25th July 2023. (**e**) As for (**b**), but for Asia during 15th–22nd April, 2023. Source: Perkins-Kirkpatrick et al. (2024) Extreme terrestrial heat in 2023. Nat Rev. Earth Environ (2024). https://doi.org/10.1038/s43017-024-00536-y. Used with permission under Springer Nature License Number 5764200772919

heatwave seasons, identified by cumulative intensity, predominantly occur after the year 2000. Notwithstanding the influence of climate variability on trends, examination of regional trends spanning various starting points between 1950 and 2000 reveals that the upward trends in historical heatwave frequency, duration, and cumulative intensity have accelerated across many regions (Perkins-Kirkpatrick & Lewis, 2020). Moreover, alongside the accelerated trend in heatwave frequency, the twenty-first century stands out for witnessing significant occurrences of simultaneous large-scale heatwaves. For these events, recent years have shown a markedly higher likelihood of occurrence compared to 40 years ago with such heatwaves having become more intense and impacting larger geographical areas, with particularly notable increases observed across Europe, the eastern United States, and Asia (Rogers et al., 2022). Accompanying the increase in dry heatwaves—heatwaves defined using air temperature criteria alone—are upward trends in the frequency of 'wet/humid' heatwaves or 'stealth heat events' (Bekris et al., 2023; Ivanovich et al., 2024; Wang et al., 2022; Willett, 2023) characterised by conditions of moderate to high levels of warmth and high humidity. These can pose serious challenges to

Table 1.1 List of 25 'hottest' heatwaves (Based on Appendix C form Cattiaux et al. (2024)[a]. Rank is by return period—return period is in years[b]

Rank	Region[c]	Year	Start date	Finish date	Return period
1	Chad	1998	24 May	29 May	2222
2	West Canada	2021	29 June	01 July	833
3	Southwest Russia	2010	21 July	14 August	769
4	Southeast ECO	1998	17 June	16 July	769
5	Libya	1988	16 June	05 July	769
6	South EEA	2003	03 August	14 August	714
7	Iran	1995	13 July	17 July	714
8	Southeast Australia	2019	14 January	17 January	667
9	South SADC	2016	01 January	07 January	625
10	Southwest USA	1994	25 June	01 July	588
11	Polar Antarctica	1978	27 December	28 December	588
12	Saudi Arabia	1998	03 August	14 August	556
13	North EEA	1994	27 July	05 August	526
14	East Canada	1996	03 August	08 August	500
15	North Andean Community	1964	08 February	19 February	476
16	West North Brazil	1963	09 October	20 October	435
17	Mongolia	1990	14 July	28 July	435
18	Central North Brazil	1963	13 October	01 November	435
19	Queensland Australia	1972	24 December	26 December	417
20	Northwest Russia	1972	0707	0714	400
21	Pacific Russia—South	2011	0717	0728	385
22	Nunavut	1962	0715	0726	385
23	North India	1995	0529	0617	385
24	Far East Arab League	1998	0603	0610	345
25	Manitoba and Ontario	1983	0709	0708	345

[a]Events less than 3 days in length have been removed from the output from Appendix C in Cattiaux et al., 2024. [b]Return period has been calculated as the reciprocal of the probability output from Appendix C in Cattiaux et al. (2024). [c]Abbreviations for regions incorporating several countries are as follows. ECO is Economic Cooperation Organisation; EEA is European Economic Area and SADC is Southern Africa Development Community. Regions are those suggested by Stone (2019) with an area of two million km^2

human and animal health (Matthews et al., 2019). To date, notable increases have been found for parts of Eurasia, Southeast Asia, southern China, and some parts of Europe (He & Chen, 2023; Li, 2020; Willett, 2023; Yu et al., 2021). Moreover, recent increases in exposure to extreme humid-heat events disproportionately affect heavily populated regions (Rogers et al., 2021). Although parts of Africa, especially in equatorial and monsoonal regions, have experienced 'stealth heat events', such as the crushing humid heatwave across the southern coastal zone of Western Africa in February 2024, frustratingly knowledge and understanding of the longer-term occurrence of such events is limited for here. This is most likely due to data sparsity and/or lack of sharing of the sub-daily data required to characterise humid heatwaves (Willett, 2023). That said, Langue et al. (2023) report a notable upturn in the frequency, duration and intensity of heat waves in general for a range of West African cities. For marine heatwaves notable increases in their frequency and duration have also been noted over the past century (Edullantes et al., 2023; Giménez et al., 2024; Marin et al., 2021; Oliver et al., 2018).

The abundance of regionally focused and occasionally city-specific research on heatwave trends, extending beyond the limited studies referenced here, highlights evidence of shifts in heatwave patterns, particularly since the beginning of the twenty-first century. This prompts the question: " to what extent might human activities have increased the risk of major heatwaves and changes in their inherent properties?" To answer this question, methodologies for undertaking climate change attribution studies have been formulated. In general climate change attribution refers to quantifying the extent to which human activities have contributed to observed changes in the climate system. Originating in the 1990s, attribution studies gained prominence as advancements in climate modelling and data analysis techniques improved—for his contribution to the development of 'attribution science' the climate scientist Klaus Hasselmann was awarded the Nobel prize in physics in 2021! While early attribution research primarily focused on attributing global temperature rises to human-induced greenhouse gas emissions, over time, because of the crucial role attribution studies play in informing climate policy, they have expanded to encompass a wide range of climate phenomena, including extreme weather events (Stott et al., 2016; Stott & Christidis, 2023). In fact, the first ever peer-reviewed attribution study of an extreme weather event was for the impactful 2003 European summer heatwave (Stott et al., 2014). Since then, many attribution studies focusing on rising levels of heat in general, record-breaking hot summers as well as other seasons and more specifically both atmospheric and marine heatwaves have been conducted (Fig. 1.4). Some studies have even turned their attention to quantifying the extent to which observed heatwave-related mortality can be attributed to human induced climate change (Ebi et al., 2020; Mitchell, 2016; Vicedo-Cabrera et al., 2021).

The results from most climate change—heatwave attribution studies add considerable grist to the heatwave imperative mill because these clearly demonstrate that many major heatwaves that have occurred over the past 20 years or so bear a clear human footprint. For example, a climate change attribution study for the dangerous humid heatwave that occurred across West Africa's Guinea zone over the period February 11–15 in 2024 found that human-induced climate change has resulted in a measure of humid heat to be approximately 4 °C higher in today's climate compared to the past. (Pinto et al., 2024). Perhaps unsurprisingly, attribution studies conducted for the 2023 heatwaves portrayed in Fig. 1.3 all point to noteworthy influences of human induced climate change on their likelihood of occurrence and levels of heat attained (Kew et al., 2023; Zachariah, 2023a, b). Because of their rarity and scale of impacts, it is worth considering the outcomes of attribution studies for the three highest ranked twenty-first century heatwave events listed in Table 1.1, namely Western Canada (widely known as the June 2021 Pacific Northwest heatwave), Southwest Russia (generally referred to as the 2010 Russian heatwave) and South EEA (the

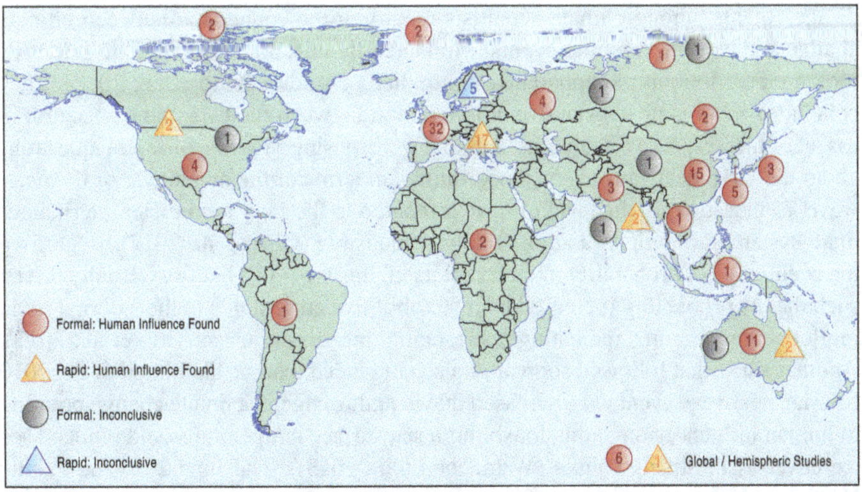

Fig. 1.4 Map showing the regions for which climate change attribution studies for heat, including heatwaves, have been conducted over the period 2011–2022. Noticeable is the low number for several regions where impactful heatwaves are known to have occurred (e.g, South America, Africa, Middle East, parts of Eurasia). Of the attribution studies conducted some are inconclusive. Formal studies are those appearing in the peer-reviewed literature. Rapid attribution studies are those performed by the World Weather Attribution initiative formed in 2015 by Dr. Geert Jan van Oldenborgh and Dr. Friederike Otto (https://www.worldweatherattribution.org/). Figure adapted from 'Mapped: How climate change affects extreme weather around the world' Carbon Brief'. https://www.carbonbrief.org/mapped-how-climate-change-affects-extreme-weather-around-the-world/ (Accessed 10/04/2024)

European 2003 heatwave event). Up until its time of occurrence, the summer of 2003 was probably the hottest in Europe since at least AD 1500. The heatwave that occurred across Europe in August of that year resulted in a devasting toll on society and environment. In some ways it spurred on the development of extreme weather attribution science. By setting a mean summer temperature threshold surpassed only in 2003 and not observed in any other year since instrumental records began in 1851, Stott et al. (2014) calculated, with high confidence (> 90%), that the human influence on the climate system substantially increased the likelihood by two-fold of the 2003 heatwave surpassing this threshold. In 2011 two studies were published related to the possible role that human-induced climate change may have played in the occurrence of the devasting 2010 Russian heatwave (Dole et al., 2011; Rahmstorf & Coumou, 2011). There is no doubt that this was an exceptional event lasting from early July until mid-August with temperatures considerably above the long-term average for this time of the year with widespread impacts reported. On the face of it, the two studies appeared to be contradictory. One concluded that natural variability was the primary driver of the heatwave event (Dole et al., 2011) while the other (Rahmstorf & Coumou, 2011) suggested that there was an 80% probability that the record July heat would not have occurred without human-related climate warming. Naturally, this sparked a flurry of activity in the media, igniting discussions about disagreements among climate scientists regarding the connection between climate change and extreme weather events. Motivated by this, a team of climate scientists took a closer look at the apparently contradictory studies (Otto et al., 2012). They concluded there is no substantive contradiction between the two studies because in essence, the respective teams of authors were posing slightly different questions about the 2010 heatwave event. Accordingly, in terms of the magnitude of the heatwave, as measured by its level of heat compared to the long term mean internal climate system variability was most likely responsible (Otto et al., 2012). Regarding the occurrence or probability of the event itself, this was most likely externally driven such that its probability increased by a factor of five attributable to the external trend in global temperatures indicating a substantial human influence (Otto et al., 2012). Another study that followed soon after also concluded that the likelihood of the 2010 Russian heatwave event was increased due to anthropogenic climate change pointing to human influences on anomalously high sea surface temperatures and reduced sea ice cover as the likely causes (Watanabe et al., 2013)). The June 2021 the Pacific Northwest heatwave across the US and Canada smashed many temperature records by several degrees. Not only was this a record shattering climate event its severe impacts extended beyond human populations to marine, terrestrial and freshwater ecosystems. Through a comprehensive attribution analysis encompassing observations, modelling, and traditional statistical methods, Philip et al. (2022) determined the Pacific Northwest heatwave would have been nearly impossible without the influence of human-induced climate change. While acknowledging the challenges associated with the statistical modelling of this event, because its maximum temperatures were far beyond the historical record, an estimated frequency of approximately once in 1000 years under present climate conditions was arrived at. This is in the ballpark of the return period given in Table 1.1 but somewhat different to that worked out by (Malinina & Gillett, 2024) based on newly available historical climate data. The attribution study also revealed that an event like the June 2021 heatwave would occur at least 150 times less frequently without human-induced climate change and what's more it was approximately 2 °C hotter than an equivalent heatwave would have been during the period from 1850 to 1900 when the global mean temperature was 1.2 °C lower. In a complimentary analysis of the same event Lucarini et al. (2023) note that while the Pacific Northwest heatwave event could be viewed as an unlikely but possible manifestation of climate variability, its probability of occurrence was greatly amplified by ongoing human-related climate change.

Although attributing major heatwaves to climate change remains a formidable methodological challenge (Van Oldenborgh et al., 2022) it would be difficult to deny that heatwave attribution studies consistently reveal a robust signal linked to human-induced climate change. This signal of course is not in the form of a one-to-one relationship in which climate change causes extreme events. Rather it is in the form of tipping the balance of probabilities such that human induced climate change can increase the likelihood of extreme heatwaves occurring, because climate variability, even if constant, rides upon an ever-increasing elevation of global base temperatures. Despite the complexities and uncertainties inherent in attribution, the findings from attribution studies are alarming. In many ways they underscore the urgent need to address the heatwave imperative not only in relation to understanding heatwaves as a physical phenomenon but also enhancing our comprehension of the lived experiences entwined with this perilous natural hazard.

1.3 Heatwave Science

To address the heatwave imperative a new type of science is required. Here this is referred to as heatwave science. This is defined as the interdisciplinary study of heatwaves, focusing on understanding their causes, characteristics, impacts (e.g., social, economic, political, infrastructural and environmental), and the potential adaptation and mitigation strategies for heatwave risk reduction. The suggested aim for a 'human-centric' heatwave science is to unravel the complex interactions between climatic processes, societal dynamics in its broadest sense, and human vulnerability, exposure and response, with the aim of informing effective strategies for heatwave risk reduction and general resilience-building. Of course, human systems should not be prioritized over natural systems in the development and application of heatwave science, given the crucial role of natural systems in maintaining planetary health. Rather than being a multidisciplinary science its ideal form should be that of an interdisciplinary science (Table 1.1) that draws together the expertise and methods from various scientific fields in a truly integrated and interactive fashion. Requisite disciplines might include for example, meteorology, climatology, oceanography, engineering, environmental science, ecology, epidemiology, public health, psychology, sociology, and political science. With such a combination of disciplines the potential for achieving a holistic view and understanding of heatwaves will be increased. Clearly the definition of heatwave science offered here is far broader than what a traditional definition of heatwave science might be, referring strictly to the study of the physical drivers and dynamics of atmospheric and marine heatwaves as physical phenomenon involving only a limited number of disciplines, residing as a sub-discipline within climate science.

While it's commonly recognized how researchers in meteorology and climatology and epidemiology can contribute to understanding heatwaves, the potential roles of other disciplines in developing heatwave science is not often articulated. In fact, despite the massive growth in the number of publications on heatwaves since 2000, most of the literature still predominantly focuses on their physical aspects either in relation to specific heatwave occurrences or more commonly climate change related heatwave projections. Papers on the current or future health impacts of heatwaves also make up a considerable portion of the total heatwave literature (Marx et al., 2021). In recognition of the narrowness of the topics typically addressed Table 1.2 aims to outline in a non-exhaustive, non-exclusive, non-hierarchical and non-independent way, how some disciplinary fields could contribute to advancing an integrated and more holistic understanding of heatwaves and thus broaden the topics addressed in the heatwave literature. The assumption here is that the boundaries between disciplines are fuzzy as might be more aligned with an interdisciplinary as opposed to a multidisciplinary approach (Table 1.3).

Table 1.2 Contrasts between an interdisciplinary and multidisciplinary approach to understanding heatwaves

Characteristic	Interdisciplinary approach	Multidisciplinary approach
Integration	The integration of methodologies, theories, and perspectives from multiple disciplines throughout the entire research process	Each discipline maintains its own methodologies, theories, and approaches throughout the research process, with limited integration between them
Integration depth	The integration of disciplines is deeper and more sustained with an emphasis on co-creating new frameworks, theories, or methodologies that transcend disciplinary boundaries	The emphasis is on leveraging diverse perspectives and expertise to address complex problems, but the integration of different disciplines may be relatively shallow with little co-creation
Who	Researchers from different disciplines actively collaborate from the outset, working together to develop research questions, design methodologies, analyse data, and interpret results. The potential for uncovering the complex multidimensional nature of heatwaves and associated feedbacks between the determinants of heatwave risk is increased	Research, experts from different disciplines work independently within their own fields and then combine their findings at the end of the research process. The potential for shedding light on how the various physical, environmental and social dimensions of heatwaves interact is lessened compared to an interdisciplinary approach
Where collaboration occurs	Collaboration occurs across disciplinary boundaries such that researchers co-produce research findings. From the outset a mutual learning process develops with a high level of reciprocity	Collaboration occurs primarily at the boundaries of the disciplines, where researchers share their findings and insights with each other rather than co-produce them. Learning occurs towards the end of the research with limited scope for a paced and reciprocal learning process
Culture and literacy	Researchers will often develop a shared language and conceptual framework to facilitate communication and collaboration across disciplines. An upshot will be an increase in heatwave science literacy	There is little development of a shared language or holistic framework. This may stymie conceptual developments and preserve the status quo in terms of heatwave science literacy

Source: Author

Table 1.3 Disciplinary contributions to heatwave science

Discipline	Contribution to heatwave science
Meteorology	Physics and dynamics of heatwaves—The atmospheric processes associated with specific heatwave events. Implications for heatwave forecasting at the short to medium timescales
Climatology	Association between large scale modes of climate variability and heatwave occurrence at intra-seasonal to decadal timescales. Climatology of heatwave characteristics (mean, variability, extremes, trends, spatial and temporal patterns). Climate mechanisms associated with compounding. Analysis of surface-atmosphere interactions and response of urban heat island to heatwaves. Implications for extended to long range heatwave forecasting, land management and urban heat risk reduction
Oceanography	Analysis of the physical oceanic and ocean-atmosphere processes associated with marine heatwave development. Implications for management of marine environments and marine-based industries such as fisheries
Atmospheric chemistry	Physical chemistry of heatwaves especially impacts on air quality. Implications for air quality management during heatwaves
Biometeorology	Study of the association between heatwaves and living organisms especially in relation to heat stress and strain and animal and plant phenology. Cross over with environmental science and ecology. Implications for managing heat stress in animal and plant production systems (e.g., dairy farming or grape growing), timing of agricultural and related activities and management of eco-tourism (e.g., cherry blossom viewing)
Environmental science and ecology	Impacts of heatwaves on natural systems, including marine, terrestrial, and aquatic ecosystems, and ecosystem services. Plant and animal heat stress. Impact of heatwaves on food chains, habitats (loss and degradation), biodiversity, soil health and population dynamics of disease vectors. Implications for environmental (land, freshwater and marine) management, habitat and biodiversity protection and vector borne disease prediction and management
Hydrology	Impact of heatwaves of freshwater processes and resources (water quality and quantity). Implications for freshwater resources management (supply and demand) as well as quality which bears implications for health and water borne diseases
Architecture	Design of shelter for heat protection and heatwave related sensitivity of indoor climates for different shelter designs
Urban Planning and landscape design	The planning and design of built environment spaces that ameliorate heatwave risks and understanding the impact of existing urban morphologies on the spatial and temporal characteristics of heat in towns and cities
Engineering	Physical reaction of critical infrastructure to extreme heat and associated impacts on infrastructure function. Implications for infrastructure dependent human systems such as health, transport and energy systems

(continued)

Table 1.3 (continued)

Discipline	Contribution to heatwave science
Sociology	Analysis of heat inequity, exposure, socio-economic vulnerability, access to adaption and mitigation options, determinants of community resilience, and effectiveness of public awareness campaigns. Implications for understanding health and other social outcomes arising from heatwaves and broader environmental justice issues as input into policy
Psychology	Understanding heatwave related human behaviour and psychological factors that influence reactions to heatwave awareness and warning information. Implications for managing mental health and tipping points for offensive or self-destructive behaviour and for informing heatwave risk communication strategies
Economics and finance	The analysis of the economic impacts of heatwaves (e.g., retail, financial markets). Implications for the development of heatwave related fiscal management plans and risk transfer options in the form of insurance products (e.g. parametric insurance and weather derivatives)
Epidemiology	Analysis of the patterns, distribution and determinants (physical health and socio-economic), of heatwave-related illnesses and deaths. Implications for informing the development of syndromic surveillance systems, health related intervention strategies and the formulation of Heat-Health Warning Systems operating within wider heat health action plans
Public health	Evaluation of the public health burden associated with heatwaves and the broader public health determinants of heatwave vulnerability. Development of syndromic surveillance systems and public health heatwave intervention strategies including education and risk communication programmes. Lead on the development and evaluation of Heat Health Action Plans. Implications for the management of health burden associated with heatwaves
Political science	Heatwave related policy, advocacy, governance, decision making, stakeholder engagement and public opinion and political behaviour analysis. The local to national to international politics of heat and heatwaves. Implications for political visibility and actions associated with heatwaves as a serious threat to society, economy and the environment
History	Using documentary sources to seek out and investigate past events, interpretation of significant environmental and societal events in relation to heatwave occurrence. Implications for context setting, expanding our historical understanding of the gravity of heatwaves and general human interest in the history of heatwaves
Anthropology	Cultural perceptions, beliefs and practices in relation to heatwaves, nature and formulation of heatwave related indigenous knowledge. Working with sociologists to understand heatwave related social vulnerability and inequity and resilience in rural and urban settings. Implications for a culturally sensitive approach to heatwave related policies and actions
The humanities	Heatwave representation in literature, fine art and popular culture (e.g., entertainment, fashion, music, sports, technology, mass media, manga and consumer goods), heatwaves in global languages and linguistics, heatwaves seen through various philosophical lenses (e.g., epistemological enquiry, metaphysical perspectives, ethical considerations). Implications for wider public understanding of heatwaves as more than mere meteorological phenomenon

Source: Author

1.4 Conceptual Framing

Under the influence of human-induced climate change the nature of heatwaves has changed. The results of attribution studies referred to earlier attest to this. Further if climate change continues unabated it is expected that heatwaves will become increasingly common, last longer and grow more intense with a significant proportion of the world's future population exposed to the deleterious effects of heatwaves (Maurya et al., 2024; Nishant et al., 2022; Wang et al., 2023). Already an overwhelming body of research is essentially telling us the same thing. Put very simply, we are in trouble if we continue down a road for which social, economic, and technological trends do not shift markedly from historical patterns, something akin to the so-called IPCC Shared Socioeconomic Pathway 2 (SSP2). While investing in climate change projections of heatwave futures is important, a much-needed imperative is to enhance our understanding of the present regarding what are the physical causes of heatwaves, what are the consequences arising from their occurrence and how has society so far responded to the ever-increasing perils associated with heatwaves. This is because to appreciate what potential heatwave futures might look like, we need a better understanding of present circumstances. Without this, and it may seem heretical to say so, heatwave related climate change projections might be conceived as somewhat speculative, especially by those less convinced about the contributions science in its widest sense can make to bettering society, Accordingly, while acknowledging the importance of human-related external forcing of the climate system for future heatwave processes and their impacts, the conceptual framework and the contents of this book give precedence to contemporary climate system internal variability and the associated societal consequences and responses arising from this. The hope is that such a focus will lend support to endeavours to reduce heatwave related risk. Figure 1.5 presents the conceptual framework around which the contents of the book are shaped.

Heatwaves are considered as the outcome of a cascade of processes from the large to regional and local scales within the climate system with the climate system comprising the five interacting components of the atmosphere, hydrosphere (including the oceans), the cryosphere, biosphere, and the lithosphere between which there are flows of energy and mass (IPCC, 2014). The dynamic relationship between climate system components as manifest by interactions and feedbacks gives rise to a range of modes of climate variability that occur across varying space and time scales. El Niño-Southern Oscillation (ENSO) is but one example of a mode of climate variability. Associated with modes of climate variability are alterations, usually at the intra-seasonal to decadal timescales, in the broad surface climate boundary conditions (e.g., sea and land surface temperatures, sea ice extent, soil moisture, state of plant ecosystems). Such 'slow' variations may provide the requite atmospheric process conditions at the 'faster' synoptic time scale (5–10 days and horizontal length scale of 1000 km or more) conducive to heatwave development, for

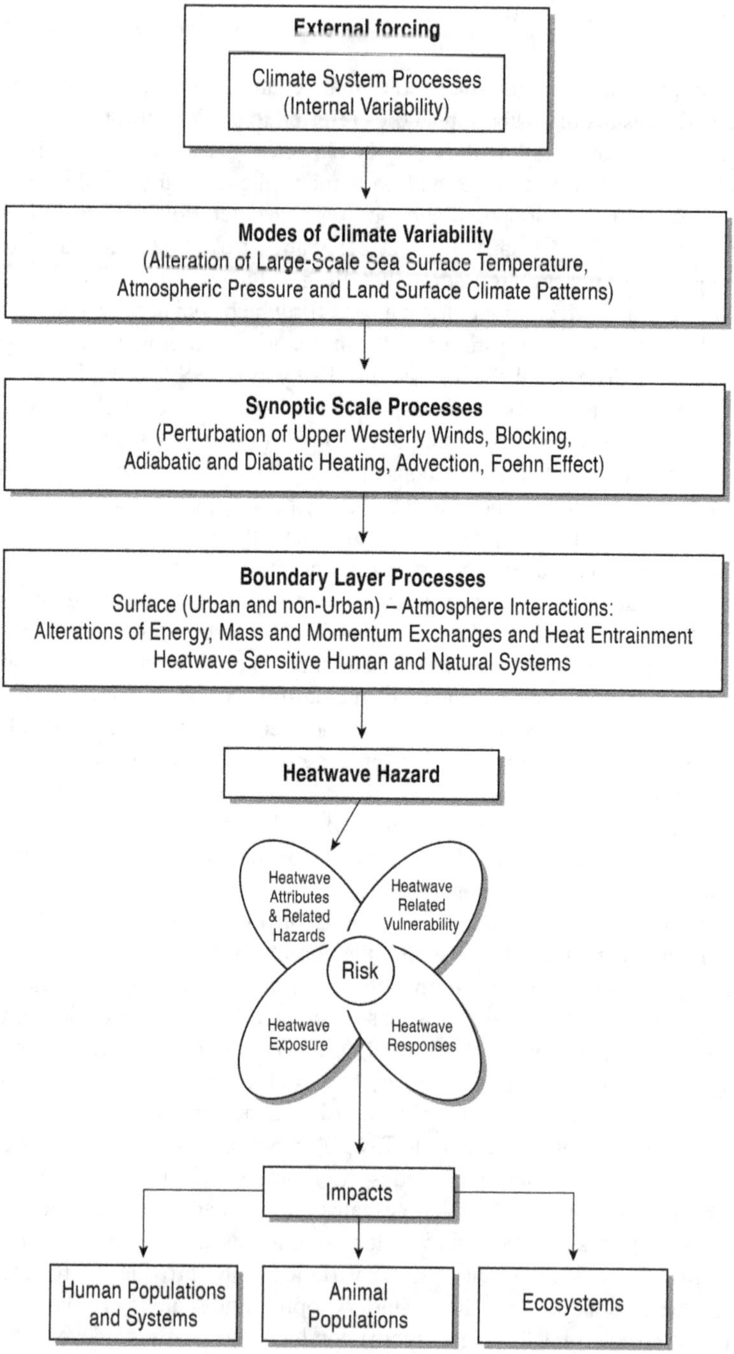

Fig. 1.5 Conceptual framework applied in this book representing a cascade of climate system processes that result in heatwave occurrence, how this is transformed to heatwave risk and ultimately impact (consequences) in human and natural systems. (Source: Author)

example through perturbation of the pattern of westerly winds in the upper atmosphere, the development of 'blocking' and enhanced solar radiation inputs to the surface. Working at the synoptic scale, these processes may tip the exchange of energy, mass and momentum between urban/non-urban surfaces and air within the boundary layer such that the heat associated with synoptic scale processes is enhanced through surface-atmosphere interactions and feedbacks with the outcome of this cascade of processes a period of multi-day anomalous heat usually codified as a heatwave. It is at this point a heatwave as a naturally occurring meteorological event transitions to a potential hazard if it intersects with human systems. Of course, the same applies to heatwave sensitive ecosystems and natural inanimate objects which have tended to be overlooked because of our generally human-centric view of heatwaves as a threat to living systems. That aside, the level of societal risk associated with a heatwave will be determined by the nature of the interaction between the heatwave hazard and other determinants of heatwave risk (Fig. 1.5).

Here the broad framing of risk as set out by the IPCC is adopted. In the case of heatwaves, risk is defined as the possibility of negative outcomes for human or natural systems arising from heatwave occurrence. Originally conceptualised as the outcome of the interaction between hazard, exposure, and vulnerability, the IPCC revised its risk framework to encompass response as a fourth determinant of risk, with response defined as any adaptation or mitigation action implemented to reduce risk (IPCC, 2022). This move was stimulated by the recognition that some responses to climate events might lead to new risks and impacts because the responses or actions put in place are either insufficient or even worse, might exacerbate the level of risk and impact. Acknowledging this but recognising that responses can have both positive and negative outcomes for risk via feedback, heatwave risk in this volume is treated as the interaction between hazard, exposure, vulnerability, and response with the realisation of risk manifest in the form of heatwave related impacts or consequences (Fig. 1.5) as expanded on in Chap. 9 (see Fig. 9.1). As for other natural hazards, the impact of heatwaves on human systems is potentially wide ranging including how people's lives, livelihoods, health, and well-being are affected, as well as heatwave related outcomes for sectors of the economy, social and cultural assets, investments, infrastructure, and services. In the case of natural systems, heatwave impacts potentially span ecosystems and their associated ecosystem services, as well as individual plant and animal species, plus inanimate components of the natural environment such as the land surface/soil, water, and air.

1.5 Structure and Overview of Book's Subject Matter

The principal subject matter of this book is atmospheric heatwaves over land with the focus skewed to human systems. However, heatwaves in marine and freshwater environments are occasionally touched upon, with the intent of sparking the reader's interest in the topics of marine, lake and stream heatwaves. These subjects of course deserve a more comprehensive exploration in a standalone book. This book is organised into four parts. Part I comprising Chaps. 2, 3 and 4, provides background material on heatwaves as weather and climate phenomenon. Parts II–IV speak to the topics referred to in the book's title, namely heatwave causes (Chaps. 5, 6, 7 and 8), heatwave consequences (Chaps. 9, 10 and 11) and heatwave responses (Chaps. 12, 13 and 14). Chapter 15 as the closing chapter offers a synthesis of the material presented in the foregoing chapters and outlines possible directions for future enquiry with regards to improving our understanding of heatwave causes, consequences, and responses.

1.5.1 *Part I: Heatwaves as Weather and Climate Phenomenon*

Chapter 2 addresses the topic of defining heatwaves, an ongoing fervent topic of discussion in the academic literature and amongst the operational community in meteorological services. In addition to outlining a range of heatwave definitions and offering comments on these, the topic of heatwave characteristics is also introduced with new material presented on heatwave typology. Chapter 3 considers heatwave analysis largely from a statistical perspective. Included in this chapter are matters related to the requisite data for heatwave analysis and associated challenges in assembling that data. Also included is an outline of the nature of Extreme Value Theory and its application in the statistical analysis of extreme heat and heatwaves. Although the next chapter on the nature and measurement of heat could well have followed the introductory chapter, as it describes the defining physical property of heatwaves, it forms Chap. 4. First the physical nature of energy and its relation to heat is described. This is followed by an introduction to the concepts of the radiation and energy balances which are referred to extensively throughout this work. Chapter 4 finishes with a description of some of the ways in which heat is measured with a range of the more commonly used thermal indices presented.

1.5.2 *Part II: Heatwave Causes*

Chapter 5 tackles the association between modes of climate variability and heatwave occurrence in a probabilistic as opposed to a deterministic framing. It begins with a general description of some of the more prominent modes of climate variability before focusing on the research evidence for associations between these and heatwave occurrence. Although not explicit there is a geographical flavour to the material presented in the second half of this chapter. Chapter 6 occupies a step along the cascade chain outlined in Fig. 1.5. It focuses on synoptic scale atmospheric processes first outlining the main physical mechanisms associated with heatwave development before going on to present case studies of several noteworthy heatwave events. The purpose of the case studies, presented using a geographical perspective, is to highlight the diversity of heatwave formation processes making the point there is no singularly dominant explanation for heatwaves. Rather the flavour or interplay of synoptic processes varies geographically. Chapter 7 focuses on the role of surface—atmosphere interactions in the development of heatwaves. Both terrestrial and marine surfaces are touched upon in relation to atmospheric and marine heatwaves respectively. The chapter begins with an attempt to define the surface as the active layer that binds the surface and the atmosphere. It then describes the nature of the boundary layer where the majority of surface—atmosphere interactions occur. A consideration of the controls on the surface–atmosphere exchanges of heat, mass and momentum and their implications for heatwave development then follows. The latter part of the chapter focuses specifically on the evidence for the role of surface-atmosphere interactions in the development of atmospheric and marine heatwaves and uniquely heatwaves in terrestrial water bodies such as lakes and streams. Because nowhere else on the planet are changes in 'the surface' more obvious than in urban settlements, Chap. 8 in the wider context of surface—atmosphere interactions, is dedicated to the matter of what happens when regional scale heatwaves interact with the urban surface. The chapter begins by introducing the concept of the urban surface, followed by an overview of the urban boundary layer. To lay the groundwork for understanding interactions between heatwaves and urban surfaces, the concept of the urban energy balance is then explored, focusing particularly on the interplay of non-advective fluxes between the urban surface and the overlying atmosphere during heatwaves. Next, it delineates the distinct characteristics of the four types of urban heat islands as the term urban heat island is often used indiscriminately. The final section of the chapter addresses the widely debated topic of whether heatwaves may amplify the urban heat island effect.

1.5.3 Part III: Heatwave Consequences

Chapter 9 begins with a reflection on the conceptualisation of heatwave impacts. Noting that a fundamental requirement for managing the impacts that arise from heatwaves is an understanding of their nature, the rest of the Chapter presents a possible five step methodology for undertaking an *ex-poste* heatwave impact assessment. To the author's knowledge, this may be one of the first attempts to initiate a discussion about the form that a systematic assessment of a specific heatwave event might take. Chapter 10 explores exposure and vulnerability as moderators of heatwave impacts. It not only addresses these topics in relation to human populations but also animal populations and marine, terrestrial, and freshwater systems. Chapter 11 turns its attention to observed heatwave impacts and in doing so deliberately avoids speculation about possible future impacts. Structured similarly to the previous chapter, impacts on humans, economy and society and infrastructure systems, along with animal populations and a range of ecosystems are outlined.

1.5.4 Part IV: Heatwave Responses

Heatwave responses, taken to mean intentional short to long-term and sustained actions and measures to reduce heatwave related risk and thus impact is covered in three chapters. Chapter 12 focuses on adaptation and starts to build the case for making a clear distinction between heatwave adaptation and mitigation options. The chapter commences with an overview of the fundamental principles of adaptation, followed by a broad discussion on various options for adapting to heatwaves. Instead of exhaustively exploring the wide array of heatwave adaptation options, the focus here is on a select few. These encompass heatwave risk governance, heatwave insurance, the appointment of Chief Heat Officers, and the creation of specialized knowledge networks dedicated to advancing heatwave risk reduction. These topics have been singled out due to their rapid emergence within the broader subject of heatwave adaptation. The final section of Chap. 12 attempts something relatively new and explores some of the theories that might explain what precipitates heatwave responses. Although included in a chapter on adaptation the material in this section should be read with heatwave mitigation options in mind as well. Because of the enormity of the subject Chap. 13 is dedicated to early warning systems as an adaptation option. Firstly, this chapter outlines the basic characteristics of Early Warning Systems (EWS). The nature of Heatwave Early Warning Systems (HEWS) and their health variant, Heat Health Warning Systems (HHWS) is then described. This is followed by material on HEWS and HHWS evaluation. Recognizing that a dependable heatwave prediction system is indispensable for effective HEWS and HHWS, the methodologies and advancements in heatwave forecasting across various timeframes is addressed next. In the context of the communication and dissemination dimension of EWS, the topic of heatwave naming is tackled in the latter parts of the chapter. Mitigation options for reducing heatwave related risk is the subject matter of Chap. 14. The focus is mainly on urban heat. From the outset the Chapter makes a case for differentiating between actions aimed at mitigating heat and those categorised as adaptation, It suggests that measures designed to alter the urban surface radiative and energy balances, as well as modify the exchange of mass and momentum between the urban surface and the overlying atmosphere, with the goal of reducing heat, should be classified as urban heat mitigation strategies. The Chapter then proceeds to describe and evaluate heatwave mitigation options under the three broad headings of modifications to the facets or elements of urban surfaces, the application of nature-based solutions (NBS) and changes to urban morphology and built form.

Lastly before proceeding, a short statement about the word related to the general subject matter of this book.

1.6 Heatwave or Heat Wave?

Although it is probably obvious from the title of this book, the compound noun "heatwave" is preferred here over the two-word alternative "heat wave". Both are grammatically correct. Although speculative and not supported by any quantitative analysis it would appear heatwave tends to occur more in academic and related outputs using British English, whereas the opposite is true for American English. Heatwave has been settled upon here for a variety of reasons. These include clarity and precision, consistency, linguistic tradition, visual representation and more pragmatically, efficiency. Using the compound word heatwave creates a single, unified term that directly conveys at the very least the concept of an extended period of abnormally high temperatures or unusual heat. In this way the use of heatwave may enhance clarity and precision in communication with the ideal the intended meaning is immediately apparent. Compound nouns are often used to describe specific phenomena or concepts in scientific discourse. By adopting heatwave as a standardized term, consistency in terminology, setting aside how a heatwave is defined of course, can be promoted across disciplines, policy making arenas and geographic regions thus facilitating effective communication. Interestingly in some scientific reports (e.g., IPCC 6AR) heat wave and heatwave are used interchangeably indicating a lack of consistency. Further as discovered while preparing this book, when research data bases such as the Web of Science are queried with the terms heatwave and heat wave, quite different listings of research papers result. In English, compound nouns are commonly used to create new terms by combining two or more words to form a single concept. Heatwave follows this linguistic convention and conforms to the pattern of other compound nouns used to describe meteorological phenomena, such as snowstorms or thunderstorms. The compound noun heatwave may conjure up a more vivid and evocative mental image than the two separate words heat and wave conveying the sense of an event characterized by a pattern of heat spreading across a region. This may assist people visualize and understand heatwave as a phenomenon more intuitively. Lastly and from a strictly editorial point of view, compound nouns can be more efficient in terms of space and word count compared to phrases composed of multiple words. Instead of heat wave, the use of heatwave may facilitate more streamlined writing and speech, particularly in situations where brevity is valued as is the case for scientific reports, journal articles, and public communications. Of course, the reasons proffered here for the use of heatwave are open to debate and in many ways could form the basis of some sociolinguistic exploration of the usage of heatwave versus heat wave. All that said it is worth mentioning the term 'cold wave 'as two separate words to describe the antithesis of a period of anomalous heat. While 'coldwave' might seem like a logical compound noun to describe a period of anomalously cold weather, it is not commonly used in meteorological terminology. Although it merits testing, this might be due to historical usage patterns and the dwindling occurrence of cold waves as a meteorological event with their warm counterpart attracting much more attention in the popular and scientific press with an associated need to use a single word as an eye-catching term.

Setting aside the issue of heatwave or heat wave, of interest is the etymology of this term. Originally the term was used to describe a wave of hot weather passing over a location or from one location to another. This is broadly akin to the modern-day understanding of the term. Alternative uses referred to an electromagnetic wave in physics or the shimmering or wave-like visual effect caused by heat convection from a hot road or land surface. According to the Oxford English Dictionary, the term heatwave, actually originally hyphenated as heat-wave, was first used in around 1842 in a paper published in the Proceedings of the Geological Society (UK) with reference to the propensity for cold waves and heat-waves to occur in continental interior regions in winter and summer respectively. Further heat-wave appears to have supplanted hot wave as the twentieth century progressed with eventually the hyphen dropped with and then, at least in British English, the compound noun of heatwave being preferred. However internationally it appears that heat wave continues to dominate according to the output from a Google Books Ngram Viewer analysis that shows how the terms 'heat wave' and 'heatwave' have occurred in a corpus of books over the period 1800 to 2019 (Fig. 1.6). Setting aside the earlier upturn and the greater occurrence of 'heat wave' (Fig. 1.6a), of interest is the rapid growth in the occurrence of the compound noun heatwave post 2000 perhaps reflecting, notwithstanding the aforementioned inconsistency, its use in IPCC reports, and the occurrence of a number of impactful heatwaves especially across Europe, Russia and Australia with an increased number of publications originating from Commonwealth countries such as Australia and Canada (Fig. 1.6b). The rapid increase in occurrence of 'heat wave' in the 1930s and again in the 1950s (Fig. 1.6a) relates to extended high impact drought periods across the United States. Similarly, an uptick in the occurrence of 'heat wave' in the 1980s relates to publications regarding extended periods of drought and high temperatures across vast tracts of the US in the early and later stages of this decade. Whether heatwave eventually

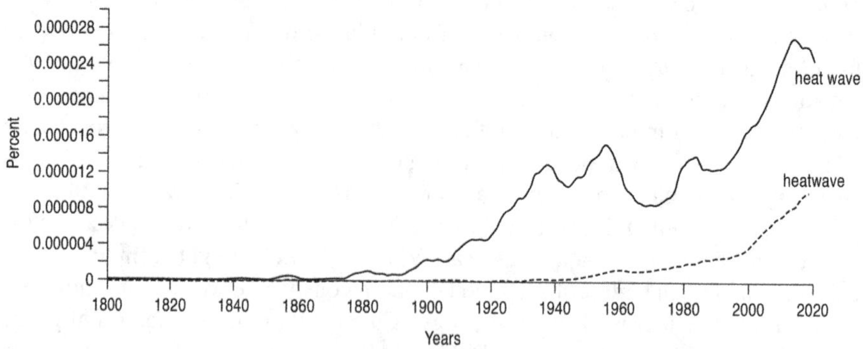

Fig. 1.6 Occurrence of (a) 'heat wave' and (b) 'heatwave' represented as a percentage of all words contained within the Google book corpus. Adapted from original analysis provided by Professor Michihisa Tsukamoto, Aichi University, Japan

supplants heat wave or vice versa remains to be seen. Of course, some might say that in the context of more important and yet to be resolved 'cultural wars', whether it is heatwave or heat wave may not matter.

References

Allen, M., Dube, O., Solecki, W., Aragón-Durand, F., Cramer, W., Humphreys, S., Kainuma, M., Kala, J., Mahowald, N., & Mulugetta, Y. (2018). Global warming of 1.5 C. An IPCC special report on the impacts of global warming of 1.5 C above pre-industrial levels and related global greenhouse gas emission pathways, in the context of strengthening the global response to the threat of climate change, sustainable development, and efforts to eradicate poverty. Sustainable development, and efforts to eradicate poverty.

Bekris, Y., Loikith, P. C., & Neelin, J. D. (2023). Short warm distribution tails accelerate the increase of humid-heat extremes under global warming. *Geophysical Research Letters, 50*(11). https://doi.org/10.1029/2022gl102164

C3S. (2024). European state of the climate 2023 summary copernicus climate change service. https://doi.org/10.24381/bs9v-8c66

Camenisch, C., Brázdil, R., Kiss, A., Pfister, C., Wetter, O., Rohr, C., Contino, A., & Retsö, D. (2020). Extreme heat and drought in 1473 and their impacts in Europe in the context of the early 1470s. *Regional Environmental Change, 20*(1), 19. https://doi.org/10.1007/s10113-020-01601-0

Cattiaux, J., Ribes, A., & Thompson, V. (2024). Searching for the most extreme temperature events in recent history. *Bulletin of the American Meteorological Society, 105*(1), E239–E256. https://doi.org/10.1175/bams-d-23-0095.1

Diffenbaugh, N. S., & Barnes, E. A. (2023). Data-driven predictions of the time remaining until critical global warming thresholds are reached. *Proceedings of the National Academy of Sciences of the United States of America, 120*(6), e2207183120. https://doi.org/10.1073/pnas.2207183120

Dole, R., Hoerling, M., Perlwitz, J., Eischeid, J., Pegion, P., Zhang, T., Quan, X. W., Xu, T. Y., & Murray, D. (2011). Was there a basis for anticipating the 2010 Russian heat wave? *Geophysical Research Letters, 38*. https://doi.org/10.1029/2010gl046582

Ebi, K. L., Astrom, C., Boyer, C. J., Harrington, L. J., Hess, J. J., Honda, Y., Kazura, E., Stuart-Smith, R. F., & Otto, F. E. L. (2020). Using detection and attribution to quantify how climate change is affecting health. *Health Affairs, 39*(12), 2168–2174. https://doi.org/10.1377/hlthaff.2020.01004

Edullantes, B., Concolis, B. M. M., Quilestino-Olario, R., Atup, D. P. D., Cortes, A., & Yñiguez, A. T. (2023). Characteristics of marine heatwaves in The Philippines. *Regional Studies in Marine Science, 62*, 102934. https://doi.org/10.1016/j.rsma.2023.102934

Giménez, L., Boersma, M., & Wiltshire, K. H. (2024). A multiple baseline approach for marine heatwaves. *Limnology and Oceanography, 69*(3), 638–651. https://doi.org/10.1002/lno.12521

He, W., & Chen, H. (2023). More extreme-heat occurrences related to humidity in China. *Atmospheric and Oceanic Science Letters, 16*(5), 100391.

Huntingford, C., Cox, P. M., Ritchie, P. D., Clarke, J. J., Parry, I. M., & Williamson, M. S. (2024). Acceleration of daily land temperature extremes and correlations with surface energy fluxes. *npj Climate and Atmospheric Science, 7*(1), 84.

IPCC. (2014). *Climate change 2013′the Physical science basis: Working group I contribution to the Fifth assessment report of the intergovernmental panel on climate change*. Cambridge University Press. https://doi.org/10.1017/CBO9781107415324

IPCC. (2022). *Climate change 2022: Impacts, adaptation, and vulnerability. Contribution of working group II to the sixth assessment report of the intergovernmental panel on cli-*

mate change. Cambridge books online, (1st ed.), Cambridge University Press. https://doi.org/10.1017/9781009325844.

Ivanovich, C. C., Horton, R. M., Sobel, A. H., & Singh, D. (2024). Subseasonal variability of humid heat during the South Asian summer monsoon. *Geophysical Research Letters, 51*(6). https://doi.org/10.1029/2023gl107382

Kew, S.F., Pinto, I., Alves, L., Santos, D., Libonati, R., Liberato, M.L.R., Philip, S., Zachariah, M., Barnes, C., Vahlberg, M., Otto, F., Clarke, B.J., & Kimutai, J. (2023). *Strong influence of climate change in uncharacteristic early spring heat in South America*. https://doi.org/10.25561/106753.

Langue, C. G. N., Lavaysse, C., Vrac, M., & Flamant, C. (2023). Heat wave monitoring over West African cities: Uncertainties, characterization and recent trends. *Natural Hazards and Earth System Sciences, 23*(4), 1313–1333. https://doi.org/10.5194/nhess-23-1313-2023

Lee, J. Y., Marotzke, J., Bala, G., Cao, L., Corti, S., Dunne, P., Engelbrecht, F., Fischer, E., Fyfe, J. C., Jones, C., Maycock, A., Mutemi, J., Ndiaye, O., Panickal, S., & Zhou, T. (2021). Future global climate: Scenario-based projections and near term information. In V. Masson-Delmotte, P. Zhai, A. Pirani, et al. (Eds.), *Climate change 2021: The physical science basis*. Contribution of working group I to the sixth assessment report of the intergovernmental panel on climate change (pp. 553–672). Cambridge University Press. https://doi.org/10.1017/9781009157896.006

Lenssen, N. J., Schmidt, G. A., Hansen, J. E., Menne, M. J., Persin, A., Ruedy, R., & Zyss, D. (2019). Improvements in the GISTEMP uncertainty model. *Journal of Geophysical Research: Atmospheres, 124*(12), 6307–6326.

Li, X. X. (2020). Heat wave trends in Southeast Asia during 1979–2018: The impact of humidity. *Science of the Total Environment, 721*, 137664. https://doi.org/10.1016/j.scitotenv.2020.137664

Lipfert, L., Hand, R., & Brönnimann, S. (2024). A global assessment of heatwaves since 1850 in different observational and model data sets. *Geophysical Research Letters, 51*, 3. https://doi.org/10.1029/2023gl106212

Lucarini, V., Galfi, V. M., Riboldi, J., & Messori, G. (2023). Typicality of the 2021 Western North America summer heatwave. *Environmental Research Letters, 18*(1). https://doi.org/10.1088/1748-9326/acab77

Malinina, E., & Gillett, N. P. (2024). The 2021 heatwave was less rare in Western Canada than previously thought. *Weather and Climate Extremes, 43*, 100642. https://doi.org/10.1016/j.wace.2024.100642

Marin, M., Feng, M., Phillips, H. E., & Bindoff, N. L. (2021). A global, multiproduct analysis of coastal marine heatwaves: Distribution, characteristics, and long-term trends. *Journal of Geophysical Research-Oceans, 126*(2). https://doi.org/10.1029/2020jc016708

Marx, W., Haunschild, R., & Bornmann, L. (2021). Heat waves: A hot topic in climate change research. *Theoretical and Applied Climatology, 146*(1–2), 781–800. https://doi.org/10.1007/s00704-021-03758-y

Matthews, T., Wilby, R. L., & Murphy, C. (2019). An emerging tropical cyclone-deadly heat compound hazard. *Nature Climate Change, 9*(8), 602–606. https://doi.org/10.1038/s41558-019-0525-6

Maurya, H. K., Joshi, N., & Suryavanshi, S. (2024). Projected changes in the heatwave's characteristics and associated population exposure over India under 1.5–3°C warming levels. *Stochastic Environmental Research and Risk Assessment., 38*, 2521–2538. https://doi.org/10.1007/s00477-024-02695-2

Mitchell, D. (2016). Human influences on heat-related health indicators during the 2015 Egyptian heat wave. *Bulletin of the American Meteorological Society, 97*(12), S70–S74. https://doi.org/10.1175/bams-d-16-0132.1

Nishant, N., Ji, F., Guo, Y. M., Herold, N., Green, D., Di Virgilio, G., Beyer, K., Riley, M. L., & Perkins-Kirkpatrick, S. (2022). Future population exposure to Australian heatwaves. *Environmental Research Letters, 17*(6). https://doi.org/10.1088/1748-9326/ac6dfa

Oliver, E. C. J., Donat, M. G., Burrows, M. T., Moore, P. J., Smale, D. A., Alexander, L. V., Benthuysen, J. A., Feng, M., Sen Gupta, A., Hobday, A. J., Holbrook, N. J., Perkins-Kirkpatrick, S. E., Scannell, H. A., Straub, S. C., & Wernberg, T. (2018). Longer and more frequent marine heatwaves over the past century. *Nature Communications, 9*, 1324. https://doi.org/10.1038/s41467-018-03732-9

Otto, F. E. L., Massey, N., van Oldenborgh, G. J., Jones, R. G., & Allen, M. R. (2012). Reconciling two approaches to attribution of the 2010 Russian heat wave. *Geophysical Research Letters, 39*. https://doi.org/10.1029/2011gl050422

Perkins-Kirkpatrick, S. E., & Lewis, S. C. (2020). Increasing trends in regional heatwaves. *Nature Communications, 11*(1). https://doi.org/10.1038/s41467-020-16970-7

Perkins-Kirkpatrick, S., Barriopedro, D., Jha, R., Wang, L., Mondal, A., Libonati, R., Wang, L., & Kornhuber, K. (2024). Extreme terrestrial heat in 2023. *Nature Reviews Earth and Environment, 5*, 244–246. https://doi.org/10.1038/s43017-024-00536-y

Philip, S. Y., Kew, S. F., van Oldenborgh, G. J., Anslow, F. S., Seneviratne, S. I., Vautard, R., Coumou, D., Ebi, K. L., Arrighi, J., Singh, R., van Aalst, M., Marghidan, C. P., Wehner, M., Yang, W. C., Li, S. H., Schumacher, D. L., Hauser, M., Bonnet, R., Luu, L. N., Lehner, F., Gillett, N., Tradowsky, J. S., Vecchi, G. A., Rodell, C., Stull, R. B., Howard, R., & Otto, F. E. L. (2022). Rapid attribution analysis of the extraordinary heat wave on the Pacific coast of the US and Canada in June 2021. *Earth System Dynamics, 13*(4), 1689–1713. https://doi.org/10.5194/esd-13-1689-2022

Pinto, I., Odoulami, R.C., Lawal, K.A., Olaniyan, E., Ibrahim, W.A., Guigma, K., Vahlberg, M., Heinrich, D., Marghidan, C.P., Vogel, M., Arrighi, J., Barnes, C., Otto, F., Philip, S., Mistry, M., Sengupta, S., Kew, S., & Kimutai, J. (2024). *Dangerous humid heat in southern West Africa about 4°C hotter due to climate change*. https://doi.org/10.25561/110082.

Rahmstorf, S., & Coumou, D. (2011). Increase of extreme events in a warming world. *Proceedings of the National Academy of Sciences of the United States of America, 108*(44), 17905–17909. https://doi.org/10.1073/pnas.1101766108

Rogers, C. D. W., Ting, M. F., Li, C. H., Kornhuber, K., Coffel, E. D., Horton, R. M., Raymond, C., & Singh, D. (2021). Recent increases in exposure to extreme humid-heat events disproportionately affect populated regions. *Geophysical Research Letters, 48*(19). https://doi.org/10.1029/2021gl094183

Rogers, C. D. W., Kornhuber, K., Perkins-Kirkpatrick, S. E., Loikith, P. C., & Singh, D. (2022). Sixfold increase in historical northern hemisphere concurrent large heatwaves driven by warming and changing atmospheric circulations. *Journal of Climate, 35*(3), 1063–1078. https://doi.org/10.1175/jcli-d-21-0200.1

Stone, D. A. (2019). A hierarchical collection of political/economic regions for analysis of climate extremes. *Climatic Change, 155*(4), 639–656. https://doi.org/10.1007/s10584-019-02479-6

Stott, P. A., & Christidis, N. (2023). Operational attribution of weather and climate extremes: What next? *Environmental Research: Climate, 2*. https://doi.org/10.1088/2752-5295/acb078

Stott, P. A., Stone, D. A., & Allen, M. R. (2014). Human contribution to the European heatwave of 2003. *Nature, 432*(7017), 610–614. https://doi.org/10.1038/nature03089

Stott, P. A., Christidis, N., Otto, F. E. L., Sun, Y., Vanderlinden, J. P., van Oldenborgh, G. J., Vautard, R., von Storch, H., Walton, P., Yiou, P., & Zwiers, F. W. (2016). Attribution of extreme weather and climate-related events. *Wiley Interdisciplinary Reviews-Climate Change, 7*(1), 23–41. https://doi.org/10.1002/wcc.380

Thompson, V., Mitchell, D., Hegerl, G. C., Collins, M., Leach, N. J., & Slingo, J. M. (2023). The most at-risk regions in the world for high-impact heatwaves. *Nature Communications, 14*(1). https://doi.org/10.1038/s41467-023-37554-1

Van Oldenborgh, G. J., Wehner, M. F., Vautard, R., Otto, F. E. L., Seneviratne, S. I., Stott, P. A., Hegerl, G. C., Philip, S. Y., & Kew, S. F. (2022). Attributing and projecting heatwaves is hard: We can do better. *Earths Futures, 10*(6). https://doi.org/10.1029/2021ef002271

Vicedo-Cabrera, A. M., Scovronick, N., Sera, F., Roye, D., Schneider, R., Tobias, A., Astrom, C., Guo, Y., Honda, Y., Hondula, D. M., Abrutzky, R., Tong, S., Coelho, M., Saldiva, P. H. N.,

Lavigne, E., Correa, P. M., Ortega, N. V., Kan, H., Osorio, S., Kysely, J., Urban, A., Orru, H., Indermitte, E., Jaakkola, J. J. K., Ryti, N., Pascal, M., Schneider, A., Katsouyanni, K., Samoli, E., Mayvaneh, F., Entezari, A., Goodman, P., Zeka, A., Michelozzi, P., de Donato, F., Hashizume, M., Alahmad, B., Diaz, M. H., Valencia, C. D., Overcenco, A., Houthuijs, D., Ameling, C., Rao, S., Di Ruscio, F., Carrasco-Escobar, G., Seposo, X., Silva, S., Madureira, J., Holobaca, I. H., Fratianni, S., Acquaotta, F., Kim, H., Lee, W., Iniguez, C., Forsberg, B., Ragettli, M. S., YLL, G., Chen, B. Y., Li, S., Armstrong, B., Aleman, A., Zanobetti, A., Schwartz, J., Dang, T. N., Dung, D. V., Gillett, N., Haines, A., Mengel, M., Huber, V., & Gasparrini, A. (2021). The burden of heat-related mortality attributable to recent human-induced climate change. *Nature Climate Change, 11*(6), 492–500. https://doi.org/10.1038/s41558-021-01058-x

Wang, X. S., He, L., Ma, X. H., Bie, Q., Luo, L., Xiong, Y. C., & Ye, J. S. (2022). The emergence of prolonged deadly humid heatwaves. *International Journal of Climatology, 42*(16), 8607–8618. https://doi.org/10.1002/joc.7750

Wang, Y. W., Zhao, N., Wu, C. Y., Quan, J. L., & Chen, M. X. (2023). Future population exposure to heatwaves in 83 global megacities. *Science of the Total Environment, 888*, 164142. https://doi.org/10.1016/j.scitotenv.2023.164142

Watanabe, M., Shiogama, H., Imada, Y., Mori, M., Ishii, M., & Kimoto, M. (2013). Event attribution of the August 2010 Russian heat wave. *Scientific Online Letters On The Atmosphere, 9*, 65–68. https://doi.org/10.2151/sola.2013-015

Wetter, O., Pfister, C., Werner, J. P., Zorita, E., Wagner, S., Seneviratne, S. I., Herget, J., Grünewald, U., Luterbacher, J., Alcoforado, M. J., Barriendos, M., Bieber, U., Brázdil, R., Burmeister, K. H., Camenisch, C., Contino, A., Dobrovolny, P., Glaser, R., Himmelsbach, I., Kiss, A., Kotyza, O., Labbé, T., Limanówka, D., Litzenburger, L., Nordl, O., Pribyl, K., Retsö, D., Riemann, D., Rohr, C., Siegfried, W., Söderberg, J., & Spring, J. L. (2014). The year-long unprecedented European heat and drought of 1540-a worst case. *Climatic Change, 125*(3–4), 349–363. https://doi.org/10.1007/s10584-014-1184-2

Willett, K. M. (2023). HadISDH. Extremes part II: Exploring humid heat extremes using wet bulb temperature indices. *Advances in Atmospheric Sciences, 40*(11), 1968–1985. https://doi.org/10.1007/s00376-023-2348-7

WMO. (2024). *State of the global climate 2023* (Vol. WMO-No. 1347). World Meteorological Organisation.

Yu, S., Tett, S. F. B., Freychet, N., & Yan, Z. W. (2021). Changes in regional wet heatwave in Eurasia during summer (1979–2017). *Environmental Research Letters, 16*(6). https://doi.org/10.1088/1748-9326/ac0745

Zachariah, M., Philip, S., Pinto, I., Vahlberg, M., Singh, R., Otto, F., Barnes, C., & Kimutai, J. (2023a). *Extreme heat in North America, Europe and China in July 2023 made much more likely by climate change*. Grantham Institute for Climate Change. https://doi.org/10.25561/105549

Zachariah, M., Vautard, R., Chandrasekaran, R., Chaithra, S. T., Kimutai, J., Arulalan, T., AchutaRao, K., Barnes, C., Singh, R., Vahlberg, M., Arrgihi, J., Raju, E., Sharma, U., Ogra, A., Vaddhanaphuti, C., Bahinipati, C. S., Tschakert, P., Pereira Marghidan, C., Mondal, A., Schwingshackl, C., Philip, S., & Otto, F. (2023b). *Extreme humid heat in South Asia in April 2023, largely driven by climate change, detrimental to vulnerable and disadvantaged communities*. Grantham Institute for Climate Change. https://doi.org/10.25561/104092

Chapter 2
Defining Heatwaves

Abstract This Chapter considers the complexities of establishing a universal definition for heatwaves, highlighting the associated challenges and implications. It begins by contrasting two possible framings for defining heatwaves—the structuralist and phenomenological approaches—and outlines the criteria for a "good" heatwave definition. Also addressed is the distinction between heatwaves and warm spells. The chapter then describes the use of absolute and relative criteria for defining atmospheric heatwaves along with the type of variables typically used to define heatwaves. A distinction is made between hazard-based and impact-based heatwave definitions. Following this background material, an overview of heatwave definitions is presented organised under the four broad headings of National Meteorological and Hydrological Services (NMHS), research literature, international working groups and sector based. Unsurprisingly a vast range of definitions are revealed with varying degrees of precision and falsifiability. Setting aside the diversity of heatwave definitions, a general description of heatwave attributes (e.g., duration and intensity) is then presented along with a consideration of their interaction. Building on the idea that not all heatwaves are the same the chapter introduces the idea of heatwave form and presents an example of a heatwave typology for Tokyo, Japan. Further to atmospheric heatwaves, the chapter considers the case of defining marine heatwaves and their categorisation in relation to intensity. The chapter concludes by acknowledging the multifaceted nature of heatwaves and the practical challenges of implementing an interdisciplinary definition that incorporates both physical and societal dimensions.

Keywords Atmospheric heatwave definition · Heatwaves and warm spells · Relative and absolute thresholds · Hazard-based and impact-based definitions · Heatwave attributes · Heatwave typology · Marine heatwave definition

G. McGregor, *Heatwaves*, Biometeorology 6,
https://doi.org/10.1007/978-3-031-69906-1_2

2.1 Introduction

Despite vocal calls from various sectors for the development of a universal heatwave definition, such a call remains unanswered. Frustrated by this, a common reaction from many in the world of climate science and climate-related risk management has been to emphasise how this situation hinders progress regarding the investigation of historical and future changes in heatwave characteristics, renders comparisons between heatwave studies ineffective and makes the assessment of heatwave impacts difficult. In brief because all conclusions and actions related to heatwaves are definition contingent, definition standardisation is considered the best way forward. Of course, counter to this claim is 'one size may not fit all' with the idea of overarching, objective truths or solutions strongly rejected. This is because heatwaves are physically complex and messy objects, and a universal definition may overlook vulnerable populations and communities that are disproportionately affected by extreme heat. Further a one-size-fits-all approach may undermine adaptation strategies and resilience measures tailored to locally specific climate conditions and heatwave experiences.

Rather than solve what appears to be an intractable issue, especially with regards to producing a universal heatwave definition, the purpose of this chapter is to present an overview of the definition and description problem for heatwaves as physical events beyond them being generally understood as periods of exceptional summer heat lasting more than 2–3 days. The chapter is structured as follows: Firstly, it presents two broad framings for defining heatwaves, followed by an examination of the generic criteria for a 'good' heatwave definition. Subsequently, the criteria and types of variables used in defining heatwaves are outlined. The chapter proceeds with a presentation and analysis of common heatwave definitions currently in use in academic and operational settings. It then considers the characterisation of heatwaves in terms of various attributes. While the primary focus remains on atmospheric heatwaves, the penultimate section addresses the definition of marine heatwaves. Finally, the chapter concludes by synthesizing its contents.

2.2 Framings for Defining Heatwaves

In defining heatwave events, two broad framings are possible. A structuralist framing would resolve heatwaves as objects defined based on clear and objective criteria, specifying for example a threshold value for a key physical variable such as temperature and a minimum length of time for that threshold to be exceeded. In contrast a phenomenological framing would give attention to how individuals and communities perceive and respond to periods of extreme heat including cultural interpretations and societal impacts. Of these two very broad possibilities, the structuralist approach to heatwave definition, or in other words the definition of heatwaves as physical as opposed to societal events, using quantitative measures of key heat variables is the approach adopted here.

Definitions which draw on physical variables are sometimes referred to as hazard based definitions. Heatwaves are also normally defined for the warm season only—the meteorological summer months of the northern or southern hemisphere of June, July, August and December, January, February respectively. Alternatively, heatwaves may be defined for the heat season if this has been identified for a specific location or region as a period extending beyond the traditional summer months. It is also important to remember that heatwaves are different to warm spells. In contrast to heatwaves, warm spells can occur any time of the year and relate to unseasonably warm weather. Heatwaves also distinguish themselves from warm spells in terms of the absolute temperatures or levels of heat achieved with such absolutes likely to engender a clear impact on human and natural systems. Phenomenologically speaking, a warm spell may be enjoyed whereas a heatwave, because of the level of heat, may induce feelings of discomfort and distress and cause significant disruption to daily life and well-being.

In contrast to the view that heatwaves are scientifically definable as pure physical events is the contention heatwaves are complex and messy thermal objects that in many ways defy definition or more positively lend themselves to being defined as multiple objects. This is because heatwaves are not just epistemologically distinct as they are comprehended through diverse forms of knowledge creation—e.g., climatology versus epidemiology—but are also ontologically varied, signifying they represent different concepts or realities to different individuals—e.g., youth versus the elderly (Boni et al., 2023). Seen in this way, rather than a definition grounded in the objectivity of climate science an interdisciplinary definition might be more appropriate recognising that heatwaves are simultaneously individual experiences, biophysical events, and socio-political phenomena (Boni et al., 2023). From a philosophical perspective this alternative view of heatwaves is attractive and deserves further exploration. However, while recognizing the multifaceted nature of heatwaves is valuable, practical implementation of an interdisciplinary based heatwave definition that incorporates phenomenological dimensions may prove quite a challenge. For this reason, this chapter continues along the lines of a structuralist approach to defining and describing heatwaves.

2.3 Criteria for a 'Good' Heatwave Definition

While the title of this section may appear to be 'a hostage to fortune', it is worth considering what the purpose of a definition is, plus some of the criteria that may render a definition, whether it be context specific or not, useful in a practical sense. First, in relation to the word heatwave, what do we mean by definition?

A heatwave definition, or any definition for that matter, is a set of words to describe its meaning and thus create its existence as an event. In essence, the set of words comprising a heatwave definition specifies the necessary properties for an event to 'qualify' as a heatwave. A corollary of this of course, is a heatwave definition will clearly communicate the properties associated with the use of the word heatwave. Using a set of words in this way to specify meaning leads to a definition of a term in a formal language context, or a formal definition. In the case of the word heatwave, this would mean that the definition of the term precedes its use. This contrasts with a 'natural' language definition of heatwaves because in this case, use of the word heatwave would precede its definition. Given this, any natural language-based definition would have to capture the actual everyday usage of the word. This is no mean feat. To obtain a natural language-based definition, a researcher would need to turn to something like a text corpus containing a collection of linguistic data related to word and expression usage. Examples of such English language corpora include the American National Corpus (ANC) and the British National Corpus (BNC). Once having identified which corpus to work with, the researcher would then undertake concordancing, aided by computer software, to locate every recorded occurrence of the word heatwave or a phrase containing it. Based on this, a natural language definition would be formed.

Of course, there are implications associated with the use of formal and natural language-based definitions. In the case of a 'formal' definition, the use of the word heatwave in spoken or written language must adhere strictly to its definition. If not, then the user of the word would be incorrect. In contrast, a natural language-based definition in failing to encapsulate the natural or everyday usage of the word, would be incorrect. Unsurprisingly a likely upshot of the contrasting approaches to defining a heatwave is confusion. This is because potentially, two definitions could be in circulation with the likelihood that 'experts' and the 'public', despite using the same words, will be speaking two different languages. Added to this is the potential that heatwave, as an English word, may not translate easily into a different language with the prospect that in some locations, especially those with a harsh hot climate, there may be no equivalent word. Although beyond the scope of this book, a furtive area of research could well be the exploration of how anomalous periods of heat are expressed in a range of languages and whether heatwave becomes a so-called borrowed word and socialised into non-English languages. One might also ask whether heatwave and the conditions it describes, has distinct geographical origins with a mainly mid-latitude temperate climate provenance, as it is in such climates that many of the impactful heatwaves have occurred and when they do occur are considered a bit of a surprise.

Having outlined the broad nature of what a definition is, the question arises as to what constitutes a good definition. It is probably useful to think of a definition as an asset as this can be applied to make conversations meaningful and assist with decision-making and assessment of situations in relation to risk management. Primarily, a good definition needs to be precise. An ambiguous or overly elaborate definition with too many elements or moving parts will lead to confusion and uncertainty. Added to this, a good definition should be falsifiable. How robust a definition will be when 'stress tested', will depend on its level of precision. The more precise a definition, the easier it will be for the identification of situations when it is falsified, thus paving the way for a revision of the definition. Ideally, a good definition should also be generative, creating value beyond its immediate intended use. In the case of a heatwave definition, the benefit might be in the form of a wider public understanding of the nature of this phenomenon and associated actions required to mitigate heatwave impacts, a good definition with generative qualities will create wider conversations.

Given the above, what might be the elements of a good heatwave definition? First to consider are the constituents of the definition and the values these must exceed to be considered 'unusual' in a statistical sense. In terms of the public communication of risk, the simpler the constituent variables and the critical threshold the better, as this will avoid scepticism arising from the lack of public understanding of an elaborate definition and the methodology underlying its derivation. Following identification of the defining variable(s) and an associated threshold value, a minimum duration for a heatwave needs to be specified. While this is usually measured in days, for sectors where sustained high temperatures at the sub-daily scale are critical, hourly data may be appropriate for definition development with a duration minimum above a sector relevant threshold specified. In sum, a good definition will need to be precise, falsifiable, based on physical variables or indices that represent heat or heat stress, have clearly specified variable or index threshold values and minimum duration criteria for exceedance of the nominated threshold. Such criteria of course are suited more to a physical definition of a heatwave event with alternative criteria necessary if a phenomological framing is preferred as a basis for definition.

2.4 Defining Heatwaves as a Physical Phenomenon

As for most extreme events, two broad approaches exist in relation to the definition of heatwaves as a physical phenomenon, namely those based on absolute criteria or relative criteria. Absolute criteria refer to the choice of a specific value for a climate variable, whereas relative criteria are values of climate variables associated with a particular percentile from a statistical distribution of values (e.g., 95th percentile). Definitions based on absolute criteria, such as a specific maximum temperature value (e.g., 35 °C) generally assume that the absolute value represents the tipping point beyond which there is a discernible physiological, ecosystem or societal response. For example, 35 °C as a threshold temperature may represent the point at which there is a thermo-physiological response amongst humans, with the heat stress associated with this temperature value engendering heat related strain measured by an increase in skin temperature or heart rate. Clearly, adopting such a fixed threshold as the criteria for defining a heatwave will always preclude the identification of heatwaves in climates where the absolute temperature never or rarely reaches the threshold level, despite anomalously warm periods being a feature of a location's climate.

The application of percentile values offers a good alternative to absolute values for defining heatwaves. This is because percentiles identify values that are extreme in a local context. Given this, a heatwave definition based on percentile values will always allow the identification of periods of anomalous heat regardless of the season and for any location. In the case of heatwaves, once chosen, the percentile is applied to the identification of all exceedances of that percentile value; absolute values are also applied in the same way. The exceedances may be single days or a sequence of successive days with the latter constituting a heatwave, if at least 2 days above a threshold value is set as the minimum duration criteria. Choosing which percentile forms the relative threshold is of course not straightforward. This is because selection should strike a balance between a percentile that allows the identification of events that persist beyond a single day, and events that are unrealistically long and exceed the timescale of the physical processes that may have generated the event; this is usually the synoptic time scale (see Chap. 6). In fact, when considering the range of possible relative threshold and duration criteria, the number of possible heatwave definitions could be considerable.

In applying percentiles as criteria, the time scale for which the percentile is calculated is critical. This is because a percentile value calculated at the seasonal or monthly time scale will be different to that for the daily scale. This is important if the heatwave definition needs to be time sensitive and consider adjustment of biophysical or social systems to the seasonally evolving pattern of heat. Calculating percentiles for annual or seasonal, or indeed monthly timescales and then applying these to the identification of heatwaves comprised of sequences of anomalously hot days will not capture any possible acclimatisation effect. Ideally, calculation of percentile values for the sub-monthly time scale should be undertaken. This could include, for example, 5-day (pentads) or 10-day (decads) periods, or ideally a moving multi-day window of say 15 days with the percentile value for the 15 day period centred on the calculation period—day 8 in the case of a 15 day window (Fig. 2.1) with possible de-seasoning of the temperature time series prior to percentile calculation to avoid possible biases (Brunner & Voigt, 2024). Equally important the percentile chosen should equate with a meaningful response in terms of the sensitivity of the biophysical or socio-economic system of interest. Therefore, a percentile value above which a clear trend in an impact variable value is detectable is best.

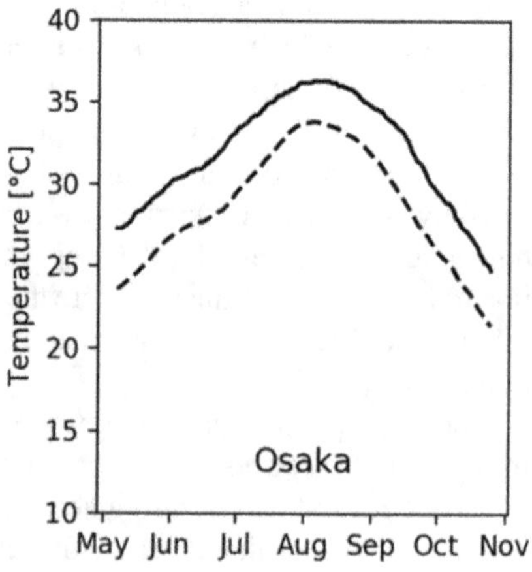

Fig. 2.1 Fifteen day running 90th percentile (solid line) and mean (dashed line) maximum temperature, May to November for Osaka, Japan. (Source: Author)

Ideally, the identification of critical threshold values derives from robust analyses of the historical relationship between an independent 'forcing' variable, such as maximum temperature, and a response variable that represents changing societal or ecosystem circumstances. Notwithstanding this and especially in the case of little data for a response variable, standard percentiles are applied as relative thresholds. Most commonly, these are the 90th and 95th percentiles although occasionally some analysts might apply values in the 97th to the 99th percentile range. Clearly the higher the percentile value the more limited will be the sample of potential heatwave days. For example, consider the number of potential heatwave days for a standard 3-month summer period of 92 days. The 95th percentile dictates that only 5% of days, equivalent to approximately 5 days, makes up the pool of potential heatwave days, which may well restrict identification of heatwave events to no more than two, assuming at least a 2-day duration requirement. At worst, no heatwave days may be identified for a relative threshold of greater than the 95th percentile if temperature extremes occur on isolated days with temperatures immediately before or after the temperature peak falling just below the 95th percentile value. Given this, selecting a less harsh 90th percentile value would seem sensible as this increases the number of potential candidate days to approximately nine. It also manages against the prospect of too many heatwave days or too long a sequence of days resulting from the application of an 80th or 85th percentile value as found by some researchers (Della-Marta et al., 2007; Perkins & Alexander, 2013). Although there is a tendency to adopt either an absolute or relative value approach to defining heatwaves, with the latter more predominant as we will discover later, definitions that use other statistical criteria might be worth contemplating if they turn out to be convincing determinants of the anomalous behaviour of any given response variable. Such alternatives might include the number of standard deviations from the long-term mean, as defined by the magnitude of standardised anomalies (e.g. z scores with units of standard deviation), for a fixed time window (a day, pentad or decad), or departure from the mean of a fixed absolute value (e.g. 5 °C above the mean). Robinson (2001) suggests that such alternatives represent departure from 'normal' social or economic conditions and may effectively connote in a definition the 'sociological components' of a heatwave.

2.5 Variables for Defining Heatwaves

Maximum and minimum temperatures alone or in combination often appear as the variables for which absolute or relative values are calculated. While such 'simple' variables have become the default for heatwave definition, there is an array of alternative heat related biometeorological indices as outlined in Chap. 4. Varying from simple empirical indices based on two or more climate variables, to complex numerically based ones that derive from a solution of the human heat balance, there are both advantages and disadvantages of using biometeorological indices for defining heatwaves. One of the advantages relates to representing the interaction between two or more climate variables that might be critical for determining levels of human thermal stress. For example, the heat index and apparent temperature, which combine measurements of temperature and atmospheric moisture, might be good candidates in this regard. A level of sophistication well beyond this are the complex human heat balance based or rational indices such as the Universal Thermal Climate Index (UTCI) or the Physiological Equivalent Temperature (PET), which yield values of a physiologically equivalent temperature or values of heat strain (e.g. skin temperature). Further information regarding these thermal indices is provided in Chap. 4.

Attractive as these alternatives may be, a good deal of pragmatism is required in choosing a climate variable or biometerological index as a basis for a heatwave definition. Such pragmatism relates to data availability as many of the alternative heat related indices are 'data hungry', ingesting variables that may not make up the standard set of climate variable measurements for a given location. As outlined by Davis et al. (2016) and McGregor and Vanos (2018), the choice of any one variable for modelling heat stress very much depends on the goal of the assessment process, with the upshot that careful consideration is required in selecting a variable or biometeorological index for heatwave definition. This is especially so if the intention of a researcher is to assess the impact of climate change on heatwave frequency. For such an exercise, the simplest of variables for defining heatwaves will work best given the uncertainty associated with estimating future values of non-thermal variables such as atmospheric moisture, solar radiation and cloud cover, all of which may be key input variables for 'complex' numerical biometeorological indices. The same could be said in relation to applying re-analysis data for calculating thermal indices used for defining heatwaves. Of course, the safest variables for defining heatwaves may well be maximum and minimum temperature as implied above. Indeed, for some locations, these variables may be the only ones available, and in relation to re-analysis products and the output from climate change models (see Chap. 3), the ones with the least uncertainty (greatest predictability). In the case of marine heatwaves measures of daily sea surface temperature are used either from in situ observations, satellite remote sensing platforms or model output data.

2.6 Heatwave Definitions

That there is a plethora of heatwave definitions is by no means an understatement. Setting aside the challenges associated with establishing how many definitions are in circulation, perhaps an impossible task for even the hardiest and persistent detectorists amongst us, heatwave definitions fall into two broad categories, namely meteorological or hazard-based and sector or impact-based. The former makes use of physical variables or indices selected without any specific impact variable in mind (e.g., number of heat related hospitalisations). They are therefore generic in nature and may take on the form of a formal or natural language definition. In contrast, sector-based heatwave definitions are designed with a specific impact variable in mind. Such sector or impact-based indices may draw on heuristics or the outcome of targeted analyses of the association between a heatwave variable or index and a chosen impact variable. Generally, sector-based heatwave definitions are of the formal type.

2.6.1 Meteorological Based Heatwave Definitions

Meteorological based heatwave definitions derive mainly from three sources. These are meteorological glossaries and the web sites of National Hydrological and Meteorological Service (NHMS) websites, the scientific literature focused on describing heatwave characteristics and international working groups tasked with developing consistent measures of heatwaves or periods of anomalous heat.

2.6.1.1 Glossary and National Meteorological Service Heatwave Definitions

Often the first port of call for someone hoping to find a precise definition of what constitutes a heatwave is a meteorological glossary or the definition of terms link on a National Meteorological and Hydrological Service's (NMHS) web site. Table 2.1 presents a non-exhaustive list of some heatwave definitions that can be found in meteorological glossaries and on NMHS websites.

Frustratingly, many of these definitions are generic in nature and subjective, lacking precise quantitative criteria. Such definitions would fail 'a good definition test' in relation to preciseness and falsifiability. For example, it is uncertain what abnormally and uncomfortably hot and unusually humid weather might mean for a specific location, without the specification of some defining values if such descriptors are used for heatwave definition. This of course relates to the wider problem of settling on a general definition of what an extreme event is, as discussed in the next chapter. In the case where specific heatwave criteria are used, these mainly take the form of absolute values (e.g. > 35 °C) only, or absolute values combined with a fixed departure of a meteorological variable from a point of reference (e.g. a Tmax 5 °C above a given climate statistic for a specified time of year or location). Further to absolute departure criteria, the specification of minimum duration for a heatwave demonstrates some variation between meteorological agencies. In some cases, an extended period is referenced but with no articulation of how long such a period might be. Some agencies do specify duration criteria but there is little inter-agency consistency. Interestingly, the meteorological service of the Netherlands adjusts the duration criteria downwards as the absolute value of the temperature departure criteria increases.

Associated with several of the definitions listed in Table 2.1 is additional information, the purpose of which is to qualify the generic elements of a definition or lend greater preciseness to the stated criteria. The IPCC definition, which states that

Table 2.1 Glossary and National Meteorological and Hydrological Services (NMHS) heatwave definitions (Source. Compiled by Author- all websites access 01/08/2023)

Definition Origin	Heatwave Definition	Source
American Meteorological Society Glossary	A period of abnormally and uncomfortably hot and usually humid weather	https://glossary.ametsoc.org/wiki/Welcome
United States National Oceanographic and Atmosphere Administration (NOAA) National Weather Service (NWS) Glossary	A period of abnormally and uncomfortably hot and unusually humid weather. Typically a heat wave lasts two or more days.	https://w1.weather.gov/glossary/
Intergovernmental Panel on Climate Change (IPCC) Glossary	A period of abnormally hot weather.	https://www.ipcc.ch/sr15/chapter/glossary/
World Meteorological Organisation	Marked unusual hot weather (Max, Min and daily average) over a region persisting at least two consecutive days during the hot period of the year based on local climatological conditions, with thermal conditions recorded above given thresholds.	https://www.wmo.int/pages/prog/wcp/ccl/opace/opace2/documents/DraftversionoftheGuidelinesontheDefinitionandMonitoringofExtremeWeatherandClimateEvents.pdf
WMO Regional Climate Centre on Climate Monitoring for Europe	Daily maximum temperature exceeds the 98th percentile over at least 3 consecutive days with maximum values of at least 28 °C	https://rcccm.dwd.de/DWD-RCCCM/EN/products/productdesc/productdesc_node.html
Australian Bureau of Meteorology	Three or more days in a row when both daytime and night-time temperatures are unusually high—In relation to the local long-term climate and the recent past.	http://www.bom.gov.au/australia/heatwave/about.shtml
Argentina National Meteorological Service	Both maximum and minimum temperatures above the corresponding 90th percentiles based on 1961–2010 reference period	https://www.smn.gob.ar/estadisticas
China Meteorological Administration	Maximum temperature greater than 35 °C for 2 consecutive days	http://www.cma.gov.cn/en2014/weather/Warnings/WarningSignals/201409/t20140919_261774.html
Danish Meteorological Institute	A period of at least 3 consecutive days when the average maximum temperature across more than fifty percent of the country exceeds 28 °C; a "warm wave" is when the same criteria are met for 25 °C	https://web.archive.org/web/20080723170544/http://www.dmi.dk/dmi/danmark_faar_varme-_og_hedeboelge

(continued)

Table 2.1 (continued)

Definition Origin	Heatwave Definition	Source
European Environment Agency	Periods of at least two consecutive days during summer (June to August) with a maximum apparent temperature and a minimum air temperature exceeding a local reference threshold derived from the 90th percentile in a particular region and at a particular time of year	https://www.eea.europa.eu/publications/europes-changing-climate-hazards-1/heat-and-cold/heat-and-cold-extreme-heat
Government of Canada	Period of more than 3 consecutive days with maximum temperature at or above 32 °C	https://www.canada.ca/en/environment-climate-change/services/sky-watchers/glossary.html
Korea Meteorological Administration	Period of 2 days or more when the daily maximum temperature exceeds 33 °C. Threshold of 33 °C is the 96th percentile of July and August daily maximum temperature averaged across all stations in Korea for the period 1973–2017	https://www.weather.go.kr/weather/warning/safetyguide_heat.jsp
Indian Meteorological Department	Heat Wave is not declared until maximum temperature of a station reaches at least 40 °C for the plains, 37 and 30 °C for coastal and hilly regions respectively. If temperature is less than these thresholds then a heatwave occurs if Tmax reaches a set value as a departure from the average Tmax for that time of the year	http://www.imdpune.gov.in/Weather/Reports/glossary.pdf
Meteo France	Three consecutive days of consistently very high temperatures throughout the day and night, with the levels varying depending on the region—Paris day 31 °C, night 21 °C; Marseille day 36 °C, night 24 °C	https://lameteorologie.fr/issues/2016/94/meteo_2016_94_45
Netherlands Meteorological Institute	At least 5 days with Tmax above 25 °C, at least 3 days with Tmax of >30 °C. Same criteria are used in Belgium and Luxembourg	https://www.knmi.nl/kennisen-datacentrum/uitleg/hittegolf
Sweden	At least five consecutive days with a daily high that exceeds 30 °C	Swedish Meteorological and Hydrological Institute[a]

(continued)

Table 2.1 (continued)

Definition Origin	Heatwave Definition	Source
Lithuania	A 'dangerous' heat wave event is a day when the daily maximum temperature is equal to or exceeds 30 °C between May and September	Lithuanian Hydrometeorological Service[a]
Latvia	At least 6 consecutive days when the daily maximum temperature exceeds the 90th percentile of daily maximum temperature calculated for the season between May and August in the 30-yr period 1961–1990.	Latvian meteorological Service[a]
Estonia	A dangerous heat wave. Event, daily maximum temperature is equal to and exceeds 30 °C between May and August for two consecutive days and covers 30% of the territory.	Estonian Environment Agency[a]
Finland	Heat warning rather than a heatwave defined as when the, daily mean temperature exceeds 20 °C and the daily maximum temperature exceeds 27 °C for a day.	Finnish Meteorological Institute[a]
Hellenic Meteorological Service	Three consecutive days at 39 °C or more. In the same period the minimum temperature is 26 °C or more.	http://www.emy.gr/emy/el/
South African Weather Service	At least 3 days with maximum temperature 5 °C above the average maximum temperature of "the hottest month"	http://www.weathersa.co.za/home/weatherques
United Kingdom Met Office (UKMO)	A period of at least three consecutive days with daily maximum temperatures meeting or exceeding the heatwave temperature threshold. The threshold varies by UK county	https://www.metoffice.gov.uk/weather/learn-about/weather/types-of-weather/temperature/heatwave

[a]Sourced from: https://datastore.copernicus-climate.eu/documents/sis-european-health/C3S_Health_Application_PUG_Heatwave_days_ERA5_v2.pdf

a heatwave is a period of abnormally hot weather, caveats this by pointing out that heatwaves and warm spells possess overlapping definitions. Further, the IPCC directs the reader to the IPCC definition for an extreme weather event. This implies that 'abnormally hot', might mean something extreme or rare for a particular place or time of the year, with rare considered as occurring less than 10% of the time, equated with the 90th percentile value of a variable from an observation-based probability density function. So, by inference, the IPCC heatwave definition is a relative one that considers a heatwave as being a period of unspecified length, presumably at least 2 days, for which the level of heat reached is greater than the recorded 90th percentile value of a 'heat variable'.

The Australian Bureau of Meteorology (BoM) is quite open about the challenges of settling on a heatwave definition. This assists with managing some of the frustration associated with understanding the more generic of the NMHS heatwave definitions. In relation to thresholds, the BoM states unequivocally there is no single temperature threshold for a heatwave in Australia. Rather a more regional and time relevant approach is adopted based on comparing the forecast maximum and minimum temperatures for each 3-day period in the coming week (e.g. Monday–Wednesday, Tuesday–Thursday) to the 'normal' temperatures expected for that location at that time of year, and to observed temperatures over the last 30 days. Armed with this information, the BoM calculates the Excess Heat Factor (EHF) (see Chap. 4) and attempts to classify a potential heatwave situation into one of three classes based on intensity, by considering the forecast temperature change, relative to the previous 30 days. In addition, preparedness and the potential impacts on people and infrastructure are considered when deciding on the nature of an impending heatwave and how to communicate this to the public. Accordingly, heatwaves might be classified as either low-intensity (most frequent type of event which most people can cope with), severe (less frequent, poses a challenge to the vulnerable) or extreme (rarest kind affecting reliability of infrastructure and posing a high level of threat to the less precautious in the population).

The UKMO's heatwave definition, like many others, is quite generic. However, information provided in a recent publication assists with clarification, in quantitative terms, of what constitutes the official UKMO heatwave definition (McCarthy et al., 2019). For the UK a heatwave is specified as occurring when a location exceeds a specified maximum temperature for at least three consecutive days with the threshold varying by broad geographic region from 25–28 °C, with northern regions possessing the lowest threshold (Fig. 2.2). To inform the identification of thresholds, a relative percentile approach was adopted with 'unusual' defined as exceeding the 90th percentile maximum temperature value for the warmest part of the year (late July to early August). When the thresholds were originally derived based on the reference period 1981–2010, the value for London was set above that for the rest of the country at a Tmax of 28 °C, presumably to reflect the 'warmer' urban climate of London. Importantly, and acknowledging that heatwave thresholds should be sensitive to the impacts of climate change, in 2022 the UK Met Office updated the climatological reference period from 1981–2010 to 1991–2020, with the upshot being six counties were moved up into the 28 °C threshold category alongside London with two other counties also entering higher categories (Fig. 2.2). As one might guess, for a mid-latitude temperate country such as the UK, the lowest threshold of 25 °C does not equate with the local 90th percentile maximum temperature value in northern regions such as Scotland, where such values are generally below 21 °C. Given this, the UKMO included the constraint that for regions with 'low' 90th percentile maximum temperature values, a heatwave occurs only when

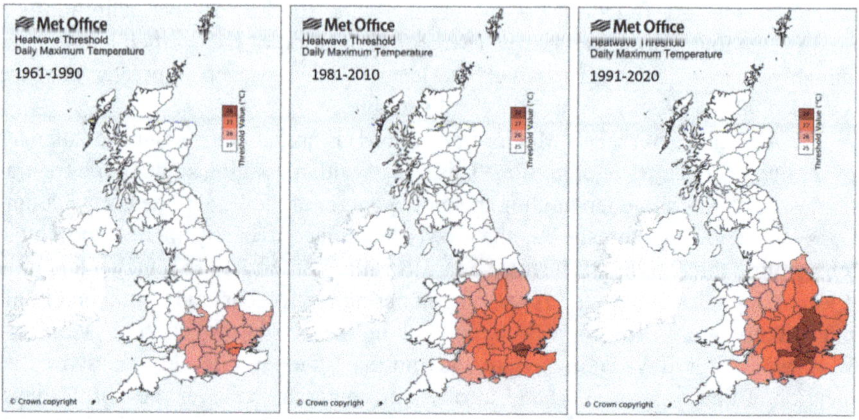

Fig. 2.2 Variation of heatwave thresholds over time. The far-right figure shows the geographical distribution of the heatwave thresholds as a result of the 2022 update to the climatological reference period. The middle figure shows the original thresholds based on the 1981–2010 reference period while the figure on the left portrays the spatial distribution of thresholds if the updated thresholds of 2022 were applied to a 1961–1990 reference period. Source: UK Met Office. Available at: https://www.metoffice.gov.uk/about-us/press-office/news/weather-and-climate/2022/heatwave-threshold-changes. Used under the UK open government licence (OGL) for public sector information https://www.nationalarchives.gov.uk/doc/open-government-licence/version/3/

maximum temperatures are greater than 25 °C, the lower limit of a 'summer day' as defined in the climatological literature (McCarthy et al., 2019). Interestingly for a mixture of reasons, the UKMO has chosen not to include any minimum temperature criteria in their official heatwave definition. These include simplicity, the fact that high minimum temperatures generally lag the same for maximum temperature and the public interest appears directed towards maximum temperature (McCarthy et al., 2019).

The situation regarding heatwave definition in India is interesting. Here, the Indian Meteorological Department (IMD) and the Indian Institute of Tropical Meteorology (IITM), both of which belong to the Indian Ministry of Earth Sciences, have developed different criteria. The IMD have developed a range of heatwave criteria based on absolute maximum temperature thresholds and departures from 'normal' maximum temperature, with normal referring to the pre-monsoon warm season (March to June) for the period 1970–1999. Acknowledging the diverse physical geography of the country, the IMD only consider heatwave conditions to have occurred if maximum temperatures have reached 40 °C for the plains and 37 °C and 30 °C for coastal and hilly regions respectively. Based on the departure from normal maximum temperature and when maximum temperature of a station is <40 °C, the IMD consider a heatwave to be associated with a maximum temperature departure of 5–6 °C, while a severe heatwave is associated with a departure from normal of 7 °C or more. In the case of a location with a normal maximum temperature > 40 °C, a departure of 4–5 °C from normal is referred to as a heatwave, with a severe heatwave occurring when the departure from normal is >6 °C. Interestingly, the reduction of the level of maximum temperature departure for a higher reference temperature is similar to the approach adopted by the Netherlands meteorological service. For the Indian plains, a heatwave is declared when actual maximum temperature is greater than 45 °C with a severe heatwave related to an actual maximum temperature greater than 47 °C for at least 2 days, with the requirement that the defining temperatures are recorded at two meteorological stations or more within a meteorological subdivision. In contrast to most of the definitions presented in Table 2.1, the IMD heatwave definition permits a single day of high temperature to be declared as a heatwave, in addition to multiple consecutive days of high temperature (Nori-Sarma et al., 2019). This is like the heatwave definition of the complementary IITM, which also grants a single day as a heatwave. However, unlike the IMD definition, that of the IITM combines relative thresholds and absolute departures to identify hot days, and heatwave and severe heatwave conditions. A further contrast relates to the calculation of temperature statistics for heatwave identification. Whereas the IMD use instrumental data for the pre-monsoon months of March to June based on the reference years 1970–1999 for calculating the normal maximum temperature, the IITM calculates daily percentiles of gridded maximum and minimum temperature for the period 1981–2010. No doubt, the contrasting reference periods result in the IITM reference values being higher than those of the IMD for equivalent temperature statistics, because of recorded increases in temperature across the two reference periods. According to the IITM, a hot day occurs if the maximum temperature is ≥39 °C and minimum temperature is ≥ the 90th percentile of the observed minimum temperature for a given day. If maximum temperature is ≥ the 95th percentile of the observed daily maximum temperature with the actual maximum temperature ≥ 39 °C, and the maximum temperature departure from normal is ≥3.5 °C, this constitutes a heatwave situation. For a severe heatwave, the maximum temperature must be ≥ the 99th percentile of the observed maximum temperature for a given day with the actual maximum temperature ≥ 39 °C and the maximum temperature departure from normal exceeding 5.5 °C. In addition to these rather complex definitions and irrespective of any relative threshold criteria, the IITM offers two simple absolute criteria for heatwave and severe heatwave conditions, those being the maximum temperature ≥ 44 and 46 °C respectively (Mandal et al., 2019). While the IITM in developing its set of heatwave criteria have taken into consideration those of the IMD, the fact that there are two agencies with slightly different heatwave definitions is perhaps not ideal. From a public messaging point of view, having two definitions in circulation may lead to some confusion amongst the public, notwithstanding the contrasting short-term and extended range forecasting time scales prioritised by the IMD and IITM respectively.

Like the UKMO heatwave definition, that for France considers the impact of the within country variation in the climatology of very high temperatures with heatwave threshold levels varying depending on the region. However, this is where the similarity stops as the regional threshold levels for France have been set with reference to the climatological relationship between various temperature metrics and mortality (Pascal et al., 2013). These thresholds are used as triggers for action in the context of the French heatwave warning system and therefore differ from heatwave definitions based on a statistics of extremes approach using relative or absolute thresholds for a given meteorological variable (see Chap. 13). This distinction is important, especially as heatwave thresholds are likely to vary from sector to sector, or even within a sector thus contrasting with those set on a purely meteorological basis. In this regard, the situation for the UK is pertinent as here there is a clear distinction between the official UKMO heatwave thresholds described above and those set using epidemiological evidence (Masato et al., 2015) and applied in the Heatwave Plan for England (see Chap. 13). Preference for a clear distinction between purely meteorologically (hazard) based heat wave criteria and those set with a specific sector (impact-based) in mind is evident for Canada. Here there appears to be a clear shift in the direction of using health data for identifying meteorological thresholds to identify heat events as opposed to heatwaves (McLean et al., 2018). Perhaps for reasons of pragmatism, or less politely nervousness, some NHMS have not declared an official heatwave definition. Japan and Germany, where there is clear evidence that heatwaves exact a heavy societal toll, are good examples of this. Currently in Japan, the Japan Meteorological Agency (JMA) only issues outlooks for extreme high temperature and extreme-high temperature-forecasts using a 35 °C maximum temperature threshold for the next day. Furthermore, this service is restricted purely to the warm season. Complementing the physical threshold is the impact based WBGT maximum temperature of 33 °C which has been identified as the tipping point for heat stroke weather (Oka et al., 2023). As yet, the German Weather Service (DWD) has not officially defined what a heatwave is. Rather the DWD has focused its efforts on developing a heat health warning system based on exceedance criteria of a biometeorological index (Matzarakis et al., 2020).

There is some evidence that definitions have changed over time as revealed by a comparison of the WMO definition listed in Table 2.1, with a previous definition. Formerly the WMO defined a heatwave as occurring when the daily maximum temperature exceeds the average maximum temperature, for the reference period 1961–1990, by 5 °C for five consecutive days or more. The shift of the WMO definition from a specific criteria-based definition to a more generic one represents an acknowledgement that being prescriptive does not recognize the possible role of regional/local and sector context factors. Relevant also is the move away in the more recent definition from just maximum temperature and the inclusion of the 'unusualness' of average and minimum temperature conditions, perhaps a recognition that heatwaves are not strictly a daytime phenomenon. Added to this, the updated definition mentions thresholds without specifying any particular values or types, a possible nod to the fact that thresholds can be situation specific requiring identification based on contextual information. Further, the former WMO definition states a climate normal reference period of 1961–1990 with no such specification in the recent definition. This is no doubt testament to the definition challenges associated with a non-stationary climate system as acknowledged by the UK Met Office in relation to its updated heatwave thresholds, with former reference periods no longer appropriate given the observed increase in global and regional temperatures. Having said that, the problem remains of what constitutes unusual hot weather, especially whether defined by applying a statistics of extremes approach to one or several climate variables, or whether the specification is in terms of societally relevant 'hot' weather. Interestingly, the WMO Regional Climate Centre on Climate Monitoring for Europe has produced an objective criteria-based heatwave definition (Table 2.1). While such contrasting approaches at the international and regional levels may be by design, reflecting a greater understanding of what constitutes a heatwave at the regional level of Europe, it does point to some inconsistency in terms of a whole of agency approach to defining a meteorological phenomenon.

2.6.1.2 Research Literature Based Heatwave Definitions

Spurred on by significant events such as the 1995 Chicago and the 2003 European heatwaves, as well as the prospect of greater heatwave frequency under climate change, the research literature on the analysis of heatwave characteristics and trends has grown substantially over the last couple of decades. Unsurprisingly, the associated literature is rich in heatwave definitions, or heatwave indices, as defining what constitutes a heatwave is the first problem any climatological study focused on this phenomenon needs to solve; the age-old 'definition of terms' issue in research (Schultz, 2009). To give the reader a sense of the diversity of heatwave definitions in circulation, a selection of meteorological based definitions from the post-2000 literature is presented in Table 2.2. Notably a number of these definitions (Anderson & Bell, 2011; Della-Marta et al., 2007; Fischer & Schar, 2010; Meehl & Tebaldi, 2004) have been applied widely to climatological studies of heatwaves. In contrast to many of the definitions provided by NMHS (Table 2.1), those applied in the scientific literature are quantitative, specifying clear absolute or relative exceedance

Table 2.2 A sample of heatwave definitions from the post-2000 climate science literature

Study	Heatwave definition	Type of threshold	Reference period	Heatwave aspect measured
Huth et al. (2000)	HW is defined based on two thresholds (T1 and T2) being met with T1 and T2 set at 30 and 25 °C respectively. Tmax $\geq$ T1 for $\geq$3 consecutive days; mean Tmax for the entire period $\geq$ T1; Tmax $\geq$ T2 each day.	Absolute	All year round	Frequency, intensity, duration
Robinson (2001)	Defined based on minimum or maximum daily temperature. Tmin > 26.7 °C or Tmax > 40.6 °C for 2 days of more.	Absolute	All year round	Intensity, duration

(continued)

Table 2.2 (continued)

Study	Heatwave definition	Type of threshold	Reference period	Heatwave aspect measured
Frich et al. (2002)	HW defined as ≥6 consecutive days with Tmax > 5 K above each calendar day's mean Tmax. Referred to widely in the literature as the heatwave duration index (HWDI).	Mixed	All year round	Duration
Meehl and Tebaldi (2004)	Based on three criteria using two location-specific maximum temperature thresholds (T1 and T2). T1 = 97.5th percentile of observed and modelled maximum temperatures. T2 = 81st percentile of maximum temperature. Given T1 and T2 a heat wave = longest period of consecutive days that jointly satisfies: (i) daily maximum temperature > T1 for at least 3 days, (ii) the average daily maximum temperature > T1 for the entire period, and (iii) the daily maximum temperature > T2 for every day of the entire period. Similar in form to Huth et al. (2000) but applies percentiles as T1 and T2.	Relative	Seasonal (e.g. June—August)	Duration
Meehl and Tebaldi (2004)	"Worst heat event" defined as 3 consecutive nights with the highest mean minimum temperature per year.	Absolute	All year round	Intensity

(continued)

Table 2.2 (continued)

Study	Heatwave definition	Type of threshold	Reference period	Heatwave aspect measured
Della-Marta et al. (2007)	HW is 3 consecutive days >80th percentile of daily maximum temperature. (1906–2003). (5 day centred calendar window)		June–August	Frequency
Fischer and Schar (2010)	Maximum temperature > the calendar day 90th percentile (15 day centred calendar window) for at least 6 days.	Relative	All year round	Frequency, intensity, duration
Anderson and Bell (2011)	Local daily mean temperature ≥ 95th percentile for more than 2 days for the period May 1 to September 30, 1987–2005.	Relative	May–September	Intensity, duration, timing
Pezza et al. (2012)	Monthly 90th percentile climatology. Maximum temperature ≥ 90th percentile of the maximum temperature for the month in which the heatwave begins for a minimum of 3 consecutive days. Minimum temperature ≥ 90th percentile of the minimum temperature for that month on the second and third days of the heatwave.	Relative	January–March	Frequency, intensity, duration
Stefanon et al. (2012)	Temperature maximum >95th percentile value of the local probability density function for the period 1950–2009 for at least four consecutive days. (21 day centred calendar window)	Relative	All year round	Duration, spatial extent

(continued)

Table 2.2 (continued)

Study	Heatwave definition	Type of threshold	Reference period	Heatwave aspect measured
Perkins and Alexander (2013)	≥ 3 consecutive days with calendar day Tmax ≥ calendar day 90th percentile (15 day centred calendar window); the same for Tmin. The EHF of Narin and Fawcett (see below) was also used.	Relative	All year round	Frequency, intensity, duration
Russo et al. (2014)	≥ 3 consecutive days with Tmax ≥90th percentile (31 day centred calendar window)	Relative	All year round	Frequency, intensity
Nairn and Fawcett (2015)	≥ 3 consecutive days with positive Excess Heat Factor (EHF) based on year-round data	Absolute	All year round	Intensity, duration, spatial extent
Schoetter et al. (2015)	Tmax ≥98th percentile for ≥3 consecutive days.	Relative	May–October	Intensity, duration, spatial extent, severity
Barbier et al. (2018)	Tmax or Tmin anomaly exceeds the 90th percentile of the distribution over the period 1950–2010 for at least 3 consecutive days. Tmax and Tmin data a prefiltered using a 90-day high pass filter (baseline period 1950–2010)	Relative	March–July	Frequency, intensity, duration, spatial extent
Mazdiyasni et al. (2019)	Heatwave defined as 2–10 hottest consecutive days per year using daily mean temperature (Tmax and Tmin not included in HW definition).	Absolute	All year round	Frequency, intensity, duration

(continued)

Table 2.2 (continued)

Study	Heatwave definition	Type of threshold	Reference period	Heatwave aspect measured
Su and Dong (2019)	Three heatwave types defined namely: Compound: ≥ 3 consecutive days with simultaneous occurrence of Tmax ≥90th percentile and Tmin ≥90th percentile). Daytime: ≥ 3 consecutive days with Tmax ≥90th percentile Night time: ≥ 3 consecutive nights Tmin ≥90th percentile (15 day centred calendar window, baseline period of 1964–81)	Relative	All year round	Frequency, intensity, spatial extent
Ullah et al. (2019)	Two thresholds (T1 and T2) for daily Tmax and Tmin distributions were used to define daytime and night time HW events, respectively. T1 = 90th percentile; T2 = 75th percentile of the daily Tmax and Tmin distributions. A daytime/night time HW event is defined as the longest period of successive days/nights fulfilling the following three conditions: (1) the daily Tmax/Tmin > T1 for a minimum of 3 consecutive days/nights, (2) the average daily Tmax/Tmin > T1 for the entire period, and (3) the daily Tmax/Tmin > T2 for each day/night of the whole event. 1980–2016	Relative	May–September	Frequency, intensity, duration, timing

Source: Compiled by Author
Severity = (product of duration (D), mean extent (E) and mean intensity (I)).

criteria for the warm season or all year round. A feature of the definitions in Table 2.2 is diversity. While such diversity is testament to the inventiveness of the heatwave science community, in some ways it is a bit of a curse because very few studies adopt a common heatwave definition. This poses great challenges if inter-comparisons of temporal or spatial heatwave climatologies are required, even for the same region, let alone for different countries or continents. The same applies to studies focused on establishing the impacts of climate change on heatwave trends. Such an 'apples and oranges' situation makes scientific assessments, as undertaken under the auspices of the IPCC, difficult with assessors needing to make a judgement call balancing evidence and agreement. That aside, there are some broad commonalities across the definitions presented in Table 2.2. Almost all use values of maximum and/or minimum temperature, no doubt due to their wide availability as standard climate variables. Further, of these two 'heat' descriptors', Tmax appears to be the primary variable with the reason often proffered that most impacts arise from 'daytime' heat—a position largely adopted in a period when the importance of nocturnal heat as a feature of heatwave climates has been overlooked.

Most definitions listed in Table 2.2 stipulate a minimum of 3 consecutive days as the duration criteria, that is, exceedance of the stated threshold must persist for 3 days, because this length of time is commonly regarded as the period over which heat-related impacts may manifest (Perkins & Alexander, 2013). This is certainly the case in the health sector where impacts emerge during extreme heat events only lasting a few days. Interestingly, two of the definitions (Fischer & Schar, 2010; Frich et al., 2002) specify a minimum duration of 6 days. This is perhaps a little harsh, limiting any heatwave climatology to a consideration of only relatively prolonged events and at the same time, discounting the climatological and societal importance of short sharp or acute periods of anomalous heat. Such unforgiving duration criteria could also produce several zero heatwave years in any climatological description, thus downplaying the possible climate and societal importance of this phenomenon and creating statistical challenges for any study concerned with establishing trends in heatwave characteristics.

Commonly temporal studies of heatwave frequency, intensity and duration have been the target of heatwave definitions. However, some definition developers have added criteria (not shown in Table 2.2) related to spatial extent. This is because their focus is on either geographically large events, or the construction of spatial heatwave climatologies (Barbier et al., 2018; Stefanon et al., 2012; Su & Dong, 2019). Overall, definitions based on a relative threshold tend to dominate. Perhaps at a stretch, the chronology of definitions in Table 2.2 indicates a shift over the last decade to definitions based on percentile as opposed to absolute values for the exceedance criteria. Generally, justifications of definitions based on absolute criteria relate to the chosen exceedance value representing a critical threshold beyond which noticeable societal impacts occur, regardless of location. In contrast, the justification of relative thresholds is with regard to relevance at the local level in addition to their portability and possible universal application.

In calculating the relative threshold value, some definitions employ a time-varying moving window approach. Fenner et al. (2019) refer to these as 'dynamic thresholds'. That aside, of interest is the spread of centred calendar day windows used. These range from relatively short 5-day (Della-Marta et al., 2007) through medium 15-day (Fischer & Schar, 2010; Perkins & Alexander, 2013) to longer 31-day windows (Russo et al., 2014). To clarify, the percentile value for the day at the centre of a specified window of time, is calculated based on a set period before and after the calendar day of interest. For example, the calculation of a 15-day calendar window 90th percentile Tmax value for August 20 is based on the daily Tmax values for the period August 13–27. If Tmax data were available for a 30-year period (e.g.1971–2000), then the sample size for calculating the 90th percentile would be 450 days (15 days × 30 years). Once calculated from the 450-day sample, the Tmax value is assigned to August 15 as the climatological 90th percentile value for that calendar day. To calculate the same for August 16, the 15-day window is moved forward by one calendar day and the process repeated.

As noted above, a range of time-varying moving windows for calculating seasonally dependent percentiles appear in heatwave studies. Usefully, Barbier et al. (2018), in analysing the nature of heatwaves across the Sahel region, provide an assessment of the impact of the nature of the moving window on analysis results. They compared analysis outcomes for a 90th percentile relative threshold based on 15-day moving window using 90-day pre-filtered temperature data with an unfiltered 15-day window value of the 90th percentile and a fixed-percentile threshold for the reference period of March to July. In doing so, they found that the pre-filtered 15-day window approach revealed a reasonably constant frequency of heatwaves throughout March to July with little difference in heatwave characteristics (e.g., intensity) across these months. In contrast, the 90th percentile defined using 'raw' non-filtered temperature data and a 15-day window approach, produced longer lasting and larger heatwaves. Further for these events, the temperature anomalies were smaller, while their mean Tmax, Tmin, and Tmean values were higher. Lastly, by using raw temperatures and a fixed non-seasonally dependent 90th percentile threshold for March to July, the authors found that heatwave occurrence is largely a product of the annual temperature cycle such that Tmax (Tmin) heatwaves reach a peak in April–May (May—June) with few heatwaves in March and July, before and after the spring temperature peak respectively.

Further to the absolute and relative criteria presented in Table 2.2, the literature includes a range of other heatwave definition criteria in the form of threshold values for various climate indices. However, as noted by Perkins and Alexander (2013), many of these are unsuitable as the building blocks for heatwave definitions—see Table 2.3.

2.6.1.3 International Working Groups

As noted earlier, the spectre of climate change has focused attention on the importance of climate extremes as a feature of a location's or region's climate, with a large number of scientific papers on this subject evidence of this. The antecedents of heightened publication activity most likely trace back to the explicit recognition in

Table 2.3 Weakness of some common climate indices for heatwave definition

Climate index	Definition of threshold	Reason for unsuitability
Tmax95pct	The calendar day of the 95th percentile of the maximum temperature (Tmax) calculated for a 15-day period with the period centred on the day of interest; each day has a different percentile	The 95th percentile threshold is high and likely reveal a low number of heatwaves a year especially if defined as 3 days or more; the number of events could be less than one per year if measured on the climatological scale. Setting a high threshold is also likely to lessen the inclusion of consecutive days in a heatwave with resultant events limited to very short periods, especially if the data record is short (e.g. for summer the 95th percentile is equivalent to 4.6 days)
TMin95pct,	The calendar day of the 95th percentile of the minimum temperature (Tmin) calculated for a 15-day period with the period centred on the day of interest; each day has a different percentile	Same as for TX95pct
TX35deg (TX40deg),	Threshold fixed at 35 °C (40 °C) for Tmax	This fixed threshold index is climate regime sensitive producing heatwave events for arid subtropical to tropical climates but only rarely for mid-latitude temperature climates. Cannot effectively represent heatwave occurrence across a range of climate regimes. Best used for a fixed location or single climate regime. Fixed threshold for a particular sector (e.g. health) ignores possible acclimatisation/adaptation/vulnerability effect
TN20deg (TN25deg)	Threshold is fixed at 20 °C (25 °C) for Tmin	As for TX35 and TX40 but mainly for subtropical to tropical climates and only rarely for mid-latitude temperature climates
CHT	Threshold set as combination of hot day (Tmax = 35 °C) followed by a tropical night (Tmin = 20 °C)	While the concept of combining a hot day and a tropical night is attractive this index suffers from the same issues as for other indices (e.g. Tx35, Tx40) that use fixed thresholds
Ta90pct (Ta95pct),	The calendar day of the 90th (95th) percentile of apparent temperature (Ta) calculated for a 15-day period with the period centred on the day of interest; each day has a different percentile. Apparent temperature is calculated according to Fischer and Schar (2010)	While the 95th percentile threshold might be high, the 90th percentile threshold stands more chance of identifying heatwaves. Apparent temperature requires measures of relative humidity that may need to be derived from other variables if this measure of atmospheric moisture is not available

(continued)

Table 2.3 (continued)

Climate index	Definition of threshold	Reason for unsuitability
WSDI	Warm spell duration index. Warm spell occurs when Tmax exceeds the 90th percentile of Tmax (calculated based on a 5 day window) for at least 6 consecutive days	The specified duration threshold of 6 days is very long and not likely to reveal many warm spell events
HWDI	Heatwave duration index. Heatwave occurs when Tmax exceeds the mean Tmax by 5 °C for at least 6 consecutive days	Same as for WSDI with concern that the 5 °C exceedance of Tmax seems arbitrary plus the index yields incomparable results for different climate zones because the 5 °C exceedance is fixed and not location/region specific. Its application in regions where the annual amplitude of temperature and inter-daily temperature variability is small, such as in equatorial/tropical latitudes, is problematic

Adapted from Perkins and Alexander 2013

the IPCC's 1995 Second Assessment Report (SAR) that data on climate extremes was inadequate. Such a paucity of data made responding to policy relevant questions such as, 'is there evidence for an increase in climate extremes' and 'how best to develop indicators and indices to measure climate extremes using observational and model data', difficult. This situation served as an impetus for the organisation in 1997 of a joint CLIVAR/GCOS/WMO workshop on indices and indicators for climate extremes. The principal aspiration of the workshop was that its outputs would lay the basis for a systematic analysis of the question 'has the climate become more variable or extreme', with the IPCC's Third Assessment Report (TAR) being the ideal recipient of workshop outcomes (Karl et al., 1999; Nicholls & Alexander, 2007). With similar aims to those of the CLIVAR/GCOS/WMO workshop, the Asia-Pacific Network (APN) for Global Change Research also convened workshops on climate extremes, but with a specific focus on Southeast Asia and the South Pacific (Manton et al., 2001).

While the APN workshop did not, the CLIVAR/GCOS/WMO workshop proposed a Heat Wave Duration Index (HWDI) for assessing changes in heatwave characteristics. Defined in two parts, the HWDI comprised a measure of the proportion of consecutive days in a month/season for which Tmax exceeded a specified percentile value, and the number of times sets of consecutive days occurred, with the constraint of a 2-day duration minimum (Folland et al., 1999). Frich et al. (2002) subsequently built on the idea of a HWDI, however, unlike that recommended by the CLIVAR/GCOS/WMO working group it was based on absolute criteria (Table 2.2). Although the Frich et al. (2002) HWDI was eventually superseded by other indices, the situation of two heatwave indices in circulation 'early on', possessing the same name, but constructed in different ways, was not ideal. In many ways this 'tribulation' captures what lies at the heart of the defining heatwaves problem, that is diversity of approaches, lack of comparability and potential confusion amongst heatwave information stakeholders and the public.

Perhaps frustrated by the potpourri of heatwave and climate extreme indices in general, a group of climate scientists assembled under the auspices of the joint World Meteorological Organization Commission for Climatology (CCl)/World Climate Research Programme (WCRP) project on Climate Variability and Predictability (CLIVAR), to form the Expert Team on Climate Change Detection, Monitoring and Indices (ETCCDMI). Rebranded as the ETCCDI and now known as the Expert Team on Sector-Specific Climate Indices (ET-SCI), its primary goal was the international coordination and development of a suite of climate change indices, including metrics of temperature extremes (Alexander et al., 2006; Donat et al., 2013). Of the 27 climate indices developed by ETCCDI, 17 were specific to temperature extremes (Table 2.4). However, as appetising as such an unprecedented menu of indices might have been, most if not all were, in their original form, unsuitable for defining heatwaves. This is because the ETCCDI indices measure extremes at the annual level, as opposed to individual events; note many of the indices have percentage of time or number of days annually as the unit of measurement. That aside, some of the ETCCDI temperature indices (e.g. SU, TR, TXx, TNx, TX90p, TN90p, WSDI) have been adopted singularly or in combination to form heatwave indices (Perkins & Alexander, 2013). For example, the relative threshold of the 90th percentile of maximum and/or minimum temperature (TX90p, TN90p), along with duration criteria, have been applied widely to fashion heatwave indices (see Table 2.2) and descriptors for various heatwave aspects/characteristics (Sect. 2.7).

Table 2.4 Temperature indices recommended by the ETCCDI calculated based on GHCN (Global Historical Climatology Network)-Daily station data. For the calculation of percentile threshold values the base period 1961–90 was used

Identifier	Indicator name	Indicator definiton	Units
TXx	Hottest day	Monthly maximum value of daily max temperature	°C
TNx	Warmest night	Monthly maximum value of daily min temperature	°C
TXn	Coldest day	Monthly minimum value of daily max temperature	°C
TNn	Coldest night	Monthly minimum value of daily min temperature	°C
TN10p	Cool nights	Percentage of time when daily min temperature < 10th percentile	%
TX10p	Cool days	Percentage of time when daily max temperature < 10th percentile	%
TN90p	Warm nights	Percentage of time when daily min temperature > 90th percentile	%

(continued)

Table 2.4 (continued)

Identifier	Indicator name	Indicator definiton	Units
TX90p	Warm days	Percentage of time when daily max temperature > 90th percentile	%
DTR	Diurnal temperature range	Monthly mean difference between daily max and min temperature	°C
GSL	Growing season length	Annual (1st Jan to 31st Dec in NH, 1st July to 30th June in SH) count between first span of at least 6 days with TG > 5 °C and first span after July 1 (January 1 in SH) of 6 days with TG < 5 °C (where TG is daily mean temperature)	Days
ID	Ice days	Annual count when daily maximum temperature <0 °C	Days
FD	Frost days	Annual count when daily minimum temperature <0 °C	Days
SU	Summer days	Annual count when daily max temperature > 25 °C	Days
TR	Tropical nights	Annual count when daily min temperature > 20 °C	Days
WSDI	Warm spell duration index	Annual count when at least six consecutive days of max temperature > 90th percentile	Days
CSDI	Cold spell duration index	Annual count when at least six consecutive days of min temperature < 10th percentile	Days

Adapted from: Donat et al. 2013

2.6.1.4 Inter-comparisons of Meteorological Based Heatwave Definitions

Contemplating the vast range of heatwave definitions, the adage of 'you get what you asked for' comes to mind, as outputs from any analysis will depend on the heatwave metric chosen. For this reason, researchers have assessed the dependence of research results on the heatwave definition chosen. In a study that applied 15 heatwave indices to an assessment of the trend in heatwave frequency across the United States, Smith et al. (2013) found the frequency, trends, and geographical patterns of heatwaves are definition dependent. For example, heatwave definitions using absolute thresholds produced a greater range in the average number of heatwave days across the US compared to those based on relative criteria. Further, not all heatwave indices indicated the same results in terms of the geographical pattern of heatwave frequency; seven indices denoted the highest frequency of heatwave occurrence for the Northwest region while seven different indices signalled the same for the Southeast region. Added to this, indices including both temperature and humidity only showed upward trends in heatwave frequency for high humidity regions of the conterminous US. Studies in China also indicate heatwave index choice is important for conclusions drawn. You et al. (2017) demonstrate that of 16 heatwave indices, three reveal annual heatwave frequencies less than 3 days while six point to annual averages greater than 10 days for the period 1961–2014. Worryingly, two indices indicate opposites in terms of the annual heatwave frequency with a difference of close to 90 heatwaves. You et al. (2017) also found no statistical association between the frequency of heatwaves defined using relative and absolute criteria, indicating little consistency in the climatologies produced under different heatwave definitions. Similarly, a consideration of heatwave trends exposed the dependency of the results on the type of definition criteria. In a similar vein, Wang et al. (2017) have considered the impact of heatwave definition on research findings. In their study, the absolute value 35 °C, (value used by the Chinese Meteorological Administration for heatwave definition), along with the local 90th percentile of Tmax as the relative threshold, were applied to the development of a China wide heatwave climatology. Perhaps unsurprisingly, 'absolute heatwaves' were found to occur mainly in the Southeast and Xinjiang regions of China where temperatures often exceed 35 °C, with 'relative heatwaves' occurring evenly across most of China. Given this and considering China's vast range of regional climates, they conclude that heatwave definitions based on relative criteria are best for the identification of trigger values for heat related early warning systems. Not wanting to labour the point, Fenner et al. (2019) in evaluating the nature of heatwaves for two German cities, also demonstrate how conclusions about long-term trends in heatwaves are definition dependent. Like the aforementioned studies for the US and China, Fenner et al. (2019) found large inter-definition differences in trend magnitude. Also revealing is the lack of temporal matching of heatwave events with only 15 heatwave episodes across the 125 years of analysis identified by all ten heatwave indices. These researchers also make the point that heatwave definitions applying criteria related to the Tmin and Tmean reveal noteworthy contrasts in heatwave trends for urban compared to less urban areas, whereas those based on Tmax do not, a finding that bears implications for the type of definition used for establishing the nature of urban heatwave trends. Similarly, Sulikowska and Wypych (2020) note the dependency of conclusions about the climatology of extreme temperature events across Europe on the definition used. Lastly, in a careful consideration of the best definition for defining the nature of heatwaves and their trend in Australia, Perkins and Alexander (2013) settle on three indices namely the 90th percentile for maximum temperature, the 90th percentile for minimum temperature and positive values of the Excess Heat Factor (EHF). While they find all heatwave definitions indicate an overall similar trend in heatwave aspects, subtle inter-definition differences in the rate of change in frequency and other aspects were revealed.

2.6.2 *Sector Based Heatwave Definitions*

An alternative tactic for settling on a heatwave definition is to adopt a sector specific or impact-based approach that focuses on a particular outcome or risk. Some obvious candidate sectors for specific impact-based heatwave definitions include agriculture, forestry, water, health, transport and infrastructure. For example, the agricultural sector might be interested in the risk heatwaves present to various stages of crop growth or livestock productivity. Similarly, infrastructure engineers may want to know if there is a particular combination of extreme heat conditions which induce a response in a key element of a region's infrastructure (e.g. a bridge, rail track, or tarmac road). Understandably, heatwaves defined based on sector- or impact-based criteria are likely to take many different forms. For some sectors, a period of 'slow burn' may be more important such that extreme heat distributed over a prolonged period is critical for the operation of a system. In such a case the heatwave threshold criteria, whether it be relative or absolute, might need to be lower than that specified for a sector in which a 'fast burn' is more critical with extreme heat delivered over a short period. Further, sector-based definitions or indices are likely to use a range of variables that best represent the drivers of climate sensitivity in a particular sector. Conscious of the need for sector-specific climate indices that could be applied to the measurement and monitoring of extremes such as heatwaves, the WMO Expert Team on Sector-Specific Climate Indices (ET-SCI) (see above) have developed several indices for the health, agriculture and water sectors. Although designed to be calculated at the annual and monthly timescales, which makes them generally inappropriate for heatwave definition, some indices can be employed in the characterisation of individual heatwave events.

Of a range of sectors, probably the health sector has made the most progress towards acknowledging the difficulty of settling on a sector-specific all-embracing heatwave definition. This sector has recognised that a wide range of heatwave definitions is possible, depending on the meteorological variable used, the specific health outcome considered and the location. Diversity manifests in the form of the *a priori* application of both absolute and relative thresholds for temperature and several thermal indices and a broad range of minimum duration criteria (Xu et al., 2016, 2018) with the result many health-related heatwave definitions diverge from the standard ones used in the meteorological community. As well as *a priori* definitions, researchers in the health sector have adopted what is very much an inductive approach and searched amongst a range of possible heatwave definitions for the one that best models a specific health risk. For example, from amongst 15 heatwave definitions, combining a range of relative thresholds and durations, Yang et al. (2019) found a heatwave definition based on the daily maximum temperature greater or equal to the 92.5th percentile and a duration of at least 3 days, produced the best fit to observed mortality amongst the general population in China. In a similar vein, but specifically for the case of infants in Brisbane, Australia, the heatwave definition that explains significant increases in infant mortality is a period more than 2 days when the average temperature rises above its 97th percentile value (Xu et al., 2017). Interestingly some researchers have found measures of heat stress such as the Wet Bulb Globe Temperature and the Universal Thermal Climate Index (UTCI) to be superior to air temperature indices for developing health outcome-based heatwave definitions (Akatsuka et al., 2016; Di Napoli et al., 2019; Heo et al., 2019). While to date most studies in the quest for the 'best' health related heatwave definition have applied statistical techniques to the analysis of climate and mortality time series, alternative approaches such as artificial intelligence are now emerging (Park & Kim, 2018).

Occasionally, climate and health researchers have adopted a bottom-up approach in defining heatwaves, whereby they use a specific extreme heat event, resulting in a high death toll, as a framing for derivation of physical heatwave descriptors and comparison of health outcomes with non-heatwave periods. However, if such an approach applies strict heatwave or period of excess mortality commencement and termination rules, some issues emerge, as illustrated best by an example. Figure 2.3 shows a plot of maximum temperature and mortality departures for the heatwave that struck London (UK) in August 2003. If we define the heatwave based on the period when maximum temperature exceeds the 90th percentile for August, then the heatwave start and end dates are August 3 and 12 respectively; see the period of positive 90th percentile departure values in Table 2.5. Alternatively, considering the days mortality is above the mean daily mortality for August (based on a 1998–2002 reference period), then the start and end dates of the period of excess mortality are August 5 and 16 respectively. This simple comparison of dates shows there is a clear offset between the temperature- and excess mortality-defined heatwaves (Fig. 2.3 and Table 2.5).

While there are some good epidemiological reasons for the offset noted above, such as a lag between temperature and associated mortality anomalies, defining a heatwave in the two different ways described here bears implications for estimation

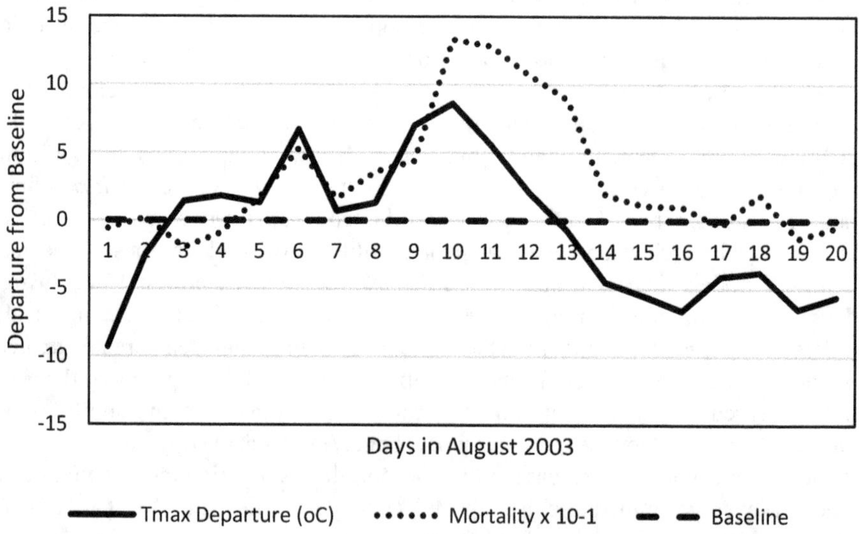

Fig. 2.3 Daily maximum temperature (Tmax) and mortality for London, August 1–20, 2003 plotted as departures from a baseline. Tmax (solid line) is the departure from the daily 90th percentile value of Tmax for August. The daily mortality departure (dotted line) is expressed as mortality x 10^{-1}, that is, the original departure value has been divided by 10. The baseline represents the reference values for Tmax (90th percentile of daily Tmax, 1998–2002) and mortality (mean daily mortality for August 1998–2002). (Source: Author)

Table 2.5 Departures for maximum temperature and mortality for the August 2003 London heatwave. Shaded values indicate the period for the 'temperature' and 'excess mortality' defined heatwaves

Day	Month	Year	Tmax Departure from Tmax 90th Percentile	Mortality Departure x 10^{-1}	Cumulative Mortality Departure X 10^{-1}
1	8	2003	-9.3	-.60	
2	8	2003	-2.5	.20	
3	8	2003	1.4	-2.00	
4	8	2003	1.8	-.90	
5	8	2003	1.3	1.70	1.7
6	8	2003	6.7	5.30	7.0
7	8	2003	.7	1.60	8.6
8	8	2003	1.3	3.60	12.2
9	8	2003	7.0	4.40	16.6
10	8	2003	8.6	13.30	29.9
11	8	2003	5.6	12.80	42.7
12	8	2003	2.1	10.70	53.4
13	8	2003	-.7	8.90	62.3
14	8	2003	-4.5	1.90	64.2
15	8	2003	-5.5	1.10	65.3
16	8	2003	-6.6	1.00	66.30
17	8	2003	-4.1	-.40	
18	8	2003	-3.8	1.90	
19	8	2003	-6.5	-1.30	
20	8	2003	-5.6	-.50	

of the immediate and delayed health burden, as demonstrated by the cumulative sum of mortality (Table 2.5). For the temperature-defined heatwave period August 3–12, the cumulative mortality count reaches 534. Although this would be less if the below average mortality from the beginning of the heatwave was considered, it nevertheless gives an idea of the immediate impact of the heatwave. In contrast, the mortality count obtains a figure of 663 for the heatwave period defined by excess mortality, lasting August 5–16 (Table 2.5). This totals 129 deaths more than the temperature defined heatwave period, a figure that may represent the delayed deaths arising from the 10-day period of extreme heat. A similar observation can also be made in relation to the physical characteristics of the 'two' heatwaves. For the temperature-based heatwave, the cumulative sum of the Tmax departures above the 90th percentile Tmax threshold value reaches 36.5 °C, while for the period of excess mortality this equates to a lower figure of 21.5 °C mainly because the temperature-based heatwave 'terminated' 4 days in advance of the excess mortality defined heatwave. For duration, there are also contrasts with positive anomalies of temperature and mortality lasting for 10 and 12 days respectively. Of course, these physical descriptors only refer to the case of the August 2003 London 'maximum temperature heatwave', and importantly, applying a different physical or heat stress descriptor may well result in revealing subtly different onset/cessation dates and cumulative statistics for this event. Clearly, defining a heatwave strictly in meteorological terms, notwithstanding a range of possible physical variables, or based on an outcome variable such as mortality, will influence conclusions drawn regarding, not only its physical characteristics, but its immediate and delayed, or total severity. This is especially so in the case of any response variable (e.g., mortality) that sporadically or habitually lags a forcing variable (e.g., temperature).

The health sector also serves as a good example of the conundrum faced when taking an impacts-based approach to heatwave definition. This is aptly demonstrated by the association between heatwave definition and estimates of the relative risk of heatwave related mortality. Table 2.6 presents the results from several studies, the general aim of which were to establish the increase in mortality risk associated with heatwave conditions and consider the influence of heatwave definition on risk estimates. A couple of important things emerge from the studies presented in Table 2.6. Clear is the dependency of the relative risk estimates on heatwave definition, with in some cases the relative risk increasing semi-systematically as the temperature threshold does for a fixed minimum duration. For example, Anderson and Bell (2009) demonstrate for a heatwave with a duration of 2 days or more, the relative risk increases from 1.04 to 1.07 as the temperature threshold progresses from the 95th to 99th percentile value for mean temperature. Similarly, the study by Gasparrini and Armstrong (2011) shows an elevation of relative risk with progressively higher relative threshold values for a heatwave with a duration of 4 days or more. For the Philippines, a continual increase in relative risk as the heatwave temperature threshold increases is also apparent (Seposo et al., 2017). Further to the value of the heatwave threshold criteria determining the estimated relative risk, is heatwave duration. For Brisbane, as heatwave duration increases from 3 to 5 days, for a constant threshold value (95th percentile of mean temperature), there is a steady climb of the relative risk of mortality (Tong et al., 2010). This also appears to be the case for Beijing, China for a heatwave exceeding the 95th percentile value

Table 2.6 Studies that consider the impact of heatwave definition on the estimates of the relative risk of heatwave related mortality. Relative risk is represented in the form of 1.XX with XX referring to the percentage increase of mortality above a reference state (e.g., non-heatwave conditions). Tmean is mean temperature, Tmax is maximum temperatiure, Tmin is minimum temperature. Numbers in subscripts refer to percentile values

Study	Location	Duration (≥)	Temperature threshold (≥)	Relative risk (95% confidence interval)
Anderson and Bell (2009)	United States	2	$Tmean_{95}$	1.04 (1.03, 1.06)
		2	$Tmean_{98}$	1.03 (1.02, 1.04)
		2	$Tmean_{99}$	1.07 (1.04, 1.10)
		2	$Tmean_{99.5}$	1.06 (1.04, 1.08)
		4	$Tmean_{98}$	1.04 (1.02, 1.06)
		4	$Tmean_{99}$	1.07 (1.03, 1.11)
		4	$Tmean_{99.5}$	1.06 (1.04, 1.08)

(continued)

Table 2.6 (continued)

Study	Location	Duration (≥)	Temperature threshold (≥)	Relative risk (95% confidence interval)
Tong et al. (2010)	Brisbane, Australia	2	35 °C	1.26 (1.10, 1.43)
		2	37 °C	1.53 (1.27, 1.84)
		3	$Tmax_{95}$	1.20 (1.02, 1.41)
		4	$Tmax_{95}$	1.14 (1.05, 1.24)
		5	$Tmax_{95}$	1.24 (1.13, 1.36)
		2	$Tmax_{97}$	1.16 (1.06, 1.27)
		3	$Tmax_{97}$	1.21 (1.08, 1.35)
Gasparrini and Armstrong (2011)	United States	2	$Tmean_{97}$	1.05 (1.03, 1.07)
		4	$Tmean_{97}$	1.05 (1.04, 1.07)
		4	$Tmean_{98}$	1.06 (1.05, 1.08)
		4	$Tmean_{99}$	1.08 (1.05, 1.10)
Dong et al. (2016)	Beijing, China	2	$Tmean_{90}$	1.02 (0.96, 1.09)
		3	$Tmean_{90}$	0.98 (0.91, 1.04)
		4	$Tmean_{90}$	1.01 (0.94, 1.08)
		5	$Tmean_{90}$	1.02 (0.95, 1.10)
		2	$Tmean_{95}$	1.00 (0.92, 1.08)
		3	$Tmean_{95}$	1.08 (0.98, 1.19)
		4	$Tmean_{95}$	1.14 (1.03, 1.26)
		5	$Tmean_{95}$	1.14 (1.03, 1.27)
Seposo et al. (2017)	Philippines	2	$Tmean_{90}$	1.12 (1.04, 1.21)
		2	$Tmean_{95}$	1.14 (1.04, 1.25)
		2	$Tmean_{97}$	1.22 (1.05, 1.43)
		4	$Tmean_{90}$	1.13 (1.03, 1.25)
		4	$Tmean_{95}$	1.14 (1.02, 1.26)
		4	$Tmean_{97}$	1.26 (1.09. 1.46)

Source: Compiled by Author

for mean temperature, as the relative risk progresses from 1.0 to 1.14 as heatwave duration increases from 2 to 4 days (Dong et al., 2016).

Corroborating the location or region-specific findings of the researchers listed in Table 2.6 is the heatwave-mortality risk meta-analysis of Xu et al. (2016). They find for heatwaves, with a duration of at least 2 days and increasingly severe temperature thresholds set at the 95th and 99th percentile values for mean temperature, that the relative risk rises from 1.04 to 1.07. Furthermore, it attains a value of 1.16 for a heatwave of at least 5 days with a threshold matching the 97th percentile value for mean temperature (Xu et al., 2016). Similarly, Nori-Sarma et al. (2019), in an assessment of the association between 13 different heatwave definitions and mortality for developing country settings, found that the relative risk of mortality varied significantly depending on how heatwaves were defined. These findings and those from the studies reviewed in Table 2.6, attest to a clear impact of the choice of heatwave definition criteria (temperature and minimum duration thresholds) on estimates of the relative risk of mortality and understandably raises alarm bells about the dependency of heatwave impact analyses on the heatwave definition used, not just for the health sector. Moreover, there is an ethical dimension to this contention. For example, if triggers for implementing heat health action plans are to be risk based (see Chap. 13), that is a trigger is set at a specific level of relative risk (e.g., 1.10 for mortality) then the heatwave alert criteria need to match this. In other words, the risk variable for a warning system should not be set independently of the heat alter system or heatwave definition criteria. Similarly, settling on a given threshold value, whether it be absolute or relative, as the trigger for an alert system, without reference to a specific risk variable, will be akin to taking a shot in the dark as the outcome for which the alarm is raised is unknown. Having said that, developing a sector-based heatwave definition may remain aspirational for many countries or locations. This is because the paucity of data on specific risks or impacts may well constrain efforts to establish risk forcing or dose-response functions. Notwithstanding the previous 'shot in the dark' remark, there may be only two options in the absence of impact data for developing heatwave definition criteria, assuming meteorological data are available. These are the application of a heuristic rule of thumb approach or a specific *a priori* index. In the case of the former, the extreme events literature could be drawn on to establish what in general constitutes an extreme value (e.g., exceedance of 90th or 95th percentile value for temperature for a given duration), while the latter may well be informed by expert judgement as embodied in specialist working groups (e.g., the ET-SCI).

2.7 Heatwave Characteristics

Setting aside the actual way in which a heatwave is defined, once events have been identified the characterisation of these becomes of interest. This is because biophysical and societal responses to heatwaves may vary depending on the nature of the heatwave, beyond how extreme the constituent temperature or other physical variable values might be.

Essentially, not all heatwaves are the same, with each having a different 'flavour'. Such flavours can best be summarised by characterising heatwaves, at either the meteorological or climatological time scales, with reference to several heatwave attributes (Perkins & Alexander, 2013) as set out in Table 2.7. Of course, descriptions of heatwave attributes need not be confined to the observational period of heatwave variables. Equally, weather forecast data could be used to present information about the characteristics of an impending heatwave event, either in deterministic or probabilistic terms. Indeed, the same applies to using re-analysis products and climate model output to illustrate the likely nature of heatwave characteristics where observational records do not exist or for future time horizons.

Table 2.7 Heatwave attributes

Aspect	Description	Comment
Magnitude	Peak value of variable used to define heatwave or variable value anomaly relative to historical reference period	Maximum temperature is commonly used but a range of thermal indices could be used (e.g. apparent temperature, WBGT). If 'nocturnal' heatwaves are of interest then minimum temperature is the critical variable. Anomalies could be expressed as departures in degrees Celsius, or standardised anomalies (units of standard of deviation)
Frequency	The number of times a heatwave occurs for a set period of analysis	Setting a 'harsh' absolute or percentile heatwave threshold will result in the recording of only a few heatwaves. Frequency can apply to all heatwaves or heatwaves of different durations
Duration	Number of days that heatwave definition criteria are exceeded	In most cases, heatwaves will have a duration of a minimum of 2 days if this is set as the duration threshold in a heatwave definition. For fine scale descriptions of a heatwave, thresholds based on hourly data may be appropriate so that duration is reported in hours. The lower the absolute or percentile threshold value set for a heatwave, the longer the heatwave duration will be
Spatial extent	The geographical area over which heatwave threshold criteria are exceeded	Heatwaves could either be localised or extensive/widespread depending on the meteorological or land surface climate circumstances. Criteria for extensive heatwaves could be based on heatwave threshold criteria being exceeded for more than one local government administration unit such as a county. Clearly the density of the weather observation network will determine how well the geographical characteristics of an observed heatwave can be defined. If gridded data are used to analyse how widespread a heatwave is then extent could be described in square kilometres or number of grid boxes
Severity	Scale of impact	Potentially many social and economic metrics exist for severity characterisation. These might include level of damage to agricultural crops, extent of algal blooms, number of heat-related deaths or hospitalisations or retail figures. The issuance of heatwave alerts or warnings could also constitute measures of severity, notwithstanding that these are usually based on exceedance of some physical threshold
Timing	Time of the year which heatwave occurs	Information on timing may shed light on whether heatwaves over time have shifted to earlier and/or latter in the year, or certain characteristics are associated with timing, e.g. are high magnitude events associated with a specific period in the warm season (do singularities exist?)

Adapted from Perkins and Alexander 2013

2.7.1 *Metrics of Heatwave Attributes*

For climatological descriptions of the possible changing nature of heatwave attributes over long periods, including into the future, Perkins and Alexander (2013) have proposed a number of metrics based on absolute values of the Excess Heat Factor (EHF) and relative values for Tmax and Tmin. While the Perkins and Alexander (2013) heatwave metrics relate in general terms to the broadly defined heatwave attributes presented in Table 2.7, it is worthwhile highlighting them here, because of their increasingly ubiquitous application in heatwave studies and facilitation of heatwave climatology inter-comparisons (Table 2.8). The metrics refer to summer only with summer defined as May—September in the Northern Hemisphere and November to March in the Southern Hemisphere. Importantly the metrics use a set of well-researched climate indices with exceedance duration criteria of three or more days. Once a set of heatwave metrics have been developed these can used to develop heatwave climatologies for specific locations or form the basis of heatwave analysis toolboxes, such as the global heatwave and warm-spell data record and analysis toolbox—referred to as the GHWR—(Raei et al., 2018) for the analysis of heatwave events.

Having stated that there is a general relationship between the characteristics in Table 2.7 and Table 2.8, some qualification is required, especially in relation to the 'magnitude' characteristic and the widespread use of 'intensity' in the heatwave literature. There appears to be some confusion with intensity defined in different ways. For example, Mazdiyasni et al. (2019) define heatwave intensity as the average of mean daily temperature over the duration of a heatwave. This is like the HWM metric in Table 2.8, albeit it relates to all heatwaves in a given time period. Alternatively, intensity has been defined as the hottest day of a heatwave. At the seasonal scale intensity is viewed as the average intensity of all heatwave events. There are also cumulative versions of intensity (cumulative intensity) such as the

Table 2.8 Perkins and Alexander (2013) Heatwave Metrics for climatological descriptions. Each Metric definition is in relation to either 90th percentile values of maximum or minimum temperature or positive values of the Extreme Heat Factor (EHF). A researcher specified base period is used for percentile calculation

Metric	Definition	Unit
Heatwave number (HWN)	Number of individual heatwaves that occur each summer.	Count
Heatwave frequency (HWF)	The number of days that contribute to heatwaves, as defined by HWN	Days
Heatwave duration (HWD)	The length of the longest heatwave identified by HWN	Days
Heatwave magnitude (HWM)	The mean temperature of all heatwaves identified by HWN	°C
Heatwave amplitude (HWA)	The peak daily value in the hottest heatwave (defined as the heatwave with the highest HWM	°C

cumulative exceedance of a specified temperature threshold across all events. This conception of intensity is used as part of the French heatwave early warning system with intensity defined as the maximum of the daily severities during the heatwave with severity measured as the difference between the daily observed temperature and a predefined threshold value (Pascal et al., 2021). To add to the confusion, Xu et al. (2016) use intensity to refer to the threshold value applied in heatwave definitions. Further, magnitude as defined in Table 2.7 appears to equate with HWA in Table 2.8, yet a different terminology is used. Clearly, the interchangeable use of some key heatwave characteristic terms is not ideal and may lead to researchers speaking at cross-purposes. Acknowledging the significance of intensity as a heatwave characteristic, especially in the context of understanding impacts, Nairn and Fawcett (2015) have attempted to develop a standardised measure of intensity with global applicability (see Chap. 4).

To demonstrate the calculation of the five heatwave metrics given in Table 2.8 an example for Tokyo is presented. Figure 2.4 plots the daily maximum temperature for the summer period May 1 to September 30 with the 90th percentile value of Tmax imposed. Applying the minimum duration criteria of 3 days or more, for which the Tmax must exceed the 90th percentile value, reveals the occurrence of three heatwave events, with durations varying between 4 and 11 days (events 2 and 3 respectively, Fig. 2.4). Following (Perkins & Alexander, 2013), the heatwave metrics based on maximum temperature are: HWN = 3, HWF = 21, HWD = 11, HWM = 35.3 °C and HWA = 37.2 °C. From these statistics, derivation of further information is possible. For example, HWF/HWN gives the average heatwave length for 2010 as 7 days, albeit the long duration of event 3 (11 days) influences this statistic. Further, event 3 accounts for around 52 percent ((HWD/HWF)/100) of the heatwave days in 2010. Moreover, knowing the 90th percentile Tmax values helps put the HWM and HWA metrics into an 'anomaly context' such that HWM and HWA values of 35.3 °C and 37.2 °C respectively, in turn represent anomalies of 1.5 °C and 2.9 °C above the 90th percentile for Tmax. Of course, slackening the minimum duration criteria to a value of 2 days would allow a number of 'near-miss' events to enter into the calculation of the heatwave metrics for 2010 – 4 events in fact. Moreover, and setting aside the inclusion of 2-day events, applying a definition for a compound heatwave definition (see Chapter—Baldwin et al. (2019)) would also influence the statistics as two compound events occurred in 2010. Specifying a heatwave in terms of the (Perkins & Alexander, 2013) criteria for Tmax can also reveal information on heatwave timing. For the case of the Tokyo 2010 summer, the first and last heatwaves start and end on calendar days 200 and 250 respectively (Fig. 2.4). Borrowing from the literature on the climatology of the onset and cessation of the wet or dry season, such start and end dates translate into a heatwave season duration of 51 days; exercising leniency with a 2-day heatwave definition would extend this to 141 days (calendar days 125–265). Interestingly, and despite the burgeoning literature on heatwave characteristics, few if any studies attempt to define the heatwave season and describe its characteristics, although allied studies on human thermal comfort have attempted to tackle the problem of the definition of the 'heat season' (McGregor et al., 2002) and its drivers (Bartzokas et al., 2013).

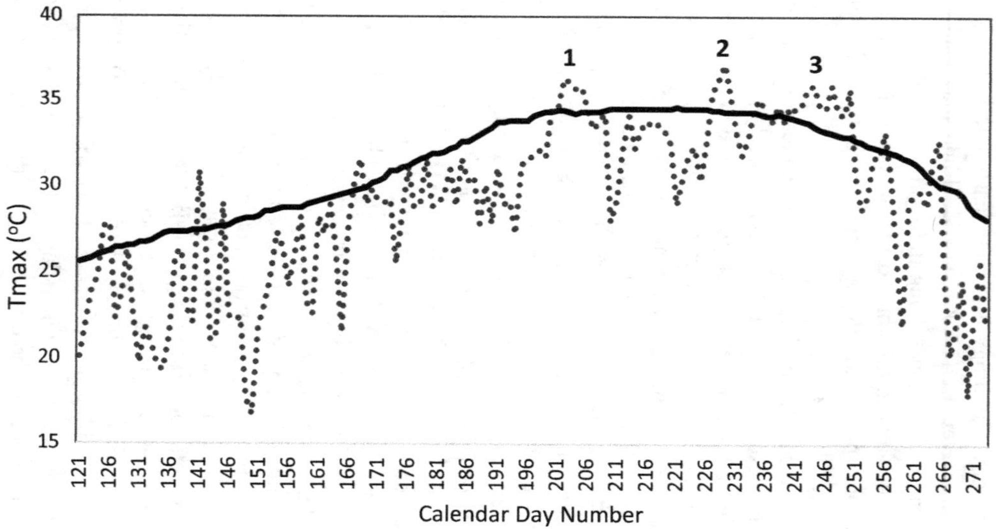

Fig. 2.4 Time plot of climatological value of 90th percentile (solid line) for base period 1972–2018 and observed maximum temperature (Tmax, dotted line) for Tokyo, summer 2010 (Day 121 = May 1; Day 273 = September 30). Numbers 1, 2 and 3 refer to heatwave events. (Source: Author)

2.7.2 Interaction Between Heatwave Attributes

While a systematic consideration of heatwave attributes as presented above might imply their independence, intuitively we know from an impacts or risk perspective, certain combinations of heatwave characteristics could lead to 'the perfect storm' in terms of severity. In the world of hydrometeorology, the amplification of hydrological impacts arising from the concurrent occurrence of, for example, high magnitude short duration or medium magnitude long duration precipitation events, has been understood for some time, with the joint occurrences of intensity, frequency and duration often represented in the form of intensity-duration-frequency (IDF) curves (Devi et al., 2022; Lanciotti et al., 2022). Recognising this, Mazdiyasni et al. (2019) made one of the first attempts to present heatwave IDF curves for six cities in the United States (Fig. 2.5). For each city, the IDF plots depict the non-exceedance probabilities for different combinations of heatwave duration and intensity. In addition to providing city-specific information on heatwave frequency, the IDF plots facilitate inter-city comparisons of the probability of occurrence of different combinations of heatwave characteristics. For example, a Chicago heatwave with a duration of up to 6-days and an associated non-exceedance probability of 50 percent (2-year return period) has an average temperature (intensity) of 38 °C or less, whereas a Phoenix heatwave with the same duration and probability characteristics has a markedly higher intensity of 47.5 °C (Mazdiyasni et al., 2019). Further to this type of information, the plots also allow establishment of duration and intensity combinations that are equally likely. Taking Los Angeles and a non-exceedance probability of 0.5 (2-year return period), a 6-day duration heatwave with an intensity of 39 °C is equally as likely as a 10-day heatwave with an intensity of around 35.8 °C (Fig. 2.5).

As noted by Mazdiyasni et al. (2019), instead of non-exceedance probabilities, the heatwave IDFs can also be described based on the idea of joint exceedance probabilities for different combinations of heat wave intensity and duration. To illustrate this, they present an associated IDF plot for Chicago, noting as an example that joint exceedance probabilities of a heatwave event lasting 6 days or more with an average temperature exceeding 38 °C, equates with an exceedance probability of less than 0.01, or a return period greater than 100 years (Fig. 2.6). Recalling the Chicago 1995 heatwave, which had a duration of approximately 4 days and an average temperature of around 38.5 °C, according to the IDF in Fig. 2.6, this often-cited event possesses an exceedance probability of close to 0.01 and thus a return period of just under 100 years. Complementing the IDF framework are intensity-area-duration (IAD) analyses of heatwaves. These consider the temporal continuity and spatial dynamics of extreme events to identify the characteristics of regional scale heatwave events (Becker et al., 2022; Ren et al., 2023). However, the situation portrayed in terms of the spatio-temporal heatwave characteristics for a region is very much dependent on the heatwave index chosen (Becker et al., 2022).

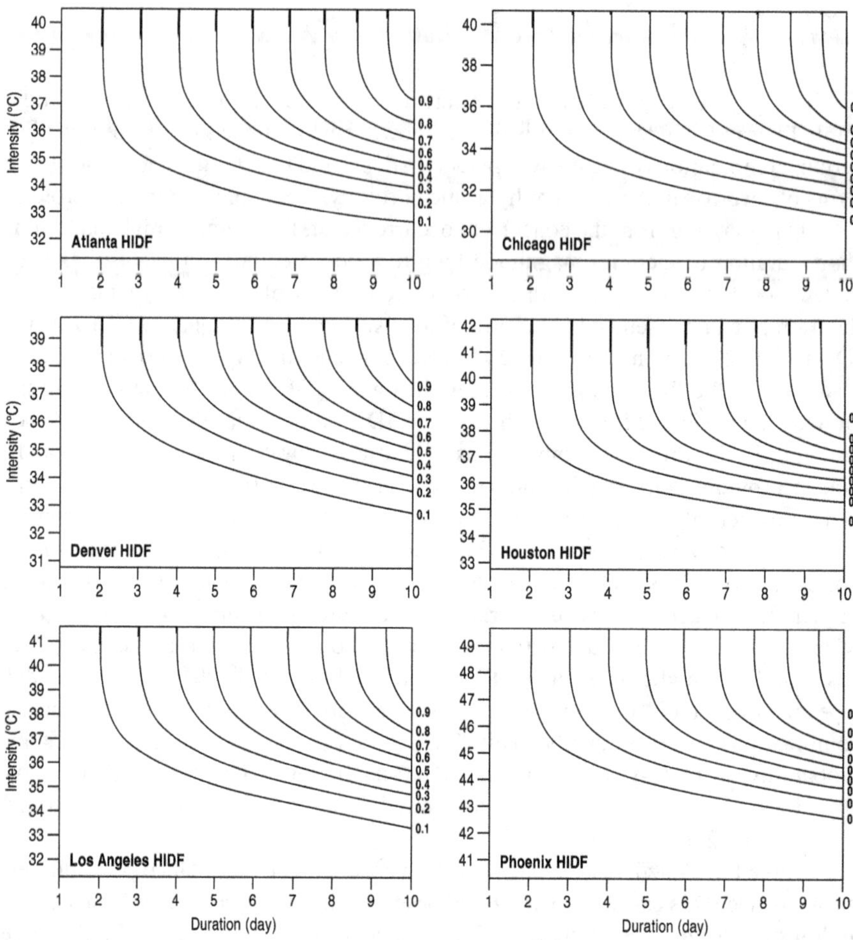

Fig. 2.5 Heat wave intensity-duration-frequency (HIDF) curves for six US cities, using non-exceedance probabilities as represented by the re red values on the right axis. Source: Mazdiyasni, O., Sadegh, M., Chiang, F. et al. Heat wave Intensity Duration Frequency Curve: A Multivariate Approach for Hazard and Attribution Analysis. Sci Rep 9, 14,117 (2019). https://doi-org.ezphost.dur.ac.uk/10.1038/s41598-019-50643-w Used under Creative Commons Attribution 4.0 International License CC BY 4.0 https://creativecommons.org/licenses/by/4.0/

2.7.3 *Heatwave Form*

Building on the idea that not all heatwaves are the same, one can imagine the time distribution of heat during a heatwave might vary from event to event, or from location to location, such that a variety of heatwave forms exist. Despite this obvious characteristic of heatwaves, heatwave form appears to be a little explored characteristic in the literature. To illustrate this consider the time plots of Tmax for the 1995 Chicago and 2003 Paris heatwaves (Fig. 2.7). Applying the type of terminology used to describe hydrograph shape (Hannah et al., 2000), it is apparent that the

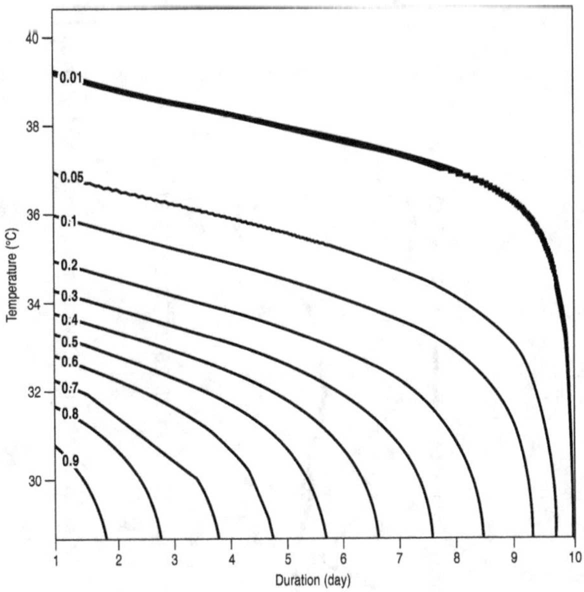

Fig. 2.6 Heat wave intensity-duration-frequency (HIDF) curves for Chicago using exceedance probabilities. The red values on the left axis represent exceedance probabilities. Source: Mazdiyasni, O., Sadegh, M., Chiang, F. et al. Heat wave Intensity Duration Frequency Curve: A Multivariate Approach for Hazard and Attribution Analysis. Sci Rep 9, 14117 (2019). https://doi-org.ezphost.dur.ac.uk/10.1038/s41598-019-50643-w Used under Creative Commons Attribution 4.0 International License CC BY 4.0 https://creativecommons.org/licenses/by/4.0/

rising and recession limbs as well as the 'peakedness' of the Chicago and Paris thermographs are quite different (Fig. 2.7). Whereas the increase in Tmax is gradual for the Paris event, it is rapid for the Chicago event. Further, the Chicago event possesses a clear short duration period of high temperatures while for Paris, two Tmax peaks 'bookend' a plateau of high maximum temperatures lasting 6–7 days. Turning attention to the recession limbs, a much more rapid decrease in Tmax is evident for the Paris event. Such simple comparisons and thinking about the possible controlling factors on thermograph shape, such as state of the land surface and nature of synoptic scale weather patterns at the time of a heatwave event, raises the question whether different heatwave forms may exist, as determined by how atmospheric processes interact with land surface conditions (see Chap. 7). Heatwave form for any given location may also very depending on its distance from the centre of action or heatwave core, where temperature anomalies are greatest. For example, Stefanon et al. (2012) present a classification of the spatial pattern of major heatwaves across Europe with clear centres of action in relation to temperature anomalies and strong spatial gradients of these. Given this, one might speculate that locations lying on the margins of a heatwave may possess a more leptokurtic or peaky thermograph, whereas those in the heatwave core might be more platykurtic (less leptokurtic) in shape, like that of the Paris 2003 event, reflecting a longer duration of anomalous heat possibly arising from sustained positive land-atmosphere feedbacks.

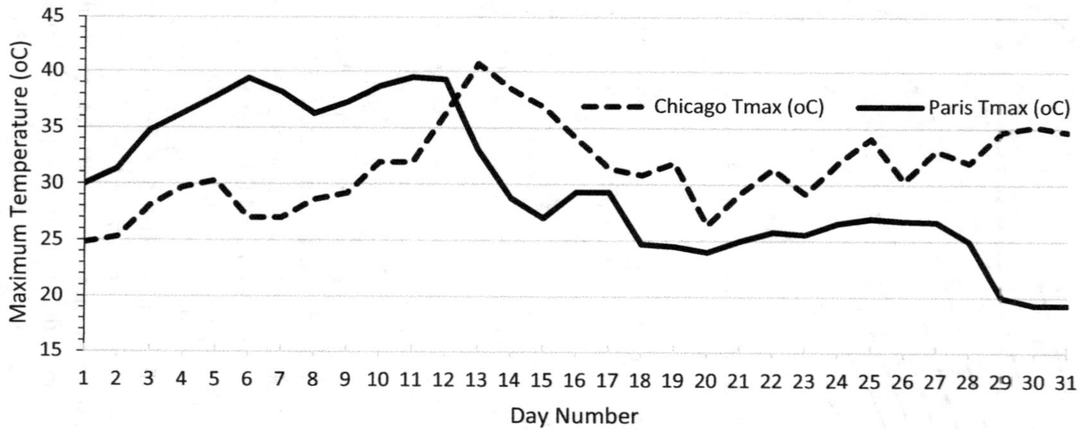

Fig. 2.7 Time plot of maximum temperature (°C) for the July 1995 Chicago and August 2003 Paris heatwave events. Day number refers to the day from the beginning of July (Chicago) and August (Paris). (Source: Author)

To lend support to the contention that diversity may be a characteristic of heatwave form, plots of standardised maximum temperature (Tmax) anomalies are presented for several heatwave events in Tokyo. Such forms (Fig. 2.8), hint at a heatwave typology for Tokyo consisting of at least five classes of thermograph shape. Without deriving quantitative metrics related to the rising and recession limbs and peak, it is clear the five heatwave events possess quite different thermal forms. For example, the September 2, 2000 event (Fig. 2.8) is comprised of a single 1 day Tmax anomaly peak. Also noticeable is the steep rise and subsequent fall of the Tmax anomaly away from the Tmax mean (see Fig. 2.8 caption for explanation). Although appearing broadly similar in form, the August 4, 2005, event (Fig. 2.8), is somewhat different to that which occurred on September 2, 2000. This is because it has significant Tmax anomalies over 2 days and the rising and recession limbs of the thermograph are gentle in comparison. The August 28, 2010, event is interesting as it occurs in a period of above average Tmax and possesses a double Tmax anomaly peak, the first being longer than the second (Fig. 2.8). Further the gradient of the rising limb for this event is gentle with peak temperatures only reaching one unit of standard deviation, much lower than that attained over a shorter period in the case of the September 2000 and August 2005 events. Lastly, the August 2010 event does not possess a rapid cessation with this event embedded in a period of relative warmth with intermittent pulses of heat as manifest by the double Tmax anomaly peak.

That heatwave onset can be abrupt, with positive Tmax anomalies developing quickly from a period of relatively cool weather is demonstrated well by the June 24, 2011 event (Fig. 2.8). An additional feature of this event is the sudden curtailment of a 2-day period of anomalous heat, followed by the rapid onset of another period of anomalous heat after a short cool interlude (days 8 and 9 in the plot). In fact, the June 24, 2011 event was followed by another heatwave event that reached an even greater Tmax anomaly on June 30 (day 12 not shown in the plot). Such Tmax anomaly peaks occurring in quick succession may well constitute a compound type of heatwave event. The same may ring true for the August 2010 event described above, albeit for this event the sharpness of the anomaly peaks is less spectacular than the June 2011 event. The form associated with the heatwave event in and around August 2, 2014 (Fig. 2.8) contrasts strongly with the other four events considered here. A broad plateau of Tmax anomalies lasting 5 days is a noteworthy feature of this event. Further, this persistent period of heat has evolved from a preceding period of above average Tmax conditions. While in some ways the double Tmax peak evident for this event resembles that described for August 2010 (Fig. 2.8), the rapid cessation of the period of anomalous warmth following the double peak sets it apart from that of the August 2010 event. Importantly, this description of heatwave physical form begs the question 'for any given location is there a spectrum of heatwave forms', a question that beckons future research. Other tantalising questions might include 'does the form of heatwave related impacts follow that of the heatwave'. Clearly, heatwave form is a furtive area for future research.

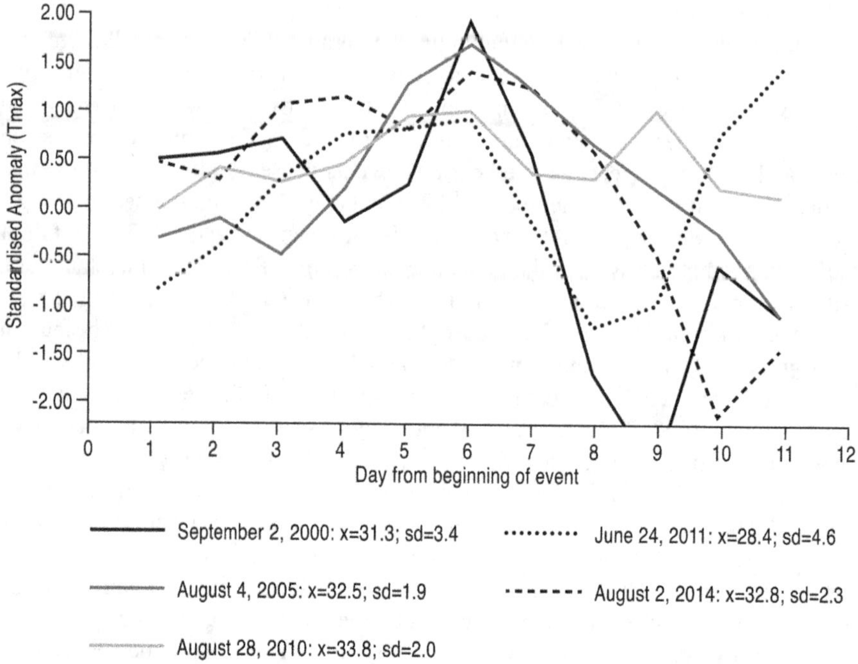

Fig. 2.8 Possible heatwave forms for Tokyo as represented by five specific heatwave events as given by date. The heatwave forms have been identified with the assistance of a Principal Components Analysis (PCA) of daily maximum temperature (Tmax). A 21-day period centred on the peak Tmax for 37 heatwave events over the period of 2000–2015 was used as input into the PCA, The standardised anomalies of Tmax (units of standard deviation) have been calculated with reference to the 21 day period within which the heatwave event occurred. A standardised anomaly value of zero represents the Tmax mean for the 21-day period. The plots presented here are only for the 11-day period centred on the Tmax peak, that is, 5 days either side of the peak. x and sd next to the dates are the mean and standard deviation of Tmax for the 21-day period in degrees Celsius (°C); in the case of the September 2, 2000 event, this would mean that the peak temperature was almost 6.8 °C above the mean of 31.3 °C, that is the peak Tmax reached 37.9 °C. (Source: Author)

2.8 Marine Heatwaves

Because of the impact of anomalously warm ocean waters on marine ecosystems, oceanographers and marine ecologists have drawn attention to the ever-increasing incidence of marine heatwaves. In doing so, they have noted the need to move away from the plethora of metrics applied in the assessment of the impacts of extreme thermal stress on marine flora and fauna and the general vagueness of definitions used for defining marine heatwaves. The preference amongst this community is to settle upon a definition that could form the basis for establishing a globally consistent detection framework to study marine heatwave drivers and variability. Acknowledging these issues Hobday et al. (2016) have proposed that marine heatwaves are "identifiable events with start and end dates, a persistent duration of at least 5 days and anomalous warmer sea surface water relative to a threshold (90th or 99th percentile) in a 30-year baseline climatology". With reference to what might constitute a good definition, this definition meets several of the criteria. This is because it is clear about the level beyond which anomalous thermal conditions exist through identifying a relative threshold. Further a minimum duration is specified which facilitates the identification of discrete events with clear onset and cessation dates. In formulating this definition, Hobday et al. (2016) state that the climatological mean and threshold are to the calculated based on a 11-day moving window for a suitably long reference period, as determined by data availability from in situ or satellite-based observations of sea surface temperature. Their definition also allows for short breaks of 2 days or less such that if five anomalously warm days are followed by 2 days less than the threshold which are then followed by another five anomalously warm days this would constitute a 12-day marine heatwave event. Conversely five warm days followed by three cool days, which are then followed by six warm days would be defined as two separate marine heatwave events. The general parameters associated with this definition resonate with the definition of compound atmospheric heatwave events put forward by Baldwin et al. (2019), as outlined in the next chapter. While clear in terms of its definition parameters, it is unclear whether marine heatwaves are restricted to the traditional summer months as for the case of atmospheric heatwaves. In other words, can a distinction be made between marine warm spells (any time of the year) and marine heatwaves which reach absolute levels of heat far beyond those of warm spells and are restricted to summer only?

Having proffered the above definition for marine heatwaves, Hobday et al. (2018) subsequently developed an intensity-based marine heatwave categorisation scheme (Fig. 2.9). This is based on multiples of the value represented by the local difference between the climatological mean and the climatological 90th percentile (Hobday et al., 2018). That both the marine heatwave definition proffered by Hobday et al. (2016) and its associated intensity scale (Hobday et al., 2018) have been met with general approval amongst the physical oceanography community is evident from the large number of studies that have applied this description of marine heatwaves. Agencies such as NOAA have also applied this definition to create marine heatwave products (Fig. 2.10). That said the Hobday et al. (2018) definition is fundamentally a hazard-based definition offering only a general description of oceanic heat stress broadly relevant to marine life. Accordingly, if there is a strong desire amongst the marine ecology community to develop impact-based marine heatwave definitions, then attention will need to be turned to applying species specific thresholds that model critical heat-related life cycle changes in marine organisms (see Chaps. 10 and 11).

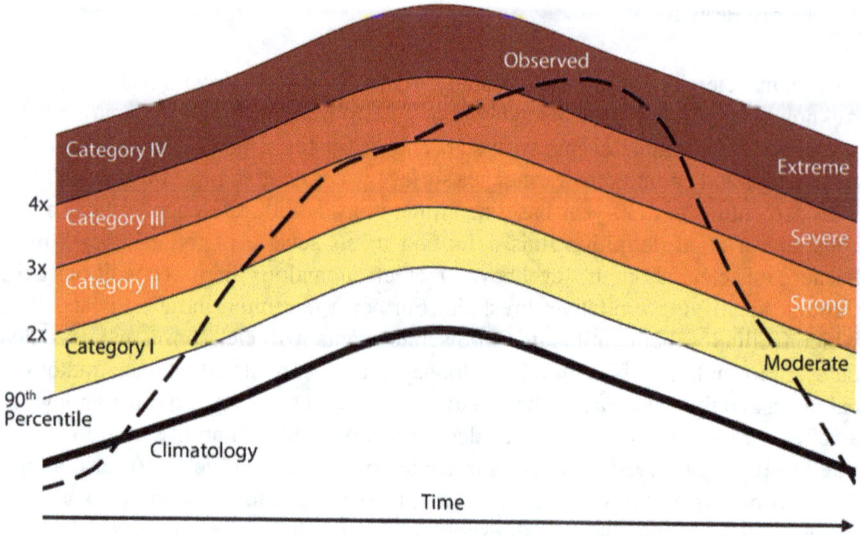

Fig. 2.9 Magnitude of scale descriptors defined by the difference between the local climatological mean and the climatological 90th percentile. Four categories are defined based on the scale of difference. Source Hobday, A.J., E.C.J. Oliver, A. Sen Gupta, J.A. Benthuysen, M.T. Burrows, M.G. Donat, N.J. Holbrook, P.J. Moore, M.S. Thomsen, T. Wernberg, and D.A. Smale. 2018. Categorizing and naming marine heatwaves. Oceanography 31(2):162–173, https://doi.org/10.5670/oceanog.2018.205. Used under Creative Commons Attribution 4.0 International (CC BY 4.0) Licence https://creativecommons.org/licenses/by/4.0/

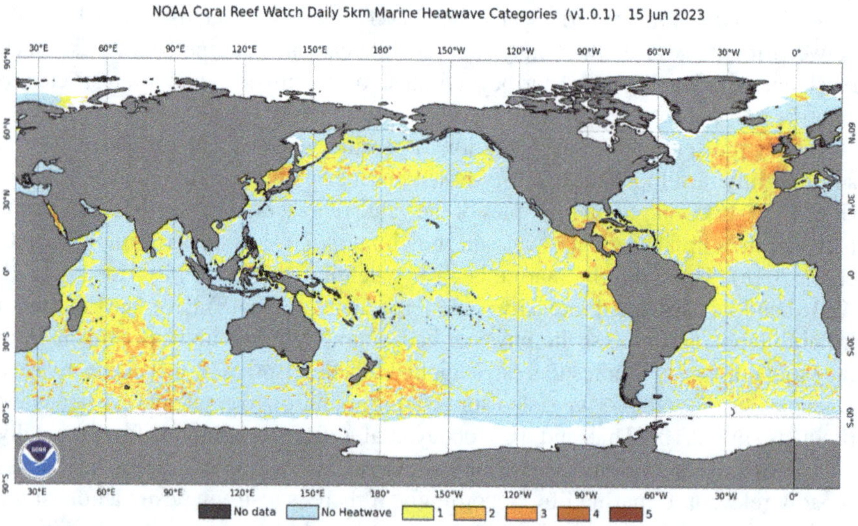

Fig. 2.10 NOAA's Marine Heatwave Watch (MHW) product based on the application of the Hobday et al. (2018) definition of heatwave categories to NOAA's Coral Reef Watch (CRW) daily global 5 km satellite CoralTemp satellite SST data product. Although the name might imply otherwise, this product provides a general description of oceanic heat stress relevant to a broad range of marine life. The example shown is for June 15, 2023. The extensive strong to severe marine heatwave experienced in the eastern North Atlantic in 2023 is clear. Several other regions also demonstrate the occurrence of marine heatwaves of varying intensity and spatial extent. (Source: NOAA, 2021 https://coralreefwatch.noaa.gov/product/marine_heatwave/)

2.9 Conclusion

Heatwaves are a meteorological phenomenon restricted to the summer months, as opposed to warm spells which relate to unusually warm weather in any season and do not reach the absolute levels of heat achieved by heatwaves. From the outset, the intent of this chapter has not been to offer a 'universal' heatwave definition. Rather the chapter has presented some basic principles for what constitutes a 'good' definition. While a universal heatwave definition may seem appealing for consistency and comparability, the complexities and context-specific nature of heatwaves suggest a more flexible approach is warranted. Hazard-based definitions like those proposed by Perkins and Alexander (2013) or Nairn and Fawcett (2015) could serve as a general benchmark for climatological analyses, facilitating cross-study comparisons. That said these definitions would need to be modified for the case of humid heatwaves. For understanding impacts or developing warning systems, bespoke impact-based definitions tailored to specific sectors, regions, or populations are likely necessary. The diversity of existing definitions reflects the inventiveness of the heatwave science community, but also highlights the challenges in developing a one-size-fits-all solution. Different definitions potentially lead to conflicting results and public confusion, underscoring the need for clear communication and transparency regarding the criteria and limitations of any given definition. Ultimately, a balanced approach that combines structuralist and phenomenological perspectives may be most appropriate but, in reality and steered by pragmatism, especially in an operational setting, this may be difficult to achieve.

References

Akatsuka, S., Uno, T., & Horiuchi, M. (2016). The relationship between the heat disorder incidence rate and heat stress indices at Yamanashi prefecture in Japan. *Advances in Meteorology, 2016*, 1–11. https://doi.org/10.1155/2016/9492815

Alexander, L. V., Zhang, X., Peterson, T. C., Caesar, J., Gleason, B., Tank, A., Haylock, M., Collins, D., Trewin, B., Rahimzadeh, F., Tagipour, A., Kumar, K. R., Revadekar, J., Griffiths, G., Vincent, L., Stephenson, D. B., Burn, J., Aguilar, E., Brunet, M., Taylor, M., New, M., Zhai, P., Rusticucci, M., & Vazquez-Aguirre, J. L. (2006). Global observed changes in daily climate extremes of temperature and precipitation. *Journal of Geophysical Research-Atmospheres, 111*(D5). https://doi.org/10.1029/2005jd006290

Anderson, B. G., & Bell, M. L. (2009). Weather-related mortality how heat, cold, and heat waves affect mortality in the United States. *Epidemiology, 20*(2), 205–213. https://doi.org/10.1097/EDE.0b013e318190ee08

Anderson, G. B., & Bell, M. L. (2011). Heat waves in the United States: Mortality risk during heat waves and effect modification by heat wave characteristics in 43 U.S. *Communities. Environmental Health Perspectives, 119*(2), 210–218. https://doi.org/10.1289/ehp.1002313

Baldwin, J. W., Dessy, J. B., Vecchi, G. A., & Oppenheimer, M. (2019). Temporally compound heat wave events and global warming: An emerging hazard. *Earths Future, 7*(4), 411–427. https://doi.org/10.1029/2018ef000989

Barbier, J., Guichard, F., Bouniol, D., Couvreux, F., & Roehrig, R. (2018). Detection of Intraseasonal large-scale heat waves: Characteristics and historical trends during the Sahelian spring. *Journal of Climate, 31*(1), 61–80. https://doi.org/10.1175/jcli-d-17-0244.1

Bartzokas, A., Lolis, C. J., Kassomenos, P. A., & McGregor, G. R. (2013). Climatic characteristics of summer human thermal discomfort in Athens and its connection to atmospheric circulation. *Natural Hazards and Earth System Sciences, 13*(12), 3271–3279. https://doi.org/10.5194/nhess-13-3271-2013

Becker, F. N., Fink, A. H., Bissolli, P., & Pinto, J. G. (2022). Towards a more comprehensive assessment of the intensity of historical European heat waves (1979–2019). *Atmospheric Science Letters, 23*(11). https://doi.org/10.1002/asl.1120

Boni, Z., Bienkowska, Z., Chwalczyk, F., Jancewicz, B., Marginean, I., & Serrano, P. Y. (2023). What is a heat(wave)? An interdisciplinary perspective. *Climatic Change, 176*(9). https://doi.org/10.1007/s10584-023-03592-3

Brunner, L., & Voigt, A. (2024). Pitfalls in diagnosing temperature extremes. *Nature Communications, 15*(1), 2087–2087. https://doi.org/10.1038/s41467-024-46349-x

Davis, R. E., McGregor, G. R., & Enfield, K. B. (2016). Humidity: A review and primer on atmospheric moisture and human health. *Environmental Research, 144*, 106–116. https://doi.org/10.1016/j.envres.2015.10.014

Della-Marta, P. M., Haylock, M. R., Luterbacher, J., & Wanner, H. (2007). Doubled length of western European summer heat waves since 1880. *Journal of Geophysical Research-Atmospheres, 112*(D15). https://doi.org/10.1029/2007jd008510

Devi, R., Gouda, K. C., & Lenka, S. (2022). Temperature-duration-frequency analysis over Delhi and Bengaluru city in India. *Theoretical and Applied Climatology, 147*(1–2), 291–305. https://doi.org/10.1007/s00704-021-03824-5

Di Napoli, C., Pappenberger, F., & Cloke, H. L. (2019). Verification of heat stress thresholds for a health-based heat-wave definition. *Journal of Applied Meteorology and Climatology, 58*(6), 1177–1194. https://doi.org/10.1175/jamc-d-18-0246.1

Donat, M. G., Alexander, L. V., Yang, H., Durre, I., Vose, R., Dunn, R. J. H., Willett, K. M., Aguilar, E., Brunet, M., Caesar, J., Hewitson, B., Jack, C., Tank, A., Kruger, A. C., Marengo, J., Peterson, T. C., Renom, M., Rojas, C. O., Rusticucci, M., Salinger, J., Elrayah, A. S., Sekele, S. S., Srivastava, A. K., Trewin, B., Villarroel, C., Vincent, L. A., Zhai, P., Zhang, X., & Kitching, S. (2013). Updated analyses of temperature and precipitation extreme indices since the beginning of the twentieth century: The HadEX2 dataset. *Journal of Geophysical Research-Atmospheres, 118*(5), 2098–2118. https://doi.org/10.1002/jgrd.50150

Dong, W. T., Zeng, Q., Ma, Y., Li, G. X., & Pan, X. C. (2016). Impact of heat wave definitions on the added effect of heat waves on cardiovascular mortality in Beijing, China. *International Journal of Environmental Research and Public Health, 13*(9). https://doi.org/10.3390/ijerph13090933

Fenner, D., Holtmann, A., Krug, A., & Scherer, D. (2019). Heat waves in Berlin and Potsdam, Germany – Long-term trends and comparison of heat wave definitions from 1893 to 2017. *International Journal of Climatology, 39*(4), 2422–2437. https://doi.org/10.1002/joc.5962

Fischer, E. M., & Schar, C. (2010). Consistent geographical patterns of changes in high-impact European heatwaves. *Nature Geoscience, 3*(6), 398–403. https://doi.org/10.1038/ngeo866

Folland, C. K., Miller, C., Bader, D., Crowe, M., Jones, P., Plummer, N., Richman, M., Parker, D. E., Rogers, J., & Scholefield, P. (1999). Workshop on indices and indicators for climate extremes, Asheville, NC, USA, 3-6 June 1997 – Breakout group C: Temperature indices for climate extremes. *Climatic Change, 42*(1), 31–43. https://doi.org/10.1023/a:1005447712757

Frich, P., Alexander, L. V., Della-Marta, P., Gleason, B., Haylock, M., Tank, A., & Peterson, T. (2002). Observed coherent changes in climatic extremes during the second half of the twentieth century. *Climate Research, 19*(3), 193–212. https://doi.org/10.3354/cr019193

Gasparrini, A., & Armstrong, B. (2011). The mmpact of heat waves on mortality. *Epidemiology, 22*(1), 68–73. https://doi.org/10.1097/EDE.0b013e3181fdcd99

Hannah, D. M., Smith, B. P. G., Gurnell, A. M., & McGregor, G. R. (2000). An approach to hydrograph classification. *Hydrological Processes, 14*(2), 317–338. https://doi.org/10.1002/(sici)1099-1085(20000215)14:2<317::aid-hyp929>3.0.co;2-t

Heo, S., Bell, M. L., & Lee, J. T. (2019). Comparison of health risks by heat wave definition: Applicability of wet-bulb globe temperature for heat wave criteria. *Environmental Research, 168*, 158–170. https://doi.org/10.1016/j.envres.2018.09.032

Hobday, A. J., Alexander, L. V., Perkins, S. E., Smale, D. A., Straub, S. C., Oliver, E. C. J., Benthuysen, J. A., Burrows, M. T., Donat, M. G., Peng, M., Holbrook, N. J., Moore, P. J., Scannell, H. A., Sen Gupta, A., & Wernberg, T. (2016). A hierarchical approach to defining marine heatwaves. *Progress in Oceanography, 141*, 227–238. https://doi.org/10.1016/j.pocean.2015.12.014

Hobday, A. J., Oliver, E. C. J., Sen Gupta, A., Benthuysen, J. A., Burrows, M. T., Donat, M. G., Holbrook, N. J., Moore, P. J., Thomsen, M. S., Wernberg, T., & Smale, D. A. (2018). Categorizing and naming marine heatwaves. *Oceanography, 31*(2), 162–173. https://doi.org/10.5670/oceanog.2018.205

Huth, R., Kysely, J., & Pokorna, L. (2000). A GCM simulation of heat waves, dry spells, and their relationships to circulation. *Climatic Change, 46*(1–2), 29–60. https://doi.org/10.1023/a:1005633925903

Karl, T. R., Nicholls, N., & Ghazi, A. (1999). CLIVAR/GCOS/WMO workshop on indices and indicators for climate extremes – Workshop summary. *Climatic Change, 42*(1), 3–7. https://doi.org/10.1023/a:1005491526870

Lanciotti, S., Ridolfi, E., Russo, F., & Napolitano, F. (2022). Intensity-duration-frequency curves in a data-rich era: A review. *Water, 14*(22). https://doi.org/10.3390/w14223705

Mandal, R., Joseph, S., Sahai, A. K., Phani, R., Dey, A., Chattopadhyay, R., & Pattanaik, D. R. (2019). Real time extended range prediction of heat waves over India. *Scientific Reports, 9*, 9008. https://doi.org/10.1038/s41598-019-45430-6

Manton, M. J., Della-Marta, P. M., Haylock, M. R., Hennessy, K. J., Nicholls, N., Chambers, L. E., Collins, D. A., Daw, G., Finet, A., Gunawan, D., Inape, K., Isobe, H., Kestin, T. S., Lefale, P., Leyu, C. H., Lwin, T., Maitrepierre, L., Ouprasitwong, N., Page, C. M., Pahalad, J., Plummer, N., Salinger, M. J., Suppiah, R., Tran, V. L., Trewin, B., Tibig, I., & Yee, D. (2001). Trends in extreme daily rainfall and temperature in Southeast Asia and the South Pacific: 1961–1998. *International Journal of Climatology, 21*(3), 269–284. https://doi.org/10.1002/joc.610

Masato, G., Bone, A., Charlton-Perez, A., Cavany, S., Neal, R., Dankers, R., Dacre, H., Carmichael, K., & Murray, V. (2015). Improving the health forecasting alert system for cold weather and heat-waves in England: A proof-of-concept using temperature-mortality relationships. *PLoS One, 10*(10), e0137804. https://doi.org/10.1371/journal.pone.0137804

Matzarakis, A., Laschewski, G., & Muthers, S. (2020). The heat health warning system in Germany-application and warnings for 2005 to 2019. *Atmosphere, 11*(2). https://doi.org/10.3390/atmos11020170

Mazdiyasni, O., Sadegh, M., Chiang, F., & AghaKouchak, A. (2019). Heat wave intensity duration frequency curve: A multivariate approach for Hazard and attribution analysis. *Scientific Reports, 9*, 14117. https://doi.org/10.1038/s41598-019-50643-w

McCarthy, M., Armstrong, L., & Armstrong, N. (2019). A new heatwave definition for the UK. *Weather, 74*(11), 382–387. https://doi.org/10.1002/wea.3629

McGregor, G. R., & Vanos, J. K. (2018). Heat: A primer for public health researchers. *Public Health, 161*, 138–146. https://doi.org/10.1016/j.puhe.2017.11.005

McGregor, G. R., Markou, M. T., Bartzokas, A., & Katsoulis, B. D. (2002). An evaluation of the nature and timing of summer human thermal discomfort in Athens, Greece. *Climate Research, 20*(1), 83–94. https://doi.org/10.3354/cr020083

McLean, K. E., Stranberg, R., MacDonald, M., Richardson, G. R. A., Kosatsky, T., & Henderson, S. B. (2018). Establishing heat alert thresholds for the varied climatic regions of British Columbia, Canada. *International Journal of Environmental Research and Public Health, 15*(9). https://doi.org/10.3390/ijerph15092048

Meehl, G. A., & Tebaldi, C. (2004). More intense, more frequent, and longer lasting heat waves in the 21st century. *Science, 305*(5686), 994–997. https://doi.org/10.1126/science.1098704

Nairn, J. R., & Fawcett, R. J. B. (2015). The excess heat factor: A metric for heatwave intensity and its use in classifying heatwave severity. *International Journal of Environmental Research and Public Health, 12*(1), 227–253. https://doi.org/10.3390/ijerph120100227

Nicholls, N., & Alexander, L. (2007). Has the climate become more variable or extreme? Progress 1992-2006. *Progress in Physical Geography-Earth and Environment, 31*(1), 77–87. https://doi.org/10.1177/0309133307073885

Nori-Sarma, A., Benmarhnia, T., Rajiva, A., Azhar, G. S., Gupta, P., Pednekar, M. S., & Bell, M. L. (2019). Advancing our understanding of heat wave criteria and associated health impacts to improve heat wave alerts in developing country settings. *International Journal of Environmental Research and Public Health, 16*(12). https://doi.org/10.3390/ijerph16122089

Oka, K., Honda, Y., & Hijioka, Y. (2023). Launching criteria of 'Heatstroke Alert' in Japan according to regionality and age group. *Environmental Research Communications, 5*(2). https://doi.org/10.1088/2515-7620/acac03

Park, J., & Kim, J. (2018). Defining heatwave thresholds using an inductive machine learning approach. *PLoS One, 13*(11), e0206872. https://doi.org/10.1371/journal.pone.0206872

Pascal, M., Le Tertre, A., & Laaidi, K. (2013). The 2003 heatwave: Impacts, public health adaptation and response in France. *Natural Disasters and Adaptation to Climate Change*, 121–127.

Pascal, M., Wagner, V., Alari, A., Corso, M., & Le Tertre, A. (2021). Extreme heat and acute air pollution episodes: A need for joint public health warnings? *Atmospheric Environment, 249*, 118249. https://doi.org/10.1016/j.atmosenv.2021.118249

Perkins, S. E., & Alexander, L. V. (2013). On the measurement of heat waves. *Journal of Climate, 26*(13), 4500–4517. https://doi.org/10.1175/jcli-d-12-00383.1

Pezza, A. B., van Rensch, P., & Cai, W. J. (2012). Severe heat waves in southern Australia: Synoptic climatology and large scale connections. *Climate Dynamics, 38*(1–2), 209–224. https://doi.org/10.1007/s00382-011-1016-2

Raei, E., Nikoo, M. R., AghaKouchak, A., Mazdiyasni, O., & Sadegh, M. (2018). Data descriptor: GHWR, a multi-method global heatwave and warm-spell record and toolbox. *Scientific Data, 5*. https://doi.org/10.1038/sdata.2018.206

Ren, Y. Q., Liu, J. P., Zhang, T. C., Shalamzari, M. J., Arshad, A., Liu, T., Willems, P., Gao, H. R., Tao, H., & Wang, T. L. (2023). Identification and analysis of heatwave events considering temporal continuity and spatial dynamics. *Remote Sensing, 15*(5). https://doi.org/10.3390/rs15051369

Robinson, P. J. (2001). On the definition of a heat wave. *Journal of Applied Meteorology, 40*(4), 762–775. https://doi.org/10.1175/1520-0450(2001)040<0762:otdoah>2.0.co;2

Russo, S., Dosio, A., Graversen, R. G., Sillmann, J., Carrao, H., Dunbar, M. B., Singleton, A., Montagna, P., Barbola, P., & Vogt, J. V. (2014). Magnitude of extreme heat waves in present climate and their projection in a warming world. *Journal of Geophysical Research-Atmospheres, 119*(22), 12500–12512. https://doi.org/10.1002/2014jd022098

Schoetter, R., Cattiaux, J., & Douville, H. (2015). Changes of western European heat wave characteristics projected by the CMIP5 ensemble. *Climate Dynamics, 45*(5–6), 1601–1616. https://doi.org/10.1007/s00382-014-2434-8

Schultz, D. M. (2009). *Eloquent science*. American Meteorological Society.

Seposo, X. T., Dang, T. N., & Honda, Y. (2017). Exploring the effects of high temperature on mortality in four cities in The Philippines using various heat wave definitions in different mortality subgroups. *Global Health Action, 10*(1). https://doi.org/10.1080/16549716.2017.1368969

Smith, T. T., Zaitchik, B. F., & Gohlke, J. M. (2013). Heat waves in the United States: Definitions, patterns and trends. *Climatic Change, 118*(3–4), 811–825. https://doi.org/10.1007/s10584-012-0659-2

Stefanon, M., D'Andrea, F., & Drobinski, P. (2012). Heatwave classification over Europe and the Mediterranean region. *Environmental Research Letters, 7*(1). https://doi.org/10.1088/1748-9326/7/1/014023

Su, Q., & Dong, B. W. (2019). Recent decadal changes in heat waves over China: Drivers and mechanisms. *Journal of Climate, 32*(14), 4215–4234. https://doi.org/10.1175/jcli-d-18-0479.1

Sulikowska, A., & Wypych, A. (2020). Summer temperature extremes in Europe: How does the definition affect the results? *Theoretical and Applied Climatology., 141*, 19–30. https://doi.org/10.1007/s00704-020-03166-8

Tong, S. L., Wang, X. Y., & Barnett, A. G. (2010). Assessment of heat-related health impacts in Brisbane, Australia: Comparison of different heatwave definitions. *PLoS One, 5*(8), e12155. https://doi.org/10.1371/journal.pone.0012155

Ullah, S., You, Q. L., Ullah, W., Hagan, D. F. T., Ali, A., Ali, G., Zhang, Y. Q., Jan, M. A., Bhatti, A. S., & Xie, W. X. (2019). Daytime and nighttime heat wave characteristics based on multiple indices over the China-Pakistan economic corridor. *Climate Dynamics, 53*(9–10), 6329–6349. https://doi.org/10.1007/s00382-019-04934-7

Wang, P. Y., Tang, J. P., Sun, X. G., Wang, S. Y., Wu, J., Dong, X. N., & Fang, J. (2017). Heat waves in China: Definitions, leading patterns, and connections to large-scale atmospheric circulation and SSTs. *Journal of Geophysical Research-Atmospheres, 122*(20), 10679–10699. https://doi.org/10.1002/2017jd027180

Xu, Z. W., FitzGerald, G., Guo, Y. M., Jalaludin, B., & Tong, S. L. (2016). Impact of heatwave on mortality under different heatwave definitions: A systematic review and meta-analysis. *Environment International, 89-90*, 193–203. https://doi.org/10.1016/j.envint.2016.02.007

Xu, Z. W., Crooks, J. L., Black, D., Hu, W. B. A., & Tong, S. L. (2017). Heatwave and infants' hospital admissions under different heatwave definitions. *Environmental Pollution, 229*, 525–530. https://doi.org/10.1016/j.envpol.2017.06.030

Xu, Z. W., Cheng, J., Hu, W. B., & Tong, S. L. (2018). Heatwave and health events: A systematic evaluation of different temperature indicators, heatwave intensities and durations. *Science of the Total Environment, 630*, 679–689. https://doi.org/10.1016/j.scitotenv.2018.02.268

Yang, J., Yin, P., Sun, J. M., Wang, B. G., Zhou, M. G., Li, M. M., Tong, S. L., Meng, B. H., Guo, Y. M., & Liu, Q. Y. (2019). Heatwave and mortality in 31 major Chinese cities: Definition, vulnerability and implications. *Science of the Total Environment, 649*, 695–702. https://doi.org/10.1016/j.scitotenv.2018.08.332

You, Q. L., Jiang, Z. H., Kong, L., Wu, Z. W., Bao, Y. T., Kang, S. C., & Pepin, N. (2017). A comparison of heat wave climatologies and trends in China based on multiple definitions. *Climate Dynamics, 48*(11), 3975–3989. https://doi.org/10.1007/s00382-016-3315-0

Chapter 3
Heatwaves as Extreme Events

Abstract Heatwaves are significant extreme events characterized by their low frequency but high magnitude, making them noticeable due to their deviation from normal climate conditions and their substantial societal and economic impacts. This chapter opens with material on the conceptualization of heatwaves as extreme events, exploring their statistical properties and the broader context of climate extremes. It highlights the complexity of defining extremes, noting that their impact is context-specific, influenced by local climatological means, atmospheric circulation variability, and societal vulnerability. The chapter discusses the multifaceted nature of heatwaves and explores their physical complexity, distinguishing between simple and complex events as well as providing an overview of heatwaves as compound and cascading events. The discussion also extends to the temporal and spatial clustering of heatwaves and introduces the idea of heatwaves as 'whiplash' events. Also sketched out are some principles for understanding the origin of heatwaves as extreme events. The challenges associated with data preparation for the statistical analysis of heatwaves as extreme events is outlined with emphasis given to the need for long-term, high-resolution climate data for understanding heatwaves as statistical entities. An overview of the various data sources for the statistical analysis of heatwaves is provided touching upon instrumental records, climate reanalysis products, and climate model outputs. The chapter also presents an overview of Extreme Value Theory and the Generalized Extreme Value (GEV) and Generalized Pareto (GP) approaches to analysing extreme values along with a discussion of the uncertainties and challenges associated with extreme value analysis.

Keywords Conceptualisation of extremes · Heatwaves as extreme events · Compound and cascading events · Whiplash events · Statistical analysis of heatwaves · Instrumental records · Climate reanalysis products · Extreme value theory

G. McGregor, *Heatwaves*, Biometeorology 6,
https://doi.org/10.1007/978-3-031-69906-1_3

3.1 Introduction

Heatwaves are one of a range of extreme climate events. By being extreme, heatwaves do not occur often, but when they do, they are noticeable because of their unusualness relative to 'normal' climate conditions, or impact on society or economy. In statistical speak, heatwaves are low frequency high magnitude events. As such heatwave researchers are interested in understanding the statistical properties and likelihood of occurrence of heatwaves related to their frequency, duration, and intensity as well as their spatial distribution, and temporal changes. The same applies to the variables that characterise the physical nature of heatwaves such as temperature and atmospheric moisture. Such analyses provide valuable information for risk managers, policymakers, urban planners, public health officials, and other stakeholders to make informed decisions related to the management of heatwave risk. While the previous chapter focused on defining heatwaves it did not address why heatwaves are considered extreme events. Given this, part of the purpose of this chapter is to place heatwaves in a climate extremes context and consider ways in which heatwaves as extreme events may be conceptualised. To understand the statistical characteristics of heatwaves the requisite climate data needs to be assembled. While on the surface this might appear a simple task a range of matters need to be considered in preparing data for statistical analysis. Accordingly, this chapter provides an overview of these matters before proceeding to introduce the main approaches to the statistical analysis of heatwaves as extreme events.

3.2 Conceptualising Heatwaves as Extremes Climate Events

Heatwaves are often casually accepted as extremes events with little consideration given to why they might be so in a sense broader than their statistical properties. In brief heatwaves are not as simple as they might appear. Accordingly, this section presents a wider view of the nature of extremes and invites the reader to think about heatwaves in that context.

3.2.1 Extreme as a Concept

The English word extreme has a deep etymology deriving from the Latin *extremus* and before that the Proto-Italic *eksterisemos*. *Extremus* meaning 'outermost', is also a superlative of the Latin *extra*, which indicates something 'outside of' or 'beyond'. Adopting a human-centric position for the meantime, extreme refers to something that is outside our usual cultural assumptions or experience. However, while extremes may fall into the category of human experience that is beyond the norm, they are easily recognisable. This is captured well by the phrase 'I know it when I see it', uttered by the US Justice Potter Stewart when adjudicating on a court case related to unsavoury material contained in a film (Gewirtz, 1996). Despite their recognisability, extremes remain difficult to define in both a statistical and societal sense. Consequently, there are wide ranging statistical definitions of extremes (Alexander et al., 2006; Katz, 2010; Stephenson, 2008). Further, how extreme an event is (its 'extremeness') is context specific, being dependent upon a location's climatological mean for a particular period of the year, as well as an atmospheric circulation variability component. Extremeness described in this way is of course purely from the physical perspective. Alternatively, we can think about the 'societal extremeness' of an event and the extent to which this might be contingent on the level of population vulnerability at the time of an event; a moderately sized physical event may well engender a heavy societal toll because of the societal state at the time and longer-term adaptation factors. As we see will in Chap. 10, vulnerability is an important moderator of the impact of heatwaves.

Added to the challenge of defining what an extreme event means is the interchangeable use of the word extreme with other adjectives such as severe, rare and high-impact when describing noticeable events. This may lead to some confusion, as each of these adjectives connote their own distinct meaning in relation to the nature of the event. As pointed out by Stephenson (2008) in a weather or climate context, an extreme event refers to an occasion when one of a range of meteorological (daily maximum temperature) or climatological (summer rainfall deficit) variables attains an extreme value that is either a maximum value or above some pre-defined critical threshold value. In the case of heatwaves, this might be an extreme value of maximum temperature or a combination of extreme values for two or more variables (e.g. temperature and wind speed). Therefore, in saying a heatwave event is extreme, the physical meteorological variable is the point of reference. In a statistical sense, extreme values are rare. A rare event thus refers to a situation with a low probability of occurrence, usually expressed in percentage terms (1% probability of occurrence—probability = 0.01) or using concepts such as return period (100-year return period). Of course, an extreme and rare event can exact a heavy toll on society making it a severe event, with severe referring to the level of loss. There are several ways of quantifying loss or severity, with some of the most common being financial costs or number of deaths. The expected loss from an event is often articulated as the level of risk, which is contingent upon the interaction between the occurrence probability of a weather or climate hazard (e.g. a heatwave), and the exposure and vulnerability of the so-called exposure unit (a specific population group, community, industry or ecosystem service) as well as whether various response strategies are in place (e.g., adaptation and mitigation measures). As noted by (Magnan et al., 2023), the notions of physical, ecological and social thresholds leading to transformational and possibly abrupt changes, the irreversibility of these changes, and the cascading effects within and across systems are often neglected when considering what constitutes severe or dangerous climate risk. Lastly, a high-impact event refers to a short-term weather system or a medium to longer-term mode of atmospheric circulation or climatic variability that provides the conditions for attaining extreme values of a particular meteorological or climatological variable (Stephenson, 2008). For example, a tropical cyclone (hurricane), as a high-impact weather system, might generate extreme wind speed values. Similarly, the El Niño Southern Oscillation (ENSO) phenomenon as a high impact mode of climatic variability may provide the requisite conditions for prolonged periods of drought, via, for example, altering the trajectory of rainfall bearing weather systems.

Further to distinguishing between extreme, severe, rare, and high-impact events, Stephenson (2008) has suggested borrowing the terms chronic and acute from the field of medicine, which are used to describe the temporality of an illness, as this may assist with differentiating rapid onset and short-lived (acute) from prolonged and/or repeating (chronic) weather and climate events. However, heatwaves, defined generically as periods of unusual heat, may not sit comfortably into this simple binary. This is because observation tells us that heatwaves can occur on a range of timescales. For example, and notwithstanding what strictly constitutes rapid and short-lived, the 1995 Chicago heatwave had a rapid onset and short duration—an acute event. Contrasting with this is the 2010 Russian event. In the case of this event, strong temperature anomalies built over several days with a period of unprecedented temperature extremes lasting nearly 6 weeks from early July into August (see Chap. 6), clearly an event that is beyond what might be considered short lived. Whether heatwaves as a general category of extreme events are acute or chronic may depend on the climate context. One might postulate that for climate regimes, in which heatwaves are not part of the 'normal' climate expectation, then a short, or even prolonged, almost one-off heatwave event could be considered acute. Counter to this, for some climate regimes heatwaves might be a repeating feature with a relatively high probability of annual recurrence making them chronic events. Indeed, in the context of climate change, heatwaves may become a chronic problem for some land and ocean regions of the world. Complementing the acute/chronic taxonomy of extreme events is also the possibility of events being placed into the categories of rare and severe and rare and non-severe (Stephenson, 2008). While such a taxonomy may be useful, the broad category of rare and non-severe in many ways is human-centric as it implies that non-human systems (e.g., a forest ecosystem), including the inanimate (e.g., ocean waters), although experiencing rare events, will suffer no loss. Given recent developments in philosophy, related to new materialism (Coole & Frost, 2010), planetarity (Elias & Moraru, 2015), and object-oriented ontology (Harman, 2018), it is perhaps worth reflecting on whether privileging the human over the non-human is advisable in creating such taxonomies that identify non-severe (non-loss) events. There is always the possibility that an extreme will affect something, whether it be directly or indirectly or be noticed or unnoticed.

3.2.2 *Extremes as Statistical Entities*

Because a measured value of one or several meteorological variables is used to describe the physical nature of a weather or climate extreme (e.g., temperature), extremes are generally referred to as events that generate a maximum value or exceed a historical value of significance. This implies that extremes are events that occupy a unique position amongst a range of historically recorded values. Given this, extreme events have distinct statistical properties (see next section). Such a strict statistical view of extremes does however negate consideration of these as societal events because it places emphasis on statistical properties as opposed to the impacts arising from anomalous events. This is important because the societal outcome associated with an extreme statistical event will very much depend on other factors at play at the time of the event, including for example how well a community may be prepared. The prospect therefore exists that an event with a 'lesser' statistical significance may generate a significant impact because of societal conditions at the time. So why when undertaking the analysis of extremes do we default to a consideration of the statistical distribution of variable values? The reason is that while long records of variable values exist, such as those from temperature monitoring instruments, the requisite data for extreme event analysis from a societal perspective is poor in comparison, or simply does not exist. Added to this are inhomogeneities and non-stationarities (see Sect. 3.2.3) in social impact data. Such issues arise from improved reporting of events with the result many observations of impacts are biased towards recent years. The fact that many reports of societal impacts use quite different metrics for describing the scale of an impact also does not help, especially if a goal is comparison between locations. Given this, much effort has been invested in exploring the requirements for a global disaster database that is consistent in terms of its thresholds, spatial resolution, and data quality measures. In many ways the adoption of the Sendai Framework for Disaster Risk Reduction in 2015 by many governments and the need for an internally consistent database of disaster events amenable to a range of analyses has catalysed such a quest (Huggel et al., 2015; Moriyama et al., 2018; Tanaka et al., 2018). So, until issues associated with climate related societal impact databases are addressed with confidence, most analyses of extremes will by default, focus on the statistical distribution of values recorded for physical variables such as temperature, precipitation, wind and so on.

3.2.3 *Multi-faceted Nature of Extremes*

Adding to the challenges of defining extreme events is their multi-faceted nature. Because extreme events possess several attributes, including rate of occurrence (probability per unit time), magnitude (intensity), temporal duration and timing, spatial scale (footprint) and multivariate dependencies (Stephenson, 2008), the possibility arises that an extreme heat event could be classified and therefore ranked in a variety of ways in terms of its relationship with other heat events. For example, while a particular heatwave event might be ranked as the highest in terms of magnitude this may not be so for spatial extent.

3.2.4 *Physical Complexity*

Another way to think about heatwaves is their physical make up with some being 'physically simple', with only one atmospheric property being critical (e.g., temperature) in terms of the events severity or the degree of loss incurred, whereas others may be 'complex' with two or more variables (e.g., high temperatures and strong winds) acting conjointly to exact heavy losses. Interestingly, and from a conceptual point of view, the extremeness of meteorological variables comprising a complex severe event may not necessarily need to be great. This is because their synergistic effects could bring about significant losses. For example, the combined effects of clear skies, moderately high but not extreme temperatures and medium strength dry winds lasting a few days over a large area will favour high inputs of solar radiation, moderately strong surface heating and heat advection. Such a combination of circumstances could have devastating impacts on agricultural crops because of the resulting exceptionally high evaporation rates arising from this combination of circumstances leading to soil moisture depletion and the mechanical battering of crops by what essentially is a prolonged blast of warm dry air.

3.2.5 *Compound, Cascading and Multi-hazard Events*

While the term 'complex' describes events arising from the interaction of two or more variables, they are really a sub-category of what is a compound event (Hao et al., 2018; Leonard et al., 2014; Liu & Huang, 2014; Visser-Quinn et al., 2019; Zscheischler et al., 2020). There is some debate however as to whether compound, multi-hazard or cascading events are one in the same (Pescaroli & Alexander, 2018; Tilloy et al., 2019). For example, cascading events are generally ones that follow one after another (Toscano et al., 2022; Zaidi, 2018) or propagate across space (Carter et al., 2021)—see Chap. 11 for material on cascading impacts. Multi-hazard refers to several events occurring either at the same time, or more typically at a specific location over time such that a characteristic of some locations is a mixture of hazards. The IPCC has defined a compound event as: (1) two or more extreme events occurring simultaneously or successively, (2) combinations of extreme events with underlying conditions that amplify the impact of the events, or (3) combinations of events that are not themselves extremes but lead to an extreme event or impact when combined (Seneviratne et al., 2012). In relation to heatwaves, examples of these three sub-categories might include:

(i) Two successive heatwaves, a heatwave and a wildfire, multiple simultaneous heatwaves,
(ii) Dry or moist heatwave conditions arising from coinciding extreme temperature and wind, or high temperatures, high humidity, and low wind speeds, respectively, and
(iii) Prolonged moderate temperatures over a region already experiencing low soil moisture levels, leading to drought and land-atmosphere feedbacks that increase the probability of a heatwave.

Although the IPCC definition appears to be straight forward, it contains several ambiguities making compound events difficult to define. Leonard et al. (2014) have identified these to be:

(i) Whether background meteorological conditions amplify an existing event or are an integral part of a compound event,
(ii) Ill-definition of the temporal and spatial scales associated with successive and simultaneous events,
(iii) Whether a compound event is a combination of multiple variables (extreme or otherwise), or the result of two or more distinct events, and
(iv) Whether the combination of events leading to an impact is restricted to non-extremes.

As a remedy to these issues and those associated with the categorisation of compound events into discrete classes (e.g., heatwave and fire, drought, and heatwave) that are independent of the physical system underlying their generation, a more general definition of compound events has been suggested (Leonard et al., 2014). This emphasises the event's impact or extremeness instead of the variables comprising the event, the dependency of the impact on multiple variables or events, and the fact that there is a physically meaningful dependence between the impact-forcing variables or events. Given this set of criteria, an example of a compound heatwave event would be one in which the impact, in terms of an exceptionally high level of the maximum temperature, depends on a blocking event (see Chap. 6) and depleted soil moisture in the middle of the summer (see Chap. 7). This is because, the maximum temperature (the impact metric) reaches an extreme level, and the impact depends on multiple processes including the blocking event and land surface boundary conditions. Further, the maximum temperature has a physical dependence on the blocking event and land surface conditions due to clear sky conditions and large solar radiation inputs to the surface. Added to this, stable anticyclonic conditions will be conducive to subsidence warming and heat accumulation in the atmospheric boundary layer with land-atmosphere feedbacks favouring the partitioning of net radiation into sensible heat flux (see Chap. 7). Timing within the season is also important as low soil moisture levels are likely to be obtained by mid- to late-summer.

Despite the alternative approach to defining compound events as outlined above compound heatwave events are probably best understood as successive heat events with short breaks between them. This is like the idea of cascading events although cascading usually refers to a different type of hazard following on closely after the occurrence of the initial hazard. In perhaps the first attempt to identify the occurrence of compound heatwave events in current and future climates, Baldwin et al. (2019) have proposed several definition criteria related to individual heatwave onset, an intermittent break, re-initiation, and cessation. Their motivation stems from the observation that many heatwave definitions largely assume periods of uninterrupted anomalous heat. Although there are some exceptions, with brief usually 1-day periods with temperatures below a stated threshold permitted, most definitions terminate a heatwave if continuous exceedance-threshold and duration criteria are not met. For example, and as aptly illustrated by Baldwin et al. (2019), strictly applying the heatwave definition of (Fischer & Schar, 2010) (see Table 2.2), to several 'deadly' post-1980 European and US heatwaves (Fig. 3.1), results in only the 2003 European and 2010 Russian events meeting the continuous 6-day exceedance criteria. Consequently, the impactful 1995 Chicago event and the temporally complex 1994 Belgium event, as well as other events depicted in Fig. 3.1, would not be classed as heatwaves.

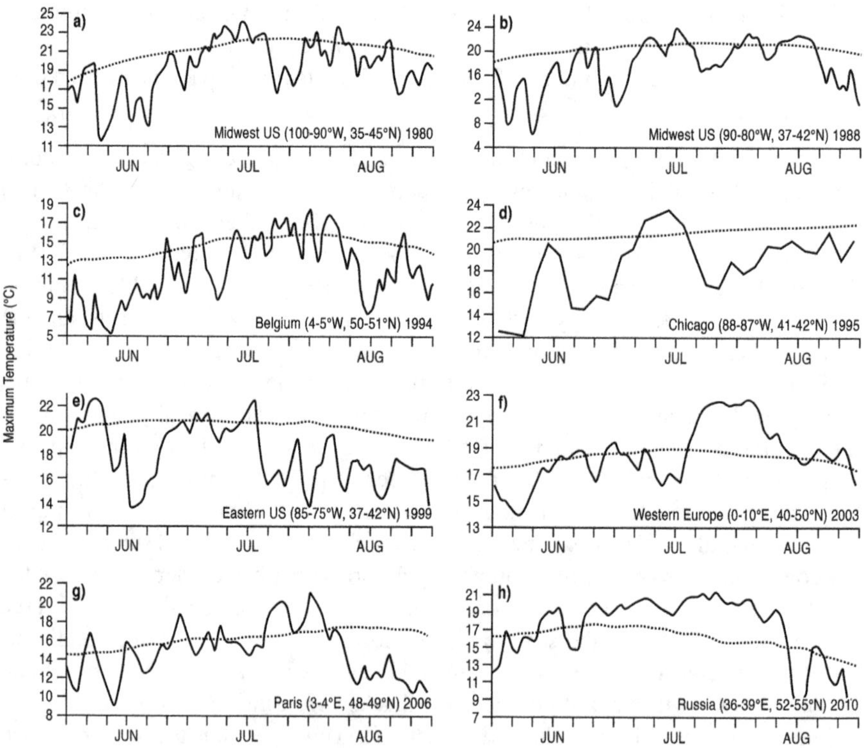

Fig. 3.1 Plot of daily minimum temperature (black) and corresponding seasonally varying heatwave thresholds (red) for eight severe heatwaves across the US and Europe. Periods for which the minimum temperature exceeds the threshold temperature are potential heatwave days. Source: Baldwin, J. W., Dessy, J. B., Vecchi, G. A., & Oppenheimer, M. (2019). Temporally compound heat wave events and global warming: An emerging hazard. Earth's Future, 7, 411–427. https://doi.org/10.1029/2018EF000989. Used under Creative Commons Attribution-Non Commercial-No Derivatives 4.0 International (CC BY-NC-ND 4.0) Licence https://creativecommons.org/licenses/by-nc-nd/4.0/

Acknowledging this shortcoming and noting that climate change could well generate more temporally complex heatwave events, Baldwin et al. (2019) proposed a modification to the Fischer and Schar (2010) definition (also referred to as the WSDI), which facilitates the identification of temporally compound heatwaves. To achieve this they dropped the requirement of exceedance criteria meeting six consecutive days and instead proposed three outline conditions namely (1) a minimum number of hot (above threshold) days occurring consecutively to start a heatwave, (2) a maximum number of cooler (below threshold) days that can occur consecutively for the heatwave to continue, and (3) a minimum number of hot days occurring consecutively that can add onto a heatwave after a break (Baldwin et al., 2019). Also permitted are multiple breaks in a single heatwave provided the additional days follow the conditions specified in 2 and 3 above. Based on the three conditions, Baldwin et al. (2019) present a range of options related to the temporal structure of compound heatwaves (Table 3.1) and apply a three number system to denote the value associated with each condition respectively. For example, a 321-compound event indicates a minimum of three hot days to initiate the event, a maximum break length of two cooler days, and at least one subsequent day that forms the compounding hot day following the 'cool' break (Baldwin et al., 2019). According to their example definitions, compound heatwave events could be as short as 3 days (a minimally sufficient '111' event) or in the case that multiple breaks are permitted and the cool break following a 6-day compound event is greater than 3 days (that is the event does not continue) then a compound event could be at least as long as 15 days (a '636' event).

In some ways, the Baldwin et al. (2019) definition of compound heatwaves is an attempt to put a stake in the ground and secure this definition as one widely accepted as the basis for identifying and analysing these important events. Undoubtedly the aspiration of all definition developers, the reality is, as with heatwave definitions in general, other definitions of compound heatwaves are likely to emerge and 'compete' with this seminal definition. No doubt, some will resonate with the general conception that compounding describes successive events linked by short 'non-event periods', depending on how event/non-events are defined. However, there is always the potential that other conceptions of compounding will be proffered. For example, some researchers use a combination of day and night temperatures to define a compound heatwave event as a period when both daily Tmax and Tmin exceed the corresponding calendar day's 90th percentile values for at least 3 days with the 90th percentile of a calendar day calculated based on a 15-day moving window around the calendar day for the reference period (Jiang et al., 2024; Li et al., 2017; Wang et al., 2020). While statistically speaking this may not appear problematic, from a physical perspective it is as if this type of approach implies that day and nighttime conditions are somehow separate which of course is not that case as heatwaves are 24-h events with daylight and nocturnal processes inextricably linked.

Setting aside the subtleties of definitions, from an impact perspective, compound, cascading or clustered events are likely to be more important than single independent events because of their cumulative effects (AghaKouchak et al., 2018; Manning

Table 3.1 Options for a temporally flexible compound heatwave definition

Definition parameter	Options
Temperature data	Daily minimum or daily maximum
Threshold percentile	90th or 95th
Minimum initial heat event duration	1, 3 or 6 days
Maximum break duration	1, 2 or 3 days
Minimum subsequent heat event duration	1, 3 or 6 days

Adapted from (Baldwin et al., 2019)

et al., 2019; Mazdiyasni & AghaKouchak, 2015). Consequently, there is now a tendency to place heatwaves within a multi-hazard framework (Baklanov et al., 2018; Depietri et al., 2018; Forzieri et al., 2016; Johnson et al., 2016; Vitolo et al., 2019), because such an approach facilitates the identification of potential disaster hotspots (Dilley et al., 2005) that arise due to the joint hazardousness of a place. Despite the attractiveness of this concept, some challenges exist including multi-hazard identification and mapping, data demands, and conflicting results from different methods of multi-hazard assessment, (Barrantes, 2018; Kappes et al., 2012; Liu et al., 2016).

3.2.6 *Temporal and Spatial Clustering*

A further characteristic of extreme events is their temporal or spatial clustering (Telesca, 2007). This is distinct from the ideas of compound and cascading events. In the case of heatwaves, clustering may manifest as a tendency for these to occur in a specific region, climate zone or climate type or the predominance of heatwaves for a set of years such that heatwave 'rich' and 'poor' periods occur. In comparison to landslides (Benz & Blum, 2019), lightning (Gou et al., 2018), wildfires (Parente et al., 2016), tornadoes (Malamud et al., 2016) and floods (Merz et al., 2018), the literature on heatwave clustering is meagre or couched more in terms of compound or cascading hazards than temporal and spatial clustering in a strict sense (Sutanto et al., 2020). For some regions it appears that clustering of heatwave events may occur. In this regard, Portugal is a good example. Here, the highest frequency of heatwaves and the heatwaves with the greatest duration tend to occur over the north-eastern and southern parts of the country, while those with the greatest amplitude occur over the central regions of Portugal (Parente et al., 2018). Similarly for Senegal, possible spatial clustering of heatwave events may occur related to the positioning of positive pressure anomalies over western parts of the Mediterranean Sea (Sambou et al., 2020). In a climate change context, there is some evidence from modelling that consistent geographical patterns of heatwaves will emerge, with heatwaves mainly clustered over southern Europe (Fischer & Schar, 2010). As well as frequency, there is the possibility that clustering might apply to other characteristics such that while one region may experience the highest frequency of heatwaves another region may be characterised by long duration events, or indeed, there could be teleconnections between regions. Added to this, clustering may vary with event magnitude or threshold. As for other natural hazards, the clustering behaviour of heatwaves is likely to relate to the land and atmosphere characteristics and processes that predispose a region, a particular time of the year or run of years to the occurrence of heatwaves. Establishing the clustering characteristics of heatwave appears to be a wide-open research area with climate type, land surface memory, blocking and the inter-annual variability of modes of atmospheric circulation being good candidates for exploring the spatial and temporal clustering of heatwave events.

3.2.7 *Heatwaves as Weather Whiplash Events*

Weather whiplash refers to a rapid and significant shift in weather conditions, often characterized by abrupt changes in temperature, precipitation patterns, or other meteorological factors within a relatively short period. Such significant fluctuations in weather conditions can cause disruptions to ecosystems, infrastructure, and human activities (Creed et al., 2023). Perhaps unsurprisingly, the social-ecological context in which whiplash events occur, for example the degree of emergency and response preparedness and capability, can moderate their level of socio-economic impact (Das et al., 2018). Although weather whiplash has been applied to describe transitions in hydrometeorological conditions at a range of timescales, from the daily to the inter-annual (He & Sheffield, 2020; Lee, 2022; Liu et al., 2023; Madrigal et al., 2024; Swain et al., 2018), and often applied to winter weather events, such as rapid snowmelt or river-ice breakup brought on by sudden warming, it would seem the term in many ways aptly describes heatwaves. This is because despite being considered as pervasive events, the global record of severe heatwave events indicates that many have evolved and decayed quite rapidly, well within the synoptic timescale of 5–10 days (see Chap. 6). In terms of atmospheric dynamics, weather whiplash occurs when there's a sudden transition from a sustained circulation pattern to a new pattern, which may or may not persist over time (Francis et al., 2022, 2023). This resonates with the type of atmospheric circulation regime shifts associated with heatwave development in the mid-latitudes, specifically the rapid transition from a non-blocking to a blocking (heatwave causing) pattern in the upper westerlies with a switch back to a non-blocking situation (see Chap. 6). Although not specific to heatwaves, Lee (2022) in a study of rapid temperature changes as a form of weather whiplash, found that over a 70-year period the ranges (variability) in 7-day and 1-day temperature have increased substantially in all seasons across the globe, with significant upward shifts since 1985. Akin to the timescale of heatwaves, the results of this particular study invites investigation of whether some of the recently experienced highly impactful heatwave events could be considered as whiplash events.

3.2.8 Origin of Extremes

Conceptually extremes and thus heatwaves can also be considered in terms of their origins. In this respect the guiding principles and the suite of underlying generic processes suggested by Stephenson (2008) (Table 3.2) offers an avenue for understanding the processes that underlie the generation of extreme events. For example, by considering the role of land-atmosphere feedbacks in mega-heatwave generation (see Chap. 7), climate scientists have shed light on the processes leading to the development of past events. This has informed the advancement of sub-seasonal to seasonal (S2S) forecasting of mega-heatwave events.

Table 3.2 Principles and processes for understanding the origin of heatwaves as extreme events (Adapted from Stephenson (2008)

Principles
The evolutionary principle: Heatwaves do not arise spontaneously. Rather they evolve continuously from less extreme events with a lifecycle of growth and decay driven by competing positive and negative feedback processes with the magnitude of an extreme heatwave dependent on this competition. In this way, heatwaves represent local heat maxima as a result of a continuous process of heat growth and decay.
The stationary principle: There is a balance between forcing and dissipation tendencies for heatwave events. Given this, local temperature maxima and minima are quasi-stationary states in which the rate of change of their amplitude is zero. A corollary of this is an imbalance of heatwave forcing and dissipation processes will manifest in a change in the heatwave climatology. For example, the process of urbanisation, via the heat island effect, may unfavourably tilt the probability distribution of temperature values to the high end.
Processes
Rapid growth due to instabilities caused by positive feedbacks: For example, the rapid growth of heatwave events due to Rossby wave folding and associated thermo-dynamical feedbacks (see chap. 6).
Displacement of a weather system: A region may start to experience a set of 'new' weather conditions. For example, extension of a subtropical ridge over an area, otherwise not affected by this circulation feature, or temporal expansion of heatwave generation conditions beyond the 'normal' season for a particular region.
Simultaneous coincidence: The cumulative effect of non-extreme conditions. For example, the successive occurrence of a number of 'warm' spells, or the synergistic effects of several meteorological variables that create impactful conditions such as moderately high temperatures in combination with high humidity.
Localization of activity into intermittent regions: A result of the regular occurrence of a set of conditions tied to the seasonal cycle or a particular weather singularity. For example, hot spells during monsoon 'breaks' associated with the movement of the monsoon trough over northern India.
Persistence or frequent recurrence of atmospheric conditions: Slow variations in the climate system related to the evolution of surface boundary conditions such as sea surface temperature or soil moisture, which lead to prolonged ocean- and/or land-based heatwaves due to positive ocean/land-atmosphere feedbacks.
Natural stochastic/chaotic variation: The chronicling of heatwave extremes is dependent on climate record length such that as the length of climate records increases the likelihood of measuring or recording extreme heatwave events increases.

3.3 Data Sources for Extreme Event Analysis

To understand the nature of extremes for a particular location or region reliable data sources are necessary. Sources of data generally comprise digitised instrumental records, reconstructed climate records, reanalysis products and the output from climate models.

3.3.1 *Instrumental Records*

Traditionally, the analysis of extremes has drawn on manual or automatic measurements of variables from weather and climate observing stations. Such observations may be available for a range of time scales. Clearly, the temporal resolution of observations will constrain conclusions drawn about extreme event occurrence. Weekly (once a week) manual observations of temperature for example will be of little use if a picture about extremes at the daily timescale is the desired outcome of an analysis. In the case of heatwaves, which occur over consecutive days and at the sub-monthly and sometimes sub-weekly time scales, continuous daily data are essential. For high-resolution analyses of heatwaves, even hourly data may be required.

By definition extremes are rare events. Accordingly, as long a record as possible for a particular climate variable or event is required to maximise the possibility of capturing the occurrence of extremes. In the case of heatwaves, often described in relation to extreme values of maximum or minimum temperature, the length of temperature records is critical. Ideally, records comprising daily data with a minimum length of 30 years are required for the analysis of extremes. Despite the importance of high-resolution data sets and the need for sets of critical climate variables for extreme event analysis, in some cases the length of climate records or indeed their spatial coverage may not be sufficient to establish with any confidence the climatology of extreme events for some regions of the world. This is usually because historical, political or economic factors (Fellous, 2008; Gunasekera et al., 2018; Strangeways, 2015), have or continue to work against the maintenance or establishment of climate monitoring networks.

A casual look at the distribution of land-based temperature stations included in the Global Historical Climate Network (GHCN) demonstrates clearly that while the global coverage of temperature observations has improved over time (Lawrimore et al., 2011), it remains skewed to regions with strong economies (Fig. 3.2). In contrast to North America and Europe temperature observation networks in regions such as Africa, South America and Southeast Asia are less well developed. Indeed, some regions of the earth's land surface lack observations. Despite this worrying situation, efforts at several levels to rectify the problem of inadequate data for climate extreme event analysis are underway. For example, in 2011, fourteen West African countries collaborated to construct a data set for the analysis of climate

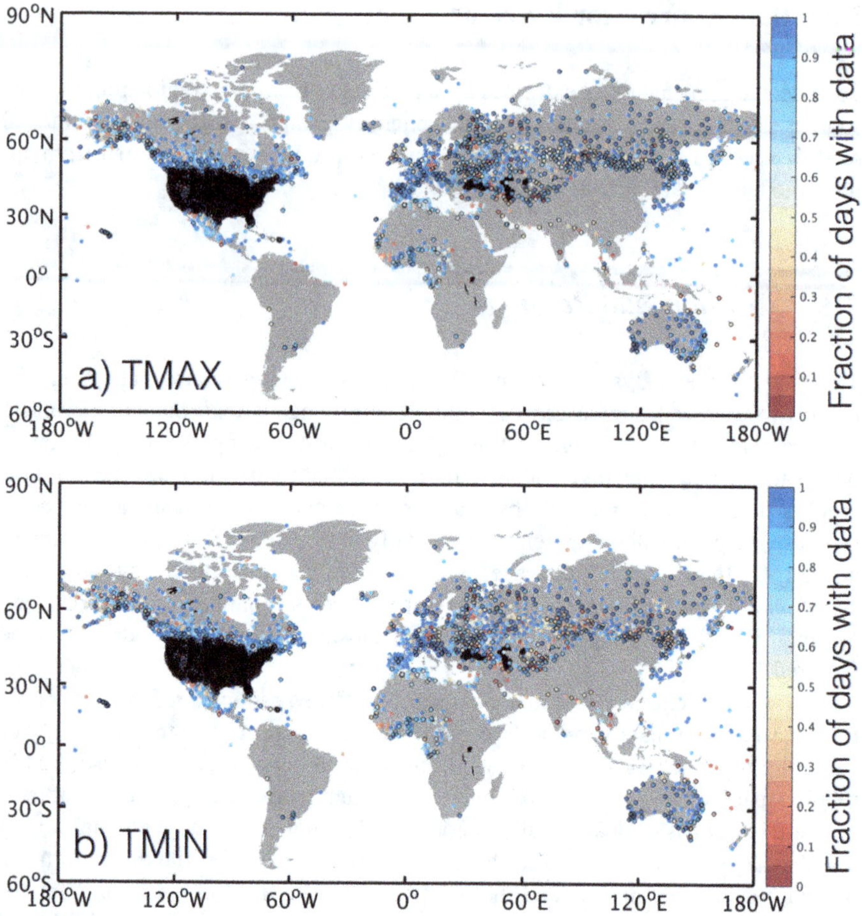

Fig. 3.2 Fraction of days with data within the record of each GHCN-D station for (**a**) maximum temperature (Tmax) and (**b**) minimum temperature (Tmin). Stations are shown if they have records beginning no later than 2015. Source: K McKinnon and National Centre for Atmospheric Research Staff (Eds). 'The Climate Data Guide: GHCN-D: Global Historical climatology Network daily temperatures'. https://climatedataguide.ucar.edu/climate-data/ghcn-d-global-historical-climatology-network-daily-temperatures. (Accessed 3/03/2024)

extremes (Barry et al., 2018). At the global level various agencies have invested significantly in compiling daily climate data that lend themselves to the analysis of extremes and thus the establishment of heatwave climatologies. A good example is the Global Historical Climate Network's daily data set (GHCN—Daily). This contains records from over 100,000 stations in 180 countries and territories and is updated daily from a variety of data streams (Menne et al., 2012). Other global data sets that bear relevance to the analysis of heatwave occurrence and their processes include a land surface specific humidity data set (Willett, Dunn, et al., 2014b; Willett, 2023), a near global gridded data set of temperature and precipitation extremes (Donat et al., 2013) and a global data set of diurnal temperature range (Thorne et al., 2016). Equally important as land-based observations are those for the world's ocean regions, especially in relation to the monitoring of sea surface temperatures (Kent et al., 2019; O'Carroll et al., 2019) and understanding the origin and extent of marine heatwaves (Frolicher & Laufkotter, 2018).

In addition to data gaps for some land and ocean regions, is the case of cities where large numbers of potentially vulnerable people live. Regrettably, detailed climate data for cities for researching both the temporal and spatial characteristics of heatwaves is largely lacking. This has constrained the development of a full understanding of the complexities of urban temperature fields and associated extremes (Chen et al., 2012; McGregor et al., 2017). This situation has arisen because weather and climate stations have been located to gather measurements representative of atmospheric processes over regions unaffected by human activity. Consequently, there are few urban weather stations with long records. Accordingly, as discussed in Chap. 8, attention is now being turned to the development of urban climate networks and information systems with satellite-based observations of surface temperature and the output from urban climate numerical models now contemplated as suitable data sources for understanding heatwaves at the urban scale.

3.3.2 Data Rescue, Reconstruction and Generation

Because of the incompleteness of climate data in both space and time, some climate scientists now invest considerable effort in historical climate data rescue and climate reconstruction (McGregor, 2015). Rescued or reconstructed data can serve a number of climate research purposes and importantly in the case of heatwave analysis extend the climate record for a single location or region allowing analyses of extremes (McGregor, 2015). Data rescue efforts at both the national and international levels include the Mediterranean Data Rescue (MEDARE) (Brunet & Jones, 2011), the Atmospheric Circulation Reconstructions over the Earth (ACRE) (Allan et al., 2011; Allan et al., 2016), the Uncertainties in Ensembles of Regional ReAnalyses (UERRA) (Ashcroft et al., 2018) and Data Rescue Archive Weather (DRAW) (Park et al., 2018) initiatives. Understandably because of the increasing importance of the role of rescued and newly constructed data sets for extreme event analysis, there has been a call for the global implementation of a set of climate data standards and protocols to ensure data quality (Muller et al., 2013; WMO, 2023b). Further sources of climate information pertinent to heatwave analysis are personal or institutional diaries, ships logbooks, expedition and other types of historical documents (McGregor, 2015). These can provide a rich resource for chronicling the occurrence of extremes (Endfield & Veale, 2018). Complimenting data rescue efforts are techniques for data generation based on historical values. For example, at the regional to local scale, stochastic weather generators, which draw on the statistical characteristics of observed meteorological time series to simulate weather time series have been developed to infill and estimate meteorological variables in data sparse regions (Chen & Brissette, 2014; Rahmani et al., 2016; Semenov, 2008; Wilks, 2012).

3.3.3 Climate Reanalysis Products

Increasingly extreme event analyses make use of climate reanalysis products. Reanalysis products are the output from numerical weather prediction models run with historical data. Unlike weather analysis and prediction models, which must be run as soon as possible to produce weather predictions and are often initialized with incomplete observational data, reanalysis has the advantage of using as complete an observational data set as possible including land/ocean surface and subsurface, upper air and satellite-based data. This is because the reanalysis modelling system can be run 'at leisure'. In some cases, reanalyses are run years after the original weather observations have been made. A further attraction of reanalysis systems is that a fixed modelling system is used. In this way, the production of an internally consistent set of weather variables is possible for the chosen period of reanalysis, thus facilitating analysis of how a range of variables may be changing on several time scales. This contrasts with weather prediction models. This is because these are often updated with new versions based on constantly evolving research with the result that outputs from model to model are not internally consistent over time, constraining their application to the statistical analysis of extreme events because stationary data are required.

Unlike observational data, which exist for fixed locations, reanalyses produce output for a fixed grid with no missing values (Fig. 3.3) usually at a latitude/longitude scale of 0.5 degrees and up to four times daily. While this might seem the idyll for climate analysis, it must be borne in mind that reanalysis data is not 'real', rather it is a numerical expression of atmospheric conditions at the point of temporal or spatial interest, based on observational data with an uneven geographical distribution. Therefore, for points on the earth's surface where there is no observational data, reanalysis models make the best guess as to what the values are. Consequently, variable values in data sparse regions of the world are highly dependent on the reanalysis model used. For some variables, for which there are no original observations, the reanalysis system derives these directly from other variables. Further reanalysis systems, as do weather prediction models, do a better job at estimating the values for some variables compared to others. For example, reanalysis estimates of temperature generally outperform those for precipitation, because reanalysis systems do not do a perfect job of modelling precipitation processes (Valmassoi et al., 2023). For these reasons, it is incumbent on any analyst to check whether reanalysis products, such as values of daily Tmax or humidity, match with reality. It is also a good idea to assess the relative performance of the various reanalysis systems, as one system may outperform others in terms of variable value accuracy and portrayal of climate phenomenon (Chen et al., 2019; Sheridan et al., 2020; Tesfaye et al., 2017).

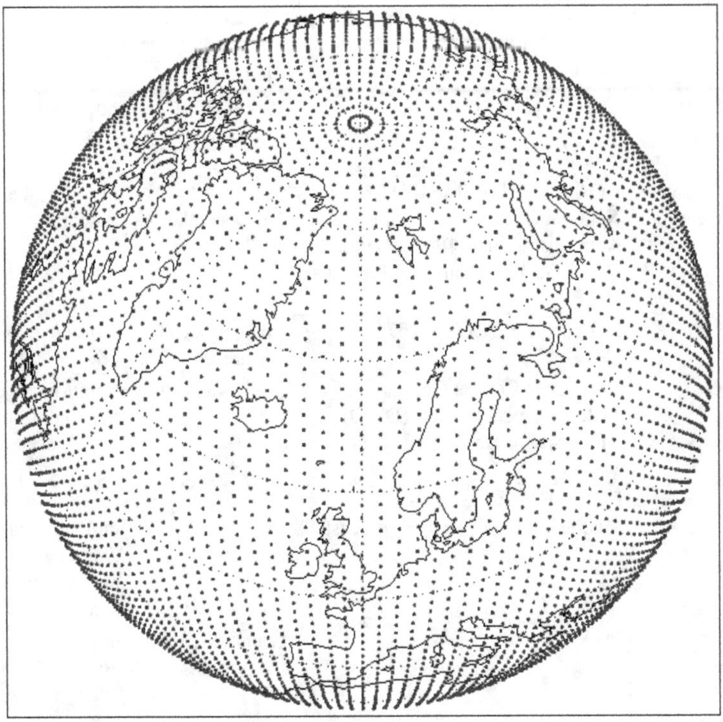

Fig. 3.3 Example of surface Gaussian grid points for which variables are available from European Centre for Medium Range Weather Forecasting (ECMWF) climate re-analysis products. The same grid points provide data in the vertical dimension as well. The grid displayed here is an O80 octahedral reduced Gaussian grid (Adapted from: ECMWF Gaussian Grids. https://confluence.ecmwf.int/display/FCST/Gaussian+grids Accessed 10/06/2024)

Several global reanalysis data sets are available for the analysis of extremes and therefore heatwaves. Examples of recently updated reanalyses include the:

(i) US National Centre for Environmental Prediction reanalysis II (NCEP-2) (Kanamitsu et al., 2002);
(ii) NCEP Climate Forecast System Reanalysis (CFSR) (Saha et al., 2010);
(iii) European Centre for Medium-Range Weather Forecasts (ECMWF) Interim Reanalysis (ERA-5) (Hersbach et al., 2020);
(iv) Japanese 55-year Reanalysis Project (JRA-55); (Kobayashi et al., 2015); and
(v) National Aeronautics and Space Administration (NASA) Modern Era Reanalysis for Research and Applications, Version-2 (MERRA-2) (Gelaro et al., 2017).

Table 3.3 lists several reanalysis products. As well as global reanalysis products, several regional products have emerged over recent times to facilitate analyses of heatwaves at finer spatial scales (Table 3.3).

Table 3.3 Global and regional reanalysis products

Reanalysis System	Origin	Period	Website
Arctic System Reanalysis (ASR)	The Ohio State University	2000–2012	https://rda.ucar.edu/datasets/ds631.1/#
CFSR Climate Forecast System (CFS) Reanalysis	NOAA/NCEP(CDAS-T382) National Oceanic and Atmospheric Administration (NOAA), National Centre for Environmental Prediction (NCEP)	January 1979 to December 2010	http://cfs.ncep.noaa.gov/cfsr
CFSv2	NOAA/NCEP(CDAS-T574)	January 2011 to present	http://cfs.ncep.noaa.gov/
COSMO regional Reanalyses Europe	Hans-Ertel-Centre, German Meteorological Service (DWD) University of Bonn	1995-present	http://reanalysis.meteo.uni-bonn.de/?Overview
ERA-20Ca, ERA40, ERA5	ECMWF European Centre for Medium-Range Weather Forecasts (ECMWF)	Various periods	https://www.ecmwf.int/en/forecasts/datasets/browse-reanalysis-datasets
JRA-55b Japanese 55-year reanalysis	JMA	January 1958 to present	https://jra.kishou.go.jp/JRA-55/index_en.html
MERRA-2	NASA/GMAO National Aeronautics and Space Administration (NASA) and Global Modelling and Assimilation Office (GMAO)	January 1980 to present	https://gmao.gsfc.nasa.gov/reanalysis/MERRA-2/
NCEP-DOE R2	NOAA/NCEP and the DOE AMIP-II project(R2) Department of Energy (DOE)	January 1979 to present	https://www.esrl.noaa.gov/psd/data/gridded/data.ncep.reanalysis2.html
NCEP-NCAR R1,	NOAA/NCEP and NCAR(R1)	January 1948 to present	https://www.esrl.noaa.gov/psd/data/gridded/data.ncep.reanalysis.html
NCEP North American Regional Reanalysis (NARR):	NCEP	1979-near present	http://www.emc.ncep.noaa.gov/mmb/rreanl/
NOAA-CIRES 20CR v2d,	NOAA and the University of Colorado CIRES(20CR) Cooperative Institute for Research in Environmental Sciences (CIRES), University of Colorado	November 1851 to December 2014	https://www.esrl.noaa.gov/psd/data/gridded/data.20thC_ReanV2c.html

All websites accessed 10/06/2024

Because reanalyses contain variables at four dimensions (3 space and one time dimension) they facilitate detailed analysis of heatwaves. For example, Yang et al. (2019) obtained geopotential height, sea level pressure, surface winds, precipitation, evaporation and soil moisture, latent and sensible heat flux data from the MERRA-2 (Gelaro et al., 2017) and the North America Regional Reanalysis (NARR) (Mesinger et al., 2006) reanalyses to study the physical mechanisms associated with US heatwaves. Similarly, for the US, Schoof et al. (2017) used four reanalyses to study the changing character of heatwaves over the period 1981 to 2015. Respectively for Argentina and Africa, Collazo et al. (2018) and Ceccherini et al. (2017) used the ECMWF ERA-Interim reanalysis to study heatwaves. While the availability of reanalysis products for heatwave analyses opens up a wide range of possibilities for developing heatwave climatologies and exploring heatwave development processes, researchers need to bear in mind the possibility of divergence of results arising from using several different reanalysis data sets as found by (Jiang et al., 2024) when considering the nature of compound heatwave events. Usefully several assessments of reanalysis products in relation to their applicability to describing or evaluating heat exposure are beginning to emerge in the literature (Di Napoli et al., 2021; Royé et al., 2020; Spangler et al., 2019; Yan et al., 2021).

3.3.4 *Climate Model Output*

A further source of data for the analysis of climate extremes and heatwaves is the output from climate models. There is a vast range of climate models available and like reanalysis products, their output is organised on a regular grid for several levels from the surface to the upper atmosphere with data reported for at least the daily scale. The Climate Model Intercomparison Project 5 (CMIP5) data set is widely used for the analysis of extremes for current and future conditions. CMIP5 and its associated models provides a framework for coordinated climate change experiments aimed at evaluating climate simulations of the recent past, providing projections of climate change, and quantifying climate feedbacks (Taylor et al., 2012). Recently CMIP6 has superseded CMIP5 (Eyring et al., 2016). Examples of CMIP5 based analyses of 'heatwave futures' are Rohini et al. (2019) for India, Palipane and Grotjahn (2018) for California, Guo et al. (2017) for China, Gibson et al. (2017) for Australia and Schoetter et al. (2015) for Europe.

3.4 Analysing Extreme Events

Heatwaves are extreme events because of their rarity. They manifest as periods of extreme heat because one or more meteorological variables reach unusually high or extreme values for a period that can range from a few days to many days in exceptional cases. As well as the meteorological variables that comprise the physical event being extreme values, the number of heatwave events when compared to the frequency of lesser 'warm events' or 'warm spells' over a long period of time, say 30 years of more, also constitute an extreme value. Because the values of the meteorological variables or the number of heatwave events are rare or extreme, they appear as very small or very large values in a probability distribution. Such extreme values reside in the extremities or the so-called tails of a probability distribution (Fig. 3.4).

Taking temperature as one constituent, and possibly the most impactful physical variable comprising a heatwave event, extreme temperature values can be treated as the highest temperature for a unit of time or exceedances above a threshold occurring at irregular times. In statistical speak the attainment or exceedance of a critical value on a continuous scale is referred to as the outcome of a stochastic process known as a marked point process. This is because the extreme temperature values are considered as random magnitude 'marks' occurring at random 'points' in time (Stephenson, 2008). Characterising the statistical properties of such 'marks' has become an important area of climate research, especially from a hazard management perspective, because understanding the relationship between event frequency, magnitude, duration and timing can provide insights into not only the processes underlying physical climate extremes but how adaptation and mitigation strategies should be formed.

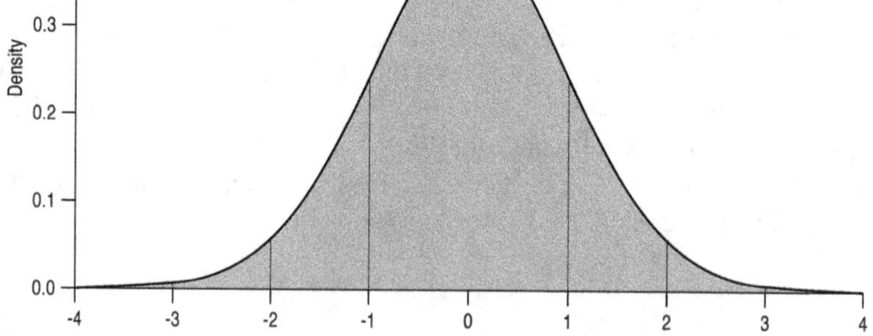

Fig. 3.4 Probability density for normally distributed values. Horizontal axis is units of standard deviation—zero represents the mean. The '68-95-99.7% rule' is often used to refer to the normal distribution given that the three percentages represent the proportion of values that lie within +/– 1, 2 and 3 standard deviations respectively (horizontal axis values). Extremes occur in the 'tails' of the distribution. Extreme values of high temperature would lie beyond 2 standard deviations, where in a normal distribution 2.5% of the total number of observed values are found. (Source: Author)

3.4.1 Extreme Value Theory

When data analysts in a wide range of fields such as science, engineering, economics and medicine mention the statistical analysis of extreme values and the modelling or mathematical representation of a marked point process, they are generally referring to Extreme Value Theory (EVT) (Coles, 2001). In the context of climate science and specifically heatwaves and their associated physical variables, the purpose of EVT is to predict the likelihood or probability of extremes greater than values previously recorded. Having said that and before proceeding, it is worth mentioning the difference between EVT and Central Limit Theorem (CLT) another often used method for studying random variables. While both EVT and CLT are 'limiting theorems' concerned with modelling how large or small data values might reach and deal with random variables, EVT only focuses on the behaviour of random variables in the tails of a distribution, that is, a set of maximum and minimum values. Therefore, an extreme value distribution is a limiting model for maximum and minimum values in a data set. In contrast, CLT concerns itself with the analysis of the complete distribution of random variables (Haan & Ferreira, 2006). Further, in the case of the CLT, as the sample size increases the sampling distribution approaches the normal distribution while with the EVT the distribution converges to quite a different form.

A term used in discussing the distribution of values of random variables is the probability density function or PDF for short. The relationship between observations and their probability is the probability density with the shape of the probability density called the probability distribution. There is a wide range of probability distributions with the uniform, normal, log-normal and exponential being reasonably common. In the case of extremes, the thing of interest is the density of probabilities for values of a random variable and thus the distribution's shape and its 'moments' such as the mean and variance. However, often for a sample of extreme values the probability distribution is unknown. This is because we usually only possess a sample of all possible values; there will always be unknown values. Given this, the challenge is to resolve the nature of the probability distribution that best describes the variable of interest. Such a process is referred to as probability density estimation, with the outcome the identification of a mathematical function that models a variable's probability distribution. The upshot of all of this is that once successfully identified, the PDF can be used to calculate the probability of the unknown values or outcomes of a given random variable.

In the case of the analysis of extreme values, there are two approaches for compiling a data set. These are the block maxima and threshold approaches. Block maxima are a collection of maximum values from a block of data with the block being a block of time or set period. For example, the annual daily maximum temperature (Tmax) could be extracted from a 30-year block to produce 30 values for analysis. Such a series is called the annual maximum series. One drawback of a strict block maxima approach is that the sample size is limited to the number of time units making up the block, years in the case of annual maxima. The block maxima approach may also miss other significant occurrences of extremes in any 1 year that exceed those of other years. Imagine for an individual year a Tmax value for a July day falling just below an independent Tmax value in the following month of August. In this case, the July Tmax value would not be included in the analysis and accordingly important information might be lost. The threshold approach, sometimes referred to as peaks over threshold (POT) compiles a time series of events by taking values above a predetermined threshold that may be of some physical significance such as the road surface temperature required for asphalt/bitumen to begin to soften (about 50 °C). In this way, several Tmax values from the same year, along with those from other years could enter into the analysis to increase the sample size, assuming all values are independent. Partial duration series is the term for a series of values compiled in this way. Because a partial duration series comprises a higher number of values for analysis it typically produces higher values for a given frequency than that yielded from the analysis of an annual maximum series. This is most obvious for events with a relatively high probability of occurrence (e.g., 20–50%). However, as the probability of occurrence decreases (e.g., less than 5%) there is a tendency for the difference in values between the partial and annual maximum series to be small. While this might seem inconsequential such differences in approach bear implications for the information generated and highlight the dependency of results on the method adopted.

Block maxima values are usually modelled by applying the Generalised Extreme Value (GEV) distribution (Coles, 2001; Katz, 2010) while generally exceedances or excesses above a threshold are modelled by the Generalized Pareto distribution (GPD). Figure 3.5 lays out the fundamental differences between these two models. The following subsections expand on the GEV and GPD modelling approaches.

3.4.1.1 Generalised Extreme Value (GEV) Distribution and Modelling Based on Block Maxima/Minima Approach

The GEV is a combination of three extreme value distribution (EVD) types, namely the Gumbel, Fréchet and Weibull distributions. A common shorthand for these extreme value distributions is EVD Type I, II and III respectively. The beauty of these is for any data set comprising a collection of maximum or minimum extreme values, one of the three types will model the probability density distribution. However, there are three caveats those being, the variable of interest has a continuous distribution, the distribution possesses an inverse distribution (distribution of the reciprocal of a random variable) and the distribution has no strong trends in it or in other words it comprises independent, identically distributed random variables. Given the ability of EVD types I, II and III to capture the distribution of extreme values, a natural question is why combine them into one model? The reason for this is that usually at the beginning of any analysis the nature of the parent distribution of extreme values is unknown. In short, what is happening in the extremities of the parent distribution remains a mystery. Given this establishing which of EVD Types I, II or III might be suitable for modelling the PDF for the variable of interest

	GEV	GP
Description	Distribution function of standardized maxima (or minima) – block maxima/minima approach	Probability of exceeding pre-determined threshold – peaks over threshold approach
Parameters	**Location μ:** position of the GEV mean	**Threshold u:** Reference value for which GP excesses are calculated
	Scale σ: multiplier that scales functions	
	Shape ξ: Parameter that describes the relative distribution of the probabilities	
General function (CDF)	For extreme values z, $G(z) = \exp\left[-\left\{1+\xi\left(\frac{z-\mu}{\sigma}\right)\right\}_+^{-1/\xi}\right]$	For threshold excess x, $H(x) = 1-\left[1+\xi\left(\frac{x-u}{\sigma_u}\right)\right]_+^{-1/\xi}$
Limit as $\xi \to 0$	Gumbel: $G(z) = \exp\left[-\exp\left\{-\left(\frac{z-\mu}{\sigma}\right)\right\}\right]$	Exponential: $H(x) = 1-\exp\left(-\frac{x-u}{\sigma}\right)$
$\xi > 0$	Fréchet	Pareto
$\xi < 0$	Weibull	Beta
Interpretation of results	**Return level:** value z_p that is expected to be exceeded on average once every $1/p$ periods, where $1-p$ is the probability associated with the quantile. Find z_p such that $G(z_p) = 1-p$.	**Return level:** value x_m that is exceeded every m times. Begin by estimating ζ_u, the probability of exceeding the threshold. Then, x_m is $x_m = \begin{cases} u + \frac{\sigma_u}{\xi}\left[(m\zeta_u)^{\xi}-1\right], & \xi \neq 0 \\ u + \sigma_u \ln(m\zeta_u), & \xi = 0 \end{cases}$

Fig. 3.5 The two basic types of extreme value distributions. GEV is Generalised Extreme Value and GP is Generalised Pareto. Adapted from: Pinheiro and Grotjahn (2015): An Introduction to Extreme Value Statistics. Available at: http://grotjahn.ucdavis.edu/EWEs/extremes_primer_v9_22_15.pdf (Accessed 23/03/2024)

becomes a process of trial and error. Combining the three distributions to produce the GEV distribution circumvents this problem.

As one might expect, each of the three individual EVD that make up the GEV model have slightly different characteristics. The Gumbel type is one of the most common EVD with a limiting distribution for data that has no upper or lower limits. It possesses two forms, one for minima and one for maxima both of which contain location and scale parameters. Interestingly the Gumbel distribution has one fixed shape that moves according to the location parameter. As the size of the location parameter increases (decreases) the Gumbel distribution moves to the left (right). The Fréchet distribution applies in the case of a distribution bounded on the lower end. Employed in the modelling of maximum values it contains individual parameters for shape, scale, and location. In contrast to the other two distributions, the Weibull is a collection of distributions, all bounded on the upper end. The Weibull EVD has the same three parameters as the Fréchet but can change its shape depending on the parameters and associated values chosen. Hence, the Weibull distribution can take on a variety of forms. For example, if the shape and location parameter in a Weibull distribution are held constant and the scale parameter is increased (decreased) then the shape elongates (protracts) and moves to the right (left) with the height of the distribution and thus the probability of the most common values decreasing (increasing). Examples of the three EVD are presented in Fig. 3.6.

Like the Fréchet (Type II) and Weibull (Type III) EVD distributions, the GEV distribution has three parameters as can be seen in Fig. 3.5, where μ, σ and ξ are the location, scale and shape parameters respectively. Each parameter refers to a characteristic of the distribution such that the location parameter denotes the centre of the distribution, the scale parameter governs how much deviation or spread there is around the location parameter while the shape parameter defines how rapidly the upper tail decays. For the GEV distribution, σ and $1 + \xi(x-\mu)/\sigma$ must be greater than zero. Further ξ and μ can have any real value. The role of the shape parameter ξ is the definition of which distribution of the three EVD distributions the GEV assumes. When ξ is zero, the GEV is equivalent to the Gumbel distribution (EVD Type I). The Fréchet distribution (EVD Type II) applies in the case when ξ is greater than zero, whereas if ξ is less than zero the GEV matches the Weibull distribution (EVD Type III) (Fig. 3.6).

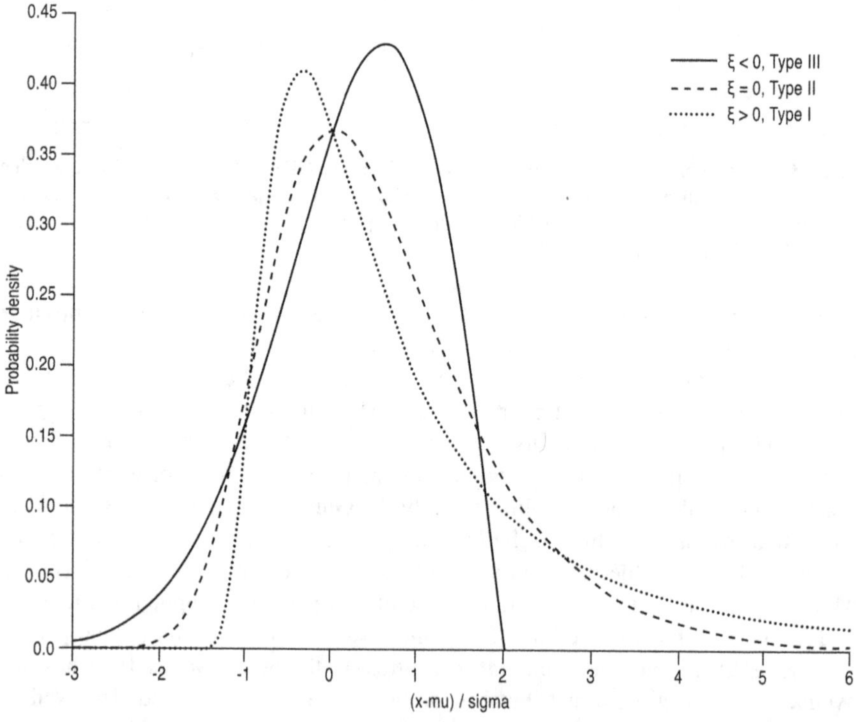

Fig. 3.6 Three types of Extreme Value Distribution (EVD) that comprise the Generalised Extreme Value (GEV) distribution. The Gumbel distribution (Type I) has a light upper tail and is positively skewed. The Frechet distribution (Type II) possesses a heavy upper tail and infinite higher order moments. The Weibull (Type III) distribution has a bounded upper tail. (Adapted from: https://www.mathworks.com/help/stats/examples/modelling-data-with-the-generalized-extreme-value-distribution.html Accessed 10/06/2024)

3.4.1.2 Generalised Pareto Distribution (GPD) and Modelling Based on Peaks over Threshold Approach

Once a data series comprised of values more than a threshold (the exceedances) has been assembled, the Generalised Pareto (GP) distribution is usually applied to modelling the probability density. Like the GEV distribution, the GP distribution is a generalization model comprised of constituent models. These are the exponential and the Pareto distributions, which form the basis of a large family of possible distributions the result being that a continuous range of shapes is possible. Although the GP distribution is similar to the GEV distribution in that it is a composite of pre-existing distributions, it contrasts with the GEV in that it is a right skewed distribution composed of only two-parameters. The dimensionless scale parameter σ defines the scale of the excesses while the shape parameter ξ indicates the overall shape of the tail. Importantly the shape parameter ξ, occasionally referred to as the tail index parameter, can take on values that are negative, positive or zero, the result being that the GP has three basic forms (Fig. 3.7). If the shape parameter is zero this indicates that the tail of a distribution of exceedances has an exponential form,

Fig. 3.7 Three types of Generalised Pareto distribution. The exponential distribution possesses a light tail with a "memoryless" property ($\xi = 0$). The Pareto is a heavy tailed distribution, occasionally referred to as the 'power law' $\xi > 0$. The Beta is a bounded distribution ($\xi < 0$). The horizontal axis is exceedances above a reference/threshold value. (Adapted from: https://www.mathworks.com/help/stats/generalized-pareto-distribution.html Accessed 10/06/2024)

values in the tail decrease exponentially. For the case of a positive shape parameter, the fall in the values in the tail aligns with that of a polynomial distribution whereas a negative shape parameter is indicative of a finite tail with an upper limit as found in a beta distribution. To determine the values of the GP scale and shape parameters, the most widely used methods include the Maximum Likelihood Estimator, the Method of Moments and the Probability Weighted Moments (Bermudez & Kotz, 2010). However, as noted by Hosking & Wallis (1987), the choice of parameter determination method is important because of sensitivity to threshold choice and the generation of unrealistic estimates.

Because the POT approach produces a series of exceedances, they can be treated as a Poisson process. Essentially a Poisson process describes a series of discrete events (e.g., exceedances) for which the average time between their occurrences is known. However, what is unknown is the exact timing of event 'arrival' because this is random. Other Poisson process characteristics are events occur independently of each other (there is no memory in the system), and two events cannot occur at the same time so that for any sub-interval (e.g., a day) or a period (e.g., 1000 days) the event can be defined as occurring or not occurring (a success or failure or a Bernoulli Trial process). What the Poisson process allows us to establish is the likely frequency of yet unknown events. In the case of temperature extremes derived from a POT approach it is the non-homogenous Poisson process which is used to reveal the number of events to be expected up to a given time period (Coles, 2001; Stephenson, 2008; Visser & Petersen, 2012). It is a non-homogenous Poisson process because the rate at which events (exceedances) occur (events per time-period) is not constant and varies with time. In contrast, if the average rate of an event were constant, then this would be a homogenous Poisson process.

3.4.1.3 Return Period

The term return period (T) appears frequently in both the scientific literature and popular press when reporting on extreme events. Often used interchangeably with recurrence interval, it is the average time until the next occurrence of a defined event, or in other words, the mean time (T) expected between a specific event and its recurrence. Return period is calculated using block maxima or minima data. Examples of its use include the Chicago 1995 and European 2003 heatwaves. In the former case, the return period of the peak apparent temperature (see Chap. 4) has been estimated to be less than 23 years (Karl & Knight, 1997). For night time temperatures in Paris during the 2003 European heatwave, the return period has been estimated at around 35–50 years (Charpentier, 2011), although other estimates using different variables indicate a much longer return period for the European heatwave event at large (Schär et al., 2004; Stott et al., 2004). As well as based on observed block maxima values, return periods can be estimated for extreme temperature events under global warming (Dosio et al., 2018; Zwiers et al., 2011). In this case, block maxima values, such as the annual daily maximum temperature for the 31-year period 2050–2080, are assembled from climate model output and the return period calculated in relation to a reference period.

Seemingly a straightforward concept, there is sometimes confusion regarding the meaning of return periods, especially amongst the public. Most likely, confusion arises from the definition stating 'average time until the next event'. This is often mistaken to mean once an event with say a 100-year return period and an associated magnitude has occurred, it will occur again in 100 years' time. This is incorrect, as at any one time the occurrence of a particular event will have a specific probability associated with it. Therefore, it is possible that two 1 in 100-year events could occur in successive years, or even in the same year. Usefully a probability can be attached to an event with a given return period. This is because return periods are inversely proportional to the annual exceedance probability (P) or the likelihood in percent of an event in any year exceeding a given magnitude (P = 1/T and T = 1/P). Therefore, a 1 in 100-year heatwave event (100-year return period) will have a 1% probability of occurrence in any 1 year. This would mean that there is around a 37% chance of 100 years passing without at least one such heatwave event occurring $((1-0.01)^{100}) \times 100) = 37\%$). In relation to probability distributions derived using GEV methods as described above, inverting the resultant distributions, once the three parameters of location, shape and scale have been estimated, produces the return period.

A simple example of the relationship between return period T and exceedance probability P is shown in Fig. 3.8 for the annual January maximum daily temperature maximum (Tmax) for Melbourne Australia based on the period 1970–2013. This example assumes a parent probability density function for Tmax equivalent to the normal distribution. In this case the exceedance probability (P) has been calculated using the simple formula P = m/(n + 1), where m is the rank number of the Tmax values with the highest Tmax value set as m = 1 and n = the number of years (44). Similarly, the return period T is calculated as T = (n + 1)/m with the same notation as for calculating P—in fact calculating one of these is only necessary given their inverse association. Here we can see a return period of 25 years is associated with a January maximum daily Tmax of around 44 °C (Fig. 3.8), equivalent to an exceedance probability of 4% (Fig. 3.8). The inverse relationship between T and P is also shown in Fig. 3.8.

Lastly a cautionary note about return period estimation. Care is required in stating with confidence the magnitude of an event with a return period well beyond the length of observational data used for estimation. This is because of insufficient knowledge about the exact shape of the tail of a distribution of exceptionally large extreme values. Given this and despite the temptation to make estimates of return levels or return periods for 1000 years, it is best to restrict estimates to return periods no more than twice the length of the original data. The situation is different if the analyst is making estimates of return period based on model output data generated for long periods into the future assuming low levels of uncertainty in model projections.

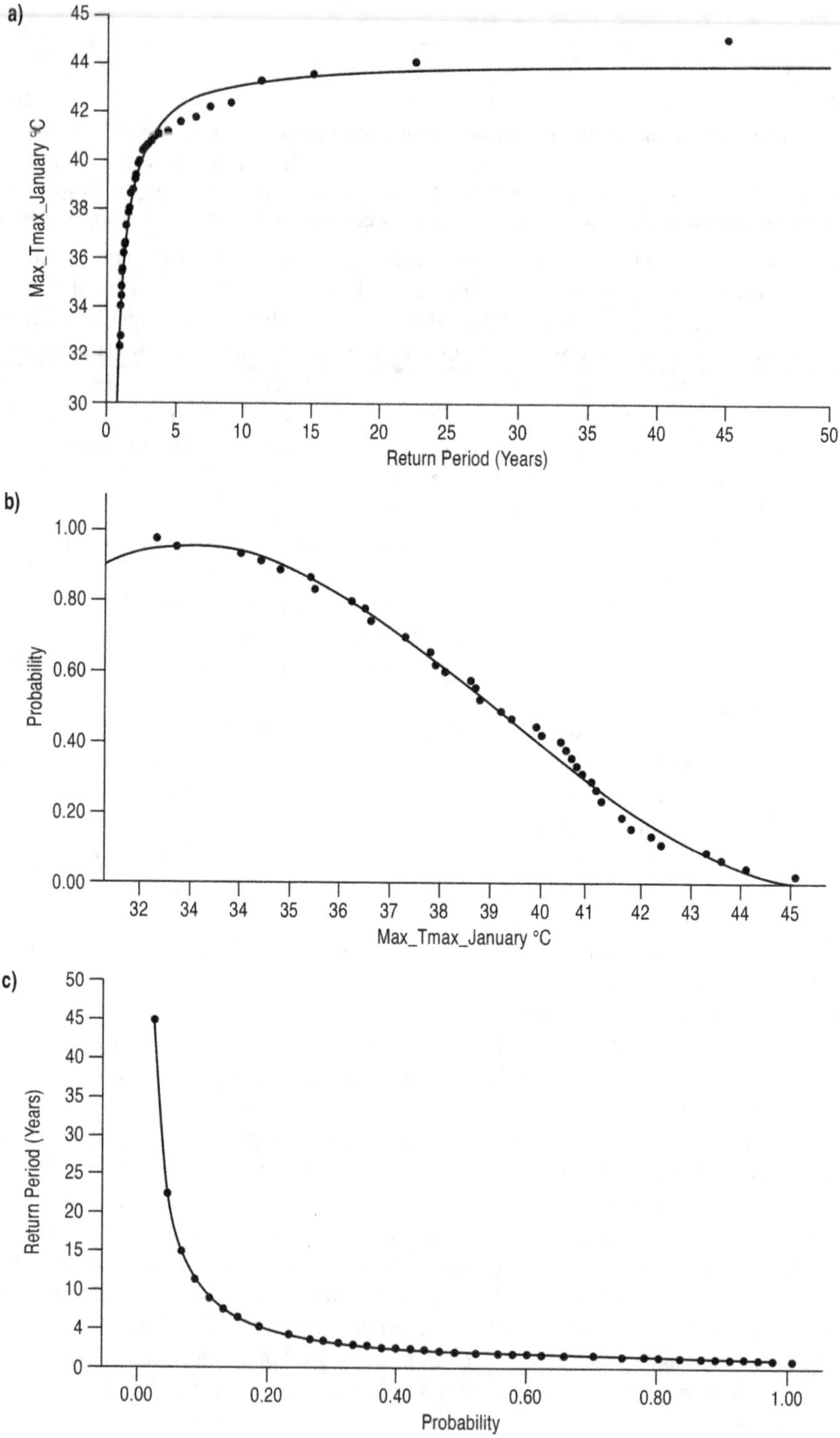

Fig. 3.8 Plots of January maximum daily maximum temperature probability and return period for Melbourne, Australia based on the period 1970–2013. January maximum daily temperature maximum and return period (**a**), January maximum daily temperature maximum and probability (**b**), and the inverse association between return period and probability (**c**). The solid line in each plot models the relationship between the variables. (Source: Author)

3.4.2 *Sources of Uncertainty in Extreme Value Statistics*

Because climate risk analysts and managers use the output from climate related extreme value analyses, whether for current or future climates, they need to be confident that the data and methods used are fit for purpose. Cumulative experience within the climate extremes community indicates in undertaking the analysis of extreme values, critical decisions are necessary given there is a range of sources of uncertainty, which can affect probability and return period estimates. The various data sources available for the analysis of extremes were described in Sect. 3.3, so the purpose of this section is to outline potential sources of uncertainty in extreme value analysis. Generally, these stem from the climate observations themselves, data inhomogeneity, non-stationarity, the robustness of reanalysis or climate model outputs or the choice of the extreme value analysis method or extreme value model.

3.4.2.1 Quality Control of Climate Observations

As a first step in any analysis of extremes, simple but important quality control procedures related to data completeness, errors and outliers should be contemplated. This applies to observational and model-based data. The importance of record length has been mentioned previously. Next to this is record completeness as serial completeness is essential for some analyses. A common criterion for completeness is four or less missing observations per year (e.g., 4 missing daily Tmax observations) (WMO, 2023a). Assuming the data series meets this stipulation then methods such as interpolation are useful for replacing missing values. Further bias corrected reanalysis data is also a possibility for replacing missing observational data (Lompar et al., 2019; Toreti et al., 2019). Data errors may arise for several reasons due to faulty instrumentation, observer error and data input and processing related mistakes (Hunziker et al., 2017). Sampling errors arising from extracting or mixing data from wrong or various sources may also occur. Typical errors might be obviously wrong values, such as Tmax being less than Tmin, or wrong dates. Related to errors is the issue of outlier values. These could be genuine trusted observations, or truly erroneous. Identification of outliers and the decision to exclude these is an important one as this could affect the outcomes of any ensuing analysis. This is because outliers are extreme values.

A variety of methods exist for identifying outliers, with identification dependent on the nature of the outlier definition, in the first place. In some cases, the experience of the analyst might be sufficient for the identification of outliers. Alternatively reference to information sources such as metadata or the media may help establish whether a value is to be trusted. In other cases, heuristics such as setting an upper or lower limit for tolerable values based on the mean of the day of interest, plus or minus four times the standard deviation, may suffice. Validation of potential outliers against immediately preceding or following values to see if they are part of a physically plausible trend is also a possibility. Where there is a decent station network, with little spatial variability in the variable of interest, then comparison of potential outliers with values recorded at nearby recording stations could be revealing. Also relevant to the identification of outliers is the visual inspection of data plots such as histograms and time series as well as stem and leaf diagrams, all of which facilitate a holistic view of the data (Wilks, 2011, 2019). Complementing the less technical assessment methods outlined above are a range of advanced statistical techniques for outlier detection (Gupta et al., 2014; Ma et al., 2017).

3.4.2.2 Data Homogeneity

Ideally, assessment of data homogeneity follows on from data quality control. Homogeneity refers to the internal consistency of data, or in other words ensuring that fluctuations or changes in data values arise from actual climate variability or climate change, that is, natural processes. Checking and controlling for homogeneity guards against non-climatic step changes or data drift affecting the analysis of extremes. Homogeneity testing involves several steps. In addition to quality control, the creation of a reference series and the detection of break points precedes the final stages of data adjustment and validation of results (WMO, 2020).

Typically, non-homogeneous data arises because of changes in climate station location, instrumentation, exposure or observing practices (Trewin, 2010; Trewin et al., 2020; WMO XE "World Meteorological Organization (WMO)", 2020). For example, the move of a climate station from its original location could result in a 'different climate' being recorded. Table 3.4 shows a good example of this for Port Macquarie in Australia. Here a move of the climate station from its original Hill Street site in the town to the airport 5 km inland occurred in 1995. While the distance of the move does not seem overly large this resulted in the climate station being situated in quite a different environment as the original Hill Street station was coastal, in contrast to the airport site located on an inland floodplain away from any marine influence. To understand the impact of the station move, the Australian Bureau of Meteorology (BoM) maintained measurements at the original site over the period 1996 to 2002, eventually closing it in 2003. Various maximum and minimum temperature statistics for the 7-year period of parallel observations are quite revealing in terms of the impact of the location move. In relation to extreme maximum temperatures, it is clear the move to an inland site bears important implications for extreme analysis (Table 3.4). Noticeably the airport station records a 99th percentile temperature 3.3 °C above that for the original site, a difference most likely due to marine related moderating effects at Hill Street, which are most pronounced on clear sky days when high maximum temperatures manifest. As noted by the BoM, the situation is not as simple as it appears with a consistently large difference between the new and original locations. This is because on the very hottest days, the coastal location of the Hill Street climate station is affected by hot dry air advection from central Australia which supresses the marine moderating influence and thus the temperature differential with the airport. For example, for the hottest day of the 7 year parallel observation period (January 2, 2002), Hill Street recorded a maximum temperature of 38.3 °C while at the airport it was 40.2 °C, a temperature differential of only 1.9 °C. Given this, simply applying a constant annual or even

Table 3.4 Comparison of maximum and minimum temperatures between the original Hill Street and new Airport locations at Port Macquarie, Australia, 1996–2002. (Source: http://etccdi.pacificclimate.org/docs/Classic_Examples.pdf) (Accessed 31/05/2024)[a]

Parameter	Maximum temperature (°C)			Minimum temperature (°C)		
	Hill St.	Airport	Difference	Hill St.	Airport	Difference
Annual mean	22.9	23.5	−0.6	14.3	12.8	1.5
Summer mean (Dec—Feb	26.1	27.0	−0.9	19.0	17.6	1.4
Winter mean (Jun—Aug)	19.1	19.3	−0.2	8.9	7.5	1.4
Lowest in period	12.5	11.3	1.2	0.0	−2.7	2.7
1st percentile	15.7	15.8	−0.1	3.9	1.2	2.7
5th percentiale	17.2	17.5	−0.3	6.2	3.9	2.3
10th percentile	18.1	18.3	−0.2	7.6	5.8	1.8
90th percentile	27.2	28.4	−1.2	20.1	19.1	1.0
95th percentile	28.0	29.8	−1.8	21.2	20.2	1.0
99th percentile	30.0	33.3	−3.3	22.4	21.9	0.5
Highest in period	38.3	40.2	−1.9	24.2	23.7	0.5

[a]*Classic examples of Inhomogeneities in climate datasets* (prepared by CCl/CLIVAR/JCOMM expert team (ET) on Climate Change Detection and Indices (ETCCDI))

monthly adjustment to bring the original temperature series at Hill Street in line with the new airport record to achieve homogeneity of extreme temperature values would be inappropriate seeing that such extremes are weather situation dependent.

Updating instrumentation, such as installing new thermometers or switching from manual to automatic measurements can also influence a data record, as would an alteration of recording times (Acquaotta et al., 2019). Instruments may also drift out of calibration such that a given measurement may no longer be accurate, possessing a warm or cold bias in the case of temperature. Exposure of a climate observation site and its instruments could also change over time due to tree growth thus introducing a shelter effect or simply grass growth could change the nature of the surface, affecting air and soil temperature and humidity measurements. Given this, the upkeep of the environment surrounding a climate station is very important and for this reason, the World Meteorological Organisation has produced strict guidelines for the maintenance of meteorological enclosures and instruments (WMO, 2023b). Because there is a raft of ways in which data could become non-homogeneous, climate station metadata, or simply data about the data, is extremely important. Such metadata can assist with understanding climate station history in terms of location and instrument changes and inform the data adjustment strategy and the removal of non-climatic influences from time series of climate data. If an analyst is interested in unadulterated climate extremes, then urban influences arising from the urban heat island effect, which may cause a warm bias in a data series, will also need consideration. Inhomogeneities can also appear in numerical model output. This is especially when blended data sets are constructed from a series of model runs with each model run being a slight improvement over the previous, because of a programme of continual model updates. Blended data sets from different models can also contain inhomogeneities because the contributing models may possess a different model architecture or ways of representing atmospheric processes. Such model related inconsistencies in data is a reason why single numerical models are preferred for producing reanalysis products.

Beyond simple tests for homogeneity mostly based on visual inspection of data plots such as double mass curves, originally developed for hydrological applications (Gao et al., 2017), quite an industry has developed in relation to the production of sophisticated methods for testing homogeneity of climate time series. Notwithstanding their eloquence or temporal focus, generally homogeneity tests can be categorised as either absolute or relative tests (Peterson et al., 1998; Venema et al., 2012; Yozgatligil & Yazici, 2016). Absolute tests only use the data from the climate station under investigation, with the idea that erroneous trends or jumps in the data, relative to the preceding behaviour in the times series, represent inhomogeneities. In many ways, this is a process of 'self-reflection' with the data reference series being itself. Relative tests on the other hand use time series of variable values from nearby stations to form a reference series. The assumption here is that the reference station has been unaffected by any possible biases arising from things that might cause inhomogeneities but has been under the influence of the same general climate conditions as the station for which homogeneity testing is to be conducted (the candidate station). Either a single reference station, or information from multiple highly correlated neighbouring stations, may form the reference series, with the candidate station ultimately compared with the reference station. The basis of relative testing is the examination of the difference or ratio of values between the candidate station and the reference station(s) (Domonkos & Coll, 2018). In the case of temperature time series, differences between the candidate and reference stations are calculated. This is because temperature is an interval scale variable. For meteorological variables measured on a proportional scale, such as precipitation, which has an absolute minimum value of zero, ratios are normally calculated. Once differences or ratios have been calculated and plotted against time, it will become apparent whether the candidate station is homogenous or not. If the difference or ratio values vary around a constant level and do not demonstrate any abrupt changes or trends, then it is likely that the candidate station is inhomogeneity free. In contrast, noticeable single or multiple jumps, or a drift in the values of differences and ratios suggest a non-homogenous time series at the candidate station (Fig. 3.9). When statistical testing of homogeneity is undertaken, it is important to keep in mind the statistical distribution of the values of interest. For example, temperature data usually follows a normal 'bell-shaped' distribution so parametric based homogeneity tests will most likely suffice. This contrasts with variables such as precipitation, which usually has a skewed distribution, with non-parametric homogeneity tests most appropriate in this case.

Examples of some climate specific and non-climate specific homogeneity tests as well as software packages that have been designed specifically for homogeneity testing of climate series and applicable at the daily scale are presented in Table 3.5. Many of the listed methods have undergone systematic assessments as to their efficacy for detecting inhomogeneities (Domonkos, 2013; Ribeiro et al., 2016; Yozgatligil & Yazici, 2016; Venema et al., 2012; Willett, Williams, et al., 2014a). While Table 3.5 presents an attractive array of homogeneity testing possibilities, an important point to note is that most homogeneity tests have been developed for

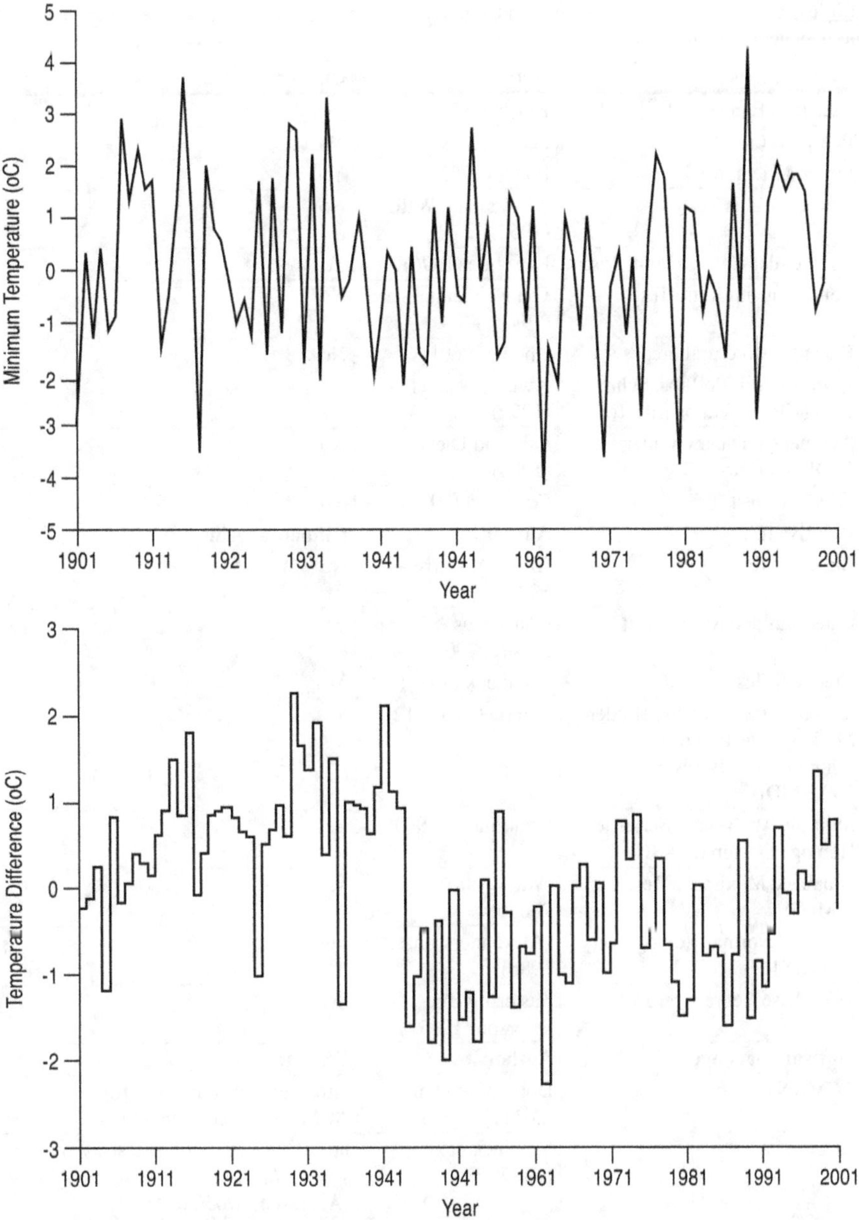

Fig. 3.9 (**a**) Monthly average of daily minimum temperature for December at Burgos, Spain, (**b**) difference of temperature values between the candidate station at Burgos and nearby reference stations which shows a clear inhomogeneity in 1941, arising from the relocation of the climate station at Burgos. (Adapted from: Aguilar et al., 2003)

Table 3.5 Some common absolute and relative homogeneity tests and software packages for homogeneity testing

Absolute tests	Author	Climate specific
Buishand Range Test	Buishand (1982)	Yes
Pettitt Test	Pettitt (1979)	No
Potter's Method	Potter (1981)	Yes
Kruskal Wallis Test	Kruskal and Wallis (1952)	No
Rank Order Change Point Test	Lanzante (1996)	Yes
Von Neumann Ratio Test	von Neumann (1941)	No
F-test for structural breaks[a]	Chow (1960)	No
Kwiatkowski, Phillips, Schmidt, Shin (KPSS) Stationarity Test	Kwiatkowski et al. (1992)	No
Augmented Dickey-Fuller (ADF) Unit Root Test	Said and Dickey (1984)	No
Friedman Test	Friedman (1937)	No
Relative tests	**Author**	**Climate specific**
Bayesian Test	(Barry & Hartigan, 1993)	No
Caussinus and Mestre Test	(Caussinus & Mestre, 2004)	Yes
Craddock Test	Craddock (1977)	Yes
Genetic Algorithm and Hidden Markov Detection of Inhomogeneities Test (GAHMDI)	Toreti et al. (2012)	Yes
Multiple Analysis of Series for Homogenisation (MASH)	Szentimrey (1999)	Yes
Penalized Maximal F Test (PMFT)	(Wang, 2008a, 2008b)	Yes
Standard Normal Homogeneity Test (SNHT)	(Alexandersson, 1986)	Yes
Two Phase Regression Test	Easterling and Peterson (1995)	Yes
Software packages	**Author**	**Website**
ACMANT	Domonkos et al. (2011)	https://github.com/dpeterfree/ACMANT (Accessed 31/05/2024)
AnClim	Stepánek et al. (2009)	http://www.climahom.eu/software-solution/anclim (Accessed 31/05/2024)
Climatol	(Guijarro, 2016; Guijarro et al., 2023)	http://www.climatol.eu/index.htm (Accessed 31/05/2024)

(continued)

Table 3.5 (continued)

Absolute tests	Author	Climate specific
GAHMDI HOMAD	Toreti et al. (2012)	Contact author
MASH	Szentimrey (1999)	https://www.met.hu/en/omsz/rendezvenyek/homogenization_and_interpolation/software/ (Accessed 31/05/2024)
RHtestsV4	Wang and Feng (2013)	http://etccdi.pacificclimate.org/software.shtml (Accessed 31/05/2024)

Complied by Author
[a]Can be used as a relative or absolute test

climate data at the monthly or annual time scale. As pointed out by Aguilar et al. (2003) homogeneity testing at the sub-monthly time scale presents some challenges. This is because detecting inhomogeneities is essentially about identifying extreme values in a candidate series that are beyond what might be expected in a stable internally consistent time series. However, as extremes are rare events there is often few extreme data values on which an assessment can be based thus creating challenges for how a time series might be adjusted (Aguilar et al., 2003). Further adjusting a daily time series based on homogeneity information for mean values may not represent what is required for the so-called higher order moments (e.g., variance and high or low percentile values) of a distribution of values. As noted in the above Port Macquarie example for location change, the differences between a candidate and reference station for extreme values may be weather situation dependent and much larger than the difference in mean values. Although the jury is still partially out in terms of the best approach to homogeneity testing at the daily time scale, several researchers have adjusted daily time series based on the information revealed by carefully selected homogeneity tests. Approaches used for homogeneity testing and adjustment at the daily time scale can be found in Aguilar et al. (2003), Chimani et al. (2018), Dumitrescu et al. (2020), Hewaarachchi et al. (2017), Squintu et al. (2019), Trewin (2013), Vincent et al. (2002) and Wijngaard et al. (2003). Aware of the importance of homogenous climate series at the daily time scales efforts have been made via the International Surface Temperature Initiative to develop global scale daily temperature series, especially as these are imperative for the assessment of extremes and weather variability at the sub-monthly time scale (Rennie et al., 2014; Thorne et al., 2013).

There is certainly a wide range of objective methods available for homogeneity testing. The fact that they are termed objective may however, lead to misplaced confidence in the final time series that emerges from homogeneity testing and adjustment. This is because the choice of method is quite subjective. As a result, two analysts attempting to homogenise the same temperature series may produce different outcomes in terms of the adjusted time series of values. To manage against this, it is best to apply at least two to three different methods for homogeneity testing to establish whether these produce convergence in terms of the results. Equally important of course is the maintenance of a detailed record of all the steps involved, so that any users of adjusted climate series can understand its derivation. Lastly because of the nature of homogeneity testing and adjustment, this can prove a contentious matter as demonstrated by claims and counterclaims about the faithfulness of adjusted time series to reality (Mullan et al., 2018; Trewin & Jones, 2015; Way et al., 2017).

3.4.2.3 Stationarity

Stationarity refers to no change over time of the low and higher order moments of a distribution of values such that the mean and variance for example is stable over the length of a time series. Another way of describing this is the statistical properties of a process that generates a time series is time invariant. This means that while a time series of values can change over time, the way in which it changes does not. There are some pragmatic reasons why stationarity is important. One is that a stationary time series is relatively easy to analyse. This is especially pertinent in the case of time series based forecasting, because the behaviour of a stationary time series is more predictable compared to one that is non-stationary. Allied to this is stationarity is an underlying assumption of many statistical techniques such as trend estimation and inferential statistics and most importantly traditional methods used for extreme value analysis. Therefore, to have confidence in the outcome of extreme event analysis, a time series needs to be stationary, and it is best to test for this characteristic, even if stationarity is assumed (Katz, 2010; Serinaldi & Kilsby, 2015; Visser & Petersen, 2012).

A range of methods for checking stationarity of climate related time series exist. For rapid appraisal of stationarity, a visual determination based on the inspection of time series plots may be effective. For example, consider the time series plots in Fig. 3.10.

The marked seasonality of the time series in Figs. 3.10d, h and i, and the strong trends in Figs. 3.10a, c, e, f and i make all these non-stationary. Further, the increase in variance over time evident in Fig. 3.10i adds to the non-stationarity of this time series. Given this, only time series in Fig. 3.10b and g are stationary. Arguably, the time series in Fig. 3.10g is potentially non-stationarity, because of its strong cycles (e.g. a region with large inter-annual variability of rainfall), however the timing of the years with high and low values is not predictable. Therefore, the process generating time series g is stochastic making this series stationary. So, when inspecting time series plots for the purpose of establishing evidence for stationarity the basic question to ask is ‘do the statistical properties depend on the time at which the series is observed’. Given this, a time series with trends or seasonality will result in values being dependent on the time they are measured. In contrast, a so-called white noise process, which is a serially uncorrelated stochastic process with a mean of zero and a constant and finite variance, is stationary. This means that at any observation in a time series arising from a stochastic process should look similar to all other

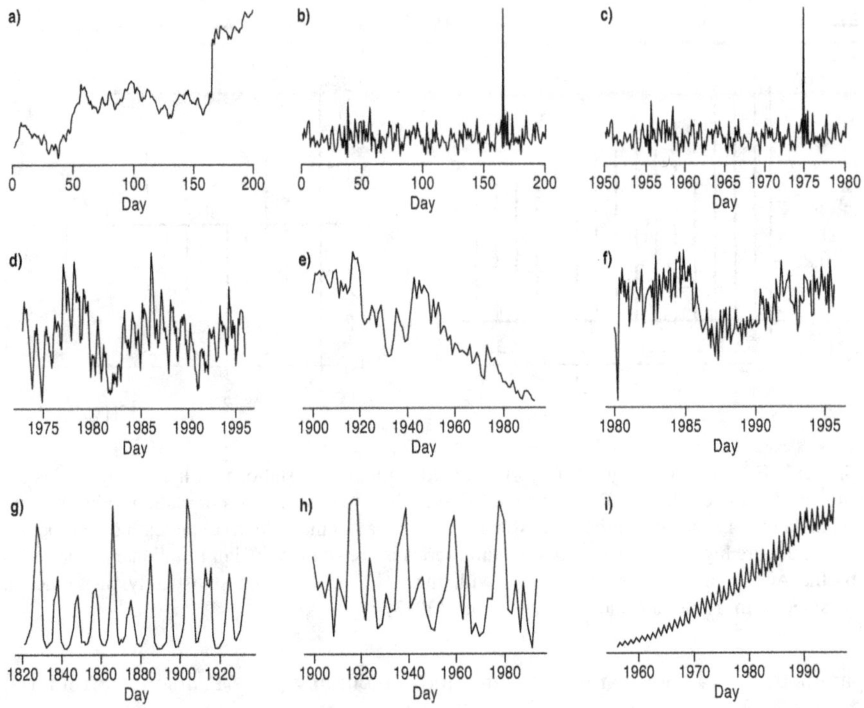

Fig. 3.10 Time series demonstrating a range of characteristics, some of which are stationary (**b** and **g** only). (Adapted from: Hyndman & Athanasopoulos, 2013)

observations. As a result, when plotted over time and allowing for some cyclical behaviour, a stationary series will have evidence of no predictable patterns in the long term, possessing a constant mean and variance (Hyndman & Athanasopoulos, 2013). Further to inspecting time series, considering the autocorrelation behaviour in a time series can reveal evidence of non-stationary behaviour. Essentially the correlation of a time series with itself at various lags, a plot of the autocorrelation function (ACF), called the correlogram, will demonstrate a slow decay of statistically significant values over a large number of lags. In contrast for a stationary time series, the ACF value will decrease quickly to zero with this being almost immediately statistically insignificant (Fig. 3.11).

While inspection of time series plots or correlograms will provide hints of a stationary state or otherwise, it is best to undertake a quantitative assessment using parametric or non-parametric tests, depending on whether the data is normally distributed. Examples of testing for trend abound in the literature with the ordinary least squares approach used most widely in the climate literature for the assessing trends in indicators of climate extremes (Visser & Petersen, 2012). To test for trend in a time series of river flow Diermanse et al. (2010) applied a number of simple

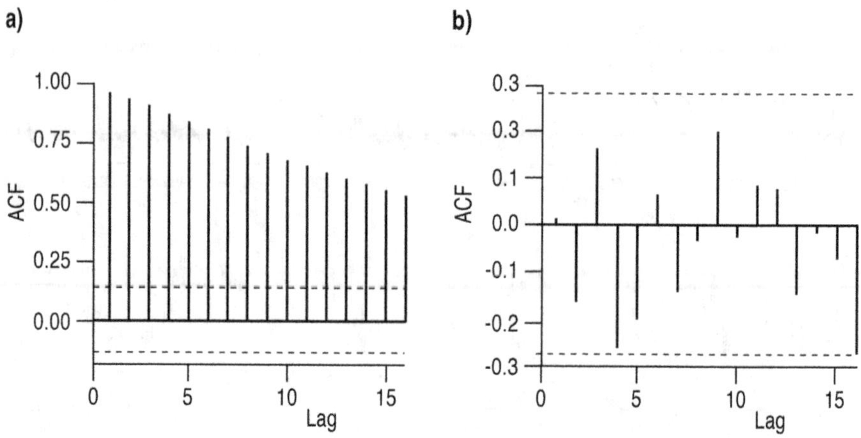

Fig. 3.11 Example correlograms for (**a**) non-stationary and (**b**) stationary time series. ACF is the autocorrelation coefficient. The horizontal dashed line represents the correlation value beyond which there is statistical significance. Note the slow decay of the ACF in (**a**) indicating that successive values are highly correlated, and the time series is non-stationary. For a stationary time series (**b**) the ACF rapidly decreases to zero with the ACF values being statistically insignificant. (Adapted from: Hyndman and Athanasopoulos (2013))

parametric and non-parametric methods including the Pearson correlation, Spearman Rank correlation, Mann-Kendall and Wilcoxon-Mann-Whitney (also known as the Wilcoxon ranks sum test or the Mann-Whitney U) tests. Other general tests for stationarity include the parametric tests of Dickey-Fuller (Said & Dickey, 1984) plus its various extensions (Hamilton, 1994), including the KPSS test (Kwiatkowski et al., 1992), and the non-parametric alternatives of Basu et al. (2009), van Delft and Eichler (2018) and Vogt and Dette (2015). Examples of the application of tests for stationarity related to trend and associated adjustments of climate time series, in preparation for extreme event analysis, include Feng et al. (2007), Fowler et al. (2010), Furio and Meneu (2011), Rea et al. (2011), Villarini et al. (2012) and Visser et al. (2018). As well as techniques for trend testing, others tests for identifying breaks in time series, similar to those applied in homogeneity testing, have also been developed (Gil-Alana, 2015; Yaya & Akintande, 2019; Zivot & Andrews, 1992).

Further to adjusting a time series to achieve stationarity, there is the option of applying a range of non-stationary models to the analysis of extremes (Cheng et al., 2014; Ouarda & Charron, 2019). As noted by Visser and Petersen (2012) non-stationary approaches generally fall into the categories of allowing the location parameter of a GEV distribution or the threshold in a GPD analysis to vary with time (Coelho et al., 2008; Katz et al., 2002). Examples of climate times series to which non-stationary models, with time-vary parameters, have been applied to the analysis of extremes, include heatwaves, climate-fire relations, precipitation, and temperature (Abaurrea et al., 2018; Davies et al., 2017; Hamdi et al., 2018; Hao et al., 2019; Ouarda & Charron, 2019; Turco et al., 2018; Vanem, 2015).

3.4.2.4 Model Choice

While several GEV models are available for analysing extremes, experience shows that model choice can have a strong influence on the conclusions drawn. For example, in an assessment of the implications of model choice, Visser & Petersen (2012) note the GEV and log-normal PDF distributions effectively model the distributions for a range of extreme indicators, with the normal PDF best suited to temperature extremes. In fact, for very good reasons, they note that the Gumbel distribution, which is more suited to skewed distributions, is not applied to temperature data. The impact of adopting a stationary or non-stationary analysis framework is clear in several studies. For example, Begueria et al. (2011) and (Carney et al., 2020) demonstrate how stationary and non-stationary approaches applied to GEV distributions affect the estimate of return periods for temperature extremes. Complementing these examples, Xie et al. (2019) show how different change-point detection models can influence conclusions drawn about the behaviour of hydroclimate time series.

Increasingly the output from numerical models forms the input to the analysis of climate extremes. For the evaluation of extremes up to the present, reanalysis data often forms a proxy for instrumental data, especially for data sparse regions or, where the observational record is less than the reanalysis period. Given that reanalysis products come from a modelling system that does not change with time, apart from major updates and the production of entirely new reanalysis data sets, the main source of uncertainty will be the faithfulness of reanalysis values to reality. Added to this is the potential uncertainty arising from analyses based on just one reanalysis product. Given there are several reanalysis producing agencies, ideally due diligence would include an inter-comparison of reanalysis products in order to establish whether these render extreme climate statistics in the same way.

With the prospect that anthropogenic climate change will increase the number of extreme heat events, there is interest in establishing what the probability distribution of extreme temperature values might look like for the future. To this end, climate scientists run climate models for a different representative concentration pathways (RCP) and shared socio-economic pathways (SSP) and apply extreme event analysis techniques to the output. While this seems quite straightforward, there are several sources of uncertainty associated with climate models, which may affect the confidence in extreme value analyses based on climate model output. Broadly, three major sources of uncertainty exist namely, scenario uncertainty, model uncertainty, and natural variability (Hawkins & Sutton, 2009). Scenario uncertainty refers to not knowing exactly what the future level of greenhouse gas emissions will be. Understanding this is important as the level of atmospheric GHG gas concentrations determines the so-called radiative forcing or the amount of radiation absorbed by the atmosphere and eventually atmospheric heating. Because each modelling group that undertakes climate change projections represents climate change processes, such as radiative forcing, in their own way, differences in model output can arise (Soden et al., 2018). Referred to as model uncertainty, this can result in a range of projections for important climate variables such as temperature. Accordingly, the results from the analysis of temperature extremes from one model might differ from that of another model. The third source of uncertainty is due to internal climate variability. This uncertainty arises because of the natural fluctuations in climate that occur in the absence of any human induced changes to climate. For example, the Pacific Decadal Oscillation, a large-scale mode of natural climate variability, may change the relationship between some critical climate variables as it moves in and out of distinct warm and cold phases (see Chap. 5). Building knowledge on how such fluctuations create changes in climate over time scales less than that of climate change is important, especially in the context of formulating climate risk management strategies based on climate model output.

To complicate matters, the three sources of uncertainty just described are not constant in time nor space, as the relative importance of these varies with prediction lead-time and spatial and temporal averaging scales (Hawkins & Sutton, 2009). While quantifying uncertainty at large spatial and temporal scales is a significant challenge this is exacerbated by the need to produce climate statistics for the local scale via the process of downscaling large scale model output to much smaller 'human scales'. Such a downscaling processes adds an additional layer of uncertainty mainly related to the type of downscaling techniques used, whether it be statistical or dynamical (Wootten et al., 2017). Clearly uncertainty bears implications for policymaking (Smith & Stern, 2011), and for this reason a huge effort continues to be invested in quantifying the various sources of uncertainty, especially in relation to their impact on the statistics of climate extremes (Visser et al., 2018; Wehner, 2010).

3.5 Conclusion

This chapter has established heatwaves as less than simple extreme events, being potentially complex and multifaceted. The chapter has also invited the reader to think of heatwaves not merely as physical events. Heatwaves are also societal events especially when described as severe, although the requisite data for portraying and analysing them as such is largely lacking. Although the availability of physical data, in the form of temperature and other variable values, are generally not a constraint when it comes to the statistical analysis of heatwaves, close attention needs to be paid to whether the data at hand is fit for purpose. This applies not only to instrumental data, but also to that derived from a range of climate reanalysis products and climate models. Of particular concern is whether record incompleteness, inhomogeneities and non-stationarity may compromise the estimation of the likelihood or probability of occurrence of values of heatwave variables reaching levels greater than previously recorded.

Under the chapeau of Extreme Value Theory, extreme values of heatwave related variables are modelled by applying the Generalised Extreme Value (GEV) or Generalised Pareto (GP) distributions, with the former using so-called block maxima data and the later drawing on data series compiled via a peak over threshold procedure. Equipped with well-established models for the statistical analysis of extreme events and good quality data, the heatwave analyst might feel the job is done. However, to be completely confident, due consideration needs to be given to other potential sources of uncertainty, including the choice of the statistical model (e.g., stationary versus non-stationary models) and in the case of data drawn from reanalysis or climate model data products, the structural variability of reanalysis and climate models.

References

Abaurrea, J., Asin, J., & Cebrian, A. C. (2018). Modelling the occurrence of heat waves in maximum and minimum temperatures over Spain and projections for the period 2031–60. *Global and Planetary Change, 161*, 244–260. https://doi.org/10.1016/j.gloplacha.2017.11.015

Acquaotta, F., Fratianni, S., Aguilar, E., & Fortin, G. (2019). Influence of instrumentation on long temperature time series. *Climatic Change, 156*(3), 385–404. https://doi.org/10.1007/s10584-019-02545-z

AghaKouchak, A., Huning, L. S., Mazdiyasni, O., Mallakpour, I., Chiang, F., Sadegh, M., Vahedifard, F., & Moftakhari, H. (2018). How do natural hazards cascade to cause disasters? *Nature, 561*(7724), 458–460. https://doi.org/10.1038/d41586-018-06783-6

Aguilar, E., Auer, I., Brunet, M., Peterson, T., & Wieringa, J. (2003). *Guidelines on climate metadata and homogenization* (Vol. WMO/TD No. 1186). World Meteorological Organization.

Alexander, L. V., Zhang, X., Peterson, T. C., Caesar, J., Gleason, B., Tank, A., Haylock, M., Collins, D., Trewin, B., Rahimzadeh, F., Tagipour, A., Kumar, K. R., Revadekar, J., Griffiths, G., Vincent, L., Stephenson, D. B., Burn, J., Aguilar, E., Brunet, M., Taylor, M., New, M., Zhai, P., Rusticucci, M., & Vazquez-Aguirre, J. L. (2006). Global observed changes in daily climate extremes of temperature and precipitation. *Journal of Geophysical Research-Atmospheres, 111*(D5). https://doi.org/10.1029/2005jd006290

Alexandersson, H. (1986). A homogeneity test applied to precipitation data. *Journal of Climatology, 6*(6), 661–675. https://doi.org/10.1002/joc.3370060607

Allan, R., Brohan, P., Compo, G. P., Stone, R., Luterbacher, J., & Brönnimann, S. (2011). The international atmospheric circulation reconstructions over the earth (ACRE) initiative. *Bulletin of the American Meteorological Society, 92*(11), 1421–1425. https://doi.org/10.1175/2011bams3218.1

Allan, R., Endfield, G., Damodaran, V., Adamson, G., Hannaford, M., Carroll, F., Macdonald, N., Groom, N., Jones, J., Williamson, F., Hendy, E., Holper, P., Arroyo-Mora, J. P., Hughes, L., Bickers, R., & Bliuc, A. M. (2016). Toward integrated historical climate research: The example of atmospheric circulation reconstructions over the earth. *Wiley Interdisciplinary Reviews-Climate Change, 7*(2), 164–174. https://doi.org/10.1002/wcc.379

Ashcroft, L., Coll, J. R., Gilabert, A., Domonkos, P., Brunet, M., Aguilar, E., Castella, M., Sigro, J., Harris, I., Unden, P., & Jones, P. (2018). A rescued dataset of sub-daily meteorological observations for Europe and the southern Mediterranean region, 1877-2012. *Earth System Science Data, 10*(3), 1613–1635. https://doi.org/10.5194/essd-10-1613-2018

Baklanov, A., Grimmond, C. S. B., Carlson, D., Terblanche, D., Tang, X., Bouchet, V., Lee, B., Langendijk, G., Kolli, R. K., & Hovsepyan, A. (2018). From urban meteorology, climate and environment research to integrated city services. *Urban Climate, 23*, 330–341. https://doi.org/10.1016/j.uclim.2017.05.004

Baldwin, J. W., Dessy, J. B., Vecchi, G. A., & Oppenheimer, M. (2019). Temporally compound heat wave events and global warming: An emerging hazard. *Earths Future, 7*(4), 411–427. https://doi.org/10.1029/2018ef000989

Barrantes, G. (2018). Multi-hazard model for developing countries. *Natural Hazards, 92*(2), 1081–1095. https://doi.org/10.1007/s11069-018-3239-6

Barry, D., & Hartigan, J. A. (1993). A BAYESIAN-analysis for change point problems. *Journal of the American Statistical Association, 88*(421), 309–319. https://doi.org/10.1080/01621459.1993.10594323

Barry, A. A., Caesar, J., Tank, A., Aguilar, E., McSweeney, C., Cyrille, A. M., Nikiema, M. P., Narcisse, K. B., Sima, F., Stafford, G., Touray, L. M., Ayilari-Naa, J. A., Mendes, C. L., Tounkara, M., Gar-Glahn, E. V. S., Coulibaly, M. S., Dieh, M. F., Mouhaimouni, M., Oyegade, J. A., Sambou, E., & Laogbessi, E. T. (2018). West Africa climate extremes and climate change indices. *International Journal of Climatology, 38*, E921–E938. https://doi.org/10.1002/joc.5420

Basu, P., Rudoy, D., & Wolfe, P. J., (2009, April 19–24). A nonparametric test for stationarity based on local Fourier analysis. In: IEEE International Conference on Acoustics, Speech and Signal Processing, Taipei, International Conference on Acoustics Speech and Signal Processing ICASSP. pp 3005–3008. https://doi.org/10.1109/icassp.2009.4960256.

Begueria, S., Angulo-Martinez, M., Vicente-Serrano, S. M., Lopez-Moreno, J. I., & El-Kenawy, A. (2011). Assessing trends in extreme precipitation events intensity and magnitude using non-stationary peaks-over-threshold analysis: A case study in Northeast Spain from 1930 to 2006. *International Journal of Climatology, 31*(14), 2102–2114. https://doi.org/10.1002/joc.2218

Benz, S. A., & Blum, P. (2019). Global detection of rainfall-triggered landslide clusters. *Natural Hazards and Earth System Sciences, 19*(7), 1433–1444. https://doi.org/10.5194/nhess-19-1433-2019

Bermudez, P. Z., & Kotz, S. (2010). Parameter estimation of the generalized Pareto distribution-Part I. *Journal of Statistical Planning and Inference, 140*(6), 1353–1373. https://doi.org/10.1016/j.jspi.2008.11.019

Brunet, M., & Jones, P. (2011). Data rescue initiatives: Bringing historical climate data into the 21st century. *Climate Research, 47*(1–2), 29–40. https://doi.org/10.3354/cr00960

Buishand, T. A. (1982). Some methods for testing the homogeneity of rainfall records. *Journal of Hydrology, 58*(1–2), 11–27. https://doi.org/10.1016/0022-1694(82)90066-x

Carney, M., Azencott, R., & Nicol, M. (2020). Nonstationarity of summer temperature extremes in Texas. *International Journal of Climatology, 40*(1), 620–640. https://doi.org/10.1002/joc.6212

Carter, T. R., Benzie, M., Campiglio, E., Carlsen, H., Fronzek, S., Hilden, M., Reyer, C. P. O., & West, C. (2021). A conceptual framework for cross-border impacts of climate change. *Global Environmental Change-Human and Policy Dimensions, 69*. https://doi.org/10.1016/j.gloenvcha.2021.102307

Caussinus, H., & Mestre, O. (2004). Detection and correction of artificial shifts in climate series. *Journal of the Royal Statistical Society Series C-Applied Statistics, 53*, 405–425. https://doi.org/10.1111/j.1467-9876.2004.05155.x

Ceccherini, G., Russo, S., Ameztoy, I., Marchese, A. F., & Carmona-Moreno, C. (2017). Heat waves in Africa 1981–2015, observations and reanalysis. *Natural Hazards and Earth System Sciences, 17*(1), 115–125. https://doi.org/10.5194/nhess-17-115-2017

Charpentier, A. (2011). On the return period of the 2003 heat wave. *Climatic Change, 109*(3–4), 245–260. https://doi.org/10.1007/s10584-010-9944-0

Chen, J., & Brissette, F. P. (2014). Comparison of five stochastic weather generators in simulating daily precipitation and temperature for the Loess Plateau of China. *International Journal of Climatology, 34*(10), 3089–3105. https://doi.org/10.1002/joc.3896

Chen, L., Ng, E., An, X. P., Ren, C., Lee, M., Wang, U., & He, Z. J. (2012). Sky view factor analysis of street canyons and its implications for daytime intra-urban air temperature differentials in high-rise, high-density urban areas of Hong Kong: A GIS-based simulation approach. *International Journal of Climatology, 32*(1), 121–136. https://doi.org/10.1002/joc.2243

Chen, S., Gan, T. Y., Tan, X. Z., Shao, D. G., & Zhu, J. Q. (2019). Assessment of CFSR, ERA-interim, JRA-55, MERRA-2, NCEP-2 reanalysis data for drought analysis over China. *Climate Dynamics, 53*(1–2), 737–757. https://doi.org/10.1007/s00382-018-04611-1

Cheng, L. Y., AghaKouchak, A., Gilleland, E., & Katz, R. W. (2014). Non-stationary extreme value analysis in a changing climate. *Climatic Change, 127*(2), 353–369. https://doi.org/10.1007/s10584-014-1254-5

Chimani, B., Venema, V., Lexer, A., Andre, K., Auer, I., & Nemec, J. (2018). Inter-comparison of methods to homogenize daily relative humidity. *International Journal of Climatology, 38*(7), 3106–3122. https://doi.org/10.1002/joc.5488

Chow, G. C. (1960). Tests of equality between sets of coefficients in 2 linear regressions. *Econometrica, 28*(3), 591–605. https://doi.org/10.2307/1910133

Coelho, C. A. S., Ferro, C. A. T., Stephenson, D. B., & Steinskog, D. J. (2008). Methods for exploring spatial and temporal variability of extreme events in climate data. *Journal of Climate, 21*(10), 2072–2092. https://doi.org/10.1175/2007jcli1781.1

Coles, S. (2001). *An introduction to statistical modeling of extreme values* (Springer Series in Statistics). Springer-Verlag.

Collazo, S., Lhotka, O., Rusticucci, M., & Kysely, J. (2018). Capability of the SMHI-RCA4 RCM driven by the ERA-interim reanalysis to simulate heat waves in Argentina. *International Journal of Climatology, 38*(1), 483–496. https://doi.org/10.1002/joc.5190

Coole, D., & Frost, S. (2010). *New materialisms: Ontology, agency and politics*. Duke University Press. https://doi.org/10.1215/9780822392996

Craddock, J. M. (1977). Homogeneous record of monthly rainfall totals for Norwich for years 1836 to 1976. *Meteorological Magazine, 106*(1262), 267–278.

Creed, I. F., Hewitt, C. M., Casson, N. J., Contosta, A. R., Campbell, J. L., Lutz, D., & Morzillo, A. T. (2023). Coupled human-natural system impacts of a winter weather whiplash event. *Ecology and Society, 28*(2). https://doi.org/10.5751/es-14174-280230

Das, A., Reed, M., & Lindenschmidt, K.-E. (2018). Sustainable Ice-Jam Flood Management for socio-economic and socio-ecological systems. *Water, 10*(2). https://doi.org/10.3390/w10020135

Davies, G., Callaghan, D. P., Gravois, U., Jiang, W., Hanslow, D., Nichol, S., & Baldock, T. (2017). Improved treatment of non-stationary conditions and uncertainties in probabilistic models of storm wave climate. *Coastal Engineering, 127*, 1–19. https://doi.org/10.1016/j.coastaleng.2017.06.005

Depietri, Y., Dahal, K., & McPhearson, T. (2018). Multi-hazard risks in New York City. *Natural Hazards and Earth System Sciences, 18*(12), 3363–3381. https://doi.org/10.5194/nhess-18-3363-2018

Di Napoli, C., Barnard, C., Prudhomme, C., Cloke, H. L., & Pappenberger, F. (2021). ERA5-HEAT: A global gridded historical dataset of human thermal comfort indices from climate reanalysis. *Geoscience Data Journal, 8*(1), 2–10. https://doi.org/10.1002/gdj3.102

Diermanse, F., Kwadijk, J., Beckers, J., & Crebas, J. (2010). Statistical trend analysis of River Rhine and Meuse discharges. doi:https://doi.org/10.7558/bhs.2010.ic9.

Dilley, M., Chen, R. S., Deichmann, U., LernerLam, A. L., & Arnold, M. (2005). Natural disaster hotspots: A global risk analysis. In *Natural disaster hotspots: A global risk analysis* (Vol. 5). World Bank.

Domonkos, P. (2013). Measuring performances of homogenization methods. *Idojaras, 117*(1), 91–112.

Domonkos, P., & Coll, J. (2018). Time series homogenisation of large observational datasets: Impact of the number of partner series on efficiency. *Climate Research, 74*(1), 31–42. https://doi.org/10.3354/cr01488

Domonkos, P., Poza, R., & Efthymiadis, D. (2011). Newest developments of ACMANT. *Advances in Science and Research, 6*, 7–11. https://doi.org/10.5194/asr-6-7-2011

Donat, M. G., Alexander, L. V., Yang, H., Durre, I., Vose, R., Dunn, R. J. H., Willett, K. M., Aguilar, E., Brunet, M., Caesar, J., Hewitson, B., Jack, C., Tank, A., Kruger, A. C., Marengo, J., Peterson, T. C., Renom, M., Rojas, C. O., Rusticucci, M., Salinger, J., Elrayah, A. S., Sekele, S. S., Srivastava, A. K., Trewin, B., Villarroel, C., Vincent, L. A., Zhai, P., Zhang, X., & Kitching, S. (2013). Updated analyses of temperature and precipitation extreme indices since the beginning of the twentieth century: The HadEX2 dataset. *Journal of Geophysical Research-Atmospheres, 118*(5), 2098–2118. https://doi.org/10.1002/jgrd.50150

Dosio, A., Mentaschi, L., Fischer, E. M., & Wyser, K. (2018). Extreme heat waves under 1.5 degrees C and 2 degrees C global warming. *Environmental Research Letters, 13*(5). https://doi.org/10.1088/1748-9326/aab827

Dumitrescu, A., Cheval, S., & Guijarro, J. A. (2020). Homogenization of a combined hourly air temperature dataset over Romania. *International Journal of Climatology, 40*(5), 2599–2608. https://doi.org/10.1002/joc.6353

Easterling, D. R., & Peterson, T. C. (1995). A new method for detecting undocumented discontinuities in climatological time series. *International Journal of Climatology, 15*(4), 369–377.

Elias, A., & Moraru, C. (2015). *The planetary turn: Relationality and geoaesthetics in the twenty-first century*. Northwestern University Press.

Endfield, G., & Veale, L. (2018). *Cultural histories memories and extreme weather: A historical geography perspective*. Routledge.

Eyring, V., Bony, S., Meehl, G. A., Senior, C. A., Stevens, B., Stouffer, R. J., & Taylor, K. E. (2016). Overview of the coupled model Intercomparison project phase 6 (CMIP6) experimental design and organization. *Geoscientific Model Development, 9*(5), 1937–1958. https://doi.org/10.5194/gmd-9-1937-2016

Fellous, J. L. (2008). Towards a global climate observing system. *Interdisciplinary Science Reviews, 33*(1), 83–94. https://doi.org/10.1179/030801808x259952

Feng, S., Nadarajah, S., & Hu, Q. (2007). Modeling annual extreme precipitation in China using the generalized extreme value distribution. *Journal of the Meteorological Society of Japan, 85*(5), 599–613. https://doi.org/10.2151/jmsj.85.599

Fischer, E. M., & Schar, C. (2010). Consistent geographical patterns of changes in high-impact European heatwaves. *Nature Geoscience, 3*(6), 398–403. https://doi.org/10.1038/ngeo866

Forzieri, G., Feyen, L., Russo, S., Vousdoukas, M., Alfieri, L., Outten, S., Migliavacca, M., Bianchi, A., Rojas, R., & Cid, A. (2016). Multi-hazard assessment in Europe under climate change. *Climatic Change, 137*(1–2), 105–119. https://doi.org/10.1007/s10584-016-1661-x

Fowler, H. J., Cooley, D., Sain, S. R., & Thurston, M. (2010). Detecting change in UK extreme precipitation using results from the climateprediction.net BBC climate change experiment. *Extremes, 13*(2), 241–267. https://doi.org/10.1007/s10687-010-0101-y

Francis, J. A., Skific, N., Vavrus, S. J., & Cohen, J. (2022). Measuring "Weather Whiplash" events in North America: A new large-scale regime approach. *Journal of Geophysical Research-Atmospheres, 127*(17). https://doi.org/10.1029/2022jd036717

Francis, J. A., Skific, N., & Zobel, Z. (2023). Weather whiplash events in Europe and North Atlantic assessed as continental-scale atmospheric regime shifts. *Npj Climate and Atmospheric Science, 6*(1). https://doi.org/10.1038/s41612-023-00542-9

Friedman, M. (1937). The use of ranks to avoid the assumption of normality implicit in the analysis of variance. *Journal of the American Statistical Association, 32*(200), 675–701. https://doi.org/10.2307/2279372

Frolicher, T. L., & Laufkotter, C. (2018). Emerging risks from marine heat waves. *Nature Communications, 9*. https://doi.org/10.1038/s41467-018-03163-6

Furio, D., & Meneu, V. (2011). Analysis of extreme temperatures for four sites across peninsular Spain. *Theoretical and Applied Climatology, 104*(1–2), 83–99. https://doi.org/10.1007/s00704-010-0324-5

Gao, P., Li, P. F., Zhao, B. L., Xu, R. R., Zhao, G. J., Sun, W. Y., & Mu, X. M. (2017). Use of double mass curves in hydrologic benefit evaluations. *Hydrological Processes, 31*(26), 4639–4646. https://doi.org/10.1002/hyp.11377

Gelaro, R., McCarty, W., Suarez, M. J., Todling, R., Molod, A., Takacs, L., Randles, C. A., Darmenov, A., Bosilovich, M. G., Reichle, R., Wargan, K., Coy, L., Cullather, R., Draper, C., Akella, S., Buchard, V., Conaty, A., da Silva, A. M., Gu, W., Kim, G.-K., Koster, R., Lucchesi, R., Merkova, D., Nielsen, J. E., Partyka, G., Pawson, S., Putman, W., Rienecker, M., Schubert, S. D., Sienkiewicz, M., & Zhao, B. (2017). The modern-era retrospective analysis for research and applications, version 2 (MERRA-2). *Journal of Climate, 30*(14), 5419–5454. https://doi.org/10.1175/jcli-d-16-0758.1

Gewirtz, P. (1996). On "'I know it when I see it"'. *Yale Law Journal, 105*(4), 1023–1047. https://doi.org/10.2307/797245

Gibson, P. B., Perkins-Kirkpatrick, S. E., Alexander, L. V., & Fischer, E. M. (2017). Comparing Australian heat waves in the CMIP5 models through cluster analysis. *Journal of Geophysical Research-Atmospheres, 122*(6), 3266–3281. https://doi.org/10.1002/2016jd025878

Gil-Alana, L. A. (2015). Linear and segmented trends in sea surface temperature data. *Journal of Applied Statistics, 42*(7), 1531–1546. https://doi.org/10.1080/02664763.2014.1001328

Gou, X. Q., Chen, M. L., & Zhang, G. S. (2018). Time correlations of lightning flash sequences in thunderstorms revealed by fractal analysis. *Journal of Geophysical Research-Atmospheres, 123*(2), 1351–1362. https://doi.org/10.1002/2017jd027206

Guijarro, J.A. (2016). Package 'climatol', version 3.0.

Guijarro, J. A., López, J. A., Aguilar, E., Domonkos, P., Venema, V. K. C., Sigró, J., & Brunet, M. (2023). Homogenization of monthly series of temperature and precipitation: Benchmarking results of the MULTITEST project. *International Journal of Climatology, 43*(9), 3994–4012. https://doi.org/10.1002/joc.8069

Gunasekera, D., Manton, M., & Zillman, J. (2018). Threat to data integrity and international exchange. *Bulletin of the American Meteorological Society, 99*(11), 2206–2207. https://doi.org/10.1175/bams-d-18-0163.1

Guo, X. J., Huang, J. B., Luo, Y., Zhao, Z. C., & Xu, Y. (2017). Projection of heat waves over China for eight different global warming targets using 12 CMIP5 models. *Theoretical and Applied Climatology, 128*(3–4), 507–522. https://doi.org/10.1007/s00704-015-1718-1

Gupta, M., Gao, J., Aggarwal, C. C., & Han, J. W. (2014). Outlier detection for temporal aata: A survey. *IEEE Transactions on Knowledge and Data Engineering, 26*(9), 2250–2267. https://doi.org/10.1109/tkde.2013.184

Haan, L., & Ferreira, A. (2006). *Extreme value theory*. Springer.

Hamdi, Y., Duluc, C. M., & Rebour, V. (2018). Temperature extremes: Estimation of on-stationary return levels and associated uncertainties. *Atmosphere, 9*(4). https://doi.org/10.3390/atmos9040129

Hamilton, J. D. (1994). *Time series analysis*. Princeton University Press. https://doi.org/10.2307/j.ctv14jx6sm

Hao, Z. C., Hao, F. H., Singh, V. P., & Zhang, X. (2018). Changes in the severity of compound drought and hot extremes over global land areas. *Environmental Research Letters, 13*(12). https://doi.org/10.1088/1748-9326/aaee96

Hao, W. L., Shao, Q. X., Hao, Z. C., Ju, Q., Baima, W., & Zhang, D. W. (2019). Non-stationary modelling of extreme precipitation by climate indices during rainy season in Hanjiang River Basin, China. *International Journal of Climatology, 39*(10), 4154–4169. https://doi.org/10.1002/joc.6065

Harman, G. (2018). *Object-oriented ontology*. Penguin.

Hawkins, E., & Sutton, R. (2009). The potential to narrow uncertainty in regional climate predictions. *Bulletin of the American Meteorological Society, 90*(8), 1095. https://doi.org/10.1175/2009bams2607.1

He, X. G., & Sheffield, J. (2020). Lagged compound occurrence of droughts and Pluvials globally over the past seven decades. *Geophysical Research Letters, 47*(14). https://doi.org/10.1029/2020gl087924

Hersbach, H., Bell, B., Berrisford, P., Hirahara, S., Horányi, A., Muñoz-Sabater, J., Nicolas, J., Peubey, C., Radu, R., Schepers, D., Simmons, A., Soci, C., Abdalla, S., Abellan, X., Balsamo, G., Bechtold, P., Biavati, G., Bidlot, J., Bonavita, M., De Chiara, G., Dahlgren, P., Dee, D., Diamantakis, M., Dragani, R., Flemming, J., Forbes, R., Fuentes, M., Geer, A., Haimberger, L., Healy, S., Hogan, R. J., Hólm, E., Janisková, M., Keeley, S., Laloyaux, P., Lopez, P., Lupu, C., Radnoti, G., de Rosnay, P., Rozum, I., Vamborg, F., Villaume, S., & Thépaut, J. N. (2020). The ERA5 global reanalysis. *Quarterly Journal of the Royal Meteorological Society, 146*(730), 1999–2049. https://doi.org/10.1002/qj.3803

Hewaarachchi, A. P., Li, Y. B., Lund, R., & Rennie, J. (2017). Homogenization of daily temperature data. *Journal of Climate, 30*(3), 985–999. https://doi.org/10.1175/jcli-d-16-0139.1

Hosking, J. R. M., & Wallis, J. R. (1987). Parameter and quantile estimation for the generalized Pareto distribution. *Technometrics, 29*(3), 339–349. https://doi.org/10.2307/1269343

Huggel, C., Raissig, A., Rohrer, M., Romero, G., Diaz, A., & Salzmann, N. (2015). How useful and reliable are disaster databases in the context of climate and global change? A comparative case study analysis in Peru. *Natural Hazards and Earth System Sciences, 15*(3), 475–485. https://doi.org/10.5194/nhess-15-475-2015

Hunziker, S., Gubler, S., Calle, J., Moreno, I., Andrade, M., Velarde, F., Ticona, L., Carrasco, G., Castellon, Y., Oria, C., Croci-Maspoli, M., Konzelmann, T., Rohrer, M., & Bronnimann, S. (2017). Identifying, attributing, and overcoming common data quality issues of manned station observations. *International Journal of Climatology, 37*(11), 4131–4145. https://doi.org/10.1002/joc.5037

Hyndman, R. J., & Athanasopoulos, G. (2013). *Forecasting: Principles and practice*. University of Western Australia.

Jiang, L., Zhang, J., Meng, X., Yang, S., Wang, J., & Shi, L. (2024). Identification and characterization of global compound heat wave: Comparison from four datasets of ERA5, Berkeley earth. *CHIRTS and CPC. Climate Dynamics, 62*(1), 631–648. https://doi.org/10.1007/s00382-023-06940-2

Johnson, K., Depietri, Y., & Breil, M. (2016). Multi-hazard risk assessment of two Hong Kong districts. *International Journal of Disaster Risk Reduction, 19*, 311–323. https://doi.org/10.1016/j.ijdrr.2016.08.023

Kanamitsu, M., Ebisuzaki, W., Woollen, J., Yang, S. K., Hnilo, J. J., Fiorino, M., & Potter, G. L. (2002). NCEP-DOE AMIP-II reanalysis (R-2). *Bulletin of the American Meteorological Society, 83*(11), 1631–1643. https://doi.org/10.1175/bams-83-11-1631(2002)083<1631:Nar>2.3.Co;2

Kappes, M. S., Keiler, M., von Elverfeldt, K., & Glade, T. (2012). Challenges of analyzing multi-hazard risk: A review. *Natural Hazards, 64*(2), 1925–1958. https://doi.org/10.1007/s11069-012-0294-2

Karl, T. R., & Knight, R. W. (1997). The 1995 Chicago heat wave: How likely is a recurrence? *Bulletin of the American Meteorological Society, 78*(6), 1107–1119. https://doi.org/10.1175/1520-0477(1997)078<1107:tchwhl>2.0.co;2

Katz, R. W. (2010). Statistics of extremes in climate change. *Climatic Change, 100*(1), 71–76. https://doi.org/10.1007/s10584-010-9834-5

Katz, R. W., Parlange, M. B., & Naveau, P. (2002). Statistics of extremes in hydrology. *Advances in Water Resources, 25*(8–12), 1287–1304. https://doi.org/10.1016/s0309-1708(02)00056-8

Kent, E. C., Rayner, N. A., Berry, D. I., Eastman, R., Grigorieva, V. G., Huang, B., Kennedy, J. J., Smith, S. R., & Willett, K. M. (2019). Observing requirements for long-term climate records at the ocean surface. *Frontiers in Marine Science, 6*. https://doi.org/10.3389/fmars.2019.00441

Kobayashi, S., Ota, Y., Harada, Y., Ebita, A., Moriya, M., Onoda, H., Onogi, K., Kamahori, H., Kobayashi, C., Endo, H., Miyaoka, K., & Takahashi, K. (2015). The JRA-55 reanalysis: General specifications and basic characteristics. *Journal of the Meteorological Society of Japan, 93*(1), 5–48. https://doi.org/10.2151/jmsj.2015-001

Kruskal, W. H., & Wallis, W. A. (1952). Use of ranks in one-criterion variance analysis. *Journal of the American Statistical Association, 47*(260), 583–621.

Kwiatkowski, D., Phillips, P. C. B., Schmidt, P., & Shin, Y. C. (1992). Testing the null hypothesis of stationarity against the alternative of a unit-root – How sure are we that economic time-series have a unit-root. *Journal of Econometrics, 54*(1–3), 159–178. https://doi.org/10.1016/0304-4076(92)90104-y

Lanzante, J. R. (1996). Resistant, robust and non-parametric techniques for the analysis of climate data: Theory and examples, including applications to historical radiosonde station data. *International Journal of Climatology, 16*(11), 1197–1226. https://doi.org/10.1002/(sici)1097-0088(199611)16:11<1197::aid-joc89>3.0.co;2-1

Lawrimore, J. H., Menne, M. J., Gleason, B. E., Williams, C. N., Wuertz, D. B., Vose, R. S., & Rennie, J. (2011). An overview of the global historical climatology network monthly mean temperature data set, version 3. *Journal of Geophysical Research-Atmospheres, 116*. https://doi.org/10.1029/2011jd016187

Lee, C. C. (2022). Weather whiplash: Trends in rapid temperature changes in a warming climate. *International Journal of Climatology, 42*(8), 4214–4222. https://doi.org/10.1002/joc.7458

Leonard, M., Westra, S., Phatak, A., Lambert, M., van den Hurk, B., McInnes, K., Risbey, J., Schuster, S., Jakob, D., & Stafford-Smith, M. (2014). A compound event framework for understanding extreme impacts. *Wiley Interdisciplinary Reviews-Climate Change, 5*(1), 113–128. https://doi.org/10.1002/wcc.252

Li, Y., Ding, Y. H., & Li, W. J. (2017). Observed trends in various aspects of compound heat waves across China from 1961 to 2015. *Journal of Meteorological Research, 31*(3), 455–467. https://doi.org/10.1007/s13351-017-6150-2

Liu, M., & Huang, M. (2014). *Compound disasters and compounding processes* (Global Assessment Report on Disaster Risk Reduction). United Nations Office for Disaster Risk Reduction.

Liu, B., Siu, Y. L., Mitchell, G., & Xu, W. (2016). The danger of mapping risk from multiple natural hazards. *Natural Hazards, 82*(1), 139–153. https://doi.org/10.1007/s11069-016-2184-5

Liu, W., Dong, F., Hu, B. X., You, R., Li, J., & Shao, G. (2023). Exploring two-decadal risk variability of drought-flood abrupt alternation in a high-plateau basin. *Ecological Indicators, 156*. https://doi.org/10.1016/j.ecolind.2023.111106

Lompar, M., Lalic, B., Dekic, L., & Petric, M. (2019). Filling gaps in hourly air temperature data using debiased ERA5 data. *Atmosphere, 10*(1). https://doi.org/10.3390/atmos10010013

Ma, L., Gu, X. D., & Wang, B. W. (2017). Correction of outliers in temperature time series based on sliding window prediction in meteorological sensor network. *Information, 8*(2). https://doi.org/10.3390/info8020060

Madrigal, C., Bedri, R., Piechota, T., Li, W., Tootle, G., & El-Askary, H. (2024). Water whiplash in Mediterranean regions of the world. *Water, 16*(3). https://doi.org/10.3390/w16030450

Magnan, A. K., O'Neill, B. C., & Garschagen, M. (2023). Further understanding "severe" climate risk. *Climate Risk Management, 42*. https://doi.org/10.1016/j.crm.2023.100538

Malamud, B. D., Turcotte, D. L., & Brooks, H. E. (2016). Spatial-temporal clustering of tornadoes. *Natural Hazards and Earth System Sciences, 16*(12), 2823–2834. https://doi.org/10.5194/nhess-16-2823-2016

Manning, C., Widmann, M., Bevacqua, E., Van Loon, A. F., Maraun, D., & Vrac, M. (2019). Increased probability of compound long-duration dry and hot events in Europe during summer (1950–2013). *Environmental Research Letters, 14*(9). https://doi.org/10.1088/1748-9326/ab23bf

Mazdiyasni, O., & AghaKouchak, A. (2015). Substantial increase in concurrent droughts and heatwaves in the United States. *Proceedings of the National Academy of Sciences of the United States of America, 112*(37), 11484–11489. https://doi.org/10.1073/pnas.1422945112

McGregor, G. (2015). Climatology in support of climate risk management: A progress report. *Progress in Physical Geography, 39*(4), 536–553. https://doi.org/10.1177/0309133315578941

McGregor, G. R., Bone, A., & Pappenberger, F. (2017). Meteorological risk: Extreme temperatures. In K. Poljanšek (Ed.), *Science for disaster risk management 2017: Knowing better and losing less* (Vol. 1, pp. 257–270). European Union. https://doi.org/10.2788/842809

Menne, M. J., Durre, I., Vose, R. S., Gleason, B. E., & Houston, T. G. (2012). An overview of the global historical climatology network-daily database. *Journal of Atmospheric and Oceanic Technology, 29*(7), 897–910. https://doi.org/10.1175/jtech-d-11-00103.1

Merz, B., Dung, N. V., Apel, H., Gerlitz, L., Schroter, K., Steirou, E., & Vorogushyn, S. (2018). Spatial coherence of flood-rich and flood-poor periods across Germany. *Journal of Hydrology, 559*, 813–826. https://doi.org/10.1016/j.jhydrol.2018.02.082

Mesinger, F., DiMego, G., Kalnay, E., Mitchell, K., Shafran, P. C., Ebisuzaki, W., Jovic, D., Woollen, J., Rogers, E., Berbery, E. H., Ek, M. B., Fan, Y., Grumbine, R., Higgins, W., Li, H., Lin, Y., Manikin, G., Parrish, D., & Shi, W. (2006). North American regional reanalysis. *Bulletin of the American Meteorological Society, 87*(3), 343–360. https://doi.org/10.1175/bams-87-3-343

Moriyama, K., Sasaki, D., & Ono, Y. (2018). Comparison of global databases for disaster loss and damage data. *Journal of Disaster Research, 13*(6), 1007–1014. https://doi.org/10.20965/jdr.2018.p1007

Mullan, B., Salinger, J., Renwick, J., & Wratt, D. (2018). Comment on "a reanalysis of long-term surface air temperature trends in New Zealand". *Environmental Modeling & Assessment, 23*(3), 249–262. https://doi.org/10.1007/s10666-018-9606-6

Muller, C. L., Chapman, L., Grimmond, C. S. B., Young, D. T., & Cai, X. M. (2013). Toward a standardized metadata protocol for urban meteorological networks. *Bulletin of the American Meteorological Society, 94*(8), 1161–1185. https://doi.org/10.1175/bams-d-12-00096.1

O'Carroll, A. G., Armstrong, E. M., Beggs, H. M., Bouali, M., Casey, K. S., Corlett, G. K., Dash, P., Donlon, C. J., Gentemann, C. L., Hoyer, J. L., Ignatov, A., Kabobah, K., Kachi, M., Kurihara, Y., Karagali, I., Maturi, E., Merchant, C. J., Marullo, S., Minnett, P. J., Pennybacker, M., Ramakrishnan, B., Ramsankaran, R., Santoleri, R., Sunder, S., Picart, S. S., Vazquez-Cuervo, J., & Wimmer, W. (2019). Observational needs of sea surface temperature. *Frontiers in Marine Science, 6*. https://doi.org/10.3389/fmars.2019.00420

Ouarda, T., & Charron, C. (2019). Changes in the distribution of hydro-climatic extremes in a non-stationary framework. *Scientific Reports, 9*. https://doi.org/10.1038/s41598-019-44603-7

Palipane, E., & Grotjahn, R. (2018). Future projections of the large-scale meteorology associated with California heat waves in CMIP5 models. *Journal of Geophysical Research-Atmospheres, 123*(16), 8500–8517. https://doi.org/10.1029/2018jd029000

Parente, J., Pereira, M. G., & Tonini, M. (2016). Space-time clustering analysis of wildfires: The influence of dataset characteristics, fire prevention policy decisions, weather and climate. *Science of the Total Environment, 559*, 151–165. https://doi.org/10.1016/j.scitotenv.2016.03.129

Parente, J., Pereira, M. G., Amraoui, M., & Fischer, E. M. (2018). Heat waves in Portugal: Current regime, changes in future climate and impacts on extreme wildfires. *Science of the Total Environment, 631-632*, 534–549. https://doi.org/10.1016/j.scitotenv.2018.03.044

Park, E. G., Burr, G., Slonosky, V., Sieber, R., & Podolsky, L. (2018). Data rescue archive weather (DRAW): Preserving the complexity of historical climate data. *Journal of Documentation, 74*(4), 763–780. https://doi.org/10.1108/jd-10-2017-0150

Pescaroli, G., & Alexander, D. (2018). Understanding compound, interconnected, interacting, and cascading risks: A holistic framework. *Risk Analysis, 38*(11), 2245–2257. https://doi.org/10.1111/risa.13128

Peterson, T. C., Easterling, D. R., Karl, T. R., Groisman, P., Nicholls, N., Plummer, N., Torok, S., Auer, I., Boehm, R., Gullett, D., Vincent, L., Heino, R., Tuomenvirta, H., Mestre, O., Szentimrey, T., Salinger, J., Forland, E. J., Hanssen-Bauer, I., Alexandersson, H., Jones, P., & Parker, D. (1998). Homogeneity adjustments of in situ atmospheric climate data: A review. *International Journal of Climatology, 18*(13), 1493–1517. https://doi.org/10.1002/(sici)1097-0088(19981115)18:13<1493::aid-joc329>3.0.co;2-t

Pettitt, A. N. (1979). A non-parametric approach to the change-point problem. *Journal of the Royal Statistical Society: Series C (Applied Statistics), 28*(2), 126–135. https://doi.org/10.2307/2346729

Pinheiro, M., & Grotjahn, R. (2015). An introduction to extreme value statistics. Tutorial at University of California at Davis. Available at: http://grotjahn.ucdavis.edu/EWEs/extremes_primer_v9_22_15.pdf (Accessed 23/03/2024)

Potter, K. W. (1981). Illustration of a new test for detecting a shift in mean in precipitation series. *Monthly Weather Review, 109*(9), 2040–2045. https://doi.org/10.1175/1520-0493(1981)109<2040:ioantf>2.0.co;2

Rahmani, E., Friederichs, P., Keller, J., & Hense, A. (2016). Development of an effective and potentially scalable weather generator for temperature and growing degree days. *Theoretical and Applied Climatology, 124*(3–4), 1167–1186. https://doi.org/10.1007/s00704-015-1477-z

Rea, W., Reale, M., & Brown, J. (2011). Long memory in temperature reconstructions. *Climatic Change, 107*(3–4), 247–265. https://doi.org/10.1007/s10584-011-0068-y

Rennie, J. J., Lawrimore, J. H., Gleason, B. E., Thorne, P. W., Morice, C. P., Menne, M. J., Williams, C. N., de Almeida, W. G., Christy, J. R., Flannery, M., Ishihara, M., Kamiguchi, K., Klein-Tank, A. M. G., Mhanda, A., Lister, D. H., Razuvaev, V., Renom, M., Rusticucci, M., Tandy, J., Worley, S. J., Venema, V., Angel, W., Brunet, M., Dattore, B., Diamond, H., Lazzara, M. A., Le Blancq, F., Luterbacher, J., Machel, H., Revadekar, J., Vose, R. S., & Yin, X. (2014). The international surface temperature initiative global land surface databank: Monthly temperature data release description and methods. *Geoscience Data Journal, 1*(2), 75–102. https://doi.org/10.1002/gdj3.8

Ribeiro, S., Caineta, J., & Costa, A. C. (2016). Review and discussion of homogenisation methods for climate data. *Physics and Chemistry of the Earth, 94*, 167–179. https://doi.org/10.1016/j.pce.2015.08.007

Rohini, P., Rajeevan, M., & Mukhopadhay, P. (2019). Future projections of heat waves over India from CMIP5 models. *Climate Dynamics, 53*(1–2), 975–988. https://doi.org/10.1007/s00382-019-04700-9

Royé, D., Iñiguez, C., & Tobías, A. (2020). Comparison of temperature-mortality associations using observed weather station and reanalysis data in 52 Spanish cities. *Environmental Research, 183*. https://doi.org/10.1016/j.envres.2020.109237

Saha, S., Moorthi, S., Pan, H.-L., Wu, X., Wang, J., Nadiga, S., Tripp, P., Kistler, R., Woollen, J., Behringer, D., Liu, H., Stokes, D., Grumbine, R., Gayno, G., Wang, J., Hou, Y.-T., Chuang, H.-Y., Juang, H.-M. H., Sela, J., Iredell, M., Treadon, R., Kleist, D., Van Delst, P., Keyser, D., Derber, J., Ek, M., Meng, J., Wei, H., Yang, R., Lord, S., Van den Dool, H., Kumar, A., Wang, W., Long, C., Chelliah, M., Xue, Y., Huang, B., Schemm, J.-K., Ebisuzaki, W., Lin, R., Xie, P., Chen, M., Zhou, S., Higgins, W., Zou, C.-Z., Liu, Q., Chen, Y., Han, Y., Cucurull, L., Reynolds, R. W., Rutledge, G., & Goldberg, M. (2010). The NCEP climate forecast system reanalysis. *Bulletin of the American Meteorological Society, 91*(8), 1015–1057. https://doi.org/10.1175/2010bams3001.1

Said, S. E., & Dickey, D. A. (1984). Testing for unit roots in autoregressive-moving average models of unknown order. *Biometrika, 71*(3), 599–607.

Sambou, M. J. G., Janicot, S., Pohl, B., Badiane, D., Dieng, A. L., & Gaye, A. (2020). Heat wave occurrences over Senegal during spring: Regionalization and synoptic patterns. *International Journal of Climatology*. https://doi.org/10.1002/joc.6220

Schär, C., Vidale, P. L., Lüthi, D., Frei, C., Häberli, C., Liniger, M. A., & Appenzeller, C. (2004). The role of increasing temperature variability in European summer heatwaves. *Nature, 427*(6972), 332–336. https://doi.org/10.1038/nature02300

Schoetter, R., Cattiaux, J., & Douville, H. (2015). Changes of western European heat wave characteristics projected by the CMIP5 ensemble. *Climate Dynamics, 45*(5–6), 1601–1616. https://doi.org/10.1007/s00382-014-2434-8

Schoof, J. T., Ford, T. W., & Pryor, S. C. (2017). Recent changes in US regional heat wave characteristics in observations and Reanalyses. *Journal of Applied Meteorology and Climatology, 56*(9), 2621–2636. https://doi.org/10.1175/jamc-d-16-0393.1

Semenov, M. A. (2008). Simulation of extreme weather events by a stochastic weather generator. *Climate Research, 35*(3), 203–212. https://doi.org/10.3354/cr00731

Seneviratne, S. I., Nicholls, N., Easterling, D., Goodess, C. M., Kanae, S., Kossin, J., Luo, Y. L., Marengo, J., McInnes, K., Rahimi, M., Reichstein, M., Sorteberg, A., Vera, C., Zhang, X. B., Rusticucci, M., Semenov, V., Alexander, L. V., Allen, S., Benito, G., Cavazos, T., Clague, J., Conway, D., Della-Marta, P. M., Gerber, M., Gong, S. L., Goswami, B. N., Hemer, M., Huggel, C., van den Hurk, B., Kharin, V. V., Kitoh, A., Tank, A., Li, G. L., Mason, S., McGuire, W., van Oldenborgh, G. J., Orlowsky, B., Smith, S., Thiaw, W., Velegrakis, A., Yiou, P., Zhang, T. J., Zhou, T. J., Zwiers, F. W., & Intergov Panel Clim C. (2012). Changes in climate extremes and their impacts on the natural physical environment. In C. B. Field, V. Barros, T. F. Stocker, D. Qin, D. J. Dokken, K. L. Ebi, M. D. Mastrandrea, K. J. Mach, G.-K. Plattner, S. K. Allen,

M. Tignor, & P. M. Midgley (Eds.), *Managing the risks of extreme events and disasters to advance climate change adaptation* (pp. 109–230). Cambridge University Press.

Serinaldi, F., & Kilsby, C. G. (2015). Stationarity is undead: Uncertainty dominates the distribution of extremes. *Advances in Water Resources, 77*, 17–36. https://doi.org/10.1016/j.advwatres.2014.12.013

Sheridan, S. C., Lee, C. C., & Smith, E. T. (2020). A comparison between station observations and reanalysis data in the identification of extreme temperature events. *Geophysical Research Letters, 47*(15). https://doi.org/10.1029/2020gl088120

Smith, L. A., & Stern, N. (2011). Uncertainty in science and its role in climate policy. *Philosophical Transactions of the Royal Society a-Mathematical Physical and Engineering Sciences, 369*(1956), 4818–4841. https://doi.org/10.1098/rsta.2011.0149

Soden, B. J., Collins, W. D., & Feldman, D. R. (2018). Reducing uncertainties in climate models implementing accurate calculations of radiative forcing can improve climate projections. *Science, 361*(6400), 326–327. https://doi.org/10.1126/science.aau1864

Spangler, K. R., Weinberger, K. R., & Wellenius, G. A. (2019). Suitability of gridded climate datasets for use in environmental epidemiology. *Journal of Exposure Science and Environmental Epidemiology, 29*(6), 777–789. https://doi.org/10.1038/s41370-018-0105-2

Squintu, A. A., van der Schrier, G., Brugnara, Y., & Tank, A. K. (2019). Homogenization of daily temperature series in the European climate Assessment & Dataset. *International Journal of Climatology, 39*(3), 1243–1261. https://doi.org/10.1002/joc.5874

Stepánek, P., Zahradnícek, P., & Skalák, P. (2009). Data quality control and homogenization of air temperature and precipitation series in the area of The Czech Republic in the period 1961–2007. In: 8th annual meeting of the European-Meteorological-Society (EMS) /7th European conference on applied climatology, Sep 29-Oct 03 2008. pp. 23–26.

Stephenson, D. B. (2008). Definition, diagnosis, and origin of extreme weather and climate events. *Climate Extremes and Society.* https://doi.org/10.1017/cbo9780511535840.004

Stott, P. A., Stone, D. A., & Allen, M. R. (2004). Human contribution to the European heatwave of 2003. *Nature, 432*(7017), 610–614. https://doi.org/10.1038/nature03089

Strangeways, I. (2015). A global climate reference network. *Weather, 70*(4), 124–129. https://doi.org/10.1002/wea.2460

Sutanto, S. J., Vitolo, C., Di Napoli, C., D'Andrea, M., & Van Lanen, H. A. J. (2020). Heatwaves, droughts, and fires: Exploring compound and cascading dry hazards at the pan-European scale. *Environment International, 134*. https://doi.org/10.1016/j.envint.2019.105276

Swain, D. L., Langenbrunner, B., Neelin, J. D., & Hall, A. (2018). Increasing precipitation volatility in twenty-first-century California. *Nature Climate Change, 8*(5), 427. https://doi.org/10.1038/s41558-018-0140-y

Szentimrey, T. (1999). Multiple analysis of series for homogenization (MASH). In *Second seminar for homogenization of surface climatological data, Geneva* (pp. 27–46). World Meteorological Organisation.

Tanaka, H., Sasaki, D., & Ono, Y. (2018). Proposed requirement definition method for developing global disaster database based on various means of data collection. *Journal of Disaster Research, 13*(6), 1015–1023. https://doi.org/10.20965/jdr.2018.p1015

Taylor, K. E., Stouffer, R. J., & Meehl, G. A. (2012). An overview of CMIP5 and the experimental design. *Bulletin of the American Meteorological Society, 93*(4), 485–498. https://doi.org/10.1175/bams-d-11-00094.1

Telesca, L. (2007). Time-clustering of natural hazards. *Natural Hazards, 40*(3), 593–601. https://doi.org/10.1007/s11069-006-9023-z

Tesfaye, T. W., Dhanya, C. T., & Gosain, A. K. (2017). Evaluation of ERA-interim, MERRA, NCEP-DOE R2 and CFSR reanalysis precipitation data using gauge observation over Ethiopia for a period of 33 years. *AIMS Environmental Science, 4*(4), 596–620. https://doi.org/10.3934/environsci.2017.4.596

Thorne, P.W., Lawrimore, J. H., Willett, K. M., Allan, R., Chandler, R. E., Mhanda, A., de Podesta, M., Possolo, A., Revadekar, J., Rusticucci, M., Stott, P. A., Strouse, G. F., Trewin, B., Wang,

X. L., Yatagai, A., Merchant, C., Merlone, A., Peterson, T. C., & Scott, E. M. (2013). The international surface temperature initiative (invited paper). In: Technol, Conf MS, Flukc, C. ct al. (eds) 9th international temperature symposium on temperature – Its measurement and control in science and industry, (2013 Mar 19–23 2012. AIP conference proceedings. pp. 1020–1029). https://doi.org/10.1063/1.4821418.

Thorne, P. W., Donat, M. G., Dunn, R. J. H., Williams, C. N., Alexander, L. V., Caesar, J., Durre, I., Harris, I., Hausfather, Z., Jones, P. D., Menne, M. J., Rohde, R., Vose, R. S., Davy, R., Klein-Tank, A. M. G., Lawrimore, J. H., Peterson, T. C., & Rennie, J. J. (2016). Reassessing changes in diurnal temperature range: Intercomparison and evaluation of existing global data set estimates. *Journal of Geophysical Research-Atmospheres, 121*(10), 5138–5158. https://doi.org/10.1002/2015jd024584

Tilloy, A., Malamud, B. D., Winter, H., & Joly-Laugel, A. (2019). A review of quantification methodologies for multi-hazard interrelationships. *Earth-Science Reviews, 196*. https://doi.org/10.1016/j.earscirev.2019.102881

Toreti, A., Kuglitsch, F. G., Xoplaki, E., & Luterbacher, J. (2012). A novel approach for the detection of inhomogeneities affecting climate time series. *Journal of Applied Meteorology and Climatology, 51*(2), 317–326. https://doi.org/10.1175/jamc-d-10-05033.1

Toreti, A., Maiorano, A., De Sanctis, G., Webber, H., Ruane, A. C., Fumagalli, D., Ceglar, A., Niemeyer, S., & Zampieri, M. (2019). Using reanalysis in crop monitoring and forecasting systems. *Agricultural Systems, 168*, 144–153. https://doi.org/10.1016/j.agsy.2018.07.001

Toscano, B., Fernandes, A. D., da Silva, M. M., & Santoro, F. M. (2022). A domain ontology on cascading effects in critical infrastructures based on a systematic literature review. *International Journal of Critical Infrastructures, 18*(1), 79–103.

Trewin, B. (2010). Exposure, instrumentation, and observing practice effects on land temperature measurements. *Wiley Interdisciplinary Reviews-Climate Change, 1*(4), 490–506. https://doi.org/10.1002/wcc.46

Trewin, B. (2013). A daily homogenized temperature data set for Australia. *International Journal of Climatology, 33*(6), 1510–1529. https://doi.org/10.1002/joc.3530

Trewin, B. C., & Jones, D. A. (2015). Comment on Boretti (2013)). Statistical analysis of the temperature records for the Northern Territory of Australia. *Theoretical and Applied Climatology, 120*(1–2), 69–74. https://doi.org/10.1007/s00704-014-1158-3

Trewin, B., Braganza, K., Fawcett, R., Grainger, S., Jovanovic, B., Jones, D., Martin, D., Smalley, R., & Webb, V. (2020). An updated long-term homogenized daily temperature data set for Australia. *Geoscience Data Journal, 7*(2), 149–169. https://doi.org/10.1002/gdj3.95

Turco, M., Rosa-Canovas, J. J., Bedia, J., Jerez, S., Montavez, J. P., Llasat, M. C., & Provenzale, A. (2018). Exacerbated fires in Mediterranean Europe due to anthropogenic warming projected with non-stationary climate-fire models. *Nature Communications, 9*. https://doi.org/10.1038/s41467-018-06358-z

Valmassoi, A., Keller, J. D., Kleist, D. T., English, S., Ahrens, B., Duran, I. B., Bauernschubert, E., Bosilovich, M. G., Fujiwara, M., Hersbach, H., Lei, L., Loehnert, U., Mamnun, N., Martin, C. R., Moore, A., Niermann, D., Ruiz, J. J., & Scheck, L. (2023). Current challenges and future directions in data assimilation and reanalysis. *Bulletin of the American Meteorological Society, 104*(4), E756–E767. https://doi.org/10.1175/bams-d-21-0331.1

van Delft, A., & Eichler, M. (2018). Locally stationary functional time series. *Electronic Journal of Statistics, 12*(1), 107–170. https://doi.org/10.1214/17-ejs1384

Vanem, E. (2015). Non-stationary extreme value models to account for trends and shifts in the extreme wave climate due to climate change. *Applied Ocean Research, 52*, 201–211. https://doi.org/10.1016/j.apor.2015.06.010

Venema, V. K. C., Mestre, O., Aguilar, E., Auer, I., Guijarro, J. A., Domonkos, P., Vertacnik, G., Szentimrey, T., Stepanek, P., Zahradnicek, P., Viarre, J., Muller-Westermeier, G., Lakatos, M., Williams, C. N., Menne, M. J., Lindau, R., Rasol, D., Rustemeier, E., Kolokythas, K., Marinova, T., Andresen, L., Acquaotta, F., Fratianni, S., Cheval, S., Klancar, M., Brunetti, M., Gruber, C., Duran, M. P., Likso, T., Esteban, P., & Brandsma, T. (2012). Benchmarking

homogenization algorithms for monthly data. *Climate of the Past, 8*(1), 89–115. https://doi.org/10.5194/cp-8-89-2012
Villarini, G., Smith, J. A., Serinaldi, F., Ntelekos, A. A., & Schwarz, U. (2012). Analyses of extreme flooding in Austria over the period 1951-2006. *International Journal of Climatology, 32*(8), 1178–1192. https://doi.org/10.1002/joc.2331
Vincent, L. A., Zhang, X., Bonsal, B. R., & Hogg, W. D. (2002). Homogenization of daily temperatures over Canada. *Journal of Climate, 15*(11), 1322–1334. https://doi.org/10.1175/1520-0442(2002)015<1322:Hodtoc>2.0.Co;2
Visser, H., & Petersen, A. C. (2012). Inferences on weather extremes and weather-related disasters: A review of statistical methods. *Climate of the Past, 8*(1), 265–286. https://doi.org/10.5194/cp-8-265-2012
Visser, H., Dangendorf, S., van Vuuren, D. P., Bregman, B., & Petersen, A. C. (2018). Signal detection in global mean temperatures after "Paris": An uncertainty and sensitivity analysis. *Climate of the Past, 14*(2), 139–155. https://doi.org/10.5194/cp-14-139-2018
Visser-Quinn, A., Beevers, L., Collet, L., Formetta, G., Smith, K., Wanders, N., Thober, S., Pan, M., & Kumar, R. (2019). Spatio-temporal analysis of compound hydro-hazard extremes across the UK. *Advances in Water Resources, 130*, 77–90. https://doi.org/10.1016/j.advwatres.2019.05.019
Vitolo, C., Di Napoli, C., Di Giuseppe, F., Cloke, H. L., & Pappenberger, F. (2019). Mapping combined wildfire and heat stress hazards to improve evidence-based decision making. *Environment International, 127*, 21–34. https://doi.org/10.1016/j.envint.2019.03.008
Vogt, M., & Dette, H. (2015). Detecting gradual changes in locally stationary processes. *Annals of Statistics, 43*(2), 713–740. https://doi.org/10.1214/14-aos1297
von Neumann, J. (1941). Distribution of the ratio of the mean square successive difference to the variance. *Annals of Mathematical Statistics, 12*, 367–395. https://doi.org/10.1214/aoms/1177731677
Wang, X. L. L. (2008a). Accounting for autocorrelation in detecting mean shifts in climate data series using the penalized maximal t or F test. *Journal of Applied Meteorology and Climatology, 47*(9), 2423–2444. https://doi.org/10.1175/2008jamc1741.1
Wang, X. L. L. (2008b). Penalized maximal F test for detecting undocumented mean shift without trend change. *Journal of Atmospheric and Oceanic Technology, 25*(3), 368–384. https://doi.org/10.1175/2007jtecha982.1
Wang, X., & Feng, Y. (2013). *RHtests V4 user manual*. Climate Research Division, Atmospheric Science and Technology Directorate Science and Technology Branch.
Wang, J., Chen, Y., Tett, S. F. B., Yan, Z. W., Zhai, P. M., Feng, J. M., & Xia, J. J. (2020). Anthropogenically-driven increases in the risks of summertime compound hot extremes. *Nature Communications, 11*(1). https://doi.org/10.1038/s41467-019-14233-8
Way, R. G., Oliva, F., & Viau, A. E. (2017). Underestimated warming of northern Canada in the Berkeley earth temperature product. *International Journal of Climatology, 37*(4), 1746–1757. https://doi.org/10.1002/joc.4808
Wehner, M. (2010). Sources of uncertainty in the extreme value statistics of climate data. *Extremes, 13*(2), 205–217. https://doi.org/10.1007/s10687-010-0105-7
Wijngaard, J. B., Tank, A., & Konnen, G. P. (2003). Homogeneity of 20th century European daily temperature and precipitation series. *International Journal of Climatology, 23*(6), 679–692. https://doi.org/10.1002/joc.906
Wilks, D. S. (2011). *Statistical Methods in the Atmospheric Sciences*. Oxford. Academic Press. 467pp.
Wilks, D. S. (2012). Stochastic weather generators for climate-change downscaling, part II: Multivariable and spatially coherent multisite downscaling. *Wiley Interdisciplinary Reviews-Climate Change, 3*(3), 267–278. https://doi.org/10.1002/wcc.167
Wilks, D. S. (2019). *Statistical Methods in the Atmospheric Sciences* (Statistical methods in the atmospheric sciences) (4th ed., pp. 1–818). https://doi.org/10.1016/c2017-0-03921-6

Willett, K. M. (2023). HadlSDH.Extremes Part II: Exploring humid heat extremes using wet bulb temperature indices. *Advances in Atmospheric Sciences, 40*(11), 1968–1985. https://doi.org/10.1007/s00376-023-2348-7

Willett, K., Williams, C., Jolliffe, I. T., Lund, R., Alexander, L. V., Bronnimann, S., Vincent, L. A., Easterbrook, S., Venema, V. K. C., Berry, D., Warren, R. E., Lopardo, G., Auchmann, R., Aguilar, E., Menne, M. J., Gallagher, C., Hausfather, Z., Thorarinsdottir, T., & Thorne, P. W. (2014a). A framework for benchmarking of homogenisation algorithm performance on the global scale. *Geoscientific Instrumentation Methods and Data Systems, 3*(2), 187–200. https://doi.org/10.5194/gi-3-187-2014

Willett, K. M., Dunn, R. J. H., Thorne, P. W., Bell, S., de Podesta, M., Parker, D. E., Jones, P. D., & Williams, C. N. (2014b). HadISDH land surface multi-variable humidity and temperature record for climate monitoring. *Climate of the Past, 10*(6), 1983–2006. https://doi.org/10.5194/cp-10-1983-2014

WMO. (2020). *Guidelines on homogenization* (Vol. WMO 1245). World Meteorological Organisation.

WMO. (2023a). *Guide to climatological practices* (Vol. WMO-No. 100). World Meteorological Organisation.

WMO. (2023b). *Guide to instruments and methods of observation volume I—Measurement of meteorological variables* (Vol. WMO-No. 8). World Meteorological Organisation.

Wootten, A., Terando, A., Reich, B. J., Boyles, R. P., & Semazzi, F. (2017). Characterizing sources of uncertainty from global climate models and downscaling techniques. *Journal of Applied Meteorology and Climatology, 56*(12), 3245–3262. https://doi.org/10.1175/jamc-d-17-0087.1

Xie, P., Gu, H. T., Sang, Y. F., Wu, Z. Y., & Singh, V. P. (2019). Comparison of different methods for detecting change points in hydroclimatic time series. *Journal of Hydrology, 577*. https://doi.org/10.1016/j.jhydrol.2019.123973

Yan, Y. C., Xu, Y. Y., & Yue, S. P. (2021). A high-spatial-resolution dataset of human thermal stress indices over South and East Asia. *Scientific Data, 8*(1). https://doi.org/10.1038/s41597-021-01010-w

Yang, Z., Dominguez, F., & Zeng, X. B. (2019). Large and local-scale features associated with heat waves in the United States in reanalysis products and the NARCCAP model ensemble. *Climate Dynamics, 52*(3–4), 1883–1901. https://doi.org/10.1007/s00382-018-4414-x

Yaya, O. S., & Akintande, O. J. (2019). Long-range dependence, nonlinear trend, and breaks in historical sea surface and land air surface global and regional temperature anomalies. *Theoretical and Applied Climatology, 137*(1–2), 177–185. https://doi.org/10.1007/s00704-018-2592-4

Yozgatligil, C., & Yaziçi, C. (2016). Comparison of homogeneity tests for temperature using a simulation study. *International Journal of Climatology, 36*(1), 62–81. https://doi.org/10.1002/joc.4329

Zaidi, R. Z. (2018). Beyond the Sendai indicators: Application of a cascading risk lens for the improvement of loss data indicators for slow-onset hazards and small-scale disasters. *International Journal of Disaster Risk Reduction, 30*, 306–314. https://doi.org/10.1016/j.ijdrr.2018.03.022

Zivot, E., & Andrews, D. W. K. (1992). Further evidence on the great crash, the oil-price shock, and the unit-root hypothesis. *Journal of Business & Economic Statistics, 10*(3), 251–270. https://doi.org/10.2307/1391541

Zscheischler, J., Martius, O., Westra, S., Bevacqua, E., Raymond, C., Horton, R. M., van den Hurk, B., AghaKouchak, A., Jezequel, A., Mahecha, M. D., Maraun, D., Ramos, A. M., Ridder, N. N., Thiery, W., & Vignotto, E. (2020). A typology of compound weather and climate events. *Nature Reviews Earth & Environment, 1*(7), 333–347. https://doi.org/10.1038/s43017-020-0060-z

Zwiers, F. W., Zhang, X. B., & Feng, Y. (2011). Anthropogenic influence on long return period daily temperature extremes at regional scales. *Journal of Climate, 24*(3), 881–892. https://doi.org/10.1175/2010jcli3908.1

Chapter 4
The Physical Nature of Heat and Its Measurement

Abstract Heat is the defining physical entity of heatwaves. Accordingly, this chapter provides an overview of the physical nature of heat and its measurement. It begins by establishing the relationship between energy and heat, describing heat as a form of kinetic energy. The concepts of the surface radiation and energy balances are introduced, highlighting the various radiative, turbulent, and conductive heat fluxes that contribute to understanding heatwave processes. The chapter then examines the mechanisms of atmospheric heating, drawing a distinction between adiabatic processes and diabatic processes. Significant attention is given to the measurement and monitoring of heat, which is crucial for assessing heat stress levels during heatwaves. Direct measurements using temperature sensors and indirect methods like thermal mapping are discussed for infrastructure applications. For human health, the chapter highlights the complexity of heat stress, which arises from synergistic effects of temperature, humidity, radiation, and other factors. Because the quantification of heat stress is a priority for heatwave related studies focused on exposure and impact relationships, an overview of thermal climate indices used to model heat stress is presented noting that these range from simple empirical indices to complex rational models that solve the human heat balance equation. The strengths and limitations of some of the more commonly used thermal climate indices is discussed with the importance of selecting appropriate indices based on the research problem or application at hand emphasised.

Keywords Heat and energy · Atmospheric heating processes · Surface radiation and energy balances · Sensible heat · Latent heat · Heat stress · Thermal climate indices

4.1 Introduction

Heat is the defining physical entity of heatwaves as an extreme weather phenomenon. Often considered as the sensation of warmth its precise scientific definition is more than that. Having defined heatwaves and considered these as extreme events in the previous two chapters, the purpose of this chapter is to outline the nature of heat

G. McGregor, *Heatwaves*, Biometeorology 6,
https://doi.org/10.1007/978-3-031-69906-1_4

as the defining property of heatwaves. The chapter begins with a description of the relationship between energy and heat. It then introduces the concepts of the surface radiation and energy (heat) balances and describes their components as represented by a range of radiative, and turbulent and conductive heat fluxes and outlines why an appreciation of these concepts helps understand heatwave processes. The chapter then moves on to sketch out the main atmospheric heating processes drawing a distinction between adiabatic and diabatic processes, noting that both contribute to the formation of heat extremes associated with heatwaves. While the first half of the chapter addresses the subject of heat at a more theoretical level, the latter half focuses on the more practical aspects of the measurement of heat and its constituent variables and the associated development of thermal climate indices for assessing levels of heat in the context of human related heat stress associated with heatwaves.

4.2 Energy and Heat

According to the First Law of Thermodynamics, also referred to as the Law of the Conservation of Energy, energy can be neither created nor destroyed. Rather all energy can do is change form. Energy can be divided into two broad groups namely kinetic and potential energy. Kinetic energy arises from movement. Heat or thermal energy is just one form of kinetic energy as it has its origins in the motion of atoms that make up gaseous, solid or liquid matter. Other types of kinetic energy include radiant, electrical, magnetic, and sound energy. In contrast, potential energy is any type of stored energy such as gravitational, nuclear, chemical, or elastic energy. In the context of heatwaves, it is energy in its radiant and sensible heat and latent heat forms which are important.

Before describing the nature of radiant energy and sensible and latent heat it is worth introducing some terminology related to energy. In most climate related literature heat energy is described in terms of watts (W), which indicates how much energy is flowing from or to an object; photons make up the energy flow. Referred to as an energy flux, 1 watt is equivalent to 1 joule per second. Because climatologists are interested in how much energy may be flowing to or from a given area, the term energy flux density, as represented by watts per square metre (Wm^{-2}) is used. This provides a measure of the rate of transfer of energy per square metre per second. For example, at the top of the atmosphere on an imaginary one square metre surface perpendicular to the sun's rays the energy flux density is 1376 Wm^{-2}. This value is also known as the solar constant because currently it is the amount of energy constantly received at the top of the earth's atmosphere every second.

Radiant energy or simply radiation, is the energy emitted from objects with a temperature greater than absolute zero. A huge source of radiant energy for the earth is that which flows from the sun. Solar energy basically powers all natural systems on the earth and makes life possible. The radiative flux from an object is called the irradiance. The amount of solar irradiance received at the earth's surface is controlled by a range of orbital characteristics which in turn exert a fundamental control on the spatial distribution of energy and thus temperature across the earth's surface over a range of timescales from the daily to the millennial. Radiant energy travels from the place it is emitted to where it is eventually absorbed in the form of waves of photons that have both electrical and magnetic properties, hence the term electromagnetic waves. As for ocean waves arriving at the shoreline, electromagnetic waves and thus radiation, can have different lengths as illustrated in Fig. 4.1. The range of wavelengths that make up the flow of radiation from an object is referred to as the electromagnetic spectrum. In the case of the sun this is wide covering electromagnetic waves from the very short to very long. Helpfully there are some basic laws (Wein's Law, Stefan-Boltzman Law) which allow us to predict the amount and relative importance of the range of radiation types emitted from an object with a basic rule of thumb being that 'hot' ('cool') objects (as measured in degrees Kelvin (K)) emit shortwave (longwave) radiation. Considering the 'hot' sun (6000 K) and its electromagnetic spectrum, something like 88 percent of radiation is emitted at short wavelengths less than 1.5 micrometres with maximum emission at 0.5 micrometres which falls in the range of wavelengths that can be detected by the human eye or visible range (0.4–0.7 micrometres). In contrast, radiation emitted by the 'cool' earth (255 K) is in the form of longwave radiation between 5 and 25 micrometres with maximum emissions at around 10 micrometres. So, in relation to heatwaves, essentially the amount of energy absorbed at the earth's surface in the form of shortwave radiation over the duration of a few days has significant implications for surface heating and thus the heating of the overlying atmosphere through sensible heat transfer and to a lesser extent upward longwave radiative flux.

Sensible heat is the heat arising from the vibration of atoms in an object whether it be gaseous, liquid, or solid. Sensible heat manifests itself in the form of temperature which can be sensed by humans, hence its name. The flow or flux of sensible heat is determined by the temperature gradient, with heat flowing spontaneously from places of high to low temperature, for example between a warm urban surface

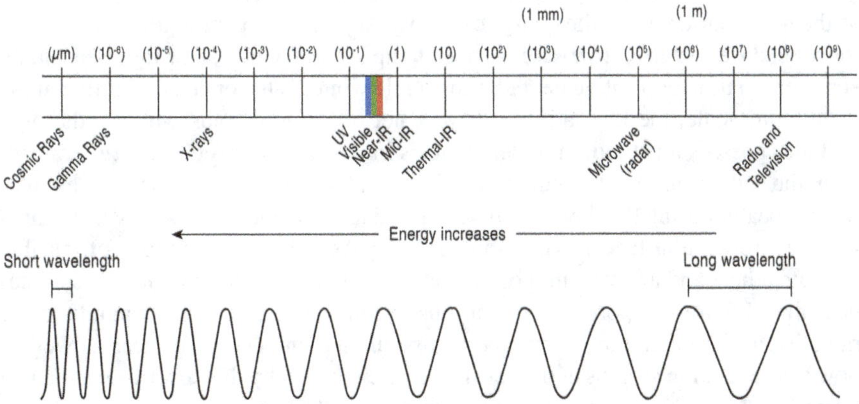

Fig. 4.1 Electromagnetic spectrum with schematic representation of wavelength. One micrometre (um) is a millionth of a metre, or a thousandth of a millimetre. (Adapted from various sources)

and an overlying cooler atmosphere. The purpose of this heat flow is to try and attain thermal equilibrium such that in the case of the urban example the same temperature for the air and urban surface is reached. In many ways the temperature gradient between places or objects is determined by their ability to heat up and thus undergo temperature change. A fundamental property of materials is their heat capacity or thermal capacitance which describes how much energy is required to raise a given volume of material by 1 °C.

Latent heat is 'hidden' heat contained within water molecules and is related to phase changes of water. For example, when water is evaporated off a surface, sensible heat is removed from the overlying atmosphere and stored as latent heat in the evaporated water molecules. This results in a cooling of the air; increasing evaporation rates is one approach to managing heat which will be discussed in Chap. 14. Latent heat stored in the water molecules or water vapour may be carried away from the site of evaporation by vertical currents of air. Eventually condensation may occur with the result that the latent heat is released back into the atmosphere as water changes from its gaseous to liquid form which causes warming. In essence whenever water changes its state or phase there is either energy absorption or release in the form of latent heat. Water phase changes that result in cooling of the air include liquid to vapour (vapourisation), ice to liquid (melting) and ice to vapour (sublimation). In contrast, phase changes such as vapour to liquid (condensation), liquid to ice (fusion/freezing) and vapour to ice (deposition) result in energy release and warming of the air.

4.3 Surface Radiation and Energy (Heat) Balances

Daily the earth's surface receives and emits radiation. The balance of radiation gains and losses is known as the radiation balance with the amount of radiation available at the earth's surface, once all gains and losses have been accounted for, referred to as the net radiation. Over the course of a day or a year, the net radiation at a location must be close to zero; otherwise, surface temperatures will rise or fall. Seasonally summer is a time of positive net radiation while winter is the opposite. Further at the annual timescale, the low latitudes have a net radiation surplus whereas the high latitudes possess a net radiation deficit. This imbalance is offset by energy transfer from the equatorial to polar regions via the global atmospheric and oceanic circulations. Locally during the day a positive net radiation balance is offset by heat transfer away from the surface into the atmosphere by the turbulent heat fluxes of sensible and latent heat and away from the surface to the subsurface by ground (subsurface) heat flux. Figure 4.2 portrays the various avenues and associated magnitudes of radiative heat gain and loss averaged across all land and ocean surfaces as well as how a net radiation surplus at the earth's surface is offset by the flow of sensible and latent heat from the surface into the atmosphere. Of the two turbulent heat fluxes, latent heat, which represents the energy consumed in evaporation, dominates. In simple terms this is because the oceans which provide an unlimited supply of water for evaporation, occupy a greater surface area, than land where the relative balance of latent and sensible heat fluxes is determined by water availability.

As can be determined from Fig. 4.2, the fluxes of sensible and latent heat to the atmosphere from the surface are driven by the flow of energy from the sun in the form of shortwave radiation. However, there are also flows of energy to the earth's surface in the form of longwave radiation from 'warm' gases in the atmosphere. Sometimes referred to as atmospheric radiation, this radiation is the result of the absorption by atmospheric gases of the outgoing longwave or terrestrial radiation emitted from the earth's surface; absorption of the terrestrial radiation warms the atmospheric gases, and this heat is subsequently emitted both upwards to space and downwards to the earth as atmospheric radiation. As mentioned above, the balance between all the flows of incoming and outgoing radiation, or radiative fluxes, is referred to as the net radiation. This represents the total amount of energy available at the earth's surface for apportionment between sensible heating, or raising the air or subsurface temperature, and latent heating, or evaporating water. The net

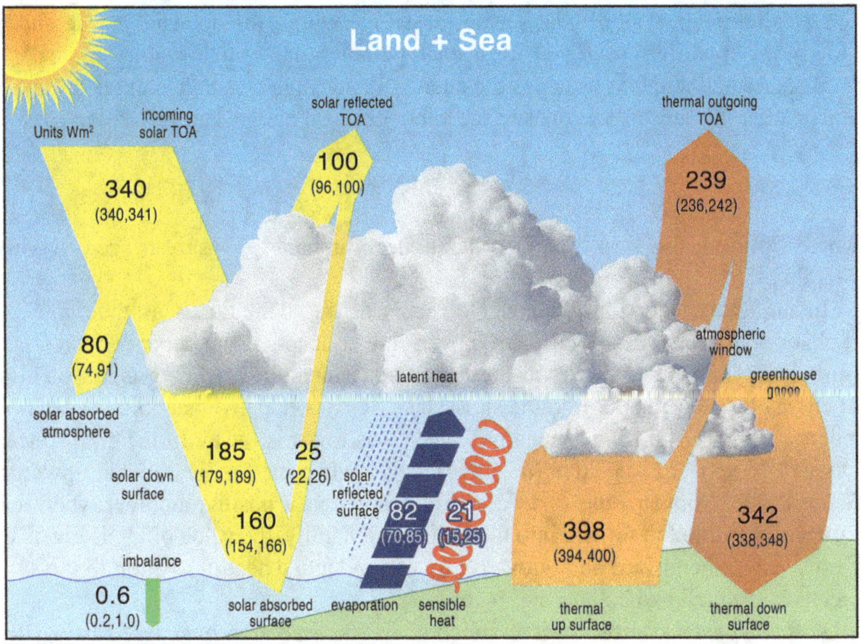

Fig. 4.2 Schematic representation of the global (land and ocean) annual mean energy balance of the Earth. Numbers indicate best estimates for the magnitudes of the globally averaged radiation (solar and thermal) and energy (sensible and latent heat) balance components; the uncertainty ranges appear in parentheses. The yellow arrows represent the flows associated with solar radiation while the orange arrows depict the flows of radiative heat emitted by the earth's surface or the atmosphere. Units are Wm^{-2}. Adapted from: Wild et al. (2015). The energy balance over land and oceans: an assessment based on direct observations and CMIP5 climate models. Clim Dyn 44, 3393–3429 (2015). https://doi-org.ezphost.dur.ac.uk/10.1007/s00382-014-2430-z Used under the terms of the Creative Commons CC BY license 4.0

radiation balance and ultimately the surface energy balance is best described using the following equation:

$$Q^* = K^* + L^*$$

where K* = K↓ – K↑ and L* = L↓ – L↑ and Q* = net radiation, K* = net shortwave radiation, K↓ = incoming shortwave (solar) radiation, K↑= reflected shortwave (solar) radiation, L* = net longwave radiation, L↓ = incoming longwave (atmospheric) radiation and L↑ = outgoing longwave (terrestrial) radiation.

Considering the above, the net radiation balance over a surface is dependent on more than just the shortwave radiation flux from the sun. In the case of the eventual magnitude of K* the amount of shortwave radiation loss (K↑) plays an important role. This is determined by surface reflectivity or albedo with light (dark) coloured surfaces possessing high (low) albedo resulting in high (low) K↑. In relation to L*, the fluxes of L↓and L↑ are dependent on the temperature of the atmosphere and the surface respectively. At night when there is no K*, air temperature near the surface will depend on whether L* is positive or negative; clear skies can lead to high rates of L↑ and the rapid fall of near surface air temperatures; the opposite is true under cloudy sky conditions such that surface temperatures can remain stable.

Because net radiation provides the energy for sensible heat and latent heat fluxes and heating the subsurface, the surface energy or heat budget can be written as.

$$Q^* = Qh + Qe + Qg$$

where Qh = sensible heat flux, Qe = latent heat flux and Qg = subsurface (ground) heat flux.

In the above equation Qh and Qe represent turbulent fluxes as their rate of exchange between the surface and the overlying atmosphere are dependent on wind (turbulence) and the actual vertical temperature or moisture gradients. Qg is the loss of energy by heat transfer into the subsurface; it's a conductive heat flux. In energy balance eqs. Q* is positive (negative) when it is towards (away from) the surface. Qh and Qe appear as positive (negative) values when the flux is directed upwards (downwards) from (to) the surface. Qg is denoted as a positive number when the flux is away from the surface into the ground. Generally, the value of Qg is given as zero because the inward and outward flux of conductive heat in relation to the surface is both small and balances over time.

A fundamental control on how much of the net radiation goes into raising the temperature via sensible heat flux (Qh) is the availability of water and thus the magnitude of the latent heat flux (Qe) or rate of evaporation (Donat et al., 2017; Gallego-Elvira et al., 2016). In the case of a dry surface during the day, most of Q* will go into Qh, whereas the opposite is true for a very moist or wet surface with Qe dominating the turbulent fluxes (Fig. 4.3). For a moist surface at night, Q* is directed upwards as there is no incoming shortwave radiation (K↓) with the nocturnal radiation balance dominated by outgoing longwave radiation (L↑). Because atmospheric moisture and temperature are higher than their surface counterparts, Qe and Qh are directed towards the surface. Similarly, there is an outflow of heat (Qg) from the subsurface to the surface at night because subsurface temperatures exceed those at the surface, a result of cooling due to the upward flux of longwave radiation.

A useful dimensionless indicator of the relative apportionment of Q* between Qh and Qe is the Bowen Ratio (*B*) which is defined as follows:

$$B = \mathrm{Qh} / \mathrm{Qe}$$

In the situation where water is freely available *B* is small, with most of the available energy (Q*) transferred to the atmosphere in the form of latent heat (Qe) (Fig. 4.3). In contrast for dry surfaces, *B* is large because the majority, or in extreme cases all, of Q* is transported to the atmosphere in the form of sensible heat (Qh) (Fig. 4.3). Some illustrative examples of values for *B* are shown in Table 4.1. These clearly demonstrate how water availability and vegetation can influence *B*. For example, the Bowen Ratio for deserts indicates that ten times more of Q* goes into sensible compared to latent heat flux where the complete opposite is true for tropical oceans. Interestingly the *B* values for London, which are typical for dense urban areas, approach that for deserts as cities climatologically are dry with vast tracts of built-up impervious surfaces with few 'wet' or vegetated surfaces. Clearly the relative partitioning of net radiation into sensible and latent heat is an extremely important determinant of surface climate. The implications of how Q* is apportioned between Qh and Qe with respect to the generation of heatwaves will be discussed in Chap. 7. How Qh and Qe influence the development of the so-called urban heat island, as

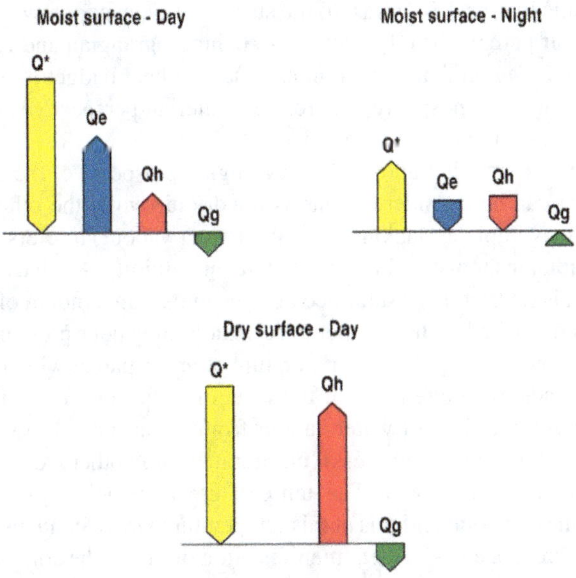

Fig. 4.3 Apportionment of Q* between turbulent (Qh and Qe) and conductive (Qg) fluxes for three different situations. (Adapted from various sources)

Table 4.1 Example Bowen Ratios for various surfaces

Type of surface	Range of Bowen Ratios
Deserts	>10.0
Semi-arid landscapes	2.0–6.0
Temperate forests and grasslands	0.4–0.8
Tropical rainforests	0.1–0.3
Tropical oceans	<0.1
London	6.0–7.0

Compiled by Author

well as the potential importance of another 'non-natural' source of heat called anthropogenic heat flux for increasing heat levels in cities will be discussed in Chap. 8.

4.4 Heating of Air

In the previous section by considering the nature of the surface energy budget it was established that the balance between various radiative fluxes and ultimately turbulent and conductive fluxes determines how much energy is available for raising the temperature of air, evaporation and increasing subsurface temperatures. As for the surface, the concept of the energy or heat budget can be applied to a parcel of air. Let's consider a parcel of air to be a very large bubble of air that has a diameter something like a large city block of about 150 m with an associated volume of about 1.8 million cubic metres. In contrast to the surface, we can adopt two viewpoints for examining an air parcel's heat budget, these being Lagrangian and Eulerian (Stull, 2011). In brief, a Lagrangian viewpoint assesses the heat budget of an air parcel as it moves about in the atmosphere, whereas an Eulerian perspective investigates the heat budget for a stationary parcel of air.

For an air parcel moving about (the Lagrangian perspective) there are two possible cases that can be considered. One is for dry air while the other is moist air. While dry air may imply air lacking moisture (water vapour) it refers to unsaturated air or air that contains moisture below 100 percent humidity. Air that contains all the moisture it can is referred to as saturated air. The maximum amount of moisture that air can hold is described by the temperature saturation vapour pressure relationship (Fig. 4.4) such that a change in the temperature of an air parcel with a given level of water vapour leads to an alteration of the level of saturation. For a fixed moisture amount (no addition or loss of water vapour from the air parcel) a warming (cooling) air parcel will become more (less) 'unsaturated' or in other words the saturation deficit will increase (decrease). The temperature at which saturation occurs is referred to as the dewpoint and it is at this temperature where condensation of moisture in the air parcel occurs—a clear manifestation of this in the atmosphere is cloud formation.

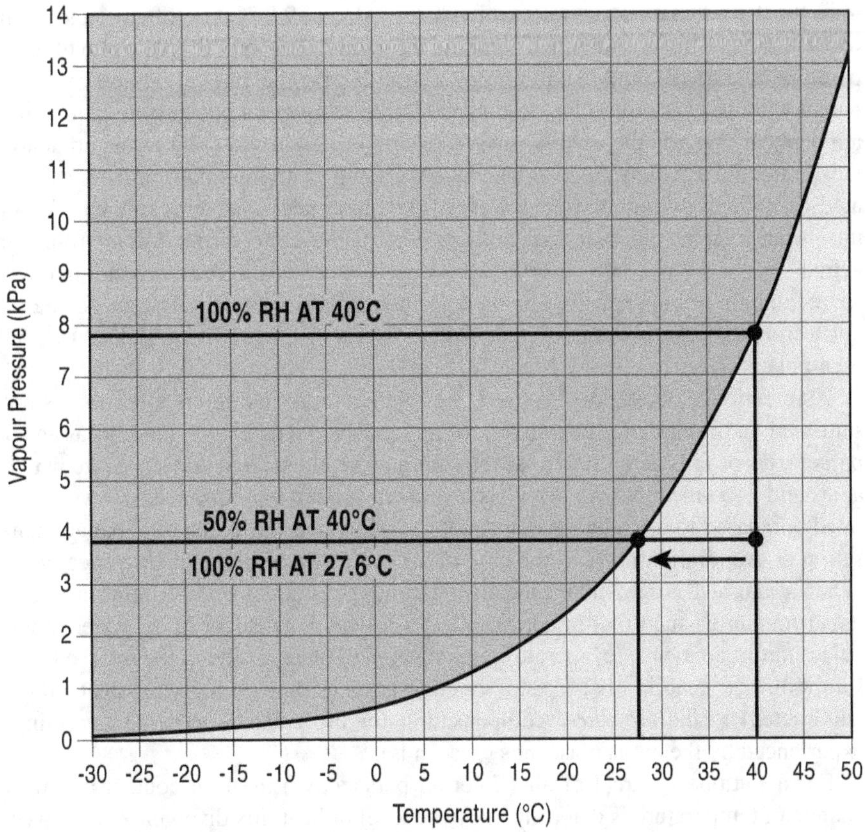

Fig. 4.4 Saturation vapour pressure as a function of temperature. For air at 40 °C the saturation vapour pressure is 7.4 kilo Pascals (kPa), that is, air at 40 °C with a water vapour pressure of 7.4 kPa would be saturated. If an air parcel at 40 °C has a vapour pressure of 3.7 kPa it would be unsaturated with a relative humidity (RH) of 50 percent, that is, it possesses half its water vapour capacity at 40 °C. To achieve a RH of 100 percent, an air parcel at 40 °C with a vapour pressure of 3.7 kPa would need to be cooled down to its dew point temperature which is 27.6 °C. The shaded zones in the figure indicate areas where the air would either be supersaturated (darker tone) or unsaturated. The solid curve represents the saturated state (Adapted from Stull, 2011)

In the case of dry or unsaturated air that does not experience any heat gain or loss through its 'sides', the first law of thermodynamics or conservation of heat describes how the temperature of an air parcel can change as it rises or falls through the atmosphere. This is because as an air parcel ascends (descends) and experiences lower (higher) atmospheric pressure it expands (contracts); in physics terms work is done by or on the air. Consequently, the gaseous molecules that make up the air parcel drift further apart (come closer together), the result being a fall (rise) of air parcel temperature. This alteration of temperature due to vertical motion without direct heat gain or loss is called an adiabatic process. The rate at which the temperature decreases (increases) in an ascending (subsiding) unsaturated parcel of air is referred to as the dry adiabatic lapse rate, which has a value of 9.8 °C/km (often rounded up to 10 °C/km). This is a so-called 'process lapse rate' as it applies to the rate of temperature change resulting from a process, in this case the vertical motion of an air parcel without heat gain or loss to that air. The descent of air parcel from high in the atmosphere towards the earth's surface results in the air warming at the adiabatic lapse rate of 10 °C/km. Such subsidence warming is an important process associated with the development of heatwaves. The theoretical adiabatic lapse rate for an unsaturated air parcel rising or falling through the atmosphere differs from the actual observed lapse rate observed on a day-to-day basis in the surrounding environmental air or atmosphere. On average the rate at which temperature decreases away from the earth's surface is 6.5 °C/km. This is often referred to as the environmental lapse rate.

As mentioned above 'dry' is used in a relative sense and refers to air that is not saturated. In the case of saturated air, the rate at which temperature changes as an air parcel rises or falls is referred to as the saturated or moist adiabatic lapse rate which at around 5–6 km above sea level has a value of around 6-7 °C/km. The lower saturated compared to dry adiabatic lapse rate is accounted for by the role that condensation or evaporation plays in the case of ascending or descending air respectively. When a saturated parcel of air ascends (descends) it is no longer in equilibrium with its surroundings such that in an ascending (descending) parcel of air water molecules start to condense (evaporate). This change of state results in the release (consumption) of heat in the case of condensation (evaporation) with heat added (subtracted) to the air parcel compensating for the adiabatic cooling (warming) experienced by the air as it ascends (descends).

For a stationary parcel of air (Eulerian perspective) the heat content and thus change of temperature is determined by the level of heat flux divergence or convergence, or in other words the amount of heat entering or leaving the parcel of air which is determined by the flux gradient or the change of flux across a given distance. Heat flux divergence is associated with more heat leaving an air parcel than entering, with a flux convergence applying in the opposite situation. Ignoring the role of internal heat generation, as might arise from condensation of moisture in a fixed parcel of air, there are four possible ways that heat can enter or leave a stationary air parcel, namely advection, conduction, turbulence, and radiation. Advection refers to transport of a property (heat or moisture) by the wind. In the case of temperature advection this means heat being transported from one place to another, for example across the boundary of an air parcel, with the rate of advection linearly dependent on the temperature and wind speed. Conduction is the process of heat transfer due to the direct contact between molecules. In the atmosphere this is most important in the very thin layer of air in direct contact with a surface whether it be the ground or for example a building surface such as a wall or roof. Conduction is the main way in which heat is moved from the surface into the atmosphere as well as from the surface to the subsurface. However, in the case of the transfer of heat from the surface into the atmosphere turbulence plays a critical role as this is the way in which heat in the very thin layers of air at the surface is carried to greater heights. In combination conduction and turbulence are often referred to as effective surface turbulent heat flux. Because turbulence, which manifests itself in the form of eddies, transports heat from one place to another it is very effective in mixing warm with cool air such that it tends to induce temperature uniformity in air. An air parcel can theoretically gain heat by molecules of air directly absorbing incoming shortwave radiation from the sun or long wave radiation emitted from the earth's surface or gaseous molecules in the atmosphere. In a cloud, haze, or smoke free atmosphere there is virtually no direct heating of the air by shortwave radiation. Notwithstanding this, shortwave radiation of course plays a critical indirect role in atmospheric heating via influencing the heat flux at the earth's surface. In contrast to shortwave radiation, the atmosphere is quite opaque to longwave radiation. Therefore, longwave or infrared radiation emitted by atmospheric gases, or the surface of the earth can be absorbed within the atmosphere. Once absorbed, the heat can be re-emitted vertically to space or towards the earth's surface. In fact, it is the absorption by atmospheric gases of outgoing longwave radiation from the earth's surface and its re-emission back to the surface which gives rise to the natural and enhanced greenhouse effect, enhanced because the concentration of heat absorbing greenhouse gases has increased over the last 150 years due to human activities. The heating of air as described above due to solar and terrestrial radiation, latent and sensible surface heat flux absorption are generally referred to as diabatic processes because these result in the addition or loss of heat from an air parcel. This contrasts with an adiabatic process.

4.5 Measurement and Monitoring of Heat

Much effort has been invested in finding ways to best measure and monitor heat because the anomalous heat associated with heatwaves is impactful and heat risk managers are interested in the relationship between the scale and duration of heat and associated impact levels. For infrastructure such as roads, rail, bridges and building walls and roofs, heat is usually assessed with reference to either direct measurements of surface temperature using a range of temperature sensors (Bevilacqua et al., 2017; Chapman et al., 2008; Yang & Bou-Zei, 2019; Yarnold & Moon, 2015), or inferred from weather variables from nearby meteorological stations or thermal mapping using remote sensing methods (Gaitani et al., 2017; Mandanici et al., 2016; Pardyjak & Stoll, 2017). Such direct and indirect measurements (Fig. 4.5) usually form the basis of attempts to model the stress and strains resulting from thermal loads on infrastructure which can be significant during heatwave events (see Chap. 11).

In the case of human health, temperature and heat are often conflated to mean the same thing. However, heat in a human health context is a complex phenomenon resulting from the synergistic effects of air temperature, humidity and ventilation levels, radiation loads, metabolic activity, and clothing (McGregor & Vanos, 2018). While this is only partially recognised in some research communities addressing climate and health issues, the complexity of heat as a physical driver of heat stress amongst humans and animals has been fully embraced by the biometeorological

Fig. 4.5 Sensors for measuring thermal response of infrastructure. Temperature sensors installed on rail track sections in an experimental rail research facility (left) and infrared temperature sensor for monitoring road surface temperature (right). Images courtesy of Professor Lee Chapman, the University of Birmingham

community. Because of the impact of anomalous and prolonged heat on human and animal health, a plethora of thermal climate indices have been developed to monitor heat stress (also referred to as heat stress indices) which is generally defined as the negative effect of the thermal energy (heat) environment on individuals. The response of exposure to heat stress or heat load is heat strain. In the case of humans and animals, this manifests as increased skin and/or core temperature and sweat rate. Cardiovascular strain in the form of increased heart rate also occurs as the heart works harder to pump blood to the skin surface to dissipate heat built up in the body core. Such manifestations of heat strain arise because of an imbalance between heat gains and heat loses to and from the body respectively as illustrated in Fig. 4.6 and portrayed by the so-called human energy/heat balance equation which takes the form of:

$$0 = \Delta S + \mathrm{M} + \mathrm{K} + \mathrm{R} \pm \mathrm{C} - \mathrm{E}$$

where heat gains (+) include metabolic heat flux (M), conductive heat flux (K) and net radiation (R) as well as convection (C) if air temperature rises above that of the skin. Losses (−) include convective heat loss (C), and evaporative heat loss

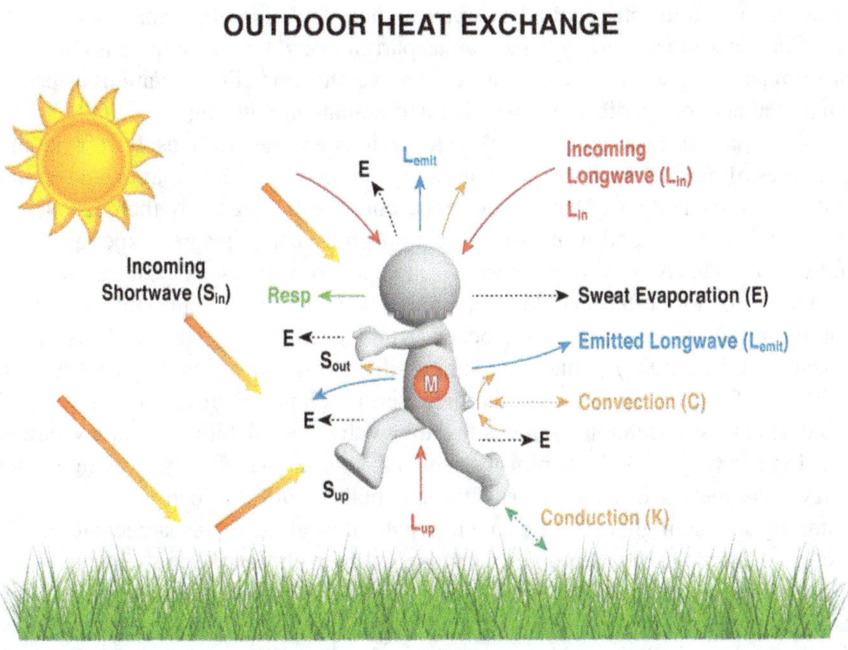

Fig. 4.6 Heat exchange between the environment and human body in an outdoor environment. In normal conditions, a human's energy balance will become positive (>0) due to metabolic heat generation (M) and radiation in the form of shortwave (Sin and Sup) and longwave (Lin and Lup) from the sky and ground. A human will normally lose heat through convection (C), evaporation (E), emitted longwave radiation (Lemit) and respiration (Resp). Source: (McGregor & Vanos, 2018). Used under Elsevier License 5605430862832

(E) respectively, if the skin temperature is higher than the air temperature and the skin surface is moist compared to the surrounding atmosphere. ΔS is change in heat storage. For no gain in body core temperature there must be a balance between heat gains and losses as represented by zero (0) in the human heat balance equation (McGregor & Vanos, 2018).

To monitor and assess the level of heat, often in the form of human or animal related heat stress, a range of thermal climate indices have been developed. Generally, the purpose of these is to assess thermal comfort, physical well-being or environmental heat risk. The production of thermal indices has almost turned into an industry with over 150 thermal climate indices appearing in the literature over the last 100 years (de Freitas & Grigorieva, 2017; Havenith & Fiala, 2016; Basarin et al., 2020). Many of these indices are two-parameter indices with the implicit assumption that they capture the complex conditions of heat exchange between the body and its thermal environment as represented by the human heat balance. In the case of indices relevant to the monitoring and analysis of heatwaves, two parameter heat stress indices usually consist of combinations of dry-bulb temperature and different measures of humidity. More complex indices based on solving the human heat balance are necessarily demanding in terms of the requisite input variables (McGregor & Vanos, 2018). Table 4.2 presents some of the common instruments used for collecting data required as inputs to the calculation of thermal climate indices. Such instruments are either set up as part of special field campaigns (Fig. 4.7) or form part of instrument arrays installed at weather and climate stations as part of a national network dedicated to weather and climate monitoring.

While data from weather and climate stations are generally useful for making estimates of heat stress at the population level, notwithstanding that many weather stations are located away from major population centres, recently there has been a surge in interest related to measuring the specific heat exposures experienced by individuals. This is because measuring heat exposure at this scale provides valuable insights into how often people or animals encounter heat conditions and when, where and why these encounters occur (Kuras et al., 2017; Hass & Ellis, 2019; Nazarian & Lee, 2021). While it was impossible to imagine measuring heat stress at the level of the individual when the first thermal climate indices were developed, rapid advances in technology now allow this with the availability of small wearable data loggers (Fig. 4.8). Such mini-instrument packages, which may be no more than a few centimetres in size, sense environmental conditions such as temperature, humidity, and sunlight and store data internally thus allowing researchers to build a picture of individual heat profiles, as modelled by a chosen thermal climate index, during anomalous heat events such as heatwaves. Small compact sensors also allow establishment of indoor climate conditions during heatwaves, the output from which facilitates the calculation of a range of thermal indices (Oetomo et al., 2022).

Building on previous reviews (Blazejczyk et al., 2012; Driscoll, 1992; Fanger, 1970; Landsberg, 1972; Parsons, 2007) an insightful appraisal of a comprehensive range of human thermal climate indices has been provided by de Freitas and Grigorieva (2017) who identified eight classes of indices in terms of what they attempt to measure namely: (A) simulation device based indices for integrated

Table 4.2 General Instruments used to collect data for the calculation of thermal climate indices. Source: McGregor G and Vanos J, 2018 Heat: a primer for public health researchers, Public Health, 161, 138–146, https://doi.org/10.1016/j.puhe.2017.11.005[a]

Variable	Instrument(s)	Modern technology	Considerations
Air temperature (T_a) (dry bulb temperature)	Thermometer ('of heat meter')	Mercury thermometer, tthermistors, thermocouples, bimetallic thermometer	Relatively low cost. Ventilated radiation shield required
Relative humidity (RH)	Hygrometers ('water meter')	Psychrometer (wet-and-dry-bulb thermometer) Wet bulb has bulb wrapped in wet muslin with air flow moving over. Also measured based on changes in thermal conductivity of air due to vapour, or change in electrical resistance of (used by Kestrel WBGTs)	Ventilation important, as is the simultaneous accurate measurement of T_a. Thus, RH and T_a often measured together. RH is rarely the correct value to use in public health and epidemiological studies, and can be converted to proper, independent parameters (see Davis et al., 2016)
Longwave (LW) radiation	Pyrgeometer ('fire-earth meter')	"Thermal": Silicon dome (transmits infrared only), black disk heats up proportional to LW radiation	Often high cost and complex. Calibrations required. Low cost option for outgoing LW is to obtain a surface temperature and apply Boltzmann's law (e.g. Campbell & Norman, 1998).
Shortwave (SW) radiation	Pyranometer ('fire meter')	"Thermal": Glass dome with black disk changing temperature proportional to SW radiation, producing a voltage output (e.g., Kipp & Zonen) "Photocell"—converts specific wavelengths of light into electrical energy (Li-COR)	Thermal: Often high cost and complex. Photocell: Are less expensive and less complex, but only valid under 'open sky' conditions. Calibrations required for both. Portab and less complex options are made by Huskeflux
Net Radiation (NR)	Net radiometer (upward and downward pyranometer and pyrgeometer)	Use pyranometer and pyrgemoeter technologies (e.g., Kipp & Zonen CNR4)	Expensive, calibrations required, complex
Wind speed and direction	Anemometer ('wind meter')	Cups or propeller rotate proportional to wind speed. Direction obtained through the use of a wind vein. Hot wire anemometers: Wind cools wire proportionate to speed	Height of measurement is crucial. Most windspeeds come from 10 m height, which can be brought to human height with log-wind equations

(continued)

Table 4.2 (continued)

Variable	Instrument(s)	Modern technology	Considerations
Radiant temperature	Globe thermometer (ISO 7726 1998), cylindrical radiation thermometer (Kenny et al., 2008) grey globe thermometer (Thorsson et al., 2007); 3D or related models integrating SW and LW radiation (Johansson et al., 2014)	Temperature sensor, such as thermistor or thermocouple, embedded in hallow globes or in conductive material (epoxy) to transfer both sensible and radiant heat to temperature sensor. Net radiometers are required for 3D setup	Slow response time of larger sensors is an issue (Budd, 2008); colour and shape inaccuracies for representing a human Very costly to have three net radiometers

[a]Used under Elsevier Licence Number 5610760305882

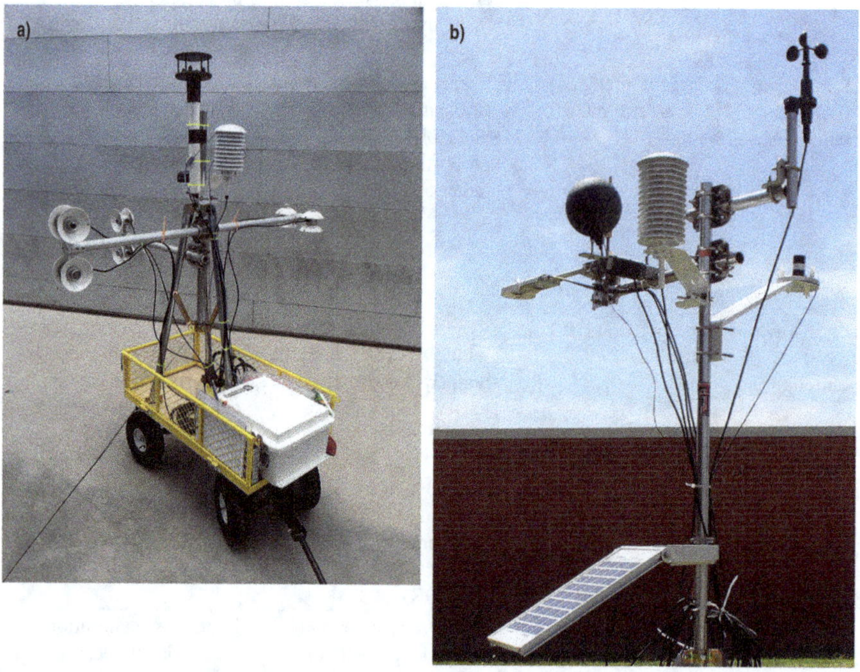

Fig. 4.7 Campaign specific instrument arrays composed of various temperature, humidity, wind speed and radiation sensors for measuring heat stress related variables. (**a**) mobile MRT cart with three net radiometers pointing each cardinal direction to get mean radiant temperature (MRT) plus temperature, humidity and wind sensors, (**b**) fixed weather station with temperature, humidity and wind sensors along with black globe thermometer for measuring radiant load. Images courtesy of Professor Ariane Midde Arizona State University (**a**), and Professor J Vanos (**b**), Arizona State University

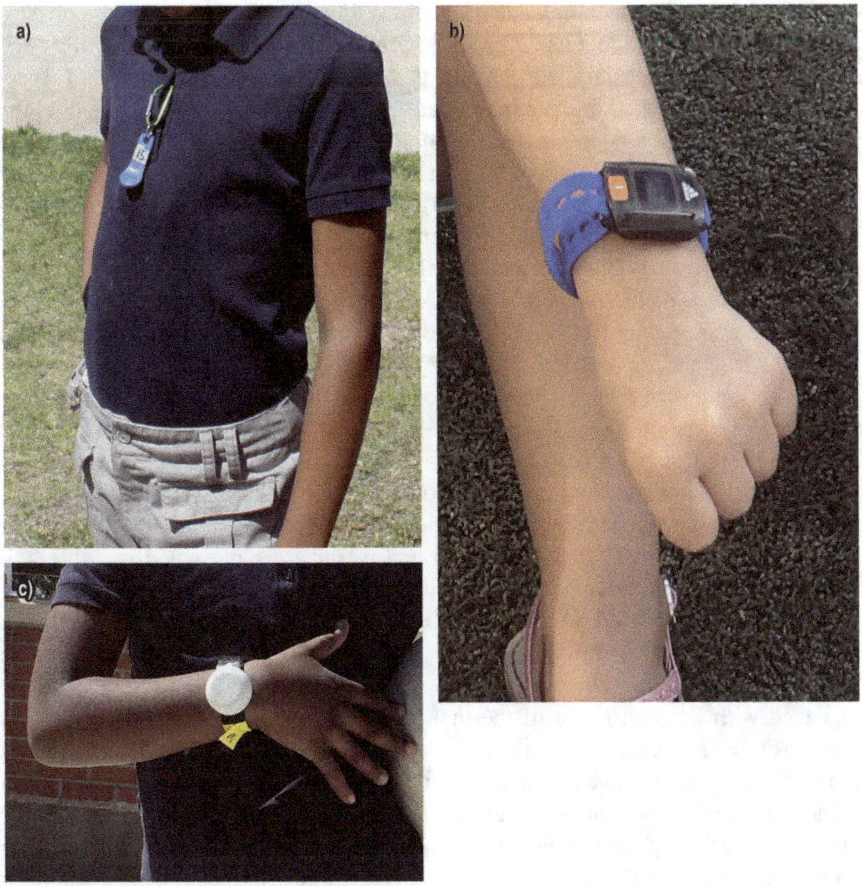

Fig. 4.8 Personal heat exposure related monitoring devices. (**a**) temperature and humidity sensor, (**b**) heart rate watch, (**c**) ultraviolet dosimeter. Images courtesy of Professor J Vanos, Arizona State University

measurement (B) single sensor (single-parameter) indices, (C) indices based on algebraic or statistical models, (D) proxy thermal strain indices, (E) proxy thermal stress indices, (F) energy balance strain indices, (G) energy balance stress indices and (H) special purpose indices. In a follow up analysis, they evaluated and assigned a score to each individual index belonging to the eight categories based on several criteria such as validity, usability, transparency, sophistication, completeness, and scope with each of these scored between a minimum of 1 and a maximum of 5 (de Freitas & Grigorieva, 2017).

The eight types of thermal climate indices identified by de Freitas and Grigorieva (2017) generally fall into three broad structural categories as noted by McGregor and Vanos (2018), namely empirical, direct and rational. Empirical indices are based on measured and verified human physiological responses to various factors of metabolic and environmental loads. In contrast direct indices do not assess physiological responses but rather use simple and practical measurements of weather variables to infer the thermal environment. In comparison with empirical and direct indices, rational thermal climate indices are complex. These numerically model the interaction of environmental and physiological variables and try and directly solve the human heat balance equation. While rational indices are based on eloquent and sophisticated solutions of the human energy balance, they are data demanding usually requiring meteorological measurements beyond the basic set available at standard weather stations and at spatial scales which standard weather observation networks cannot deliver on. For this reason, they are not generally preferred for large scale epidemiological studies of population response during heatwaves and indeed as triggers for heatwave warning systems (see Chap. 13). Rather rational indices are better suited to specific applications in estimating indoor/outdoor thermal comfort, urban design, occupational heat exposure, exertional heat stress and even the design of bespoke personal heat warnings and thus the achievement of population-to-person scaling of the response to heat (McGregor & Vanos, 2018). Some of the more frequently used empirical, direct and rational indices applied in general human thermal climate research and practice are presented in Table 4.3. While this is an impressive list, the number of indices that form the basis of hazard- or impact-based heatwave warning systems (see Chap. 13) or have been applied in heatwave analyses, is more limited. Details for some of these are expanded on below in Table 4.4.

Faced with the vast range of thermal climate index possibilities, the question arises as to which is the best in the context of not only defining heatwaves but as a predictor of a heat-related impact. While the ranking of indices presented in Table 4.3 based on the work of de Freitas and Grigorieva (2017) might help answer this question, this was never the intention of these researchers. Rather they emphasised that indices should be selected based on the purpose of the problem at hand for example whether an evaluation of environmental stress or a consequence of that stress (strain) is required. Similar sentiments have also been expressed by Grundstein and Vanos (2021) in relation to sports performance under challenging heat conditions.

A clear research problem that has emerged over the years is how to best quantify the impacts of heat on human health. This is especially pertinent in the case of the push to develop heatwave and heat-health warning systems (see Chap. 13) with an associated scramble to identify which descriptor of environmental heat stress best predicts heat-related mortality and morbidity (Barnett et al., 2010; Staiger et al., 2019). Accordingly, there has been a growth in the number of studies that try and answer this question explicitly. For example, Barnett et al. (2010) analysed the relationship between three measures of temperature with and without humidity, apparent temperature and the Humidex, and mortality for 107 US cities. No single measure of heat was found to be superior for predicting mortality with the conclusion reached that the best thermal index is that with the least amount of missing data, good spatial coverage using input data from meteorological measurement sites across the region of concern and is easily understood by the public. In a similar

Table 4.3 Common thermal climate indices grouped according to whether direct, empirical, or rational with the thermal climate index class type and evaluation score (maximum 30) of de Freitas and Grigorieva (2017) also listed where applicable—see text for explanation of class type

Index	Index type & Main inputs*	Scale/Units	Main applications	Class type and (score)
Direct & Empirical				
Heat index	Direct, yet empirically derived in its conception. T_a, RH.	°F	Heatwave warnings and guidance, USA.	C (16)
Humidex	Direct index, yet empirically derived. T_a, VP	°C	Heatwave warnings and guidance, Canada	C (15)
NET Effective Temperature (NET)	Direct. T_a, RH, V	°C	Formerly heatwave warning and guidance, Hong Kong	C (23)
Wet Bulb Globe Temperature (WBGT)	Direct. T_w, T_a, T_g	°F/°C	Exertional heat stress. Heat stroke warning and guidance in Japan	C (20)
Simplified Wet-bulb Globe Temperature Approximation (sWBGT)	Direct	°C	Applied in the Australian horse racing industry	Not considered
Wet bulb temperature (T_w)	Direct. T_w.	°F/°C	T_w > 35 °C indicated as limits of habitability for human adaptation to heat.	B (14)
Physiological Strain Index (PSI)	Empirical. Requires physiological inputs of heart rate and BCT.	Strain(0–10)	Clinical studies, exercise.	D (19)
Environmental Stress Index (ESI)	Empirical. SR, RH, T_a	°C	Exercise (athletic, military, occupational).	C (22)
Discomfort Index (DI)	Direct. T_w, T_a	°F	Human (dis)comfort required for air conditioning, sedentary activity.	C (18)
Excess Heat Factor (EHF)	Direct and partly empirical	$°C^2_L$, 'L' refers to the local nature of the threshold	Heatwave warning and guidance Australia	Not considered
Hong Kong Heat Index (HKHI)	Direct—Modified WBGT index	°C	Work safety indicator produced by Hong Kong observatory	Not considered

(continued)

Table 4.3 (continued)

Index	Index type & Main inputs*	Scale/Units	Main applications	Class type and (score)
Rational (complex heat budget models)				
Physiological Equivalent Temperature (PET)	Rational. T_a, T_{mrt}, RH, V, M_{act}	°C	Thermal comfort, urban design.	G (26)
Perceived Temperature	Rational T_a, T_{mrt}, RH, V, M_{act}, I_{cl}	°C	Thermal sensation and related heat stress	G (27)
COMfort FormulA (COMFA)/ COMFA*	Rational. T_a, NR, VP, V, M_{act}, I_{cl}. Utilizes Eq. 1.	W m^{-2}	Thermal comfort, urban design, heat stress prediction sedentary/active (COMFA*).	A (13) specified as cylinder index
Man-ENvironment heat EXchange model (MENEX)	Rational. T_a, NR, RH, V, M_{act}, I_{cl}. Utilizes Eq. 1.	W m^{-2}	Thermal comfort, Urban Design, heat stress during exercise.	F (27)
Heat Stress Index (HSI)	Rational. VP, T_a, V, M_{act}. Ratio of required (E_{req}) to maximum (E_{max}) evaporation.	Scale 0–100 (HSI = E_{req}/ E_{max})	Heat stress prediction, classic and exertional.	F (26)
Standard Effective Temperature (SET)	Rational. T_a, T_{mrt}, RH, V, M_{act}, I_{cl}.	°C	Two-node method, represents skin temperature and wettedness.	G (26)
Thermal Work Limit (TWL)	Rational. T_a, NR, VP, V.	W m^{-2} or METs	Occupational and exertional heat stress.	F (24)
Universal Thermal Comfort Index (UTCI)	Rational. T_a, T_{mrt}, RH, V, M_{act}, I_{cl}	°C	Physiologically-based thermal comfort.	G (27)
Apparent Temperature	Rational. T_a, T_{mrt}, RH, V, M_{act}, I_{cl}	°C	Assessment of hot/ humid weather; Thermal comfort; clothing	G (23)

Modified from: McGregor & Vanos, 2018

[a] air temperature (T_a), globe thermometer temperature (T_g), natural wet bulb thermometer temperature (T_w), relative humidity (RH), vapour pressure (VP), windspeed/ventilation (V), mean radiant temperature (T_{mrt}), net radiation (NR), clothing insulation (I_{cl}), metabolic activity (M_{act}), solar radiation (SR)

study for 105 US cities, Bobb et al. (2011)) also found no dominant single measure of heat for predicting mortality but did highlight considerable inter-city variation in the best predictor of mortality. These general findings for the US have been corroborated by Spangler et al. (2023) but importantly, these researchers note that maximum values of the thermal indices investigated are superior to their minimum counterparts in relation to assessing the relative risk of heat-related morbidity and all-cause mortality. Beyond the individual country level, Lo et al. (2023) have undertaken the first systematic multi-country assessment of the association between

Table 4.4 Details for a selection of thermal climate indices applied in heatwave and heat health warning systems

Indicator or index	Details
Heat Index (HI) °F	The Heat Index (HI) combines air temperature and relative humidity (RH) to determine an apparent or equivalent temperature—how hot it feels given the actual temperature and the relative humidity of air. The National Weather Service of the United States uses a Heat Index chart to portray the outcomes of temperature and humidity combinations (https://www.weather.gov/ffc/hichart) and as a basis for its public messages about heat safety (Hawkins et al., 2017). Dangerous conditions that are likely to result in sunstroke, muscle cramps, and/or heat exhaustion with heat stroke possible with prolonged exposure and/or physical activity are associated with HI values greater than 41 °C. The HI does not consider solar load and assumes a constant wind and walking speed. The original formula is given below. Although based on Steadman's original conceptualisation of the Apparent Temperature (Steadman, 1979), there are numerous approximations (Anderson et al., 2013). Because the original index is not suited to extremely high values of temperature and humidity it has been extended beyond its original range. Importantly application of the extended version of the HI demonstrates that the original HI underestimates heat stress for severe heatwaves (Romps & Lu, 2022; Lu & Romps, 2023). The original formula is: $$Heat\ Index(HI) = -42.379 + 2.04901523(T_f) + 10.14333127(RH) - 0.22475541(T_f)(RH) - (6.83783*10^{-3})(T_f^2) - (5.481717*10^{-2}) RH^2 + (1.22874*10^{-3})(T_f^2)RH + (8.5282*10^{-4})(T_f)(RH^2) - (1.99*10^{-6})(T_f^2)(RH^2)$$ Where: T_f air temperature in degrees Fahrenheit RH relative humidity expressed as a whole number

(continued)

Table 4.4 (continued)

Indicator or index	Details
Humidex °C	Humidex value is an apparent or equivalent temperature similar in principle to the Heat Index. Originally formulated by Anderson (1965), Masterton et al. (1979) promoted it as a measure of thermal discomfort but in the absence of an empirical basis. Underlying the Humidex are the assumptions a thermoregulation "neutral point" of the human body (no heat gains or losses) occurs in the air temperature range of 27–30 °C for standard conditions (nude male, still air, resting), and the body cannot continue to maintain equilibrium when air temperature and relative humidity >32 °C and 75%, respectively equivalent to a psychrometric wet bulb temperature of 28 °C. Such assumptions render the Humidex sensitive to changes in human behaviour, expcsure to the sun and wind. Generally computed using hourly data approximation methods have been developed to facilitate its calculation using daily data (Diaconescu et al., 2023) (Diaconescu et al., 2023). Dangerous conditions are considered to occur with Humidex values >45 °C with heat stroke imminent at values >54 °C. Humidex = (air temperature) + h where, h = (0.5555)*(e – 10.0); and e = 6.11 * exp.(5417.7530 * ((1/273.16) – (1/dewpoint))) A simplified version that only uses temperature and the relative humidity is: Humidex = T + 5/9 * (e-10) where: e = vapour pressure(6.112*10^(7.5*T/(237.7 + T))*H/100) T = air temperature (degrees Celsius) H = humidity (%)

Excess Heat Factor (EHF)	Excess Heat Factor (EHF) measures heatwave intensity. It is a location specific percentile based index that incorporates an additional component to account for acclimatisation. It is therefore made up of two components that represent (1) heat in excess of the local climatic heatwave threshold (EHI_{sig}), and (2) heat in excess of the recent experience of heat (EHI_{accl}) such that: $EHF = EHI_{sig} \times \max(1, EHI_{accl})$ where $EHI_{sig} = DMT_{3\text{-}day} - DMT_{95}$ and $EHI_{accl} = DMT_{3\text{-}day} - DMT_{30\text{-}day}$ and $DMT_{3\text{-}day}$ = average daily temperature of the next 3 day period (t, t + 1, t + 2) DMT_{95} = the percentile of daily temperatures for a 30 year reference period $DMT_{30\text{-}day}$ = the average daily temperature of the previous 30 day period Using the EHF values a heatwave severity index can be calculated as: Severity = EHF/EHF_{85} Where EHF_{85} = the local 85th percentile value of the EHF (Nairn & Fawcett, 2015; Nairn et al., 2018) As the EHF is only based on temperature as a measure of heat stress, in humid (tropical) environments the EHF may not capture all heatwave days compared to an EHF calculated with an alternative temperature and humidity-based measure of heat stress (Nairn et al., 2022). This relates to the assumption that DMT, as the simple average between daily maximum and minimum temperature, is implicitly affected by relative humidity changes, which may not necessarily hold true. Some studies have directly compared the performance of the EHF as a predictor of mortality with other thermal indices (e.g., Morais et al. (2020)).

(continued)

Table 4.4 (continued)

Indicator or index	Details
Wet Bulb Globe Temperature (WBGT) °C	Originally developed for application for military training exercises (Budd, 2008; Minard, 1961), the wet-bulb globe temperature (WBGT) is a heat stress index recognised by the International Standards Organisation (ISO) with the identifier ISO 7243 2017. Its value represents the thermal environment to which an individual is exposed during their total exposure over the working day (up to 8 h) and therefore does not apply to very short exposures to heat. The WBGT is reported as a single value based on measurements from three devices namely (1) A black-globe thermometer. This usually consists of a 150-mm black globe with a thermometer located at the centre. The black-globe temperature (Tg) represents the integrated effects of radiation and wind, (2) A natural wet-bulb thermometer. This is a thermometer with its bulb covered with cotton wick irrigated with distilled water from a reservoir. Evaporation from the wet bulb cools the thermometer with the amount of evaporation dependent on the level of atmospheric moisture. The greater the evaporation the greater the depression of the natural wet-bulb temperature (Tnwb). Like the black-globe thermometer, the wet-bulb thermometer is not shielded from wind or radiation, thus representing the integrated effects of humidity, wind and radiation, and (3) A thermometer shielded from radiation. Usually located in a weather screen (e.g., Stevenson Screen) or radiation shield, it records the standard dry bulb air temperature (Ta). The standard formulation for the WBGT is: $WBGT = 0.7T_{nwb} + 0.2T_g + 0.1T_a$ Its application requires the estimation of metabolic heat and the effects of clothing with the WBGT thresholds varying with level of metabolic rate/activity. A standard clothing ensemble equivalent to 0.6clo (a measure of clothing insulation) is assumed. For a resting individual sitting at ease—a metabolic rate of 115 watts—the WBGT reference limit for a person acclimatised (unacclimatised) to heat is 33 °C (32 °C). For a person engaged in strenuous activity such as intense digging or walking at a speed greater than 7 km/h—very high metabolic rate of 520 Watts—the WBGT reference for an individual acclimatised (unacclimatised) to heat is 25 °C (20°C). The Hong Kong Observatory have produced a modified version of the original WBGT where WBGT = 0.8Tnwb + 0.05Tg + 0.15Ta. Because of the absence of Tg measures in many locations, several attempts have been made to produce simplified versions of the WBGT (Armstrong et al., 2007; Brimicombe et al., 2023; Grundstein & Cooper, 2018; Kong & Huber, 2022; Liljegren et al., 2008). However, excluding the Tg value violates its intended use (Vanos et al., 2020).

Universal Thermal Climate Index (UTCI)	The UTCI measures the human physiological response to the thermal environment based on solving the human heat/energy balance considering physiology, clothing and a representative activity (Błażejczyk, 2021; Bröde et al., 2012; Fiala et al., 2012; Jendritzky et al., 2012). Four general variables are required to calculate the UTCI: air temperature and dew point temperature (or relative humidity) and wind speed at 2 and 10 m above ground level respectively and mean radiant temperature (MRT). The output from the UTCI model is an equivalent temperature related to a strain index that represents the synergistic behaviour of metabolic rate, core temperature, skin wettedness, blood flow and sweat rate (Bröde et al., 2012). The UTCI thermal stress categories and associated UTCI value ranges that align with conditions expected during heatwaves are >46, extreme heat stress (>46); very strong heat stress (38–46) and strong heat stress (32–38). Possible limitations of the UTCI relate to the standard person for which the UTCI is modelled—74 kg person walking at 1.1 m/s—which is not representative of all people and activities. Further it is undefined for very high temperatures and humidities. Because the UTCI is a complex index attempts have been made to derive a simplified version. The UTCI has been applied to the analysis of heat-related health risk and the identification of heat-health thresholds (Di Napoli et al., 2018, 2019) and has been put forward as a candidate for the global forecasting of thermal health hazards and developing a global heatwave warning system (Pappenberger et al., 2015).
Perceived Temperature (PT) °C	The Perceived Temperature (PT) is an equivalent temperature. The PT, like the UTCI it is based on a human heat budget model but is structured differently. The Predicted Mean Vote (PMV) of Fanger (1970) is the general basis for the Klima-Michel model, the operational thermal assessment procedure model of the German Weather Service from which PT emerges. Being an equivalent temperature, the PT is defined as the air temperature of a reference environment in which the thermal perception would be the same as in the actual environment. A reference subject with an internal heat production of 135 Wm^{2} (walking at 4 km/h on flat ground) is used for the calculation. For the reference environment, the mean radiant temperature equals the air temperature and wind velocity is equivalent to a slight draught (Staiger et al., 2012). The PT forms the basis of the German Weather Service heat health warning system (Matzarakis et al., 2020) with a built in acclimatisation component based on the HeRate approach (Koppe & Jendritzky, 2005). Dangerous heat episodes are identified when the PT value exceeds 32 °C such that the thermal perception of hot is matched with strong heat stress and PT values of 32—38 °C, while very hot is associated with extreme heat stress and PT values >38 °C.

Compiled by Author

thermal indices (heat stress metrics) and warm season mortality. Using eight metrics for 604 locations across 39 countries they found considerable inter-country variability in the optimal heat stress metric for modelling mortality. For 40 percent of the countries, temperature metrics with no or little humidity modification associate best with mortality while apparent temperature dominated in a further 40 percent of the 39 countries. While one might expect that heat stress metrics that include humidity were superior in countries with warm and humid summers this was not the case; dry-bulb temperature and humidity-based heat stress metrics were found to perform similarly in estimating heat-related mortality.

There is no doubt that the role of humidity in determining heat-health outcomes and how this variable is handled in heatwave studies has become a hotly debated topic (Davis et al., 2016). Generally epidemiological studies of heatwaves, using a range of measures of heat stress, have found little or no association between humidity and heat-health outcomes. As outlined by Baldwin et al. (2023) a variety of reasons may exist for this apparent weak association. Amongst others, these include little influence of humidity on heat-related health conditions at high temperatures typical of heatwaves, a bias towards heat-health studies in cool and less humid extra-tropical regions where temperature and humidity are temporally correlated, an improper consideration of temperature and humidity relationships in the interpretation of epidemiological model results and the combination of the high levels of temperature and humidity required for humidity to engender heat strain are seldom seen in the historical record (Baldwin et al., 2023). Added to this, Cvijanovic et al. (2023) note spatial and temporal mismatches between maximum daily values of temperature and temperature and humidity-based thermal indices for recent heatwaves across Europe, the US and Asia which bears implications for the saliency of heat risk communications if temperature maps alone are used.

In the wider context of the debate about the role of humidity in exacerbating heatwave related heat stress and/or temperature extremes (Kang & Eltahir, 2018; Fang et al., 2023; Li et al., 2022; Mishra et al., 2020; Rachid & Qureshi, 2023; Russo et al., 2017; Wouters et al., 2017; Zhang et al., 2023), Simpson et al. (2023) make the powerful point that many comparisons of heat stress indices are fraught with problems. This relates to the assumption of a one-to-one mapping between different heat stress indices. This of course is not correct because many indices that are typically compared do not operate on a common scale even if they use the same unit of measurement such as degrees Celsius. This renders direct comparisons impractical. In recognition of this, Simpson et al. (2023) have proposed a 'calculus of variations' approach which facilitates a comparison of the sensitivity of heat stress indices to temperature and humidity that is independent of index scale. To measure the degree of change in a heat stress index value due to a change in humidity, Simpson et al. (2023) calculated a measure of the marginal temperature equivalent change, or in other words how much of a change in humidity produces a change in the heat stress index value equivalent to a unit change in temperature. The sensitivity of seven heat stress indices to changes in humidity were assessed for three broad climate regimes. The results summarised in Table 4.5 show considerable discrepancies in the sensitivity between AT, HI and UTCI, and WBT for the low temperature

Table 4.5 General degree of sensitivity of heat stress indices[a] to changes in humidity for three climate regimes. Based on Simpson et al. (2023)

Climate regime	Relative degree of sensitivity of heat stress index to change in humidity		
	Low	Medium	High
Low temperature	AT, HI, UTCI	Humidex, WBGTin, sWBGT	WBT
Hot-Humid			AT, HI, UTCI, WBT, Humidex, WBGTin, sWBGT
Hot-Dry	HI, UTCI		AT, WBT, Humidex, WBGTin, sWBGT

[a]*AT* is apparent temperature, *HI* is heat index, *UTCI* is universal thermal climate index, *WBT* is wet bulb temperature, Humidex, *WBGT* in is wet bulb globe temperature for indoors (no radiation) and *sWBGT* is simplified Wet bulb globe temperature

case and between the HI and UTCI, and the AT, WBT, Humidex, WBGTin, and sWBGT for the hot-dry regime (see below Table 4.5 for translation of index acronyms). For the hot-humid regime, all seven indices demonstrate relatively high sensitivity to changes in humidity and thus respond similarly. Building on these results Simpson et al. (2023) demonstrate how the use of different indices with contrasting sensitivities could lead to contradictory results in heatwave related heat-health studies that focus on the importance of humidity under heatwaves conditions. Accordingly, they caution against the blind use of indices and encourage researchers to critically reflect upon the appropriateness of preferred indices for the research problem at hand. This eloquent study certainly serves as a reality check for those not only undertaking heat stress index intercomparisons but also applying heat stress indices in the analysis of heatwave/heat-health associations, especially for hot-humid and hot-dry conditions.

4.6 Conclusion

This chapter has provided an overview of some basic concepts related to heat, a defining characteristic of heatwaves and introduced the nature of the radiation and energy balances. The partitioning of net radiation into sensible and latent heat fluxes influenced by surface moisture availability, has profound implications for temperature extremes during heatwaves. While latent heat flux can provide a cooling effect, increased atmospheric moisture can exacerbate humidity-driven heat stress, underscoring the complex interplay between these processes. The heating of air can occur via a range of processes including heat flux from the earth's surface, heat advection, direct absorption of radiative energy and both adiabatic and diabatic processes. Heatwaves because of their extreme levels of heat cause heat stress. In the context of human and animal health, heat stress is a multifaceted phenomenon resulting from the synergistic effects of air temperature, humidity, ventilation, radiation load, metabolic activity, and factors like clothing and body cover. Direct and indirect measures of atmospheric and surface temperature and other heat-related variables assist with understanding the nature of heatwave processes and heat stress. While a range of heat relevant variables are measured at standard weather and climate stations, special field studies that use complex constellations of instrumentation and 'wearable sensors' are often deployed to measure heat stress at the individual scale. To represent heat stress, thermal climate indices have been developed. These range from simple univariate or bivariate indices to complex rational models that try and numerically solve the balance between heat gains to and heat losses from the body. A burgeoning number of intercomparison studies of thermal climate indices clearly indicate there is no best one size fits all index for understanding the heat stress characteristics and heat-related impacts of heatwaves. Rather the choice of index needs to fit the purpose of the heatwave research problem or application at hand.

References

Anderson, S. (1965). *Humidex calculation*. Atmosferic Environment Service CDS (24–65).

Anderson, G. B., Bell, M. L., & Peng, R. D. (2013). Methods to calculate the Heat Index as an exposure metric in environmental health research. *Environmental Health Perspectives, 121*(10), 1111–1119. https://doi.org/10.1289/ehp.1206273

Armstrong, L. E., Casa, D. J., Millard-Stafford, M., Moran, D. S., Pyne, S. W., Roberts, W. O., & Amer Coll Sports, M. (2007). Exertional heat illness during training and competition. *Medicine and Science in Sports and Exercise, 39*(3), 556–572. https://doi.org/10.1249/MSS.0b013e31802fa199

Baldwin, J. W., Benmarhnia, T., Ebi, K. L., Jay, O., Lutsko, N. J., & Vanos, J. K. (2023). Humidity's role in heat-related health outcomes: A heated debate. *Environmental Health Perspectives, 131*(5), 55001. https://doi.org/10.1289/ehp11807

Barnett, A. G., Tong, S., & Clements, A. C. A. (2010). What measure of temperature is the best predictor of mortality? *Environmental Research, 110*(6), 604–611. https://doi.org/10.1016/j.envres.2010.05.006

Basarin, B., Lukic, T., & Matzarakis, A. (2020). Review of biometeorology of heatwaves and warm extremes in Europe. *Atmosphere, 11*(12). https://doi.org/10.3390/atmos11121276

Bevilacqua, P., Mazzeo, D., Bruno, R., & Arcuri, N. (2017). Surface temperature analysis of an extensive green roof for the mitigation of urban heat Island in southern mediterranean climate. *Energy and Buildings, 150*, 318–327. https://doi.org/10.1016/j.enbuild.2017.05.081

Błażejczyk, K. (2021). UTCI—10 years of applications. *International Journal of Biometeorology, 65*(9), 1461–1462. https://doi.org/10.1007/s00484-021-02174-1

Blazejczyk, K., Epstein, Y., Jendritzky, G., Staiger, H., & Tinz, B. (2012). Comparison of UTCI to selected thermal indices. *International Journal of Biometeorology, 56*(3), 515–535. https://doi.org/10.1007/s00484-011-0453-2

Bobb, J. F., Dominici, F., & Peng, R. D. (2011). A Bayesian model averaging approach for estimating the relative risk of mortality associated with heat waves in 105 U.S. cities. *Biometrics, 67*(4), 1605–1616. https://doi.org/10.1111/j.1541-0420.2011.01583.x

Brimicombe, C., Lo, C. H. B., Pappenberger, F., Di Napoli, C., Maciel, P., Quintino, T., Cornforth, R., & Cloke, H. L. (2023). Wet bulb globe temperature: Indicating extreme heat risk on a global grid. *Geohealth, 7*(2), e2022GH000701. https://doi.org/10.1029/2022gh000701

Bröde, P., Fiala, D., Błażejczyk, K., Holmér, I., Jendritzky, G., Kampmann, B., Tinz, B., & Havenith, G. (2012). Deriving the operational procedure for the Universal Thermal Climate Index (UTCI). *International Journal of Biometeorology, 56*, 481–494.

Budd, G. M. (2008). Wet-bulb globe temperature (WBGT) – Its history and its limitations. *Journal of Science and Medicine in Sport, 11*(1), 20–32. https://doi.org/10.1016/j.jsams.2007.07.003

Chapman, L., Thornes, J. E., Huang, Y., Cai, X., Sanderson, V. L., & White, S. P. (2008). Modelling of rail surface temperatures: A preliminary study. *Theoretical and Applied Climatology, 92*(1–2), 121–131. https://doi.org/10.1007/s00704-007-0313-5

Campbell, G. S., & Norman, J. M. (1998). Radiation fluxes in natural environments. In: *An Introduction to Environmental Biophysics*, 167–184. https://doi.org/10.1007/978-1-4612-1626-1_11

Cvijanovic, I., Mistry, M. N., Begg, J. D., Gasparrini, A., & Rodo, X. (2023). Importance of humidity for characterization and communication of dangerous heatwave conditions. *Npj Climate and Atmospheric Science, 6*(1). https://doi.org/10.1038/s41612-023-00346-x

Davis, R. E., McGregor, G. R., & Enfield, K. B. (2016). Humidity: A review and primer on atmospheric moisture and human health. *Environmental Research, 144*, 106–116. https://doi.org/10.1016/j.envres.2015.10.014

de Freitas, C. R., & Grigorieva, E. A. (2017). A comparison and appraisal of a comprehensive range of human thermal climate indices. *International Journal of Biometeorology, 61*(3), 487–512. https://doi.org/10.1007/s00484-016-1228-6

Di Napoli, C., Pappenberger, F., & Cloke, H. L. (2018). Assessing heat-related health risk in Europe via the Universal Thermal Climate Index (UTCI). *International Journal of Biometeorology, 62*(7), 1155–1165. https://doi.org/10.1007/s00484-018-1518-2

Di Napoli, C., Pappenberger, F., & Cloke, H. L. (2019). Verification of heat stress thresholds for a health-based heat-wave definition. *Journal of Applied Meteorology and Climatology, 58*(6), 1177–1194. https://doi.org/10.1175/jamc-d-18-0246.1

Diaconescu, E., Sankare, H., Chow, K., Murdock, T. Q., & Cannon, A. J. (2023). A short note on the use of daily climate data to calculate Humidex heat-stress indices. *International Journal of Climatology, 43*(2), 837–849. https://doi.org/10.1002/joc.7833

Donat, M. G., Pitman, A. J., & Seneviratne, S. I. (2017). Regional warming of hot extremes accelerated by surface energy fluxes. *Geophysical Research Letters, 44*(13), 7011–7019. https://doi.org/10.1002/2017gl073733

Driscoll, D. M. (1992). Thermal comfort indexes. Current uses and abuses. *National Weather Digest, 17*(4), 33–38.

Fang, W., Li, Z. X., Gao, J. H., Meng, R. L., He, G. H., Hou, Z. L., Zhu, S., Zhou, M. G., Zhou, C. L., Xiao, Y. Z., Yu, M., Huang, B., Xu, X. J., Lin, L. F., Xiao, J. P., Jin, D. H., Qin, M. F., Yin, P., Xu, Y. Q., Hu, J. X., Liu, T., Huang, C. R., & Ma, W. J. (2023). The joint and interaction effect of high temperature and humidity on mortality in China. *Environment International, 171*, 107669. https://doi.org/10.1016/j.envint.2022.107669

Fanger, P. O. (1970). *Thermal comfort. Analysis and applications in environmental engineering*. Danish Technical Press.

Fiala, D., Havenith, G., Brode, P., Kampmann, B., & Jendritzky, G. (2012). UTCI-Fiala multi-node model of human heat transfer and temperature regulation. *International Journal of Biometeorology, 56*(3), 429–441. https://doi.org/10.1007/s00484-011-0424-7

Gaitani, N., Burud, I., Thiis, T., & Santamouris, M. (2017). Aerial survey and in-situ measurements of materials and vegetation in the Urban fabric. *Procedia Engineering, 180*, 1335–1344. https://doi.org/10.1016/j.proeng.2017.04.296

Gallego-Elvira, B., Taylor, C. M., Harris, P. P., Ghent, D., Veal, K. L., & Folwell, S. S. (2016). Global observational diagnosis of soil moisture control on the land surface energy balance. *Geophysical Research Letters, 43*(6), 2623–2631. https://doi.org/10.1002/2016gl068178

Grundstein, A., & Cooper, E. (2018). Assessment of the Australian Bureau of Meteorology wet bulb globe temperature model using weather station data. *International Journal of Biometeorology, 62*(12), 2205–2213. https://doi.org/10.1007/s00484-018-1624-1

Grundstein, A., & Vanos, J. (2021). There is no 'Swiss Army Knife' of thermal indices: The importance of considering 'why?' and 'for whom?' when modelling heat stress in sport. *British Journal of Sports Medicine, 55*(15), 822–824. https://doi.org/10.1136/bjsports-2020-102920

Hass, A. L., & Ellis, K. N. (2019). Using wearable sensors to assess how a heatwave affects individual heat exposure, perceptions, and adaption methods. *International Journal of Biometeorology, 63*(12), 1585–1595. https://doi.org/10.1007/s00484-019-01770-6

Havenith, G., & Fiala, D. (2016). Thermal indices and Thermophysiological modeling for heat stress. *Comprehensive Physiology, 6*(1), 255–302. https://doi.org/10.1002/cphy.c140051

Hawkins, M. D., Brown, V., & Ferrell, J. (2017). Assessment of NOAA National Weather Service methods to warn for extreme heat events. *Weather Climate and Society, 9*(1), 5–13. https://doi.org/10.1175/wcas-d-15-0037.1

Jendritzky, G., de Dear, R., & Havenith, G. (2012). UTCI-Why another thermal index? *International Journal of Biometeorology, 56*(3), 421–428. https://doi.org/10.1007/s00484-011-0513-7

Johansson, E., Thorsson, S., Emmanuel, R., & Krüger, E. (2014). Instruments and methods in outdoor thermal comfort studies–The need for standardization. *Urban Climate, 10*, 346–366.

Kang, S., & Eltahir, E. A. (2018). North China plain threatened by deadly heatwaves due to climate change and irrigation. *Nature Communications, 9*(1), 2894.

Kenny, N. A., Warland, J. S., Brown, R. D., & Gillespie, T. G. (2008). Estimating the radiation absorbed by a human. *International Journal of Biometeorology, 52*, 491–503.

Kong, Q. Q., & Huber, M. (2022). Explicit calculations of wet-bulb globe temperature compared with approximations and why it matters for labor productivity. *Earths Futures, 10*, 3. https://doi.org/10.1029/2021ef002334

Koppe, C., & Jendritzky, G. (2005). Inclusion of short-term adaptation to thermal stresses in a heat load warning procedure. *Meteorologische Zeitschrift, 14*(2), 271–278. https://doi.org/10.1127/0941-2948/2005/0030

Kuras, E. R., Richardson, M. B., Calkins, M. M., Ebi, K. L., Hess, J. J., Kintziger, K. W., Jagger, M. A., Middel, A., Scott, A. A., Spector, J. T., Uejio, C. K., Vanos, J. K., Zaitchik, B. F., Gohlke, J. M., & Hondula, D. M. (2017). Opportunities and challenges for personal heat exposure research. *Environmental Health Perspectives, 125*(8), 085001. https://doi.org/10.1289/ehp556

Landsberg, H. E. (1972). *The assessment of human bioclimate, a limited review of physical parameters* (Vol. WMO-No. 331). World Meteorological Organization.

Li, H. Z., Lo, M. H., Ryu, D., Peel, M., & Zhang, Y. Q. (2022). Possible increase of air temperature by irrigation. *Geophysical Research Letters, 49*(20). https://doi.org/10.1029/2022gl100427

Liljegren, J. C., Carhart, R. A., Lawday, P., Tschopp, S., & Sharp, R. (2008). Modeling the wet bulb globe temperature using standard meteorological measurements. *Journal of Occupational and Environmental Hygiene, 5*(10), 645–655. https://doi.org/10.1080/15459620802310770

Lo, Y. T. E., Mitchell, D. M., Buzan, J. R., Zscheischler, J., Schneider, R., Mistry, M. N., Kysely, J., Lavigne, E., da Silva, S. P., Roye, D., Urban, A., Armstrong, B., Gasparrini, A., Vicedo-Cabrera, A. M., & Multi-Country Multi-City, M. C. C. (2023). Optimal heat stress metric for modelling heat-related mortality varies from country to country. *International Journal of Climatology, 43*, 5553–5568. https://doi.org/10.1002/joc.8160

Lu, Y. C., & Romps, D. M. (2023). Predicting fatal heat and humidity using the heat index model. *Journal of Applied Physiology, 134*(3), 649–656. https://doi.org/10.1152/japplphysiol.00417.2022

Mandanici, E., Conte, P., & Girelli, V. A. (2016). Integration of aerial thermal imagery, LiDAR data and ground surveys for surface temperature mapping in urban environments. *Remote Sensing, 8*(10). https://doi.org/10.3390/rs8100880

Masterton, J. M., Service CAE, Richardson, F. A., & Canada, C. E. (1979). *Humidex: A method of quantifying human discomfort due to excessive heat and humidity*. Environment Canada, Atmospheric Environment.

Matzarakis, A., Laschewski, G., & Muthers, S. (2020). The heat health warning system in Germany-application and warnings for 2005 to 2019. *Atmosphere, 11*(2). https://doi.org/10.3390/atmos11020170

McGregor, G. R., & Vanos, J. K. (2018). Heat: A primer for public health researchers. *Public Health, 161*, 138–146. https://doi.org/10.1016/j.puhe.2017.11.005

Minard, D. (1961). Prevention of heat casualties in marine corps recruits – Period of 1955–60, with comparative incidence rates and climatic heat stresses in other training categories. *Military Medicine, 126*(4), 261–272. https://doi.org/10.1093/milmed/126.4.261

Mishra, V., Ambika, A. K., Asoka, A., Aadhar, S., Buzan, J., Kumar, R., & Huber, M. (2020). Moist heat stress extremes in India enhanced by irrigation. *Nature Geoscience, 13*(11), 722–728. https://doi.org/10.1038/s41561-020-00650-8

Morais, L., Lopes, A., & Nogueira, P. (2020). Which heatwave measure has higher predictive power to prevent health risks related to heat: EHF or GATO IV? – Evidence from modelling Lisbon mortality data from 1980 to 2016. *Weather and Climate Extremes, 30*, 100287. https://doi.org/10.1016/j.wace.2020.100287

Nairn, J. R., & Fawcett, R. J. B. (2015). The excess heat factor: A metric for heatwave intensity and its use in classifying heatwave severity. *International Journal of Environmental Research and Public Health, 12*(1), 227–253. https://doi.org/10.3390/ijerph120100227

Nairn, J., Ostendorf, B., & Bi, P. (2018). Performance of excess heat factor severity as a global heatwave health impact index. *International Journal of Environmental Research and Public Health, 15*(11). https://doi.org/10.3390/ijerph15112494

Nairn, J. R., Moise, A., & Ostendorf, B. (2022). The impact of humidity on Australia's operational heatwave services. *Climate Services, 27*, 100315. https://doi.org/10.1016/j.cliser.2022.100315

Nazarian, N., & Lee, J. K. W. (2021). Personal assessment of urban heat exposure: A systematic review. *Environmental Research Letters, 16*, 3. https://doi.org/10.1088/1748-9326/abd350

Oetomo, A., Jalali, N., Costa, P. D. P., & Morita, P. P. (2022). Indoor temperatures in the 2018 heat wave in Quebec, Canada: Exploratory study using Ecobee smart thermostats. *Jmir Formative Research, 6*(5), e34104. https://doi.org/10.2196/34104

Pappenberger, F., Jendritzky, G., Staiger, H., Dutra, E., Di Giuseppe, F., Richardson, D. S., & Cloke, H. L. (2015). Global forecasting of thermal health hazards: The skill of probabilistic predictions of the Universal Thermal Climate Index (UTCI). *International Journal of Biometeorology, 59*(3), 311–323. https://doi.org/10.1007/s00484-014-0843-3

Pardyjak, E. R., & Stoll, R. (2017). Improving measurement technology for the design of sustainable cities. *Measurement Science and Technology, 28*(9), 092001. https://doi.org/10.1088/1361-6501/aa7c77

Parsons, K. (2007). *Human thermal environments: The effects of hot, moderate, and cold environments on human health, comfort and performance*. CRC Press.

Rachid, A., & Qureshi, A. M. (2023). Sensitivity analysis of heat stress indices. *Climate, 11*(9). https://doi.org/10.3390/cli11090181

Romps, D. M., & Lu, Y. C. (2022). Chronically underestimated: A reassessment of US heat waves using the extended heat index. *Environmental Research Letters, 17*(9). https://doi.org/10.1088/1748-9326/ac8945

Russo, S., Sillmann, J., & Sterl, A. (2017). Humid heat waves at different warming levels. *Scientific Reports, 7*, 7477. https://doi.org/10.1038/s41598-017-07536-7

Simpson, C. H., Brousse, O., Ebi, K. L., & Heaviside, C. (2023). Commonly used indices disagree about the effect of moisture on heat stress. *Npj Climate and Atmospheric Science, 6*(1). https://doi.org/10.1038/s41612-023-00408-0

Spangler, K. R., Adams, Q. H., Hu, J. K., Braun, D., Weinberger, K. R., Dominici, F., & Wellenius, G. A. (2023). Does choice of outdoor heat metric affect heat-related epidemiologic analyses in the US Medicare population? *Environmental Epidemiology, 7*(4). https://doi.org/10.1097/ee9.0000000000000261

Staiger, H., Laschewski, G., & Gratz, A. (2012). The perceived temperature – A versatile index for the assessment of the human thermal environment. Part A: Scientific basics. *International Journal of Biometeorology, 56*(1), 165–176. https://doi.org/10.1007/s00484-011-0409-6

Staiger, H., Laschewski, G., & Matzarakis, A. (2019). Selection of appropriate thermal indices for applications in human biometeorological studies. *Atmosphere, 10*(1). https://doi.org/10.3390/atmos10010018

Steadman, R. G. (1979). Assessment of sultriness. 1. Temperature-humidity index based on human physiology and clothing science. *Journal of Applied Meteorology, 18*(7), 861–873. https://doi.org/10.1175/1520-0450(1979)018<0861:taospi>2.0.co;2

Stull, R. (2011). *Meteorology for scientists and engineers* (3rd ed.). University of British Columbia.

Thorsson, S., Lindberg, F., Eliasson, I., & Holmer, B. (2007). Different methods for estimating the mean radiant temperature in an outdoor urban setting. *International Journal of Climatology: A Journal of the Royal Meteorological Society, 27*(14), 1983–1993.

Vanos, J. K., Baldwin, J. W., Jay, O., & Ebi, K. L. (2020). Simplicity lacks robustness when projecting heat-health outcomes in a changing climate. *Nature Communications, 11*(1). https://doi.org/10.1038/s41467-020-19994-1

Wild, M., Folini, D., Hakuba, M. Z., Schar, C., Seneviratne, S. I., Kato, S., Rutan, D., Ammann, C., Wood, E. F., & Konig-Langlo, G. (2015). The energy balance over land and oceans: An assessment based on direct observations and CMIP5 climate models. *Climate Dynamics, 44*(11–12), 3393–3429. https://doi.org/10.1007/s00382-014-2430-z

Wouters, H., De Ridder, K., Poelmans, L., Willems, P., Brouwers, J., Hosseinzadehtalaei, P., Tabari, H., Vanden Broucke, S., van Lipzig, N. P. M., & Demuzere, M. (2017). Heat stress increase under climate change twice as large in cities as in rural areas: A study for a densely populated midlatitude maritime region. *Geophysical Research Letters, 44*(17), 8997–9007. https://doi.org/10.1002/2017gl074889

Yang, J. C., & Bou-Zei, E. (2019). Designing sensor networks to resolve spatio-temporal urban temperature variations: Fixed, mobile or hybrid? *Environmental Research Letters, 14*(7). https://doi.org/10.1088/1748-9326/ab25f8

Yarnold, M. T., & Moon, F. L. (2015). Temperature-based structural health monitoring baseline for long-span bridges. *Engineering Structures, 86*, 157–167. https://doi.org/10.1016/j.engstruct.2014.12.042

Zhang, K. R., Cao, C., Chu, H. R., Zhao, L., Zhao, J. Y., & Lee, X. H. (2023). Increased heat risk in wet climate induced by urban humid heat. *Nature, 617*, 738–742. https://doi.org/10.1038/s41586-023-05911-1

Part II
Heatwave Causes

Chapter 5
Modes of Climate Variability

Abstract The chapter highlights the growing interest in how modes of climate variability influence temperature extremes and heatwaves and explores the relationship between climate variability and the occurrence of heatwaves. It begins by defining modes of climate variability as quasi-periodic variations in ocean and atmospheric circulation patterns and outlines what is meant by climate teleconnections. The Chapter adopts a probabilistic frame. It posits that certain atmospheric and oceanic circulation patterns may tip the balance of probability towards heatwave occurrence or non-occurrence because the associated atmospheric, ocean and land surface conditions either enhance or diminish the probability of extreme heat events in different regions. Several key modes of climate variability are discussed, including the El Niño Southern Oscillation (ENSO), North Atlantic Oscillation (NAO), Madden Julian Oscillation (MJO), Indian Ocean Dipole (IOD), Southern Annular Mode (SAM), Pacific Decadal Oscillation (PDO), and Atlantic Multidecadal Oscillation (AMO). Each mode's characteristics, geographical impacts, and their potential to influence heatwave occurrences are examined. The chapter underscores the complexity of the association between modes of climate variability and heatwave occurrence, especially as multiple modes may amplify or dampen the likelihood of heatwave occurrence and calls for a deeper understanding of the physical mechanisms that underlie the cascade of processes that link modes of climate variability with atmospheric and marine heatwave occurrence.

Keywords Intra-seasonal · Inter-annual and decadal modes of climate variability · Teleconnections · Atmospheric heatwaves · Marine heatwaves

5.1 Introduction

Modes of climate variability can be generally defined as quasi-periodic variations in ocean and atmospheric circulation patterns that possess an oscillatory behaviour. Associated with modes of climate variability is the concept of climate teleconnections. A climate teleconnection means an association in the variation of climate between locations or regions separated by a great distance. For example, regions on

G. McGregor, *Heatwaves*, Biometeorology 6,
https://doi.org/10.1007/978-3-031-69906-1_5

opposite sides of an ocean basin may experience very different summer conditions such as hot and dry in one region while in a region on the opposite side of the ocean basin cool wet conditions occur and vice versa. This seesaw like connection in climate arises because an atmosphere or ocean-atmosphere 'bridge' links two regions through atmospheric and/or ocean circulation patterns being coupled. Climate teleconnections, which are most obvious at the intra-seasonal to inter-annual to decadal time scales, express themselves most obviously through spatial variations in atmospheric pressure, air and/or sea surface temperature, cloud, wind, or precipitation patterns.

Over the past decade or so there has been mounting interest in establishing how modes of climate variability might play a role in the occurrence of temperature extremes and heatwaves, or perhaps more accurately, the conditions that create the opportunity for the development of heatwaves. This interest stems from the assumption that heatwave occurrence is the outcome of a cascade of processes that link the large-scale physical state and dynamics of the ocean and the atmosphere with land and marine surfaces on time scales beyond that of an actual heatwave event. In essence, certain configurations of atmospheric and ocean climate patterns or modes of climate variability may set the scene for an increased likelihood of heatwave occurrence. This idea will be explored in this chapter. Prior to outlining the evidence for a connection between modes of climate variability and the frequency of temperature extremes and heatwaves, first an overview of the general characteristics and climate implications associated with these modes will be provided.

5.2 Modes of Climate Variability: Nature and Climate Characteristics

Many modes of climate variability have been identified (Kucharski et al., 2010; McGregor, 2017; Sheridan & Lee, 2012; de Viron et al., 2013; Williams et al., 2017), all of which could be considered as potential drivers of intra-seasonal to inter-decadal variability of heatwave occurrence (Table 5.1). Further, a plethora of teleconnection indices have been constructed to measure, monitor and summarise the status of modes of climate variability. Falling broadly into atmospheric, oceanic and blended categories, teleconnection indices are in effect, statistical constructs comprising single numbers which attempt to capture a range of often complex ocean and/or atmospheric process interactions. A criticism often levelled at teleconnection indices is that they cannot be expected to encapsulate the intricate coupled processes that give rise to multifaceted modes of climate variability. Given this it is important to make a distinction between teleconnection indices as statistical constructs and the complex climate phenomena which they attempt to represent. Notwithstanding this potential weakness, teleconnection indices have been applied widely in the representation and analysis of modes of climate variability.

Table 5.1 Modes of climate variability and associated acronyms.[a]

Arctic Oscillation (AO)/Northern Annual Mode (NAM)	El Niño Southern Oscillation (ENSO)
North Atlantic Oscillation (NAO)	East Atlantic Pattern (EA)
West Pacific Pattern (WP)	East Pacific/ North Pacific Pattern (EP/NP)
Pacific/ North American Pattern (PNA)	East Atlantic/West Russia Pattern (EA/WR)
Scandinavia Pattern (SCA)	Northern Hemisphere Pattern (TNH)
Polar/ Eurasia Pattern (POL)	Pacific Transition Pattern (PT)
Pacific South American Pattern (PSA)	Southern Annular Mode (SAM)/ Antarctic Oscillation (AO)
Indian Ocean Dipole (IOD)	South Pacific Wave Pattern (SPW)
Quasi Biennial Oscillation (QBO)	Madden Julian Oscillation (MJO)
Pacific Decadal Oscillation (PDO)	Atlantic Multidecadal Oscillation AMO

Compiled by Author

[a] While there might be a commonly accepted acronym for each teleconnection there may be many different versions of a particular index. This is because different methods, data sets, atmosphere and ocean variables, criteria and sampling periods might be used in their construction. The modes presented in this table are a non-exhaustive list

In relation to heatwave occurrence several modes of climate variability have been identified as potentially important. These mainly include El Niño Southern Oscillation (ENSO), North Atlantic Oscillation (NAO), Madden Julian Oscillation (MJO), Indian Ocean Dipole (IOD), Southern Annular Mode (SAM), Pacific Decadal Oscillation (PDO) and the Atlantic Multidecadal Oscillation (AMO) and will be described here in ascending order of timescale. The geographical regions from which climate impacts are spawned by these modes are portrayed in Fig. 5.1. Although other modes of climate variability listed in Table 5.1 are not outlined here, this does not mean they may not influence the propensity for heatwave development. Rather their absence is an indication of the infancy of research on how heatwave development may be predisposed to certain atmosphere and ocean climate configurations. No doubt as research progresses in this field, modes of variability other than those described here will be revealed as important.

5.2.1 *Intra-seasonal*

5.2.1.1 Madden Julian Oscillation

The Madden–Julian Oscillation (MJO) is an important intra-seasonal mode of tropical variability operating at time scales of 30–90 days. It is the outcome of a close coupling between atmospheric circulation and tropical deep convection. Unlike other modes of variability, the MJO is a travelling as opposed to a standing pattern which moves eastwards over parts of the Indian and Pacific Oceans. Its movement is characterised by successive periods of enhanced and supressed rainfall in equatorial regions stretching from the western regions of the Indian Ocean to the western

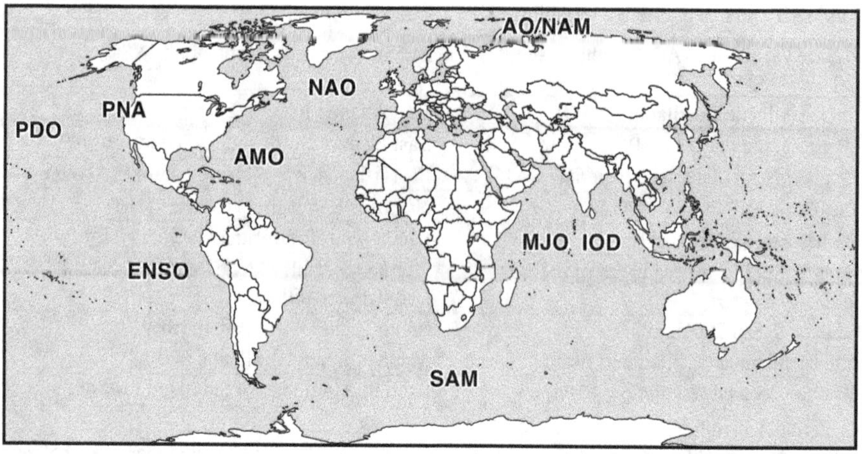

Fig. 5.1 Regions associated with various modes of climate variability. In terms of climate impacts, the modes of climate variability can affect regions far beyond their point of origin or so-called centres of action. AO/NAM is Arctic Oscillation/Northern Annular Mode, AMO is Atlantic Multidecadal Oscillation, ENSO is El Niño Southern Oscillation, IOD is Indian Ocean Dipole, MJO is Madden Julian Oscillation, NAO is North Atlantic Oscillation, PDO is Pacific Decadal Oscillation, PNA is Pacific North America pattern, SAM is Southern Annular Mode. (Source: Author)

and central Pacific. The MJO is composed of eight phases which have been identified by the way in which rainfall, because of enhanced convection, propagates eastward from the Indian Ocean (Fig. 5.2).

As portrayed in Fig. 5.2, phase 1 (F) begins with enhanced convection over the western Indian Ocean. In Phases 2 and 3 (G and H) this moves eastwards out over the Indian Ocean and parts of the Indian subcontinent. During Phases 4 and 5 (A and B), enhanced convection has reached the broad region of Indonesia and the far western Pacific (Maritime Continent). Subsequently, enhanced convection moves further eastward over the western Pacific during Phases 6 to 8 (C, D, E) with enhanced rainfall ultimately fading away over the central Pacific. With the decay of Phase 8 the next MJO cycle begins in the western Indian Ocean. Because the MJO operates along the equatorial regions a clear dipole of convection and thus rainfall anomalies occur over the MJO cycle such that enhanced convection over the western Pacific and suppressed convection over the Indian Ocean takes place during Phase 4–6 while the opposite is true during Phase 2–3.

In terms of its influence on heatwave occurrence, the MJO and its successive periods of enhanced convection has the potential to influence tropical atmospheric (diabatic) heating patterns which may lead to perturbations in the subtropical jet stream (Seo & Son, 2012) with associated knock-on effects for Rossby wave patterns and blocking in the mid latitudes. Further, the suppressed convection phase of the MJO in the Indian Ocean region could delay the onset of the summer monsoon, hence extending the pre-monsoon season period of hot dry conditions thus increasing the likelihood of heatwave occurrence. Similarly for regions where heatwave

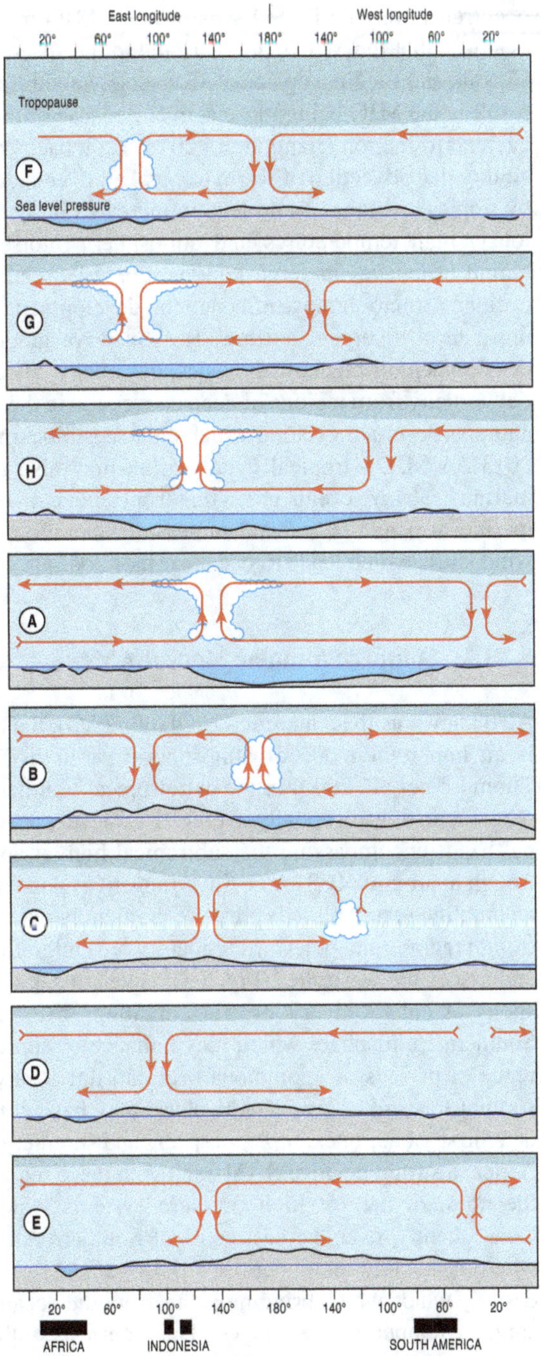

Fig. 5.2 MJO phases—F to E showing the eastwards progression of a zone of enhanced convection (clouds) originating on the western side of the Indian Ocean off the east coast of Africa and finally dissipating over the eastern Pacific in the vicinity of South America. The arrows refer to the zonal-along equator overturning circulation with upward (downward) arrows indicating ascending (descending) air. The lines at the top and bottom of each panel represent upper tropospheric and sea level pressure anomalies, respectively. Areas of ascending air are matched with negative surface (positive tropospheric) pressure anomalies. Redrawn from https://physicstoday.scitation.org/do/10.1063/PT.5.4014/full/ and adapted from Madden and Julian (1972)

occurrence may be ENSO sensitive, the MJO may be important as there is some evidence that the MJO can contribute to the speed of development and intensity of El Niño and La Nina episodes. Conversely neutral and weak ENSO years appear to result in the MJO being more active. Because of the impact of the MJO on tropical cyclone formation (Jiang et al., 2018; Klotzbach & Oliver, 2015), there is the possibility that advection of warm tropical air over subtropical to mid-latitude locations by tropical cyclones could lead to increases in the likelihood of periods of anomalously high temperatures and atmospheric moisture levels giving rise to warm humid heatwaves in some regions. For Australia, MJO related enhanced tropical cyclone formation may influence the development of heatwaves indirectly through upper-level divergent outflow, Rossby wave guide perturbation and downstream development of blocking episodes, or directly, via advection of anticyclonic potential vorticity from regions of deep convection in the vicinity of tropical cyclones into an upper-level anticyclone which enhances blocking and subsidence (Parker et al., 2013). A MJO—tropical cyclone link may also be important for the demise of marine heatwave events given that tropical cyclones drastically increase the surface mixing of ocean layers and increase evaporation rates through increased surface wind stress which leads to ocean surface cooling.

5.2.1.2 Southern Annular Mode (SAM)

Also known as the Antarctic Oscillation (AO), the Southern Annular Mode (SAM) is an important mode of atmospheric variability (Thompson & Wallace, 2000; Thompson et al., 2000) in the extratropical Southern Hemisphere. It involves periodic fluctuations in zonally averaged sea level pressure (SLP) between two latitudinal bands encompassing the subtropical high at around 40°S and the circumpolar trough at around 60°S. A positive (negative) phase of the SAM is characterized by an amplified (suppressed) pressure gradient between the subtropical high and polar trough regions and poleward (equatorward) migration of the mid-latitude jet stream and associated westerlies (Fig. 5.3). Given this, the SAM has the potential to create spells of enhanced dry or wet conditions over the mid-latitude regions of the Southern Hemisphere which may enhance or supress the likelihood of the occurrence of periods of anomalous heat and thus heatwaves. While this sounds quite straight forward, a given SAM phase may have contrasting climate effects dependent on season. For example, in spring and summer for southern Australia and South Africa, a strong positive SAM can mean these regions come under the influence of the northern half of high pressure systems which have been displaced south. Consequently, over the east coast of Southern Africa and south eastern Australia where heatwaves are endemic, a positive SAM phase results in easterly to southeasterly winds which bring moist air from the oceans to the east and suppression of summer temperature extremes. This contrasts with a SAM negative phase when over Australia there is a tendency for the subtropical high pressure system to be displaced north with the result that hot dry northerly air is advected on its western flank over southern Australia thus increasing the likelihood of temperature extremes

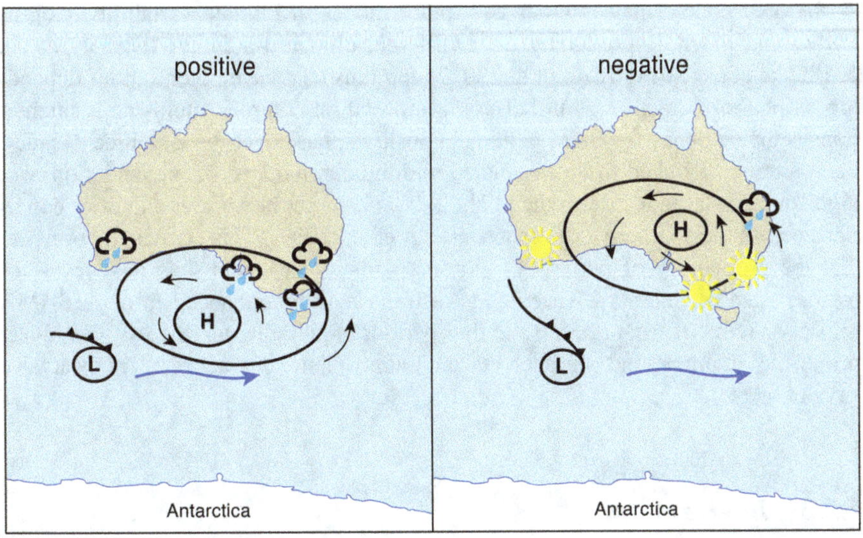

Fig. 5.3 Schematic representation of summer impacts of positive (left) and negative (right) phases of SAM. H and L indicate high- and low-pressure systems respectively. Blue arrows represent the westerly wind flow. Note how under a SAM negative phase southeast Australia, where the impactful 2009 heatwave event occurred, experiences increased clear sky conditions/radiative heat input with high pressure dominating. Under a positive SAM the opposite is true with wet conditions owing to onshore cool moist winds. Adapted from: Australian Broadcasting Corporation News https://www.abc.net.au/news/2018-08-14/southern-annular-mode-and-how-it-affects-our-weather/10106134 (Accessed 26/05/24)

and dry conditions (Meneghini et al., 2007). On the southwest coast of South Africa where Cape Town is located, a positive phase of SAM can enhance the probability of drought and anomalously high temperatures (Sousa et al., 2018).

While early SAM research focused on its climate impacts across the Southern Hemisphere, it has emerged that its effects can extend well beyond its provenance. For example, Dou et al. (2017) have shown how SAM in the austral autumn can influence summer monsoon rainfall over the Indian sub-continent and adjacent areas. This occurs via a cross-equatorial ocean-atmosphere bridge in that a strong (weak) boreal autumn SAM anomaly can prompt (halt) a positive South Indian Ocean dipole sea surface temperature anomaly which in turn enhances (suppresses) rainfall in June to July. This bears implications for heatwave generation given the criticality of the timing and intensity of the summer monsoon for heatwave occurrence over the Indian subcontinent such that a weakened SAM in the Southern Hemisphere autumn could increase the probability of heatwave promoting conditions. As well as its effect on the Indian subcontinent, the boreal autumn SAM also appears to impact summer monsoon rainfall delivery over the South China Sea and Korea (Liu et al., 2018; Prabhu et al., 2017) thus creating the possibility of SAM-heatwave relationships in this region.

SAM of course may interact with other modes of climate variability such as ENSO and the IOD producing quite different climate outcomes (Blamey et al., 2018; Cai et al., 2011; Tozer et al., 2017) and thus heatwave potentialities depending on phase interactions. Furthermore, imposed on complex multi-mode interactions is the tendency for SAM to adopt a positive phase over the last three decades. Such a trend, believed to be associated with human induced ozone depletion over high southern latitudes (Perlwitz et al., 2008) has brought warmer and drier conditions across the southern extra-tropics (Fogt et al., 2009). This creates the prospect that the probability of periods of extreme summer heat and thus heatwaves for regions such as Southern Africa and Australia may be enhanced through SAM's impact on winter/spring rainfall and thus periods of pre-summer dryness which may prime the summer land surface, via land-atmosphere interactions, for heatwave development.

5.2.2 *Inter-annual*

5.2.2.1 El Niño Southern Oscillation (ENSO)

The ENSO phenomenon refers to variations in sea surface temperature and atmospheric pressure conditions across the tropical Pacific Ocean. Comprised of two major components, namely the El Niño or ocean, and the Southern Oscillation or atmospheric components, ENSO is a major driver of inter-annual climate variability at a range of geographical scales. El Niño events are associated with anomalously high sea surface temperatures in the eastern Pacific, and accordingly are referred to as "warm events". The "cool event" counterpart carries the name of La Niña. Both El Niño and La Niña are part of the ENSO cycle that lasts on average from 12–18 months with a periodicity of 2–7 years. El Niño (La Niña) or warm (cool) event conditions first begin to manifest as positive (negative) sea surface temperature (SST) anomalies in the central and eastern Pacific around July. These continue to develop as the ENSO cycle progresses, reaching a peak around January to February of the following year. Abatement of SST anomalies occurs in the subsequent months of March to August with cessation of the El Niño (La Niña) event normally by the end of the Northern Hemisphere summer. A La Niña phase does not necessarily immediately follow an El Niño and vice versa as "neutral" conditions can intervene when SST conditions in the eastern and central Pacific are in and around "normal".

SST anomalies and other ocean and atmospheric indices are used to monitor ENSO (McGregor & Ebi, 2018). Amongst a wide range of ENSO indices two ocean-based indices, namely the Niño 3.4 and Ocean Nino Index (ONI) and the multivariate ENSO index (MEI) have been used most widely in explorations of ENSO related climate impacts, including heatwaves. The Niño 3.4 and ONI indices measure Pacific equatorial SST anomalies from around the dateline to the South American coast (5 °N-5 °S, 120°–170 °W). They are the two 'official' ENSO indices produced by the US National and Oceanic and Atmospheric Administration (NOAA). Both are used in subtly different ways for classifying ENSO events (McGregor & Ebi, 2018). The MEI is a blended index as it amalgamates the ocean and atmosphere variables of sea level pressure, zonal and meridional surface wind components, SST, surface air temperature and total cloudiness fraction of the sky into a single index (Wolter & Timlin, 2011). A plot of the ONI since 1950 clearly demonstrates when El Niño and La Niña events of varying strengths have occurred (Fig. 5.4). Noteworthy are the very strong El Niño events of 1997–98 and 2015–16 with the recent 2023–24 event falling within the upper bound of the strong category. Interestingly since about the mid-1970s strong El Niño events have dominated compared to strong La Niña events, most likely due to the modulation of ENSO phase strength by the Pacific Decadal Oscillation and Inter-decadal Pacific Oscillation.

The climate impacts of El Niño and La Niña can be significant. Strong La Niña events generally result in an exaggeration of the "normal" climate pattern; warm and wet (cool and dry) locations become warmer and wetter (cooler and drier) than usual. In contrast El Niño events largely result in a reversal of the normal climate pattern such that warm and wet (cool and dry) locations experience cool and dry (warm and wet) conditions over the duration of the event. Despite their origin in the tropical South Pacific, the geographical extent of intense El Niño and La Niña events can be considerable with climate anomalies apparent across large parts of not only the tropics/subtropics but mid-latitudes (Figs. 5.5 and 5.6). Through the ability of the two phases of ENSO to alter weather and climate patterns across vast tracts of the globe, for example the Rossby wave pattern, storm tracks, atmospheric blocking and thus temperature, precipitation and cloud conditions, there is considerable potential for ENSO to have an impact on the likelihood of heatwave occurrence.

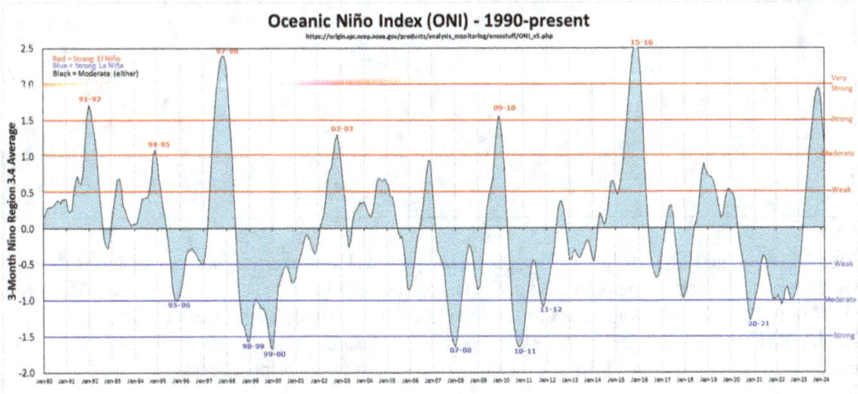

Fig. 5.4 Plot of the NOAA ONI with El Niño and La Nina events identified. To qualify as an 'event' five consecutive overlapping 3-month periods must attain a departure of plus (El Niño) or minus (La Nina) 0.5 °C or more. The size of the ONI departure is also used to classify event strength as follows: weak = 0.5 to 0.9; moderate = 1.0 to 1.4; strong = 1.5 to 1.9; very strong ≥2.0. Present in the title is January 2024. (Source: Golden Gate Weather Services https://www.ggweather.com/enso/oni.htm). Used with permission from Jan Null, CCM, Golden Gate Weather Services

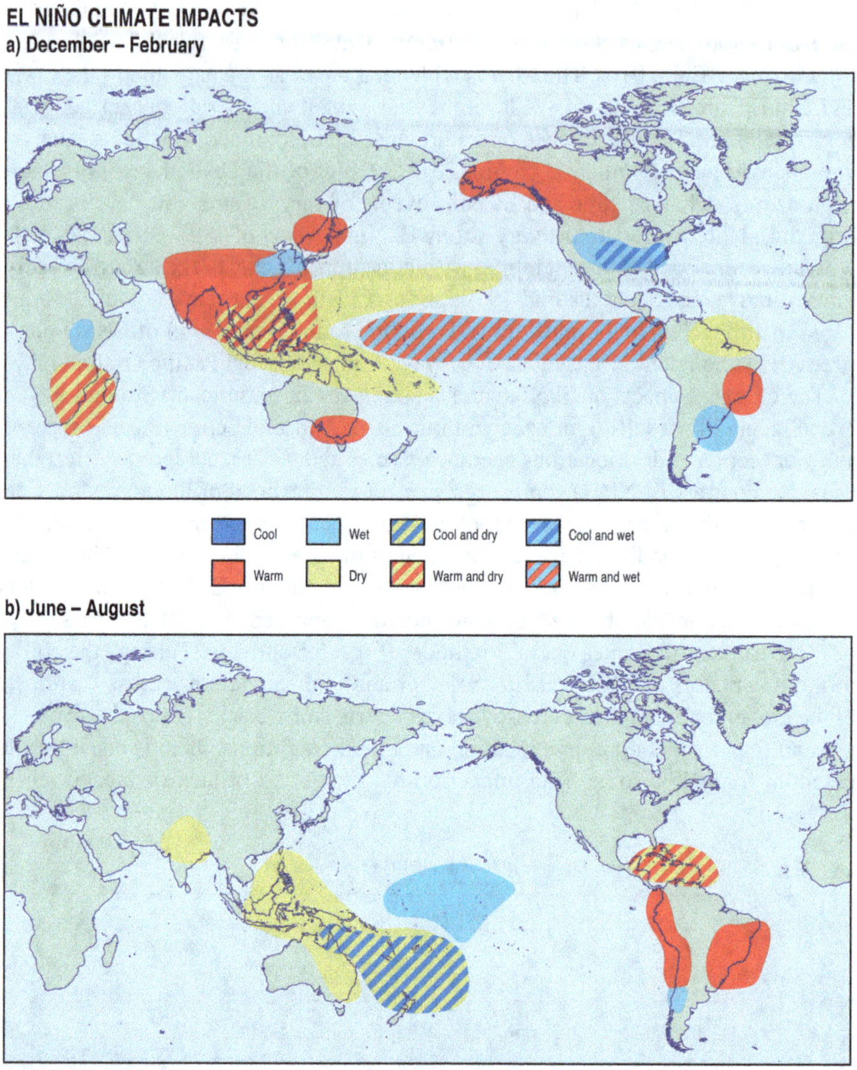

Fig. 5.5 Average El Niño climate impacts for the periods December to January and June to August which broadly match the peak (**a**) and decay (**b**) periods of El Niño respectively Source: McGregor and Ebi (2018) El Niño Southern Oscillation (ENSO) and Health: An Overview for Climate and Health Researchers. Atmosphere. 2018; 9(7):282. https://doi.org/10.3390/atmos9070282. Used under Creative Common CC BY Licence https://www.mdpi.com/openaccess

Given the type of impacts illustrated in Fig. 5.5, one might surmise that the likelihood of heatwave occurrence may increase in those areas that experience warmer and/or drier or more humid conditions during summer as an outcome of an ENSO warm or cold phase. Importantly, as well as concurrent associations, a given ENSO phase may prime the land surface, via reduced rainfall and thus soil moisture

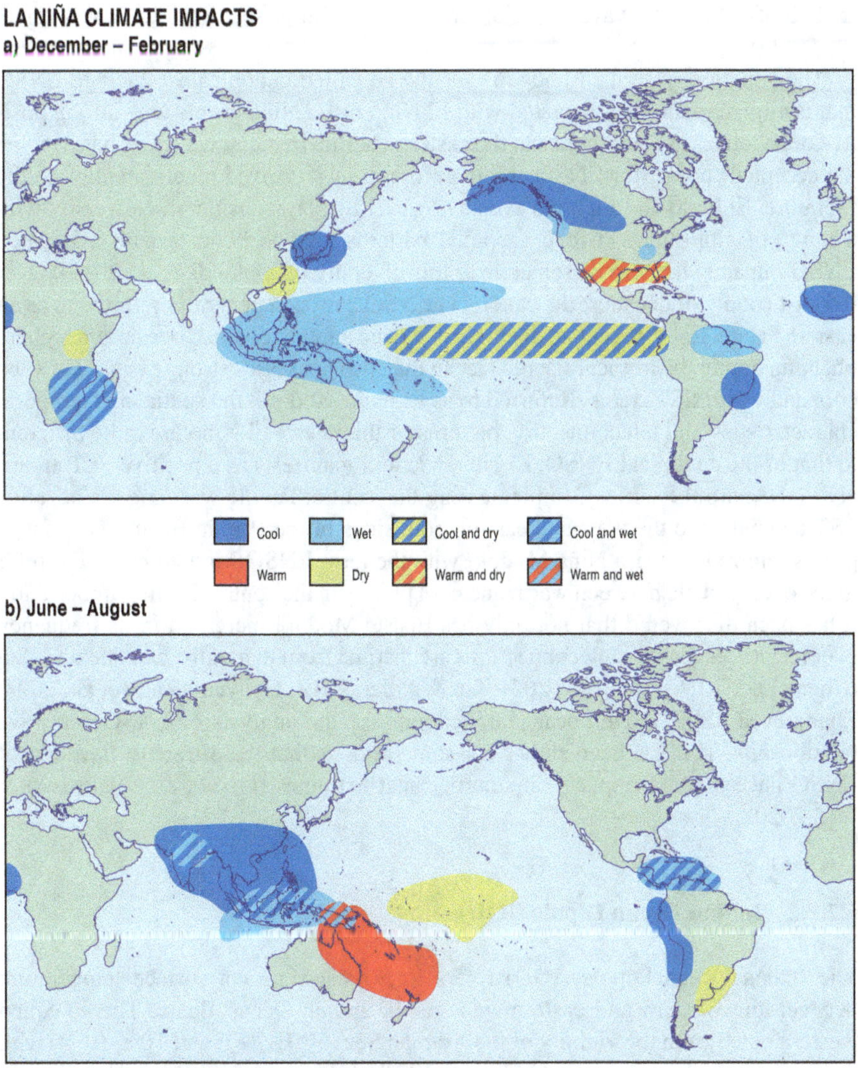

Fig. 5.6 Average La Niña climate impacts for the periods December to January and June to August which broadly match the peak (**a**) and decay (**b**) periods of La Niña respectively Source: McGregor and Ebi (2018). El Niño Southern Oscillation (ENSO) and Health: An Overview for Climate and Health Researchers. Atmosphere. 2018; 9(7):282. https://doi.org/10.3390/atmos9070282 Used with permission under Creative Common CC BY Licence https://www.mdpi.com/openaccess

deficits, in the months preceding the heatwave season thus increasing the likelihood of periods of extreme heat via land-atmosphere feedbacks (see Chap. 7). Because neither of the ENSO phases occurs in isolation, ENSO's interaction with other modes of climate variability may determine the climate outcome and therefore the probability for heatwave development at the intra-seasonal to inter-annual timescales.

The situation described above is for the so-called canonical or average ENSO that is climate patterns associated with El Niño or La Niña events with strong positive or negative SST anomalies in the eastern Pacific. In many ways this has become the accepted 'norm' for ENSO and therefore, it has received much attention in the literature. El Niño and La Niña events however display considerable diversity with a range of amplitudes, triggers, spatial patterns, and life cycles with associated ENSO impacts likely to be sensitive to this diversity (Capotondi et al., 2015). Over the past couple of decades, the canonical ENSO has been generally referred to as an eastern Pacific ENSO event, to distinguish it from another coupled ocean-atmosphere phenomenon in the tropical Pacific Ocean that also possesses strong swings between cool and warm SST states. Referred to as El Niño Modoki, the spatial and temporal characteristics and teleconnection patterns of this 'new' El Niño are quite different to that of the canonical ENSO. El Niño Modoki manifests as a positive SST anomaly in the central Pacific. Complementing this central Pacific warming are negative SST anomalies to the west and east on both sides of the Pacific Basin. Its counterpart is referred to as La Niña Modoki with the term ENSO Modoki coined to refer to the overall cycle between warm and cool phases in the central Pacific. Importantly it has been discovered that not only has ENSO Modoki increased in its frequency over recent decades, its teleconnections are distinct from that of the canonical ENSO (Chen et al., 2016; Fan et al., 2021; Gu & Adler, 2019; Nguyen-Thanh et al., 2023; Zhang et al., 2022). This bears implications for the analysis of ENSO heatwave relationships as it has been shown for some regions that the different 'flavours' of ENSO have distinct impacts on summer heat extremes (Gao et al., 2020; Luo & Lau, 2020).

5.2.2.2 Indian Ocean Dipole (IOD)

The Indian Ocean Dipole (IOD) refers to a seesaw in sea surface temperature between the western and eastern sides of the Indian Ocean Basin. The so-called western pole lies in the vicinity of the Arabian Sea (50°E-70°E and 10°S-10°N) with the eastern pole found south of Indonesia (90°E-110°E and 10°S-0°N). The climatological pattern of warm waters in the eastern parts of the Indian Ocean Basin and cooler waters in the west of the basin means that air rises in the east, moves west and then subsides over the western regions of the basin, the result being a surface westerly wind flow along the equator. When SST are anomalously high in the western Indian Ocean an IOD positive phase is said to occur (Fig. 5.7a) with a negative phase associated with unusually high SST in the eastern Indian Ocean (Fig. 5.7b). A neutral phase refers to SST conditions being near normal in the region of the eastern pole. Positive or negative phases of the IOD tend to develop in and around the equatorial region of the Indian Ocean in the northern hemisphere spring, reaching a peak in the autumn. Being a coupled ocean-atmosphere phenomenon, the atmospheric response of the opposite phases of the IOD is clearly seen in the pattern

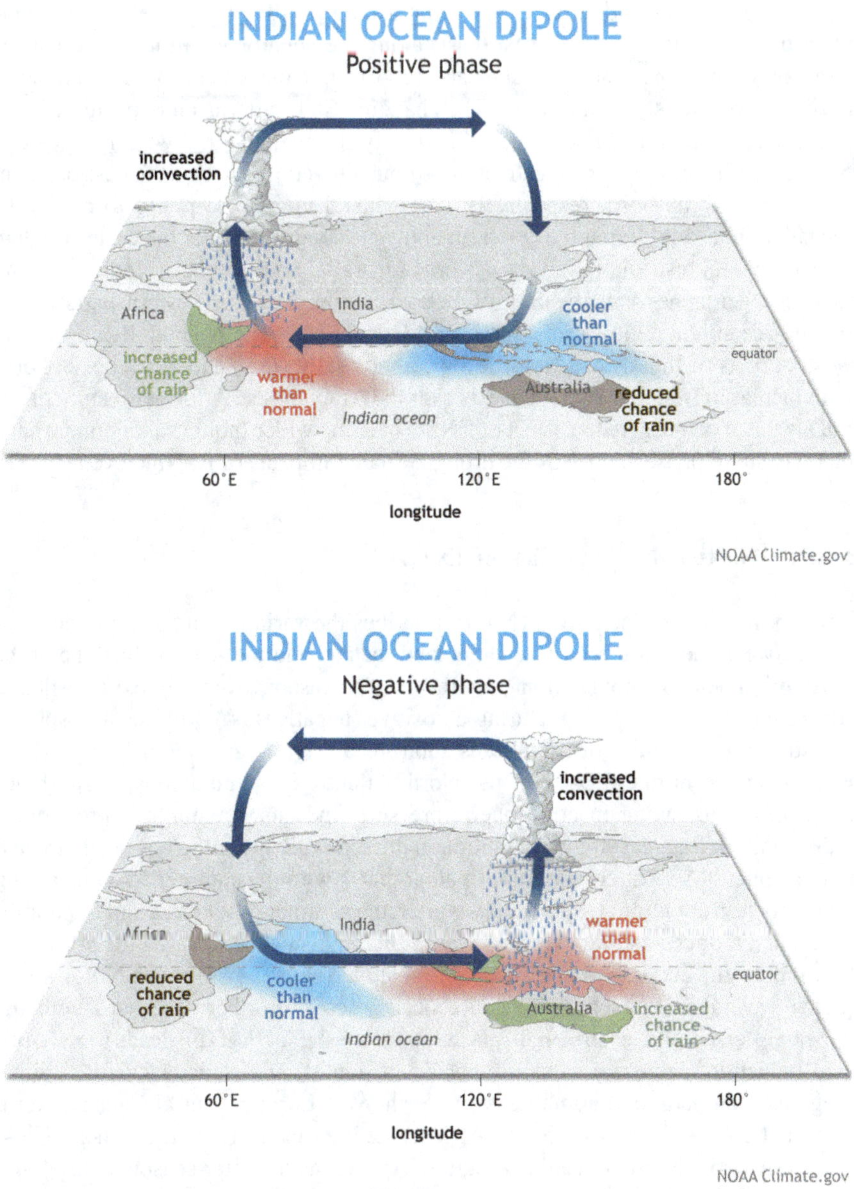

Fig. 5.7 The positive (top) and negative (bottom) phases of the Indian Ocean Dipole (IOD). The intervening neutral phase (not shown) is a 'weaker' version of the negative IOD phase. Adapted from: ENSO Blog Team at climate.gov https://www.climate.gov/news-features/blogs/enso/meet-enso%E2%80%99s-neighbor-indian-ocean-dipole#comments

of winds across the Indian Ocean. During a IOD positive (negative) phase winds blow from the east (west) to west (east) along the equator with the result that the Arabian Sea becomes anomalously warm (cool) while the waters south of Indonesia become anomalously cool (warm). The changing SST patterns on opposite sides of the Indian Ocean are related to the anomalously strong equatorial westerly (positive phases) or easterly winds (negative phase) pushing cool water to the east or warm water west of the basin respectively. The general climate response to a positive (negative) phase is cool and dry (warm and wet) conditions over the eastern Indian Ocean region, especially north-western Australia and southern Indonesia and reduced cyclogenesis in the Bay of Bengal, while over the western parts of the Indian Ocean in a region stretching from Madagascar to the Arabian Peninsula and western parts of India, conditions become abnormally warm and wet (cool and dry) with increased (reduced) cyclogenesis over the Arabian Sea. A further feature of the IOD is that it can negate the effect of ENSO over the wider Indian subcontinent such that a positive phase supresses the drought related impacts of El Niño events.

5.2.2.3 North Atlantic Oscillation (NAO)

The North Atlantic Oscillation (NAO) describes the variation in atmospheric pressure between the broad region of Greenland/Iceland and the Azores. It is one of the most prominent teleconnection patterns in all seasons over the North Atlantic (Barnston & Livezey, 1987). During a positive (negative) NAO phase atmospheric pressure in the vicinity of the Azores subtropical high is anomalously high (low) while over the northern parts of the North Atlantic it is anomalously low (high) (Fig. 5.8). This swing in atmospheric pressure finds its maximum expression in winter and for this reason most attention in the scientific literature has been directed to the winter NAO pattern and its climate impacts with the summer NAO receiving relatively less attention. Compared to winter, the summer NAO has a smaller amplitude and a reduced spatial extent as well as being displaced towards the north-east (Hurrell et al., 2003). For the case of a positive (negative) summer NAO, above (below) average atmospheric pressure occurs across northern latitudes with the upper westerlies and jet stream displaced to the north (south). This leads to a significant reduction (increase) in cloud cover and rainfall and a tendency for positive (negative) temperature anomalies over North-West Europe, central Canada, North Central USA, and parts of North Africa (Mediterranean region, southern USA, North Mexico, and the Greenland region). As well as the inter-seasonal variability in the NAO, considerable inter-annual variability can also occur with positive and negative phases lasting several years. For example, the mid-1950s to 1960s and 2000 to 2020 were dominated by a negative summer NAO with the intervening period characterised by considerable inter-annual variability (Fig. 5.9).

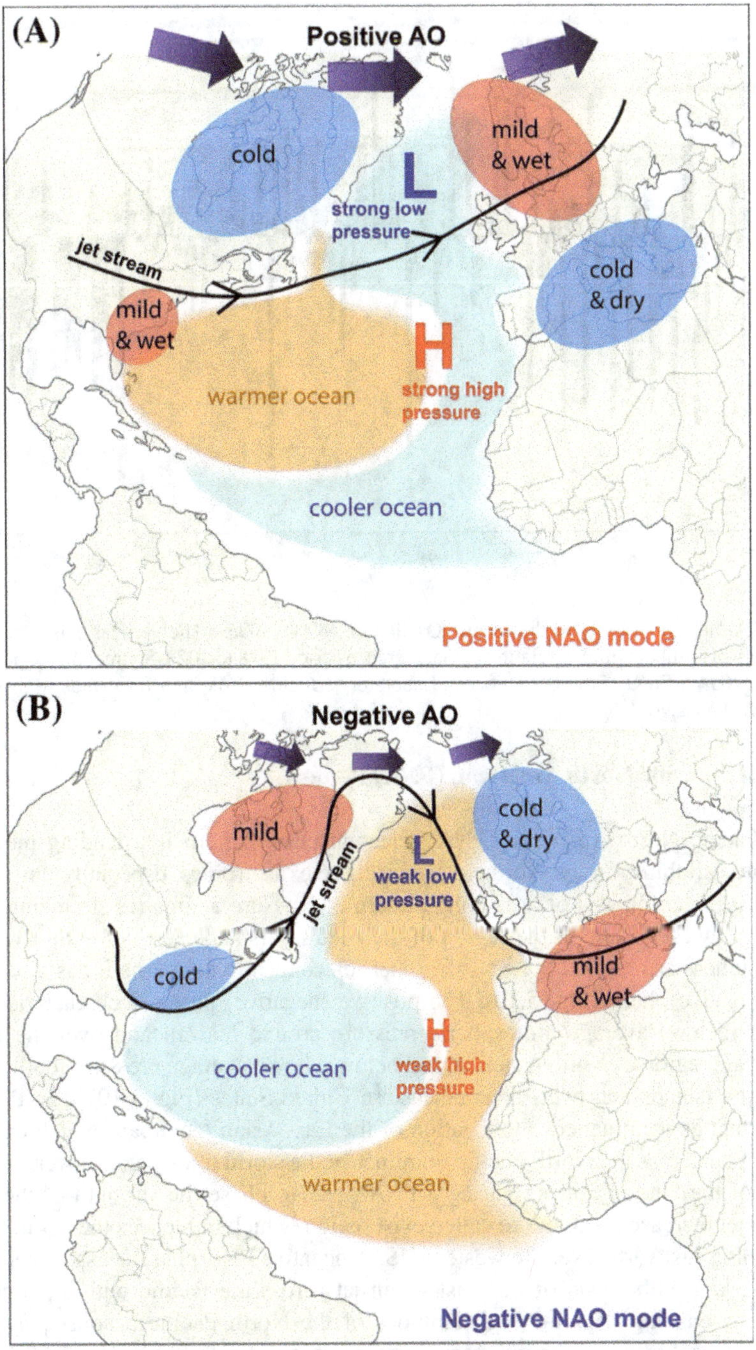

Fig. 5.8 Two phases (modes) of the North Atlantic Oscillation. (**a**) positive NAO phase, (**b**) negative NAO phase. (Source Báez et al. (2021) North Atlantic Oscillation and fisheries management during global climate change. Rev. Fish Biol Fisheries 31, 319–336 (2021). https://doi.org/10.1007/s11160-021-09645-z. Used with permission under Springer Licence 5523390774596)

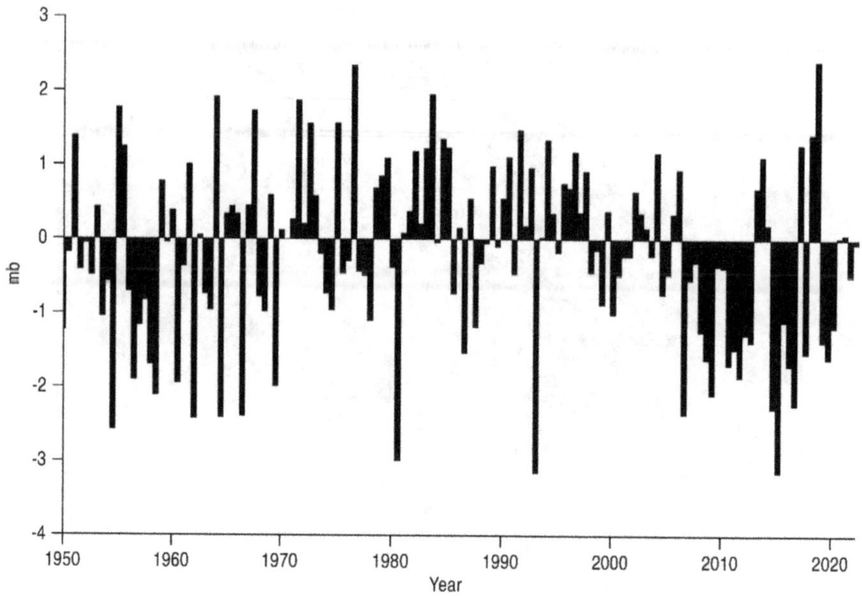

Fig. 5.9 July to August North Atlantic Oscillation (NAO) Index. The vertical axis is pressure anomaly in millibars (mb) in relation to the reference period 1948 to 2023. Source: Image provided by the NOAA/ESRL Physical Sciences Laboratory, Boulder Colorado from their web site at https://psl.noaa.gov/

5.2.2.4 Pacific North American (PNA) Pattern

The Pacific North American (PNA) teleconnection pattern is a leading mode of climate variability in the Northern Hemisphere extra-tropics, especially during the winter. It is composed of a tri-polar pattern of pressure anomalies stretching from the Northwest Pacific in the region of the Aleutian Islands across the mid-west to the southeastern United States. Like other teleconnection patterns it has a positive and negative phase (Fig. 5.10). The positive (negative) phase is characterised by above (below) average atmospheric pressure around Hawaii and over the inter-mountain region of North America, and below (above) average pressure south of the Aleutian Islands and over the southeastern United States (Fig. 5.10). The PNA is important as it influences the location of the East Asian jet stream which extends from Asia across the North Pacific to influence the storm track pattern over western parts of the United States. During a PNA positive phase the strength of the East Asian jet increases with the so called exist region, which influences the storm track, extending eastwards over the western US. A negative PNA phase is associated with a westward withdrawal of the East Asian jet across the Pacific with a pattern of blocking ensuing over the high latitudes of the North Pacific a consequence of which is a split flow of the jet stream over the central North Pacific. While the PNA pattern is largely absent in the summer and likely to have little effect on summer heatwave occurrence, a positive PNA increases the likelihood of anomalously high

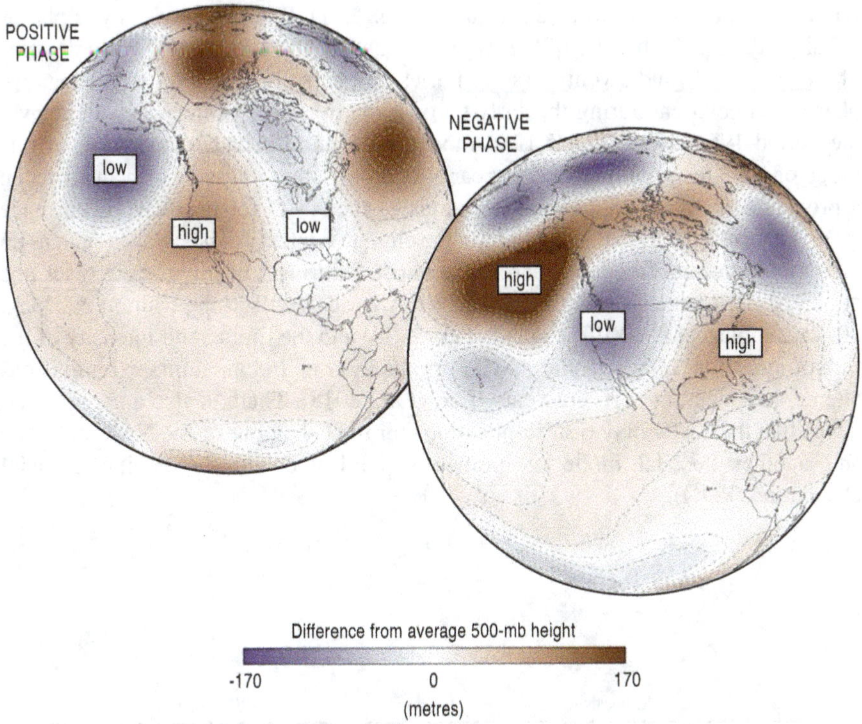

Fig. 5.10 Positive (left) and negative (right) phases of the Pacific North American (PNA) pattern represented by atmospheric pressure anomalies compared to the 1981–2010 average. (Adapted from NOAA Global Climate Dashboard at https://www.climate.gov/climatedashboard)

winter temperatures over western Canada and the far west of the United States and thus the possibility of winter warm spell events. As for other teleconnection patterns the PNA does not operate independently with positive (negative) phases of the PNA tending to be associated with El Niño (La Niña) events.

5.2.3 Decadal

5.2.3.1 Atlantic Multidecadal Oscillation (AMO)

The Atlantic Multidecadal Oscillation (AMO) refers to the slow variation of sea surface temperature across the North Atlantic Basin with the North Atlantic transitioning between generally 'warm' and 'cool' phases on a time scale of approximately 30 years (Kerr, 2000). This decadal SST pattern differs from the tri-polar SST pattern associated with the NAO at the inter-annual timescale. Positive (negative) SST anomalies are referred to as the AMO positive (negative) or warm (cool) phase with the strongest anomalies occurring in the subpolar region compared to relatively weak ones in tropical latitudes (Fig. 5.11). Warm positive AMO phases occurred during the middle of the twentieth century and the late and early decades of the twentieth and twenty-first centuries respectively, while cool or negative phases were evident during the early parts of the twentieth century as well as over the period 1964–1995 (Fig. 5.12). Although the multi-decadal time scale of this mode of climate variability may appear at odds with the timescale of heatwaves, in a probabilistic framework the implications of the AMO for heatwave occurrence manifest through the AMO's impact on the climate context for heatwaves such as an increase in drought frequency across the mid-US during a positive phase (McCabe et al., 2004), warmer summers over North America and Europe (Sutton & Dong, 2012; Zampieri et al., 2017; Vigaud et al., 2018) and the timing and intensity of the Indian monsoon, although the AMO's modulation of Indian summer rainfall has altered somewhat over recent years (Luo et al., 2018). There is also emerging evidence that the AMO may bear implications for the behaviour of the NAO and PNA and another decadal mode of climate variability called the Pacific Decadal Oscillation (PDO).

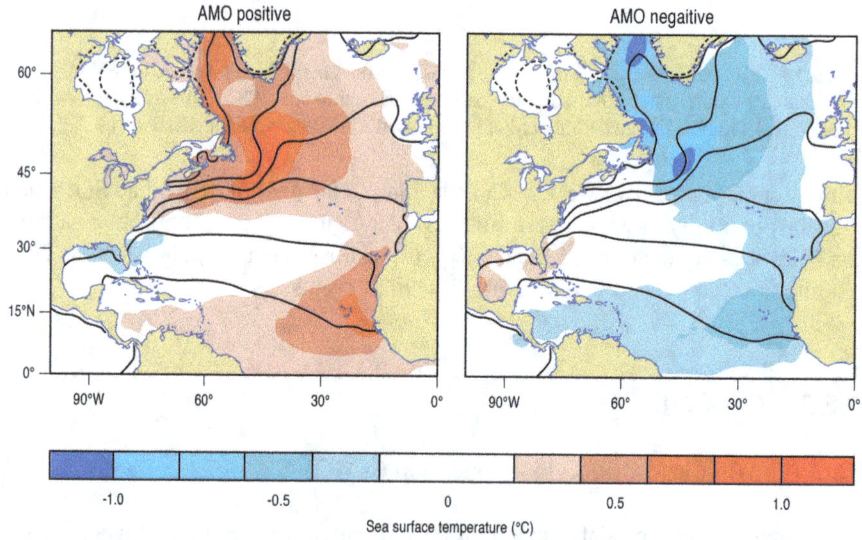

Fig. 5.11 Two phases of the Atlantic Multidecadal Oscillation (AMO) as represented by sea surface temperature (SST) anomalies. The black contours represent the January mean field for SST whereas the shading depicts the SST anomaly compared to a 72 year climatology for the positive phase (left panel) and negative phase (right panel). (Source: Yamamoto and Palter (2016). The absence of an Atlantic imprint on the multidecadal variability of wintertime European temperature. Nat Commun 7, 10,930 (2016). https://doi.org/10.1038/ncomms10930. Used under Creative Commons Attribution 4.0 International License http://creativecommons.org/licenses/by/4.0/)

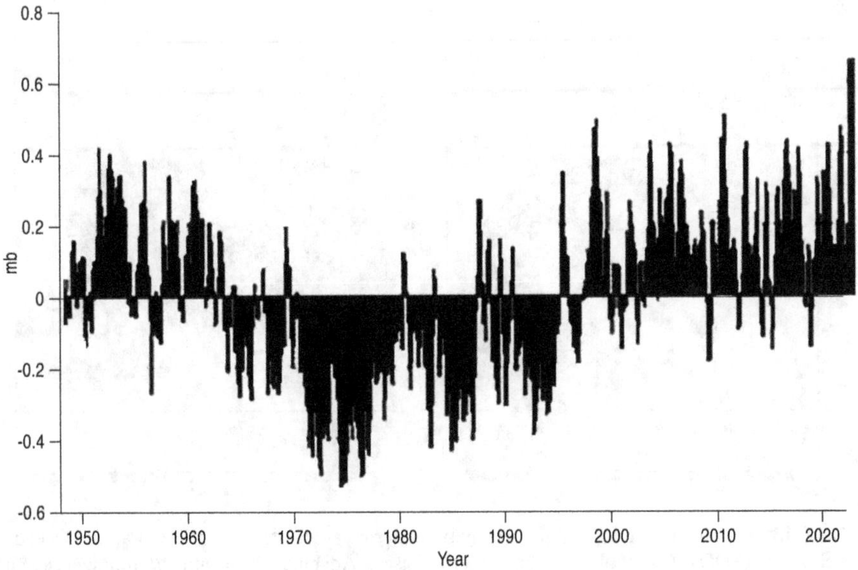

Fig. 5.12 Time series of Atlantic Multidecadal Oscillation (AMO) index January to December 1948–2023. The vertical axis is sea surface temperature anomaly in relation to the reference period 1948 to 2023. The zero-reference line distinguishes between warm (positive values) and cool (negative values) phases of the AMO. (Source: Image provided by the NOAA/ESRL Physical Sciences Laboratory, Boulder Colorado from their web site at https://psl.noaa.gov/. Source data set is Kaplan (SST))

5.2.3.2 Pacific Decadal Oscillation (PDO) and Inter-decadal Pacific Oscillation (IPO)

Like the AMO, the Pacific Decadal Oscillation (PDO) refers to multi-decadal variations of SST conditions but for the north Pacific Basin, typically north of 20°N (Mantua et al., 1997). A warm (cool) phase of the PDO corresponds with positive (negative) SST anomalies off the west coast of North America in the northeast Pacific in tandem with negative (positive) SST anomalies extending from Japan out into the central and western North Pacific. Positive (negative) SST anomalies across the central and eastern tropical Pacific are also a feature of the PDO warm (cool) phase. (Fig. 5.13). This spatial pattern of SST anomalies resembles that of ENSO with the PDO warm (cool) phase like the SST pattern associated with El Niño (La Niña). Periods of a predominantly warm or positive PDO phase during the twentieth century include 1926–1946, 1977–1997 with cold phases evident for 1905–1925, 1947–1976 and 1998–2018 (Fig. 5.14). The implications of the PDO for heatwave occurrence manifest through indirect SST related alterations of atmospheric circulation/weather regimes at the inter-annual time scale. Having said that, the impact of PDO phase on the likelihood of heatwave occurrence in any 1 year will be very much dependent on its interaction with other modes of climate variability such that the impacts associated with El Niño (La Niña) are enhanced during a PDO warm

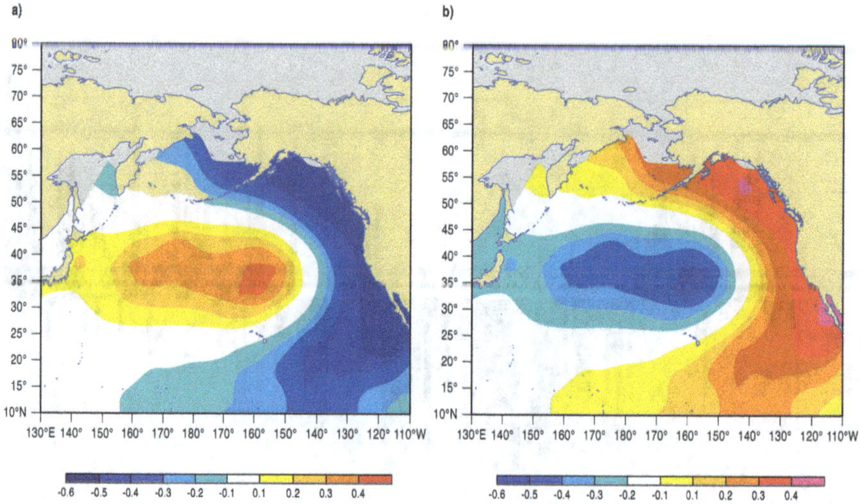

Fig. 5.13 Sea surface temperature anomaly patterns associated with the Pacific Decadal Oscillation (PDO) (**a**) cool and (**b**) warm phases. Adapted from https://optimizeias.com/cyclone-frequency-may-rise-over-indian-coast-from-the-warming-of-pacific-study/

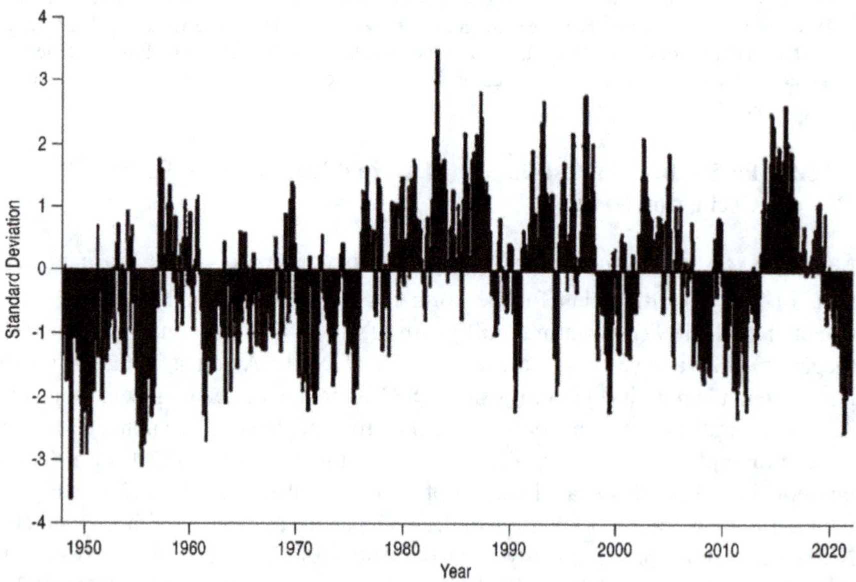

Fig. 5.14 Time series of Pacific Decadal Oscillation (PDO), January to December 1948–2022. The zero-reference line distinguishes between warm (positive values) and cool (negative values) phases of the PDO. Vertical axis is units of standard deviation. (Source: Image provided by the NOAA/ESRL Physical Sciences Laboratory, Boulder Colorado from their web site at https://psl.noaa.gov/)

(cool) phase (Lupo et al., 2007). As generally during a PDO warm phase, El Niño induces spatially more extensive and severe droughts relative to a cold PDO phase (Wang et al., 2014), theoretically the potential for an increased probability of heatwaves embedded in a period of drought is enhanced when an El Niño event is in phase with a PDO warm event.

In contrast to the PDO, the IPO refers to variations in SST and sea level pressure with a cycle of around 15–30 years, across both the North and South Pacific basins. Maximum SST anomalies are found in tropical latitudes of the Pacific away from the equator. The IPO should not be confused with the quasi-decadal oscillation which refers to maximum SST anomalies straddling the equator with a periodicity of around 8–12 years.

5.3 Atmospheric Heatwaves and Modes of Climate Variability

The idea put forward in this chapter is that alterations to the atmospheric circulation that lead to temperature extremes and heatwave conducive weather regimes can often be traced back to interactions between the ocean and atmosphere or modes of climate variability at the intra-seasonal to decadal time scales. There is evidence that extreme maximum temperatures can be significantly influenced by ENSO for a range of regions across the world (Arblaster & Alexander, 2012; Donat et al., 2014; Hoy et al., 2013; Kenyon & Hegerl, 2008) as well as Madden Julian Oscillation related anomalies in tropical convection (Cassou et al., 2005; Matsueda & Takaya, 2015). Similarly, the NAO has been found to influence the occurrence of high temperature extremes across Europe (Hoy et al., 2013; Kenyon & Hegerl, 2008; Scaife et al., 2008). Given these general findings the purpose of this section is to explore specific links between modes of climate variability and temperature extremes and heatwaves for several broad geographical regions.

At the global scale Kenyon and Hegerl (2008) have shown that the impact of ENSO on temperature extremes is most obvious for regions bordering the Pacific Ocean. This is particularly so for the western Pacific where temperature extremes, especially maximum and minimum temperatures, are very much dependent on the ENSO cycle through the impact of SST anomalies on surface air temperatures (Whan et al., 2014) and weather regimes that are likely to generate temperature extremes (Lorrey & Fauchereau, 2018). For Hawaii in the North Pacific, Zhu and Li (2017) have shown for the exceptionally hot 2015 summer, when there was 98 exceptional local heatwave days, that an El Niño-like pattern in the tropics and low (high) SST anomalies over the western (eastern) North Pacific, were the primary contributors to abnormally high temperatures arising from an anomalous southwesterly flow over Hawaii and an associated slackening of the normal cooling northeasterly trade winds.

For extreme temperature days and months over North America, the PNA and the Northern Annular Mode (NAM) have the strongest forcing while comparatively ENSO is less important with overall relationships strongest for winter when synoptic-scale, transient weather disturbances are noteworthy drivers of extreme temperatures. During North American summers, large scale modes of variability are less influential as the partitioning of available energy between latent and sensible heat, as controlled by soil moisture content, is more important in determining temperature extremes (Loikith & Broccoli, 2014). However, modes of climate variability may play an indirect role in the development of summer heatwaves through priming the land surface. Hoerling et al. (2013) describe how anomalously low SST in the Pacific associated with a La Niña event were important for generating precipitation deficits, soil moisture decline and the onset of the Texas 2011 drought, during which a significant heatwave event occurred. A 'cold' Pacific Ocean and the role of La Niña and a negative phase of the PDO along with a positive AMO (warm Atlantic) in generating drought over Texas have been confirmed by Verdon-Kidd et al. (2017) and Deng et al. (2018b). Given this, the potential for US heatwaves may well be enhanced due to the interaction of drought inducing modes of climate variability through providing the requisite land surface conditions for generating land-atmosphere interactions that increase the likelihood of heatwaves embedded in periods of major drought (see Chap. 7). The role of large-scale teleconnections in modulating the occurrence of heatwaves through their impact on land-atmosphere interactions has been demonstrated via modelling studies. Ruprich-Robert et al. (2018) outline for northern Mexico and the southern United States how Atlantic multidecadal variability and Atlantic warming lead to an increased number of heatwave days by about 30% compared to Atlantic cooling. A warmer Atlantic favours subsidence over this region leading to relative humidity, cloud cover, and precipitation decreases, all of which are conducive to soil drying and a change in the ratio of sensible to latent fluxes. Using modelling, Vigaud et al. (2018) also corroborate the role of multi-decadal Atlantic variability in producing periods of extreme high temperatures. Beyond the indirect role of inter-annual to decadal modes of variability increasing the likelihood of temperature extremes, clear evidence is emerging of the direct link between intra-seasonal oscillations and heatwave occurrence in North America. For example, strong Madden-Julian Oscillation (MJO) phase pairs of 8 and 1 or 2 and 3 and convection in the Indian Ocean and Southeast Asia and eastern tropical Pacific have been shown to precede extreme Californian summertime heatwaves (Grotjahn et al., 2016; Lee & Grotjahn, 2019; Wang et al., 2024), while convection anomalies associated with the western Pacific intra-seasonal oscillation generate significant surface air temperature anomalies over North America (Yang et al., 2020).

Over the Meso-Americas—Caribbean region there is emerging evidence that extreme heat events, which occur in the wet season reaching a peak in August, may be related to the La Niña phase of ENSO, but this relationship is not consistent for all years with a high frequency of heat events (Ramirez-Beltran et al., 2017). This points to the complexity of the weather and climate patterns that are likely to drive heatwave generation in this region with the interplay between a range of factors that determine the nature of the wet season such as, tropical Pacific and Atlantic SST, the seasonal progression of the ITCZ, the North Atlantic (Azores) Subtropical High, the NAO, the Caribbean Low Level Jet and the strength of the easterlies and vertical wind shear over the Caribbean, all likely to be important for explaining heatwave occurrence (Angeles et al., 2010; Jury, 2016; Moron et al., 2015; Torres-Valcarcel, 2018).

For South America there is considerable geographical dependence of the impact of modes of climate variability on temperature extremes. Across northwestern South America ENSO exerts a strong influence while for other parts of northern South America monthly extreme high temperatures appear to be related to anomalously high SST in the Atlantic equatorial region. While the AMO appears to have little influence on the extreme temperature climatology of South America, the PDO exerts a noticeable influence on temperature extremes for the months of June to August in tandem with ENSO, most likely because the PDO is closely associated with ENSO (Loikith et al., 2017). For Argentina, El Niño conditions favour high minimum temperatures throughout the year, whereas maximum extreme temperatures only appear to be forced by El Niño during the months of July to September (Rusticucci et al., 2016), a finding that resonates with that of Loikith et al. (2017) for the broader northern half of South America. With regards to specific heatwave events, El Niño is implicated in the development of the early February 2010 event in southeast Brazil. As described by Geirinhas et al. (2019), this heatwave was associated with a quasi-stationary mid-tropospheric anticyclonic anomaly over the South Atlantic which may be attributed, to a considerable extent, to the remote forcing from an anomalous SST pattern over the central Pacific Ocean associated with El Niño. For South American extra-tropical latitudes, SAM appears to be the dominant mode of variability in terms of influencing temperature extremes in association with transient weather systems and in particular its role in intensifying wave activity in the westerlies which enhance the formation of high-pressure anomalies across South America (Loikith et al., 2017; Zhang & Xie, 2023).

Across Europe and the broad Eurasian region, the NAO generally affects all year round temperature extremes (Kenyon & Hegerl, 2008). Studies that explicitly consider the influence of the summer NAO (SNAO) have provided some insight into the influence of modes of climate variability in the North Atlantic sector on temperature extremes. Although not identified as such, the patterns of sea level pressure variability identified by Della-Marta et al. (2007) as being important for European heatwaves resemble those associated with a positive SNAO phase, that is, a widespread anomalous high pressure over the Scandinavian region. Although principally concerned with the impact of the SNAO on precipitation, the work of Bladé et al. (2012) has implications for understanding the occurrence likelihood of elevated summer temperature extremes because of the fundamental impact of land surface conditions (soil moisture) on the ratio of sensible to latent heating. They found during summers with a strong positive SNAO, strong anticyclonic conditions and suppressed precipitation prevail in the region of the UK/Scandinavia, while across the Mediterranean region anomalously wet conditions predominate. This set of circumstances would increase the likelihood of elevated (suppressed) summer temperatures for northwestern (southern) Europe. This contention is supported by the observed impact of the SNAO on temperature extremes across the Iberian Peninsula (IP) where a negative SNAO leads to a lowering of the maximum temperatures in the northwestern and northern regions of the IP and to a strengthening of the north-south IP maximum temperature gradient (Favà et al., 2016). Acknowledging that no single mode of climate variability might account for summer heatwaves, Della-Marta et al. (2007) found that the concurrent states of the SNAO and AMO, the preceding SST conditions in both the North Atlantic and Mediterranean Basins and land surface conditions all interacted to explain the occurrence of heatwaves across Europe. Interestingly in a study that assesses the drivers of long-term changes in heatwaves over the UK, Sanderson et al. (2017) suggest that the summertime North Atlantic Oscillation plays a less significant role in determining heatwave numbers and length compared to the Atlantic Multidecadal Oscillation, while at the inter-annual timescale Hall and Hanna (2018) note the more straight forward role of the NAO in influencing summer temperature anomalies compared to the same for winter. Pointing to the significance of distant forcing for altering the likelihood of temperature extremes in France and Egypt, Boé et al. (2010) note the importance of inter-annual variations in the position of the Inter-Tropic Convergence Zone (ITCZ) possibly influenced by the MJO. The NAO has been found to partly influence temperature extremes over Turkey by moderating the frequency of weather systems with a tropical origin and continental tropical dry and hot air masses (Turkes & Erlat, 2018). For western Russia, where the mega-heatwave of 2010 occurred, Sun et al. (2016) note the increasing importance of an inverse association between ENSO and temperature extremes during the past three decades. Here the ENSO signal in surface temperature is via El Niño related tropical convection anomalies over the Pacific. These instigate a Rossby wave that extends towards western Russia increasing the likelihood of blocking and warm air advection in summer. For the broad Eurasian region Zhou and Wu (2016) found that an inter-decadal mode of heatwave frequency, centred around eastern Europe-central Asia and Mongolia-southwestern China, is influenced by the interplay between mega-ENSO events and the AMO, while at the inter-annual timescale a tri-polar mode of heatwave frequency, centred over northwestern Europe, northeastern Asia and East Asia, is more related to just mega-ENSO events. Other teleconnections important for Eurasian temperature extremes relate to remote forcing from the Indian Ocean-western Pacific region where a boreal summer intra-seasonal oscillation in the form of periods of enhanced and suppressed tropical convection have been found to influence extreme temperature events across the Northern Hemisphere (Dai & Bloecker, 2019). This resonates with the findings of Cassou et al. (2005) Matsueda and Takaya (2015) who point to the significance of tropical heating anomalies for generating extreme temperature events in the Northern Hemisphere mid-latitudes.

ENSO appears to play a role in the occurrence of heatwaves across the Yellow River Valley in China. For here, Chen and Zhou (2018) have shown that around 66 percent of the total variability of the July–August heatwaves can be attributed to anomalous SST forcings; the remaining third or so are due to internal variability. Of the SST-forced heatwaves, about half can be attributed to the impact of anomalously warm Pacific SSTs with the other half associated with non-ENSO related SST forcings in the North Pacific and the North Atlantic. The key influence of ENSO on heatwaves and temperature extremes over the Yellow River Valley is via the extension of the western North Pacific Subtropical High (WNPSH) westwards over the region, the likelihood of which increases during a decaying El Niño and developing La Niña along with a warm Indian Ocean and cold central Pacific, respectively. Yoon et al. (2024) have also noted the role of the WNPSH in the development of the 2013 heatwave event across southern China. The importance of the WNPSH is also visible through the influence of the Pacific-Japan (PJ) teleconnection pattern on heatwave occurrence across East Asia. For example, El et al. (2021) find a 90 percent increase in the occurrence of extreme heatwave days at locations across Korea and Japan during a positive phase of the PJ teleconnection pattern (strong WNPSH) noting that enhanced warm air advection, with increased insolation and adiabatic heating associated with a PJ positive phase drive temperatures to extreme levels. Cha et al. (2022) also note the influence of the PJ pattern on the inter-annual variability of heatwave days for Korea. For the record-breaking 2018 heatwave across Japan, Korea and northeast China it appears that an abnormally enhanced MJO over the western Pacific and its associated diabatic heating may have set into motion a wave train of atmospheric anomalies leading to an intensification of the all-important heatwave controlling WNPSH (Hsu et al., 2020).

Across the Sahel region of Africa, it appears that of a range of tropical modes of climate variability, the MJO is the most important influence on heatwave occurrence. Here, nighttime heatwaves are more impacted by the MJO over the eastern Sahel while daytime heatwaves are more affected over the western Sahel with the MJO influencing atmospheric moisture and longwave radiation and the advection of hot dry air (Guigma et al., 2021). The MJO also appears to play a role in the occurrence of Sahelian spring heatwaves but less so their intensity (Guigma et al., 2021). For the 2015–2016 South African heatwave event it is likely that ENSO related positive SST anomalies over the Indian Ocean that initiated precipitation deficits and drove down soil moisture content in concert with anomalously high atmospheric pressure were the main drivers (Wehrli et al., 2019). The importance of intense ENSO-induced drought events for high heatwave frequency over South Africa, such as during the 1982–1983, 1991–1992, and 2015–2016 ENSO events has also been noted by Mbokodo et al. (2023). Interestingly despite 1997–1998 being one of the strongest El Niño seasons over South Africa, heatwave frequency and intensity for this El Niño event was moderated by the development of an intense Angola low, which promoted the development of intermittent cloudy and wet conditions (Mbokodo et al., 2023).

For the Northern Hemisphere as a whole teleconnection patterns at the hemispheric scale in the form of cores of high pressure over eastern Europe, northeastern Asia, the eastern North Pacific, the eastern United States, and Greenland have been revealed as important drivers of temperature extremes. Resembling a wave-5 pattern in the upper westerlies, these centres of high pressure are frequently associated with surface temperature extremes across Eurasia. (Lee et al., 2017). For the 2018 boreal summer, Kornhuber et al. (2019) found a link between simultaneous heatwave occurrence in North America, Western Europe, and the Caspian Sea region and an amplified northern hemisphere-wide wavenumber-7 circulation pattern. Further they note this teleconnection pattern mirrors that associated with previous European heatwaves in 2003, 2006, and 2015, also finding a noticeable increase in the frequency of the wave-7 pattern in recent decades. Li and Ruan (2018) and Li et al. (2019) also note the importance of an upper level hemispheric-wide atmospheric pressure teleconnection pattern for the generation of temperature extremes over China as do Deng et al. (2018a) for eastern Europe and Northern China, while (Fragkoulidis et al., 2018) point to the significance of large-amplitude perturbations in the upper troposphere, in the form of Rossby wave packets, for increasing the probability of lower-tropospheric temperature extremes over the northern hemisphere mid-latitudes.

For the Australian continent where extreme summer heat has far reaching societal consequences, ENSO, the IOD, SAM as well as the MJO modulate the likelihood of summer temperature extremes. For example, anomalously drier and hotter conditions occur over northeastern Australia and the southern coast during El Niño and a positive phase of the IOD, while wetter and cooler conditions are observed during La Niña and a IOD negative phase (Min et al., 2013). Because a positive (negative) phase of the SAM brings wetter and cooler (drier and warmer) conditions over much of the eastern continent in summer, the probability of exceedance of the 90th percentile of weekly-mean maximum temperature anomalies is reduced (enhanced) during a SAM positive (negative) phase (Marshall et al., 2014). Although there appears to be little evidence for concurrent impacts of the IOD and SAM on heatwave occurrence in tandem with ENSO, analyses that have considered the impact on summer heatwaves of modes of climate variability in preceding seasons indicate that the state of the winter IOD may weakly affect heatwave occurrence in northeast Australia, while heatwave timing in central eastern Australia is affected by SAM conditions (Loughran et al., 2017a). This lag effect may well be moderated by the impact of IOD, SAM and indeed ENSO on soil moisture conditions and land atmosphere feedback mechanism that might enhance heating (Alexander, 2011). Regarding the MJO, the observed probability of heat events varies according to MJO phase and time of year, with the greatest impact of the MJO on extreme heat occurring over southern Australia in spring during MJO phases 2–3 (Marshall et al., 2014). For summer heatwaves over southeastern Australia, Parker et al. (2014) have shown that their likelihood is increased during MJO phases 3–6. As well as influencing the general probability distribution of heatwave events with La Niña (El Niño) summers possessing fewer (more) heatwave days overall, ENSO has been shown to influence a range of heatwave characteristics (Loughran et al., 2017a; Perkins et al., 2015) across Australia. In general, longer (shorter) heatwaves tend to be associated with El Niño (La Niña) in northern Australia with the events with the greatest (least) temperature anomalies occurring during El Niño (La Niña) in inner southeast Australia. Further, during El Niño (La Niña) summers, the heatwave season starts earlier (later) (Loughran et al., 2017a, 2019; Perkins et al., 2015). Over the east coast of Australia, where there are major population centres, the interplay between ENSO phase and the position of the Tasman high pressure system appear instrumental in determining heatwave outcomes such that during El Niño (La Niña) years, the Tasman High moves north (south) resulting in intense (less intense) heatwaves over southeast Australia. This kind of connection points to the link that exists between timescales of atmospheric variability such that large scale modes of climate variability may provide the atmospheric setting for heatwave-conducive weather regimes (Boschat et al., 2015; Pezza et al., 2012). Although conceptually the two types of El Niño (El Niño Modoki and Eastern Pacific El Niño) may generate different heatwave patterns and characteristics, there appears to be minimal difference for these in the case of Australian heatwaves (Loughran et al., 2017b).

Given that Indochina lies close to the western dipole of the ENSO it is perhaps no surprise that this mode of climate variability has an impact on heatwave occurrence in this region. Here the preceding winter El Niño (La Niña) conditions have been found to amplify (weaken) heatwaves across the majority of Indochina by increasing (decreasing) their frequency and prolonging (shortening) their duration, as well as elevating (reducing) their amplitude (Lin et al., 2018). Heightened heatwave activity during an ENSO warm phase is related to a regime of subsidence of anomalously dry hot air over the western Pacific resulting in suppressed precipitation and surface drying (Luo & Lau, 2018). Further to the west over the Indian subcontinent, El Niño also plays a role in increasing the likelihood of heatwave occurrence by delaying the onset of the Indian summer monsoon as manifest via a weakening of south westerlies over the Arabian Sea and the occurrence of many clear sky days. As a result, the pre-monsoon period of intense heat is prolonged (Murari et al., 2016). While El Niño 's role in heatwave occurrence in the broad region of the Indian subcontinent and Indochina appears to be quite straight forward, it is likely that El Niño—IOD interactions (Lestari & Koh, 2016) and other teleconnections will modulate this supposedly simple relationship. For example, Zheng et al. (2017) demonstrate how the positive and negative phases of an Indo-Pacific convection oscillation (IPCO) and high summer (June–August) surface air temperatures across southern Asia are related. When convection over the western North Pacific (north Indian Ocean) is enhanced (suppressed) in a positive (negative) phase, the Southeast Asian summer monsoon is vigorous (weakened) and southern Asia is cooler (warmer) than normal with this association greatly enhanced during the decaying stages of an ENSO event.

5.4 Marine Heatwaves and Modes of Climate Variability

Marine heatwaves as defined earlier refer to periods of anomalous temperature in the near surface layers of the ocean as captured by SST measurements. Given there is link between large scales modes of climate variability and atmospheric heatwaves unsurprisingly the same applies in general to the case of marine heatwaves. This is because modes of climate variability such as those listed in Table 5.1 may work to influence processes at the local scale or act as 'remote forcers' of atmosphere and ocean processes at distant places (Holbrook et al., 2019; Oliver et al., 2021; Sen Gupta et al., 2020). Modes of climate variability relevant to MHW formation, mainly reside between the intra-seasonal and inter-annual timescales and spatial scales of 10^1 km and 10^4 km (Holbrook et al., 2019). That said, decadal variability may affect the extent to which a given intra-seasonal to interannual mode may provide the requisite large-scale climate setting for MHW occurrence. While the following material hints at some of the ocean surface atmosphere exchanges involved in the generation of MHW these will be expanded upon in Chap. 7.

Using a consistent methodology for identifying MHW, Holbrook et al. (2019) identified the major modes of climate variability that increase the likelihood of MHW occurrence noting that compound effects are most likely at play in the generation of MHW. For the tropical latitudes of the Pacific and Indian Ocean basins, ENSO and its associated teleconnections are pivotal in the occurrence of MHW. While the IOD influences MHW occurrence in the Indian Ocean basin, it has little influence on those in the Pacific Ocean basin (Holbrook et al., 2019). For the western equatorial Indian Ocean, considered to be a MHW hotspot, a high incidence of MHW days associate with a positive IOD, an El Niño or a compound IOD/El Niño event (Qi et al., 2022). Anomalously high sea surface levels and anomalous equatorial easterly winds during these modes of climate variability are conducive to the westward propagation of downwelling equatorial Rossby waves that lead to upper-ocean warm water convergence all of which conspire to suppress cool water upwelling, the generation of anomalously high SST and MHW occurrence (Qi et al., 2022). Specifically for the Mozambique Channel, lying within the wider western equatorial Indian Ocean region, Mawren et al. (2022) found that long duration, intense January to March MHW are linked to a strong positive phase of the Subtropical Indian Ocean Dipole (SIOD). The importance of this mode of climate variability is exemplified by the extreme 2017 summer Mozambique Channel MHW event. Lasting for 48 days and reaching a maximum intensity of 3.44 °C above the long-term average SST, the uniqueness of this MHW was due in large part to how the strong positive SIOD weakened the Mozambique Channel trough and facilitated the westward extension of the Mascarene High, leading to a slackening of winds and decreased ocean surface layer mixing and clear sky related strong positive insolation anomalies. Further to the east in the Bay of Bengal there appears to be some seasonal dependency of the sole or joint influence of modes of climate variability on MHW occurrence (Lin et al., 2023). Here ENSO and the NAO account for 50 to 70 percent of the inter-annual variability of MHW (Gao et al., 2022). However, a closer consideration of the spatial and temporal dynamics of MHW occurrence reveals a greater influence of El Niño on MHW occurrence across southern latitudes of the Bay as El Niño waxes and wanes from the boreal autumn to summer months with only a weak signature in the northeast of the Bay over the months of April to August during the decaying phase of El Niño. In contrast, the NAO finds it maximum expression in the northeast of the Bay over the months of March to June with a strong negative phase of this mode of variability conducive to a high incidence of MHW (Gao et al., 2022). For China's marginal seas and offshore waters (Yao et al., 2020) have found MHW to be associated with strong El Niño events. Through decreased cloud cover and weaker winds, solar radiation inputs to the sea surface are enhanced and ocean heat loss is weakened respectively. An El Niño associated weakening but warmer Kuroshio current also appears to be important. Because the PDO is effective in modulating the frequency and intensity of El Niño events, the positive phase of this mode of variability can increase the likelihood of MHW occurrence in the tropical latitudes (Salinger et al., 2016; Scannell et al., 2016). At shorter timescales the MJO through its influence on air-sea coupling also appears important for MHW generation across the tropics (Holbrook et al., 2019; Maloney & Kiehl, 2002).

Multiple drivers influence MHW occurrence in the mid-latitude regions of the North Pacific. Here El Niño and positive phases of the PDO and IPO are important. In contrast, El Niño is the main determinant of the increased likelihood of MHW in the mid-latitudes of the South Pacific. For the northwest Atlantic region, there is a tendency for MHW to be associated with the negative phase of the NAO (Holbrook et al., 2019). In terms of processes in the respective ocean basins, El Niño and a negative NAO phase invoke atmospheric blocking. This precipitates subsidence warming and high surface radiative load, a decrease in surface wind stress and consequently surface layer mixing with this concoction highly conducive to rapid warming of upper ocean waters. Interestingly for regions associated with western boundary ocean currents, Holbrook et al. (2019) not only report a low incidence of MHW in these regions but only tenuous associations between MHW and modes of climate variability. Based on this finding they conclude that many MHW in these regions may be internally generated by local scale forcing. This certainly appears to be the case for MHW off the southeast coast of Queensland, Australia. Here for some ecologically important marine areas in sheltered and shallow water locations, the influence of El Niño is very weak with local air-sea exchanges associated with certain weather patterns more important than remote forcing (Heidemann & Ribbe, 2019). On the opposite side of the ocean basins where eastern boundary currents are found, positive SST anomalies in the Niño3.4 region enhance the possibility of MHW days off the coasts of North and South America. The likelihood of MHW days off the west coast of North America are also increased during a PDO positive phase (Holbrook et al., 2019). While positive Niño3.4 SST anomalies influence the occurrence of MHW off the west coasts of the Americas, negative Niño3.4 SST anomalies and a negative phase of El Niño Modoki (low SST in the central Pacific flanked by high SST on both sides of the equatorial Pacific) increase the likelihood of MHW off the west coast of Australia. A negative phase of El Niño Modoki also appears to be important for MHW occurrence off Peru as during this phase positive SST anomalies off the Peruvian coast are a feature and likely to suppress the upwelling of cool oceanic water to the surface (Varela et al., 2021). Further to the south in the waters off central Chile, while the opposite phases of the ENSO cycle generally increase the probability of intense MHW, the spatial distribution of mixed layer temperature anomalies are different for El Niño and La Niña. During El Niño, coastal waters demonstrate the greatest MHW related temperature anomalies whereas during La Niña MHW tend to occur offshore (Carrasco et al., 2023). Off western Africa MHW days are enhanced during a positive phase of the Atlantic Niño index (Holbrook et al., 2019).

Adding to the convolution of the relationship between modes of climate variability and MHW are the so-called Ningaloo Niño events. These occur during strong La Niña events when anomalously high sea levels in the western Pacific and strong trade winds cause Indonesian throughflow when warm western Pacific waters flow westward between Australia and the Indonesian archipelago. Eventually entering the eastern margins of the Indian Ocean, the throughflow joins the Indian Ocean south equatorial current or the Leeuwin current to the south. In the latter case, these anomalously warm waters 'supercharge' the heat content of the southward flowing Leeuwin current resulting in extreme positive SST anomalies as recorded for the 2000 and 2011 MHW off the coast of western Australia (Feng et al., 2013). That La Niña and associated Ningaloo Niño events are more common and persistent during the negative phase of the Interdecadal Pacific Oscillation (Ryan et al., 2021) adds to the complexity of the multiple timescales at which modes of climate variability operate to increase the likelihood of MHW in this eastern boundary current region. Although strictly an ocean mode of variability, it would appear for at least the last four decades, that the Atlantic Meridional Overturning Circulation has not influenced MHW occurrence over most of the global oceans (Ren & Liu, 2021).

5.5 Conclusion

This chapter has highlighted the critical role that modes of climate variability play in modulating the likelihood of atmospheric and marine heatwave occurrence across various timescales. By altering atmospheric and oceanic circulation patterns, these quasi-periodic oscillations can tip the balance towards conditions that either enhance or diminish the probability of extreme heat events in different regions. The strength of the association between specific modes of climate variability and heatwave incidence varies geographically, underscoring the complex interplay between local processes and remote drivers. In some instances, compounding effects from multiple modes operating in concert can further amplify or dampen heatwave probability, an area ripe for further exploration. While statistical analyses have unveiled many such linkages, a deeper understanding of the underlying physical mechanisms—the atmospheric and oceanic "bridges" that communicate changes across distant regions—remains a pressing research imperative. Ultimately, this chapter underscores the pivotal role of modes of climate variability in shaping the likelihood and characteristics of heatwaves. Unravelling how modes of climate variability substantively influence heatwave occurrence will bode well for endeavours focused on heatwave forecasting and prediction.

References

Alexander, L. (2011). Climate science: Extreme heat rooted in dry soils. *Nature Geoscience, 4*(1), 12–13. https://doi.org/10.1038/ngeo1045

Angeles, M. E., Gonzalez, J. E., Ramirez-Beltran, N. D., Tepley, C. A., & Comarazamy, D. E. (2010). Origins of the Caribbean rainfall bimodal behavior. *Journal of Geophysical Research-Atmospheres, 115*. https://doi.org/10.1029/2009jd012990

Arblaster, J. M., & Alexander, L. V. (2012). The impact of the El Nino-southern oscillation on maximum temperature extremes. *Geophysical Research Letters, 39*. https://doi.org/10.1029/2012gl053409

Báez, J. C., Gimeno, L., & Real, R. (2021). North Atlantic oscillation and fisheries management during global climate change. *Reviews in Fish Biology and Fisheries, 31*(2), 319–336. https://doi.org/10.1007/s11160-021-09645-z

Barnston, A. G., & Livezey, R. E. (1987). Classification, seasonality and persistence of low-frequency atmospheric circulation patterns. *Monthly Weather Review, 115*(6), 1083–1126. https://doi.org/10.1175/1520-0493(1987)115<1083:Csapol>2.0.Co;2

Bladé, I., Liebmann, B., Fortuny, D., & van Oldenborgh, G. J. (2012). Observed and simulated impacts of the summer NAO in Europe: Implications for projected drying in the Mediterranean region. *Climate Dynamics, 39*(3–4), 709–727. https://doi.org/10.1007/s00382-011-1195-x

Blamey, R. C., Kolusu, S. R., Mahlalela, P., Todd, M. C., & Reason, C. J. C. (2018). The role of regional circulation features in regulating El Nino climate impacts over southern Africa: A comparison of the 2015/2016 drought with previous events. *International Journal of Climatology, 38*(11), 4276–4295. https://doi.org/10.1002/joc.5668

Boé, J., Cassou, C., Terray, L., Parey, S., & Dubus, L. (2010). Link between large scale atmospheric circulation and heat waves for seasonal forecasting and climate change impact studies. *Houille Blanche-Revue Internationale De L Eau, 4*, 67–71. https://doi.org/10.1051/lhb/2010043

Boschat, G., Pezza, A., Simmonds, I., Perkins, S., Cowan, T., & Purich, A. (2015). Large scale and sub-regional connections in the lead up to summer heat wave and extreme rainfall events in eastern Australia. *Climate Dynamics, 44*(7–8), 1823–1840. https://doi.org/10.1007/s00382-014-2214-5

Cai, W. J., van Rensch, P., & Cowan, T. (2011). Influence of global-scale variability on the subtropical ridge over Southeast Australia. *Journal of Climate, 24*(23), 6035–6053. https://doi.org/10.1175/2011jcli4149.1

Capotondi, A., Wittenberg, A. T., Newman, M., Di Lorenzo, E., Yu, J. Y., Braconnot, P., Cole, J., Dewitte, B., Giese, B., Guilyardi, E., Jin, F. F., Karnauskas, K., Kirtman, B., Lee, T., Schneider, N., Xue, Y., & Yeh, S. W. (2015). Understanding ENSO diversity. *Bulletin of the American Meteorological Society, 96*(6), 921–938. https://doi.org/10.1175/bams-d-13-00117.1

Carrasco, D., Pizarro, O., Jacques-Coper, M., & Narvaez, D. A. (2023). Main drivers of marine heat waves in the eastern South Pacific. *Frontiers in Marine Science, 10*. https://doi.org/10.3389/fmars.2023.1129276

Cassou, C., Terray, L., & Phillips, A. S. (2005). Tropical Atlantic influence on European heat waves. *Journal of Climate, 18*(15), 2805–2811. https://doi.org/10.1175/jcli3506.1

Cha, Y. M., Choi, J. W., & Ahn, J. B. (2022). The interannual synchronization of the heatwave days in Korea and western North Pacific tropical cyclone genesis frequency. *International Journal of Climatology, 42*(15), 7817–7838. https://doi.org/10.1002/joc.7679

Chen, X. L., & Zhou, T. J. (2018). Relative contributions of external SST forcing and internal atmospheric variability to July-August heat waves over the Yangtze River valley. *Climate Dynamics, 51*(11–12), 4403–4419. https://doi.org/10.1007/s00382-017-3871-y

Chen, X., Ling, J., & Li, C. Y. (2016). Evolution of the Madden-Julian oscillation in two types of El Nino. *Journal of Climate, 29*(5), 1919–1934. https://doi.org/10.1175/jcli-d-15-0486.1

Dai, A., & Bloecker, C. E. (2019). Impacts of internal variability on temperature and precipitation trends in large ensemble simulations by two climate models. *Climate Dynamics, 52*(1–2), 289–306. https://doi.org/10.1007/s00382-018-4132-4

de Viron, O., Dickey, J. O., & Ghil, M. (2013). Global modes of climate variability. *Geophysical Research Letters, 40*(9), 1832–1837. https://doi.org/10.1002/grl.50386

Della-Marta, P. M., Luterbacher, J., von Weissenfluh, H., Xoplaki, E., Brunet, M., & Wanner, H. (2007). Summer heat waves over western Europe 1880-2003, their relationship to large-scale forcings and predictability. *Climate Dynamics, 29*(2–3), 251–275. https://doi.org/10.1007/s00382-007-0233-1

Deng, K. Q., Yang, S., Ting, M. F., Lin, A. L., & Wang, Z. Q. (2018a). An intensified mode of variability modulating the summer heat waves in Eastern Europe and northern China. *Geophysical Research Letters, 45*(20), 11361–11369. https://doi.org/10.1029/2018gl079836

Deng, K. Q., Yang, S., Ting, M. F., Tan, Y. H., & He, S. (2018b). Global monsoon precipitation: Trends, leading modes, and associated drought and heat wave in the northern hemisphere. *Journal of Climate, 31*(17), 6947–6966. https://doi.org/10.1175/jcli-d-17-0569.1

Donat, M. G., Peterson, T. C., Brunet, M., King, A. D., Almazroui, M., Kolli, R. K., Boucherf, D., Al-Mulla, A. Y., Nour, A. Y., Aly, A. A., Nada, T. A. A., Semawi, M. M., Dashti, H. A., Salhab, T. G., Fadli, K. I., Muftah, M. K., Eida, S. D., Badi, W., Driouech, F., Rhaz, K., Abubaker, M. J. Y., Ghulam, A. S., Erayah, A. S., Ben Mansour, M., Alabdouli, W. O., Dhanhani, J. S., & Al Shekaili, M. N. (2014). Changes in extreme temperature and precipitation in the Arab region: Long-term trends and variability related to ENSO and NAO. *International Journal of Climatology, 34*(3), 581–592. https://doi.org/10.1002/joc.3707

Dou, J., Wu, Z. W., & Zhou, Y. F. (2017). Potential impact of the may southern hemisphere annular mode on the Indian summer monsoon rainfall. *Climate Dynamics, 49*(4), 1257–1269. https://doi.org/10.1007/s00382-016-3380-4

El, N., Kim, J., Jun, S. Y., Cha, D. H., Park, M. S., Kim, J. H., & Kim, H. G. (2021). The role of the Pacific-Japan pattern in extreme heatwaves over Korea and Japan. *Geophysical Research Letters, 48*(18). https://doi.org/10.1029/2021gl093990

Fan, F. X., Lin, R. P., Fang, X. H., Xue, F., Zheng, F., & Zhu, J. (2021). Influence of the eastern Pacific and Central Pacific types of ENSO on the south Asian summer monsoon. *Advances in Atmospheric Sciences, 38*(1), 12–28. https://doi.org/10.1007/s00376-020-0055-1

Favà, V., Curto, J. J., & Llasat, M. C. (2016). Regional differential behaviour of maximum temperatures in the Iberian Peninsula regarding the summer NAO in the second half of the twentieth century. *Atmospheric Research, 182*, 319–334. https://doi.org/10.1016/j.atmosres.2016.08.006

Feng, M., McPhaden, M. J., Xie, S. P., & Hafner, J. (2013). La Nina forces unprecedented Leeuwin current warming in 2011. *Scientific Reports, 3*. https://doi.org/10.1038/srep01277

Fogt, R. L., Perlwitz, J., Monaghan, A. J., Bromwich, D. H., Jones, J. M., & Marshall, G. J. (2009). Historical SAM variability. Part II: Twentieth-century variability and trends from reconstructions, observations, and the IPCC AR4 models. *Journal of Climate, 22*(20), 5346–5365. https://doi.org/10.1175/2009jcli2786.1

Fragkoulidis, G., Wirth, V., Bossmann, P., & Fink, A. H. (2018). Linking northern hemisphere temperature extremes to Rossby wave packets. *Quarterly Journal of the Royal Meteorological Society, 144*(711), 553–566. https://doi.org/10.1002/qj.3228

Gao, T., Luo, M., Lau, N. C., & Chan TO. (2020). Spatially distinct effects of two El Nino types on summer heat extremes in China. *Geophysical Research Letters, 47*(6). https://doi.org/10.1029/2020gl086982

Gao, X., Li, G., Liu, J. W., & Long, S. M. (2022). The trend and interannual variability of marine heatwaves over the Bay of Bengal. *Atmosphere, 13*(3). https://doi.org/10.3390/atmos13030469

Geirinhas, J. L., Trigo, R. M., Libonati, R., Castro, L. C. O., Sousa, P. M., Coelho, C. A. S., Peres, L. F., & Magalhaes, M. (2019). Characterizing the atmospheric conditions during the 2010 heatwave in Rio de Janeiro marked by excessive mortality rates. *Science of the Total Environment, 650*, 796–808. https://doi.org/10.1016/j.scitotenv.2018.09.060

Grotjahn, R., Black, R., Leung, R., Wehner, M. F., Barlow, M., Bosilovich, M., Gershunov, A., Gutowski, W. J., Jr., Gyakum, J. R., Katz, R. W., Lee, Y.-Y., Lim, Y.-K., & Prabhat. (2016). North American extreme temperature events and related large scale meteorological patterns:

A review of statistical methods, dynamics, modeling, and trends. *Climate Dynamics, 46*(3–4), 1151–1184. https://doi.org/10.1007/s00382-015-2638-6

Gu, G. J., & Adler, R. F. (2019). Precipitation, temperature, and moisture transport variations associated with two distinct ENSO flavors during 1979–2014. *Climate Dynamics, 52*(12), 7249–7265. https://doi.org/10.1007/s00382-016-3462-3

Guigma, K. H., Guichard, F., Todd, M., Peyrille, P., & Wang, Y. (2021). Atmospheric tropical modes are important drivers of Sahelian springtime heatwaves. *Climate Dynamics, 56*(5–6), 1967–1987. https://doi.org/10.1007/s00382-020-05569-9

Hall, R. J., & Hanna, E. (2018). North Atlantic circulation indices: Links with summer and winter UK temperature and precipitation and implications for seasonal forecasting. *International Journal of Climatology, 38*, E660–E677. https://doi.org/10.1002/joc.5398

Heidemann, H., & Ribbe, J. (2019). Marine heat waves and the influence of El Nino off Southeast Queensland, Australia. *Frontiers in Marine Science, 6*. https://doi.org/10.3389/fmars.2019.00056

Hoerling, M., Kumar, A., Dole, R., Nielsen-Gammon, J. W., Eischeid, J., Perlwitz, J., Quan, X. W., Zhang, T., Pegion, P., & Chen, M. Y. (2013). Anatomy of an extreme event. *Journal of Climate, 26*(9), 2811–2832. https://doi.org/10.1175/jcli-d-12-00270.1

Holbrook, N. J., Scannell, H. A., Sen Gupta, A., Benthuysen, J. A., Feng, M., Oliver, E. C. J., Alexander, L. V., Burrows, M. T., Donat, M. G., Hobday, A. J., Moore, P. J., Perkins-Kirkpatrick, S. E., Smale, D. A., Straub, S. C., & Wernberg, T. (2019). A global assessment of marine heatwaves and their drivers. *Nature Communications, 10*. https://doi.org/10.1038/s41467-019-10206-z

Hoy, A., Sepp, M., & Matschullat, J. (2013). Large-scale atmospheric circulation forms and their impact on air temperature in Europe and northern Asia. *Theoretical and Applied Climatology, 113*(3–4), 643–658. https://doi.org/10.1007/s00704-012-0813-9

Hsu, P. C., Qian, Y. T., Liu, Y., Murakami, H., & Gao, Y. X. (2020). Role of abnormally enhanced MJO over the western Pacific in the formation and subseasonal predictability of the record-breaking northeast Asian heatwave in the summer of 2018. *Journal of Climate, 33*(8), 3333–3349. https://doi.org/10.1175/jcli-d-19-0337.1

Hurrell, J. W., Kushnir, Y., Ottersen, G., & Visbeck, M. (2003). An overview of the North Atlantic oscillation. *Geophysical Monograph-American Geophysical Union, 134*, 1–36.

Jiang, X. A., Xiang, B. Q., Zhao, M., Li, T., Lin, S. J., Wang, Z., & Chen, J. H. (2018). Intraseasonal tropical cyclogenesis prediction in a global coupled model system. *Journal of Climate, 31*(15), 6209–6227. https://doi.org/10.1175/jcli-d-17-0454.1

Jury, M. R. (2016). Inter-annual rainfall variability in the eastern Antilles and coupling with the regional and intra-seasonal circulation. *Theoretical and Applied Climatology, 126*(3–4), 727–737. https://doi.org/10.1007/s00704-015-1612-x

Kenyon, J., & Hegerl, G. C. (2008). Influence of modes of climate variability on global temperature extremes. *Journal of Climate, 21*(15), 3872–3889. https://doi.org/10.1175/2008jcli2125.1

Kerr, R. A. (2000). A North Atlantic climate pacemaker for the centuries. *Science, 288*(5473), 1984–1986. https://doi.org/10.1126/science.288.5473.1984

Klotzbach, P. J., & Oliver, E. C. J. (2015). Modulation of Atlantic basin tropical cyclone activity by the Madden-Julian Oscillation (MJO) from 1905 to 2011. *Journal of Climate, 28*(1), 204–217. https://doi.org/10.1175/jcli-d-14-00509.1

Kornhuber, K., Osprey, S., Coumou, D., Petri, S., Petoukhov, V., Rahmstorf, S., & Gray, L. (2019). Extreme weather events in early summer 2018 connected by a recurrent hemispheric wave-7 pattern. *Environmental Research Letters, 14*(5). https://doi.org/10.1088/1748-9326/ab13bf

Kucharski, F., Kang, I.-S., Straus, D., & King, M. P. (2010). Teleconnections in the atmosphere and oceans. *Bulletin of the American Meteorological Society, 91*(3), 381–383.

Lee, Y. Y., & Grotjahn, R. (2019). Evidence of specific MJO phase occurrence with summertime California Central Valley extreme hot weather. *Advances in Atmospheric Sciences, 36*(6), 589–602. https://doi.org/10.1007/s00376-019-8167-1

Lee, M. H., Lee, S., Song, H. J., & Ho, C. H. (2017). The recent increase in the occurrence of a boreal summer teleconnection and its relationship with temperature extremes. *Journal of Climate, 30*(18), 7493–7504. https://doi.org/10.1175/jcli-d-16-0094.1

Lestari, R. K., & Koh, T. Y. (2016). Statistical evidence for asymmetry in ENSO-IOD interactions. *Atmosphere-Ocean, 54*(5), 498–504. https://doi.org/10.1080/07055900.2016.1211084

Li, J., & Ruan, C. (2018). The North Atlantic–Eurasian teleconnection in summer and its effects on Eurasian climates. *Environmental Research Letters, 13*(2), 024007. https://doi.org/10.1088/1748-9326/aa9d33

Li, M. Y., Yao, Y., Luo, D. H., & Zhong, L. H. (2019). The linkage of the large-scale circulation pattern to a Long-lived heatwave over Mideastern China in 2018. *Atmosphere, 10*(2). https://doi.org/10.3390/atmos10020089

Lin, L. J., Chen, C. C., & Luo, M. (2018). Impacts of El Nino-southern oscillation on heat waves in the Indochina peninsula. *Atmospheric Science Letters, 19*(11). https://doi.org/10.1002/asl.856

Lin, X. Y., Qiu, Y., Wang, J. J., Teng, H., Ni, X. T., & Liang, K. M. (2023). Seasonal diversity of El Nino-induced marine heatwave increases in the bay of Bengal. *Geophysical Research Letters, 50*, 3. https://doi.org/10.1029/2022gl100807

Liu, T., Li, J. P., Li, Y. J., Zhao, S., Zheng, F., Zheng, J. Y., & Yao, Z. X. (2018). Influence of the may southern annular mode on the South China Sea summer monsoon. *Climate Dynamics, 51*(11–12), 4095–4107. https://doi.org/10.1007/s00382-017-3753-3

Loikith, P. C., & Broccoli, A. J. (2014). The influence of recurrent modes of climate variability on the occurrence of winter and summer extreme temperatures over North America. *Journal of Climate, 27*(4), 1600–1618. https://doi.org/10.1175/jcli-d-13-00068.1

Loikith, P. C., Detzer, J., Mechoso, C. R., Lee, H., & Barkhordarian, A. (2017). The influence of recurrent modes of climate variability on the occurrence of monthly temperature extremes over South America. *Journal of Geophysical Research-Atmospheres, 122*(19), 10524–10538. https://doi.org/10.1002/2017jd027561

Lorrey, A. M., & Fauchereau, N. C. (2018). Southwest Pacific atmospheric weather regimes: Linkages to ENSO and extra-tropical teleconnections. *International Journal of Climatology, 38*(4), 1893–1909. https://doi.org/10.1002/joc.5304

Loughran, T. F., Perkins-Kirkpatrick, S. E., & Alexander, L. V. (2017a). Understanding the spatio-temporal influence of climate variability on Australian heatwaves. *International Journal of Climatology, 37*(10), 3963–3975. https://doi.org/10.1002/joc.4971

Loughran, T. F., Perkins-Kirkpatrick, S. E., Alexander, L. V., & Pitman, A. J. (2017b). No significant difference between Australian heat wave impacts of Modoki and eastern Pacific El Nino. *Geophysical Research Letters, 44*(10), 5150–5157. https://doi.org/10.1002/2017gl073231

Loughran, T. F., Pitman, A. J., & Perkins-Kirkpatrick, S. E. (2019). The El Nino-southern Oscillation's effect on summer heatwave development mechanisms in Australia. *Climate Dynamics, 52*(9–10), 6279–6300. https://doi.org/10.1007/s00382-018-4511-x

Luo, M., & Lau, N. C. (2018). Synoptic characteristics, atmospheric controls, and long-term changes of heat waves over the Indochina Peninsula. *Climate Dynamics, 51*(7–8), 2707–2723. https://doi.org/10.1007/s00382-017-4038-6

Luo, M., & Lau, N. C. (2020). Summer heat extremes in northern continents linked to developing ENSO events. *Environmental Research Letters, 15*(7). https://doi.org/10.1088/1748-9326/ab7d07

Luo, F. F., Li, S. L., & Furevik, T. (2018). Weaker connection between the Atlantic multidecadal oscillation and Indian summer rainfall since the mid-1990s. *Atmospheric and Oceanic Science Letters, 11*(1), 37–43. https://doi.org/10.1080/16742834.2018.1394779

Lupo, A. R., Kelsey, E. P., Weitlich, D. R., Woolard, J. E., Mokhov, I. I., Guinan, P. E., & Akyuz, F. A. (2007). Interannual and interdecadal variability in the predominant Pacific region SST anomaly patterns and their impact on climate in the mid-Mississippi valley region. *Atmosfera, 20*(2), 171–196.

Madden, R. A., & Julian, P. R. (1972). Description of global-scale circulation cells in tropics with a 40-50 day period. *Journal of the Atmospheric Sciences, 29*(6), 1109. https://doi.org/10.1175/1520-0469(1972)029<1109:dogscc>2.0.co;2

Maloney, E. D., & Kiehl, J. T. (2002). MJO-related SST variations over the tropical eastern Pacific during northern hemisphere summer. *Journal of Climate, 15*(6), 675–689.

Mantua, N. J., Hare, S. R., Zhang, Y., Wallace, J. M., & Francis, R. C. (1997). A Pacific interdecadal climate oscillation with impacts on salmon production. *Bulletin of the American Meteorological Society, 78*(6), 1069–1079. https://doi.org/10.1175/1520-0477(1997)078<1069:apicow>2.0.co;2

Marshall, A. G., Hudson, D., Wheeler, M. C., Alves, O., Hendon, H. H., Pook, M. J., & Risbey, J. S. (2014). Intra-seasonal drivers of extreme heat over Australia in observations and POAMA-2. *Climate Dynamics, 43*(7–8), 1915–1937. https://doi.org/10.1007/s00382-013-2016-1

Matsueda, S., & Takaya, Y. (2015). The global influence of the Madden-Julian Oscillation on extreme temperature events. *Journal of Climate, 28*(10), 4141–4151. https://doi.org/10.1175/jcli-d-14-00625.1

Mawren, D., Hermes, J., & Reason, C. J. C. (2022). Marine heatwaves in the Mozambique Channel. *Climate Dynamics, 58*(1–2), 305–327. https://doi.org/10.1007/s00382-021-05909-3

Mbokodo, I. L., Bopape, M. J. M., Ndarana, T., Mbatha, S. M. S., Muofhe, T. P., Singo, M. V., Xulu, N. G., Mohomi, T., Ayisi, K. K., & Chikoore, H. (2023). Heatwave variability and structure in South Africa during summer drought. *Climate, 11*(2). https://doi.org/10.3390/cli11020038

McCabe, G. J., Palecki, M. A., & Betancourt, J. L. (2004). Pacific and Atlantic Ocean influences on multidecadal drought frequency in the United States. *Proceedings of the National Academy of Sciences of the United States of America, 101*(12), 4136–4141. https://doi.org/10.1073/pnas.0306738101

McGregor, G. (2017). Hydroclimatology, modes of climatic variability and stream flow, lake and groundwater level variability: A progress report. *Progress in Physical Geography, 41*(4), 496–512. https://doi.org/10.1177/0309133317726537

McGregor, G. R., & Ebi, K. (2018). El Nino Southern Oscillation (ENSO) and health: An overview for climate and health researchers. *Atmosphere, 9*(7), 32. https://doi.org/10.3390/atmos9070282

Meneghini, B., Simmonds, I., & Smith, I. N. (2007). Association between Australian rainfall and the southern annular mode. *International Journal of Climatology, 27*(1), 109–121. https://doi.org/10.1002/joc.1370

Min, S. K., Cai, W. J., & Whetton, P. (2013). Influence of climate variability on seasonal extremes over Australia. *Journal of Geophysical Research-Atmospheres, 118*(2), 643–654. https://doi.org/10.1002/jgrd.50164

Moron, V., Frelat, R., Jean-Jeune, P. K., & Gaucherel, C. (2015). Interannual and intra-annual variability of rainfall in Haiti (1905–2005). *Climate Dynamics, 45*(3–4), 915–932. https://doi.org/10.1007/s00382-014-2326-y

Murari, K. K., Sahana, A. S., Daly, E., & Ghosh, S. (2016). The influence of the El Nino southern oscillation on heat waves in India. *Meteorological Applications, 23*(4), 705–713. https://doi.org/10.1002/met.1594

Nguyen-Thanh, H., Ngo-Duc, T., & Herrmann, M. (2023). The distinct impacts of the two types of ENSO on rainfall variability over Southeast Asia. *Climate Dynamics*. https://doi.org/10.1007/s00382-023-06673-2

Oliver, E. C. J., Benthuysen, J. A., Darmaraki, S., Donat, M. G., Hobday, A. J., Holbrook, N. J., Schlegel, R. W., & Sen Gupta, A. (2021). Marine heatwaves. *Annual Review of Marine Science, 13*(13), 313–342. https://doi.org/10.1146/annurev-marine-032720-095144

Parker, T. J., Berry, G. J., & Reeder, M. J. (2013). The influence of tropical cyclones on heat waves in southeastern Australia. *Geophysical Research Letters, 40*(23), 6264–6270. https://doi.org/10.1002/2013gl058257

Parker, T. J., Berry, G. J., & Reeder, M. J. (2014). The structure and evolution of heat waves in southeastern Australia. *Journal of Climate, 27*(15), 5768–5785. https://doi.org/10.1175/jcli-d-13-00740.1

Perkins, S. E., Argueso, D., & White, C. J. (2015). Relationships between climate variability, soil moisture, and Australian heatwaves. *Journal of Geophysical Research-Atmospheres, 120*(16), 8144–8164. https://doi.org/10.1002/2015jd023592

Perlwitz, J., Pawson, S., Fogt, R. L., Nielsen, J. E., & Neff, W. D. (2008). Impact of stratospheric ozone hole recovery on Antarctic climate. *Geophysical Research Letters, 35*(8). https://doi.org/10.1029/2008gl033317

Pezza, A. B., van Rensch, P., & Cai, W. J. (2012). Severe heat waves in Southern Australia: Synoptic climatology and large scale connections. *Climate Dynamics, 38*(1–2), 209–224. https://doi.org/10.1007/s00382-011-1016-2

Prabhu, A., Kripalani, R., Oh, J. H., & Preethi, B. (2017). Can the southern annular mode influence the Korean summer monsoon rainfall? *Asia-Pacific Journal of Atmospheric Sciences, 53*(2), 217–228. https://doi.org/10.1007/s13143-017-0029-0

Qi, R. S., Zhang, Y., Du, Y., & Feng, M. (2022). Characteristics and drivers of marine heatwaves in the Western equatorial Indian Ocean. *Journal of Geophysical Research-Oceans, 127*(10). https://doi.org/10.1029/2022jc018732

Ramirez-Beltran, N. D., Gonzalez, J. E., Castro, J. M., Angeles, M., Harmsen, E. W., & Salazar, C. M. (2017). Analysis of the heat index in the Mesoamerica and Caribbean region. *Journal of Applied Meteorology and Climatology, 56*(11), 2905–2925. https://doi.org/10.1175/jamc-d-16-0167.1

Ren, X. L., & Liu, W. (2021). The role of a weakened Atlantic meridional overturning circulation in modulating marine heatwaves in a warming climate. *Geophysical Research Letters, 48*(23). https://doi.org/10.1029/2021gl095941

Ruprich-Robert, Y., Delworth, T., Msadek, R., Castruccio, F., Yeager, S., & Danabasoglu, G. (2018). Impacts of the Atlantic multidecadal variability on North American summer climate and heat waves. *Journal of Climate, 31*(9), 3679–3700. https://doi.org/10.1175/jcli-d-17-0270.1

Rusticucci, M., Kysely, J., Almeira, G., & Lhotka, O. (2016). Long-term variability of heat waves in Argentina and recurrence probability of the severe 2008 heat wave in Buenos Aires. *Theoretical and Applied Climatology, 124*(3–4), 679–689. https://doi.org/10.1007/s00704-015-1445-7

Ryan, S., Ummenhofer, C. C., Gawarkiewicz, G., Wagner, P., Scheinert, M., Biastoch, A., & Boning, C. W. (2021). Depth structure of Ningaloo Nino/Nina events and associated drivers. *Journal of Climate, 34*(5), 1767–1788. https://doi.org/10.1175/jcli-d-19-1020.1

Salinger, J., Hobday, A. J., Matear, R. J., O'Kane, T. J., Risbey, J. S., Dunstan, P., Eveson, J. P., Fulton, E. A., Feng, M., Plaganyi, E. E., Poloczanska, E. S., Marshall, A. G., & Thompson, P. A. (2016). Decadal-scale forecasting of climate drivers for marine applications. *Advances in Marine Biology, 74*(74), 1–68. https://doi.org/10.1016/bs.amb.2016.04.002

Sanderson, M. G., Economou, T., Salmon, K. H., & Jones, S. E. O. (2017). Historical trends and variability in heat waves in the United Kingdom. *Atmosphere, 8*(10). https://doi.org/10.3390/atmos8100191

Scaife, A. A., Folland, C. K., Alexander, L. V., Moberg, A., & Knight, J. R. (2008). European climate extremes and the North Atlantic Oscillation. *Journal of Climate, 21*(1), 72–83. https://doi.org/10.1175/2007jcli1631.1

Scannell, H. A., Pershing, A. J., Alexander, M. A., Thomas, A. C., & Mills, K. E. (2016). Frequency of marine heatwaves in the North Atlantic and North Pacific since 1950. *Geophysical Research Letters, 43*(5), 2069–2076. https://doi.org/10.1002/2015gl067308

Sen Gupta, A., Thomsen, M., Benthuysen, J. A., Hobday, A. J., Oliver, E., Alexander, L. V., Burrows, M. T., Donat, M. G., Feng, M., Holbrook, N. J., Perkins-Kirkpatrick, S., Moore, P. J., Rodrigues, R. R., Scannell, H. A., Taschetto, A. S., Ummenhofer, C. C., Wernberg, T., & Smale, D. A. (2020). Drivers and impacts of the most extreme marine heatwaves events. *Scientific Reports, 10*(1). https://doi.org/10.1038/s41598-020-75445-3

Seo, K. H., & Son, S. W. (2012). The global atmospheric circulation response to tropical diabatic heating associated with the Madden-Julian Oscillation during northern winter. *Journal of the Atmospheric Sciences, 69*(1), 79–96. https://doi.org/10.1175/2011jas3686.1

Sheridan, S., & Lee, C. C. (2012). Synoptic climatology and the analysis of atmospheric teleconnections. *Progress in Physical Geography-Earth and Environment, 36*(4), 548–557. https://doi.org/10.1177/0309133312447935

Sousa, P. M., Blamey, R. C., Reason, C. J. C., Ramos, A. M., & Trigo, R. M. (2018). The 'Day Zero' Cape Town drought and the poleward migration of moisture corridors. *Environmental Research Letters, 13*(12). https://doi.org/10.1088/1748-9326/aaebc7

Sun, C., Li, J. P., & Ding, R. Q. (2016). Strengthening relationship between ENSO and western Russian summer surface temperature. *Geophysical Research Letters, 43*(2), 843–851. https://doi.org/10.1002/2015gl067503

Sutton, R. T., & Dong, B. W. (2012). Atlantic Ocean influence on a shift in European climate in the 1990s. *Nature Geoscience, 5*(11), 788–792. https://doi.org/10.1038/ngeo1595

Thompson, D. W. J., & Wallace, J. M. (2000). Annular modes in the extratropical circulation. Part I: Month-to-month variability. *Journal of Climate, 13*(5), 1000–1016. https://doi.org/10.1175/1520-0442(2000)013<1000:Amitec>2.0.Co;2

Thompson, D. W. J., Wallace, J. M., & Hegerl, G. C. (2000). Annular modes in the extratropical circulation. *Part II: Trends. Journal of Climate, 13*(5), 1018–1036. https://doi.org/10.1175/1520-0442(2000)013<1018:Amitec>2.0.Co;2

Torres-Valcarcel, A. R. (2018). Teleconnections between ENSO and rainfall and drought in Puerto Rico. *International Journal of Climatology, 38*, E1190–E1204. https://doi.org/10.1002/joc.5444

Tozer, C. R., Kiem, A. S., & Verdon-Kidd, D. C. (2017). Large-Scale Ocean-atmospheric processes and seasonal rainfall variability in South Australia: Potential for improving seasonal hydroclimatic forecasts. *International Journal of Climatology, 37*, 861–877. https://doi.org/10.1002/joc.5043

Turkes, M., & Erlat, E. (2018). Variability and trends in record air temperature events of Turkey and their associations with atmospheric oscillations and anomalous circulation patterns. *International Journal of Climatology, 38*(14), 5182–5204. https://doi.org/10.1002/joc.5720

Varela, R., Rodriguez-Diaz, L., de Castro, M., & Gomez-Gesteira, M. (2021). Influence of eastern upwelling systems on marine heatwaves occurrence. *Global and Planetary Change, 196*. https://doi.org/10.1016/j.gloplacha.2020.103379

Verdon-Kidd, D. C., Scanlon, B. R., Tong, R. C., & Fernando, D. N. (2017). A comparative study of historical droughts over Texas, USA and Murray-Darling Basin, Australia: Factors influencing initialization and cessation. *Global and Planetary Change, 149*, 123–138. https://doi.org/10.1016/j.gloplacha.2017.01.001

Vigaud, N., Ting, M., Lee, D. E., Barnston, A. G., & Kushnir, Y. (2018). Multiscale variability in north American summer maximum temperatures and modulations from the North Atlantic simulated by an AGCM. *Journal of Climate, 31*(7), 2549–2562. https://doi.org/10.1175/jcli-d-17-0392.1

Wang, S. S., Huang, J. P., He, Y. L., & Guan, Y. P. (2014). Combined effects of the Pacific decadal oscillation and El Nino-Southern Oscillation on global land dry-wet changes. *Scientific Reports, 4*. https://doi.org/10.1038/srep06651

Wang, J., DeFlorio, M. J., Gershunov, A., Guirguis, K., Delle Monache, L., & Ralph, F. M. (2024). Association of western US compound hydrometeorological extremes with Madden-Julian oscillation and ENSO interaction. *Communications Earth & Environment, 5*(1), 314.

Wehrli, K., Guillod, B. P., Hauser, M., Leclair, M., & Seneviratne, S. I. (2019). Identifying key driving processes of major recent heat waves. *Journal of Geophysical Research-Atmospheres, 124*(22), 11746–11765. https://doi.org/10.1029/2019jd030635

Whan, K., Alexander, L. V., Imielska, A., McGree, S., Jones, D., Ene, E., Finaulahi, S., Inape, K., Jacklick, L., Kumar, R., Laurent, V., Malala, H., Malsale, P., Pulehetoa-Mitiepo, R., Ngemaes, M., Peltier, A., Porteous, A., Seuseu, S., Skilling, E., Tahani, L., Toorua, U., & Vaiimene, M. (2014). Trends and variability of temperature extremes in the tropical Western Pacific. *International Journal of Climatology, 34*(8), 2585–2603. https://doi.org/10.1002/joc.3861

Williams, P. D., Alexander, M. J., Barnes, E. A., Butler, A. H., Davies, H. C., Garfinkel, C. I., Kushnir, Y., Lane, T. P., Lundquist, J. K., Martius, O., Maue, R. N., Peltier, W. R., Sato, K., Scaife, A. A., & Zhang, C. D. (2017). A census of atmospheric variability from seconds to decades. *Geophysical Research Letters, 44*(21), 11201–11211. https://doi.org/10.1002/2017gl075483

Wolter, K., & Timlin, M. S. (2011). El Nino/Southern Oscillation behaviour since 1871 as diagnosed in an extended multivariate ENSO index (MEI.ext). *International Journal of Climatology, 31*(7), 1074–1087. https://doi.org/10.1002/joc.2336

Yamamoto, A., & Palter, J. B. (2016). The absence of an Atlantic imprint on the multidecadal variability of wintertime European temperature. *Nature Communications, 7*. https://doi.org/10.1038/ncomms10930

Yang, Y., Zhu, Z. W., Li, T., & Yao, M. Y. (2020). Effects of western Pacific intraseasonal convection on surface air temperature anomalies over North America. *International Journal of Climatology, 40*(6), 2913–2923. https://doi.org/10.1002/joc.6373

Yao, Y. L., Wang, J. J., Yin, J. J., & Zou, X. Q. (2020). Marine heatwaves in China's marginal seas and adjacent offshore waters: Past, present, and future. *Journal of Geophysical Research-Oceans, 125*, 3. https://doi.org/10.1029/2019jc015801

Yoon, A., Kim, J., Lee, J., Sung, H. M., Hong, J.-W., Min, S.-K., Lee, J., & Hong, J. (2024). Factor analysis of recent major heatwaves in East Asia. *Geoscience Frontiers, 15*(1), 101730.

Zampieri, M., Toreti, A., Schindler, A., Scoccimarro, E., & Gualdi, S. (2017). Atlantic multidecadal oscillation influence on weather regimes over Europe and the Mediterranean in spring and summer. *Global and Planetary Change, 151*, 92–100. https://doi.org/10.1016/j.gloPlacha.2016.08.014

Zhang, B., & Xie, Z. (2023). Mechanism of the record-breaking heatwave event dynamics in South America in January 2022. *Atmosphere, 14*(9). https://doi.org/10.3390/atmos14091326

Zhang, M. J., Cao, Q., Zhu, F. L., Lall, U., Hu, P., Jiang, Y. Z., & Kan, G. Y. (2022). The asymmetric effect of different types of ENSO and ENSO Modoki on rainy season over the Yellow River basin, China. *Theoretical and Applied Climatology, 149*(3–4), 1567–1581. https://doi.org/10.1007/s00704-022-04128-y

Zheng, J. Y., Li, Y. J., Li, J. P., Xue, J. Q., Guo, Y. P., Liu, T., & Wang, Q. Y. (2017). The relationship between indo-Pacific convection oscillation and summer surface air temperature in southern Asia. *Scientific Online Letters On The Atmosphere, 13*, 199–204. https://doi.org/10.2151/sola.2017-036

Zhou, Y. F., & Wu, Z. W. (2016). Possible impacts of mega-El Nino/Southern Oscillation and Atlantic multidecadal oscillation on Eurasian heatwave frequency variability. *Quarterly Journal of the Royal Meteorological Society, 142*(697), 1647–1661. https://doi.org/10.1002/qj.2759

Zhu, Z. W., & Li, T. (2017). The record-breaking hot summer in 2015 over Hawaii and its physical causes. *Journal of Climate, 30*(11), 4253–4266. https://doi.org/10.1175/jcli-d-16-0438.1

Chapter 6
Synoptic Scale Atmospheric Processes and Heatwaves

Abstract This chapter explores how synoptic scale atmospheric processes that operate on a horizontal scale of 1000–2500 km and a vertical scale of up to 10 km, with a timescale of 5–10 days, influence heatwave development. Firstly, heatwaves are contextualised within larger modes of climate variability which set the broad 'slow' boundary conditions that influence the 'faster' synoptic scale processes discussed in this chapter. Three primary atmospheric processes crucial for heatwave development are highlighted namely blocking, advection, and the foehn effect. Blocking, a significant factor, involves persistent high-pressure systems that disrupt normal weather patterns, leading to prolonged periods of atmospheric stability and extreme heat. The nature of different blocking patterns is described. The significance of advection, the transport of heat and moisture by wind, for heatwave development is also outlined along with the importance of the interaction of local and large-scale atmospheric circulation with heat or moisture sources. The foehn effect, involving warm, dry downslope winds on the lee side of mountain ranges, is also discussed. On occasions these winds are found to cause rapid temperature increases and decreased humidity, contributing to heatwave conditions in several regions. Further to outlining the general synoptic processes associated with heatwave development, the chapter provides a geographical perspective on heatwave development, examining notable events in North America, Europe, Russia, the Indian subcontinent, Australia, East Asia, South America, and Africa. It underscores the regional variability of synoptic scale processes and their complex interactions with local climate settings. The chapter concludes by emphasizing the importance of understanding these processes for improving heatwave prediction.

Keywords Heatwave development · Synoptic scale processes · Blocking · Blocking types · Advection · Fohn effect · Regional variations in heatwave development

G. McGregor, *Heatwaves*, Biometeorology 6,
https://doi.org/10.1007/978-3-031-69906-1_6

6.1 Introduction

In the previous chapter, the association between heatwaves and modes of climate variability at the intra-seasonal to decadal timescales was considered. The point was made that such modes don't directly bring about the occurrence of heatwaves, rather they shift the likelihood in one direction or another of heatwaves occurring. This is because modes of variability can influence the nature of 'heatwave relevant' atmospheric and ocean processes and patterns operating at the so-called synoptic or weather scale. Essentially the synoptic scale is nested within the spatio-temporal scales of the larger modes of climate variability. The synoptic scale spans a horizontal length scale of 1000–2500 km, a vertical scale of up to 10 km and a time scale of 5–10 days. Typical synoptic scale systems include high pressure and low pressure systems and the associated upper level Rossby wave pattern, tropical cyclones, air masses, cold and warm fronts and it could be argued heatwaves, especially in a mid-latitude context, and their counterpart cold waves. Because understanding synoptic scale atmospheric processes and patterns can assist with improving our ability to forecast and predict heatwaves, the purpose of this chapter is to present an overview of the synoptic scale situation involved in heatwave generation. The chapter begins with a consideration of the processes associated with blocking, advection and the foehn effect. Because these generic processes may manifest in different ways, depending on the climate setting, the chapter then offers a geographical perspective on atmospheric processes for heatwave generation based on a survey of the heatwave development literature for several major geographical regions along with several case studies of important heatwave events. Unlike the previous chapter in which marine heatwaves were considered, this chapter focuses solely on atmospheric heatwaves. The processes associated with marine heatwave formation are dealt with in Chap. 7 which considers land/ocean surface-atmosphere interactions and heatwave development.

6.2 Atmospheric Processes

Beyond the fundamental physical processes associated with atmospheric heating, as described in Chap. 4, there are several atmospheric processes that operate largely on the synoptic timescale which are important for heatwave generation and the attainment of temperature extremes (Barriopedro et al., 2023; Perkins, 2015; Roethlisberger & Papritz, 2023). Mainly an outcome of the weather regime or atmospheric pressure pattern in place in the lead up to, or during a heatwave event, these generic processes include blocking, advection and the foehn effect, all of which will be described in this section.

6.2.1 Blocking

Setting aside possible geographical biases in the literature, if a meta-analysis of studies focused on understanding the atmospheric processes underlying the development of heatwaves was to be conducted it would no doubt reveal that the phenomenon referred to as 'blocking' is of prime importance. This is often the conclusion drawn when the literature on the physical causes of heatwaves is summarised. Blocking is a meteorological event that can occur at any time of the year. It manifests as a persistent area of high pressure 'anchored' over a specific region for several days. This type of event literally blocks or redirects the normal west to east zonal flow of air that in mid-latitudes brings a succession of high and low pressure systems on about a 7–10 day cycle and thus the alternation of short periods of generally wet and dry conditions. On weather charts two main types of blocking patterns manifest, namely Omega and Dipole blocks (Fig. 6.1). Amplified ridges of high pressure may also be considered as blocks (Woollings et al., 2018). Omega blocks are named so because the pressure pattern they form resembles the uppercase Greek letter omega, Ω. In the case of an Omega block an area of high pressure is usually bound to the east and west by centres of low pressure (Fig. 6.1a). Dipole blocks, which are also referred to as Rex blocks, occur when the westerlies split into a diffluent flow with branches flowing eastwards, north and south of a centre of high pressure. Often associated with a Dipole blocking situation is a closed centre of low pressure on the equatorward side of the anticyclone (Fig. 6.1b). Amplified ridges are oxbow shaped areas of high pressure that extend from the subtropical latitudes into the mid-latitudes (Fig. 6.1c). Unlike Omega and Dipole blocks, amplified ridges of high pressure do not possess any closed isobars on weather charts. That aside, ridges are effective in blocking and diverting zonal flows with a strong meridional flow component a characteristic of high pressure ridges.

Common meteorological characteristics of blocking for affected regions are unusually high and persistent atmospheric pressure, very calm conditions for areas sitting directly under the area of the blocking high with little ventilation, large scale subsidence and adiabatic warming of the air as it sinks below the area of high pressure and in summer, strong shortwave radiation inputs to the surface because of clear skies. In summer this combination of processes often conspires to produce

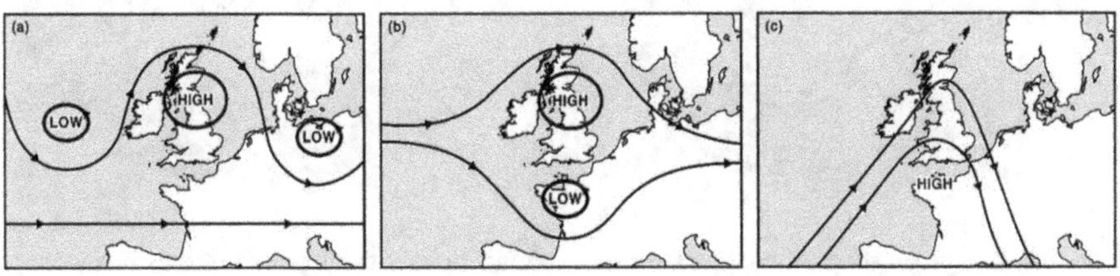

Fig. 6.1 Atmospheric blocking situations. (**a**) Omega block, (**b**) Diploe (Rex) block and (**c**) amplified ridge of high pressure. Adapted from: Met Office Learn About https://www.metoffice.gov.uk/weather/learn-about

anomalous and often long-lasting periods of extreme heat. Depending on the circumstances, a co-effect of summer blocking and heatwaves can be poor air quality (Kalisa et al., 2018; Mues et al., 2012; Ordóñez et al., 2017; Pu et al., 2017). Advection of warm air from distant locations can also occur with blocking (see next section). Despite a range of blocking theories having been advanced, currently there is no unified comprehensive dynamic theory able to account for blocking onset, maintenance, and decay (Kurgansky, 2022; Lupo, 2021). What might explain this is that the dynamic mechanisms generating periods of blocking may be geographically dependent and unique (Tibaldi et al., 1995) such that there is no reason for blocking dynamics to evolve in a similar way for any two like events (Jézéquel, Cattiaux, et al., 2018a; Jézéquel, Yiou, & Radanovics, 2018b). Notwithstanding the lack of a unified theory, it is clear periods of blocking arise because of a perturbation in the flow of the upper westerly winds or the so called Rossby wave pattern. Normally the upper westerly winds flow more or less directly from the west to the east (Fig. 6.2a) but occasionally they adopt a more meridional flow pattern meandering great distances north and south away from their mean zonal trajectory (Fig. 6.2b).

As might be apparent from the previous chapter meteorologists often refer to the 'waviness' of the Rossby wave pattern using the idea of a zonal or hemispheric wavenumber, that is the number of waves encircling the hemisphere at any one time. Each Rossby wave is comprised of a ridge, or area of high pressure oriented northwards (southwards) and trough or area of low pressure oriented southwards (northwards) in the northern (southern) hemisphere (Fig. 6.2). A high wave number refers to several ridges/troughs encircling the hemisphere and describes the situation when

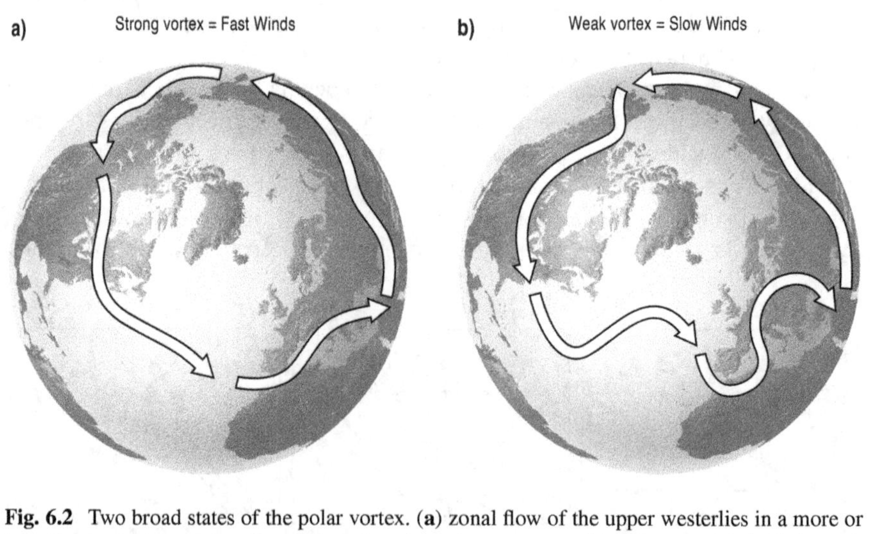

Fig. 6.2 Two broad states of the polar vortex. (**a**) zonal flow of the upper westerlies in a more or less direct west to east direction. This pattern has a low wave number. The associated windspeeds are high. (**b**) meridional flow of the upper westerlies with a highly perturbed Rossby wave pattern with winds wondering great distances north and south. This pattern has a high wave number. The associated windspeeds are low. (Source: Adapted from various sources)

the westerlies are highly perturbed with largely southerly or northerly flows of air dominating (Fig. 6.2b). In relation to heatwaves this bears implications for regions lying under the influence of an area of high pressure as this is where calm clear sky conditions are likely to exist which enhance the probability of both subsidence warming and strong surface heating and the development of a so-called heat dome. Regions lying to the west of an area of high pressure in the northern hemisphere may also experience elevated temperatures especially if the high pressure's orientation favours advection of warm air from subtropical/tropical latitudes. Further in a high (low) zonal wave number situation, with meridional (zonal) air flow dominating, the easterly progression of ridges and troughs is slow (fast). This results in prolonged periods of stable (unstable) atmospheric conditions within the vicinity of the high (low) pressure, hence the 'stalling' of weather systems and the creation of a period of blocking.

When the zonal flow becomes extremely perturbed, with the pressure contours becoming rapidly and irreversibly deformed, as if the upper westerlies and associated Rossby waves almost overturn and fold back on themselves. Rossby waves are said to 'break' (Fig. 6.3). Such Rossby wave-breaking events can result in ridges of high pressure being snapped off leading to an intense blocking event (Altenhoff et al., 2008). During such an anticyclonic wave-breaking event, the jet stream shifts north (south) in the Northern (Southern) Hemisphere; cyclonic wave-breaking events can also occur with the opposite effects on the jet stream. In relation to heatwave development, Rossby wave-breaking is potentially important as it can influence low frequency modes of atmospheric circulation. For example,

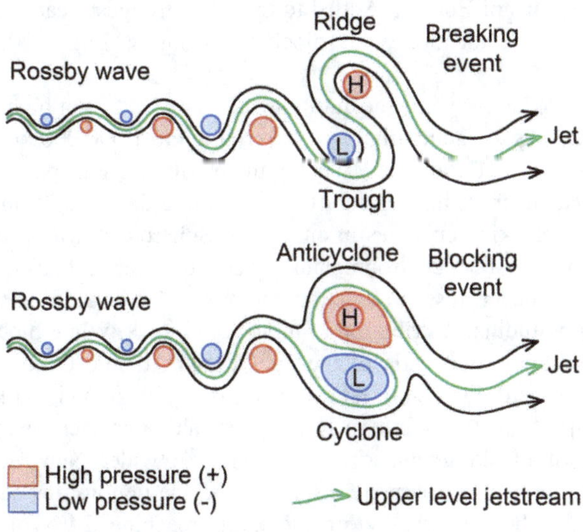

Fig. 6.3 Rossby wave breaking leading to a Dipole block (see Fig. 6.1b). (Source: Daneshvar and Freund (2019). Examination of a relationship between atmospheric blocking and seismic events in the Middle East using a new seismo-climatic index. Swiss Journal of Geosciences 112, 435–451. https://doi.org/10.1007/s00015-019-00343-4. Used with permission under Springer Nature licence number 5536560939824)

anticyclonic wave-breaking can favour a positive phase of the North Atlantic Oscillation (Benedict et al., 2004) which in the summer may enhance the likelihood, on climatological timescales, of the occurrence of heatwaves (see Sect. 5.3). Similarly large-scale modes of atmospheric circulation such as the MJO and ENSO may influence the frequency of Rossby wave-breaking on intra-seasonal to inter-annual timescales (Cassou, 2008; Moore et al., 2010). On meteorological or synoptic time scales, typically associated with the lifetime of weather events, anticyclonic wave-breaking can lead to the development of severe heatwaves and enhance the conditions for wildfires (Engel et al., 2013; Fragkoulidis et al., 2018; Kaplan et al., 2017; Sharma et al., 2022).

Beyond generally agreed qualitative descriptions of what constitutes blocking, there is a wide-ranging literature on the quantification of blocking as captured by a plethora of numerical blocking indices which vary in their level of sophistication (Doblas-Reyes et al., 2002; Barnes et al., 2012). Blocking climatologies based on such indices have revealed interesting aspects of the spatial and temporal characteristics of 'blocks' (Barriopedro et al., 2006; Dunn-Sigouin et al., 2013; Oliveira et al., 2014; Sinclair, 1996; Tyrlis & Hoskins, 2008) as well as Rossby wave-breaking (Ndarana & Waugh, 2011; Souders et al., 2014) and their association with periods of extreme temperature (Chan et al., 2019). Further there appears to be regional variation in the association between blocking and heatwaves with some regions demonstrating colocation while others do not (Chan et al., 2022; Jeong et al., 2022; Pfahl & Wernli, 2012). For example, in the northern hemisphere colocation appears to be greatest over the land compared to the ocean. Over land, colocation is strongest in the northwestern and northeastern regions of North America and the northern regions of Eurasia. Added to this, the strongest heat events appear to show the greatest colocation with blocking (Fang & Lu, 2020; Narinesingh et al., 2023).

Of relevance to heatwaves is the duration and spatial extent of blocking episodes. Atmospheric blocks are generally large covering up to 1,000,000 km^2. On average blocking episodes last for around 7 days in the Northern Hemisphere with those in the Southern Hemisphere having durations of about 2 days less (Lupo, 2021). The most persistent blocking episodes in either hemisphere can last up to 20–30 days often precipitating distinct environmental and social impacts. It has been found that blocking tends to occur in several distinct regions as can be seen in Fig. 6.4, which displays the longitudinal distribution of a blocking index. While blocking is a distinct feature of the eastern side of the Pacific and Atlantic Ocean basins in the Northern Hemisphere (Fig. 6.4a), relatively speaking, blocking is far less important across the majority of the Southern Hemisphere although there is evidence for a concentration of blocking episodes in around the wider New Zealand region (Fig. 6.4b). Of the two Northern Hemisphere blocking regions, the eastern Atlantic possesses the highest frequency of blocking, reaching 24% of days in the Scandinavian region. Only around 7% of days in the eastern Pacific region are characterised by blocking.

In addition to distinct spatial characteristics, blocking displays notable temporal features with significant seasonal variation. For example, North Pacific blocks are

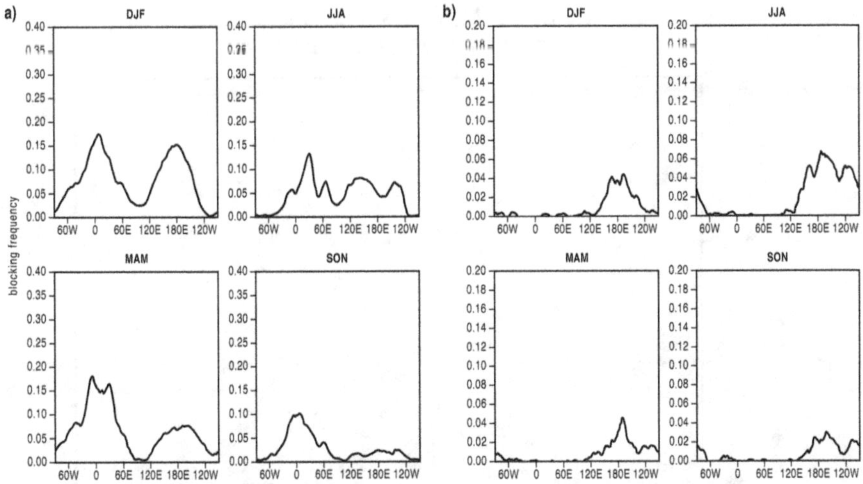

Fig. 6.4 Frequency of "blocked days" for each of the four seasons (DJF, MAM, JJA, SON) for (**a**) the northern hemisphere (1950 to 2000), and (**b**) the southern hemisphere (1979 to 2000) (Adapted from: Climate Predicition Centre of the US National Oceanic and Atmospheric Administration (NOAA). https://www.cpc.ncep.noaa.gov/products/precip/CWlink/MJO/block.shtml)

more frequent during the boreal winter while in the eastern Atlantic sector they occur more often in late winter and early spring. Although mainly a winter phenomenon, in summer, the east Atlantic sector is noteworthy for possessing blocking frequencies more than the annual climatological average. Blocking frequency also displays strong variability at the intra-seasonal to decadal time scales with mixed evidence for robust trends in blocking frequency (Barnes et al., 2014; Efe & Lupo, 2022). While trends in frequency may not be evident, some studies have found that summer blocking episodes have become longer (Wazneh et al., 2021) which bears implications for heatwave duration. Further there is mounting evidence of significant associations between summer heatwave magnitudes and the number of days influenced by atmospheric blocking in regions such as Northern Europe, Western Russia, and East Asia (An & Zuo, 2021; Choi et al., 2021; Schaller et al., 2018) and blocking frequency and heatwave frequency in general (Chan et al., 2019, 2022) (Fig. 6.5).

6.2.2 Advection

Advection refers to transport by the wind, such that an atmospheric property is transported on the wind from one place to another. In the case of heatwaves, temperature and moisture advection may be important as heat and/or moisture arriving over a location from a distant source may affect local temperature and atmospheric humidity levels. The rate of temperature and/or moisture change at a location will

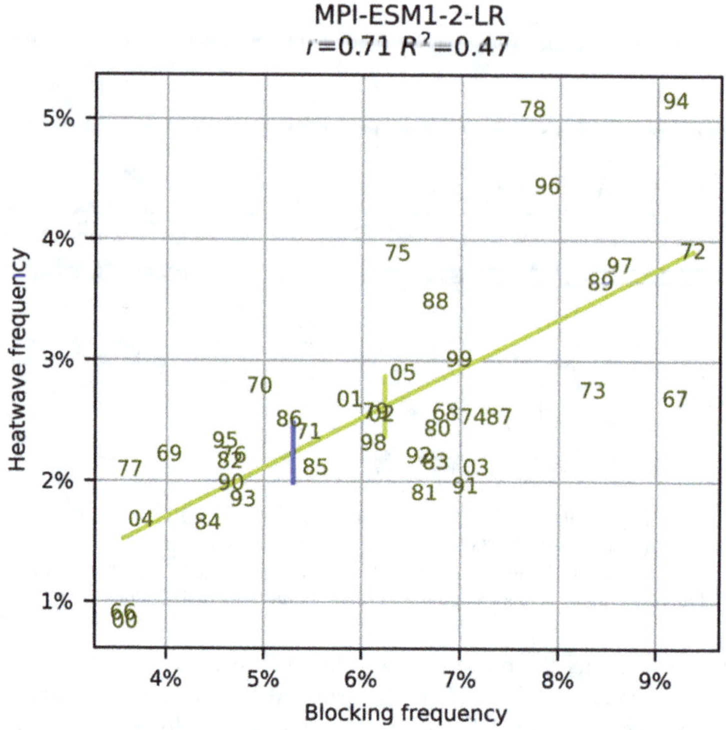

Fig. 6.5 Relationship between blocking frequency and heatwave frequency in the northern hemisphere north of 40 °N for the period 1966–2005 (correlation coefficient (r) is 0.71). Each data point represents the mean for the indicated summer. Years are identified by the associated two trailing digits such that 1966 is 66 and 2005 is 05. The solid line running through the centre of the distribution of data points is the best fit regression line. (Source: Chan et al. (2022). Heatwave–blocking relation change likely dominates over decrease in blocking frequency under global warming. npj Clim Atmos Sci 5, 68. https://doi.org/10.1038/s41612-022-00290-2. Used under Creative Commons Attribution 4.0 International License http://creativecommons.org/licenses/by/4.0/)

depend on the amount of advective heat/moisture flux. This linearly increases with the mean atmospheric temperature and/or moisture content and wind speed. The higher the temperature/humidity and stronger the wind flow the greater the advective flux. Assuming flux convergence continues over an area, with no mechanism (flux divergence) to remove the accumulating heat/moisture, then heat and moisture levels will continue to build.

Several hypothetical situations can be imagined for heat and moisture advection that might be relevant to heatwave development. These might include heat and/or moisture transport from over a hot dry or warm wet land surface or a nearby water body such as a lake, reservoir or warm ocean. Whether a potential heat or moisture source plays a role in heatwave development will depend on the interaction of the local to large scale atmospheric circulation with that source. Accordingly, how locations are positioned downwind of the heat or moisture source is important, with subtle changes in the wind speed and direction critical in determining the role of warm and/or moist air advection in heatwave development. In this regard the degree of sinuosity of the upper westerlies (zonal wave number) can exert a strong influence on the role of advection in heatwave development with heat or moisture advection occurring along the western flanks of the block in the Northern Hemisphere; a Rossby wave pattern with a high wave number will be able to tap into heat and moisture sources from sub-tropical to tropical locations much more so than a pattern with a low wave number. That warm and/or moist air advection plays a role in heatwave development and the elevation of temperature and moisture levels is apparent in several instances with heat/moisture often being transported over considerable distances (Collazo et al., 2018; Cowan et al., 2017; Darand et al., 2018; Demirtas, 2018; Na et al., 2019; Sfîca et al., 2017; Singh & Kumar, 2018) with heat advection possibly dominating over mid-latitude oceans (Roethlisberger & Papritz, 2023).

6.2.3 *Foehn Effect*

Occasionally on the lee side of mountain ranges a warm dry downslope wind is observed. Such winds were originally referred to as Foehn winds to describe the downslope winds north of the Alps that descend over Austria, Germany and Switzerland with their origins being to the south of the Alps. Regional variations on the Foehn are the Chinook on the eastern side of the US and Canadian Rocky Mountains, the Zonda that descends from the Andes over Argentina, the Aspre in France, the Austru in Romania, the Halny in the Carpathian Mountains, the Fogony in the Catalan Pyrenees and the Nor'wester in New Zealand to name a few. Common to all the regional variations of the Foehn wind is a weather pattern that generates an upstream wind flow perpendicular to the mountain range over which it ultimately descends.

The Foehn effect refers to the warming, at the adiabatic lapse rate of 10 °C/km, which the descending air on the lee side of the mountain undergoes, the result of which can be a rapid increase in temperature and a decrease of humidity downwind of the mountain range. Given this, downslope or Foehn-like winds on the lee side of major mountain ranges may play a role in generating heatwaves or short periods of excess heat. For example, there is a strong influence of downslope warming over the Cascade Mountains in the initiation of daytime heatwave events in the Pacific Northwest of the US (Bumbaco et al., 2013) with the Santa Ana winds of southern California also implicated with periods of extreme heat (Schwarz et al., 2020). For New Zealand, Tuller (1980) has commented on the importance of the so-called 'Nor'wester', for driving up temperatures across the Canterbury Plains and their significance for creating stressful human thermal comfort conditions. Over northwest China and for areas to the east of the sizeable mountains in this region, the foehn effect has a substantial influence on creating near surface heatwaves through adiabatic warming in north-westerly winds that descend from the mountains during periods of upper-level westerly flow (Chen & Lu, 2015). For Beijing the foehn winds also appear important for the generation of heatwave events (Ma et al., 2014). A foehn-like wind has been identified as the driver of record-breaking temperatures and heatwaves in Kumagaya City, Japan (Takane et al., 2014, 2017). Geirinhas et al. (2019a) also point to the role of foehn-like regional katabatic winds in heating the already warm air associated with the February 2010 Rio di Janeiro heatwave event. Similarly, foehn winds, in the form of the locally named Puelche wind, appear to be implicated in the development of heatwaves in Santiago, Chile (Barrett et al., 2016). For parts of South Korea, localised heatwaves have also been found to be associated with foehn winds (Lee, 1994; Kim & Hong, 1996; Kim & Min, 2001).

6.3 Atmospheric Processes and Heatwave Development: A Geographical Perspective

From the previous section it is clear blocking is potentially an important atmospheric process for the development of heatwaves. In many ways it has become the default atmospheric explanation for the generation of heatwaves possibly compounded by its clear role in precipitating the severe 1995 Chicago, 2003 European, 2010 Russian and 2021 Pacific Northwest heatwave events which have received considerable attention in the literature (Dole et al., 2011; Heeter et al., 2023; Matsueda, 2011; Neal et al., 2022; Trenberth & Fasullo, 2012). Blocking also appears to have been the driver of the record heatwaves experienced across the US southwest and southern Europe in 2023. In addition to blocking, other processes might be at play in heatwave generation. Given this the purpose of this section is to explore heatwave development through a wider geographical lens not solely focused on Europe and North America to provide insight into the range of observed synoptic scale atmospheric processes that generate heatwaves. This section is organised by broad geographical region given that heatwave development processes may vary depending on the regional climate setting (see Sect. 10.2).

6.3.1 North America

There is no escaping the fact that what happened in the summer of 1995 in Chicago, elevated heatwaves to a place in the minds of meteorologists and climatologists not previously given over to this type of 'unspectacular' meteorological event. With its occurrence it became immediately obvious that a forensic meteorological analysis of this devastating event, which resulted in in over 600 heat related deaths, despite only lasting for a few days, was required. Kunkel et al. (1996), and others since, have provided good descriptions of the development of the Chicago 1995 heatwave which lasted from July 12 to 15. A social autopsy of this event is also highly relevant (Klinenberg, 1999; Klinenberg, 2015).

At the height of the heatwave, around July 14, air temperatures at Chicago's Midway Airport weather station reached 41.1 °C (Kunkel et al., 1996). This is reflected clearly in the spatial pattern of near surface air temperature anomalies across the mid-west of the United States with these exceeding 4 °C in the broader Chicago/Lake Michigan region (Fig. 6.6a). The large-scale atmospheric situation which led to the Chicago heatwave involved the development of an upper-level ridge of high pressure and associated high surface temperatures over the southwestern United States around 5–7 days in advance of the Chicago heatwave onset. Over the period of July 7–12 the high pressure ridge over the southwest of the US moved north-eastwards towards the broader Chicago area bringing with it clear skies and high temperatures. On reaching the eastern half of the US the upper-level ridge stalled and remained stationary from July 12–15 with the development of a clear period of atmospheric blocking (Palecki et al., 2001). Evidence for a classic blocking high pressure is clear in a plot of 500mb geopotential height anomalies (Fig. 6.6b) and the 200mb vector wind pattern for the period of the Chicago heatwave (Fig. 6.6c).

Over the southern parts of Lake Michigan, the 500mb surface was around 120 geopotential metres (gpm) higher than its climatological height over the period of July 12–14 (Fig. 6.6b) indicative of a deep layer of very warm air centred over Chicago. Further the 500mb height anomaly pattern over the west coast, and off the east coast of the US, as well as to the north over the Canadian Arctic, where troughs or centres of low pressure were apparent, is emblematic of a typical Omega block (compare Figs. 6.1a and 6.6b) At the 200mb level the vector wind pattern for July 13–14, 1995 clearly shows the type of airflow associated with the Omega blocking pattern such that the jet stream has been diverted a considerable distance to the north over eastern Canada around the quasi-stationary area of anomalously high pressure with an almost closed circulation of flow over Illinois. Over the west coast, and off the east coast of the US, the airflow pattern is a response to the aforementioned troughs, which along with the presence of the block over Illinois, produced the convoluted pattern of upper level winds which to the west of the block resulted in very hot air advection all the way into the upper mid-west from the southwestern parts of the US (Fig. 6.6c). The very high temperatures experienced across the Chicago area can be understood as a combination of hot air advection from the southwest and high inputs of solar radiation under relatively clear skies associated with the blocking pattern. Interestingly the severity of the heat, and thus the impacts on the population, was exacerbated by unusually high dew point temperatures during the Chicago heatwave. As explained by Kunkel et al. (1996) such high dew points arose because soil moisture levels prior to the heatwave were close to capacity with high temperatures and solar radiation inputs able to drive high evaporation rates, the result of which were unusually high atmospheric water vapour content and humidity levels. Importantly the presence of the blocking high, its regime of subsidence and an attendant strong subsidence inversion restricted the high humidity levels to below 1000 metres above ground level. Consequently, while hourly air temperatures rose

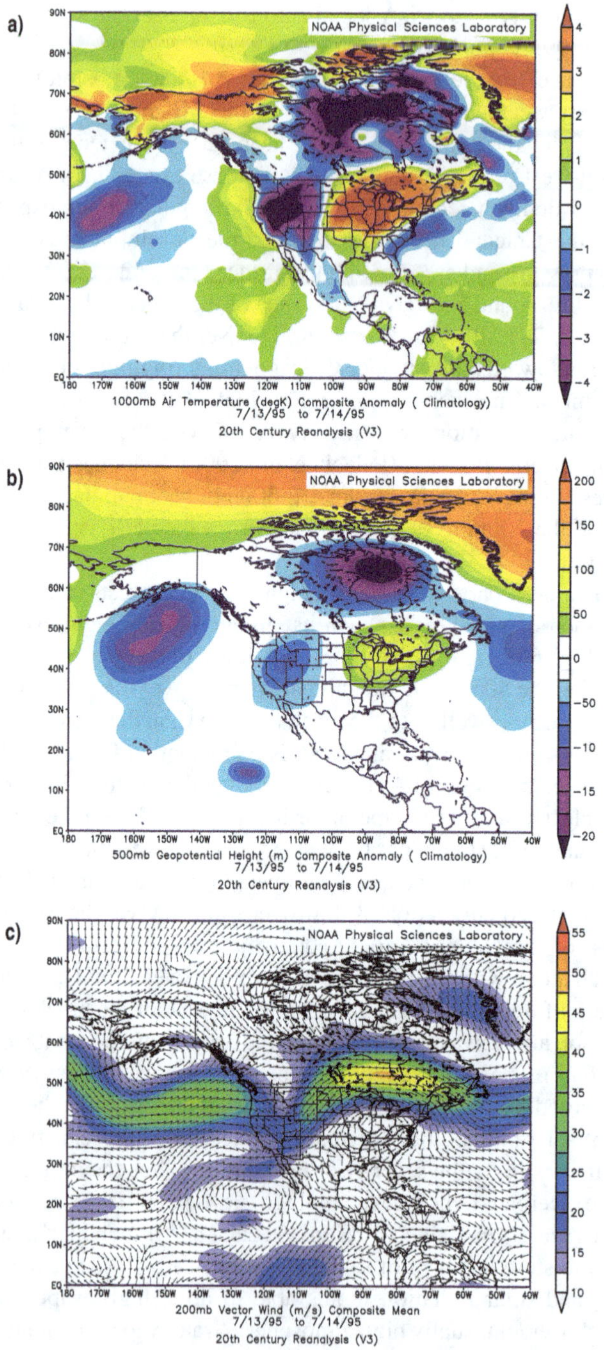

Fig. 6.6 Synoptic scale situation July 13–14, 1995 associated with the Chicago 1995 heatwave. (**a**) 1000mb temperature (°C) anomalies, (**b**) 500mb geopotential height anomalies and (**c**) composite mean wind speed (m/s) at 200mb. Image provided by the NOAA/ESRL Physical Sciences Laboratory (PSL), Boulder Colorado from their web site at https://psl.noaa.gov/ and generated based on NCEP/NCAR Reanalysis data (Kalnay et al., 1996)

to 36—38 °C around the middle of the day at Chicago's O'Hare Airport, apparent temperatures (see Chap. 4) achieved levels more than 45 °C on July 13 and 14 at the climax of the heatwave (Kunkel et al., 1996).

At the time and for some time since, the Chicago 1995 heatwave was considered exceptional. However, the 'statistically implausible' (Thompson et al., 2023) Pacific Northwest heatwave event of June 2021 superseded the Chicago event by some distance with maximum temperatures reaching 49.6 °C. Understandably the unprecedented nature of this event has generated much interest in its development processes (Loikith & Kalashinikov, 2023; Mass et al., 2024). Considerations of the synoptic scale processes associated with the Pacific Northwest heatwave reveal the occurrence of an upper-level atmospheric blocking event in the days preceding the height of the heatwave around June 28–30, 2021. This blocking event was initiated by an episode of anticyclonic wave breaking on June 24 (column (a) Fig. 6.7), an outcome of a weakened jet stream due to slack meridional temperature gradients over the North Pacific at the surface and in the upper troposphere (Neal et al., 2022). Eventually the wave break leads to the development of a classic Omega block from around June 26 (column (a) Fig. 6.7). This sequence of events June 24–28, resulted

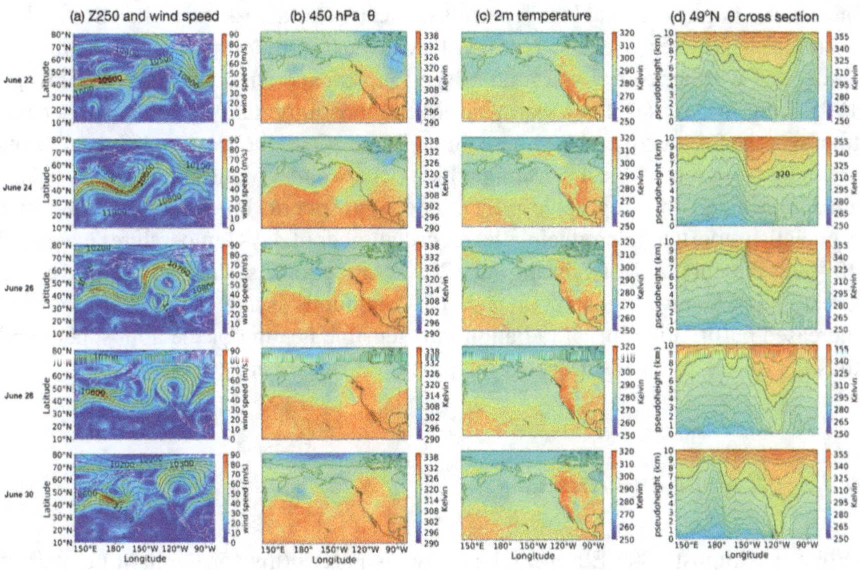

Fig. 6.7 Circulation and temperature over North Pacific and North America 22–30 June 2021. Column (**a**) 250 mb geopotential height (contours in meters) and wind speed (m/sec). Column (**b**) 450 mb potential temperature (Kelvin). Column (**c**) 2 m temperature (Kelvin). Column (**d**) vertical cross sections of potential temperature at 49 °N (contour interval = 5 Kelvin). The 320 K isentrope is highlighted in Column d. (Source: Neal et al. (2022). The 2021 Pacific Northwest heat wave and associated blocking: Meteorology and the role of an upstream cyclone as a diabatic source of wave activity. Geophysical Research Letters, 49, e2021GL097699. https://doi.org/10.1029/2021GL097699. Used under Creative Commons Licence BY-NC 4.0 https://creativecommons.org/licenses/by-nc/4.0/). The dates June 22, June 24, June 26, June 28 and June 30 have been added to the original figure for convenience of reading

in a tongue of warm air being pulled up from lower latitudes and incorporated within the blocking anticyclone (column b, Fig. 6.7). This aided stable stratification of the atmosphere at height and the development of an upper core of very warm air in the block that over time subsided towards the surface (column d, Fig. 6.7) enhancing atmospheric stability, suppression of longwave radiative cooling of the air under clear skies and any possible convection arising from strong surface heating (Neal et al., 2022). Of note is the fascinating possibility that intense latent heating of the atmosphere thousands of kilometres upwind over the North Pacific or even the Northwestern Pacific, resulting from deep cyclogenesis formed a heat source for the Pacific Northwest heatwave event through its conversion to sensible heat via subsidence (Baier et al., 2023; Neal et al., 2022; Schumacher et al., 2022; Terray, 2023). By June 30, the strong surface temperature anomalies associated with the record temperatures align with the blocking pattern suggesting that 2–3 days were required for the various synoptic scale processes, combined with shifting patterns of land surface dryness, to produce the extreme temperatures that had devastating consequences for both human and natural systems (see Chap. 11).

Beyond the detailed analysis of the controls on the development of the Chicago 1995 and the Pacific Northwest 2021 heatwaves, several studies have investigated the general large-scale and local controls on heatwaves across the United States. For example, heatwaves over the Northeast, Midwest and Northern Great Plains are associated with terrestrial warm air advection, as experienced during the Chicago 1995 event, whereas oceanic warm air advection is important for heatwave generation over the Southeast and Southern Great Plains. Over the western US, heatwaves are more a product of local warming which occurs under reduced maritime cool air advection aided by orographic effects and atmospheric blocking with the latter especially important in producing the record number of days temperatures exceeded 40 °C in Phoenix, Arizona during July 2023. At the local scale, especially over the Great Plains, heatwave generation is tied to continuous drying of soil and associated increased partitioning of net radiation into sensible heating of the atmosphere and thus the raising of local surface air temperature (Yang et al., 2019).

6.3.2 *Europe and Russia*

While the August 2003 European heatwave is noteworthy for the high number of heat-related fatalities across Western Europe it is also significant in terms of its duration, intensity and atmospheric origins. On occasions it has been referred to as a mega-heatwave and accordingly, has received much attention in the climatological literature (Beniston, 2004; Black et al., 2004; Jézéquel et al., 2018b; Russo et al., 2015; Schaller et al., 2018; Vautard et al., 2013). Several large-scale atmospheric circulation factors have been suggested as important for the development of 2003 European heatwave. These include persistent blocking, the northward displacement of the Azores anticyclone and the African ITCZ, a strong positive phase of the East Atlantic teleconnection pattern, the southward shift of the extratropical storm tracks, the strong amplitude of the summer Northern Annular Mode and anomalously clear skies and downward net radiative fluxes. In addition to these, prolonged SST anomalies in the northern North Atlantic and the Mediterranean, and land-atmosphere feedback processes (see Chap. 7) have also been suggested as forcing mechanisms (García-Herrera et al., 2010).

In terms of the synoptic climatology of the August 2003 event, what makes it quite remarkable is that it was embedded in a summer with some noteworthy anomalous atmospheric circulation characteristics. Unlike the Chicago 1995 heatwave, the 2003 European event occurred in a summer with persistent and long lasting 500mb height anomalies which had become established several weeks in advance of the August heatwave event. This is clearly seen in a time-longitude plot of 500 mb anomalies which shows a stationary ridge of unusually high pressure centred on 15°E for most of July (Fig. 6.8). This prolonged and extensive area of anomalous high pressure most likely accounts for the mid-July heatwave event that occurred over parts of Western Europe prior to the devastating August event. As July passed into August, the ridge of high pressure intensified and established itself further west over the date line remaining fixed there, and straddling a considerable latitudinal range, until mid-August. The intensity of the ridge is manifest in the magnitude of the 500mb height anomalies which reached 120 gpm over Western Europe during the height of the heatwave around August 8–12 (Fig. 6.8). The areas of anomalously low 500mb heights to the west of the high pressure ridge throughout July and August indicate the presence of a strong wave pattern in the upper level winds as demonstrated in a plot of the 200mb vector wind over Europe (Fig. 6.9a). Evidence of strong subsidence under and east of the intense ridge over central and Western Europe can be seen in the form of the velocity potential anomaly field (Fig. 6.9b). Such deep subsidence produced strong adiabatic heating in the descending air and promoted clear skies and large solar radiation inputs, as well as preventing centres of low pressure moving east off the Atlantic.

Feudale and Shukla (2011) note that the anomalous pattern of subsidence over central and southern Europe that endured throughout the majority of July and August, and contributed to the anomalous summer conditions, was most likely due to the descending branch of the African Inter-Tropical Convergence Zone (ITCZ) being further north than its usual summertime position. While most of the atmospheric evidence in the form of 500mb height and velocity potential anomalies and the pattern of upper-level winds point to a text book Omega block as being responsible for the 2003 heatwave (Ogi et al., 2005; Trigo et al., 2005; Vautard et al., 2005), García-Herrera et al. (2010) contest this based on the analysis of a blocking index calculated for successive 15 day periods throughout July and August 2003. Rather than classic Omega blocking, they suggest the northerly displacement of the Atlantic Subtropical High was more likely a contributing mechanism to the 2003 heatwave.

An interesting perspective on the 2003 heatwave event is offered using back trajectory analysis, a synoptic climatological analysis technique not often applied to the analysis of heatwaves. The purpose of back trajectory analysis is to establish the path followed by air masses arriving at a location with most back trajectory analyses

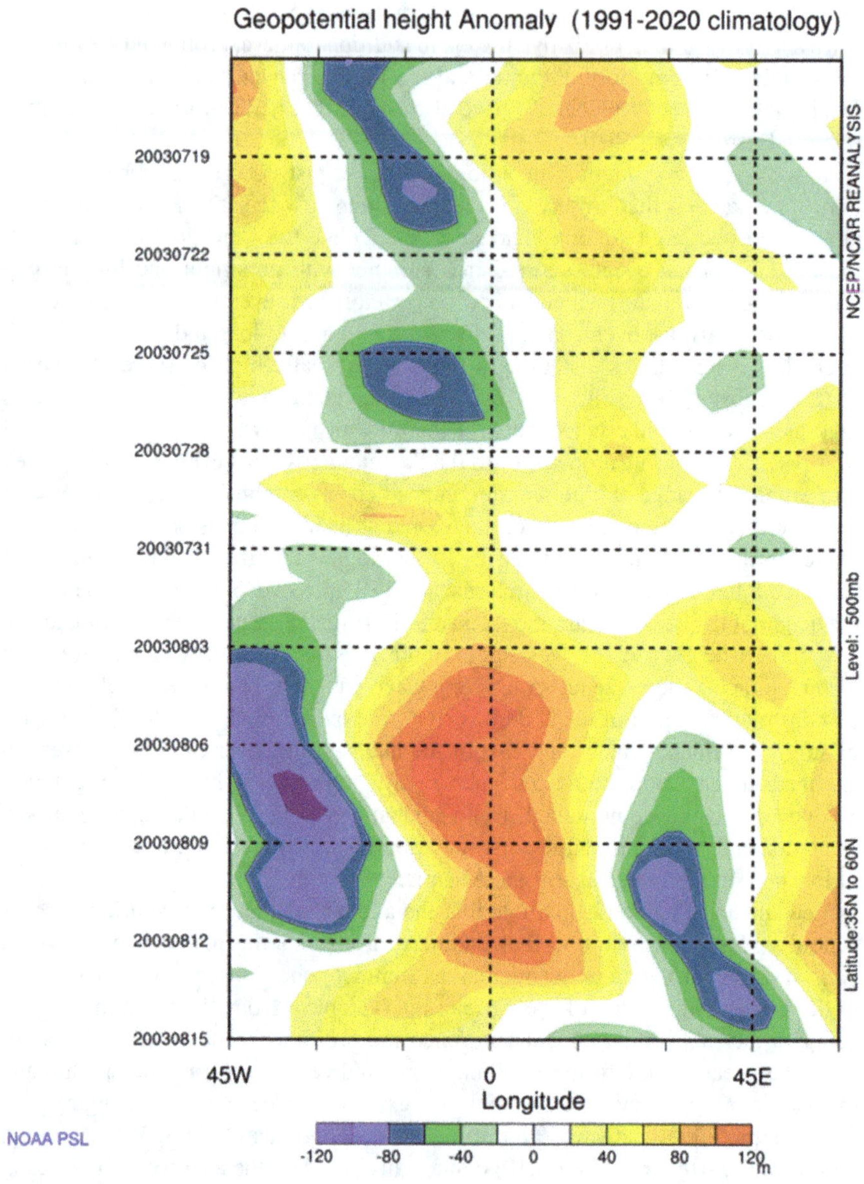

Fig. 6.8 Time longitude plot of 500 mb geopotential height anomalies (compared to climatology 1991–2020) July 16 to August 15, 2003 during the 2003 European heatwave. Image provided by the NOAA/ESRL Physical Sciences Laboratory (PSL), Boulder Colorado from their web site at https://psl.noaa.gov/ and generated based on NCEP/NCAR Reanalysis data (Kalnay et al., 1996)

directed at specific events. A plot of air mass back trajectories using London as a reference point for the period August 1–15, 2003 is shown in Fig. 6.10. This demonstrates clearly how the direction of air mass transport and possible advection of

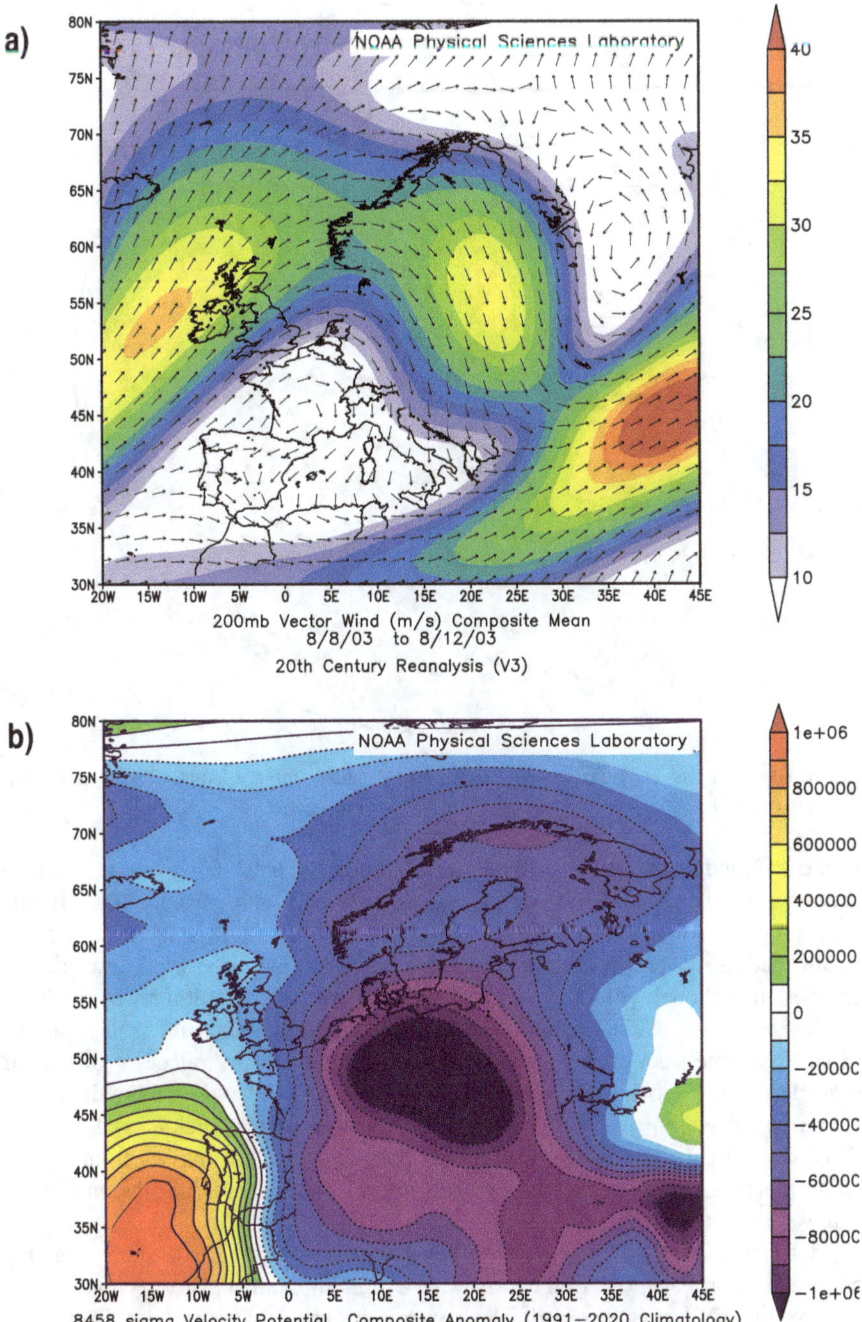

Fig. 6.9 (**a**) Mean vector wind (direction and speed) and (**b**) velocity potential anomaly (compared to climatology 1981–2010), August 8 to 12, 2003 during the 2003 European heatwave. Source: Image provided by the NOAA/ESRL Physical Sciences Laboratory (PSL), Boulder Colorado from their web site at https://psl.noaa.gov/ and generated based on NCEP/NCAR Reanalysis data (Kalnay et al., 1996)

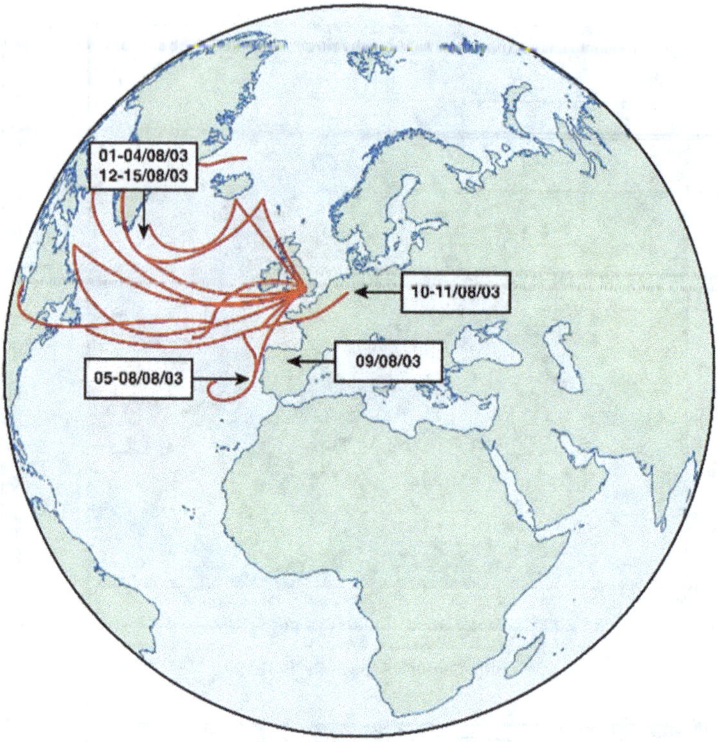

Fig. 6.10 Air mass back trajectory analysis for London for the August 2003 heatwave. (Source: Author)

heat over London changed over the duration of the August 2003 heatwave. Prior to heatwave onset, mild and moist air masses arrived over London from off the Atlantic for the period August 1–4, 2003. However, as the intense ridge of high pressure established itself over the United Kingdom around August 5, the wind flow became far more meridional with air masses transported over London all the way from the Iberian Peninsula to the south. This plume of warm air, often referred to as the 'Spanish plume', continued until August 9 after which the trajectory of the air moved to a more easterly direction during August 10–11 resulting in hot dry continental air advection over London, from central Europe. Come August 12, in the early stages of the breakdown in the high pressure ridge, air masses from off the Atlantic again started to impose themselves on London, heralding the end to the heatwave event after August 13.

In terms of human tragedy, the Russian heatwave of 2010, another mega-heatwave, is probably one of the most severe in living human history with approximately 50,000 deaths recorded over several weeks of extreme heat. The sheer magnitude of this event appears to be outside human experience to date with its incomprehensibility causing some to suggest that this is a sign of the impacts of human induced climate change on extreme events (Rahmstorf & Coumou, 2011;

Otto et al., 2012). That aside the 2010 Russian heatwave is spectacular as a meteorological event and climate anomaly not only because of its duration (Barriopedro et al., 2011) and impacts, but especially as it may have spurned severe flooding over Pakistan (Lau & Kim, 2012). Its origins, as has been the case for previous Russian heatwaves and extreme summer drought events (Atlas et al., 1993; Fischer et al., 2007), are closely tied with the pattern of upper-level atmospheric flow and the occurrence of a prolonged blocking pattern (Dole et al., 2011) which in the case of the 2010 heatwave lasted from around June 20 to August 18.

While the duration and the magnitude of the prolonged 2010 Russian heatwave was impressive, embedded in the broader 8-week long heatwave event was a period of anomalous warmth containing two distinct pulses of heat on July 26–29 and August 4–8 when surface temperature anomalies in Moscow reached a maximum of 12—14 °C and temperatures there remained above 30 °C for 33 consecutive days. Unsurprisingly air temperature records were broken (Grumm, 2011). The key feature of this period of 'extra' heat was an intense and persistent anticyclonic Omega blocking pattern (Fig. 6.11) that split the jet stream into conspicuous polar and subpolar branches, as clearly seen in a plot of the vector wind at 200 mb (Fig. 6.12). Viewed in a wider geographical context, the snaking upper level winds over western Russia, which extended well into the Arctic, belonged to a broader pattern of

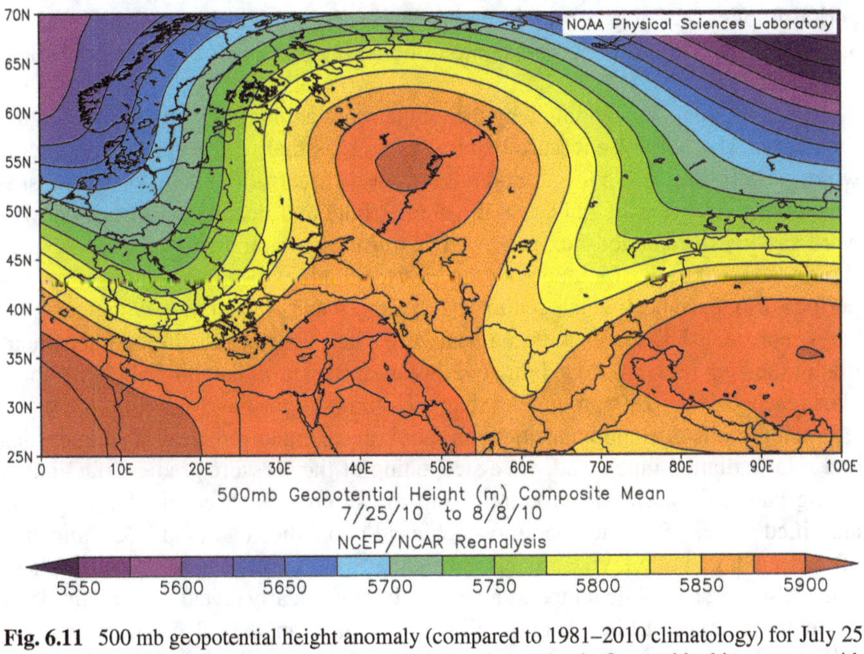

Fig. 6.11 500 mb geopotential height anomaly (compared to 1981–2010 climatology) for July 25 to August 8, 2010, during the Russian heatwave showing a classic Omega blocking pattern with troughs of low pressure on the trailing and leading edges of the ridge of high pressure. Image provided by the NOAA/ESRL Physical Sciences Laboratory (PSL), Boulder Colorado from their web site at https://psl.noaa.gov/ and generated based on NCEP/NCAR Reanalysis data (Kalnay et al., 1996)

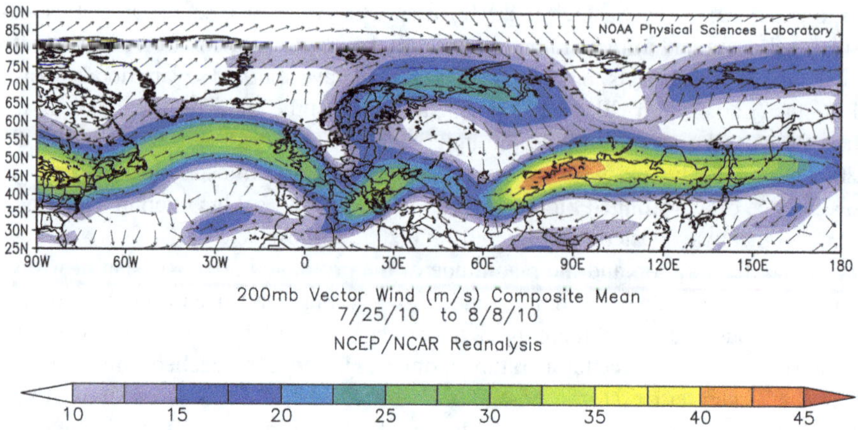

Fig. 6.12 200 mb mean vector wind pattern July 25 to August 8, 2010, during the Russian heatwave. Note split jet with branches to the north and south of the Omega block. Image provided by the NOAA/ESRL Physical Sciences Laboratory (PSL), Boulder Colorado from their web site at https://psl.noaa.gov/ and generated based on NCEP/NCAR Reanalysis data (Kalnay et al., 1996)

perturbed upper westerlies that wandered over a considerable latitudinal range (Fig. 6.12). As noted by Lau and Kim (2012), the 500 mb pressure pattern for this period implies strong geostrophic flow between 60 °N and the near-subtropics at around 30 °N, with the 5850 gpm contour indicating the division between polar and sub-tropical air.

In relation to understanding why the temperature anomalies of July 25 to August 8 occurred, the 500 mb pressure pattern, as well as that of the upper-level vector winds, is helpful. During this period not only were clear skies due to strong subsidence associated with the intense anticyclonic ridge and thus high values of downward shortwave radiation flux present, but warm air was transported over western Russia from subtropical latitudes on the western flank of the Omega block. The two periods of intense heat which occurred July 26–29 and August 4–8, which in some ways represent 'heatwaves within a heatwave', had their origins in the intensification of the high pressure ridge lying over western Russia when 500mb height anomalies reached 160—180gpm greater than the climatological average (Fig. 6.13).

Such periods of ridge intensification would have enhanced clear sky conditions and solar radiation inputs and, via a steepening of the pressure gradient, facilitated strong pulses of warm air advection from the subtropics. Accordingly, under the amplified ridge, 850mb temperatures achieved anomalies of 12—14 °C indicative of a deep layer of very warm air extending well into the middle troposphere (Fig. 6.14a and b). While temperature anomaly plots clearly reveal the magnitude of the temperature departures from the climatological mean, they also demonstrate the shifting geographical pattern of anomalous heat between July 26–29 and August 4–8 (Fig. 6.14a and b). For the latter period, the area of maximum anomalous warmth moves to the east away from the area where the July 26–29 temperature anomalies found their maximum expression. This eastward movement can be

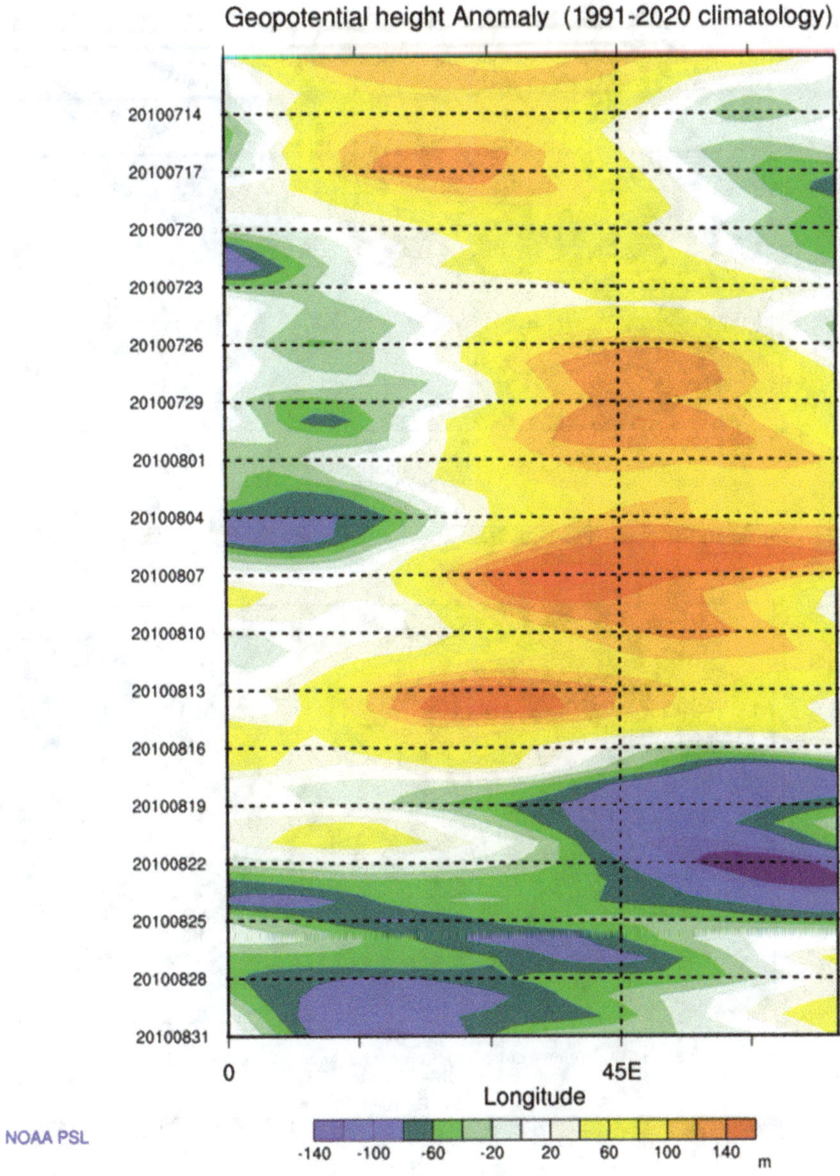

Fig. 6.13 Time longitude plot of 500 mb geopotential height anomalies (compared to 1980–2010 climatology) July 11 to August 31, 2010 during the Russian heatwave. Image provided by the NOAA/ESRL Physical Sciences Laboratory (PSL), Boulder Colorado from their web site at https://psl.noaa.gov/and generated based on NCEP/NCAR Reanalysis data (Kalnay et al., 1996)

accounted for by the changing longitudinal distribution of 500mb height anomalies which expand to the east implying that the high pressure ridge increased its area of intensification and thus impact on near surface conditions (Fig. 6.13). Over the

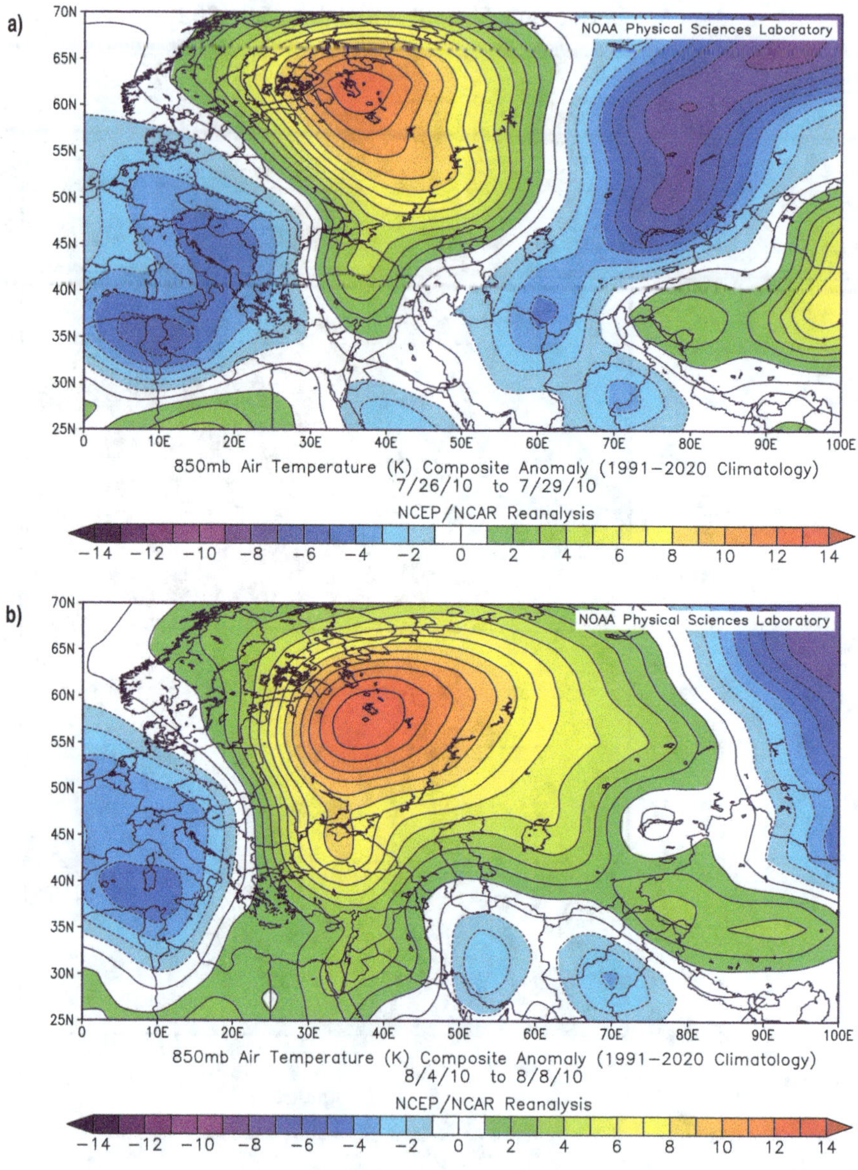

Fig. 6.14 850 mb temperature anomalies (compared to 1981–2010 climatology) for (**a**) July 26 to 29, 2010 and (**b**) August 4 to 8, 2010 during the Russian heatwave demonstrating possible heatwaves with a heatwave. Image provided by the NOAA/ESRL Physical Sciences Laboratory (PSL), Boulder Colorado from their web site at https://psl.noaa.gov/ and generated based on NCEP/NCAR Reanalysis data (Kalnay et al., 1996)

subsequent days the ridge moved back to the west. Eventually the prolonged summer blocking episode that had dominated since mid June was curtailed by the movement over western Russia of a cold front, as manifest by the below average 500mb heights, from around August 18 thus terminating the heatwave (Fig. 6.13).

6.3.3 *Indian Subcontinent*

India has a history of devastating heatwaves with the event of May 23–27, 2015 notable for causing more than 2500 deaths, many of which were concentrated in the state of Andhra Pradesh (Chandran et al., 2017; Dodla et al., 2017; Nair et al., 2017), a coastal province in India's southeast. Unlike the heatwaves described above for Chicago, Europe and Russia, the Indian May 2015 event was not associated with a classic Omega or diffluent blocking pattern. Rather a deep trough extending across northern India from the northwest to the southeast set up a flow of hot dry air that penetrated all the way to the east coast. The advection of hot dry air from the northwest was made possible by a strong pressure gradient between an intense ridge of high pressure oriented in a northwest to southeast direction and lying over Uzbekistan and parts of Kazakhstan and an area of anomalous low pressure centred on northeast Pakistan which extended across northern India. The strong southeastwards advection of hot dry desert air in the direction of Andhra Pradesh was assisted by the presence of anticyclonic flow along the northern flanks of a ridge of high pressure, visible near the surface and the 500 mb level (Fig. 6.15a and b), extending from Africa and the Arabian Peninsula over the Arabian Sea to the southwestern regions of India. Thus, the low pressure trough bound to the north by the foothills of the Himalayas and to the south by the ridge of high pressure with its attendant north-westerly flow along its north to north-eastern flank created essentially a funnel for hot air advection over the south-east coast of India, as portrayed in the 850mb vector wind field (Fig. 6.16). Zore and Landu (2024) discuss the possibility the conditions associated with this heatwave were partly influenced by the Boreal Summer Intra-Seasonal Oscillation.

Another distinct feature, which has been noted for previous heatwaves over the Andhra Pradesh region, such as that which occurred mid-May to early June in 2003 (Bhadram et al., 2005), is a trough of low pressure extending southwards along or just off the east coast of the Indian Peninsula. Such a feature enhances the inflow of hot continental air across the affected region. Critical also to the prolongation of heatwaves over the Andhra Pradesh region is a weak or absent cooling sea breeze, as occurred for both the 2003 and 2015 heatwave events, due to the atmospheric pressure and wind field configurations over the southern parts of the Bay of Bengal (Dodla et al., 2017; Singh & Kumar, 2018; Sandeep & Prasad, 2018). In addition to the damaging heatwaves over coastal eastern India, Ratnam et al. (2016) identified a north-central type as the other major heatwave category in India. These are associated with blocking over the North Atlantic. This results in an upper-level cyclonic anomaly west of North Africa and the generation of anomalous Rossby waves near

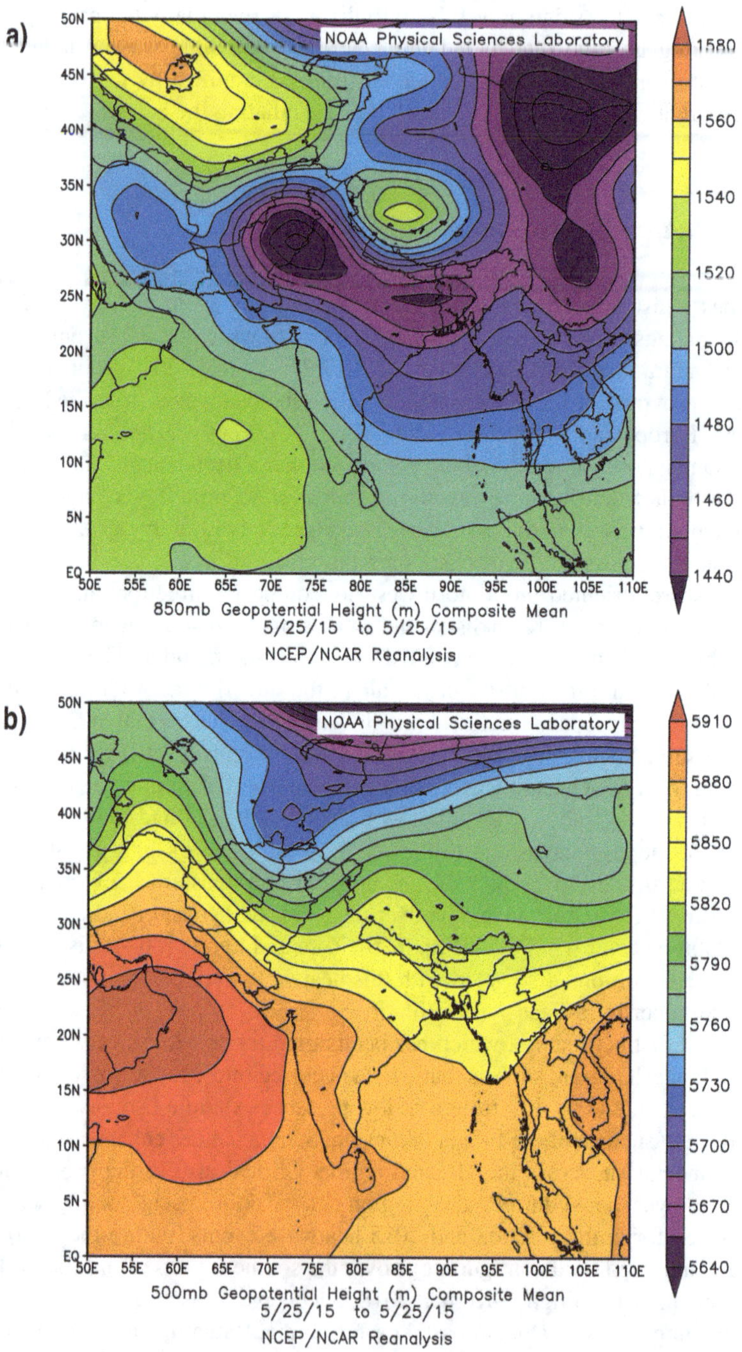

Fig. 6.15 Mean (**a**) 850 mb and (**b**) 500 mb geopotential height fields for May 25, 2015, during the Andhra Pradesh heatwave. Note deep trough of low pressure at the 850 mb level oriented in north-west south-east direction (**a**), flanked by an area of high pressure over the Arabian Sea especially at the 500 mb level (**b**). Image provided by the NOAA/ESRL Physical Sciences Laboratory (PSL), Boulder Colorado from their web site at https://psl.noaa.gov/ and generated based on NCEP/NCAR Reanalysis data (Kalnay et al., 1996; Nair et al., 2017)

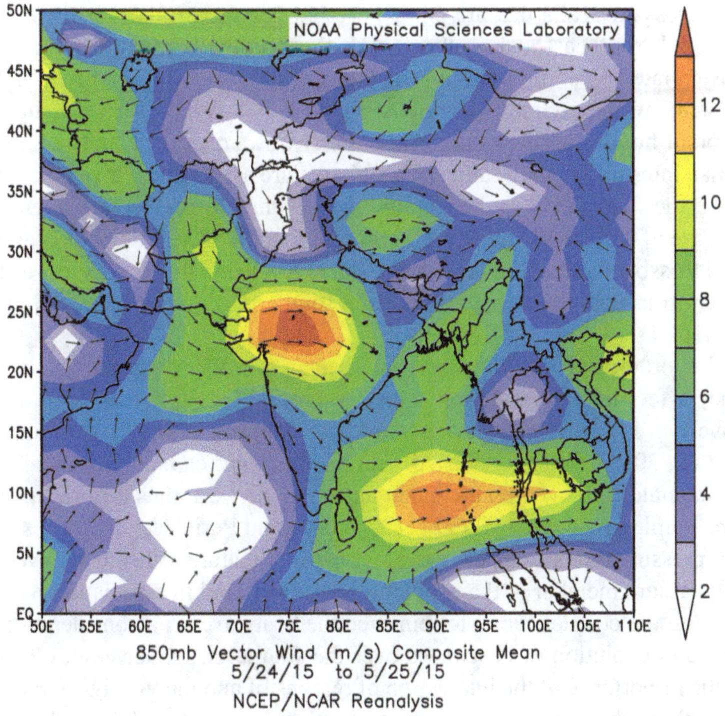

Fig. 6.16 850 mb mean vector (direction and speed) wind field May 24–25, 2015 during Andhra Pradesh heatwave. Note north-westerly flows of air directed towards the southeast coast resulting in heat advection from north-west India, the outcome of the pressure configuration described in Fig. 6.15. Image provided by the NOAA/ESRL Physical Sciences Laboratory (PSL), Boulder Colorado from their web site at https://psl.noaa.gov/ and generated based on NCEP/NCAR Reanalysis data (Kalnay et al., 1996)

the entrance of the African Jet with the resulting quasi-stationary Rossby wave-train producing intense ridging over the Indian subcontinent. In turn this causes anomalous subsidence with adiabatic warming in combination with high radiation inputs producing temperature extremes and heatwave conditions over north-central India (Ratnam et al., 2016).

6.3.4 *Australia*

Because of the significant heatwave events experienced across parts of Australia, especially in January 2009 and 2013, there is a burgeoning literature that addresses the physical dynamics associated with Australian heatwaves. While there's a tendency for the literature to focus on heatwaves in southeast Australia, where the major heatwaves of 2009 and 2013 occurred, Risbey et al. (2018) identify four regions across Australia that are predisposed to heatwaves, these being the southwest, central, southeast and northeast. For the southwest where Perth is located, heatwave related blocking situations manifest in the form of a persistent high pressure system over the Great Australian Bight. This synoptic set up results in the advection of hot dry northerly to north-easterly flows over the southwest, from the Australian interior. For the southeast, heatwaves are associated with a blocking situation over the Tasman Sea in the vicinity of the climatological feature often referred to as the New Zealand blocking region. As for the southwest, this blocking feature is tied to Rossby wave activity over the Indian Ocean and the jet stream wave-guide that steers transient weather systems over southern regions of Australia (Risbey et al., 2018).

The importance of blocking over the Tasman Sea for southeast heatwaves and how they affect major population centres such as Melbourne has been discussed extensively in the literature (Perkins-Kirkpatrick et al., 2016; Parker et al., 2014; Pezza et al., 2012). Often associated with the Tasman Sea blocking situation is an area of anomalously low pressure much further to the west, thus forming a high-low pressure couplet. As noted by Pezza et al. (2012) and Purich et al. (2014) such couplets, or pressure dipoles, appear to be a consistent feature of heatwave events with the low pressure member of the couplet formed over land in association with a heat trough, with the couplet acting to reinforce the heatwave. In a consideration of the structure and evolution of heatwaves over the southeast, Parker et al. (2014) confirmed the importance of the interaction of regions of anomalous high and low pressure such that subsidence of air between an upper level anticyclone and trough can produce warming by compression of the air with this being approximately equivalent to the warming achieved by advection at 1–2 degrees kelvin per day. The near surface alignment of anomalously high and low pressure is critical for determining the direction and intensity of advection. This is clearly seen for the case of the 2009 southeast Australian heatwave for which the position and intensity of the Tasman Sea block and areas of low pressure to the west in a couplet configuration critical in funnelling, from a parched eastern-central Australia, hot dry northerly flows of continental air over the southeast (Fig. 6.17).

In the lead up to heatwaves in the southeast region, Risbey et al. (2018) note the blocking high pressure system is amplified by Rossby wave activity upstream over the Indian Ocean. This corroborates the findings of Purich et al. (2014) who, by tracking the paths of anticyclonic systems leading to heatwave events, show that approximately 8–10 days prior to heatwave occurrence the anticyclonic system is located over the central-southern Indian Ocean. As it tracks from the Indian Ocean towards southwest Australia, transcends the Great Australian Bight and enters the Bass Straight before coming to rest over the Tasman Sea, the 'offending' anticyclonic system goes through a cycle of intensification, weakening and re-intensification respectively. Quinting and Reeder (2017) have added further understanding to the atmospheric dynamics underlying heatwave formation through analysing air mass trajectories that enhance upper-level potential vorticity anomalies and characterize heatwaves over the southeast.

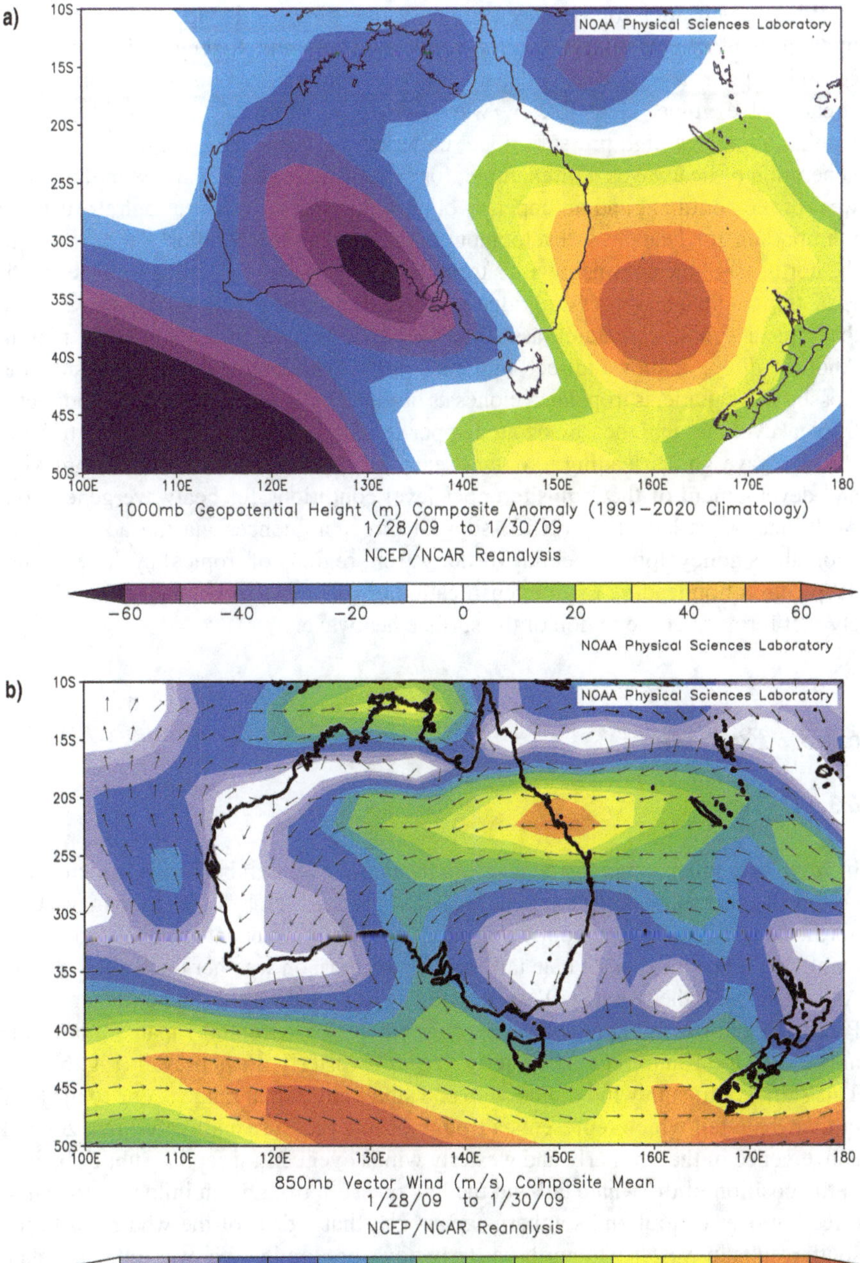

Fig. 6.17 (**a**) 1000 mb geopotential height anomaly (compared to 1981–2010 climatology) and (**b**) 850 mb mean vector wind (direction and speed), January 28 to 30, 2009 during the Australian southeast (Victoria) heatwave. Note high-low pressure anomaly couplet (**a**), and resulting northerly flows conducive to advection of hot dry continental air over the southeast (**b**). Image provided by the NOAA/ESRL Physical Sciences Laboratory (PSL), Boulder Colorado from their web site at https://psl.noaa.gov/ and generated based on NCEP/NCAR Reanalysis data (Kalnay et al., 1996)

While much attention has been drawn to the direct influence of Rossby wave patterns on blocking and heatwave occurrence in Australia, Risbey et al. (2018) note central and northeast Australian heatwaves occur in regions outside the summer jet stream wave guide. Here Rossby wave activity is indirectly responsible for the amplification of a high pressure ridge, underlain by a low pressure trough, and the generation of heatwaves. For example, for central Australia a heat low in the north-west of the continent and an upper-level high pressure ridge over central Australia is important for heatwave development and maintenance. Similarly, a heat low in the northeast of the continent with a trough of low pressure extending into the south-east, along with a high pressure ridge positioned over the east coast of Australia, are the typical synoptic features associated with heatwaves in the northeast region. Another direct and/or indirect influence on Australian heatwaves, outside the Rossby wave guide, is tropical cyclones as outlined by Parker et al. (2013). Indirectly tropical cyclones and their associated upper-level divergent outflow may perturb the Rossby wave guide, leading to downstream Rossby wave overturning or breaking and development of the requisite upper level conditions for heatwave generation. Additionally, Parker et al. (2013) suggest direct influences via the advection of anomalous anticyclonic potential vorticity from regions of tropical cyclone related deep convection leading to an intensification of the upper level anticyclonic anomaly that forms over the region of the surface heatwave.

6.3.5 East Asia

6.3.5.1 Japan

In July 2018 Japan experienced a severe heatwave event embedded in which was a period of record-breaking temperatures that reached 41.1 °C in Kumagaya City northeast of Tokyo. The synoptic origins of this heatwave event are well known and manifest as the so-called 'whale tail' pressure pattern on weather maps (Fig. 6.18a). Associated with the 'whale tail' are gentle flows of southerly warm humid air off the Pacific Ocean over Chubu and Hyogo regions in south central Japan while to the north, over the Kanto region, westerly to north-westerly flows prevail (Fig. 6.18b). This stark contrast in wind direction over a small latitudinal range is due to the kink in the whale tail which represents a trough of low pressure and therefore a zone of convergence of the southerly and westerly winds over central Japan. Subtle changes in the position of the whale tail and the low pressure trough can influence the wind direction over central and southern Japan such that a drift of the whale tail to the south results in westerly to north-westerly flows predominating over most of Japan. Although the near surface pressure pattern resembles a whale tail, with an associated complex surface wind field depending on tail positioning, the synoptic situation at the upper level resembles more an upstream blocking situation as illustrated by a mapping of 500mb height anomalies and the vector wind field for days at the peak of the July 2018 heatwave event (Fig. 6.19a and b). The 'blocking'

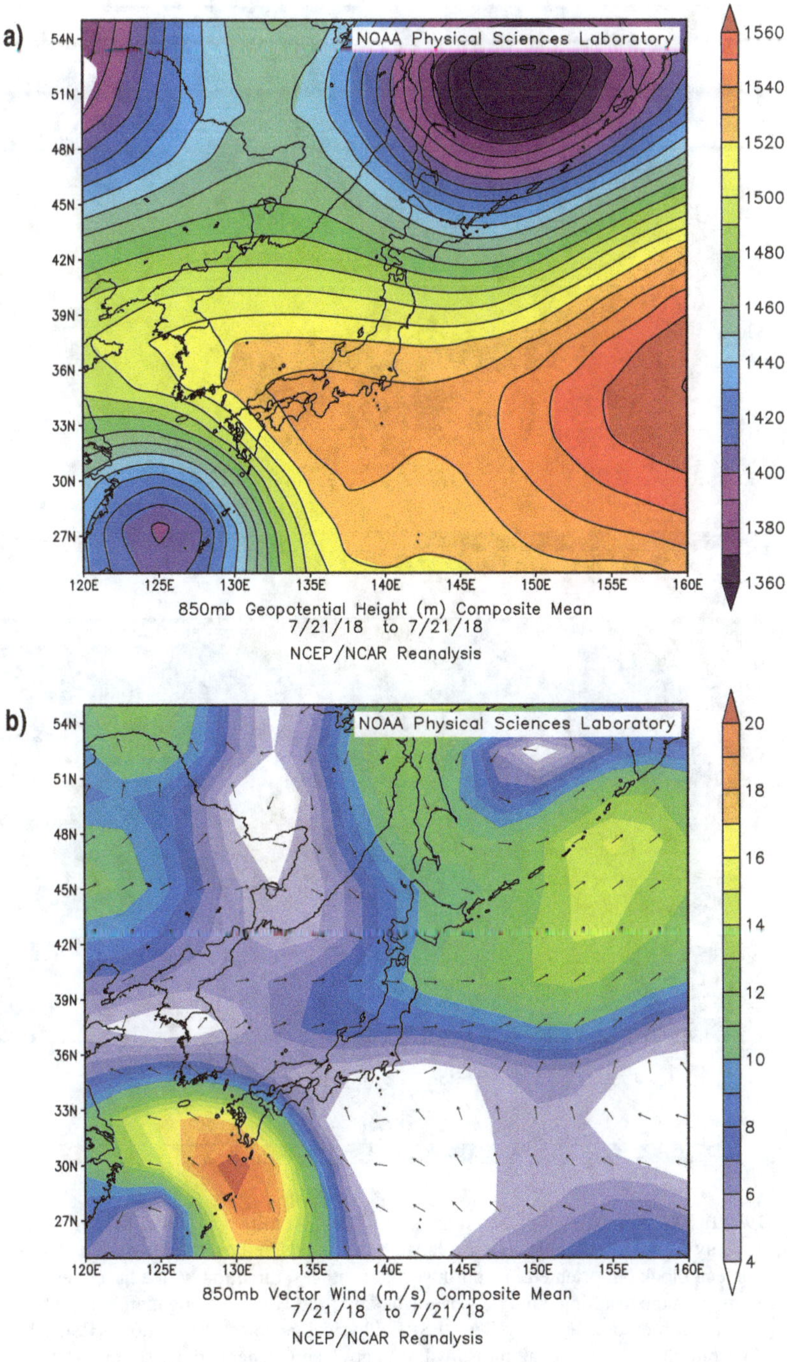

Fig. 6.18 (**a**) mean 850 mb geopotential height and (**b**) mean 850 mb vector wind field (direction and speed) for July 21, 2018 during the Japan 2018 heatwave. Note how the pressure configuration (**a**), gives rise to contrasting wind direction across a relatively small latitudinal range over central Japan (**b**). Image provided by the NOAA/ESRL Physical Sciences Laboratory (PSL), Boulder Colorado from their web site at https://psl.noaa.gov/ and generated based on NCEP/NCAR Reanalysis data (Kalnay et al., 1996)

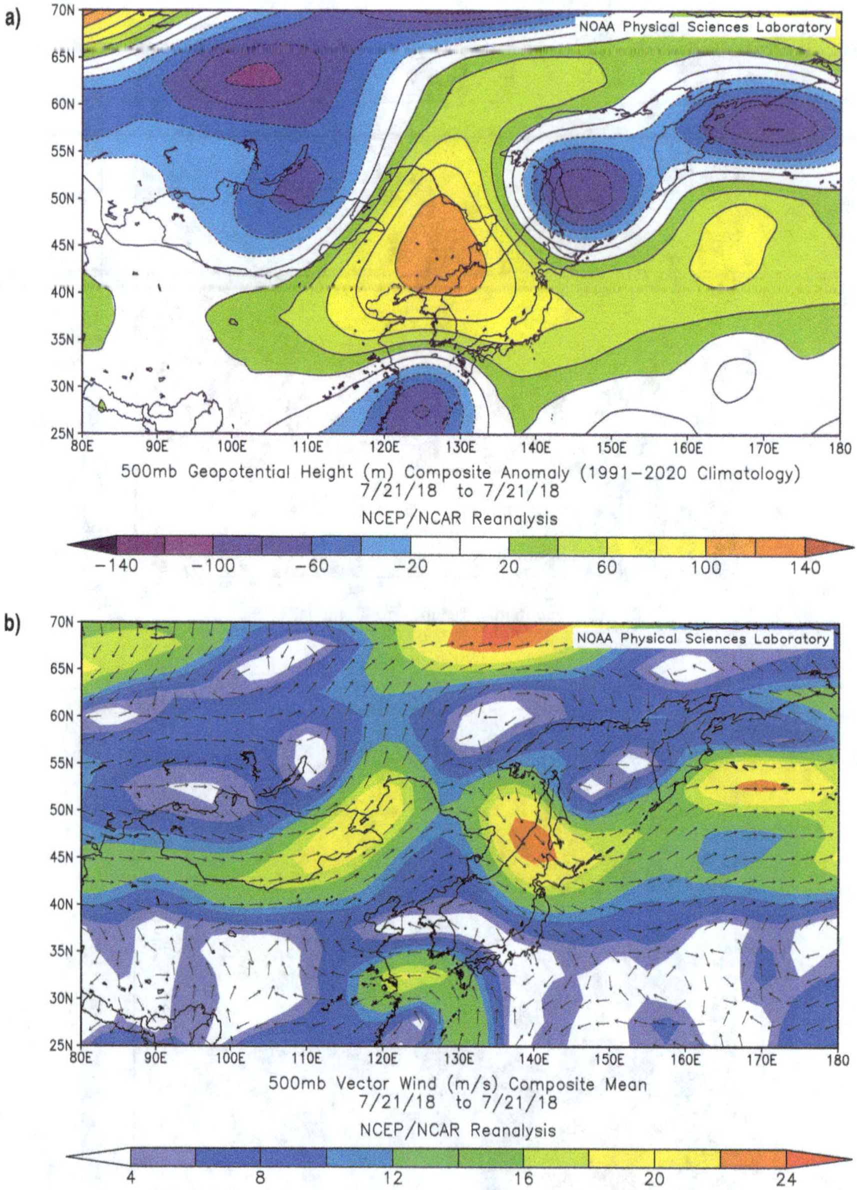

Fig. 6.19 (**a**) 500 mb mean geopotential height and (**b**) 500 mb vector wind field (direction and speed) for July 21, 2018 at the peak of the Japan heatwave. Note how the pressure pattern resembles an Omega block but oriented in a northwest to southeast direction with adjacent areas of low pressure (**a**), with the resulting wind flow over Hokkaido and Honshu being mainly from the northwest (**b**). Image provided by the NOAA/ESRL Physical Sciences Laboratory (PSL), Boulder Colorado from their web site at https://psl.noaa.gov/ and generated based on NCEP/NCAR Reanalysis data (Kalnay et al., 1996)

anomalously high pressure over the Japan Sea, which in many ways resembles an Omega type block, perturbs the upper wind field and the jet stream resulting in north-westerly flows over central Japan on the eastern flank of the block (Fig. 6.19b).

The climatological importance of the whale tail for heatwaves in Japan has been confirmed by systematic analyses of the large-scale synoptic scale conditions leading to extreme multi-day high temperature events with the whale tail pattern found to be associated with 41% of extreme high temperature days for the Tokyo metropolitan area (Takane et al., 2014, 2017). However, because of slight movements in the positioning of the whale tail, geographical contrasts in the importance of this synoptic pattern for heatwave events across Japan exist (Kitada et al., 1998; Onuma, 2003). To understand the physical mechanisms associated with heatwaves Takane & Kusaka (2011) and Takane et al. (2015) analysed the mechanisms giving rise to extreme high temperature events in the Tokyo metropolitan area. They found that there are subtle changes in the driving mechanisms throughout the day. During the morning sensible heat supply is mainly from the surface, via small-scale turbulent sensible heat transfer into the atmosphere. This drives up morning air temperature. In contrast in the afternoon most of the heat input is from advective heat transport from the west. This is associated with a large-scale synoptic pressure pattern, akin to the whale tail type, which generates a foehn-like wind and adiabatic warming originating from the Chubu Mountains to the west of greater Tokyo. This situation for Japan clearly displays the importance of the interplay between large to local scale factors in the generation of heatwaves.

6.3.5.2 South Korea

Although heatwaves in Japan appear to be related to a fairly consistent 'whale tail' synoptic pattern, in near-by South Korea Yeh et al. (2018) note that the record-breaking heatwave of August 2016 was associated with a large-scale atmospheric circulation pattern quite different to that of 'typical' August heatwaves. Rather than strong upper-level geopotential height anomalies over the Korean Peninsula conducive to surface radiative heating because of anomalously high downward shortwave radiation along with warm air advection from the tropics, as found for typical heatwaves (Lee & Lee, 2016; Yoon et al., 2018; Yoon et al., 2020), the August 2016 event was characterised by anomalously low geopotential height south of Japan and low level north-easterly winds around the Korean Peninsula. Further, persistent and anomalous high pressure at upper levels over Mongolia and the Kamchatka Peninsula were distinct features of the August 2016 heatwave. This situation caused warm air advection from a hot and dry Mongolia over South Korea with surface temperatures driven to high levels. In seeking an explanation for the contrasting synoptic features for typical heatwaves and that of August 2016, Yeh et al. (2018) suggest the location of strong convection in the tropics is critical through influencing the generation of Rossby wave trains and the positioning of regions of anomalously high and low pressure in the upper atmosphere in and around the Korean Peninsula. In an analysis of 'pure' heatwaves, as opposed to tropical nights, Hong et al. (2018) note barotropic-like atmospheric conditions, a stable lower layer and dry air due to anomalous easterly winds over the Korean Peninsula, all of which engender above-normal levels of incoming shortwave radiation and adiabatic warming, are important for heatwave generation. In contrast, anomalously high downwards longwave radiation and/or warm humid air advection from the southwest associated with baroclinic-like atmospheric conditions are associated with tropical nights. For the extreme 2018 Korean heatwave event, Yoon et al. (2024) report the importance of time varying influences including the initial importance of the West Pacific North Pacific High for creating stable conditions conducive to strong radiative heating of the surface and adiabatic warming via subsidence, followed by warm air advection from the east enhanced by typhoon passage with heatwave conditions on the lee (western) side of Taebaek Mountains the result of a foehn effect. The significance of a sustained and intense West Pacific North Pacific High for prolonging the 2018 Korean heatwave as well explaining the length of Korea's second longest heatwave in 1994 has been emphasised by Ha et al. (2020). Based on an analysis of a 40-year heatwave record, Choi et al. (2021) note a high frequency of summer blocking over the Sea of Okhotsk suppresses heatwave occurrence over the Korean Peninsula as this positively influences the strength of the East Asian summer monsoon and thus the intrusion of cool rain bearing atmospheric flows. A general situation which contrasts with this is an extension of the North Pacific subtropical high over the Korean Peninsula with a cyclonic anomaly over the East China Sea. This configuration resembles a dipole or Rex blocking pattern. The persistence of this configuration, accompanied by strong subsidence, precipitation suppression, high maximum temperatures, a weakened westerly jet stream, clear skies and increased insolation has been offered as an explanation for the 21-day long heatwave experienced across South Korea in August 2018 (Ha et al., 2020).

6.3.5.3 China

In a search to discover the main synoptic patterns associated with heatwave occurrence over China, Wang et al. (2017) used empirical orthogonal function analysis and identified 'dipole' and 'tripole' 500mb geopotential height anomaly patterns as being important. In the case of the dipole pattern, negative 500mb anomalies, centred over Siberia and covering most of the northern China matched with positive 500mb anomalies over southern China centred over the East China Sea are a feature. This anomalous pressure configuration produces cyclonic (anticyclonic) wind flow anomalies over northern (southern) China and associated below (above) average solar radiation inputs and sensible heat flux. In the case of the tripole pattern, central China experiences negative 500mb height anomalies whereas both northern and southern latitudes are characterised by 500mb height anomalies well above average. As a result, northeast China and parts of southern China are under the influence of anomalous anticyclonic flows while cyclonic wind anomalies are apparent over the Bohai Gulf (Wang et al., 2017). Of the two 500mb height anomaly patterns the dipole pattern accounts for the highest percentage of variance in heatwave occurrence at the inter-annual time scale. Bearing in mind that the patterns identified by Wang et al. (2017) are climatological, and individual heatwave events are likely to demonstrate departures from such canonical patterns, the July 2003 and August 1998 Shanghai heatwave events, which feature widely in the literature (Tan et al., 2007), serve as good examples of the dipole (Fig. 6.20) and tripole (Fig. 6.21) patterns respectively. In both cases, a significant feature is the advection of southerly air over Shanghai from deep in the tropics along the western flanks of the Western North Pacific Subtropical High. In the language of air masses, Tan et al. (2004) refer to such air masses as Maritime Tropical plus (MT+) because of their anomalous heat and moisture content.

In another attempt to establish the dynamical controls on heatwaves, but specifically for central east China where Beijing is located, Freychet et al. (2017) studied composites of a range of climate fields for predefined heatwave events. They found that mid-troposphere high pressure, as represented on 500mb height anomalies, along with enhanced subsidence under a mid-tropospheric ridge and a northward shift of the subtropical jet are conducive to the development of extended heatwave events. This set of circumstances results in enhanced shortwave radiation inputs to the surface favouring elevated surface and boundary layer temperatures. Higher than usual specific humidity was also found to be a characteristic of the events studied. The authors point to this as being important for reducing nocturnal cooling and therefore preserving anomalously high nocturnal temperatures. For a range of regions across China, Chen and Lu (2015) also applied climate field composite analysis in an attempt to uncover the synoptic scale drivers of heatwaves. While the

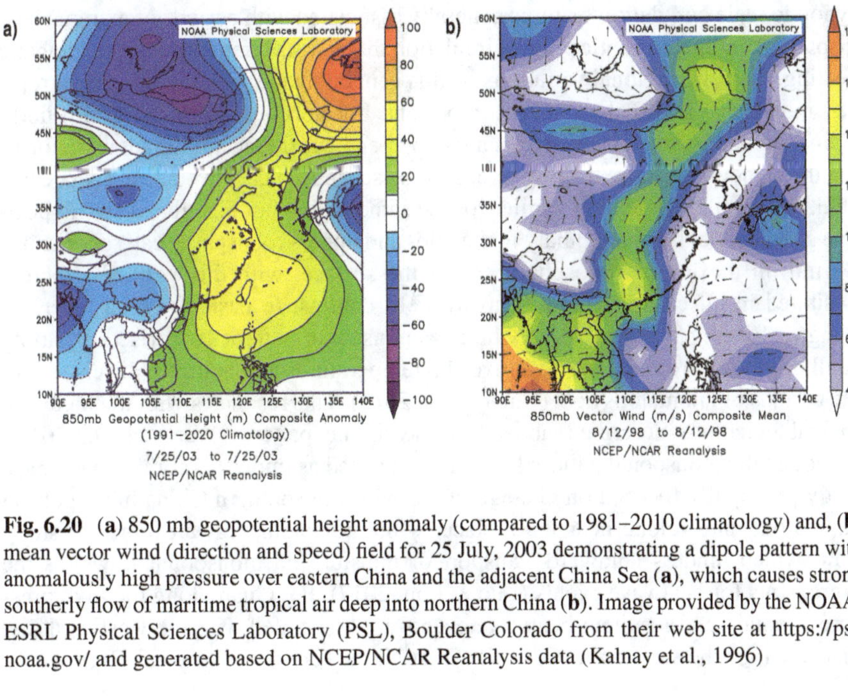

Fig. 6.20 (**a**) 850 mb geopotential height anomaly (compared to 1981–2010 climatology) and, (**b**) mean vector wind (direction and speed) field for 25 July, 2003 demonstrating a dipole pattern with anomalously high pressure over eastern China and the adjacent China Sea (**a**), which causes strong southerly flow of maritime tropical air deep into northern China (**b**). Image provided by the NOAA/ESRL Physical Sciences Laboratory (PSL), Boulder Colorado from their web site at https://psl.noaa.gov/ and generated based on NCEP/NCAR Reanalysis data (Kalnay et al., 1996)

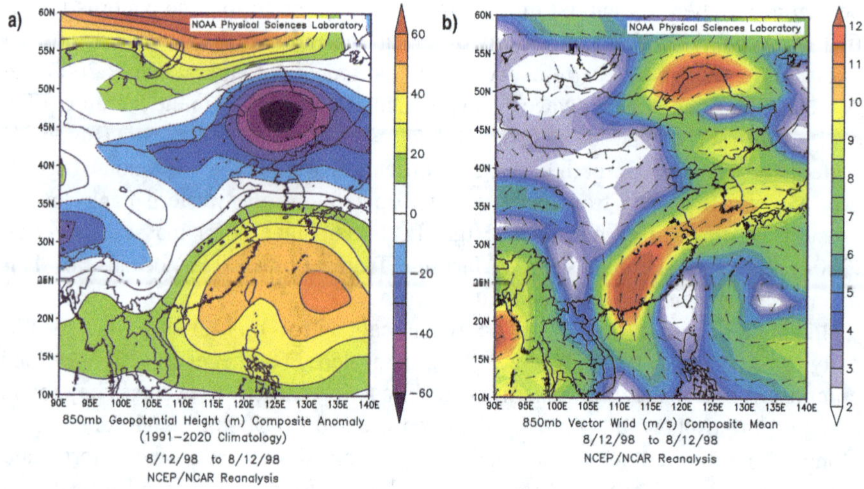

Fig. 6.21 (**a**) 850 mb geopotential height anomaly (compared to 1981–2010 climatology) and (**b**) mean vector wind (direction and speed) field for August 12, 1998 demonstrating a tripole pattern of from the north, high-low-high pressure with a noticeable area of anomalously high pressure over the western Pacific (**a**). This results in advection of maritime tropical air along the western flank for the anomalously strong west North Pacific Subtropical High and over Shanghai (**b**). Image provided by the NOAA/ESRL Physical Sciences Laboratory (PSL), Boulder Colorado from their web site at https://psl.noaa.gov/ and generated based on NCEP/NCAR Reanalysis data (Kalnay et al., 1996)

synoptic scale circulation features generally associated with periods of extreme heat display an anomalous anticyclone circulation that boosts temperatures via adiabatic heating, there are regional variations on this general theme with separate extratropical, monsoonal and foehn patterns important. For north-eastern and eastern north China, an anomalous upper-level anticyclone and subsidence, with adiabatic heating throughout the entire troposphere, sets the scene for heatwave occurrence. Chen et al. (2017) find something similar for the same region but note that a strong positive geopotential height anomaly in the mid- and upper-troposphere caps anomalous warmth in the lower troposphere and at the surface, with this vertical structure exhibited in all eastern China heatwaves. In contrast to eastern China, over the Yangtze River valley and south China, a monsoonal pattern comprised of strong southerly flows from the tropics drives up temperatures and humidity levels to create periods of anomalous heat (Chen & Lu, 2015). This resonates with the meteorological outcomes described above for the dipole pattern (Wang et al., 2017). Although the monsoonal pattern has been suggested as important for south China, a study principally focused on Guangdong province in southern China highlights an anomalous anticyclone near the surface, with anomalous seaward north-westerly flow, which hinders landward transport of moisture on monsoonal flows, as the ingredients for heatwave events (Luo & Lau, 2017). For Guangdong the high pressure anomaly and resultant heatwave events are associated with the westward displacement of the western flank of the West North Pacific Subtropical High with a westward migrating positive 500mb height anomaly over the course of the heatwave event not only over Guangdong but also parts of Indochina. For northeast China (NEC) Wang et al. (2018) found that an anticyclonic circulation anomaly prevailing over the west of NEC and a higher summer surface sea temperature in Japan Sea results in more frequent heatwaves.

As noted previously composite analyses provide an average picture of the synoptic situation associated with a particular climate event. Acknowledging this, Wang et al. (2016) undertook a synoptic climatological analysis of specific heatwaves events in 2003, 2006, and 2013 to increase the granularity of the understanding of

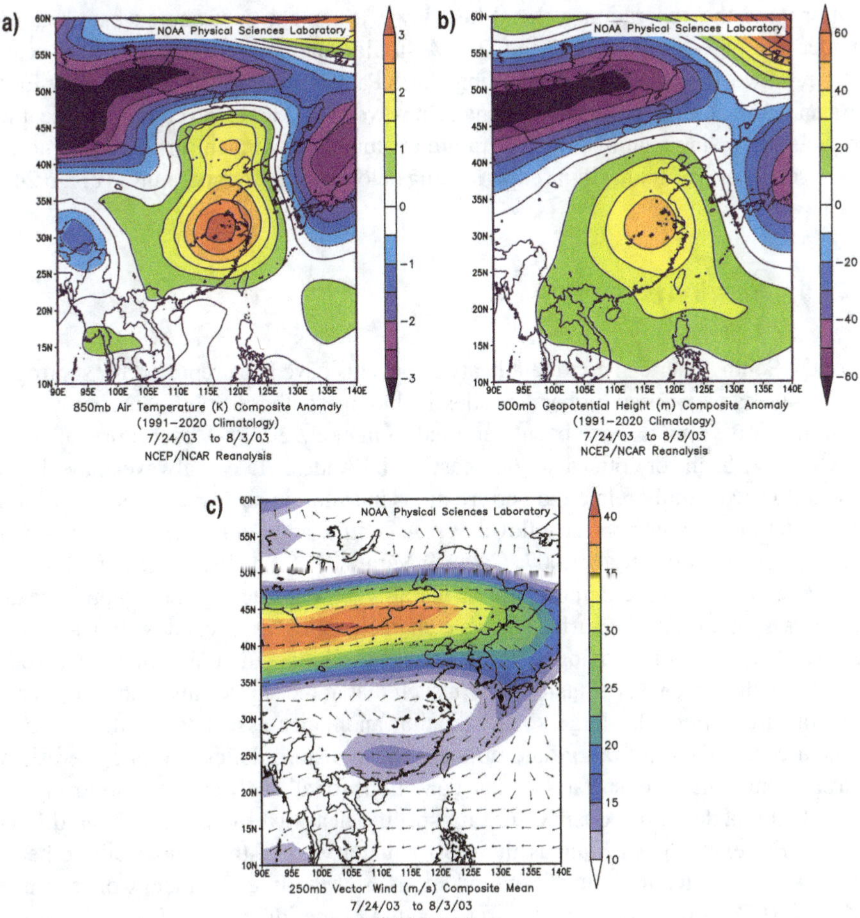

Fig. 6.22 (**a**) 850 mb temperature and (**b**) 500 mb geopotential height anomalies (compared to 1981–2010 climatology) and (**c**) mean vector wind (direction and speed) field for the July 24—August 3, 2003 heatwave event over eastern coastal China. Image provided by the NOAA/ESRL Physical Sciences Laboratory (PSL), Boulder Colorado from their web site at https://psl.noaa.gov/ and generated based on NCEP/NCAR Reanalysis data (Kalnay et al., 1996)

heatwave dynamics in China. For a heatwave event in late July early August 2003, located mainly over the south-eastern coastal provinces of China (Fig. 6.22a–c), Wang et al. (2016) found that the southward displacement of the East Asian jet stream induced anticyclonic anomalies to the south, together with a south-westward intensification of the western side of West North Pacific Subtropical High pressure system. Under this situation the heatwave event was found to be mainly restricted to south of the Yangtze River. In the case of the 2006 heatwave, mainly confined to south-west China, an anomalously northward movement of the East Asian jet stream along with a remarkable westward extension of the West North Pacific Subtropical High produced a substantive physical barrier to the penetration of the southwest monsoon over the region resulting in widespread moisture deficits, dry surface conditions and the generation of local heat lows, testament to strong surface heating under unusually clear skies (Fig. 6.23). A similar interaction, as for the 2003 heatwave, between the westward extending North Pacific Subtropical High and the East Asian jet was found for the East China heat wave of 2013. However, in this case, the region possessing significant temperature anomalies was much further to the north with some parts of south China experiencing below average temperatures (Fig. 6.24).

6.3.6 South America

For the South American continent heatwave events have been reported for subtropical latitudes as well as higher latitudes including southern South America where four heatwave regions have been identified (Suli et al., 2023). Most reports of heatwaves have been for countries lying east of the Andes. Here heatwaves have been found to be generally related to perturbations in synoptic-scale wave activity which promotes anomalous northerly flows over Argentina and Paraguay, southern Brazil and Uruguay (Cerne et al., 2007; Cerne & Vera, 2011). For Brazil, Geirinhas et al. (2018) have lent some detail to this generality by considering the synoptic scale processes, as well as the surface climate characteristics, associated with heatwaves for several large cities. Using a composite approach, they note that there is a well-defined radiative energy transfer pattern such that at the surface all heatwaves demonstrated exceptionally large solar radiation anomalies associated with clear sky conditions, and precipitation and atmospheric moisture deficits. While there is a strong commonality of surface climate conditions for all heatwaves, the atmospheric circulation patterns responsible for city specific heatwaves displayed some differences. For example, in Manaus the most equatorward of the cities studied, heatwaves are sensitive to the position and strength of the Inter-Tropical Convergence Zone (ITCZ) which in turn is tied to the nature of meridional sea surface temperature gradients over the tropical Atlantic Ocean and the position of the South Atlantic Convergence Zone (SACZ). Heatwaves occur in Manaus when the ITCZ is displaced much further north than its summer climatological position. In the upper atmosphere strong positive 500mb height anomalies promote a regime of subsidence in the descending arm of the Hadley Cell circulation and clear skies. Such a

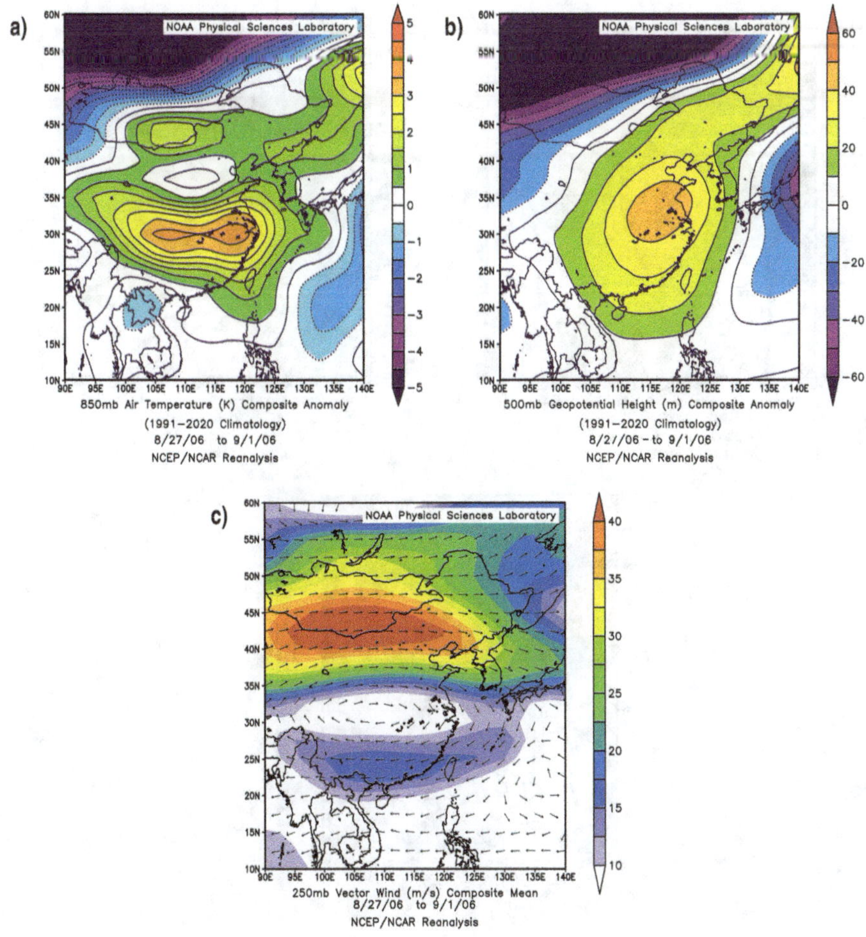

Fig. 6.23 (**a**) 850 mb temperature and (**b**) 500 mb geopotential height anomalies (compared to 1981–2010 climatology) and (**c**) mean vector wind (direction and speed) field for the August 27—September 1, 2006 southwest China heatwave event. Image provided by the NOAA/ESRL Physical Sciences Laboratory (PSL), Boulder Colorado from their web site at https://psl.noaa.gov/ and generated based on NCEP/NCAR Reanalysis data (Kalnay et al., 1996)

weakened ITCZ and the associated processes allow strong radiative forcing of high surface temperatures. For Sao Paulo and Porto Alegre, heatwaves are associated with strong positive 500mb geopotential height anomalies positioned over the southern coast of Brazil. This pressure configuration is representative of a persistent and intense quasi-stationary upper-level anticyclonic circulation pattern which promotes strong surface heating via anomalous surface radiation receipts. A key player in the atmospheric set up for heatwaves for these two cities is the westward expansion of the South Atlantic Subtropical High (SASTH) which leads to the development of the persistent upper-level high pressure ridge. The extension of the SASTH

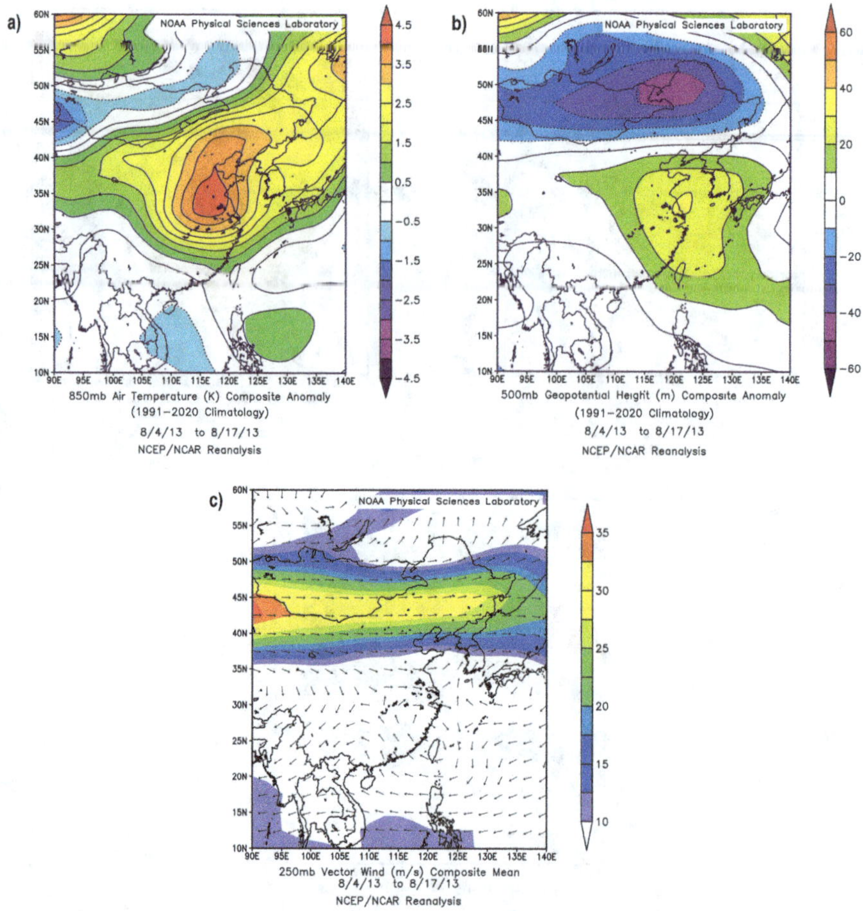

Fig. 6.24 (**a**) 850 mb temperature, (**b**) 500 mb geopotential height anomalies (compared to 1981–2010 climatology) and (**c**) mean vector wind (direction and speed) field for the August 4–17, 2013 East China heatwave event. Image provided by the NOAA/ESRL Physical Sciences Laboratory (PSL), Boulder Colorado from their web site at https://psl.noaa.gov/ and generated based on NCEP/NCAR Reanalysis data (Kalnay et al., 1996)

creates a blocking situation such that the northward movement of baroclinic waves in the form of a surface cold front with their attendant cool and moist conditions is prevented.

For Argentina, Cerne et al. (2007) have provided a detailed account of the sequence of synoptic scale processes associated with a heatwave that occurred in 2003 between January 23 and February 2. This event arose from the complex interplay between several of the components of the atmospheric circulation system over subtropical South America, including the Bolivian High over the eastern Pacific, the trough and ridge system associated with the South Atlantic Convergence Zone (SACZ), the SASTH, the Argentinian Low and the South American Low-Level Jet

(SALLJ). Leading up to the heatwave event, the Bolivian High weakened as did the SACZ as the SASTH expanded westward over eastern Argentina. This promoted a regime of strong upper-level subsidence which suppressed convective rainfall. Near the surface an extratropical front intensified the development of the so-called Argentinian Low which is essentially a summertime thermal low. The intensification of the Argentinian Low subsequently led to the strengthening of the SALLJ and the occurrence of anomalously strong northerly flows and the advection of heat over Argentina. Many of the features diagnosed for the 2003 event by Cerne et al. (2007) and other heatwaves of note (Campetella & Rusticucci, 1998; Rusticucci et al., 2016) are apparent in the synoptic fields for the extreme heatwave that occurred over Argentina at the end of 2013 when maximum temperatures in northwest Argentina reached 45 °C and remained more than 35 °C in Buenos Aires for several days. Estimated as a 1 in 100 year event (Barros et al., 2015), temperature anomalies at the 1000mb level exceeded 6 °C over the period of December 23–30, 2013 (Fig. 6.25a). Consideration of the 1000 mb geopotential height field clearly shows the near surface northern Argentinian Low stretching northwards over northern Argentina and into Paraguay (Fig. 6.25b) generated by strong surface heating. This enhanced the SALLJ and produced a strong northerly flow and the advection of warm air over northern Argentina (Fig. 6.25c). Fortuitously, Cerne et al., 2007 were able to quantify the heat sources for the 2003 Argentinian heatwave because this event occurred at the time of a major field experiment. Based on what they describe and drawing parallels between the development of the 2003 and 2013 event, there were probably time-varying sources of heat that contributed to the development of the 2013 heatwave (Alvarez et al., 2019). As for the 2003 event it is likely that diabatic warming, because of enhanced convection in the tropics, contributed to the positive tendency of temperature anomalies in the early stages of the 2013 event. Both tropical convection and the development of the intense subtropical anticyclone over the west coast of South America would have also favoured subsidence and the development of clear skies and large solar radiation inputs over northern Argentina promoting strong sensible heating of the boundary layer and the establishment of the heatwave. As the strong surface heating under clear skies progressed, intensification of the northern Argentinian Low would have facilitated strong heat advection from the Amazon Basin on northerly flows within the SALLJ with heat advection most likely supplanting local sensible heating as the major heat source for the heatwave. This set of circumstances is very similar to those described for the February 2010 Rio de Janeiro heatwave event in Brazil (Geirinhas et al., 2018; Geirinhas et al., 2019b) and for the wider record-breaking heatwave event across South America, January 9–16, 2022 (Zhang & Xie, 2023).

On the other side of the Andes, Barrett et al. (2016) have described quite a different set of synoptic scale circumstances for the development of a heatwave in March 2015 over central and southern Chile, including Santiago where a record-breaking surface air temperature of 36.8 °C was recorded. Based on an analysis of large-scale climate fields they identified the likely heatwave forcing mechanisms to be anomalous ridging in the lower troposphere and easterly downslope flow off the Andes as a response to an anomalous mid-troposphere ridge-trough pattern; the downslope

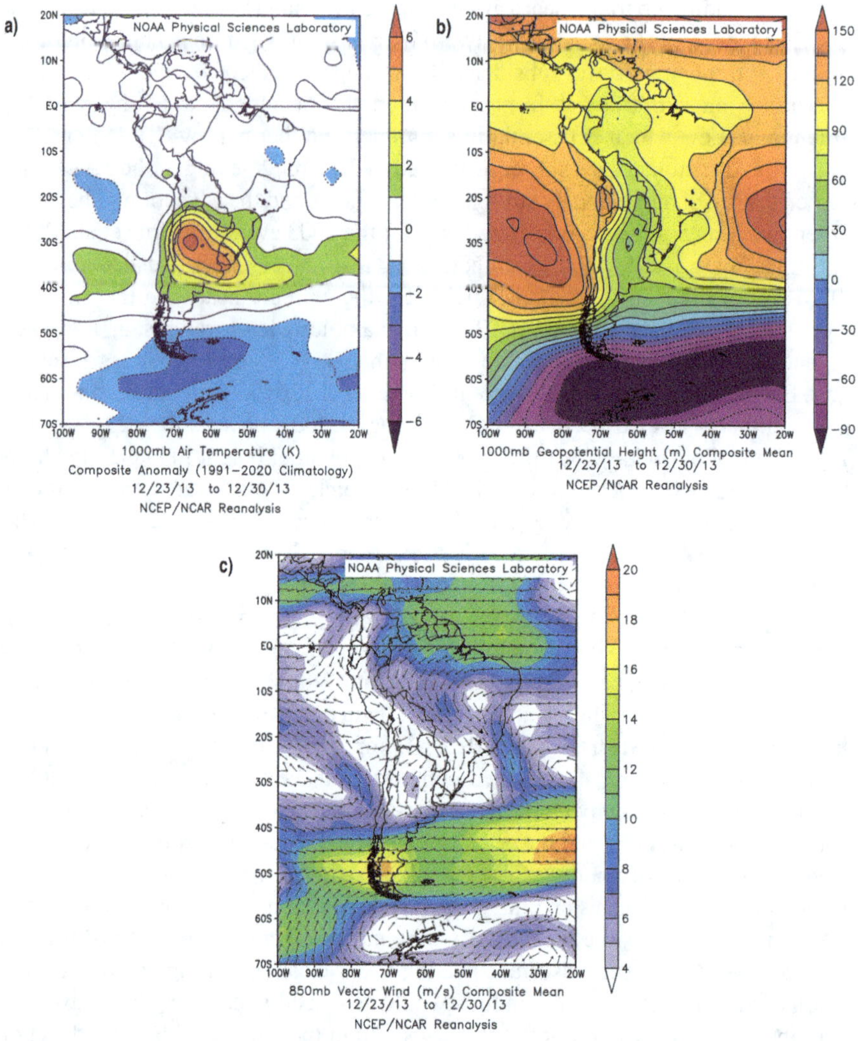

Fig. 6.25 (**a**) 1000 mb air temperature anomaly (compared to 1981–2010 climatology), (**b**) 1000 mb mean geopotential height, and (**c**) mean 850 mb vector wind (direction and speed) field associated with the Buenos Aires December 23–30, 2013 heatwave event. Image provided by the NOAA/ESRL Physical Sciences Laboratory (PSL), Boulder Colorado from their web site at https://psl.noaa.gov/ and generated based on NCEP/NCAR Reanalysis data (Kalnay et al., 1996)

flow is a foehn-like wind referred locally to as the Puelche wind. Interestingly, the heatwave event over central and southern Chile was matched by a damaging extreme precipitation event in northern Chile caused by the steering of a cut-off low which moved slowly eastwards and around the strong high pressure ridge associated with the heatwave further to the south. Such a spatial juxtaposition of extreme heat and precipitation events has also been noted for Australia (Perkins-Kirkpatrick et al.,

2016) with heatwaves in the Australian southeast being matched by anomalous precipitation events in northeast Australia. Also noteworthy is the extreme precipitation and flood events experienced over southeast Brazil during the time of the Argentinian December 2013 heatwave because of a very active SACZ due to the convergence of northerly flows from the Amazon Basin generated by a northern Argentinian Low enhanced SALLJ and easterly winds flowing around the northern flank of the SASTH. The association of the SACZ with heatwaves over southeast Patagonia has also been noted with two thirds of all heatwave events over the period of 1872–2010 linked with enhanced ascent in the SACZ (Jacques-Coper et al., 2021). This set of circumstances and that recorded in this chapter for Pakistan, Chile and Australia indicate that teleconnections may well exist with heat and precipitation extreme events occurring almost simultaneously at distant locations linked by anomalous synoptic scale conditions over large distances.

6.3.7 Africa

Most diagnostic studies of the physical mechanisms associated with heatwaves have focused on countries with developed economies where large populations concentrated in urban areas with a considerable GDP may be affected. This contrasts with the situation for Africa where the understanding of the nature and driving mechanisms of heatwaves is still in its infancy (Harrington & Otto, 2020). In response to this there has been a noticeable upturn in the number of climatological studies of heatwaves across the African continent such as for West Africa, the Sahelian region, Kenya, Mozambique, Senegal and Tanzania (Amou et al., 2021; Bouniol et al., 2021b; Engdaw et al., 2022; Gyilbag et al., 2021; Iyakaremye et al., 2022; Langue et al., 2023; Marghidan et al., 2023; Sambou et al., 2020).

To understand the large-scale drivers of heatwaves in Africa Fontaine et al. (2013) undertook a composite analysis of spring and summer heatwave episodes of at least 4 days duration across Northern Africa for the period 1979 to 2011 noting that clear evidence exists for anomalous atmospheric circulation patterns during heatwaves. For spring heatwaves, they found strong surface and lower atmosphere positive temperature anomalies over the Sahel and Sahara to be associated with a low and mid-level cyclonic rotation over Morocco. At the upper level a well-established Rossby wave pattern caused anomalous vertical ascent (descent) over the western (eastern) regions of North Africa. This configuration leads to a weakening of the Harmattan with an increase (decrease) of atmospheric moisture west (east) of the dateline. Fontaine et al. (2013) also note that in the spring months two to three heat waves propagate eastward. This points to the roles of short Rossby waves possibly associated with a distortion of the jet stream. Prior to the movement of spring heat waves in the direction of the Arabian Peninsula, anomalously warm conditions over Libya and south westerly flows over Western Sahara occur. At the same time a large trough is positioned over the North Atlantic. Downstream of this there is enhanced mid-tropospheric subsidence and anticyclonic rotation over the continent which then migrates toward the Arabian Peninsula and eventually decays (Fontaine et al., 2013). In the case of summer heatwaves, similar but weaker positive low-level temperature anomalies (less than 3 °C) are associated with a cyclonic anomaly in central Sahel (Fontaine et al., 2013). This favours the eastward progression of the West African monsoon to the dateline. Concomitantly a midlevel anticyclonic anomaly forms over the Western Sahara, which increases flux divergence associated with the African Easterly Jet in a southward direction with less atmospheric moisture westward to 10 degrees west.

Bearing in mind the description of the atmospheric circulation patterns provided for summer North African heatwaves by Fontaine et al. (2013) is based on a 30 year composite, some but not all of the aforementioned features are evident for the late June early July 2017 heatwave over Libya when temperatures reached 47 °C in Tripoli and the city was plunged into darkness due to power cuts. During this event temperature anomalies reached 4-6 °C in the lower atmosphere. In fact, a large area of anomalously high temperatures stretched from the North African coast across the Mediterranean and over Greece where temperatures reached more than 40 °C in Athens (Fig. 6.26a). Consideration of the 850mb pressure anomalies for June 28—July 1, 2017, show some distinct perturbations in the pressure field, including an intense anticyclonic anomaly over the Azores, a deep trough extending from the mid latitudes over Algeria and Morocco and to the east, a ridge of high pressure stretching southwestwards from the Black Sea region over Greece and into the eastern Mediterranean (Fig. 6.26b). This atmospheric configuration generated a weak cyclonic circulation over western North Africa and strong anticyclonic flows around the high pressure ridge over eastern North Africa (Fig. 6.26c). The upshot of this was southerly flows over the northwestern regions of Libya and advection of hot dry air off the Sahara over Tripoli. A regime of subsidence over the eastern Mediterranean and northeast Africa, associated with the anticyclonic ridging there, would have facilitated clear skies and strong downward fluxes of shortwave radiation forcing strong surface heating, adding to the advective heat source from the Sahara and the development of the 2017 Libyan heatwave.

Specifically for West Africa and the countries mainly affected by the West African monsoon, Oueslati et al. (2017) describe the atmospheric dynamics and radiative and turbulent fluxes associated with heatwaves for this region. Their analyses point to quite a different set of mechanisms compared to North Africa and the situation described above for the 2017 Libyan heatwave. They identify three main contributors to the generation of heatwaves. The direct main driver appears to be the greenhouse effect of water vapour such that enhanced downward longwave radiation from a moist atmosphere raises minimum temperatures via a water vapour feedback mechanism as outlined by Guichard et al. (2009). Critical to this is the role the atmospheric circulation plays in sustaining the positive temperature anomalies through moisture advection from the Atlantic Ocean over the Gulf of Guinea into West Africa. The third factor, which is more prevalent in the southern Sahel and along the Gulf of Guinea coast, is increased downward shortwave radiation which drives up maximum temperatures under reduced cloud cover. Many of these driving forces, especially the water vapour effect and high SST and associated moisture a)

b)

c)

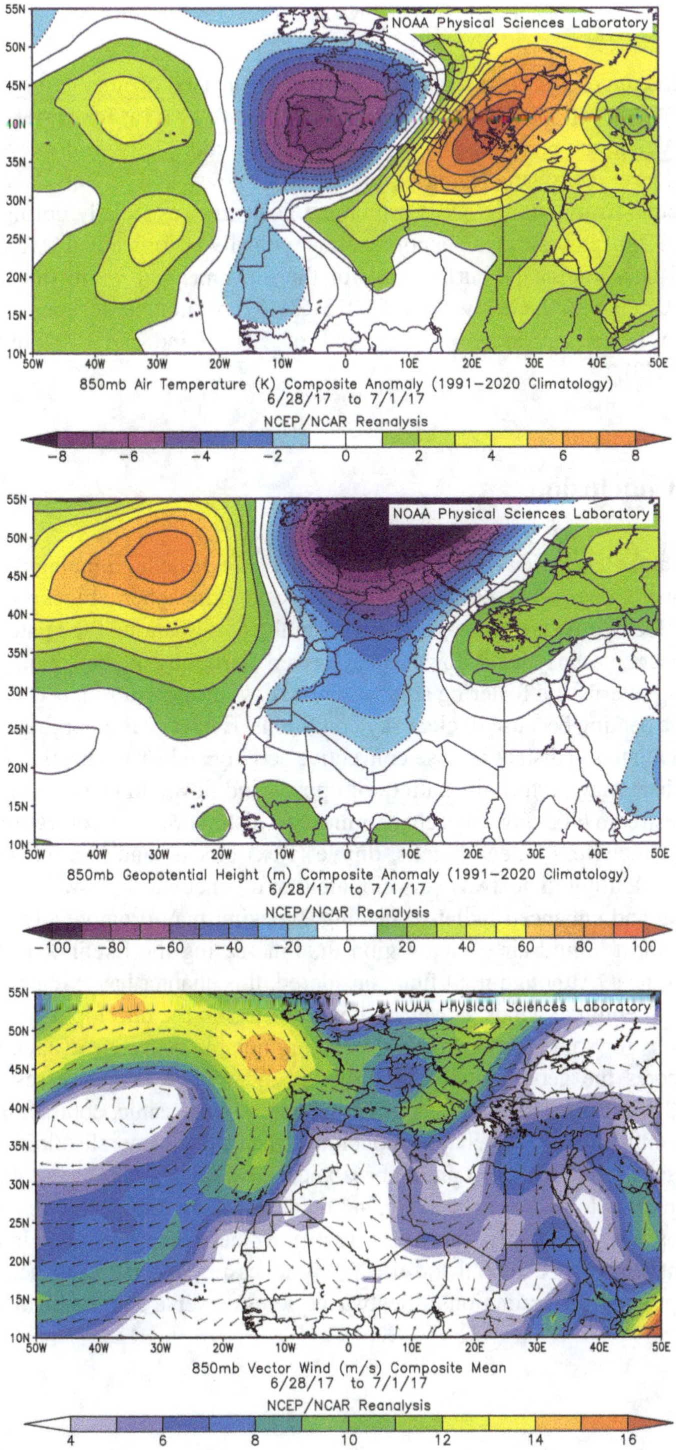

Fig. 6.26 850 mb (**a**) air temperature and (**b**) geopotential height anomalies (compared to 1981–2010 climatology) and (**c**) 850 mb mean vector wind (direction and speed) field for June 28—July 1, 2017 Libyan heatwave event. Image provided by the NOAA/ESRL Physical Sciences Laboratory (PSL), Boulder Colorado from their web site at https://psl.noaa.gov/ and generated based on NCEP/NCAR Reanalysis data (Kalnay et al., 1996)

advection appear to be associated with the occurrence of notable nighttime heatwaves across various parts of Africa (Bouniol et al., 2021a; Igun et al., 2023). For South Africa, the presence and persistence of a middle-level high-pressure system situated over Botswana/Namibia, commonly known as the Botswana High, has been found to initiate the most prolonged heatwaves especially during periods of drought. This atmospheric phenomenon is typically accompanied by a nearby continental heat low at the surface, and/or the movement of warm tropical air over South Africa with the heatwaves of 2015–2016 most notable for their intensity and duration being embedded in a period of ENSO induced drought (Mbokodo et al., 2023).

6.4 Conclusion

This chapter has reviewed the synoptic scale processes underpinning the genesis of atmospheric heatwaves, emphasizing the roles of atmospheric blocking, advection, and the foehn effect. Blocking, arising from the perturbation of the pattern of upper atmosphere westerly wind flow is a key driver, disrupting the typical progression of weather patterns and fostering adiabatic warming through subsidence and enhanced radiative heating because of clear sky conditions. Evidence is emerging that diabatic heating related to distant intense convective activity and strong heating of the upper atmosphere may perturb the pattern of upper wind flow and increase the likelihood of downstream blocking. This chapter makes a distinction between various blocking patterns, including omega blocks, dipole (Rex) blocks, and high-pressure ridges, each contributing to heatwave development in distinct ways. As well as subsidence warming and enhanced radiative heating, blocking may promote advection of hot dry or warm humid air over a region thus increasing the likelihood of enhanced levels of heat. Although not often considered, this chapter has revealed the importance of foehn like winds as well as low-level jets in the development of heatwave conditions for some locations. By examining notable heatwave events, the chapter underscores the geographical variability of synoptic scale processes, shaped by regional climate regimes and geographic factors like oceanic, continental, and orographic influences. This chapter concludes that while generalisations from well-known events aid in understanding broad heatwave related synoptic scale processes, it is important to recognise the regional nuances inherent in heatwave development and thus the complexity and contextuality of heatwave genesis. Understanding these intricacies is essential for effectively assessing and reducing heatwave risks and improving the prediction of heatwaves across diverse geographical settings.

References

Altenhoff, A. M., Martius, O., Croci-Maspoli, M., Schwierz, C., & Davies, H. C. (2008). Linkage of atmospheric blocks and synoptic-scale Rossby waves: A climatological analysis. *Tellus Series a-Dynamic Meteorology and Oceanography, 60*(5), 1053–1063. https://doi.org/10.1111/j.1600-0870.2008.00354.x

Alvarez, M. S., Cerne, B., Osman, M., & Vera, C. S. (2019). Intraseasonal and low frequency processes contributing to the December 2013 heat wave in Southern South America. *Climate Dynamics, 53*(7–8), 4977–4988. https://doi.org/10.1007/s00382-019-04838-6

Amou, M., Gyilbag, A., Demelash, T., & Xu, Y. L. (2021). Heatwaves in Kenya 1987–2016: Facts from CHIRTS high resolution satellite remotely sensed and station blended temperature dataset. *Atmosphere, 12*(1). https://doi.org/10.3390/atmos12010037

An, N., & Zuo, Z. (2021). Changing structures of summertime heatwaves over China during 1961–2017. *Science China Earth Sciences, 64*, 1242–1253.

Atlas, R., Wolfson, N., & Terry, J. (1993). The effect of SST and soil moisture anomalies on GLA model simulations of the 1988 US summer drought. *Journal of Climate, 6*(11), 2034–2048.

Baier, K., Rubel, M., & Stohl, A. (2023). The 3-week-long transport history and deep tropical origin of the 2021 extreme heat wave in the Pacific northwest. *Geophysical Research Letters, 50*(24). https://doi.org/10.1029/2023gl105865

Barnes, E. A., Slingo, J., & Woollings, T. (2012). A methodology for the comparison of blocking climatologies across indices, models and climate scenarios. *Climate Dynamics, 38*(11–12), 2467–2481. https://doi.org/10.1007/s00382-011-1243-6

Barnes, E. A., Dunn-Sigouin, E., Masato, G., & Woollings, T. (2014). Exploring recent trends in northern hemisphere blocking. *Geophysical Research Letters, 41*(2), 638–644. https://doi.org/10.1002/2013gl058745

Barrett, B. S., Campos, D. A., Veloso, J. V., & Rondanelli, R. (2016). Extreme temperature and precipitation events in March 2015 in central and northern Chile. *Journal of Geophysical Research-Atmospheres, 121*(9), 4563–4580. https://doi.org/10.1002/2016jd024835

Barriopedro, D., García-Herrera, R., Lupo, A. R., & Hernández, E. (2006). A climatology of northern hemisphere blocking. *Journal of Climate, 19*(6), 1042–1063. https://doi.org/10.1175/jcli3678.1

Barriopedro, D., Fischer, E. M., Luterbacher, J., Trigo, R. M., & García-Herrera, R. (2011). The hot summer of 2010: Redrawing the temperature record map of Europe. *Science, 332*(6026), 220–224.

Barriopedro, D., García-Herrera, R., Ordóñez, C., Miralles, D. G., & Salcedo-Sanz, S. (2023). Heat waves: Physical understanding and scientific challenges. *Reviews of Geophysics, 61*(2). https://doi.org/10.1029/2022rg000780

Barros, V. R., Boninsegna, J. A., Camilloni, I. A., Chidiak, M., Magrín, G. O., & Rusticucci, M. (2015). Climate change in Argentina: Trends, projections, impacts and adaptation. *Wiley Interdisciplinary Reviews-Climate Change, 6*(2), 151–169. https://doi.org/10.1002/wcc.316

Benedict, J. J., Lee, S., & Feldstein, S. B. (2004). Synoptic view of the North Atlantic oscillation. *Journal of the Atmospheric Sciences, 61*(2), 121–144. https://doi.org/10.1175/1520-0469(2004)061<0121:svotna>2.0.co;2

Beniston, M. (2004). The 2003 heat wave in Europe: A shape of things to come? An analysis based on Swiss climatological data and model simulations. *Geophysical Research Letters, 31*(2). https://doi.org/10.1029/2003gl018857

Bhadram, C., Amatya, B., Pant, G., & Kumar, K. K. (2005). Heat waves over Andhra Pradesh: A case study of summer 2003. *Mausam, 56*(2), 385–394.

Black, E., Blackburn, M., Harrison, G., Hoskins, B., & Methven, J. (2004). Factors contributing to the summer 2003 European heatwave. *Weather, 59*(8), 217–223.

Bouniol, D., Guichard, F., Barbier, J., Couvreux, F., & Roehrig, R. (2021a). Sahelian heat wave characterization from observational data sets. *Journal of Geophysical Research-Atmospheres, 126*(11). https://doi.org/10.1029/2020jd034465

Bouniol, D., Guichard, F., Barbier, J., Couvreux, F., & Roehrig, R. (2021b). Sahelian heat wave characterization from observational data sets. *Journal of Geophysical Research: Atmospheres, 126*(11), e2020JD034465. https://doi.org/10.1029/2020JD034465

Bumbaco, K. A., Dello, K. D., & Bond, N. A. (2013). History of Pacific northwest heat waves: Synoptic pattern and trends. *Journal of Applied Meteorology and Climatology, 52*(7), 1618–1631.

Campetella, C., & Rusticucci, M. (1998). Synoptic analysis of an extreme heat wave over Argentina in march 1980. *Meteorological Applications, 5*(3), 217–226. https://doi.org/10.1017/S1350482798000851

Cassou, C. (2008). Intraseasonal interaction between the Madden-Julian Oscillation and the North Atlantic Oscillation. *Nature, 455*(7212), 523–527. https://doi.org/10.1038/nature07286

Cerne, S. B., & Vera, C. S. (2011). Influence of the intraseasonal variability on heat waves in subtropical South America. *Climate Dynamics, 36*(11–12), 2265–2277. https://doi.org/10.1007/s00382-010-0812-4

Cerne, S. B., Vera, C. S., & Liebmann, B. (2007). The nature of a heat wave in eastern Argentina occurring during SALLJEX. *Monthly Weather Review, 135*(3), 1165–1174. https://doi.org/10.1175/mwr3306.1

Chan, P.-W., Hassanzadeh, P., & Kuang, Z. (2019). Evaluating indices of blocking anticyclones in terms of their linear relations with surface extremes. *Geophysical Research Letters, 46*(9), 4904–4912. https://doi.org/10.1029/2019gl083307

Chan, P.-W., Catto, J. L., & Collins, M. (2022). Heatwave-blocking relation change likely dominates over decrease in blocking frequency under global warming. *Npj Climate and Atmospheric Science, 5*(1). https://doi.org/10.1038/s41612-022-00290-2

Chandran, M. A. S., Rao, A., Sandeep, V. M., Pramod, V. P., Pani, P., Rao, V. U. M., Kumari, V. V., & Rao, C. S. (2017). Indian summer heat wave of 2015: A biometeorological analysis using half hourly automatic weather station data with special reference to Andhra Pradesh. *International Journal of Biometeorology, 61*(6), 1063–1072. https://doi.org/10.1007/s00484-016-1286-9

Chen, R., & Lu, R. Y. (2015). Comparisons of the circulation anomalies associated with extreme heat in different regions of eastern China. *Journal of Climate, 28*(14), 5830–5844. https://doi.org/10.1175/jcli-d-14-00818.1

Chen, Y., Hu, Q., Yang, Y. M., & Qian, W. H. (2017). Anomaly based analysis of extreme heat waves in Eastern China during 1981–2013. *International Journal of Climatology, 37*(1), 509–523. https://doi.org/10.1002/joc.4724

Choi, W., Ho, C. H., Jung, J., Chang, M., & Ha, K. J. (2021). Synoptic conditions controlling the seasonal onset and days of heatwaves over Korea. *Climate Dynamics, 57*(11–12), 3045–3053. https://doi.org/10.1007/s00382-021-05853-2

Collazo, S., Lhotka, O., Rusticucci, M., & Kysely, J. (2018). Capability of the SMHI-RCA4 RCM driven by the ERA-interim reanalysis to simulate heat waves in Argentina. *International Journal of Climatology, 38*(1), 483–496. https://doi.org/10.1002/joc.5190

Cowan, T., Hegerl, G. C., Colfescu, I., Bollasina, M., Purich, A., & Boschat, G. (2017). Factors contributing to record-breaking heat waves over the Great Plains during the 1930s dust bowl. *Journal of Climate, 30*(7), 2437–2461. https://doi.org/10.1175/jcli-d-16-0436.1

Daneshvar, M. R. M., & Freund, F. T. (2019). Examination of a relationship between atmospheric blocking and seismic events in the Middle East using a new seismo-climatic index. *Swiss Journal of Geosciences, 112*(2–3), 435–451. https://doi.org/10.1007/s00015-019-00343-4

Darand, M., Garcia-Herrera, R., Asakereh, H., Amiri, R., & Barriopedro, D. (2018). Synoptic conditions leading to extremely warm periods in Western Iran. *International Journal of Climatology, 38*(1), 307–319. https://doi.org/10.1002/joc.5177

Demirtas, M. (2018). The high-impact 2007 hot summer over Turkey: Atmospheric-blocking and heat-wave episodes. *Meteorological Applications, 25*(3), 406–413. https://doi.org/10.1002/met.1708

Doblas-Reyes, F. J., Casado, M., & Pastor, M. (2002). Sensitivity of the northern hemisphere blocking frequency to the detection index. *Journal of Geophysical Research: Atmospheres, 107*(D2), ACL 6-1-ACL 6-22.

Dodla, V. B., Satyanarayana, G. C., & Desamsetti, S. (2017). Analysis and prediction of a catastrophic Indian coastal heat wave of 2015. *Natural Hazards, 87*, 395–414.

Dole, R., Hoerling, M., Perlwitz, J., Eischeid, J., Pegion, P., Zhang, T., Quan, X. W., Xu, T. Y., & Murray, D. (2011). Was there a basis for anticipating the 2010 Russian heat wave? *Geophysical Research Letters, 38*. https://doi.org/10.1029/2010gl046582

Dunn-Sigouin, E., Son, S. W., & Lin, H. (2013). Evaluation of northern hemisphere blocking climatology in the global environment multiscale model. *Monthly Weather Review, 141*(2), 707–727. https://doi.org/10.1175/mwr-d-12-00134.1

Efe, B., & Lupo, A. R. (2022). Global trends in the occurrence and characteristics of blocking anticyclones using Sen innovative trend analysis. *Theoretical and Applied Climatology, 150*(1–2), 763–776. https://doi.org/10.1007/s00704-022-04195-1

Engdaw, M. M., Ballinger, A. P., Hegerl, G. C., & Steiner, A. K. (2022). Changes in temperature and heat waves over Africa using observational and reanalysis data sets. *International Journal of Climatology, 42*(2), 1165–1180. https://doi.org/10.1002/joc.7295

Engel, C. B., Lane, T. P., Reeder, M. J., & Rezny, M. (2013). The meteorology of Black Saturday. *Quarterly Journal of the Royal Meteorological Society, 139*(672), 585–599. https://doi.org/10.1002/qj.1986

Fang, B. J., & Lu, M. Q. (2020). Heatwave and blocking in the northeastern Asia: Occurrence, variability, and association. *Journal of Geophysical Research-Atmospheres, 125*(6). https://doi.org/10.1029/2019jd031627

Feudale, L., & Shukla, J. (2011). Influence of sea surface temperature on the European heat wave of 2003 summer. Part I: An observational study. *Climate Dynamics, 36*, 1691–1703.

Fischer, E. M., Seneviratne, S. I., Vidale, P. L., Luthi, D., & Schar, C. (2007). Soil moisture – Atmosphere interactions during the 2003 European summer heat wave. *Journal of Climate, 20*(20), 5081–5099. https://doi.org/10.1175/jcli4288.1

Fontaine, B., Janicot, S., & Monerie, P. A. (2013). Recent changes in air temperature, heat waves occurrences, and atmospheric circulation in Northern Africa. *Journal of Geophysical Research-Atmospheres, 118*(15), 8536–8552. https://doi.org/10.1002/jgrd.50667

Fragkoulidis, G., Wirth, V., Bossmann, P., & Fink, A. H. (2018). Linking northern hemisphere temperature extremes to Rossby wave packets. *Quarterly Journal of the Royal Meteorological Society, 144*(711), 553–566. https://doi.org/10.1002/qj.3228

Freychet, N., Tett, S., Wang, J., & Hegerl, G. (2017). Summer heat waves over Eastern China: Dynamical processes and trend attribution. *Environmental Research Letters, 12*(2), 024015. https://doi.org/10.1088/1748-9326/aa5ba3

García-Herrera, R., Díaz, J., Trigo, R. M., Luterbacher, J., & Fischer, E. M. (2010). A review of the European summer heat wave of 2003. *Critical Reviews in Environmental Science and Technology, 40*(4), 267–306. https://doi.org/10.1080/10643380802238137

Geirinhas, J. L., Trigo, R. M., Libonati, R., Coelho, C. A. S., & Palmeira, A. C. (2018). Climatic and synoptic characterization of heat waves in Brazil. *International Journal of Climatology, 38*(4), 1760–1776. https://doi.org/10.1002/joc.5294

Geirinhas, J. L., Trigo, R. M., Libonati, R., Castro, L. C., Sousa, P. M., Coelho, C. A., Peres, L. F., & Mônica de Avelar, F. (2019a). Characterizing the atmospheric conditions during the 2010 heatwave in Rio de Janeiro marked by excessive mortality rates. *Science of the Total Environment, 650*, 796–808.

Geirinhas, J. L., Trigo, R. M., Libonati, R., Castro, L. C. O., Sousa, P. M., Coelho, C. A. S., Peres, L. F., & Magalhaes, M. (2019b). Characterizing the atmospheric conditions during the 2010 heatwave in Rio de Janeiro marked by excessive mortality rates. *Science of the Total Environment, 650*, 796–808. https://doi.org/10.1016/j.scitotenv.2018.09.060

Grumm, R. H. (2011). The central European and Russian heat event of July–August 2010. *Bulletin of the American Meteorological Society, 92*(10), 1285–1296.

Guichard, F., Kergoat, L., Mougin, E., Timouk, F., Baup, F., Hiernaux, P., & Lavenu, F. (2009). Surface thermodynamics and radiative budget in the Sahelian Gourma: Seasonal and diurnal cycles. *Journal of Hydrology, 375*(1–2), 161–177. https://doi.org/10.1016/j.jhydrol.2008.09.007

Gyilbag, A., Amou, M., Tulcan, R. X. S., Zhang, L., Demelash, T., & Xu, Y. L. (2021). Characteristics of enhanced heatwaves over Tanzania and scenario projection in the 21st century. *Atmosphere, 12*(8). https://doi.org/10.3390/atmos12081026

Ha, K. J., Yeo, J. H., Seo, Y. W., Chung, E. S., Moon, J. Y., Feng, X., Lee, Y. W., & Ho, C. H. (2020). What caused the extraordinarily hot 2018 summer in Korea? *Journal of the Meteorological Society of Japan, 98*(1), 153–167. https://doi.org/10.2151/jmsj.2020-009

Harrington, L. J., & Otto, F. E. L. (2020). Reconciling theory with the reality of African heatwaves. *Nature Climate Change, 10*(9), 796–798. https://doi.org/10.1038/s41558-020-0851-8

Heeter, K. J., Harley, G. L., Abatzoglou, J. T., Anchukaitis, K. J., Cook, E. R., Coulthard, B. L., Dye, L. A., & Homfeld, I. K. (2023). Unprecedented 21st century heat across the Pacific northwest of North America. *Npj Climate and Atmospheric Science, 6*(1). https://doi.org/10.1038/s41612-023-00340-3

Hong, J. S., Yeh, S. W., & Seo, K. H. (2018). Diagnosing physical mechanisms leading to pure heat waves versus pure tropical nights over the Korean Peninsula. *Journal of Geophysical Research-Atmospheres, 123*(14), 7149–7160. https://doi.org/10.1029/2018jd028360

Igun, E. H., Xu, X. Y., Shi, Z. T., & Jia, G. S. (2023). Enhanced nighttime heatwaves over African urban clusters. *Environmental Research Letters, 18*(1). https://doi.org/10.1088/1748-9326/aca920

Iyakaremye, V., Zeng, G., Ullah, I., Gahigi, A., Mumo, R., & Ayugi, B. (2022). Recent observed changes in extreme high-temperature events and associated meteorological conditions over Africa. *International Journal of Climatology, 42*(9), 4522–4537. https://doi.org/10.1002/joc.7485

Jacques-Coper, M., Veloso-Aguila, D., Segura, C., & Valencia, A. (2021). Intraseasonal teleconnections leading to heat waves in Central Chile. *International Journal of Climatology, 41*(9). https://doi.org/10.1002/joc.7096

Jeong, D. I., Cannon, A. J., & Yu, B. (2022). Influences of atmospheric blocking on north American summer heatwaves in a changing climate: A comparison of two Canadian earth system model large ensembles. *Climatic Change, 172*, 1–2. https://doi.org/10.1007/s10584-022-03358-3

Jézéquel, A., Cattiaux, J., Naveau, P., Radanovics, S., Ribes, A., Vautard, R., Vrac, M., & Yiou, P. (2018a). Trends of atmospheric circulation during singular hot days in Europe. *Environmental Research Letters, 13*(5). https://doi.org/10.1088/1748-9326/aab5da

Jézéquel, A., Yiou, P., & Radanovics, S. (2018b). Role of circulation in European heatwaves using flow analogues. *Climate Dynamics, 50*(3–4), 1145–1159. https://doi.org/10.1007/s00382-017-3667-0

Kalisa, E., Fadlallah, S., Amani, M., Nahayo, L., & Habiyaremye, G. (2018). Temperature and air pollution relationship during heatwaves in Birmingham, UK. *Sustainable Cities and Society, 43*, 111–120. https://doi.org/10.1016/j.scs.2018.08.033

Kalnay, E., Kanamitsu, M., Kistler, R., Collins, W., Deaven, D., Gandin, L., Iredell, M., Saha, S., White, G., Woollen, J., Zhu, Y., Chelliah, M., Ebisuzaki, W., Higgins, W., Janowiak, J., Mo, K. C., Ropelewski, C., Wang, J., Leetmaa, A., Reynolds, R., Jenne, R., & Joseph, D. (1996). The NCEP/NCAR 40-year reanalysis project. *Bulletin of the American Meteorological Society, 77*(3), 437–471. https://doi.org/10.1175/1520-0477(1996)077<0437:tnyrp>2.0.co;2

Kaplan, M. L., Tilley, J. S., Hatchett, B. J., Smith, C. M., Walston, J. M., Shourd, K. N., & Lewis, J. M. (2017). The record Los Angeles heat event of September 2010: 1. Synoptic-scale-meso-beta-scale analyses of interactive planetary wave breaking, terrain-and coastal induced circulations. *Journal of Geophysical Research-Atmospheres, 122*(20), 10729–10750. https://doi.org/10.1002/2017jd027162

Kim, Y.-S., & Hong, S.-G. (1996). A study of quasi-Foehn in the Youngdong-district in late spring orearly summer. *Journal of the Korean Meteorological Society, 32*(4), 593–600.

Kim, S. Y., & Min, K. D. (2001). Effects of terrain and land-use on local circulation and temperature change at Taegu region in summer. *Journal of the Korean Meteorological Society, 37*, 487–512.

Kitada, T., Okamura, K., & Tanaka, S. (1998). Effects of topography and urbanization on local winds and thermal environment in the Nohbi Plain, coastal region of central Japan: A numerical analysis by mesoscale meteorological model with a κ-ε turbulence model. *Journal of Applied Meteorology, 37*(10), 1026–1046. https://doi.org/10.1175/1520-0450(1998)037<1026:eotauo>2.0.co;2

Klinenberg, E. (1999). Denaturalizing disaster: A social autopsy of the 1995 Chicago heat wave. *Theory and Society, 28*(2), 239–295. https://doi.org/10.1023/a:1006995507723

Klinenberg, E. (2015). *Heat wave: A social autopsy of disaster in Chicago*. University of Chicago press.

Kunkel, K. E., Changnon, S. A., Reinke, B. C., & Arritt, R. W. (1996). The July 1995 heat wave in the Midwest: A climatic perspective and critical weather factors. *Bulletin of the American Meteorological Society, 77*(7), 1507–1518. https://doi.org/10.1175/1520-0477(1996)077<1507:tjhwit>2.0.co;2

Kurgansky, M. V. (2022). A simple model of blocking action over a hemisphere. *Theoretical and Applied Climatology, 147*(1–2), 65–71. https://doi.org/10.1007/s00704-021-03782-y

Langue, C. G. N., Lavaysse, C., Vrac, M., & Flamant, C. (2023). Heat wave monitoring over West African cities: Uncertainties, characterization and recent trends. *Natural Hazards and Earth System Sciences, 23*(4), 1313–1333. https://doi.org/10.5194/nhess-23-1313-2023

Lau, W. K. M., & Kim, K. M. (2012). The 2010 Pakistan flood and Russian heat wave: Teleconnection of hydrometeorological extremes. *Journal of Hydrometeorology, 13*(1), 392–403. https://doi.org/10.1175/jhm-d-11-016.1

Lee, H.-Y. (1994). The Nopsae; A Foehn type wind over the Young Suh region of Central Korea. *Journal of the Korean geographical society, 29*(3), 266–280.

Lee, W. S., & Lee, M. I. (2016). Interannual variability of heat waves in South Korea and their connection with large-scale atmospheric circulation patterns. *International Journal of Climatology, 36*(15), 4815–4830. https://doi.org/10.1002/joc.4671

Loikith, P. C., & Kalashinikov, D. A. (2023). Meteorological analysis of the Pacific northwest June 2021 heatwave. *Monthly Weather Review, 151*(5), 1303–1319. https://doi.org/10.1175/mwr-d-22-0284.1

Luo, M., & Lau, N. C. (2017). Heat waves in southern China: Synoptic behavior, long-term change, and urbanization effects. *Journal of Climate, 30*(2), 703–720. https://doi.org/10.1175/jcli-d-16-0269.1

Lupo, A. R. (2021). Atmospheric blocking events: A review. *Annals of the New York Academy of Sciences, 1504*(1), 5–24. https://doi.org/10.1111/nyas.14557

Ma, H. Y., Shao, H. Y., & Song, J. (2014). Modeling the relative roles of the foehn wind and urban expansion in the 2002 Beijing heat wave and possible mitigation by high reflective roofs. *Meteorology and Atmospheric Physics, 123*(3–4), 105–114. https://doi.org/10.1007/s00703-013-0289-x

Marghidan, C. P., van Aalst, M., Blanford, J., Guigma, K., Pinto, I., Maure, G., & Marrufo, T. (2023). Heatwaves in Mozambique 1983–2016: Characteristics, trends and city-level summaries using high-resolution CHIRTS-daily. *Weather and Climate Extremes, 40*. https://doi.org/10.1016/j.wace.2023.100565

Mass, C., Ovens, D., Christy, J., & Conrick, R. (2024). The Pacific northwest heat wave of 25-30 June 2021: Synoptic/mesoscale conditions and climate perspective. *Weather and Forecasting, 39*(2), 275–291. https://doi.org/10.1175/waf-d-23-0154.1

Matsueda, M. (2011). Predictability of Euro-Russian blocking in summer of 2010. *Geophysical Research Letters, 38*(6).

Mbokodo, I. L., Bopape, M. J. M., Ndarana, T., Mbatha, S. M. S., Muofhe, T. P., Singo, M. V., Xulu, N. G., Mohomi, T., Ayisi, K. K., & Chikoore, H. (2023). Heatwave variability and structure in South Africa during summer drought. *Climate, 11*(2). https://doi.org/10.3390/cli11020038

Moore, R. W., Martius, O., & Spengler, T. (2010). The modulation of the subtropical and extratropical atmosphere in the Pacific Basin in response to the Madden-Julian Oscillation. *Monthly Weather Review, 138*(7), 2761–2779. https://doi.org/10.1175/2010mwr3194.1

Mues, A., Manders, A., Schaap, M., Kerschbaumer, A., Stern, R., & Builtjes, P. (2012). Impact of the extreme meteorological conditions during the summer 2003 in Europe on particulate matter concentrations. *Atmospheric Environment, 55*, 377–391. https://doi.org/10.1016/j.atmosenv.2012.03.002

Na, Y., Lu, R. Y., Lu, B., Chen, M., & Miao, S. G. (2019). Impact of the horizontal heat flux in the mixed layer on an extreme heat event in North China: A case study. *Advances in Atmospheric Sciences, 36*(2), 133–142. https://doi.org/10.1007/s00376-018-8133-3

Nair, S., Rao, P. C., & Pai, D. (2017). Synoptic situation associated with the heat wave condition during 17 May to 1 June 2015 over India. *Current Science*, 364–369.

Narinesingh, V., Booth, J. F., & Ming, Y. (2023). Northern hemisphere heat extremes in a warmer climate: More probable but less colocated with blocking. *Geophysical Research Letters, 50*(2). https://doi.org/10.1029/2022gl101211

Ndarana, T., & Waugh, D. W. (2011). A climatology of Rossby wave breaking on the southern hemisphere tropopause. *Journal of the Atmospheric Sciences, 68*(4), 798–811. https://doi.org/10.1175/2010jas3460.1

Neal, E., Huang, C. S. Y., & Nakamura, N. (2022). The 2021 Pacific northwest heat wave and associated blocking: Meteorology and the role of an upstream cyclone as a diabatic source of wave activity. *Geophysical Research Letters, 49*(8). https://doi.org/10.1029/2021gl097699

Ogi, M., Yamazaki, K., & Tachibana, Y. (2005). The summer northern annular mode and abnormal summer weather in 2003. *Geophysical Research Letters, 32*(4). https://doi.org/10.1029/2004gl021528

Oliveira, F. N. M., Carvalho, L. M. V., & Ambrizzi, T. (2014). A new climatology for southern hemisphere blockings in the winter and the combined effect of ENSO and SAM phases. *International Journal of Climatology, 34*(5), 1676–1692. https://doi.org/10.1002/joc.3795

Onuma, K. (2003). Abnormal high temperatures caused by foehn wind in the Nohbu Plain. *Journal of Meteorological Research, 55*, 31–55.

Ordóñez, C., Barriopedro, D., García-Herrera, R., Sousa, P. M., & Schnell, J. L. (2017). Regional responses of surface ozone in Europe to the location of high-latitude blocks and subtropical ridges. *Atmospheric Chemistry and Physics, 17*(4), 3111–3131. https://doi.org/10.5194/acp-17-3111-2017

Otto, F. E. L., Massey, N., van Oldenborgh, G. J., Jones, R. G., & Allen, M. R. (2012). Reconciling two approaches to attribution of the 2010 Russian heat wave. *Geophysical Research Letters, 39*. https://doi.org/10.1029/2011gl050422

Oueslati, B., Pohl, B., Moron, V., Rome, S., & Janicot, S. (2017). Characterization of heat waves in the Sahel and associated physical mechanisms. *Journal of Climate, 30*(9), 3095–3115. https://doi.org/10.1175/jcli-d-16-0432.1

Palecki, M. A., Changnon, S. A., & Kunkel, K. E. (2001). The nature and impacts of the July 1999 heat wave in the midwestern United States: Learning from the lessons of 1995. *Bulletin of the American Meteorological Society, 82*(7), 1353–1368.

Parker, T. J., Berry, G. J., & Reeder, M. J. (2013). The influence of tropical cyclones on heat waves in Southeastern Australia. *Geophysical Research Letters, 40*(23), 6264–6270. https://doi.org/10.1002/2013gl058257

Parker, T. J., Berry, G. J., & Reeder, M. J. (2014). The structure and evolution of heat waves in southeastern Australia. *Journal of Climate, 27*(15), 5768–5785. https://doi.org/10.1175/jcli-d-13-00740.1

Perkins, S. E. (2015). A review on the scientific understanding of heatwaves-their measurement, driving mechanisms, and changes at the global scale. *Atmospheric Research, 164*, 242–267. https://doi.org/10.1016/j.atmosres.2015.05.014

Perkins-Kirkpatrick, S. E., White, C. J., Alexander, L. V., Argueso, D., Boschat, G., Cowan, T., Evans, J. P., Ekstrom, M., Oliver, E. C. J., Phatak, A., & Purich, A. (2016). Natural hazards in Australia: Heatwaves. *Climatic Change, 139*(1), 101–114. https://doi.org/10.1007/s10584-016-1650-0

Pezza, A. B., van Rensch, P., & Cai, W. J. (2012). Severe heat waves in Southern Australia: Synoptic climatology and large scale connections. *Climate Dynamics, 38*(1–2), 209–224. https://doi.org/10.1007/s00382-011-1016-2

Pfahl, S., & Wernli, H. (2012). Quantifying the relevance of atmospheric blocking for co-located temperature extremes in the northern hemisphere on (sub-)daily time scales. *Geophysical Research Letters, 39*. https://doi.org/10.1029/2012gl052261

Pu, X., Wang, T. J., Huang, X., Melas, D., Zanis, P., Papanastasiou, D. K., & Poupkou, A. (2017). Enhanced surface ozone during the heat wave of 2013 in Yangtze River Delta region, China. *Science of the Total Environment, 603*, 807–816. https://doi.org/10.1016/j.scitotenv.2017.03.056

Purich, A., Cowan, T., Cai, W. J., van Rensch, P., Uotila, P., Pezza, A., Boschat, G., & Perkins, S. (2014). Atmospheric and oceanic conditions associated with southern Australian heat waves: A CMIP5 analysis. *Journal of Climate, 27*(20), 7807–7829. https://doi.org/10.1175/jcli-d-14-00098.1

Quinting, J. F., & Reeder, M. J. (2017). Southeastern Australian heat waves from a trajectory viewpoint. *Monthly Weather Review, 145*(10), 4109–4125. https://doi.org/10.1175/mwr-d-17-0165.1

Rahmstorf, S., & Coumou, D. (2011). Increase of extreme events in a warming world. *Proceedings of the National Academy of Sciences of the United States of America, 108*(44), 17905–17909. https://doi.org/10.1073/pnas.1101766108

Ratnam, J. V., Behera, S. K., Ratna, S. B., Rajeevan, M., & Yamagata, T. (2016). Anatomy of Indian heatwaves. *Scientific Reports, 6*. https://doi.org/10.1038/srep24395

Risbey, J. S., O'Kane, T. J., Monselesan, D. P., Franzke, C. L. E., & Horenko, I. (2018). On the dynamics of Austral heat waves. *Journal of Geophysical Research-Atmospheres, 123*(1), 38–57. https://doi.org/10.1002/2017jd027222

Roethlisberger, M., & Papritz, L. (2023). Quantifying the physical processes leading to atmospheric hot extremes at a global scale. *Nature Geoscience, 16*(3), 210. https://doi.org/10.1038/s41561-023-01126-1

Russo, S., Sillmann, J., & Fischer, E. M. (2015). Top ten European heatwaves since 1950 and their occurrence in the coming decades. *Environmental Research Letters, 10*(12). https://doi.org/10.1088/1748-9326/10/12/124003

Rusticucci, M., Kysely, J., Almeira, G., & Lhotka, O. (2016). Long-term variability of heat waves in Argentina and recurrence probability of the severe 2008 heat wave in Buenos Aires. *Theoretical and Applied Climatology, 124*(3–4), 679–689. https://doi.org/10.1007/s00704-015-1445-7

Sambou, M. J. G., Janicot, S., Pohl, B., Badiane, D., Dieng, A. L., & Gaye, A. (2020). Heat wave occurrences over Senegal during spring: Regionalization and synoptic patterns. *International Journal of Climatology*. https://doi.org/10.1002/joc.6220

Sandeep, A., & Prasad, V. S. (2018). Intra-annual variability of heat wave episodes over the east coast of India. *International Journal of Climatology, 38*, E617–E628. https://doi.org/10.1002/joc.5395

Schaller, N., Sillmann, J., Anstey, J., Fischer, E. M., Grams, C. M., & Russo, S. (2018). Influence of blocking on northern European and Western Russian heatwaves in large climate model ensembles. *Environmental Research Letters, 13*(5). https://doi.org/10.1088/1748-9326/aaba55

Schumacher, D. L., Hauser, M., & Seneviratne, S. I. (2022). Drivers and mechanisms of the 2021 Pacific northwest heatwave. *Earths Futures, 10*(12). https://doi.org/10.1029/2022ef002967

Schwarz, L., Malig, B., Guzman-Morales, J., Guirguis, K., Ilango, S. D., Sheridan, P., Gershunov, A., Basu, R., & Benmarhnia, T. (2020). The health burden fall, winter and spring extreme heat events in the in Southern California and contribution of Santa Ana winds. *Environmental Research Letters, 15*(5). https://doi.org/10.1088/1748-9326/ab7f0e

Sfîca, L., Croitoru, A. E., Iordache, I., & Ciupertea, A. F. (2017). Synoptic conditions generating heat waves and warm spells in Romania. *Atmosphere, 8*(3). https://doi.org/10.3390/atmos8030050

Sharma, A. R., Jain, P., Abatzoglou, J. T., & Flannigan, M. (2022). Persistent positive anomalies in geopotential heights promote wildfires in western North America. *Journal of Climate, 35*(19), 2867–2884. https://doi.org/10.1175/jcli-d-21-0926.1

Sinclair, M. R. (1996). A climatology of anticyclones and blocking for the southern hemisphere. *Monthly Weather Review, 124*(2), 245–263. https://doi.org/10.1175/1520-0493(1996)124<0245:acoaab>2.0.co;2

Singh, C., & Kumar, S. V. J. (2018). Meteorological conditions for development of heat wave over coastal Andhra Pradesh and Telangana. *Journal of Indian Geophysical Union, 22*(3), 349–358.

Souders, M. B., Colle, B. A., & Chang, E. K. M. (2014). The climatology and characteristics of Rossby wave packets using a feature-based racking technique. *Monthly Weather Review, 142*(10), 3528–3548. https://doi.org/10.1175/mwr-d-13-00371.1

Suli, S., Barriopedro, D., García-Herrera, R., & Rusticucci, M. (2023). Regionalisation of heat waves in southern South America. *Weather and Climate Extremes, 40*. https://doi.org/10.1016/j.wace.2023.100569

Takane, Y., & Kusaka, H. (2011). Formation mechanisms of the extreme high surface air temperature of 40.9°C observed in the Tokyo metropolitan area: Considerations of dynamic foehn and foehn-like wind. *Journal of Applied Meteorology and Climatology, 50*(9), 1827–1841. https://doi.org/10.1175/jamc-d-10-05032.1

Takane, Y., Kusaka, H., & Kondo, H. (2014). Climatological study on mesoscale extreme high temperature events in the inland of the Tokyo metropolitan area, Japan, during the past 22 years. *International Journal of Climatology, 34*(15), 3926–3938. https://doi.org/10.1002/joc.3951

Takane, Y., Kusaka, H., & Kondo, H. (2015). Investigation of a recent extreme high-temperature event in the Tokyo metropolitan area using numerical simulations: The potential role of a 'hybrid' foehn wind. *Quarterly Journal of the Royal Meteorological Society, 141*(690), 1857–1869. https://doi.org/10.1002/qj.2490

Takane, Y., Kondo, H., Kusaka, H., Katagi, J., Nagafuchi, O., Nakazawa, K., Kaneyasu, N., & Miyakami, Y. (2017). Foehn-like wind with a traditional foehn effect plus dry-diabatic heating from the ground surface contributing to high temperatures at the end of a leeward area. *Journal of Applied Meteorology and Climatology, 56*(7), 2067–2079. https://doi.org/10.1175/jamc-d-16-0257.1

Tan, J. G., Kalkstein, L. S., Huang, J. X., Lin, S. B., Yin, H. B., & Shao, D. M. (2004). An operational heat/health warning system in shanghai. *International Journal of Biometeorology, 48*(3), 157–162. https://doi.org/10.1007/s00484-003-0193-z

Tan, J. G., Zheng, Y. F., Song, G. X., Kalkstein, L. S., Kalkstein, A. J., & Tang, X. (2007). Heat wave impacts on mortality in Shanghai, 1998 and 2003. *International Journal of Biometeorology, 51*(3), 193–200. https://doi.org/10.1007/s00484-006-0058-3

Terray, L. (2023). A storyline approach to the June 2021 northwestern north American heatwave. *Geophysical Research Letters, 50*(5). https://doi.org/10.1029/2022gl101640

Thompson, V., Mitchell, D., Hegerl, G. C., Collins, M., Leach, N. J., & Slingo, J. M. (2023). The most at-risk regions in the world for high-impact heatwaves. Nature. *Communications, 14*(1). https://doi.org/10.1038/s41467-023-37554-1

Tibaldi, S., Ruti, P., Tosi, E., & Maruca, M. (1995). Operational predictability of winter blocking at ECMWF – An update. *Annales Geophysicae-Atmospheres Hydrospheres and Space Sciences, 13*(3), 305–317. https://doi.org/10.1007/s005850050166

Trenberth, K. E., & Fasullo, J. T. (2012). Climate extremes and climate change: The Russian heat wave and other climate extremes of 2010. *Journal of Geophysical Research-Atmospheres, 117*. https://doi.org/10.1029/2012jd018020

Trigo, R. M., García-Herrera, R., Díaz, J., Trigo, I. F., & Valente, M. A. (2005). How exceptional was the early August 2003 heatwave in France? -: Art. no. L10701. *Geophysical Research Letters, 32*(10). https://doi.org/10.1029/2005gl022410

Tuller, S. E. (1980). The effects of a foehn wind on human thermal exchange: The Canterbury nor'wester. *New Zealand Geographer, 36*(1), 11–19.

Tyrlis, E., & Hoskins, B. J. (2008). Aspects of a northern hemisphere atmospheric blocking climatology. *Journal of the Atmospheric Sciences, 65*(5), 1638–1652. https://doi.org/10.1175/2007jas2337.1

Vautard, R., Honoré, C., Beekmann, M., & Rouil, L. (2005). Simulation of ozone during the August 2003 heat wave and emission control scenarios. *Atmospheric Environment, 39*(16), 2957–2967. https://doi.org/10.1016/j.atmosenv.2005.01.039

Vautard, R., Gobiet, A., Jacob, D., Belda, M., Colette, A., Déqué, M., Fernández, J., García-Díez, M., Goergen, K., & Güttler, I. (2013). The simulation of European heat waves from an ensemble of regional climate models within the EURO-CORDEX project. *Climate Dynamics, 41*, 2555–2575.

Wang, W. W., Zhou, W., Li, X. Z., Wang, X., & Wang, D. X. (2016). Synoptic-scale characteristics and atmospheric controls of summer heat waves in China. *Climate Dynamics, 46*(9–10), 2923–2941. https://doi.org/10.1007/s00382-015-2741-8

Wang, P. Y., Tang, J. P., Sun, X. G., Wang, S. Y., Wu, J., Dong, X. N., & Fang, J. (2017). Heat waves in China: Definitions, leading patterns, and connections to large-scale atmospheric circulation and SSTs. *Journal of Geophysical Research-Atmospheres, 122*(20), 10679–10699. https://doi.org/10.1002/2017jd027180

Wang, L., Wang, W. J., Wu, Z. F., Du, H. B., Shen, X. J., & Ma, S. (2018). Spatial and temporal variations of summer hot days and heat waves and their relationships with large-scale atmospheric circulations across Northeast China. *International Journal of Climatology, 38*(15), 5633–5645. https://doi.org/10.1002/joc.5768

Wazneh, H., Gachon, P., Laprise, R., de Vernal, A., & Tremblay, B. (2021). Atmospheric blocking events in the North Atlantic: Trends and links to climate anomalies and teleconnections. *Climate Dynamics, 56*(7–8), 2199–2221. https://doi.org/10.1007/s00382-020-05583-x

Woollings, T., Barriopedro, D., Methven, J., Son, S. W., Martius, O., Harvey, B., Sillmann, J., Lupo, A. R., & Seneviratne, S. (2018). Blocking and its response to climate change. *Current Climate Change Reports, 4*(3), 287–300. https://doi.org/10.1007/s40641-018-0108-z

Yang, Z., Dominguez, F., & Zeng, X. B. (2019). Large and local-scale features associated with heat waves in the United States in reanalysis products and the NARCCAP model ensemble. *Climate Dynamics, 52*(3–4), 1883–1901. https://doi.org/10.1007/s00382-018-4414-x

Yeh, S. W., Won, Y. J., Hong, J. S., Lee, K. J., Kwon, M., Seo, K. H., & Ham, Y. G. (2018). The record-breaking heat wave in 2016 over South Korea and its physical mechanism. *Monthly Weather Review, 146*(5), 1463–1474. https://doi.org/10.1175/mwr-d-17-0205.1

Yoon, D., Cha, D. H., Lee, G., Park, C., Lee, M. I., & Min, K. H. (2018). Impacts of synoptic and local factors on heat wave events over southeastern region of Korea in 2015. *Journal of Geophysical Research-Atmospheres, 123*(21), 12081–12096. https://doi.org/10.1029/2018jd029247

Yoon, D., Cha, D. H., Lee, M.-I., Min, K.-H., Kim, J., Jun, S.-Y., & Choi, Y. (2020). Recent changes in heatwave characteristics over Korea. *Climate Dynamics, 55*(7–8), 1685–1696. https://doi.org/10.1007/s00382-020-05420-1

Yoon, A., Kim, J., Lee, J., Sung, H. M., Hong, J. W., Min, S. K., & Hong, J. (2024). Factor analysis of recent major heatwaves in East Asia. *Geoscience Frontiers, 15*(1). https://doi.org/10.1016/j.gsf.2023.101730

Zhang, B., & Xie, Z. (2023). Mechanism of the record-breaking heatwave event dynamics in South America in January 2022. *Atmosphere, 14*(9). https://doi.org/10.3390/atmos14091326

Zore, T., & Landu, K. (2024). Assessing the role of boreal summer intraseasonal oscillation on the severe heatwave of May 2015 over India. *Quarterly Journal of the Royal Meteorological Society*. n/a (n/a). https://doi.org/10.1002/qj.4765

Chapter 7
Surface Atmosphere Interactions and Heatwaves

Abstract This chapter considers the nature of the interactions between land and water surfaces and the overlying atmosphere involved in the development of heatwaves. It begins by defining the concept of the "active surface," the three-dimensional interface that exchanges heat, mass, and momentum with the atmosphere noting that this surface is dynamic, exhibiting significant diurnal and seasonal variations in its characteristics. The chapter then explores the structure and dynamics of the atmospheric boundary layer (ABL) where much of the exchange of heat, mass and momentum that influences heatwave development occurs, highlighting how surface conditions and atmospheric stability influence heatwave development. A significant portion of the chapter is dedicated to the controls on heat, mass, and momentum exchanges between the surface and the atmosphere and the implications of these for heatwave development. Also addressed is the impact of soil moisture on heatwave intensity, noting that dry soils can exacerbate heatwaves by increasing sensible heat flux and reducing evaporative cooling. Further to the development of atmospheric heatwaves, the role of surface atmosphere interactions for marine heatwave development is discussed with a focus on the surface mixed layer of water bodies and the processes that lead to temperature anomalies. The importance of air-sea heat fluxes, oceanic heat advection, and vertical mixing in the formation and persistence of marine heatwaves is highlighted. Uniquely the surface atmosphere interactions considered important for lake and stream heatwaves are also considered in this chapter. In conclusion, the chapter underscores the complexity of surface-atmosphere interactions and their critical role in heatwave dynamics. It calls for a nuanced understanding of these processes to improve our interpretation of the development processes associated with specific atmospheric, marine, lake or stream heatwave events.

Keywords Atmospheric heatwaves · Marine heatwaves · Lake and stream heatwaves · Surface atmosphere interactions · Sensible and latent heat · Heat, mass and momentum exchanges · Soil moisture · Atmospheric and marine boundary layers

G. McGregor, *Heatwaves*, Biometeorology 6,
https://doi.org/10.1007/978-3-031-69906-1_7

7.1 Introduction

The atmosphere and the earth's solid and liquid surfaces are directly in contact so it does not take too much imagination to deduce that there will be exchanges of energy, mass, and momentum between them. This is important as implicit in the previous chapters is the atmosphere does not act independently of the surface in the generation of heatwaves. Rather climate processes at the earth's surface and associated feedback processes may increase the likelihood of heatwave occurrence and/or exacerbation of heatwave intensity and duration. The purpose of this chapter is to provide an overview of how climate processes at the earth's surface in the form of heat, mass and momentum exchange may play a role in heatwave development. The chapter will address these processes over land and water surfaces in relation to atmospheric and marine and terrestrial water body heatwaves respectively. The role of climate processes over urban surfaces and the interaction of these with heatwaves will be addressed in the following chapter. The starting point is a consideration of what is meant by 'the surface' after which the nature of the boundary layer, where surface-atmosphere interactions take place, is described. A consideration of the controls on the surface–atmosphere exchanges of heat, mass and momentum and their implications for heatwave development then follows. The latter part of the chapter focuses specifically on the evidence for the role of surface-atmosphere interactions in the development of atmospheric and marine heatwaves and heatwaves in terrestrial water bodies such as lakes and streams.

7.2 The Surface

Notionally the surface appears a simple enough concept but when couched in a surface-atmosphere interaction context, it is more than the one-dimensional layer on which we plant our feet or float on. When it comes to climate, the term active surface is more apposite. Conceptualized in such a way, the climatically active surface is the three-dimensional surface that exchanges heat, mass and momentum with the atmosphere. At the daily time scale, the active surface, whether it be land or water based typically exhibits large diurnal swings in temperature, the result of heat absorption (gain) by day and emission (loss) at night (Fig. 7.1). For longer time scales such as the seasonal and beyond, the active surface experiences the greatest climate extremes, whether in terms of heat (highest and lowest land and water temperatures) or moisture (wettest and driest land surface materials). Examples of active surfaces include the upper most layers of a bare land surface, a tree canopy, the façade and roofs of buildings (see Chap. 8) and the uppermost layers of water bodies. Time introduces a possible fourth dimension to the active surface, as an active surface's three-dimensional properties may be non-stationary, changing over short to millennial time scales, thus influencing the nature of surface to atmosphere exchanges.

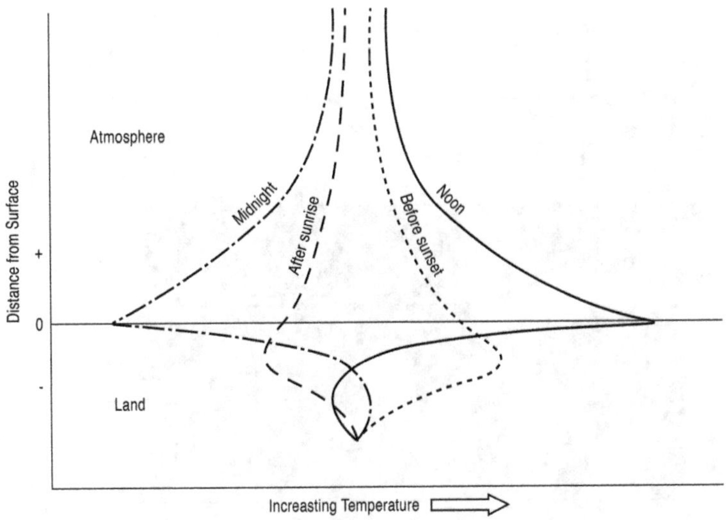

Fig. 7.1 Idealized thermal profile. Temperatures at various distances above and below the surface at four different times of day. (Adapted from Gates (1963))

In the case of the land surface, an extension of the active surface concept is the critical zone (CZO). In broad terms, this is the thin layer of the earth extending from the top of the vegetation canopy to the upper zones of the groundwater aquifer (Fig. 7.2). Displaying large geographical variations in its thickness, ranging from 0.7 to 223.5 m with an average value of 36.8 m across continental areas and attaining its greatest depths in the subtropical to temperate zones, the above ground proportion makes up on average 19.9% of its total depth (Xu & Liu, 2017). Although it has attracted a variety of reactions, including it is simply a refashioning of prior science (Giardino & Houser, 2015; Lin, 2010), the concept is a useful one. This is because it enables the idea that surface processes deeper than the shallow veneer of soil, could play a role in surface atmosphere exchanges and the development of heatwaves and in doing so opens the study of surface atmosphere interactions as an interdisciplinary enterprise. Having said that, the critical zone as originally conceived (Ashley, 1998) focused mainly on soil and regolith processes with little regard given to surface atmosphere interactions. Rather the atmosphere was viewed as providing a one-way flow of energy and mass to the surface (Fig. 7.2). Over the last decade or so however, the CZO has been recognised as a useful construct for articulating the nature of the medium that moderates the dual carriageway of land atmosphere exchanges and explaining climate outcomes at a range of scales (Quiring et al., 2015). That the more distal parts of the CZO may play a role in moderating the characteristics of heatwaves has been opened up by work undertaken on the influence of groundwater state on Australian heatwaves (Mu et al., 2022).

Unlike the land surface, water bodies are fluid and are subject to mixing processes arising from wind blowing over their surface and the merging of different water masses. The structure of water bodies is almost ubiquitous, consisting of a

Fig. 7.2 The critical zone. (Adapted from Giardino and Houser (2015)

surface mixed layer separated from a much deeper layer by a thermocline or zone of rapid temperature decrease with depth. Above the thermocline in the surface mixed layer, often referred to broadly as the ocean boundary layer in the case of the oceans, temperature, density and the chemical make-up (e.g., salinity in the case of the oceans), of the water is almost vertically uniform because of the turbulent mixing that occurs here (Fig. 7.3). This bears implications for atmospheric processes and weather and climate variations on timescales important for heatwave development. This is because the thickness of the surface mixed layer determines the heat content and mechanical inertia of that part of the water body that interacts directly with the atmosphere (de Boyer et al., 2004; Kara et al., 2003). For this reason, the surface mixed layer in both marine and lake environments is considered the climatically active layer.

A further nearly universal characteristic of water bodies is the depth of the surface mixed layer varies seasonally. In summer shallow mixed layers are the result of strong heating and weak winds, a situation that contrasts with the much deeper winter mixed layers arising from cooling and strong winds. For the global oceans, the average depth of the surface mixed layer is around 200 m, but this can vary from as little as 25–30 m in the summer and up to 250 m in the winter for some ocean

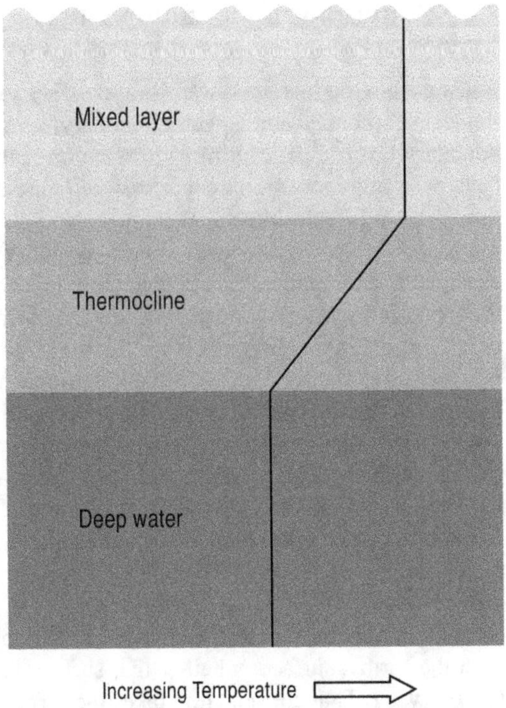

Fig. 7.3 General thermal structure of the ocean. This structure also generally applies to lakes. (Source: Author)

basins. For lakes and the like, the thickness of the surface mixed layer not only depends on seasonal influences but the size and depth of the water body. Further large lakes go through an annual cycle of turnover. In the autumn, the warm surface waters from the summer begin to cool and densify. As cooling progresses, the surface water becomes denser causing it to sink. As it does so, it forces the cool water lying at the bottom of the lake to rise to the surface. The type of processes giving rise to changes in the temperature of the near surface layers of water bodies and thus the formation of marine, lake and stream heatwaves will be outlined in associated sections later in this chapter.

While the active surface might appear an abstract concept, there are some sound pragmatic reasons that warrant a consideration of its nature as presented above. This is because understanding surface atmosphere interactions necessarily involves measurement and modelling. However, any endeavour to measure or model exchanges between all 'climatically active' surfaces and the atmosphere at the local scale and beyond may have to remain an ideal because of current measurement and modelling limitations. Given this, it is incumbent on the researcher to understand fully whether the active surface, conceptualised as a three- or four-dimensional object, is the one that actually is the focus of any attempted measurement or modelling activity.

7.3 Atmospheric Boundary Layer

Arising from the interaction of the atmosphere and the earth's surface is the formation of the so-called atmospheric boundary layer (ABL) also known as the planetary boundary layer (PBL). Lying below the free atmosphere, significant exchanges of heat, mass and momentum occur between the earth's surface and the atmosphere in this space. While the large-scale atmospheric circulation situation may set the scene for heatwave development, the interplay between the various surface atmosphere exchanges within the ABL and how these respond to antecedent surface conditions may well influence how heatwaves develop, especially over land surfaces. Before describing the general controls on the magnitude and direction of heat, mass and momentum exchanges, this section will provide a brief introduction to the nature of the ABL over land and water surfaces. For clarification although ABL is the commonly used term, for ocean and lake environments it is often prefixed with the words marine or lake respectively. Further, the subsurface 'watery' equivalents of the ABL are the ocean and lake boundary layers, more commonly referred to as the ocean or lake surface mixed layer. In the above section the ocean and lake surface mixed layers have been identified as the climatically active surface in these water bodies.

In simple terms, the ABL is that layer of the atmosphere that is in 'conversation' with the earth's surface, whether that surface be land with a particular land cover type or water. That conversation takes the form of the energy, mass and momentum exchange between the atmosphere and the land/water surface. As for the earth's atmosphere, the ABL possesses a structure. This is illustrated in Fig. 7.4 for fair weather conditions over land. During the daytime, the ABL is essentially a statically unstable mixed layer (ML). It is separated from the free atmosphere by a stable layer called the entrainment zone (EZ) where an inversion can be found. The EZ exhibits sporadic turbulence in its lower parts and is the layer through which air from the free atmosphere is assimilated (entrained) into the mixed layer, the result of which is an increase in mixed layer depth during the day. Towards the bottom of the mixed layer, lying above the surface, is the surface layer. On average, this occupies up to 10% of the total ABL depth. Sometimes referred to as the constant flux layer, because turbulent fluxes are more or less invariable with height, the SL is heavily influenced by surface related heat conduction, frictional drag and evaporation which leads to noticeable variations in wind speed, temperature, and humidity with height (Stull, 2011). At night and with the cessation of surface heating, turbulence in the lower parts of the EZ stops and a strongly stable capping inversion replaces the EZ with entrainment of air from the free atmosphere into the ABL terminated. At this point, identifiable are two clear ABL sublayers. The residual layer (RL) represents a hangover from the previous day, as its moisture and air quality characteristics are those of the previous day's ML. Beneath the RL, a very stable boundary layer (SBL) can be found. The standout characteristics of this layer are the low surface temperatures compared to those above the surface and an associated strong surface-based temperature inversion, a reversal of the daytime moisture

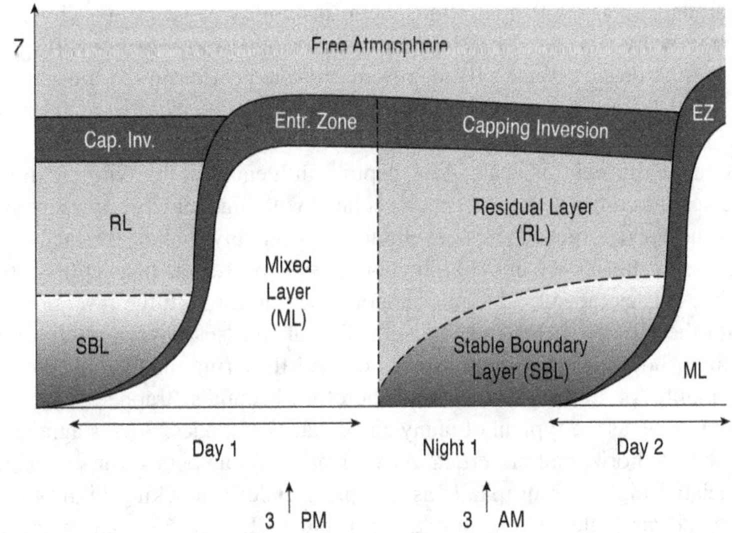

Fig. 7.4 Idealised structure of the land-based atmospheric boundary layer (ABL) during the day and night for summer fair weather conditions. White indicates nonlocally statically unstable air (as in ML), light grey (as in the RL) is neutral stability, and darker greys indicate stronger static stability. EZ is entrainment zone. (Adapted from Stull (2011))

gradient and non-turbulent calm conditions at the surface with a steep vertical gradient of wind speed. With sunrise and the reestablishment of surface heating the nocturnal structure of the ABL reverts to that typified by daytime conditions with vertical gradients of atmospheric properties changing accordingly.

A widely discussed characteristic of the ABL is its depth. Defined as the distance between the surface and the middle of the EZ (Stull, 2011), where sharp vertical gradients in temperature, humidity, turbulence, wind and the concentration of some natural and anthropogenic chemical and physical constituents of the atmosphere occur, its depth can be determined by measurements from radiosondes, and a variety of surface and satellite based remote sensors. ABL depth is neither constant in time or space, being dependent on a range of factors such as the rate of surface heating and cooling, wind strength, the advection of heat and moisture from elsewhere, large scale vertical motions and the 'roughness' of the surface.

Generally, for any one location ABL depth displays a clear diurnal cycle related to the strength of surface heating and turbulent mixing. Over land, greatest depths ranging anywhere between 200 and 5000 m are usually attained towards late afternoon around 1500 h when surface heating and mixing reaches a maximum. At around 1800–2100 h during a sunset transition period, radiative cooling commences and proceeds throughout the night, resulting in a shallowing of the ABL to depths around 10% of their daytime value accompanied by an increasingly stable ABL (Arya, 2005; Liu & Liang, 2010; Seidel et al., 2012). Following sunrise, the diurnal cycle repeats again with the ABL depth building from a night to day transition period around 0600-0900 h as atmospheric instability replaces stability attaining maximum heights of around 1200 – 1500 m in the mid-afternoon hours (Fig. 7.5) In the case of the oceans, peak ABL depths are reached earlier in the diurnal cycle at around 1200 h with a suppressed diurnal range in depth compared to over land (Fig. 7.6) (Liu & Liang, 2010).

Beyond the diurnal timescale, ABL depth is influenced at the synoptic time scale by large-scale weather developments associated with high and low pressure systems and thus surface divergence and convergence respectively, and the passage of fronts. The process of subsidence under high pressure systems results in a depression of the top of the ABL as the subsiding air cannot penetrate through the inversion associated with the EZ or CI. Instead, the subsiding air presses down on the top of the ABL. An upshot of this is the shallowing of the ML during the day and the RL and SBL at night. Air under the inversion therefore becomes 'trapped' and stagnant conditions arise, as are typical of many air pollution episodes. Such stagnant conditions with little horizontal or vertical movement of air can persist for several days if the associated high pressure stalls as is typical under a blocking high, with heat, moisture and air pollutants accumulating in the ABL with ABL air taking on the characteristics of the surface over which it lies. In the case of a low pressure system, associated surface convergence and ascent of air away from the surface (removal of mass), the ABL is forced upwards. Under very strong convergence and ascent, as found in thunderstorms, the upward forces can be so great that air breaks through and destroys the capping inversion with the result that ABL air mixes with that of

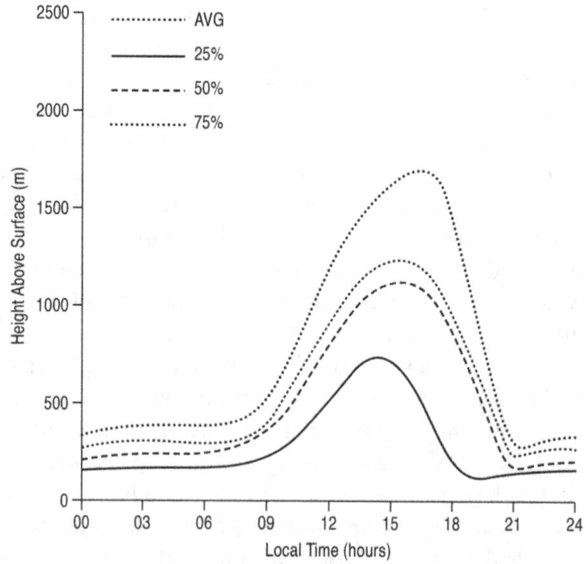

Fig. 7.5 The statistics of ABL height diurnal cycles over land. Percentage figures are quantiles with 50% being the median which tracks closely the average (AVG). (Adapted from Liu and Liang (2010)). © American Meteorological Society. Used with permission

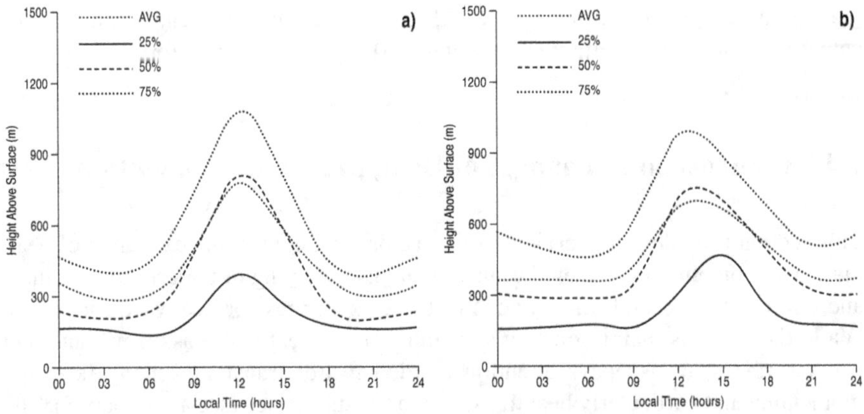

Fig. 7.6 The statistics of ABL height diurnal cycles over tropical oceans. (**a**) Western Pacific 'warm pool' and (**b**) Eastern Pacific 'cold tongue'. Percentage figures are quantiles with 50% being the median which tracks closely the average (AVG). (Adapted from Liu and Liang (2010)).

the free atmosphere. Advancing cold air, which will have its own ABL, and its associated front, can undercut and force the air contained in a warm air mass away from the surface with initially the top of the ABL forced to rise. Eventually however, the vertical forces resulting from the undercutting will create strong ascent of the warm air and a breakdown of the capping inversion. Further as the cold front advances and ABL air is detached from the surface, the ABL ceases to exist, as it is no longer associated with the surface—it becomes disconnected from its boundary. This situation is typical of the occlusion stage of a mid-latitude low pressure system.

Seasonal variations in the ABL depth mimic those at the diurnal timescale related to the times of the year when strong surface heating occurs. Accordingly, over land greatest ABL depths obtain over subtropical regions in summer reaching more than 3 km on average with clear summer to winter differences in ABL depth attained over North Africa, the Middle East, North America, and Australia (Seidel et al., 2012; von Engeln & Teixeira, 2013). In contrast to subtropical to mid-latitude regions, land surfaces dominated by tropical forests, such as the Amazon and central Africa, possess a less pronounced seasonal cycle with relatively low ABL depths associated with the cooling and stabilising effects of high rates of evapotranspiration. As for the diurnal cycle, a damped seasonal cycle typifies ABL depths over the oceans. However, a clear characteristic of the marine ABL depth is its spatial variability. From the east of major ocean basins where the ABL is supressed due to cool surface waters and stratocumulus cloud cover, ABL depth increases downwind in the trade wind regions attaining large values in the ITCZ, especially during the northern hemisphere summer. Interestingly a strong seasonal cycle is evident off the northeast coasts of continents in the northern hemisphere. Here large values are reached in the winter months associated with strong vertical mixing of the atmosphere related to mid-latitude storm systems. Because of small differences between sea and air temperatures, the summer ABL over the Mediterranean basin has low values compared to the winter (Seidel et al., 2012; von Engeln & Teixeira, 2013).

7.4 Controls on Exchanges of Heat, Mass and Momentum

This section provides an overview of the factors that control the exchange of heat, mass and momentum between the surface and its overlying or underlying mediums and outlines the general implications of these exchanges for heatwave formation. While the focus is mainly on 'upward' transfers of heat and mass from land and water surfaces, it also pays attention to 'downward' transfers, chiefly heat and momentum and particularly heat transfer into the subsurface (see also Chap. 8). The main determinant of the exchange of heat and mass between the surface and the atmosphere and vice versa is the steepness and direction of vertical gradients of temperature and moisture respectively. Wind speed and surface roughness are the main determinants of the degree of momentum exchange. For the transfer of heat from the surface to subsurface the chief control is the associated subsurface temperature gradient.

7.4.1 Sensible Heat Exchange

The magnitude of the temperature difference between a surface and the overlying air is the primary control on the rate of sensible heat transfer. The larger the temperature gradient the larger the sensible heat transfer, all other things being equal. In simple terms, if a surface is warmer (cooler) than its atmospheric surrounds then the direction of heat transfer will be from the surface (atmosphere) to the atmosphere (surface). As noted earlier a fundamental control on surface temperature is the surface energy balance, itself an outcome of the surface radiation budget and the amount of net radiation (Q^*) apportioned between sensible (Q_h), latent (Q_e) and subsurface (Q_g) heat flux. That said how a surface and hence the overlying atmosphere might respond to any surface energy forcing dependents on the thermal properties of the medium receiving and transferring the heat. These properties include density (ρ), specific heat (c), heat capacity (C), thermal conductivity (k), thermal diffusivity (α_h) and thermal admittance (μ).

The specific heat of a material, whether it be air, soil, rock, concrete, biomass, or water, is a fundamental determinant of its thermal behaviour. Defined as the amount of heat required to raise the temperature of a unit mass of material (e.g., 1 kg) by 1 °C, specific heat conditions the rate of heating or cooling of a material. Comparatively, soil and rock have a low specific heat compared to water, which accounts for why land masses heat up and cool down more rapidly than water bodies. In combination, density and specific heat determine the heat capacity ($C = \rho c$) of a unit volume of material such that a material with high heat capacity can store more heat. Heat capacity is also an indicator of the expected temperature change in a volume of material arising from the absorption of heat. Thermal conductivity (*k*) refers to the ability of a material to conduct heat along a temperature gradient; conduction is by molecule-to-molecule transfer of heat. High (low) values of *k* imply transmission of relatively large (small) amounts of heat through a material for a given temperature gradient. Thermal diffusivity (α_h) is the ratio of thermal conductivity to heat capacity ($\alpha_h = k/C$). It is a measure of the ease with which changes in temperature propagate through a material; high (low) values indicate that surface changes in temperature will penetrate rapidly (slowly) through a medium bringing temperature changes over a sizeable (shallow) depth of the layer adjacent to the surface.

While these properties are intrinsic to a volume of material, thermal admittance (μ), sometimes referred to as thermal inertia, is a state of a surface, or more specifically, the interface between two interacting mediums. In simple terms it is a measure of the ability of a surface to accept or release heat. In the case of land (water) based heatwaves, the interface is where the ground (ocean) surface meets the overlying air. For fluids, such as air or water, the fluid's thermal diffusivity, which depends on the regime of turbulence, controls the thermal admittance. For this reason, the thermal admittance of air and water take on a wide range of values being low (high) for tranquil (unsettled) conditions. For solid materials such as soil, rock or indeed building materials, thermal admittance values are quite invariable and conditioned on other thermal properties such as heat capacity and thermal conductivity. Having said that, this invariability is for a given moisture state because a saturated soil compared to an unsaturated one will have a higher heat capacity and conductivity. Hence, the thermal admittance values of wet/saturated soils can be several times larger than that for dry/unsaturated soils. In relation to heatwaves, thermal admittance is an important thermal property of interacting mediums. This importance stems from its control on the apportionment of heat between a surface and its adjacent atmospheric and subsurface layers. For example, if the thermal admittance of the material comprising the surface is large, most of the heat received at the surface will go into heating the subsurface layers whether these be soil, rock or water. In other words, a large amount of the available heat will go into subsurface storage with little remaining to heat the atmosphere. The upshot of this is a damped diurnal cycle of surface temperature. On the other hand, a large diurnal range in surface temperature, with high (low) daytime (nocturnal) temperature values is typical of surface materials with low thermal admittance—little of the available heat goes into storage with the majority given over to heating the atmosphere. A convenient way to express this is the ratio of sensible to subsurface heat (Q_h/Q_g) or alternatively, the ratio of air to surface thermal admittance (μ_a/μ_s) (Oke et al., 2017; Demuzere et al., 2008). Such ratios provide a relative indication of the thermal storage capacity or thermal mass of a material; a material with a large thermal mass stores a large amount of heat and smooths out diurnal and seasonal temperature variations.

Table 7.1 presents thermal property values for a range of substances. This reveals the strong contrasts in heat capacity and thermal conductivity for non-turbulent air

Table 7.1 Thermophysical properties of various natural mediums relevant to heatwaves (Adapted from Sharqawy et al. (2010) and Oke et al. (2017))

Material	State	Mass density (ρ)—kg m^{-3} x 10^3	Specific heat (c)—J kg^{-1} K^{-1} x 10^3	Heat capacity (C) —MJ m^{-3} K^{-1}	Thermal conductivity (k)—W m^{-1} K^{-1}	Thermal diffusivity (α_h)—m^2 s^{-1} x 10^{-6}	Thermal admittance (μ) J m—2 s$^{-1/2}$ K^{-1}
Air (Still)	10 °C	0.00124	1.006	0.0012	0.024	19.4	5
Air (Still)	20 °C,	0.00120	1.007	0.0012	0.026	20.7	5
Air (Still)	25 °C	0.00118	1.007	0.0011	0.026	21.4	5
Air (Still)	35 °C	0.00114	1.007	0.0011	0.026	22.7	5
Air (Still)	45 °C	0.00110	1.007	0.0011	0.027	24.1	5
Air (Turbulent)	20 °C	0.0012	1.007	0.0012	~125	~10 × 10^6	390
Water (Still)	4 °C	1.00	4.19	4.19	0.57	0.14	1545
Water (Still)	10 °C	0.999	4.19	4.19	0.578	0.138	1560
Water (Still)	25 °C	0.997	4.13	4.13	0.606	0.146	1580
Seawater[a]	10 °C	1.027	3.99	4.09	0.586	1.42	1530
Seawater[a]	25 °C	1.023	4.00	4.09	0.608	1.48	1560
Sandy soil	Dry	1.60	0.80	1.28	0.3	0.24	620
Sandy soil	Wet	2.00	1.45	2.96	2.2	0.74	2550
Clay soil	Dry	1.60	0.89	1.42	0.25	0.18	600
Clay soil	Wet	2.00	1.55	3.10	1.58	0.51	2201
Peat soil	Dry	0.30	1.92	0.58	0.06	0.10	190
Peat soil	Wet	1.10	3.65	4.02	0.5	0.12	1420

[a] for salinity of 35 g/kg (3.5%)

and water, with air having the lowest values of all natural materials. Air however has a high thermal diffusivity because of its low density, and this increases with temperature. Although the values of thermal properties for still air change over a temperature range from below freezing to well beyond 100 °C, for the range of air temperatures normally associated with heatwave conditions, values are quite invariant, apart from those for thermal diffusivity. A key determinant of the magnitude of thermal conductivity, diffusivity and admittance however is the motion of air or water, with values increasing markedly under turbulent conditions. For the land surface, the criticality of moisture status is readily apparent from Table 7.1. As soil moisture increases so does the heat capacity, thermal conductivity and diffusivity as water displaces air from the soil pores—water is a far more effective heat conductor and transmitter compared to air. This of course affects the thermal admittance or thermal mass of land surface materials with dry soils being poor at storing heat, the consequence of which is the significant diurnal temperature variations and thus large contributions of sensible heat to the atmosphere for dry compared to wet soils.

As for heat transfer in the land based ABL, that in the marine ABL depends on the steepness of the surface to atmosphere temperature gradient. However, over the oceans, these are small relative to those over land. Added to this is the temporal and spatial homogeneity of surface temperature over the ocean, a characteristic that clearly distinguishes sea surface temperature (SST) from land surface temperature (LST), as does the damped diurnal and seasonal temperature range due to the large thermal mass of the oceans. A range of factors affects SST and therefore the flux of heat to the atmosphere at any one time. Oceanic factors include the depth of the marine boundary layer, the degree of turbulent mixing in the marine mixed layer, the advection of heat by ocean currents and the existence of cold-water upwelling. Atmospheric factors include the net radiation budget at the ocean surface, the magnitude of latent and sensible heat fluxes as well as precipitation inputs. Amongst the atmospheric factors, latent heat flux or evaporation, a cooling process, keeps the ocean surface relatively cool. Evaporation related suppression of SST is aided by upward sensible heat flux and mixing of surface warm with cooler water at depth in the mixed layer. Together these processes, which intensify with increased wind speed, work to keep SST close to a thermal equilibrium with atmospheric temperature. Accordingly, gentle surface to atmosphere temperature gradients produce only modest levels of sensible heat flux from the oceans into the atmosphere. The exception of course is where warm poleward flowing ocean currents reach mid-high latitudes on the western boundary of the major ocean basins (e.g. Gulf Stream, Kuroshio Current, East Australian Current). Here atmospheric temperatures are considerably lower than the warm current SST hence sensible heat fluxes can be large and aided by vigorous wind flow in these 'normally' windy regions. Although the flux of sensible heat is generally from the ocean to the atmosphere, an occasional feature of some eastern regions of the ocean basins is a flow of sensible heat from the atmosphere to the ocean surface. This occurs when strong upwelling of cold water from deep in the ocean produces SST lower than the temperature of the overlying atmosphere.

For water bodies in general, heat may also be mixed down from the surface into the surface mixed layer with the degree of mixing related to surface wind stress. This is discussed in the section on momentum transfer. Under very calm atmospheric conditions mixing in the surface layer is limited. In this situation, especially with clear sky conditions, radiative heating of the water surface can lead to a build-up of heat in a thin near-surface layer of the water body and the development of strong surface temperature anomalies. An increase in wind stress will work to dissipate the near surface heat encouraging either surface to atmosphere sensible and latent heat flux or the downward mixing of warm surface waters.

7.4.2 Moisture Exchange

Surface atmosphere exchanges of mass occur largely in the form of moisture, that is, water in its liquid or gaseous form; additional exchanges may include other gases and particulate matter. Water related mass inputs to the surface occur as rainfall or snowfall. The formation of dew resulting from atmospheric moisture condensing on cool surfaces is less important. In relation to air temperature or atmospheric heatwaves, the vertical movement of moisture away from the surface, via the process of evaporation and evapotranspiration (ET), is important. This is because evaporation depletes surface and sub-surface moisture reserves with the result that increasingly large amounts of available net radiation go into the sensible heating of the atmosphere and thus raising air temperature, a process exacerbated by positive surface atmosphere feedbacks. In the case of heatwaves in water bodies, evaporation and thus latent heat flux from the surface into the atmosphere can work to moderate surface temperature extremes because of its cooling effects.

Notwithstanding the importance of moisture availability, over a non-vegetated surface the main controls on the rate of evaporation are net radiation amount, the surface to atmosphere gradient of moisture, turbulence and rate of heat and moisture advection. For a vegetated surface, the nature of the vegetation and its associated physiology adds to the aforementioned factors as a control on evapotranspiration (ET). Over water bodies, moisture supply is non-limiting. Accordingly, evaporation rates are mainly dependent on available energy and the surface to atmosphere humidity gradient as driven by the degree of mixing in the ABL and determined by the magnitude of horizontal and vertical motions both of which are critical for maintaining air above the surface in an unsaturated state. Because evaporation/evapotranspiration broadly depends on water availability, there is often a difference between actual evaporation/evapotranspiration (AET) and potential evapotranspiration (PET), or in other words, reality and what is possible. AET and PET are important concepts with PET often referred to as the climatic demand, that is, the maximum possible evaporation given the energy available at a location for undertaking evaporation; over the subtropical oceans rates for AET can approach that of PET.

Although there are a multitude of factors that control evapotranspiration rates, these can be stripped back to whether evaporation is moisture limited or energy limited (Dirmeyer et al., 2000; Seneviratne et al., 2010), (Fig. 7.7). An effective way to characterise these two evapotranspiration (ET) modes is via the notion of the evaporative fraction (EF) expressed as the ratio of latent heat flux (Qe, amount of energy consumed in evapotranspiration) to net radiation (Q*, energy available at earth's surface). If EF (Qe/Q*) = 1, then 100% of the net radiation available at the earth's surface goes into ET (Seneviratne et al., 2010). In the case of land-based ET, whether one of the two possible regimes operates depends on the exceedance of a critical soil moisture value ($\boldsymbol{\theta}_{\mathbf{CRIT}}$), equating to around 80% of field capacity (Dirmeyer et al., 2000), or the maximum amount of moisture a soil can hold as determined by its physical characteristics (e.g. texture, structure, porosity). If soil

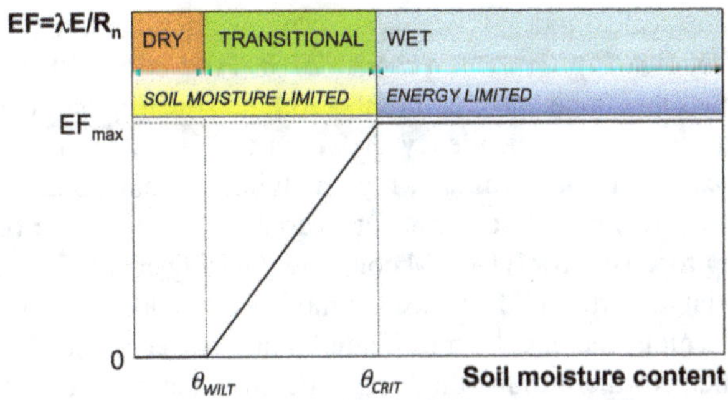

Fig. 7.7 Soil moisture and ET regimes. EF is the evaporative fraction defined as the ratio of latent heat flux (Q_e) to net radiation (Q*). EF_{max} is the maximum value for EF. $\boldsymbol{\theta}_{\mathbf{WILT}}$ and $\boldsymbol{\theta}_{\mathbf{CRIT}}$ represent critical values associated with wilting point and soil field capacity respectively. (Source: Seneviratne et al. (2010). Used under Elsevier License Number 5036470089230)

moisture is greater than $\boldsymbol{\theta}_{\mathbf{CRIT}}$, then the ET regime is said to be energy limited, or in other words EF is independent of soil moisture content and moisture availability is a non-limiting factor (this is essentially the situation over water bodies). In contrast, when soil moisture is less than $\boldsymbol{\theta}_{\mathbf{CRIT}}$, soil moisture provides a control on ET with the ET regime classed as soil moisture limited. ET does not continue endlessly until the soil is completely devoid of moisture. Rather ET stops when the so-called wilting point ($\boldsymbol{\theta}_{\mathbf{WILT}}$) is reached (Fig. 7.7). This is the situation when a plant has no water available; the soil still contains water, but it is difficult for the plant to extract it via its root system. The wilting point, which is plant type dependent, therefore sets the ultimate limit for ET as beyond this no water extraction from the soil occurs. As for field capacity, soil texture plays a role in determining the wilting point with fine textured soils retaining higher moisture contents than their coarse textured counterparts at the wilting point—essentially particles making up fine textured soils have greater surface tension and therefore can 'hang onto' their water more so than those making up course textured soils. Considering the relationship between the thresholds for determining whether ET is soil moisture or energy limited, three EF (climate)-soil moisture regimes may be identified, namely wet (soil moisture > $\boldsymbol{\theta}_{\mathbf{CRIT}}$), dry (soil moisture < $\boldsymbol{\theta}_{\mathbf{WILT}}$) and transitional ($\boldsymbol{\theta}_{\mathbf{WILT}} \leq$ soil moisture $\leq \boldsymbol{\theta}_{\mathbf{CRIT}}$). For the wet and dry regimes, ET is independent on soil moisture, whereas for the transitional regime, soil moisture strongly constrains ET and thus surface to atmosphere feedbacks (Seneviratne et al., 2010).

While the soil moisture EF relationships presented provide a conceptually attractive framework for considering the extent to which soil moisture controls ET, it is important to acknowledge there are both temporal and spatial variations in soil moisture (SM) - ET coupling (Quiring et al., 2015). For example, both observations and modelling reveal the SM—EF relationship can be non-linear and weather dependent with a non-linear response of EF under low soil moisture conditions and strong associations apparent in ideal weather comprising clear skies, low winds, and no recent precipitation (Basara & Crawford, 2002; Dirmeyer et al., 2000). At the inter-annual time scale, modes of ocean-atmosphere circulation may also moderate the SM - EF relationship (Ford & Schoof, 2016; Ford et al., 2017). Of the three EF—SM regimes mentioned above, the EF - soil moisture limited regime is the one for which there is a clear dependency of ET on SM, with strong surface—atmosphere feedbacks likely. Accordingly, regions lying in transitional zones between wet and dry climates, where soil moisture variability is a feature of the climate, show a greater propensity for ET—SM coupling (Quiring et al., 2015) and by extension the potential importance of surface—atmosphere feedbacks in heatwave development. Such regions include the mid-United States, especially the Southern Great Plains, the Sahel, Northern India, the Iberian Peninsula and some parts of Southern Europe and southern and southeastern Australia (Dirmeyer et al., 2013; Teuling et al., 2009; Zhao et al., 2019).

Compared to the land surface, the flux of moisture and thus latent heat into the atmosphere from over the oceans is significant both in absolute terms and in relative terms compared to sensible heat flux—over the oceans, latent heat flux is approximately ten times greater than that for sensible heat. As the term latent heat flux implies, the flow of heat from the ocean into the atmosphere represents not only a heat transfer but also a transfer of mass. The result of this is a decrease in ocean surface temperature and an increase in atmospheric humidity. Because evaporation is not water limited over the oceans, in the presence of a sustained surface to atmosphere humidity gradient, theoretically evaporation could take place almost continuously. However, over time, the air overlying the ocean approaches saturation—100% humidity—limiting further evaporation. For this reason, wind is critical for preserving an ocean surface to atmosphere humidity gradient as turbulence replaces moist air with drier air from elsewhere and in doing so reinstates the humidity gradient necessary for evaporation to continue.

The flux of moisture into the atmosphere from the ocean surface exerts a fundamental influence on the stability of the marine atmosphere. As moisture builds in the marine ABL from surface evaporation, heat accumulates. Further, through the process of condensation, when moist ascending air cools and condenses, latent heat is released into the atmosphere. For every kilogram of moisture that condenses around 2.5 megajoules (MJ) of heat is released. This represents a considerable diabatic heat source and serves to destabilise the atmosphere thus adding to the buoyancy of air parcels over ocean waters. Referred to as latent instability, it is boosted by the build-up of humidity in the marine ABL. Convective instability can also arise when a saturated marine ABL is overlain by relatively dry air. This is often the forerunner of explosive convection or severe thunderstorm development as increasingly buoyant air in the marine ABL breaks through the capping inversion at the base of the dry mid-level air and rises to great heights in the troposphere with instability exacerbated by large amounts of latent heat release. Such deep convection can instigate convergence of surface winds, further enhancing strong vertical updrafts of moist air, the vertical transfer of momentum and the return of mass back to the ocean surface in the form of precipitation.

7.4.3 Momentum Exchange

There are very few places on the earth where there are expansive perfectly smooth surfaces. Therefore, whenever wind flows across the earth's land surface, it is affected by its roughness elements, whether these be mountains, buildings or trees. Because the surface exerts drag on the atmosphere, the wind slows down near the surface. In doing so it dissipates its energy in the form of kinetic energy in the direction of the surface as manifest by trees bending in the wind, dust being mobilised or the generation of sea waves. An outcome of the interaction of wind flow with a 'rough' surface is the generation of mechanical turbulence, the vehicle for the transfer of momentum into the ABL. Because of the vertical exchange of horizontal momentum, a frictional force is developed. In the free atmosphere above the ABL there is no frictional force, so the frictional force has an impact on the vertical distribution of wind direction in the ABL such that in the situation of non-changing weather conditions (a barotropic ABL) winds veer with height (a veering wind turns clockwise (anticlockwise) with height in the Northern (Southern) Hemisphere).

An important impact on momentum transfer in the ABL is atmospheric stability with the result that the vertical distribution of wind speed can vary between day and night-time, especially for the case of clear sky conditions often associated with heatwave development. Under clear skies during the day, strong solar heating of the surface can generate instability and a variety of convective circulations (e.g., thermals, plumes, dust devils). The associated vertical transfer of heat from the surface into the ABL gives rise to convective or buoyancy related turbulence and the vertical mixing of momentum. This often results in the negation of vertical gradients of wind speed and direction throughout most of the ABL apart from the very lower parts of the surface layer and the EZ at the top of the ABL near the capping inversion. With the cessation of surface heating near night fall, outgoing longwave radiation leads to surface cooling with a surface-based temperature inversion ensuing. Under such conditions reduced buoyancy curtails the vertical exchange of momentum in the inversion layer, the result of which are steep gradients of wind speed and direction in the nocturnal boundary layer with a low-level jet of strong winds often present near the top of the surface-based inversion.

Comparatively speaking water body surfaces, and here the ocean is the focus, are far less aerodynamically rough than that of land surfaces. For this reason, the ocean surface has far less impact on the vertical transfer of momentum into the ABL via mechanically generated turbulence than land surfaces. That is not to say that the ocean surface is benign in this regard. Fast flowing ocean currents can exert physical drag on the overlying atmosphere, and air passing over SST fronts, where there are steep horizontal gradients of temperature and pressure, may induce turbulent mixing of momentum and thus alter marine ABL structure. However, in comparison to the atmosphere to ocean surface transfer of momentum, the magnitude of these upward directed mechanically generated momentum fluxes is largely insignificant.

The chief way in which atmosphere to ocean surface momentum transfer manifests is the generation of surface waves, currents and turbulence in the upper layers of the ocean. This is because the atmosphere, through the transfer of kinetic energy, exerts significant drag on the ocean surface thus creating so-called wind stress. As discussed earlier, momentum transfer is of fundamental importance to the depth and degree of mixing in the marine boundary layer and thus the largely even distribution of temperature in near surface waters. Consequently, reduced atmosphere to ocean surface momentum flux at the local level will decrease the downward (upward) mixing of warm (cool) water in the ocean boundary/surface mixed layer, facilitate the accumulation of near surface heat, and enhance the likelihood of marine heatwaves. Momentum transfer resulting in the generation of ocean currents may also hold implications for heat advection from distant sources and thus the development of expansive spatially coherent strong SST anomalies codified as marine heatwaves.

7.5 Atmospheric Heatwaves

The well-established agricultural adage 'drought begets drought' attests to the understanding by those who work the land that the land and atmosphere interact via positive feedback to exacerbate already dry surface conditions. While understood in an informal scientific way by the agricultural and related sectors, it wasn't until the mid-1970s that conceptual models appeared in the literature postulating the physical processes underlying land atmosphere feedback (Charney, 1975). Such models focused on the role of soil moisture in the apportionment of net radiation between sensible and latent heat fluxes. Supported by a burgeoning number of field experiments, such conceptual frameworks formed the basis for the development of mathematical models of soil vegetation atmosphere (SVAT) interactions. While there was somewhat of a revolution in the evolution of SVAT models from the 1980s into the first decade of the new millennium (Pitman, 2003), such models were often applied at time scales beyond that typical of heatwaves, especially if assessment of the impact of climate change on temperature extremes was the focus.

In many ways the seminal August 2003 European heatwave turned the heads of the land-atmosphere modelling community because the intensity and duration of this event raised suspicions about the role of strong positive feedbacks between antecedent soil moisture deficits and the overlying atmosphere in forcing temperatures to record levels over a prolonged period of time (Ferranti & Viterbo, 2006). Fischer et al. (2007) made one of the early attempts to assess the nature of soil moisture atmosphere interactions during the 2003 European heatwave. They found although there was not a surfeit of total surface net radiation, as might be expected as a driver of exceptional temperatures, surface temperature anomalies in early August nevertheless reached very high levels. This was a clear indication of the role that strong soil moisture deficits, which had accumulated over the preceding months, favoured channeling of net radiation into sensible heating of the atmosphere. Consequently, and because of very low soil moisture levels, latent heat flux was constrained and evaporative cooling ineffective in moderating heatwave development. At both the event and climatological scales, the role of antecedent soil moisture conditions and its impact on the surface energy balance and the catalyst of the soil moisture—temperature positive feedback mechanism, has been cited several times as the driving mechanism and the necessary basis for the occurrence of the 2003 European heatwave (Chiriaco et al., 2014; Hirschi et al., 2011; Quesada et al., 2012). Further Lorenz et al. (2010) view depleted soil moisture as an essential determinant of heatwave duration and persistence.

Added impetus for a focus on land atmosphere interactions in the context of heatwave generation was provided by the Russian 2010 mega-heatwave. Using both surface and upper air meteorological observations and complementary experiments using a mechanistic model of the soil-water–atmosphere column for both the 2003 European and 2010 Russian events, Miralles et al. (2014) were able to shed light on the role of soil moisture temperature feedbacks in the generation of mega-heatwaves. Notwithstanding the importance of the persistent blocking pattern which favoured warm air advection and clear skies for both the 2003 and 2010 events (see Chap. 6), their analyses pushed the understanding of the near surface physical mechanisms associated with mega-heatwave development beyond the hitherto widely accepted idea of the direct impact of soil moisture deficits on the surface energy balance and temperature extremes (Alexander, 2011). Importantly they discovered that while the high atmospheric demand for moisture intensified soil desiccation and thus strong upward fluxes of surface sensible heat (the direct surface energy balance effect), the diurnal convection that followed facilitated formation in the atmosphere of deep and warm nocturnal residual warm layers via nocturnal entrainment of warm air from near the surface. During the night the residual layers of heat, which resided at heights more than 3000 m above the surface, remained decoupled from the surface by a strong ground-based thermal inversion. With daylight, and the breakdown of the nocturnal surface-based inversion, heat from the residual layer was transported downwards and merged into the diurnal atmospheric boundary layer thus elevating surface temperatures. Because the residual layers were in effect acting as places of heat storage, the daily repetition of this process, and the repeated blending of residual layer heat into the boundary layer, caused intensification of surface temperatures. Based on these findings Miralles et al. (2014) presented a conceptual model of the development of mega-heatwaves (Fig. 7.8). This portrays mega-heatwaves as the outcome of drying soils enhancing diurnal warm air entrainment into atmospheric layers above the surface, leading to the formation of persistent residual layers of heat that through their reintegration into the boundary layer, favour the progressive increase of surface temperatures as a heatwave evolves and temperatures extremes beyond the expected range of variability are attained. In presenting this conceptual model they note that a seasonal history of rainfall deficits is not a precursor to the development of mega-heatwaves as has been suggested by others (e.g., Hirschi et al. (2011)). Rather the extreme soil drying can be obtained at the intra-seasonal timescale because of the intense atmospheric demand for moisture and the associated rapid soil dry-out in the early phases of such events. Accordingly, the European mega-heatwaves of 2003 and 2010 are viewed as an outcome of the requisite synoptic scale conditions which enabled the combination of the multi-day memory of the land surface and the atmospheric boundary layer to drive surface

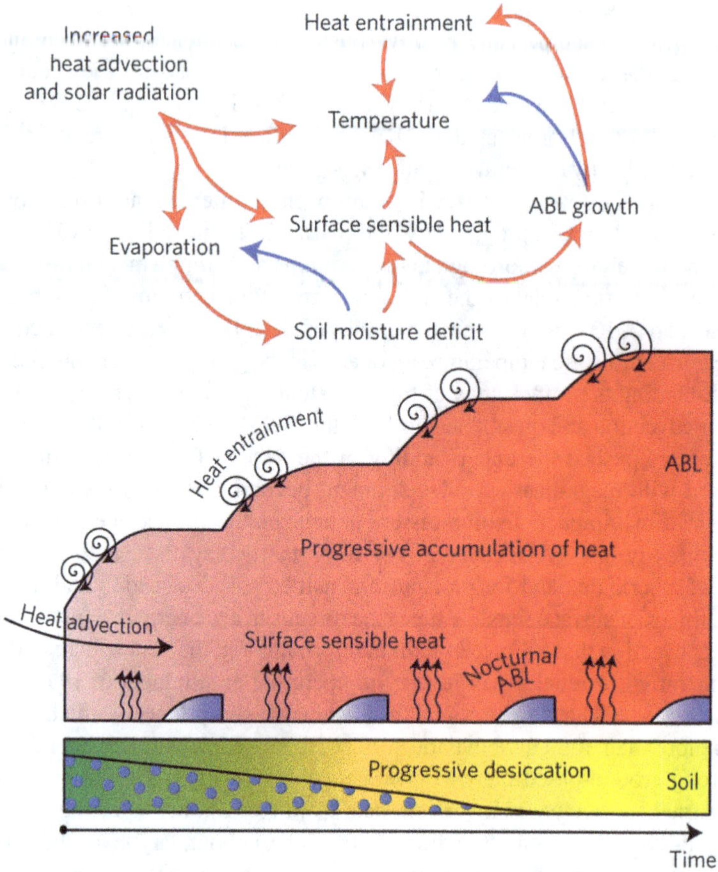

Fig. 7.8 Conceptual model explaining the development of mega-heatwaves via the combined effects of soil desiccation and atmospheric heat accumulation and the instigation of a positive soil moisture–air temperature feedback mechanism. The red and blue arrows indicate positive and negative correlations between processes respectively. Source: Miralles et al. (2014). Mega-heatwave temperatures due to combined soil desiccation and atmospheric heat accumulation. Nature Geoscience 7, 345–349. https://doi.org/10.1038/ngeo2141. Used with permission under Springer License Number 5520601250607)

temperatures to unprecedented levels. Interestingly, but without acknowledging the conceptualization of Miralles et al. (2014), Hauser et al. (2016) estimated that depleted soil moisture increased the risk of the severe 2010 Russian heatwave six-fold. For the record breaking western Mediterranean heatwave in April 2023, Lemus-Canovas et al. (2024) found strong soil moisture deficits occurring with an intense subtropical ridge were key to the amplification and duration of this record-breaking heatwave. They estimated significantly lower temperatures if the soils had been wet. At the global scale and particularly for the northern hemisphere extra-tropics, it has been found that an increase in the length of dry spells has contributed to an increase in heatwave duration and frequency (He et al., 2022) adding to the mounting evidence that precipitation anomalies, via reduced land surface moisture, can act as an important primer for heatwave development.

As the reader may have gathered, much of the literature concerning heatwave related land atmosphere interactions has originated from Europe spurred on by a desire to understand mechanisms associated with the 2003 European and 2010 Russian mega-heatwaves. However, as noted by Perkins et al. (2015), evidence for land surface conditions interacting with the large-scale synoptic environment in the generation of heatwaves is emerging for other regions. By undertaking a range of numerical modelling experiments for the western US heat event of June 2015, Philip et al. (2018) demonstrated strong coupling between temperature, soil moisture deficits and the energy balance, with mean and extreme temperatures elevated for the condition of dry soils, a consequence of the partitioning of available net radiation away from latent heat flux. Similarly, Lee et al. (2016) in a consideration of land atmosphere conditions over the southeastern US, found that sensible heat increased, and latent heat flux decreased in the case of a high frequency of heatwaves with high sensible heat flux over dry soils causing significant near surface warming in excess of 2 °C. For Illinois, despite two heatwave types possessing similar synoptic-scale origins, land surface moisture availability appears to be the critical determinant of whether 'extreme hot dry' or 'moist oppressive' heatwave events occur, with the former (latter) associated with reductions (increases) in evapotranspiration due to limited (non-limited) soil water (Ford & Schoof, 2017).

Over the last decade the volume of research on the role of land surface state in generating heat waves in Australia has increased substantially. At the climatological scale Perkins et al. (2015) demonstrate no ubiquitous continent-wide relationship between soil moisture conditions and heatwaves and report a weaker coupling between land surface conditions and heatwaves than that recounted in the literature for Europe. Building on this work, Herold et al. (2016) show statistical responses of heatwave intensity and frequency to soil moisture deficits. In the case of intensity, 'mild' heatwaves were found to be 'hotter' when antecedent soil moisture was low. Conversely, the duration of 'strong' heatwaves was found to increase with low antecedent soil moisture conditions. Noting the general contrast in land surface climate conditions between the tropical and mid-latitude regions in the eastern half of the continent, Herold et al. (2016) also observe, that where heatwave-soil moisture coupling is evident, such as in the eastern and northern parts of the country, high antecedent soil moisture produces few heatwave days and low heatwave temperatures, while elsewhere high heatwave temperatures, but not in all cases high heat wave occurrence, are associated with low antecedent soil moisture conditions. Although not focused strictly on heatwaves, Holmes et al. (2017) note the climate regime conditioning of the dependence of the number of hot days on soil moisture status. They found a strong inverse association between soil moisture and hot days for tropical savanna and temperate regimes but noted that for semi-arid and arid climate regimes decreasing moisture status produced the strongest increase in the magnitude of 90th percentile temperatures. That there are spatiotemporal variations in the importance of land atmosphere coupling for heatwave occurrence, via the role of soil moisture, is supported by Hirsch et al. (2019) and Hirsch and King (2020). For Northern Australia, they note strong dependence of temperature anomalies associated with individual heatwave events on soil moisture status and thus levels of sensible heating of the boundary layer. In contrast, for southern Australia, there is little influence of antecedent moisture conditions with heatwave development largely driven by an atmosphere (top down) as opposed to a land (bottom up) coupling regime.

Further to unearthing soil moisture heatwave associations based on empirical analyses, numerical modelling has been applied in the assessment of the nature of land atmosphere interactions such as occurred during the infamous Black Saturday bushfire event associated with the late January early February 2009 southeast Australian heatwave. Modelling of this event showed that at short lead times before the bushfire, decreasing (increasing) soil moisture increased (decreased) maximum temperatures (Kala et al., 2015). Other modelling work in Australia has demonstrated that the probability of a particular large-scale weather pattern being associated with a heatwave is strongly tempered by the partitioning of available net radiation between sensible and latent heat fluxes, as determined by soil moisture status at the local level. Further, for the situation of reduced latent heat flux under favourable synoptic conditions, the likelihood of heat wave occurrence increases several-fold (Gibson et al., 2017).

Beyond Europe, the US and Australia, analyses of heatwaves from a land atmosphere perspective confirm the importance of soil moisture temperature feedbacks, albeit focused mainly on the direct surface energy budget effect. In a diagnostic study of the 2015 Indian east coast heatwave described in Chap. 6, Ghatak et al. (2017) identified two heatwave phases with the first displaying little evidence of a soil moisture deficit related sensible heat flux anomaly in the core of the heatwave. Rather, high temperatures were driven by clear skies and a positive net radiation anomaly. However, with the onset of the second stage of the heatwave, by which time significant soil moisture deficits had accumulated, a significant sensible heat flux anomaly had developed leading Ghatak et al. (2017) to conclude that substantial local forcing of air temperature, via a soil moisture air temperature feedback mechanism, contributed to the intensity of the heatwave. For Bangladesh, Nissan et al. (2017) also point to the role of soil moisture as a prerequisite for the generation of severe heatwaves and suggest antecedent soil moisture conditions may possess some utility for heatwave forecasting.

Despite significant land atmosphere modelling activity in China, only a small literature to date has focused specifically on heatwaves. That aside evidence of heatwave characteristics driven by land atmosphere coupling exists. For northern China, Zhang et al. (2015) undertook a climatological analysis of local soil moisture forcing of 10-day maximum temperature anomalies using 50 years of observational and reanalysis data along with numerical modelling. They found, under the synoptic scale setting of positive subtropical high pressure anomalies, heatwave frequency in July to be inversely related to the soil moisture status in March to June, and as Holmes et al. (2017) found for Australia, the time scales and strength of soil moisture forcing were climate type dependent; relatively short in semi-humid and humid regions, longer in arid regions. Wu and Zhang (2015) have also found similar spring soil moisture-heatwave associations, noting that spring conditions can account for up to a third of the variance of summer hot days and heatwave occurrence across northern China.

From the above it is more than apparent that surface atmosphere interactions and associated feedback mechanisms play a key role in the self-intensification of heatwaves over land surfaces. However, some fascinating land-atmosphere interaction and heatwave questions remain. One is whether a constellation of land atmosphere feedbacks might facilitate the spatiotemporal propagation of heatwaves, that is, either the geographical expansion of a heatwave to cover a wider area or indeed the generation of a heatwave in a region at some distance from the initial heatwave. A further question relates to the role of vegetation type in exacerbating or dampening heatwave effects. Concerning the processes that might catalyse the spatial propagation of heatwaves, it is likely that local and remote land-precipitation feedbacks and associated soil evaporation and transpiration decline might provide the context for initial heatwave development, strengthening any subsequent propagation or concatenation (Miralles et al., 2019). Certainly heat advection in the form of 'sensible heat torrents' (Schumacher et al., 2019) from upwind areas under extreme drought or the onset of 'flash drought' (Christian et al., 2020) could well play a part in heatwave intensification and spatial extension (Fig. 7.9). This could well be via a tipping or bifurcation point mechanism (Wang et al., 2020), by which heat advection may tip the scales with land atmosphere feedbacks via soil desiccation hastening downwind heatwave onset and intensification. While heat advection may enhance heatwave events, advection through cool air ventilation can lead to their termination (Hirsch & King, 2020).

Generally, it is accepted that well-vegetated surfaces ameliorate the effects of heatwaves mainly through the cooling arising from evapotranspiration and the steering of available energy into latent as opposed to sensible heat flux (Forzieri et al., 2020). This generalization is especially apparent in the urban heat adaptation literature. However, not all vegetated surfaces respond in a similar way as different plant functional types may mediate the nature of land—atmosphere feedbacks and thus enhance or dampen the effects of anomalous warmth associated with heatwaves (Li et al., 2021). For example, Teuling et al. (2010) uncovered a difference in how European forest and grassland ecosystems responded to the European 2003 heatwave finding that over forests heating was twice that over grasslands. However, they note that while increased evaporation, due to elevated heatwave-related insolation and temperature, initially suppresses heating over grassland, this ultimately leads to soil moisture depletion and a shift to surface heating because of a fall in latent heat flux. Forests on the other hand, through conservative use of water throughout prolonged anomalously hot periods, mitigate the impact of extreme heat which somewhat compensates for the short-term heating effects produced in the early stages of prolonged heatwaves (Teuling et al., 2010). In the absence of moisture advection, forests may also be an important atmospheric moisture source during heatwaves (Pranindita et al., 2022).

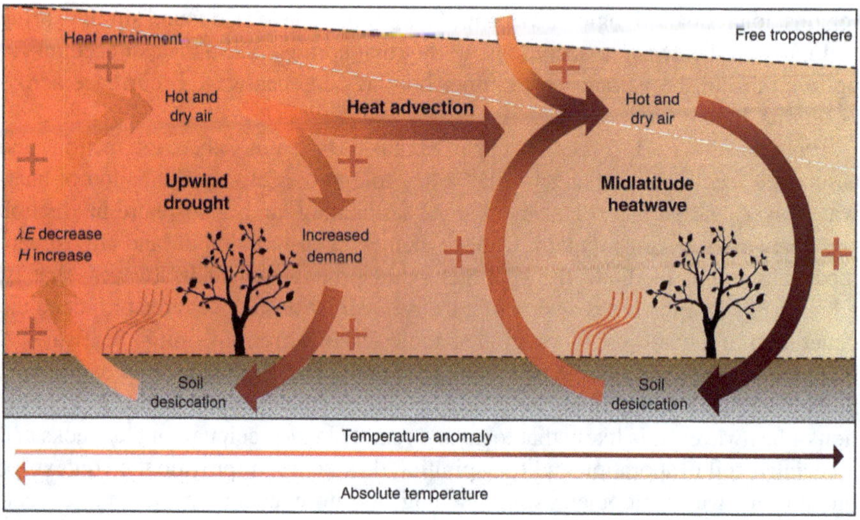

Fig. 7.9 The role of heat advection, from an area under drought, in downwind heatwave development. This conceptualisation builds on the ideas of Miralles et al. (2014) as seen in Fig. 7.8 above. As soil progressively dries throughout a drought more of the available energy is partitioned into sensible heating of the boundary layer causing this layer to warm and grow in depth. Turbulence in the deepening boundary layer may also entrain heat from the so-called entrainment zone near the overlying capping inversion. Under the right circumstances, large or regional scale winds may advect the boundary layer heat downstream thus propagating an upwind drought to a downwind heatwave. The dashed line running left to right represents the climatological height of the boundary layer, being less than that under strong sensible heat during. The darkness of the arrows portrays the relative departure of temperature from the climatological mean. (Source: Schumacher et al. (2019). Amplification of mega-heatwaves through heat torrents fuelled by upwind drought. Nature Geoscience 12, 712–717. https://doi.org/10.1038/s41561-019-0431-6 Used with permission under Springer License Number 5520610470168)

Beyond the individual event scale, Li et al. (2021) have explored how the response of different plant functional types to extreme heat may influence the interannual variation of heatwave occurrence. Using a measure of leaf area to characterize vegetation growth, they found for regions such as Central Europe and parts of southern and southeastern North America, where grasslands, temperate deciduous forests or temperate needle leaf forests tend to dominate the land cover, low leaf area tends to increase the likelihood of heatwaves. This contrasts with the situation across the northwest and northeast regions of North America. Here, where boreal evergreen forests dominate, an increase in heatwave occurrence was found to be associated with a greener landscape as represented by a higher than usual leaf area index. How different vegetation types partition available energy between fluxes of sensible and latent heat can account for these findings (Fig. 7.10). Important for determining vegetation heatwave associations is the degree of vegetation—atmosphere coupling, whether moisture supply or moisture demand limited conditions set the climate context and the nature of vegetation canopy albedo characteristics when a given vegetation type is in full leaf. For boreal evergreen forests, the climate

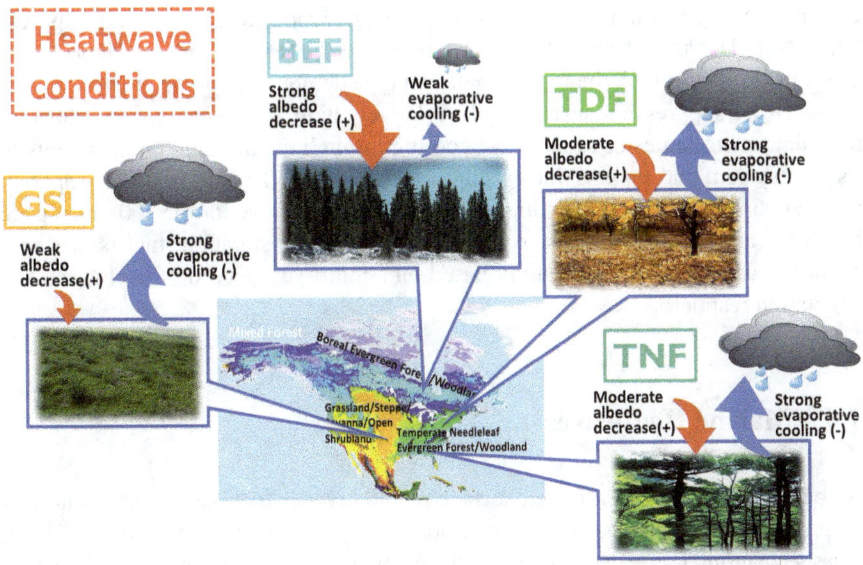

Fig. 7.10 Heatwave vegetation type interactions across North America as determined by vegetation canopy albedo (absorption of shortwave radiation) and degree of atmospheric demand and/or water supply limited evaporative cooling (latent heat flux). BEF is boreal evergreen forest, TEF is temperate evergreen forest, TDF is temperate deciduous forest, TNF is temperate needle leaf forest and GSL is grassland. Positive (+) and negative (−) signs indicate direction of influence of temperature and this heatwave occurrence (Source: Li et al. (2021). Vegetation-heatwave correlations and contrasting energy exchange responses of different vegetation types to summer heatwaves in the Northern Hemisphere during the 1982–2011 period. Agricultural and Forest Meteorology, 296, Article Number 108208. Used with Permission Elsevier License Number 5023620368685)

context is moisture demand limited because of the generally cool atmospheric conditions. When summer growth is strong and leaf area is great, the deeply green canopies of boreal evergreen forests present a low albedo surface to the atmosphere. The upshot of a strong decrease in surface albedo and limited evapotranspiration is the partitioning of considerable amounts of available energy into sensible heating of canopy air because of shortwave radiation absorption by the dark canopy. In combination with little evaporative cooling, this sets up strong positive feedbacks between forest surfaces and the overlying atmosphere that enhance heating and the prolongation and amplification of heatwave conditions. This contrasts strongly with the association between heatwaves and temperate needle leaf forests, despite these forests possessing similar canopy structure characteristics to that of boreal evergreen forests. For temperate needle leaf forests, the climate setting is neither moisture demand or moisture supply limited because of plentiful rainfall and warm atmospheric conditions. Further strong summer growth produces only a moderate fall in canopy albedo (Li et al., 2021). Under this set of circumstances large amounts of energy flow into latent as opposed to sensible heat flux with increased levels of moisture supply to the atmosphere. As a result, an evaporative cooling-cloud-precipitation feedback mechanism works to shorten and suppress heatwaves. The same feedback mechanism operates over temperate deciduous forests and grasslands with the latter vegetation type demonstrating the greatest impact on heatwave suppression because of its relatively high albedo canopy and capacity to maintain high rates of evaporative cooling through making rapid adjustments to its root system and tapping into groundwater reserves (Li et al., 2021). Apparent contradictions between this finding and that described above for the 2003 European heatwave are due to contrasts in the timescales of investigation, that is, a single event versus a climatological study based on a multi-year record of heatwaves and vegetation response.

7.6 Marine Heatwaves

As outlined earlier, the so-called surface mixed layer is that part of the ocean that directly interacts with the overlying atmosphere in terms of the exchange of heat, mass and momentum. It is in the surface mixed layer where marine heatwaves develop as manifest by anomalously high sea surface temperatures or 'hot blobs' as some have referred to them. To understand the development of marine heatwaves researchers have applied a mixed-layer temperature budget approach (Holbrook et al., 2019; Oliver et al., 2021). This describes changes in the temperature of the mixed surface layer as dependent on air-sea exchanges of heat, mass and momentum, horizontal advection of heat by ocean currents and eddies, horizontal and vertical mixing, and entrainment of water from depth into the mixed layer (Holbrook et al., 2019; Holbrook et al., 2020; Oliver et al., 2021). In the context of this framework, this section mainly focuses on air-sea or in other words ocean—atmosphere exchanges and their role in the formation, evolution and decay of what has been termed 'air-sea heat flux type' marine heatwaves (Oliver et al., 2021). Although air-sea heat exchange as a driver of marine heatwaves is the focus here, the importance of the other drivers of surface mixed layer temperature will also be touched upon mainly in relation to how they may moderate ocean—atmosphere exchanges.

The ocean surface is both a recipient and emitter of energy with the net surface heat flux an outcome of the balance between heat gains and losses as portrayed by the components of general energy balance equation (Chap. 4). In contrast to the opaque land surface, shortwave radiation can penetrate the surface layers of water bodies and achieve heating of water beyond the immediate surface. Depending on water clarity, around 75–95% of the shortwave radiation entering the water column is absorbed within 10 m of the surface making it a powerful energy source for heating of the surface mixed layer. That shortwave radiation which is not absorbed escapes to depth through the bottom of the surface mixed layer in the form of so-called penetrative radiation. On an annual timescale, there is generally a balance between solar radiative heating of the ocean and cooling via the heat loss pathways of outgoing longwave radiative and sensible and latent heat flux (Cronin et al., 2019). While the magnitude of sensible and latent heat flux is driven by ocean to atmosphere temperature and humidity differences respectively, the amount of long-wave radiative heat loss is determined by the SST. As there is a non-linear association between longwave radiation and temperature, as described by the Stefan-Boltzman Law, the higher the SST the greater the longwave radiative flux. The upshot of this is, as SST increases so does the amount of upward longwave radiative heat flux. Often referred as the thermostat mechanism, this serves as a powerful negative feedback on SST, with the increased levels of longwave radiation working to cool a warming surface mixed layer (Karnauskas, 2022). Taking all this together, air-sea heat flux type marine heatwaves can be considered as the outcome of an imbalance between heat gains and losses in the surface mixed layer. This makes increased 'downward' shortwave radiation flux, either by itself or in combination with reduced 'upward' longwave radiative flux and sensible and/or latent heat flux, a strong candidate for setting the 'air-sea exchange scene' for marine heatwave development (Fig. 7.11). Given this, it is perhaps no surprise that some of the most intense marine heatwave events are associated with persistent anticyclonic (high pressure) systems and anomalously weak winds, conditions conducive to increased insolation and reduced ocean to atmosphere heat losses (Olita et al., 2007; Sen Gupta et al., 2020; Sparnocchia et al., 2006). As posited by Oliver et al. (2021), under ideal conditions of weak winds and large insolation inputs, a background state typified by stratification and a constant mixed-layer depth, transitions to one of a shoaling mixed layer under reduced wind stress and downward heat flux, with the subsequent development of strong positive mixed layer temperature anomalies (Fig. 7.11).

In many ways what happed in and over the Tasman Sea during the 2017–18 austral summer, provides a good example of the requisite conditions for an 'air-sea heat flux type' (Oliver et al., 2021) marine heatwave. The 2017–18 Tasman Sea heatwave was impressive, in terms of not only its spatial extent, but also its duration and intensity. Covering an extensive area of the Tasman Sea basin, the event lasted for 221 days between October 2017 and April 2018. It reached a maximum intensity in the range of 2.5–3.0 °C (Perkins-Kirkpatrick et al., 2019; Oliver et al., 2021) easily surpassing the magnitude of the Tasman Sea 2015–16 event which itself was considered significant at the time (Oliver et al., 2017), albeit possessing different origins compared to the 2017–18 event (Perkins-Kirkpatrick et al., 2019). According to the marine heatwave categorisation scheme of (Hobday et al., 2018) (see Chap. 2), this event can be considered as severe, but not extreme (Oliver et al., 2021). A persistent area of high pressure over the Tasman Sea (Fig. 7.12a–b) with associated clear skies as manifest by large outgoing longwave radiation anomalies (Fig. 7.12 c–d), large insolation inputs to the sea surface and climatologically weak winds (Fig. 7.11e-f) conspired to set the scene for heatwave formation.

From November 2017 through to January 2018, significant SST anomalies developed covering vast tracts of the Tasman Sea, with a core region of exceptional warmth situated west of the South Island of New Zealand through November to January (Fig. 7.13a–c). A critical determinant of the event's intensity appears to be the very shallow surface mixed layer of around 30 m, significantly shallower than that estimated for the 2015–16 event. Given near surface temperature tendency is

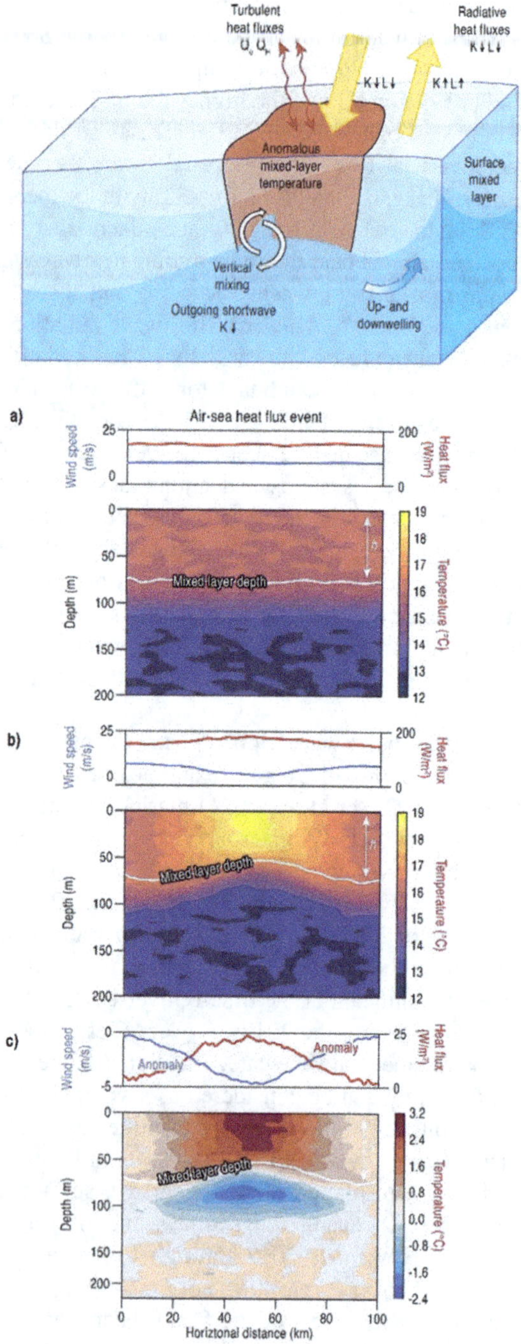

Fig. 7.11 Ocean surface—atmosphere exchanges involved in 'air-sea heat flux type' marine heatwaves. Top part of figure shows general processes at the surface and within near surface ocean layers including radiative and sensible and latent heat exchanges as well as vertical mixing and up/down-welling. The pre-MHW background state is shown in (**a**). This consists of constant mixed-layer depth with stratification, Perturbation by an atmospheric event that produces a slackening of surface winds (e.g., intense blocking situation), an increase in the magnitude of air–sea heat fluxes due to cloud free conditions and shoaling of the mixed-layer depth is portrayed in (**b**). The outcome of this anomalous situation as manifest by anomalously high temperatures in the mixed layer and anomalously low temperatures immediately below the mixed layer is represented by (**c**). (Adapted from Oliver et al. (2021))

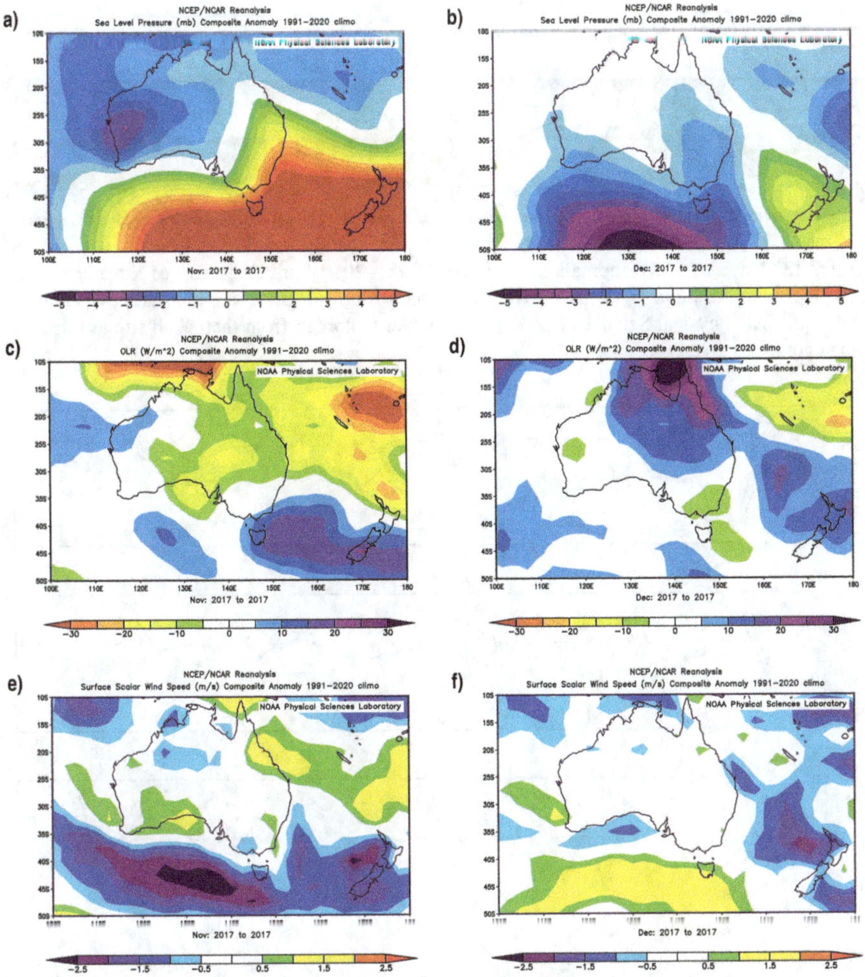

Fig. 7.12 Climate anomalies associated with the 2017–18 Tasman Sea marine heatwave. Sea level pressure in mb (**a**–**b**), outgoing longwave radiation in W/m^2 (**c**–**d**) and wind speed in m/s (**e**–**f**) for November 2017 (left) and December 2017 (right). Note the spatial alignment of strong anomalies for all three variables and their persistence November through to December. Reference period is 1991–2020. Images provided by the NOAA/ESRL Physical Sciences Laboratory, Boulder Colorado from their Web site at http://psl.noaa.gov/

dependent on the inverse association between net heat flux and mixed layer depth, the shallow mixed layer meant radiative heating was not mixed very far beyond the near surface layers, a consequence of little air to sea momentum transfer arising from climatologically weak surface winds (Fig. 7.12e–f). The upshot was little heat dissipation through either downward mixing or upward latent heat flux, with heat accumulation in the relatively shallow surface layer a result. This is clearly demonstrated in time series plots of mixed layer depth, net surface heat flux and

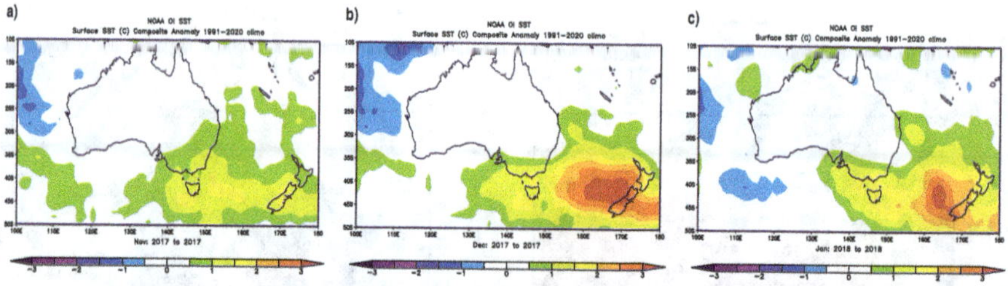

Fig. 7.13 NOAA OI SST anomalies compared to 1981–2010 climatology for (**a**) November 2017, (**b**) December 2017 and (**c**) January 2018. Reference period is 1991–2020. Images provided by the NOAA/ESRL Physical Sciences Laboratory, Boulder Colorado from their Web site at http://psl.noaa.gov/

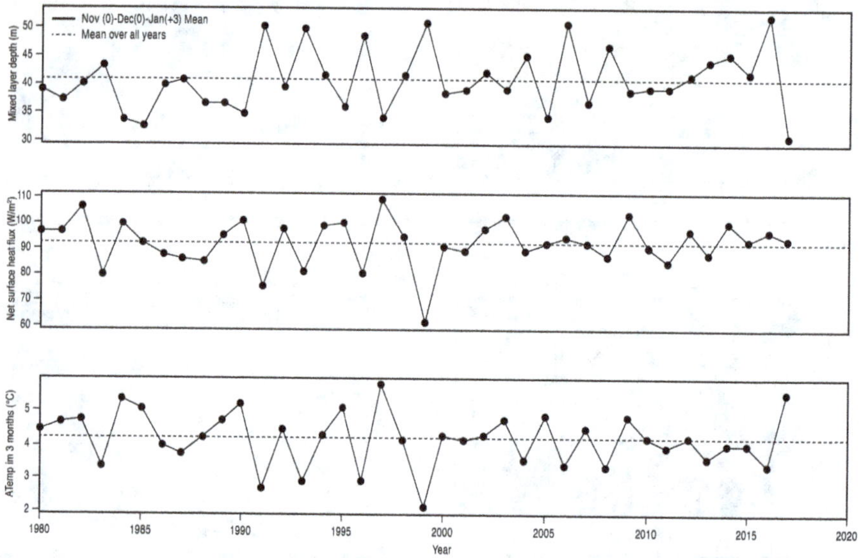

Fig. 7.14 Time series of estimated average November to January values of mixed layer depth (top), net heat flux (middle) and depth-averaged mixed-layer warming (bottom) for the period 1980–81 to 2017–18. Note the shallowness of the mixed layer depth—lowest value since 1980—and the strong warming of the mixed layer although net heat flux was only marginally above its climatological value. (Source: Perkins-Kirkpatrick et al. (2019). The role of natural variability and anthropogenic climate change in the 2017/18 Tasman Sea marine heatwave. Bulletin of the American Meteorological Society, 100, S105–S110. https://doi-org.ezphost.dur.ac.uk/10.1175/BAMS-D-18-0116.1 © American Meteorological Society, Used with permission)

temperature tendency (Fig. 7.14). As noted by Perkins-Kirkpatrick et al. (2019), although net surface heat flux was not remarkable in a climatological context, the shallowing of the mixed layer and associated containment of shortwave radiative heating near the surface was a critical determinant of the event's noteworthy temperature anomalies. The shallowness of the mixed layer along with the strong downward air-sea heat flux is likely to account for the persistence of this event, despite the event being initiated by a burst of warm water advection (Kajtar et al., 2022).

In addition to the 2017–18 Tasman Sea event described above, other MHW that could serve equally well as good examples of air-sea heat flux type marine heatwaves include the 2015–16 Indian Ocean—Coral Sea, 2003 and 2006 Mediterranean, 2017 coastal Peru/Ecuador, 2017 southwest Atlantic shelf, 2016 East China Sea, 2021 mid-latitude western North Pacific, 2017 Mozambique Channel and 1999, 2008, 2012, and 2016 Kuroshio-Oyashio Extension Region MHW events (Benthuysen et al., 2018; Bensoussan et al., 2010; Du et al., 2022; Echevin et al., 2018; Manta et al., 2018; Mawren et al., 2022; Pak et al., 2022; Tan et al., 2022, 2023). Richaud et al. (2024) also point to the importance of atmosphere—ocean heat exchange in driving MHW development in the Arctic Ocean. As well as spawning MHW over large ocean areas, air-sea heat fluxes have the potential to generate extreme localized thermal anomalies. In the context of coral reef environments these have been referred to as MHW hotspots (Fordyce et al., 2019). Here several abiotic factors conspire to drive coral reef water temperatures to extreme levels. These can prove lethal for coral and precipitate coral bleaching events. Chief amongst the various factors is low windspeed and the role it plays in reduced surface water mixing, lowered water turbidity and deeper shortwave radiation penetration, and increased humidity which reduces the effectiveness of upward latent heat flux as a heat loss pathway. If combined with low tides, clear skies, little water exchange with the open ocean and negligent windspeeds, the conditions they precipitate can, at the coral reef scale, significantly amplify regional thermal anomalies associated with a large scale MHW (Fordyce et al., 2019). Figure 7.15 aptly portrays in a conceptual sense the contrasting conditions at play between conditions that work synergistically to exacerbate or moderate extreme reef temperatures.

A number of marine heatwave studies have pointed to the possible role of a SST—cloud feedback mechanism in marine heatwave development, whereby positive SST anomalies at the outset of an event lead to a reduction of low level marine boundary layer cloud amount that permits enhanced shortwave radiative heat flux into the surface mixed layer and thus temperature increases. The simplicity of this mechanism makes it an attractive one to apply in explaining the development of air-sea heat exchange type marine heatwaves. However, as pointed out by Schmeisser et al. (2019) the actual situation may not be that simple with spatio-temporal variability a characteristic of SST—cloud feedback. Based on their analyses of the persistent extra-tropical 'Northeast Pacific Blob', they suggest the situation for marine heatwave development in the higher latitudes of the Pacific Basin is better described in terms of a SST—cloud response, rather than a SST—cloud feedback. This is because despite a net positive radiative heat flux arising from increased insolation receipt under reduced cloud conditions, cooling upward turbulent fluxes of latent and sensible heat offset radiative heat gains. These work to create a slight negative heat flux. At the same time, small anomalies in oceanic heat transport into the NE Pacific and processes at the bottom of the ocean mixed layer, countered the ocean to atmosphere heat loss, with such a combination of both atmospheric and ocean

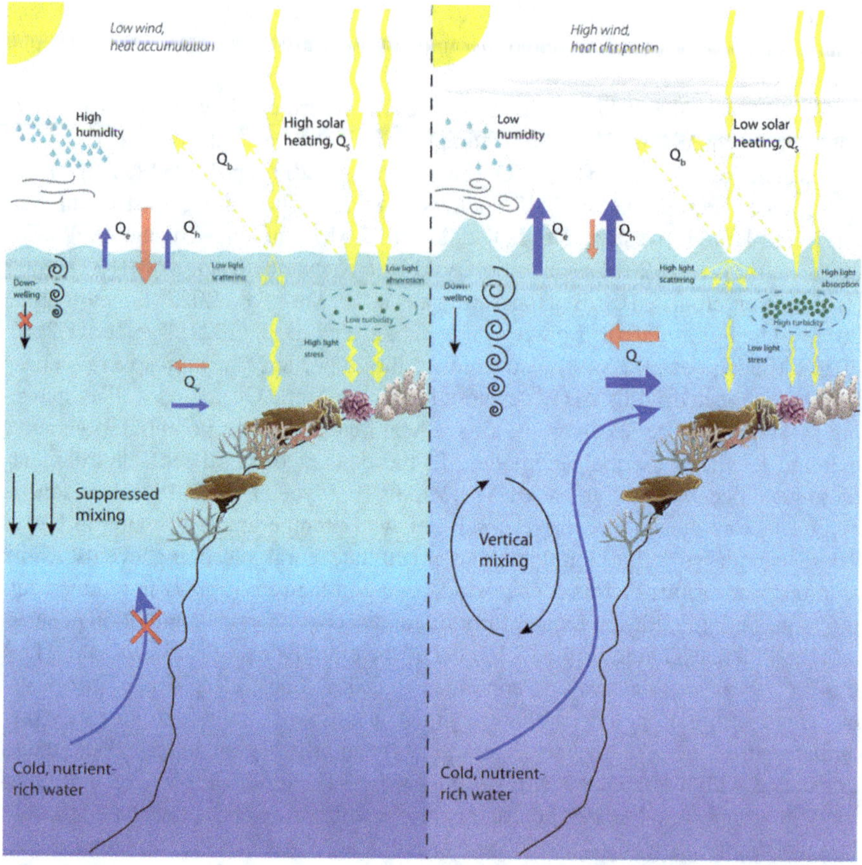

Fig. 7.15 Factors that force (left) and moderate (right) the rapid and intense heating in reef environments. Note the criticality of windspeed determined surface mixing, turbidity and vertical exchanges of the turbulent fluxes of sensible (Qe) and latent (Qe) heat as well as horizontal exchange of water between the reef and the open ocean (Qv). (Source: Fordyce et al. (2019). Marine heatwave hotspots in coral reef environments: physical drivers, ecophysiological outcomes, and impact upon structural complexity. Frontiers in Marine Science. 6:498. doi: 10.3389/fmars.2019.00498. under the terms of the Creative Commons Attribution License (CC BY) https://creativecommons.org/licenses/by/4.0/

processes maintaining this marine heatwave for over 2 years (Schmeisser et al., 2019). This contrasts with the situation for some sub-tropical locations where atmospheric processes in the form of SST—cloud feedback and associated positive heat flux anomalies have been identified as important drivers of heatwave formation as described for events off the coast of the Baja Peninsula, southern California (Myers et al., 2018), the South Atlantic (Rodrigues et al., 2019) and over the central Tasman Sea (Shi et al., 2020). Interestingly a positive low-cloud feedback mechanism may have played a role in the rapid resurgence of the 2015–16 Northeast Pacific Blob (Blob2) in the summer of 2019 (Amaya et al., 2020). Whether events that have their origin in SST—cloud feedback mechanisms form a subset of air-sea heat flux type marine heatwave events has not been mooted yet.

The 2015–16 Northeast Pacific Blob referred to above highlights the importance of oceanic processes in marine heatwave development. In fact, for some marine heatwave events, ocean processes are the dominant force. Accordingly, it is possible to recognise some marine heatwaves as 'ocean advection type' events (Oliver et al., 2021). Examples include the Tasman 2015–16, Leeuwin/Ningaloo Nina 2011, mid-Atlantic Bight 2017, and Tokai, Japan 2017 marine heatwaves (Feng et al., 2013; Gawarkiewicz et al., 2019; Oliver et al., 2017; Ryan et al., 2021; Sugimoto et al., 2020). Such events evolve because of horizontal heat advection brought about by anomalous ocean currents, the result being discernible changes in surface mixed layer temperature at a range of temporal and spatial scales. As the ocean and atmosphere for most of the time is an intimately coupled system, local air-sea heat flux interactions may interact with anomalous oceanic heat advection or other oceanic processes to strengthen or maintain anomalous mixed layer temperatures (Holbrook et al., 2019). This certainly seems to be the case for the marine heatwaves that occurred over three consecutive summers of 2016–2018 in the South China Sea and South Yellow Sea regions. Here the combination of increased shortwave radiation heating, anomalous oceanic heat advection and reduced vertical mixing and upward air-sea heat exchanges boosted by weak winds, conspired to drive up surface mixed layer temperatures (Gao et al., 2020). Enhanced atmosphere to ocean momentum transfer, because of local to regional scale wind flow anomalies, may also play a role in coastal marine heatwave development. This is because down-welling of warm surface waters may be promoted under certain wind field configurations leading to temperature anomalies stretching down to a considerable depth in the water column (Schaeffer & Roughan, 2017). Thus, any atmosphere—ocean exchange process that enhances downwelling of warm surface water and/or muffles upwelling of cold water to the surface, will work to heighten the likelihood of marine heatwave development. In this respect some modes of climate variability may be remote forcers of changes in atmosphere—ocean exchanges that prime marine heatwave development (see Chap. 5). It has also been suggested that as MHW expand in spatial scale, their primary drivers transition from oceanic to atmospheric processes with this shift in dominance being more extensive in eddy-rich regions, while in the interior of gyres, it tends to occur on a smaller scale (Bian et al., 2024).

7.7 Terrestrial Water Bodies and Heatwaves

Compared to marine heatwaves, the subject of heatwaves in terrestrial water bodies or water courses remains largely untreated. Rather than referring to heatwaves per se, most studies on climate anomalies in lakes or streams simply refer to temperature extremes with little attention given to duration characteristics, although in the case of lakes, recent attempts have been made to apply a marine heatwave definition to the analysis of the impacts of climate change on lakes (Woolway et al., 2021a;

Woolway et al., 2021b). That aside, there is a burgeoning literature on the atmospheric factors that influence lake and stream thermal extremes largely spurred on by a need to assess climate change impacts and associated ecohydrological outcomes.

While a range of factors influence the long-term climate characteristics of lakes, the atmospheric processes that generate extreme lake temperature events at the intra-seasonal (heatwave) timescale are akin to those associated with so-called air-sea heat flux type marine heatwaves. A key determinant of lake surface temperature is shortwave radiative heating with anomalous fluxes over lake surfaces accounting for a large proportion of short-term lake temperature changes (Edinger et al., 1968). As for the case of marine heatwaves, the depth over which shortwave radiative heating is distributed is a critical determinant of near surface water temperature tendency. Consequently, if there is little atmosphere to lake surface momentum transfer, mixing in the near surface water column will be inhibited, with the result the mixed surface layer shoals reducing the volume of water that interacts directly with the overlying atmosphere. Accordingly, surface waters will accumulate shortwave radiative heat quickly. This is especially the case in shallow lakes and reservoirs with subdued stratification (Toffolon et al., 2014). Although shortwave radiation in combination with air to lake momentum transfer has the potential to increase lake evaporation and therefore temper the rate of surface heating, under calm conditions as often occur during heatwaves, latent heat flux as a cooling process may have little impact as a heat loss pathway. In addition to latent heat flux, outgoing longwave radiation ($L\uparrow$) has the potential to offset the heating due to incoming shortwave ($K\downarrow$) however the flux of $L\uparrow$ may not be sufficient to offset $K\downarrow$ (Xiao et al., 2020) despite the potential offered by the so-called thermostat mechanism as described above in the case of SST. For lakes situated within complex topography, spatial contrasts in lake surface exposure may exert a noticeable influence on atmosphere to lake heat exchanges with sun shading and wind sheltering effects giving rise to spatial variations in shortwave radiative heating and sensible and latent heat flux respectively (Rahaghi et al., 2019; Sugita et al., 2020) both which have the potential to create a heterogeneous pattern of lake surface water temperature and local hot and cold spots. That atmospheric heatwaves have a clear impact on atmosphere to lake energy exchanges and thus short-term excursions of the thermal state from its long-term mean is apparent for the European heatwaves of 2003 and 2018 (Jankowski et al., 2006; Woolway et al., 2020) as well as heatwaves events across east Asia (Wang et al., 2023, 2024).

There is a large literature on stream (river) temperatures, given temperature is a key determinant of stream ecology, biogeochemistry and water quality and sets the conditions for a range of services such as industrial plant cooling, recreation and water supply. Most of the time streams are in constant motion and manage to dissipate heat quickly. However, at times, especially during periods of low flow or when streams disappear into their beds leaving pools of stranded water, they become susceptible to overheating. As for lakes there are a range of non-climatic factors that may partly influence stream temperature regimes (Caissie, 2006; Isaak et al., 2020; Laize et al., 2017). However, there is no denying that atmosphere—water exchanges of heat energy, in the form of radiative and turbulent fluxes exert a fundamental control on stream temperatures, with associated variables such as air temperature, relative humidity, cloud cover and wind speed as well as the volume of water in the stream channel being important (Evans et al., 1998). Heat may also be added via advection, rainfall and or groundwater. While there are several parallels between the atmospheric controls on lake and stream temperatures, the fact that the stream bed, in a climatic sense, may be able to 'see the atmosphere' makes streams a little different to lakes. This is because the stream bed through receiving radiative, sensible and frictional heat inputs may serve as a heat sink or heat source (Evans et al., 1998). Of the energy fluxes, shortwave radiation (K↓) dominates stream heat gains with longwave radiation (L↑) and latent heat (Q_E↑) constituting the main heat loss pathways (Caissie, 2016; Dugdale et al., 2018). In the absence of significant heat advection, K↓ and Q_E↑ are likely to be the main drivers of in stream heatwave events with the latter flux working to moderate heat gains. From measurements conducted on streams, heat gain from fluxes such as condensation, sensible heat transfer, bed conduction and friction are insignificant with the same holding true for energy losses via evaporation, sensible heat transfer and stream bed conduction (Evans et al., 1998; Laszewski, 2015). Sky view or the openness to the sky of a water course can significantly influence stream surface energy exchanges. Because sky view can vary dramatically along the length of a water course, whether it be in a natural or urban setting, many streams are characterised by a heterogeneous pattern of stream temperature (Mihalevich et al., 2020). In this regard presence or otherwise of riparian vegetation can be a critical determinant of energy exchange and thus stream temperature (Trimmel et al., 2018). For example, after tree harvest an upward shift in the value of stream temperature extremes can occur which may bear implications for the exceedance of thermal thresholds for some freshwater fauna (Reiter et al., 2020).

Urban streams are warmer than their non-urban counterparts partly because of the extra heat in cities associated with the urban heat island effect (Timm et al., 2020). Localised areas of heat storage and the engineered nature of urban water courses which characteristically have large width to depth ratios and thus large surface areas for heat exchange also contribute to higher water temperatures. Added to this, base flows in urban streams are often low and shallow, making the low volumes of water susceptible to rapid warming under large inputs of shortwave radiation. Under this set of circumstances not only does radiative heating of the water occur but heat absorption by the often engineered (e.g., concrete) channel results in heat being released back into the water from storage in the channel bed and banks. Shortwave radiation reflected off the channel bed is also likely to result in some radiative heating of the often-shallow water column. Such atmosphere—stream surface energy exchanges conspire to create harsh in-stream thermal environments in cities with extreme water temperature conditions occurring far more often in urban compared to non-urban streams (Stewart et al., 2020; Zeiger & Hubbart, 2015). Rapid changes in urban stream temperature may also occur because of the nature of atmosphere—land surface energy exchanges that occur away from the stream channel in highly built-up areas. As will be described in the next chapter, the material that makes up urban surfaces is very effective at absorbing radiative heat from the sun, for example a low albedo (dark) asphalt road surface. This makes the urban surface something akin to a hot barbeque plate compared to its rural counterpart, especially after a prolonged period of strong surface heating. Consequently, any runoff across such a surface, as might occur after a summer thunderstorm, is likely to be heated from below with heat transference raising the temperature of the runoff. On entering an urban stream, the warm runoff can produce a temperature surge or thermal shock, with stream temperatures rising quickly. Such waves of heat generated from urban runoff have been known to result in stream temperature surges of 3.5 °C in 30 min lasting up to 3 h with some exceeding 7 °C with dissipation times of 7 h (Nelson & Palmer, 2007).

Semantics aside, whether rapid changes in stream temperature can be considered as 'stream heatwaves' depends on the necessity, perhaps from an eco-hydrological perspective, to distinguish between short sharp thermal shocks at the sub-daily time scale and periods when stream temperatures are maintained for at least 2–3 days above a critical temperature threshold, the exceedance of which exacts slow cumulative thermal stress on stream biota. That thermal stress events occur in stream systems, akin to atmospheric and marine heatwaves, is apparent for several heatwave events. For example, for the European 2003, 2006, 2015 and 2018 heatwave events, exceedance of stream temperatures beyond ecologically relevant thermal thresholds lasted up to 30 days (Feng et al., 2018; Jackson et al., 2020; Piccolroaz et al., 2018). These temperature exceedances also affected the operation of thermal nuclear power plants (see Chap. 11). Tassone et al. (2022) by applying a stream heatwave definition akin to that used by Hobday et al. (2018) for marine heatwaves, identified 3984 riverine heatwave events across the United States for the period 1996 to 2021 also noting an increasing heatwave frequency in streams and rivers for the same period. That multi-day in stream heat events occur is perhaps no surprise. However, such events are not widely reported in the scientific literature. Rather most studies focus on the long-term trends of stream and river thermal sensitivity. Consequently, much work remains to be done establishing whether 'stream heatwaves' are a climatological feature of both natural and human modified water courses.

7.8 Conclusion

The type of exchanges that occur between land, and marine and terrestrial water surfaces and the atmosphere considered relevant to the development of atmospheric, marine, lake and stream heatwaves have been considered in this chapter. A key determinant of heatwave occurrence over land surfaces is how available energy is apportioned between sensible and latent heat. This makes surface moisture status critical. Exacerbation of near surface temperature anomalies during heatwave events may arise not only because of positive feedbacks between a dry surface and the overlying atmosphere but also due to entrainment of heat from the ABL or from advection of heat from upwind hot dry areas. The type of vegetation cover may also play a role in determining the likelihood of heatwave occurrence via surface albedo, canopy heating and evapotranspiration interactions. In marine and lake environments, variations in the surface mixed layer determine water temperature tendencies and thus the development of marine and lake heatwaves. While marine heatwaves stem from changes in air-sea heat exchange processes and oceanic heat advection, the development of lake heatwaves aligns more closely with air-sea heat exchange processes. For streams and rivers, water depth, flow velocity and sky view (shading) are important controls on whether radiative heating of the water column leads to anomalous water temperatures and the occurrence of stream heatwaves, a phenomenon not reported widely in the literature. Heat advection may also be important for heatwave generation in urban streams due to runoff from warm urban surfaces or discharges of warm water into streams. Setting aside the pitfalls that come with synthesising a large amount of information, it is clear from the material presented in this chapter that the configuration of surface—atmosphere interactions involved in atmosphere, marine, lake or stream heatwave formation may vary from event to event. A challenge then is to move from generalising to diagnosing the specific surface—atmosphere heating mechanisms that give rise to individual heatwave events.

References

Alexander, L. (2011). Climate science: Extreme heat rooted in dry soils. *Nature Geoscience, 4*(1), 12–13. https://doi.org/10.1038/ngeo1045

Amaya, D. J., Miller, A. J., Xie, S. P., & Kosaka, Y. (2020). Physical drivers of the summer 2019 North Pacific marine heatwave. *Nature Communications, 11*(1). https://doi.org/10.1038/s41467-020-15820-w

Arya, S. P. (2005). Micrometeorology and atmospheric boundary layer. *Pure and Applied Geophysics, 162*, 1721–1745.

Ashley, G. (1998). Where are we headed?"Soft" rock research into the new millennium. In: *Geological Society of America Abstract/Program*, vol 7. pp A–148.

Basara, J. B., & Crawford, K. C. (2002). Linear relationships between root-zone soil moisture and atmospheric processes in the planetary boundary layer. *Journal of Geophysical Research-Atmospheres, 107*(D15). https://doi.org/10.1029/2001jd000633

Bensoussan, N., Romano, J. C., Harmelin, J. G., & Garrabou, J. (2010). High resolution characterization of Northwest Mediterranean coastal waters thermal regimes: To better understand responses of benthic communities to climate change. *Estuarine Coastal and Shelf Science, 87*(3), 431–441. https://doi.org/10.1016/j.ecss.2010.01.008

Benthuysen, J. A., Oliver, E. C. J., Feng, M., & Marshall, A. G. (2018). Extreme marine warming across tropical Australia during austral summer 2015–2016. *Journal of Geophysical Research-Oceans, 123*(2), 1301–1326. https://doi.org/10.1002/2017jc013326

Bian, C., Jing, Z., Wang, H., & Wu, L. X. (2024). Scale-dependent drivers of marine heatwaves globally. *Geophysical Research Letters, 51*, 3. https://doi.org/10.1029/2023gl107306

Caissie, D. (2006). The thermal regime of rivers: A review. *Freshwater Biology, 51*(8), 1389–1406.

Caissie, D. (2016). River evaporation, condensation and heat fluxes within a first-order tributary of Catamaran Brook (New Brunswick, Canada). *Hydrological Processes, 30*(12), 1872–1883. https://doi.org/10.1002/hyp.10744

Charney, J. G. (1975). Dynamics of deserts and drought in Sahel. *Quarterly Journal of the Royal Meteorological Society, 101*(428), 193–202. https://doi.org/10.1002/qj.49710142802

Chiriaco, M., Bastin, S., Yiou, P., Haeffelin, M., Dupont, J. C., & Stéfanon, M. (2014). European heatwave in July 2006: Observations and modeling showing how local processes amplify conducive large-scale conditions. *Geophysical Research Letters, 41*(15), 5644–5652. https://doi.org/10.1002/2014gl060205

Christian, J. I., Basara, J. B., Hunt, E. D., Otkin, J. A., & Xiao, X. M. (2020). Flash drought development and cascading impacts associated with the 2010 Russian heatwave. *Environmental Research Letters, 15*(9). https://doi.org/10.1088/1748-9326/ab9faf

Cronin, M. F., Gentemann, C. L., Edson, J., Ueki, I., Bourassa, M., Brown, S., Clayson, C. A., Fairall, C. W., Farrar, J. T., & Gille, S. T. (2019). Air-sea fluxes with a focus on heat and momentum. *Frontiers in Marine Science, 6*, 430.

de Boyer, M. C., Madec, G., Fischer, A. S., Lazar, A., & Iudicone, D. (2004). Mixed layer depth over the global ocean: An examination of profile data and a profile-based climatology. *Journal of Geophysical Research-Oceans, 109*(C12). https://doi.org/10.1029/2004jc002378

Demuzere, M., De Ridder, K., & Van Lipzig, N. (2008). Modeling the energy balance in Marseille: Sensitivity to roughness length parameterizations and thermal admittance. *Journal of Geophysical Research: Atmospheres, 113*(D16).

Dirmeyer, P. A., Zeng, F. J., Ducharne, A., Morrill, J. C., & Koster, R. D. (2000). The sensitivity of surface fluxes to soil water content in three land surface schemes. *Journal of Hydrometeorology, 1*(2), 121–134. https://doi.org/10.1175/1525-7541(2000)001<0121:tsosft>2.0.co;2

Dirmeyer, P. A., Jin, Y., Singh, B., & Yan, X. Q. (2013). Evolving land-atmosphere interactions over North America from CMIP5 simulations. *Journal of Climate, 26*(19), 7313–7327. https://doi.org/10.1175/jcli-d-12-00454.1

Du, Y. Z., Feng, M., Xu, Z. H., Yin, B. S., & Hobday, A. J. (2022). Summer marine heatwaves in the Kuroshio-Oyashio extension region. *Remote Sensing, 14*(13). https://doi.org/10.3390/rs14132980

Dugdale, S. J., Malcolm, I. A., Kantola, K., & Hannah, D. M. (2018). Stream temperature under contrasting riparian forest cover: Understanding thermal dynamics and heat exchange processes. *Science of the Total Environment, 610*, 1375–1389.

Echevin, V., Colas, F., Espinoza-Morriberon, D., Vasquez, L., Anculle, T., & Gutierrez, D. (2018). Forcings and evolution of the 2017 coastal El Nino off northern Peru and Ecuador. *Frontiers in Marine Science, 5*. https://doi.org/10.3389/fmars.2018.00367

Edinger, J. E., Duttweiler, D. W., & Geyer, J. C. (1968). Response of water temperatures to meteorological conditions. *Water Resources Research, 4*(5), 1137. https://doi.org/10.1029/WR004i005p01137

Evans, E. C., McGregor, G. R., & Petts, G. E. (1998). River energy budgets with special reference to river bed processes. *Hydrological Processes, 12*(4), 575–595.

Feng, M., McPhaden, M. J., Xie, S. P., & Hafner, J. (2013). La Nina forces unprecedented Leeuwin current warming in 2011. *Scientific Reports, 3*. https://doi.org/10.1038/srep01277

Feng, M. L., Zolezzi, G., & Pusch, M. (2018). Effects of thermopeaking on the thermal response of alpine river systems to heatwaves. *Science of the Total Environment, 612*, 1266–1275. https://doi.org/10.1016/j.scitotenv.2017.09.042

Ferranti, L., & Viterbo, P. (2006). The European summer of 2003: Sensitivity to soil water initial conditions. *Journal of Climate, 19*(15), 3659–3680. https://doi.org/10.1175/jcli3810.1

Fischer, E. M., Seneviratne, S. I., Luthi, D., & Schar, C. (2007). Contribution of land-atmosphere coupling to recent European summer heat waves. *Geophysical Research Letters, 34*(6). https://doi.org/10.1029/2006gl029068

Ford, T. W., & Schoof, J. T. (2016). Oppressive heat events in Illinois related to antecedent wet soils. *Journal of Hydrometeorology, 17*(10), 2713–2726. https://doi.org/10.1175/jhm-d-16-0075.1

Ford, T. W., & Schoof, J. T. (2017). Characterizing extreme and oppressive heat waves in Illinois. *Journal of Geophysical Research-Atmospheres, 122*(2), 682–698. https://doi.org/10.1002/2016jd025721

Ford, T. W., Quiring, S. M., & Frauenfeld, O. W. (2017). Multi-decadal variability of soil moisture-temperature coupling over the contiguous United States modulated by Pacific and Atlantic Sea surface temperatures. *International Journal of Climatology, 37*(3), 1400–1415. https://doi.org/10.1002/joc.4785

Fordyce, A. J., Ainsworth, T. A., Heron, S. F., & Leggat, W. (2019). Marine heatwave hotspots in coral reef environments: Physical drivers, ecophysiological outcomes, and impact upon structural complexity. *Frontiers in Marine Science, 6.* https://doi.org/10.3389/fmars.2019.00498

Forzieri, G., Miralles, D. G., Ciais, P., Alkama, R., Ryu, Y., Duveiller, G., Zhang, K., Robertson, E., Kautz, M., Martens, B., Jiang, C. Y., Arneth, A., Georgievski, G., Li, W., Ceccherini, G., Anthoni, P., Lawrence, P., Wiltshire, A., Pongratz, J., Piao, S. L., Sitch, S., Goll, D. S., Arora, V. K., Lienert, S., Lombardozzi, D., Kato, E., Nabel, J., Tian, H. Q., Friedlingstein, P., & Cescatti, A. (2020). Increased control of vegetation on global terrestrial energy fluxes. Nature. *Climate Change, 10*(4), 356–362. https://doi.org/10.1038/s41558-020-0717-0

Gao, G. D., Marin, M., Feng, M., Yin, B. S., Yang, D. Z., Feng, X. R., Ding, Y., & Song, D. H. (2020). Drivers of marine heatwaves in the East China Sea and the South Yellow Sea in three consecutive summers during 2016–2018. *Journal of Geophysical Research-Oceans, 125*(8). https://doi.org/10.1029/2020jc016518

Gates, D. M. (1963). Energy exchange in the biosphere. *Soil Science, 96*(1), 76.

Gawarkiewicz, G., Chen, K., Forsyth, J., Bahr, F., Mercer, A. M., Ellertson, A., Fratantoni, P., Seim, H., Haines, S., & Han, L. (2019). Characteristics of an advective marine heatwave in the middle Atlantic bight in early 2017. *Frontiers in Marine Science, 6.* https://doi.org/10.3389/fmars.2019.00712

Ghatak, D., Zaitchik, B., Hain, C., & Anderson, M. (2017). The role of local heating in the 2015 Indian heat wave. *Scientific Reports, 7*, 7707. https://doi.org/10.1038/s41598-017-07956-5

Giardino, J. R., & Houser, C. (2015). Chapter 1 – Introduction to the critical zone. In J. R. Giardino & C. Houser (Eds.), *Developments in earth surface processes* (Vol. 19, pp. 1–13). Elsevier. https://doi.org/10.1016/B978-0-444-63369-9.00001-X

Gibson, P. B., Pitman, A. J., Lorenz, R., & Perkins-Kirkpatrick, S. E. (2017). The role of circulation and land surface conditions in current and future Australian heat waves. *Journal of Climate, 30*(24), 9933–9948. https://doi.org/10.1175/jcli-d-17-0265.1

Hauser, M., Orth, R., & Seneviratne, S. I. (2016). Role of soil moisture versus recent climate change for the 2010 heat wave in western Russia. *Geophysical Research Letters, 43*(6), 2819–2826. https://doi.org/10.1002/2016gl068036

He, B., Zhong, Z. Q., Chen, D. L., Liu, J. G., Chen, Y. N., Miao, C. Y., Ding, R. Q., Yuan, W. P., Guo, L. L., Huang, L., Hao, X. M., & Chen, A. F. (2022). Lengthening dry spells intensify summer heatwaves. *Geophysical Research Letters, 49*(19). https://doi.org/10.1029/2022gl099647

Herold, N., Kala, J., & Alexander, L. V. (2016). The influence of soil moisture deficits on Australian heatwaves. *Environmental Research Letters, 11*(6). https://doi.org/10.1088/1748-9326/11/6/064003

Hirsch, A. L., & King, M. J. (2020). Atmospheric and land surface contributions to heatwaves: An Australian perspective. *Journal of Geophysical Research-Atmospheres, 125*(17). https://doi.org/10.1029/2020jd033223

Hirsch, A. L., Evans, J. P., Di Virgilio, G., Perkins-Kirkpatrick, S. E., Argueso, D., Pitman, A. J., Carouge, C. C., Kala, J., Andrys, J., Petrelli, P., & Rockel, B. (2019). Amplification of Australian heatwaves via local land-atmosphere coupling. *Journal of Geophysical Research-Atmospheres, 124*(24), 13625–13647. https://doi.org/10.1029/2019jd030665

Hirschi, M., Seneviratne, S. I., Alexandrov, V., Boberg, F., Boroneant, C., Christensen, O. B., Formayer, H., Orlowsky, B., & Stepanek, P. (2011). Observational evidence for soil-moisture impact on hot extremes in southeastern Europe. *Nature Geoscience, 4*(1), 17–21.

Hobday, A. J., Oliver, E. C. J., Sen Gupta, A., Benthuysen, J. A., Burrows, M. T., Donat, M. G., Holbrook, N. J., Moore, P. J., Thomsen, M. S., Wernberg, T., & Smale, D. A. (2018). Categorizing and naming marine heatwaves. *Oceanography, 31*(2), 162–173. https://doi.org/10.5670/oceanog.2018.205

Holbrook, N. J., Scannell, H. A., Sen Gupta, A., Benthuysen, J. A., Feng, M., Oliver, E. C. J., Alexander, L. V., Burrows, M. T., Donat, M. G., Hobday, A. J., Moore, P. J., Perkins-Kirkpatrick, S. E., Smale, D. A., Straub, S. C., & Wernberg, T. (2019). A global assessment of marine heatwaves and their drivers. *Nature Communications, 10*, 2624. https://doi.org/10.1038/s41467-019-10206-z

Holbrook, N. J., Sen Gupta, A., Oliver, E. C. J., Hobday, A. J., Benthuysen, J. A., Scannell, H. A., Smale, D. A., & Wernberg, T. (2020). Keeping pace with marine heatwaves. *Nature Reviews Earth & Environment, 1*(9), 482–493. https://doi.org/10.1038/s43017-020-0068-4

Holmes, A., Rüdiger, C., Mueller, B., Hirschi, M., & Tapper, N. (2017). Variability of soilmoisture proxies and hot days across the climate regimes of Australia. *Geophysical Research Letters, 44*(14), 7265–7275. https://doi.org/10.1002/2017gl073793

Isaak, D. J., Luce, C. H., Horan, D. L., Chandler, G. L., Wollrab, S. P., Dubois, W. B., & Nagel, D. E. (2020). Thermal regimes of perennial rivers and streams in the western United States. *Journal of the American Water Resources Association, 56*(5), 842–867. https://doi.org/10.1111/1752-1688.12864

Jackson, F. L., Fryer, R. J., Hannah, D. M., & Malcolm, I. A. (2020). Predictions of national-scale river temperatures: A visualisation of complex space-time dynamics. *Hydrological Processes, 34*(12), 2823–2825. https://doi.org/10.1002/hyp.13761

Jankowski, T., Livingstone, D. M., Buhrer, H., Forster, R., & Niederhauser, P. (2006). Consequences of the 2003 European heat wave for lake temperature profiles, thermal stability, and hypolimnetic oxygen depletion: Implications for a warmer world. *Limnology and Oceanography, 51*(2), 815–819. https://doi.org/10.4319/lo.2006.51.2.0815

Kajtar, J. B., Bachman, S. D., Holbrook, N. J., & Pilo, G. S. (2022). Drivers, dynamics, and persistence of the 2017/2018 Tasman Sea marine heatwave. *Journal of Geophysical Research-Oceans, 127*(8). https://doi.org/10.1029/2022jc018931

Kala, J., Evans, J. P., & Pitman, A. J. (2015). Influence of antecedent soil moisture conditions on the synoptic meteorology of the black Saturday bushfire event in Southeast Australia. *Quarterly Journal of the Royal Meteorological Society, 141*(693), 3118–3129. https://doi.org/10.1002/qj.2596

Kara, A. B., Rochford, P. A., & Hurlburt, H. E. (2003). Mixed layer depth variability over the global ocean. *Journal of Geophysical Research-Oceans, 108*(C3). https://doi.org/10.1029/2000jc000736

Karnauskas, K. B. (2022). A simple coupled model of the wind–evaporation–SST feedback with a role for stability. *Journal of Climate, 35*(7), 2149–2160.

Laize, C. L. R., Meredith, C. B., Dunbar, M. J., & Hannah, D. M. (2017). Climate and basin drivers of seasonal river water temperature dynamics. *Hydrology and Earth System Sciences, 21*(6), 3231–3247. https://doi.org/10.5194/hess-21-3231-2017

Laszewski, M. (2015). Heat fluxes and river energy budget on the example of lowland Swider River. *Quaestiones Geographicae, 34*(1), 65–74. https://doi.org/10.1515/quageo-2015-0006

Lee, E., Bieda, R., Shanmugasundaram, J., & Richter, H. B. (2016). Land surface and atmospheric conditions associated with heat waves over the Chickasaw nation in the south Central United States. *Journal of Geophysical Research-Atmospheres, 121*(11), 6284–6298. https://doi.org/10.1002/2015jd024659

Lemus-Canovas, M., Insua-Costa, D., Trigo, R. M., & Miralles, D. G. (2024). Record-shattering 2023 spring heatwave in western Mediterranean amplified by long-term drought. *Npj Climate and Atmospheric Science, 7*(1). https://doi.org/10.1038/s41612-024-00569-6

Li, J., Tam, C. Y., Tai, A. P. K., & Lau, N. C. (2021). Vegetation-heatwave correlations and contrasting energy exchange responses of different vegetation types to summer heatwaves in the northern hemisphere during the 1982–2011 period. *Agricultural and Forest Meteorology, 296*, 108208. https://doi.org/10.1016/j.agrformet.2020.108208

Lin, H. (2010). Earth's critical zone and hydropedology: Concepts, characteristics, and advances. *Hydrology and Earth System Sciences, 14*(1), 25–45. https://doi.org/10.5194/hess-14-25-2010

Liu, S. Y., & Liang, X. Z. (2010). Observed diurnal cycle climatology of planetary boundary layer height. *Journal of Climate, 23*(21), 5790–5809. https://doi.org/10.1175/2010jcli3552.1

Lorenz, R., Jaeger, E. B., & Seneviratne, S. I. (2010). Persistence of heat waves and its link to soil moisture memory. *Geophysical Research Letters, 37*. https://doi.org/10.1029/2010gl042764

Manta, G., de Mello, S., Trinchin, R., Badagian, J., & Barreiro, M. (2018). The 2017 record marine heatwave in the southwestern Atlantic shelf. *Geophysical Research Letters, 45*(22), 12,449–412,456.

Mawren, D., Hermes, J., & Reason, C. J. C. (2022). Marine heatwaves in the Mozambique Channel. *Climate Dynamics, 58*(1–2), 305–327. https://doi.org/10.1007/s00382-021-05909-3

Mihalevich, B. A., Neilson, B. T., Buahin, C. A., Yackulic, C. B., & Schmidt, J. C. (2020). Water temperature controls for regulated canyon-bound rivers. *Water Resources Research, 56*(12). https://doi.org/10.1029/2020wr027566

Miralles, D. G., Teuling, A. J., van Heerwaarden, C. C., & de Arellano, J. V. G. (2014). Mega-heatwave temperatures due to combined soil desiccation and atmospheric heat accumulation. *Nature Geoscience, 7*(5), 345–349. https://doi.org/10.1038/ngeo2141

Miralles, D. G., Gentine, P., Seneviratne, S. I., & Teuling, A. J. (2019). Land-atmospheric feedbacks during droughts and heatwaves: State of the science and current challenges. *Annals of the New York Academy of Sciences, 1436*(1), 19–35. https://doi.org/10.1111/nyas.13912

Mu, M. Y., Pitman, A. J., Kauwe, M. G. D., Ukkola, A. M., & Ge, J. (2022). How do groundwater dynamics influence heatwaves in Southeast Australia? *Weather and Climate Extremes, 37*, 100479. https://doi.org/10.1016/j.wace.2022.100479

Myers, T. A., Mechoso, C. R., Cesana, G. V., DeFlorio, M. J., & Waliser, D. E. (2018). Cloud feedback key to marine heatwave off Baja California. *Geophysical Research Letters, 45*(9), 4345–4352. https://doi.org/10.1029/2018gl078242

Nelson, K. C., & Palmer, M. A. (2007). Stream temperature surges under urbanization and climate change: Data, models, and responses. *Journal of the American Water Resources Association, 43*(2), 440–452. https://doi.org/10.1111/j.1752-1688.2007.00034.x

Nissan, H., Burkart, K., Coughlan de Perez, E., Van Aalst, M., & Mason, S. (2017). Defining and predicting heat waves in Bangladesh. *Journal of Applied Meteorology and Climatology, 56*(10), 2653–2670.

Oke, T. R., Mills, G., Christen, A., & Voogt, J. A. (2017). *Urban climates*. Cambridge University Press.

Olita, A., Sorgente, R., Natale, S., Gabersek, S., Ribotti, A., Bonanno, A., & Patti, B. (2007). Effects of the 2003 European heatwave on the Central Mediterranean Sea: Surface fluxes and the dynamical response. *Ocean Science, 3*(2), 273–289. https://doi.org/10.5194/os-3-273-2007

Oliver, E. C. J., Benthuysen, J. A., Bindoff, N. L., Hobday, A. J., Holbrook, N. J., Mundy, C. N., & Perkins-Kirkpatrick, S. E. (2017). The unprecedented 2015/16 Tasman Sea marine heatwave. *Nature Communications, 8*. https://doi.org/10.1038/ncomms16101

Oliver, E. C. J., Benthuysen, J. A., Darmaraki, S., Donat, M. G., Hobday, A. J., Holbrook, N. J., Schlegel, R. W., & Sen Gupta, A. (2021). Marine heatwaves. *Annual Review of Marine Science, 13*(13), 313–342. https://doi.org/10.1146/annurev-marine-032720-095144

Pak, G., Noh, J., Park, Y. G., Jin, H., & Park, J. H. (2022). Governing factors of the record-breaking marine heatwave over the mid-latitude western North Pacific in the summer of 2021. *Frontiers in Marine Science, 9*. https://doi.org/10.3389/fmars.2022.946767

Perkins, S. E. (2015). A review on the scientific understanding of heatwaves-their measurement, driving mechanisms, and changes at the global scale. *Atmospheric Research, 164*, 242–267. https://doi.org/10.1016/j.atmosres.2015.05.014

Perkins, S. E., Argueso, D., & White, C. J. (2015). Relationships between climate variability, soil moisture, and Australian heatwaves. *Journal of Geophysical Research-Atmospheres, 120*(16), 8144–8164. https://doi.org/10.1002/2015jd023592

Perkins-Kirkpatrick, S. E., King, A. D., Cougnon, E. A., Grose, M. R., Oliver, E. C. J., Holbrook, N. J., Lewis, S. C., & Pourasghar, F. (2019). The role of natural variability and anthropogenic climate change in the 2017/18 Tasman Sea marine heatwave. *Bulletin of the American Meteorological Society, 100*(1), S105–S110. https://doi.org/10.1175/bams-d-18-0116.1

Philip, S. Y., Kew, S. F., Hauser, M., Guillod, B. P., Teuling, A. J., Whan, K., Uhe, P., & van Oldenborgh, G. J. (2018). Western US high June 2015 temperatures and their relation to global warming and soil moisture. *Climate Dynamics, 50*(7–8), 2587–2601. https://doi.org/10.1007/s00382-017-3759-x

Piccolroaz, S., Toffolon, M., Robinson, C. T., & Siviglia, A. (2018). Exploring and quantifying river thermal response to heatwaves. *Water, 10*(8). https://doi.org/10.3390/w10081098

Pitman, A. J. (2003). The evolution of, and revolution in, land surface schemes designed for climate models. *International Journal of Climatology, 23*(5), 479–510. https://doi.org/10.1002/joc.893

Pranindita, A., Wang-Erlandsson, L., Fetzer, I., & Teuling, A. J. (2022). Moisture recycling and the potential role of forests as moisture source during European heatwaves. *Climate Dynamics, 58*(1–2), 609–624. https://doi.org/10.1007/s00382-021-05921-7

Quesada, B., Vautard, R., Yiou, P., Hirschi, M., & Seneviratne, S. I. (2012). Asymmetric European summer heat predictability from wet and dry southern winters and springs. *Nature Climate Change, 2*(10), 736–741. https://doi.org/10.1038/nclimate1536

Quiring, S. M., Ford, T. W., & Yuan, S. (2015). Climate of the critical zone. In *Developments in earth surface processes* (Vol. 19, pp. 79–111). Elsevier.

Rahaghi, A. I., Lemmin, U., Cimatoribus, A. A., & Barry, D. A. (2019). The importance of systematic spatial variability in the surface heat flux of a large lake: A multiannual analysis for Lake Geneva. *Water Resources Research, 55*(12), 10248–10267. https://doi.org/10.1029/2019wr024954

Reiter, M., Johnson, S. L., Homyack, J., Jones, J. E., & James, P. L. (2020). Summer stream temperature changes following forest harvest in the headwaters of the Trask River watershed, Oregon Coast Range. *Ecohydrology, 13*, 3. https://doi.org/10.1002/eco.2178

Richaud, B., Hu, X. M., Darmaraki, S., Fennel, K., Lu, Y. Y., & Oliver, E. C. J. (2024). Drivers of marine heatwaves in the Arctic Ocean. *Journal of Geophysical Research-Oceans, 129*(2). https://doi.org/10.1029/2023jc020324

Rodrigues, R. R., Taschetto, A. S., Sen Gupta, A., & Foltz, G. R. (2019). Common cause for severe droughts in South America and marine heatwaves in the South Atlantic. *Nature Geoscience, 12*(8), 620–626. https://doi.org/10.1038/s41561-019-0393-8

Ryan, S., Ummenhofer, C. C., Gawarkiewicz, G., Wagner, P., Scheinert, M., Biastoch, A., & Boning, C. W. (2021). Depth structure of Ningaloo Nino/Nina events and associated drivers. *Journal of Climate, 34*(5), 1767–1788. https://doi.org/10.1175/jcli-d-19-1020.1

Schaeffer, A., & Roughan, M. (2017). Subsurface intensification of marine heatwaves off southeastern Australia: The role of stratification and local winds. *Geophysical Research Letters, 44*(10), 5025–5033. https://doi.org/10.1002/2017gl073714

Schmeisser, L., Bond, N. A., Siedlecki, S. A., & Ackerman, T. P. (2019). The role of clouds and surface heat fluxes in the maintenance of the 2013-2016 Northeast Pacific marine heatwave. *Journal of Geophysical Research-Atmospheres, 124*(20), 10772–10783. https://doi.org/10.1029/2019jd030780

Schumacher, D. L., Keune, J., van Heerwaarden, C. C., de Arellano, J. V. G., Teuling, A. J., & Miralles, D. G. (2019). Amplification of mega-heatwaves through heat torrents fuelled by upwind drought. *Nature Geoscience, 12*(9), 712–717. https://doi.org/10.1038/s41561-019-0431-6

Seidel, D. J., Zhang, Y. H., Beljaars, A., Golaz, J. C., Jacobson, A. R., & Medeiros, B. (2012). Climatology of the planetary boundary layer over the continental United States and Europe. *Journal of Geophysical Research-Atmospheres, 117*. https://doi.org/10.1029/2012jd018143

Sen Gupta, A., Thomsen, M., Benthuysen, J. A., Hobday, A. J., Oliver, E., Alexander, L. V., Burrows, M. T., Donat, M. G., Feng, M., Holbrook, N. J., Perkins-Kirkpatrick, S., Moore, P. J., Rodrigues, R. R., Scannell, H. A., Taschetto, A. S., Ummenhofer, C. C., Wernberg, T., & Smale, D. A. (2020). Drivers and impacts of the most extreme marine heatwaves events. *Scientific Reports, 10*(1), 19359. https://doi.org/10.1038/s41598-020-75445-3

Seneviratne, S. I., Corti, T., Davin, E. L., Hirschi, M., Jaeger, E. B., Lehner, I., Orlowsky, B., & Teuling, A. J. (2010). Investigating soil moisture-climate interactions in a changing climate: A review. *Earth-Science Reviews, 99*(3–4), 125–161. https://doi.org/10.1016/j.earscirev.2010.02.004

Sharqawy, M. H., Lienhard, J. H., & Zubair, S. M. (2010). Thermophysical properties of seawater: A review of existing correlations and data. *Desalination and Water Treatment, 16*(1–3), 354–380. https://doi.org/10.5004/dwt.2010.1079

Shi, J., Chen, Z. Y., Ding, S. S., & Lu, Y. Q. (2020). A hot blob eastward of New Zealand in December 2019. *Atmosphere, 11*(12). https://doi.org/10.3390/atmos11121267

Sparnocchia, S., Schiano, M. E., Picco, P., Bozzano, R., & Cappelletti, A. (2006). The anomalous warming of summer 2003 in the surface layer of the Central Ligurian Sea (Western Mediterranean). *Annales Geophysicae, 24*(2), 443–452. https://doi.org/10.5194/angeo-24-443-2006

Stewart, I. T., Engh, M. K., & Parchem, C. P. (2020). Effect of climate warming and drought on urban stream temperatures in a semiarid, seasonal system. *Journal of the American Water Resources Association, 56*(4), 551–567. https://doi.org/10.1111/1752-1688.12846

Stull, R. (2011). *Meteorology for scientists and engineers* (3rd ed.). University of British Columbia.

Sugimoto, S., Qiu, B., & Kojima, A. (2020). Marked coastal warming off Tokai attributable to Kuroshio large meander. *Journal of Oceanography, 76*(2), 141–154. https://doi.org/10.1007/s10872-019-00531-8

Sugita, M., Ogawa, S., & Kawade, M. (2020). Wind as a main driver of spatial variability of surface energy balance over a shallow 102-km^2 scale lake: Lake Kasumigaura, Japan. *Water Resources Research, 56*(8), e2020WR027173.

Tan, H.-J., Cai, R.-S., & Wu, R.-G. (2022). Summer marine heatwaves in the South China Sea: Trend, variability and possible causes. *Advances in Climate Change Research, 13*(3), 323–332.

Tan, H.-J., Cai, R.-S., Bai, D.-P., Hilmi, K., & Tonbol, K. (2023). Causes of 2022 summer marine heatwave in the East China Seas. *Advances in Climate Change Research, 14*(5), 633–641.

Tassone, S. J., Besterman, A. F., Buelo, C. D., Ha, D. T., Walter, J. A., & Pace, M. L. (2022). Increasing heatwave frequency in streams and rivers of the United States. *Limnology and Oceanography Letters., 8*, 295–304. https://doi.org/10.1002/lol2.10284

Teuling, A. J., Hirschi, M., Ohmura, A., Wild, M., Reichstein, M., Ciais, P., Buchmann, N., Ammann, C., Montagnani, L., Richardson, A. D., Wohlfahrt, G., & Seneviratne, S. I. (2009). A regional perspective on trends in continental evaporation. *Geophysical Research Letters, 36*. https://doi.org/10.1029/2008gl036584

Teuling, A. J., Seneviratne, S. I., Stockli, R., Reichstein, M., Moors, E., Ciais, P., Luyssaert, S., van den Hurk, B., Ammann, C., Bernhofer, C., Dellwik, E., Gianelle, D., Gielen, B., Grunwald, T., Klumpp, K., Montagnani, L., Moureaux, C., Sottocornola, M., & Wohlfahrt, G. (2010). Contrasting response of European forest and grassland energy exchange to heatwaves. *Nature Geoscience, 3*(10), 722–727. https://doi.org/10.1038/ngeo950

Timm, A., Ouellet, V., & Daniels, M. (2020). Swimming through the urban heat Island: Can thermal mitigation practices reduce the stress? *River Research and Applications, 36*(10), 1973–1984. https://doi.org/10.1002/rra.3732

Toffolon, M., Piccolroaz, S., Majone, B., Soja, A. M., Peeters, F., Schmid, M., & Wuest, A. (2014). Prediction of surface temperature in lakes with different morphology using air temperature. *Limnology and Oceanography, 59*(6), 2185–2202. https://doi.org/10.4319/lo.2014.59.6.2185

Trimmel, H., Weihs, P., Leidinger, D., Formayer, H., Kalny, G., & Melcher, A. (2018). Can riparian vegetation shade mitigate the expected rise in stream temperatures due to climate change during heat waves in a human-impacted pre-alpine river? *Hydrology and Earth System Sciences, 22*(1), 437–461. https://doi.org/10.5194/hess-22-437-2018

von Engeln, A., & Teixeira, J. (2013). A planetary boundary layer height climatology derived from ECMWF reanalysis data. *Journal of Climate, 26*(17), 6575–6590. https://doi.org/10.1175/jcli-d-12-00385.1

Wang, C. H., Wang, Z. H., & Sun, L. D. (2020). Early-warning signals for critical temperature transitions. *Geophysical Research Letters, 47*(14). https://doi.org/10.1029/2020gl088503

Wang, W. J., Shi, K., Wang, X. W., Wang, S. Q., Zhang, D., Peng, Y. Y., Li, N., Zhang, Y. L., Zhang, Y. B., Qin, B. Q., Woolway, R. I., & Jeppesen, E. (2023). A record-breaking extreme heat event caused unprecedented warming of lakes in China. *Science Bulletin, 68*(6), 578–582. https://doi.org/10.1016/j.scib.2023.03.001

Wang, W. J., Shi, K., Wang, X. W., Zhang, Y. L., Qin, B. Q., Zhang, Y. B., & Woolway, R. I. (2024). The impact of extreme heat on lake warming in China (vol 15, 70, 2024). *Nature Communications, 15*(1). https://doi.org/10.1038/s41467-024-44936-6

Woolway, R. I., Jennings, E., & Carrea, L. (2020). Impact of the 2018 European heatwave on lake surface water temperature. *Inland Waters, 10*(3), 322–332. https://doi.org/10.1080/20442041.2020.1712180

Woolway, R. I., Anderson, E. J., & Albergel, C. (2021a). Rapidly expanding lake heatwaves under climate change. *Environmental Research Letters, 16*(9). https://doi.org/10.1088/1748-9326/ac1a3a

Woolway, R. I., Jennings, E., Shatwell, T., Golub, M., Pierson, D. C., & Maberly, S. C. (2021b). Lake heatwaves under climate change. *Nature, 589*(7842), 402–407. https://doi.org/10.1038/s41586-020-03119-1

Wu, L. Y., & Zhang, J. Y. (2015). The relationship between spring soil moisture and summer hot extremes over North China. *Advances in Atmospheric Sciences, 32*(12), 1660–1668. https://doi.org/10.1007/s00376-015-5003-0

Xiao, W., Zhang, Z., Wang, W., Zhang, M., Liu, Q., Hu, Y. B., Huang, W. J., Liu, S. D., & Lee, X. H. (2020). Radiation controls the interannual variability of evaporation of a subtropical lake. *Journal of Geophysical Research-Atmospheres, 125*(8). https://doi.org/10.1029/2019jd031264

Xu, X. L., & Liu, W. (2017). The global distribution of Earth's critical zone and its controlling factors. *Geophysical Research Letters, 44*(7), 3201–3208. https://doi.org/10.1002/2017gl072760

Zeiger, S. J., & Hubbart, J. A. (2015). Urban Stormwater temperature surges: A central US watershed study. *Hydrology, 2*(4), 193–209. https://doi.org/10.3390/hydrology2040193

Zhang, J., Liu, Z. Y., & Chen, L. (2015). Reduced soil moisture contributes to more intense and more frequent heat waves in Northern China. *Advances in Atmospheric Sciences, 32*(9), 1197–1207. https://doi.org/10.1007/s00376-014-4175-3

Zhao, M., Zhang, H. Q., & Dharssi, I. (2019). On the soil moisture memory and influence on coupled seasonal forecasts over Australia. *Climate Dynamics, 52*(11), 7085–7109. https://doi.org/10.1007/s00382-018-4566-8

Chapter 8
The Urban Surface and Heatwaves

Abstract This chapter considers the role of urban surfaces in exacerbating heatwave conditions in urban environments. The chapter opens by exploring the morphological complexity and climatic diversity of urban surfaces, highlighting how urban materials, along with the three-dimensional structure of cities, influence the exchange of heat, mass and momentum between the urban surface and the overlying atmosphere. The characteristics of the urban boundary layer (UBL) and how this is influenced by urban surface climate processes is also outlined. The concept of the urban energy balance is used to explain how the partitioning of available energy between sensible and latent heat flux, as influenced by moisture availability, determines the thermal response of cities during heatwaves. It also addresses the role of heat storage in urban materials and its impact on maintaining high nighttime temperatures. Anthropogenic heat flux as an additional heat source and its role in the exacerbation of urban heat during heatwaves is also discussed. Because debate persists on the synergistic effects of heatwaves and urban heat islands (UHI) the nature of the four types of UHI is described before exploring whether heatwaves amplify UHI intensity. The chapter concludes that heatwaves can enhance levels of urban heat by altering energy partitioning between the turbulent and conductive heat fluxes but cautions that to understand the interaction between heatwaves and urban heat island intensity, a full consideration of the simultaneous processes in adjacent urban and rural settings is required. The specific role of air conditioning in enhancing levels of urban heat is also touched upon.

Keywords Urban surface atmosphere interactions · Heatwave development · Urban boundary layer · Urban energy balance · Anthropogenic heat flux · Amplification of urban heat · Urban heat island

8.1 Introduction

From the previous chapter it is clear surface atmosphere interactions and especially surface/sub-surface moisture status play a critical role in the generation of heatwaves. Knowing this begs the question whether long-term changes in land cover

G. McGregor, *Heatwaves*, Biometeorology 6,
https://doi.org/10.1007/978-3-031-69906-1_8

and thus surface moisture status and associated atmospheric processes, may increase the likelihood of extreme temperature events and thus heatwaves. Nowhere else on the planet are changes in land cover more obvious than in urban settlements. Here human engineered surfaces have largely replaced natural surfaces, with the result that associated alterations in the exchange of heat, moisture and momentum produce a spatial climate anomaly codified as the urban heat island (UHI). In simple terms, the UHI refers to cities being 'hotter' than their rural surrounds. This bears implications for the attainment of healthy cities and the nature of urban based ecosystem services. For this reason, the interaction of heatwaves with urban environments warrants special attention. The assumption adopted here is by itself, the urban surface does not generate heatwaves. Rather urban surfaces interact in a synergistic way with weather systems associated with heatwaves altering the intensity and duration of heatwave related temperature anomalies. The chapter commences by considering the concept of the urban surface, followed by an outline of the nature of the urban boundary layer. As a basis for understanding heatwave—urban surface interactions, the nature of the urban energy balance is then described with a particular emphasis of the interaction between the non-advective fluxes of the urban energy balance during heatwaves. Subsequently, the distinctive attributes of the four urban heat island types are delineated. Before concluding the chapter with a synthesis of its contents, a discussion of the interaction between heatwaves and the urban surface is presented with the aim to discern how heatwaves may exacerbate the UHI effect or, more accurately, urban heat.

8.2 Nature of the Urban Surface

The urban surface is both morphologically complex and climatically diverse. The former attribute relates to the range of morphological units that make up the urban surface (Fig. 8.1) while the latter arises from its assortment of radiative, thermal, moisture and aerodynamic properties. Modification of the atmosphere over urban areas is the outcome of the interaction of three key features of the urban surface namely, fabric, structure, and land cover, all of which underpin the urban surface's morphological complexity and climatic diversity. Together fabric, structure and land cover comprise urban form. Accompanying urban form's role in urban climate is urban function. Fabric, as one component of urban form, refers to the materials such as concrete, tarmac, brick, and rock to name a few, that make up the facets (e.g., building wall, street surface) of the elements (e.g., buildings and roads) of urban spaces (Fig. 8.1). Because urban materials possess quite different climate properties, urban fabric is a key control on how a surface absorbs, reflects, or emits radiation, and receives, stores, and transmits heat and moisture. Fabric is therefore a key determinant of the nature of the vast range of climatically active surfaces in urban areas.

Urban structure is defined as the three-dimensional arrangement of building elements as described by their physical dimensions and the spaces between them. The

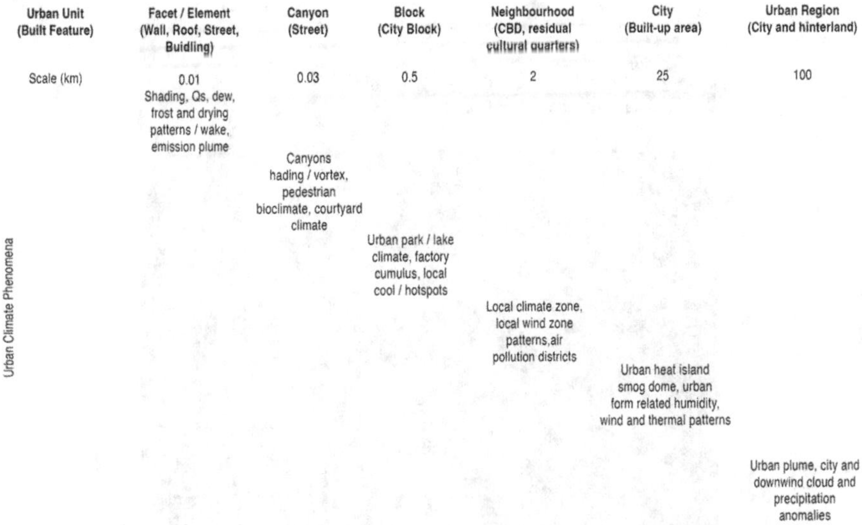

Fig. 8.1 Range of urban morphological units, associated scales, and climate urban climate phenomena. Scale values are for a city of approximately one million people. (Adapted from Oke et al. (2017))

complex three-dimensional nature of urban surfaces makes them rough in aerodynamic terms. At the city scale, structural complexity affects the way large scale wind flows associated with weather systems interact with the urban surface to create turbulence, influence the depth of the urban atmosphere, and produce downwind effects. It also controls overall urban albedo as the greater the structural complexity, the greater the likelihood that reflected shortwave radiation is intercepted on its way back to space. Similarly, at the intra-urban scale, structure affects local radiative exchange via the so-called sky view factor, or openness of urban spaces to the sky. Deep narrow streets, or urban canyons (Fig. 8.2), are likely to result in the channeling of wind flow with radiation being reflected, emitted and re-absorbed multiple times. Interestingly the aerodynamic roughness of high-density cities such as Tokyo, Shanghai and New York can impede land or sea breezes which usually bring some respite from high city temperatures (Bauer, 2020; Kusaka et al., 2000; Meir et al., 2013; Shen et al., 2019; Yamato et al., 2017; You et al., 2019).

Viewed from the air, cities and towns present a mosaic of different land covers, whether they be vast areas of suburban housing, city parks, light industrial areas, motorways, or tight high-density central business districts, all of which possess distinctive urban fabric and structural characteristics. Further to urban form as a feature of the urban surface, is urban function and urban metabolism. These reflect how a range of resources are consumed and transformed via, for example, industrial and manufacturing processes, transport, air conditioning, space heating and artificial irrigation with associated emissions of heat, moisture, and air pollutants which can impact the urban atmosphere and thus climate.

Fig. 8.2 Street canyon in downtown Chicago. (Source: Author)

Understandably, because of its convoluted and multi-faceted nature, classifying the urban surface has been a perplexing task for urban climatologists (Masson et al., 2020). Acknowledging the diverse ways in which fabric, land cover, structure and urban function may interact, the notion of local climate zones (LCZ), as homogeneous climate patches (Stewart & Oke, 2012), has emerged as a way of codifying urban surface complexity (Fig. 8.3). In terms of the range of urban morphological units, varying from facets to an entire urban region, LCZ equate with the neighbourhood scale, with horizontal dimensions in the region of 2 km × 2 km for a medium sized city (Fig. 8.1). The LCZ classification comprises ten built and seven natural surface types (Fig. 8.3) and has been shown to have general global applicability (Demuzere et al., 2019). The idea is each type of LCZ, with its own distinct surface permeability, roughness, thermal, energy and water consumption characteristics, will modify local climate in a certain way via how available energy is partitioned between the various sinks of sensible, latent, and subsurface heat flux (Stewart et al., 2014). Extending this idea further, knowing the mix of local climate zones that comprise a city assists with understanding the climate characteristics of that city. For example, one can imagine the climate contrasts for an archetypical 'garden

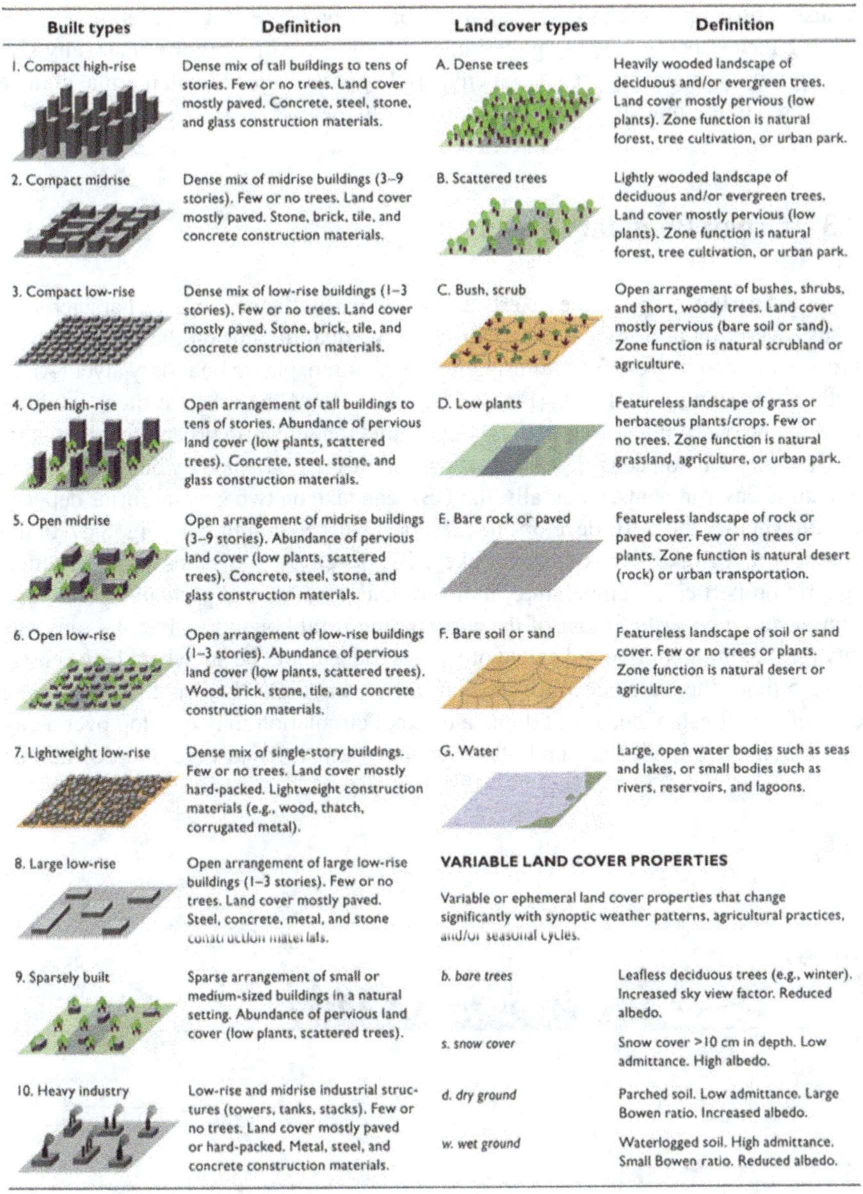

Fig. 8.3 Classification of local climate zones (LCZ). (Source: Stewart and Oke (2012); © American Meteorological Society, Used with permission)

city' as conceived by Ebenezer Howard (Howard, 2003) or indeed the Metutopia envisaged by Helmut Landsberg (Landsberg, 1973), made up of LCZ typified by open green areas with low rise dwellings, low-density commercial activity, and extensive green-blue infrastructure, and an 'industrial city' dominated by heavy industry, little open green space and a compact urban core. All other climate determining factors being equal (e.g., latitude, altitude, proximity to the coast, city size and so on), theoretically the garden city would possess a less harsh thermal climate compared to the industrial city, especially during heatwave periods.

8.3 Urban Boundary Layer

As noted earlier, 'the surface' exerts a fundamental influence on the characteristics of the overlying atmosphere such that in the vertical dimension the atmosphere is broadly divided into the free atmosphere and the atmospheric boundary layer (ABL) with its associated sub-layers (Fig. 7.4). In the case of the urban surface, its influence on the overlying atmosphere is so profound that the term urban boundary layer (UBL) has been coined to describe that part of the ABL affected by built as opposed to natural environments. Generally, the UBL can take on two general forms depending on whether the UBL develops in calm or disturbed conditions (Fig. 8.4). In the absence of regional winds, a dome like UBL develops over cities with the atmospheric properties of temperature, moisture and gaseous and particulate composition in stark contrast to those of the surrounding rural boundary layer. For any one city the maximum horizontal extent of its heat dome could be anywhere between 1.5 and 3.5 times the diameter of the city's urban area at night (Fan et al., 2017). In the case of a well-established heat dome a thermal circulation may develop over a city like a convection cell with updrafts over the warm thermal core and downdrafts

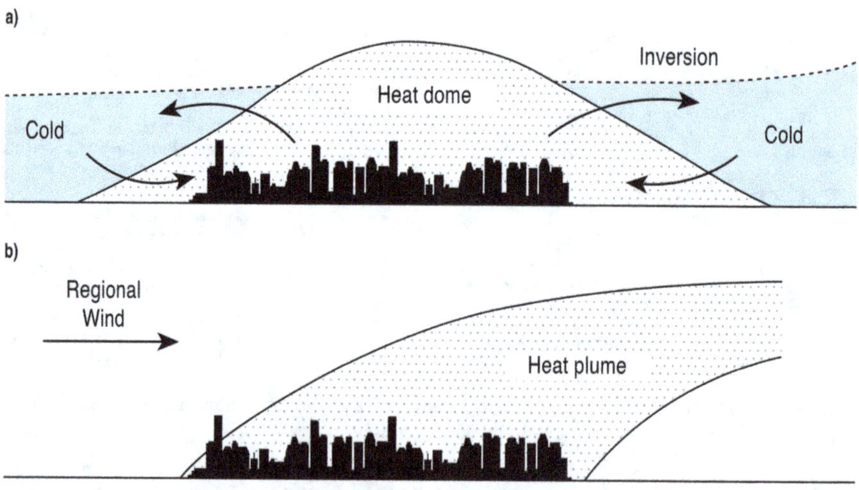

Fig. 8.4 Contrasting forms of the UBL with regards to heat during (**a**) calm conditions—heat dome and (**b**) under a gentle regional wind—heat plume. The contrasting UBL forms could also apply to other atmospheric properties, especially pollution. (Adapted from MetLink—Royal Meteorological Society: https://www.metlink.org/fieldwork-resource/microclimates /)

towards the margins of the city as well as the inflow of cool rural air drawn in along a thermally induced rural to city atmospheric pressure gradient (Varentsov, Wouters, et al., 2018b). For the situation of moist rural air converging upon an urban core from all directions, ascent of that air may result in enhanced cloud formation over urban areas. Under conditions of gentle to moderate, large- and regional-scale wind flow and the absence of strong mechanical mixing of urban air with that above it, the shape of the UBL may be distorted in a downwind direction to create an urban plume (Fig. 8.4). In case of a polluted urban atmosphere, the urban dome and urban plume are often clearly visible from outside the city.

As for the ABL, the UBL possesses a reasonably well understood structure broadly divisible into a mixed layer and local internal boundary layers (Fig. 8.5a). During calm conditions and over the diurnal cycle, a daytime convective mixed layer is replaced by a nocturnal residual layer, which lies above a stable surface layer as is the case for the classical ABL. In some cases where sensible heat flux from the surface into the atmosphere continues into the evening, a nocturnal mixed layer consisting of a shallow convective or near-neutral layer of turbulence may develop in urban areas with a weakly stable residual layer overlying this (Barlow, 2014; Barlow et al., 2015). The internal boundary layer (IBL) is composed of three layers (Fig. 8.5b) the total depth of which will vary according to the roughness of the surface, for example as defined by the nature and size of a LCZ and the degree of turbulence as moderated by the interaction of the surface with local air flow. From the surface upwards, the urban canopy layer (UCL) is the volume of internal and external air found below roof height. Here, the majority of human activity is conducted, with urban metabolic processes playing out in the form of the exchange and transformation of energy and mass. The top of the UCL is also where the upper bound of the imaginary volume of air in which the fluxes operative in the urban energy balance are found (Sect. 8.4). Lying above the UCL is the roughness sublayer (RSL) a layer that owes its existence to the degree to which the surface roughness elements thermodynamically disturb the overlying air via upward momentum and turbulent heat flux (Fig. 8.5b). The depth of the roughness sublayer can range anywhere between two to five times the average height of the surface roughness elements making up a LCZ. The inertial sub-layer (ISL) lies above the roughness sublayer and is where the turbulence is homogeneous, and fluxes vary little with height (Barlow, 2014). The vertical dimension of the ISL can range anywhere between 20—250 m. Clearly as the urban surface is heterogeneous with a vast array of LCZ juxtaposed, the IBLs may overlap with the aggregate effect being a point above the surface where surface heterogeneity no longer dominates the spatial variation in turbulence—the so-called blending height (Fig. 8.5c). Eventually the ISL gives way to the overlying mixed layer where all atmospheric properties are relatively homogeneous.

The top of the UBL is defined by the height of the capping inversion as is the case for the classical ABL. As all cities, or built-up areas, are not the same, UBL depth and its intra-city variation can vary from city to city (Pal et al., 2012). Generally, the UBL is deeper and less stable than its rural counterpart reaching depths of 2500 m with UBL depth dependent on city size, urban form and the degree of surface

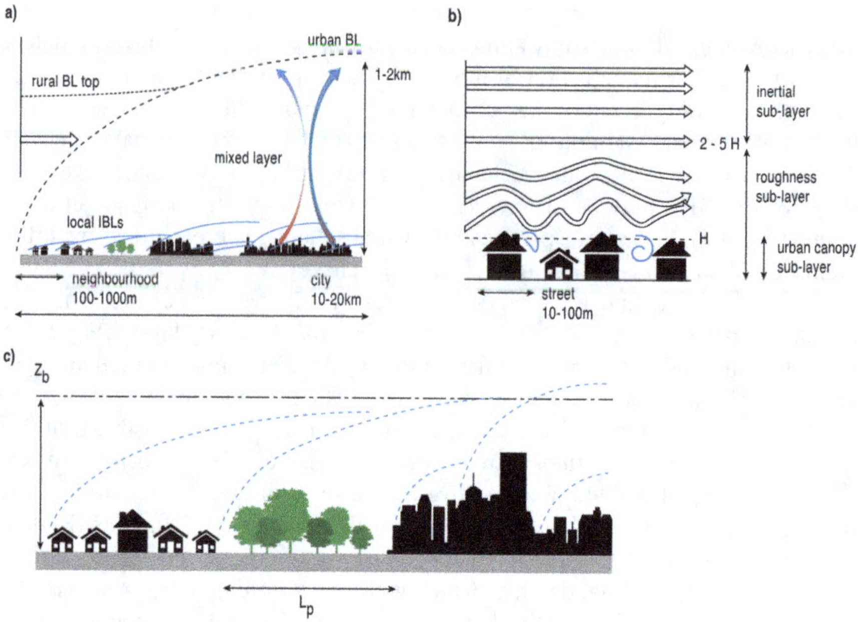

Fig. 8.5 Features of the urban boundary layer. (**a**) Illustration of the daytime convective urban boundary layer with wind blowing left to right. The dashed lines represent the top of rural and urban boundary layers. Overlapping IBLs are shown as solid lines in the lower part of the figure—see (**c**) below. Representative horizontal distances for neighbourhood and city scales are given. Note the vertical scale is exaggerated; (**b**) Detail for the IBL composed of the UCL, RSL and ISL. Arrows represent the trajectory taken by air with wind blowing left to right. H is height; (**c**) Overlapping IBLs for an assortment of urban neighbourhoods all vertically merging at the blending height (Zb). Horizontal arrows represents the length-scale of surface heterogeneity—usually ranging over distances of 500—5000 m. For urban areas, the ratio of the blending height to length scale of heterogeneity generally varies between 0.03 and 0.13. (Source: (Barlow, 2014) Progress in observing and modelling the urban boundary layer. *Urban Climate* 10, 216–240. Used with permission)

heating. Changing weather conditions may affect UBL depth. For example, in some locations a developing sea breeze has been shown to deepen the daytime mixing height within the UBL (Ado, 1992; Yoshikado, 1990), while in others the opposite may apply (Ribeiro et al., 2018). The nocturnal UBL is shallow and may only attain depths around 25% of its daytime equivalent. As noted above, mixing of the atmosphere may continue at night due to sensible heat flux and urban roughness resulting in patchy elevated inversions across a city depending on the nature and distribution of LCZ (Oke et al., 2017). The implications of nocturnal mixing are any heat residing in the nocturnal residual layer built up during a heatwave could be mixed back down to the surface thus suppressing nighttime cooling which, under normal circumstances, offers some respite from daytime heat.

8.4 Urban Energy Balance

As noted in Chap. 4, the energy balance concept is fundamental to understanding the nature of climate at a range of scales with the assumption that the sun is the primary energy source for the surface net radiation balance and thus the energy available for heating the atmosphere or the subsurface and undertaking evaporation. However, in the case of urban environments there is one additional source of energy, that being heat emitted from human activities, or in technical terms, anthropogenic heat flux comprised of both sensible and latent heat components. Heat storage and release are also important aspects of the urban energy balance. This is because cities display a propensity for capturing and releasing heat, a function of a city's urban form (fabric, structure, and land cover) and thermal mass. Interestingly, but worryingly, human-made or anthropogenic mass now exceeds 50% of the weight of all living biomass on earth with most of that mass attributable to concrete, steel, gravel, brick and asphalt, all of which form the building blocks and infrastructure of our cities. It has been estimated that if current trends continue, anthropogenic mass, including waste, is expected to be almost three times all dry biomass on earth by 2040, with much of that related to the growth of cities and their urban populations (Elhacham et al., 2020). Such projections imply that heat storage and release will become an ever increasingly important and powerful component of the urban energy balance into the future.

While for 'natural' environments the point of reference for considering the energy balance is usually a planar surface, in the case of urban landscapes it is easier to consider the energy balance in relation to a volume (Oke, 1988) with upper and lower bounds set at around roof height and just below the surface respectively (Fig. 8.6). In this way, the energy balance refers to energy fluxes through the top of the imaginary volume while acknowledging to and from the volume there will be advection of heat. Further within the volume, there will be changes of energy storage equating to the aggregate heat flux between the atmosphere and all solid surfaces within the volume (Arnfield, 2003). Accordingly, the urban energy balance is written as:

$$Q^* + Q_F = Q_H + Q_E + \Delta Q_S + \Delta Q_A \tag{8.1}$$

Where Q^*, Q_H and Q_E are as defined earlier while Q_F is anthropogenic heat flux, ΔQ_A is net advection through the sides of the volume and ΔQ_S is the storage heat flux related to heat storage by all entities composing the volume such as air, trees, building fabric and the soil (Arnfield, 2003; Barlow, 2014). Having described the general form of the urban energy balance the following subsections will outline the importance of the non-advective heat (Q^*, Q_F, Q_H, Q_E and ΔQ_S) components of the urban energy balance in the wider context of heatwaves.

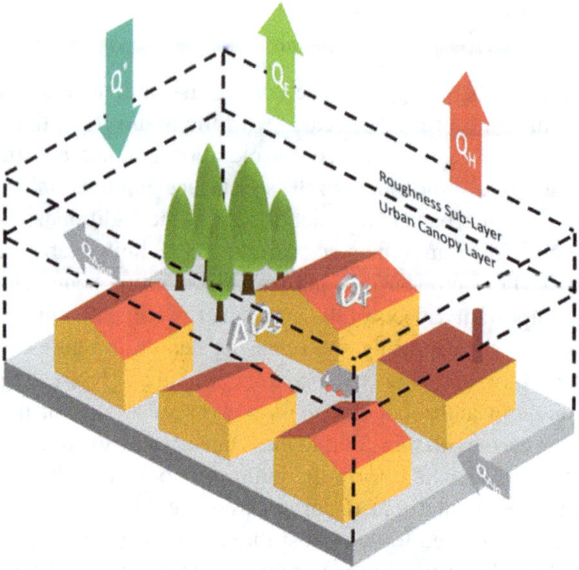

Fig. 8.6 The volume concept applied in the case of the urban energy balance. The energy balance symbols are those as appear in the urban energy balance—Eq. 8.1. Source: Chrysoulakis et al. (2018). Urban energy exchanges monitoring from space. Sci Rep 8, 11498 (2018). https://doi.org/10.1038/s41598-018-29873-x Reused under Creative Commons License 4 (https://creativecommons.org/licenses/by/4.0/)

8.4.1 The Importance of the Apportionment of Q^* Between Q_h and Q_e

During a heatwave the increase in air temperature in a city, or in other words its thermal response, will depend on the behaviour of all components of the urban energy balance. Having said that, the amount of available energy ($Q^* + Q_F$) partitioned between Q_H and Q_E, as controlled by moisture availability, is a critical determinant of thermal response. Typically, cities are dry environments. This is because city planners and engineers make great efforts to ensure excess water is removed, as quickly and efficiently as possible, from built up areas by storm water systems at the building to city scale to control urban flooding. Further, soil moisture levels are generally low in cities because of surface sealing by concrete, tarmac, and other urban surface coverings. Added to these 'urban desert-making' factors is the generally low percentage of land cover given over to substantial tracts of vegetation. These features of a city's hydroclimatological landscape conspire to direct the available energy into sensible heating of the atmosphere thus driving up air temperatures. Accordingly in cities with a high (low) proportion of LCZ classed as compact low rise to high rise and heavy industry, and low surface moisture availability (open low to high rise sparsely built with available moisture), the Bowen ratio values are generally high (low) approaching values akin to those for deserts (forests), as conceptually mapped in Fig. 8.7.

How the relationships mapped in Fig. 8.7 play out is clear from actual measurements of energy balance components for different urban land cover types. For example, consider the diurnal march of Q^*, Q_H, Q_E and ΔQ_S for three urban surfaces with contrasting housing densities in Melbourne (equating with LCZ 3, 5 and 7—Fig. 8.8a-c) and a rural site outside Melbourne Australia (Fig. 8.8d) as described by Coutts et al. (2007). Essentially Q^* was found to vary little between the urban sites with only subtle differences apparent arising from the influence on the radiation balance of albedo and the height-to-width ratio of buildings. In contrast to the rural site (Fig. 8.8d), all urban sites (Figs. 8.8 a-c) displayed low Q_E values attributable to a general summertime water balance deficit, impervious urban surfaces and little surface water following rainfall due to efficient urban drainage systems. This is reflected in high Bowen ratio values for the urban sites, with daily summertime values often more than 5 (Coutts et al., 2007) and comparable to those found for urban areas elsewhere such as Marseilles (Lemonsu et al., 2004), Mexico City (Tejeda-Martinez & Jauregui-Ostos, 2005), Ouagadougou (Offerle et al., 2005) and Vancouver (Grimmond & Oke, 1999).

For the three Melbourne urban sites Q_H and ΔQ_S were the dominant fluxes, with Q_H closely tracking Q^*. Interestingly ΔQ_S displays an asymmetric pattern leading Q^* and Q_H in the morning by around 1–2 h, most likely the result of a large

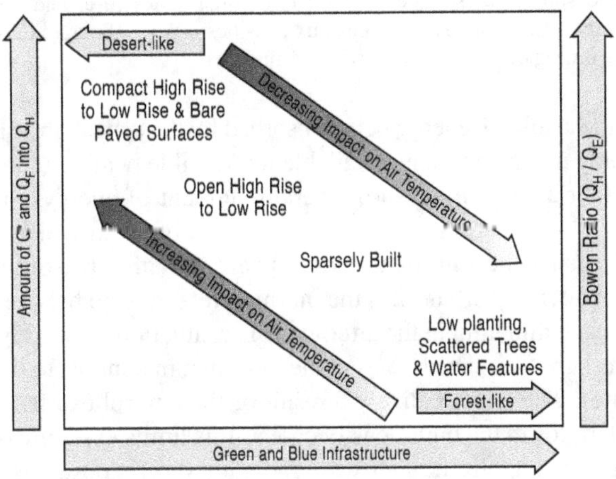

Fig. 8.7 Relationship between the relative proportion of green and blue space (a proxy of moisture availability and surface sealing) and the relative amount of available energy in the form of net radiation (Q^*) and anthropogenic heat flux (Q_F) apportioned to sensible heat (Q_H) for a range of urban surface types or LCZ as appear in Fig. 8.3. Bowen ratio is the ratio of sensible (Q_H) to latent (Q_E) heat with high (low) values akin to those for desert-like (forest-like) climates—top left-hand corner (bottom right hand corner)—see Chap. 4 for typical Bowen Ratio values. Direction of arrows on vertical and horizontal axes indicate direction of increase. Decreasing (increasing) impact on air temperature is interpreted as cooling (warming) of the urban atmosphere. (Source: Author. Inspired by Fig. 6.17 in Oke et al. (2017))

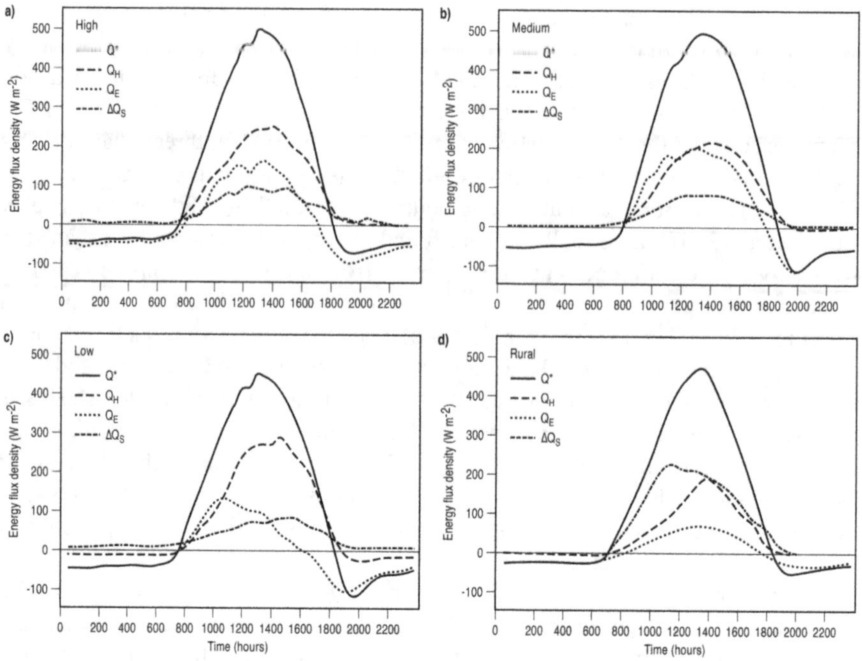

Fig. 8.8 Diurnal pattern of the flux of net radiation (Q*), sensible heat (Q_H), latent heat (Q_E), heat storage (ΔQ_S) and subsurface storage (Q_G) for three urban sites (**a-c**) and one rural site (**d**), Melbourne Australia. Values are expressed as flux densities—Wm^2. (Source: Coutts et al. (2007); © American Meteorological Society, used with permission)

proportion of the available energy being absorbed by the urban fabric and going into storage with only meagre amounts available for sensible heating of the atmosphere. As the urban fabric can only absorb a certain amount of energy, equivalent to its maximum heat storage capacity, sooner or later most of the available energy will be directed into sensible heating of the atmosphere. For this reason, in the case of medium and low-density urban sites the morning relationship between Q_H and ΔQ_S reverses with Q_H dominating in the afternoon. The situation for the high-density site is interesting as here Q_H exceeds ΔQ_S in the morning, in contrast to the medium and low-density sites. This is most likely a result of the control exerted by the greater height to width ratios in the high-density case as this limits exposure of urban facets to the morning sun and thus the shortwave radiation flux; exposure is greater for the more open medium and low density sites. This same controlling factor may also account for the noticeably earlier peak of ΔQ_S with decreasing urban density. ΔQ_S is also fascinating as its total daytime amount (as deduced from the total area under the curve) decreases in line with a fall in urban density, testament to the importance of urban mass for determining total heat storage capacity. As already noted, Q_H overshadows the other non-radiative fluxes in the afternoon. Interestingly it remains positive in the early evening for the high and medium density sites while it is conspicuously negative for the low-density site. This situation arises because of the considerable flow of heat out of storage for the high and medium density sites as indicated by the negative ΔQ_S values. As noted by Coutts et al. (2007) this occurs because of the impervious surface cover which acts as a daytime heat sponge. Further the contrasting urban canopy height to width and wall to plan area ratios affect nocturnal cooling via influencing the amount of longwave radiation loss, pointing to the importance of building density for urban climate. The upshot of a positive Q_H, as supported by ΔQ_S throughout the evening hours, is the maintenance of high air temperatures into the evening. Turning attention again to Q_E, notwithstanding its subordinate status to Q_H, it also remains as a positive flux into the evening. This is especially so for the low-density site. Here evaporation continued throughout the night with a positive Q_E flux supported by energy release from storage (Coutts et al., 2007). Evaporation would not occur in the absence of moisture availability. In this regard, the relatively high vegetation cover of the low-density site compared to its medium and high-density counterparts and overnight garden and lawn irrigation appear to play an important role in maintaining noticeable levels of positive Q_E flux. This finding attests to the importance of Q_E as an energy sink and thus the potential role moisture availability plays in moderating harsh urban temperatures. Compared to the urban sites studied by Coutts et al. (2007), the energy partitioning into the non-radiative fluxes at the rural site is somewhat different. Here Q_E dominates. Noteworthy is the steep Q_E increase in the morning with Q_E reaching a peak prior to that for Q^*, a result of available moisture being 'burned off' in the morning hours as receipts of shortwave radiation increased hence building the energy available for evapotranspiration. For most of the day at the rural site Q_E exceeds Q_H indicating sustained moisture availability for evaporation, no doubt a product of vegetation tapping into available soil moisture reserves. Although Q_E exceeds Q_H during daytime hours, Q_H does approach Q_E once midday has passed indicating a significant amount of energy is flowing into sensible heating of the atmosphere and thus raising the air temperature. Unlike the distribution of ΔQ_S at the urban sites, subsurface heat storage (Q_G) at the rural site follows the pattern of Q^* over the diurnal cycle. Also noteworthy is the low level of Q_G at the rural site compared to ΔQ_S at the urban sites, a result of the higher mass per unit volume in the urban case.

The situation described for Melbourne and hinted at in Fig. 8.6, points to the significance of the fraction of the urban surface that is built/vegetated for influencing how available energy is partitioned between the various non-radiative fluxes with associated implications for urban climate. Recognizing this, Loridan and Grimmond (2012) identified four generic urban zones that characterize energy partitioning based on how critical values of an index that describes the area of 'active' built and vegetated surfaces interact. For the 17 cities considered, they found a general inverse relationship between the total active surface area (sum of built and vegetated surfaces), and the fraction of incoming radiation ($Q\downarrow$ or $K\downarrow + L\downarrow$) lost as outgoing radiative fluxes ($Q\uparrow$ or $K\uparrow + L\uparrow$). In other words, the more 'climatically active' an area, as determined by the degree of exposure of the urban fabric to incoming radiation and the amount of vegetation, both of which vary seasonally, then the greater the likelihood the available energy will be consumed by Q_H, Q_E or

ΔQ_S,with only a small proportion lost to the atmosphere and beyond as outgoing radiation (Q↑ or K↑ + L↑). How the individual non-radiative fluxes were consumed was revealed by considering their response to the extent of active built and vegetated surfaces for each of the 17 cities. The fraction of Q* flowing into ΔQ_S was found to increase with the extent of the built surface while the Q_E fraction increased with the degree of vegetation cover.

With an eye to developing a typology of energy partitioning based on the interaction of active built and vegetated surfaces, Loridan and Grimmond (2012) defined active built and vegetated surface threshold values beyond which there were clear trends in the apportionment of incoming Q↓ to non-radiative fluxes. Their results are summarized in Table 8.1. Of special significance is the percentage of incoming radiation (Q↓) consumed by the various energy sinks Q_H, Q_E or ΔQ_S for each type of urban zone as given in the fourth column of Table 8.1 These figures bear potential significance in relation to understanding how heatwaves may interact with a spectrum of urban surfaces characterized by different proportions of climatically active built and vegetated surfaces.

Recalling the idea of a 'garden city', this matches in general terms the class 2 urban zone in Table 8.1. In this case, and reinforcing the point made earlier, for a given energy input (Q↓) a greater proportion of that energy will be dissipated by latent heat flux, a cooling process, with this in some ways offsetting warming of the urban atmosphere via release of heat from storage in the urban fabric (ΔQ_S) or sensible heating of the atmosphere (Q_H). Building on the type of associations outlined in Table 8.1, the worst-case urban zone scenario during heatwave conditions is likely to be that described by class 4. This is because the large amount of available energy that goes into storage (circa 19%) will become available for nocturnal release back into the urban atmosphere thus elevating the likelihood of strong positive nighttime temperature anomalies—note the systematic increase of ΔQ_S/ Q↓ with increasingly urban conditions. Conspiring with the high value of ΔQ_S/ Q↓ are the low Q_E/ Q↓ ratios for high density urban zones, the result of which will be little cooling of the atmosphere via evapotranspiration and thus counteraction of atmospheric heating driven by Q_H and ΔQ_S. As we will see later, the contrasting patterns of energy partitioning evident for the four Melbourne sites described above and as outlined in Table 8.1, forms the basis for a generic explanation of the development of the spatial temperature anomaly widely referred to as the urban heat island (UHI).

8.4.2 Anthropogenic Heat Flux

Anthropogenic heat flux (Q_F) is a unique feature of the urban energy balance. Although this generally refers to the emission of sensible heat associated with energy consumption, the two are not entirely synchronous in magnitude and timing as Q_F can be greater than the energy consumed in its production, and peak Q_F emissions may lag energy consumption peaks. Q_F is distinct from anthropogenic

Table 8.1 Four urban zones based on partitioning of energy fluxes in relation to the extent of climatically active built and vegetated surfaces. (Compiled based on information contained in Loridan and Grimmond (2012))

Class	Active surface combination in relation to critical threshold	General urban condition (zone)	Energy partitioning with mean ratios as percentages of incoming radiation ($Q\downarrow$)
1	Built and vegetation < threshold	Winter conditions	Most of $Q\downarrow$ radiated back to the atmosphere. $Q\uparrow/Q\downarrow \sim 76$. Little of $Q\downarrow$ goes into storage (ΔQ_S) or used up via turbulent (Q_H, Q_E) processes: $\Delta Q_S/Q\downarrow \sim 6$; $Q_H/Q\downarrow \sim 12$; $Q_E/Q\downarrow \sim 6$.
2	Built < but vegetated > threshold	Low density and most vegetated: Extensive active vegetated surface and overall increase in total active surface	Amount of $Q\downarrow$ radiated back to the atmosphere decreases: $Q\uparrow/Q\downarrow \sim 59$. Upturn in storage: $\Delta Q_S/Q\downarrow \sim 12$. Turbulent fluxes increase in line with active surface increase. Q_E becomes comparable to Q_H: $Q_H/Q\downarrow \sim 15$; $Q_E/Q\downarrow \sim 14$
3	Built and vegetated > threshold	Medium density and increasingly urban: Extent of built active surface increasing	Amount of $Q\downarrow$ radiated back to the atmosphere decreases $Q\uparrow/Q\downarrow \sim 57$. Further increase in storage of $Q\downarrow$(c.f. class 2): $\Delta Q_S/Q\downarrow \sim 16$. Total turbulent fluxes comparable to class 2 but changing balance of Q_H and Q_E in line with increase in active built surface: $Q_H/Q\downarrow \sim 17$; $Q_E/Q\downarrow \sim 10$
4	Built > but vegetated < threshold	High density and most urban: Extensive built active surface and diminished vegetation cover	Amount of $Q\downarrow$ radiated back to the atmosphere increases due to reduction of active vegetated surface $Q\uparrow/Q\downarrow \sim 59$. Further increase in storage of $Q\downarrow$ at the expense of Q_H and Q_E as relative importance of built surface increases (c.f. class 2 and 3). $\Delta Q_S/Q\downarrow$ 19. Decrease in total turbulent fluxes especially Q_E driven by diminished vegetation cover: $Q_H/Q\downarrow \sim 14$; $Q_E/Q\downarrow \sim 8$

moisture in the form of water vapour, which has its origins in the operation of evaporative cooling systems that remove heat from buildings (sometimes referred to as anthropogenic latent heating) and combustion of hydrocarbon fuels either in vehicle engines or combustion furnaces (Sailor, 2011; Sailor et al., 2015). Notwithstanding the myriad of possible individual sources of Q_F in a city it is useful to think of its origin in terms of three broad sectors namely buildings, transport, and industry with human metabolic heat output (not to be confused with urban metabolism) usually subsumed in Q_F estimates for buildings (Sailor, 2011). General controls on Q_F include the level of space heating and cooling demand and urban density and energy efficiency (Oke et al., 2017). Added to this are cycles of human activity at a range of time scales.

Whenever the heating or air conditioning in a room/building is turned on to obtain temperatures matching the required level of human thermal comfort, energy is consumed, and heat is produced. In the case of space heating, much of the heat produced to make indoor temperatures comfortable eventually escapes to the external environment leaking through areas such as windows or doors or seeping through the building walls. Heat is also flushed from buildings through drainage systems, for example, warm water from bathrooms or kitchens. When and where indoor cooling is required, because of intolerable levels of thermal discomfort, large amounts of energy can be consumed via the mechanical conditioning of air often achieved by forcing warm air through evaporative coolers or condensers. Heat removed from the indoor hot air, which can also contain considerable amounts of water vapour, is vented directly into street canyons or into the urban boundary layer from roof mounted heat exhausts. Accordingly, both building heating and cooling systems are important sources of anthropogenic heat.

On a per unit area basis the flux of anthropogenic heat, that is the anthropogenic heat flux density, is greater for areas with high population density. Simply put, the greater the concentration of people the greater the Q_F with an approximate linear relation evident between population density and Q_F (Oke et al., 2017). Even though two cities may have similar per capita energy consumption, if one of the cities is more built up than the other (e.g., concentrated versus dispersed population), Q_F can be expected to be greater for the case of the high density city, despite the greater energy efficiency of highly built up cities on a per capita basis. Q_F is not constant over time and can display significant diurnal, intra-weekly and seasonal variations. At the diurnal time scale, daytime Q_F exceeds that for night with morning and early evening peaks generally matched with the pattern of commuting. Because Q_F in cities is closely related to work activity (commuting, building lighting, cooling/heating), weekend and holiday values of Q_F are often lower than weekday/workday values. The seasonal distribution of Q_F generally equates with the broad climate setting and the seasonal march of heating/cooling requirements. In warm-humid climates with tolerable winters, Q_F in summer may exceed that for winter because the energy required for summer indoor cooling is greater than that for winter indoor heating. The opposite of course applies in the case of cool/cold climates where the energy demand for space heating during the winter months will top that consumed for summer indoor air conditioning. Unsurprisingly there appears to be some noteworthy secular trends in Q_F especially for cities with rapidly increasing populations (Jin et al., 2019; Jin, Wang, & Wang, 2020a; Lindberg et al., 2013).

Unlike Q^*, Q_H and Q_E, Q_F cannot be measured directly using instrumentation. Rather it must be estimated using methods that can be broadly classed into top-down, bottom-up and energy balance residual approaches (Oke et al., 2017; Sailor, 2011). Top-down, or alternatively inventory approaches, base Q_F estimates on a consideration of population density and energy consumption statistics from buildings and vehicles. This type of data is usually available at quite large aggregate scales and requires downscaling to smaller temporal and spatial scales. In contrast, bottom-up methods, occasionally referred to as GIS (Geographical Information System) approaches, model energy consumption at the building scale. The output from associated building energy models is then scaled-up to larger urban morphological units, such as the block or neighbourhood, to establish the level of Q_F. The energy balance residual approach involves measuring as accurately as possible, using a variety of ground and satellite-based instrumentation, all other components of the urban energy balance to find the total radiative energy sources and turbulent and conductive energy sinks. Q_F is then assumed to be the value that is required to close the energy balance equation, or in other words, the energy flux density value required to balance the energy source (Q^* and Q_F) and energy sink (Q_H, Q_E, ΔQ_S) sides of the urban energy balance equation. Essentially this is the residual value on the left-hand side of the energy balance equation. Rather than any one single approach being the silver bullet for estimating Q_F, inter-comparison studies indicate the sectors for which Q_F estimates are required (e.g., buildings, transport or industry), should lead approach choice (Sailor, 2011).

There is a burgeoning literature on Q_F estimates with the compilation of multi-city and global data bases of Q_F a recent feature (Capel-Timms et al., 2020; Chapman et al., 2016; Dong et al., 2017; Jin et al., 2019; Sailor et al., 2015; Varquez et al., 2021). In addition to the general dependence of Q_F on climate setting, season and population density, the scale at which estimates are made influences the resultant Q_F value (Gabey et al., 2019). Generally, it is at the neighborhood scale in downtown areas of large dense cities where the highest hourly values exist. Here, where building volume is large along with high daytime population density and a great demand for space cooling or heating, values of Q_F can be considerable. For example, in the densest part of Tokyo at the peak of air-conditioning demand, estimated values of Q_F exceed 1000 W m^{-2} (Ichinose et al., 1999). This contrasts with lower hourly values for local climate zones typified by low density, low-rise open urban development, where hourly values may range from 5 to 50 W m^{-2} (Oke et al., 2017). Q_F values estimated at the city scale are generally low because of the effects of spatial averaging across a wide range of land cover types or local climate zones with contrasting characteristics and thus Q_F emissions. That aside, the results from many studies, which are not necessarily comparable because of the application of different methods for estimating Q_F, indicate that anthropogenic heat emissions from cities can be significant. For example, studies using the LUCY (Large Scale Urban Consumption of Energy) model for estimating Q_F indicate Tokyo has the highest average heat emissions at around 61.0 W m^{-2} closely followed by New York (~ 59.0 W m^{-2}). Cities such as Paris, San Francisco, Shanghai, London and Vancouver were found to have Q_F values in the range of 20–35 W m^{-2}, while Manchester, Helsinki, Houston and Toulouse had averages lower than 9 W m^{-2}. Recalling the point made above about scale, the LUCY model indicates the highest heat fluxes at the intra-urban scale are in New York (577 W m^{-2}), Paris (261.5 W m^{-2}), Tokyo (178 W m^{-2}), San Francisco (173.6 W m^{-2}), Vancouver (119 W m − 2) and London (106 W m^{-2}). For these cities building related heat emissions accounted for more than 90% of total Q_F with the remainder related to almost equal contributions from human metabolic heat and vehicles (Allen et al., 2011). Of course, these values for Q_F are averages and do not match those daily estimates for heatwave events—see Sect. 8.6. Although the contribution of human metabolic heat to total Q_F may be relatively small for mid-latitude cities, this heat source can reach significant levels in high density tropical cities where it may comprise more than 20% of Q_F. Mumbai, the world's most densely populated mega-city is a good example. Here the combined heat flux density from human and animal metabolism of 6.5 W m $^{-2}$, exceeds that related to electricity use in buildings (5.8 W m $^{-2}$) and fuel combustion in vehicles (3.9 W m^{-2}) (Stewart & Kennedy, 2017).

In the context of heatwaves and knowing that cities produce significant amounts of heat, several questions come to mind. These include to what extent does Q_F play a role in raising urban temperatures and thus contribute to an enhancement of the urban heat island effect, is Q_F higher during heatwave events because of energy demand for air conditioning, and do Q_F emissions during heatwave events engender positive feedbacks precipitating further demand for air conditioning and consequently additional heat emissions? These are big questions, for which the answers are slowly emerging. A later section on the synergistic effects of urban heat and heatwaves outlines some of the current thinking about the role of Q_F as a both a 'dupe' and 'provocateur' during heatwave events.

8.4.3 *Heat Storage*

In addition to Q_F, ΔQ_S is another 'significant other' of the urban energy balance as already hinted at in the above description of the diurnal march of energy balance components for contrasting urban surfaces in Melbourne (Fig. 8.8). ΔQ_S is important because the elements making up the built environment have a propensity for the storage and release of heat, a process that helps explain some of the unique features of the climate of urban areas and one potentially critical during heatwave events in terms of sustaining positive temperature anomalies. As noted in Chap. 7, the thermal properties specific heat (c), heat capacity (C), thermal conductivity (k), thermal diffusivity (αh) and thermal admittance (μ) determine the climatic behaviour of all natural and anthropogenic materials. However, the ability of urban environments to store and release heat is not simply the result of a contrast between the thermal properties of materials found in urban and 'natural' environment. Rather it is a result of the interplay between the climatic behaviour of urban materials, urban geometric form or structure and moisture availability (Oke et al., 2017). The role of moisture has already been mentioned in relation to evapotranspiration as a cooling process and the amount of net radiation consumed in latent heat flux.

In relation to ΔQ_S, of the thermal properties thermal admittance (μ) is one of the most important for the reasons described in Chap. 7. This is because μ controls the flow of available surface energy (Q*) into the atmosphere or the substrate, that is the materials comprising the urban surface. In the case of large (small) μ, most (little) of the available heat goes into the substrate/subsurface with the result that the surface diurnal temperature range is damped (heightened). Because the jury is still out on the matter of significant long-term (climatological) differences of μ between urban and rural environments (Oke et al., 2017), and the magnitude of μ is conditioned on moisture status of a particular material, it is likely that the area of the climatically active surface plays a critical role in the amount of heat storage. For this reason, urban form/structure is important.

Comparatively speaking, the convoluted three-dimensional form of cities makes them far more complex than their relatively flat rural counterparts. Consequently, cities tend to have large total aspect ratios, that is, the surface area of all the vertical and horizontal elements making up the urban surface is far greater than the total plan area they cover. This makes the area of climatically active surfaces in cities very large meaning as the number and size of urban elements increase so does the area over which energy, moisture and pollution exchanges occur. Large aspect ratios, which generally imply a wide occurrence of deep urban canyons in a city and thus a reduced sky view, tend to increase net shortwave and net longwave radiation amounts and thus heat absorption. This is because in deep urban streets (e.g., Fig. 8.2) the opportunities for multiple reflection and thus absorption of shortwave radiation is increased as beams of sunshine 'bounce' off and 'land on' vertical surfaces thus decreasing the amount of shortwave radiation that escapes to the atmosphere. Similarly, the canyon like nature of streets in densely built-up cities work via a reduced sky view, to limit the amount of heat that escapes as outgoing longwave radiation, especially at night. As amply demonstrated for the city of Hong Kong (Chen et al., 2012; Ng et al., 2011), aspect ratio also has a fundamental impact on city ventilation. Here sheltering and reduced turbulence enhances the accumulation of heat (and pollutants) in street canyons particularly in conditions when the regional wind flow is weak or local sea or land breezes are unable to penetrate the urban core.

Although direct measurements of ΔQ_S can be attempted using heat flux plates 'buried' in urban materials, it is usually estimated by one of four methods or a combination of them (Roberts (Roberts et al., 2006; Oke et al., 2017), with critical input parameters provided by a range of measurement platforms, such as weather stations, flux towers or satellites. That heat storage is a noteworthy component of the urban energy balance is clear for several cities (Hrisko, Ramamurthy, & Gonzalez, 2021a; b; Masson et al., 2002; Sun et al., 2017). In fact, ΔQ_S can account for a major proportion of the daytime net all wave radiation (Q^*) in dense urban areas, achieving levels up to ten times greater than comparable horizontal natural surfaces such as soil, with materials such as concrete and asphalt culpable for large urban ΔQ_S values (Lindberg et al., 2020; Oliphant et al., 2018). In the case of 'old Marseille', a densely built-up district with narrow streets (large aspect ratios), buildings possessing thick limestone walls (high thermal mass) and minimal vegetation, the importance of the diurnal course of ΔQ_S for a typical summer day is clear (Fig. 8.9). Here, because of the very dry urban setting, with little vegetation and thus minimal evapotranspiration (low values of Q_E), most of the Q^* is available for sensible heating of the air or the urban substrate. Accordingly, ΔQ_S in many ways rivals Q_H, especially in the morning hours as significant amounts of heat go into storage because of the large thermal mass of building walls as they become exposed to the strong summer morning sun (Fig. 8.9). In the evening the heat that comes out of storage is substantial, as indicated by the strong negative values of ΔQ_S, the upshot being heating of air in the narrow streets and thus the maintenance of nocturnal temperatures only 3 4 °C lower than the daytime high of 26 27 °C (Lemonsu et al., 2004).

While the results described for Marseille provide an appreciation of the importance of ΔQ_S for a specific part of a city, from the perspective of heatwaves, knowing about the spatial variability of this quantity is relevant. This is because sizeable

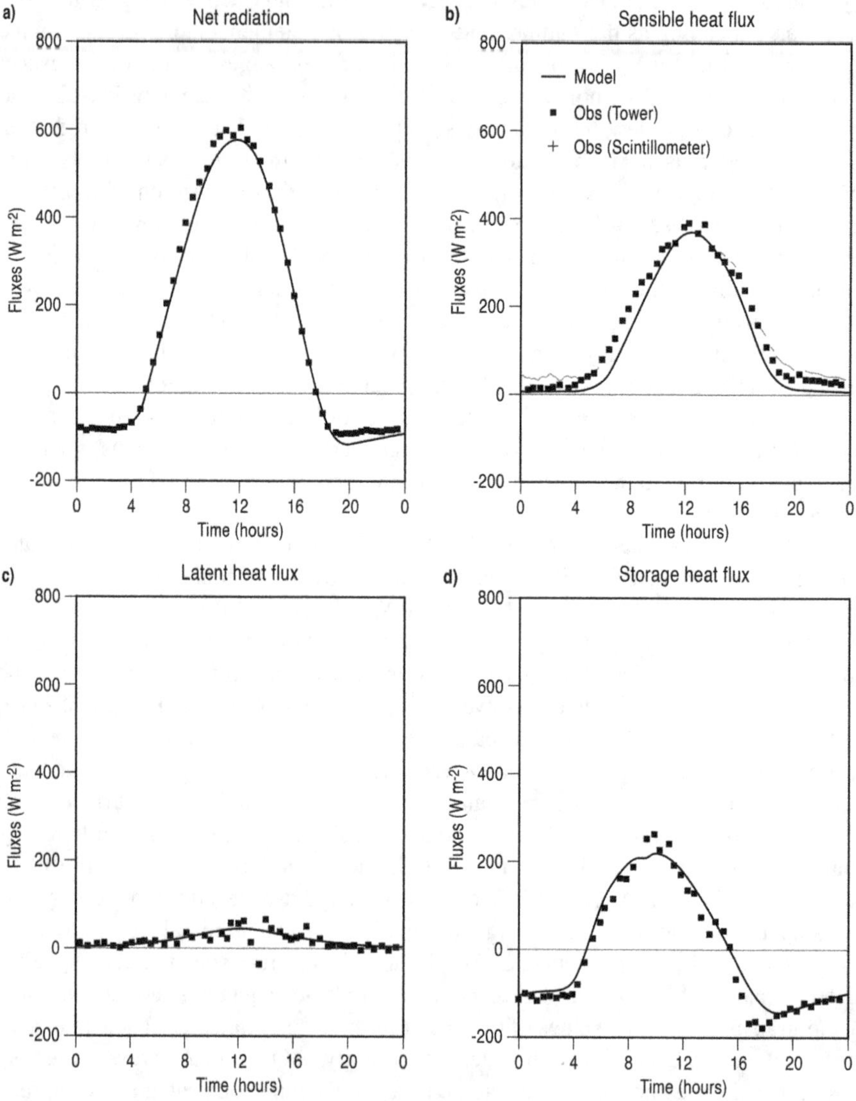

Fig. 8.9 The diurnal course of (**a**) net radiation Q*, (**b**) sensible heat Q_H, (**c**) latent heat Q_E and (**d**) storage heat ΔQ_S fluxes for 'old Marseille'. (Source: Lemonsu et al. (2004); © American Meteorological Society, used with permission)

ΔQ_S 'hotspots' at the intra-urban scale may interact with regional scale temperature anomalies associated with heatwave conditions to create nocturnal high heat risk areas due to the flow of heat into the urban atmosphere from storage. This combined with an urban form/structure typified by low sky view, will suppress nighttime cooling. Satellite based estimates of ΔQ_S for several cities have afforded a good understanding of the scale of variability of heat storage within and release from urban areas (Lindberg et al., 2020). An example for daytime and nighttime under mostly clear sky conditions for a selection of summer days is presented in Figs. 8.10 and 8.11 for New York, Los Angeles, Chicago and Phoenix (Hrisko, Ramamurthy, Melecio-Vazquez, & Gonzalez, 2021b).

Daytime large positive ΔQs values, representative of substantial amounts of heat going into storage, match areas with the highest urban densities, an indication of the large thermal mass in heavily built-up areas typical of the urban core of large cities. Of the four cities, some of the highest urban core ΔQs values (> 250–300 W m^{-2}) appear for Los Angeles and Phoenix, which possess Mediterranean and desert climates respectively and receive high values of incoming shortwave radiation that is readily absorbed by urban materials. In all cases, lower ΔQs values equate with suburban areas with a lower thermal mass compared to their urban core counterparts (Hrisko, Ramamurthy, Melecio-Vazquez, & Gonzalez, 2021b). Interestingly, New York and Chicago appear to demonstrate less of an overall urban core to hinterland spatial gradient in ΔQs compared to Los Angeles and Phoenix. This is most likely a result of contrasting climate settings. In the case of Los Angeles and Phoenix, the surrounding non-urban/more rural like areas release sensible heat back into the atmosphere because of the strong surface heating and low thermal mass. Accordingly, less heat goes into storage. Contrasts in surface albedo are also likely to play a role. The higher albedo (lighter colour) and mainly 'dry' surfaces adjacent to Los Angeles and Phoenix are likely to reflect more of the incoming shortwave radiation back to the atmosphere. This contrasts with the lower albedo ('darker') hinterlands of New York and Chicago, a result of summertime deciduous tree and plant growth because of a generally moister summer than Los Angeles and Phoenix. Comparing the four cities also reveals greater within-metropolis variability of ΔQs for New York and Chicago. This reflects the heterogeneity of the urban surface in the case of these two cities arising from the juxtaposition of contrasting local climate zones (Hrisko, Ramamurthy, Melecio-Vazquez, & Gonzalez, 2021b).

Unsurprisingly, the nocturnal pattern of ΔQs for the four cities mimics that of the daytime. Generally, areas of high nocturnal heat release (negative ΔQs values) match those with high daytime heat storage (Fig. 8.11). In many ways such areas can be considered as nocturnal hotspots or sources of intense nocturnal heat that potentially contribute to nighttime warming of the atmosphere at the local intra-urban scale and the suppression of nighttime cooling. For New York (Fig. 8.11a), stark contrasts in heat release occur between densely built-up areas with high thermal mass such as downtown Manhattan and Brooklyn and New York's famous Central Park where -ΔQs values are low (Hrisko, Ramamurthy, Melecio-Vazquez, & Gonzalez, 2021b). The same also applies in the case of Chicago where clear

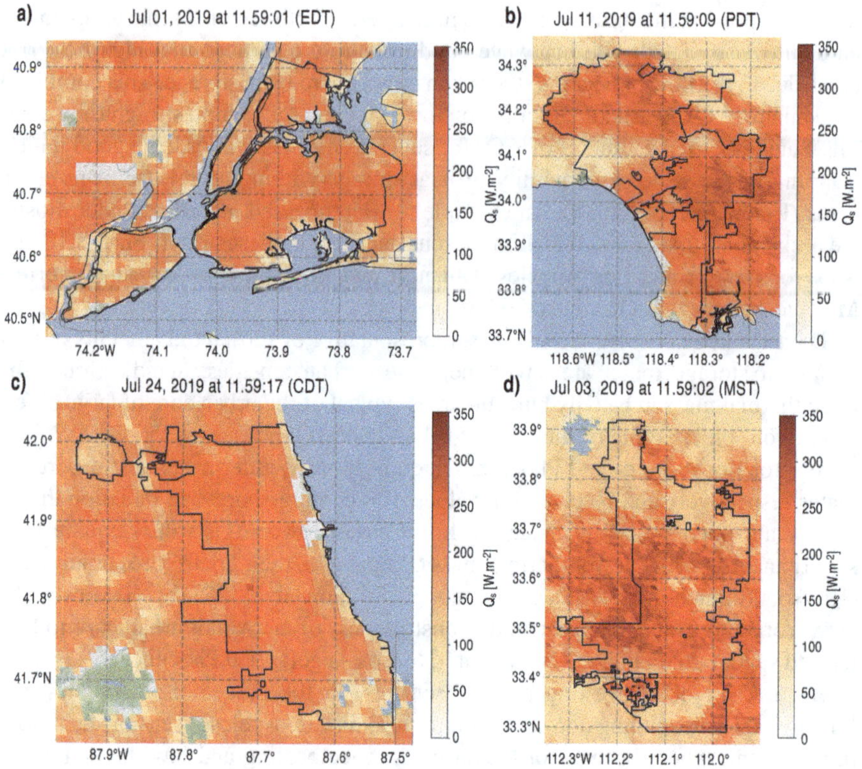

Fig. 8.10 Daytime spatial variation of ΔQs under mostly clear skies in July 2019 across four US cities: (**a**) New York City; (**b**) Los Angeles; (**c**) Chicago; and (**d**) Phoenix. (Source: Hrisko, Ramamurthy, Melecio-Vazquez, and Gonzalez (2021b). Spatiotemporal Variability of Heat Storage in Major U.S. Cities—A Satellite-Based Analysis. Remote Sensing. 13(1):59. https://doi.org/10.3390/rs13010059) Used under Creative Commons Licence 4.0 https://creativecommons.org/licenses/by/4.0/)

contrasts in heat release appear between the built-up districts in central northern Chicago and the large area of low heat release in the southeast quadrant associated the Cranberry Slough and Cap Sauers Holding Nature Reserves, the McMahon, Wolf Road and Saganashkee Woods and several lakes (Fig. 8.11c). For Chicago, the gradient in heat release between the lakeshore area and the linear front of the urban core is unambiguous. Comparatively, Los Angeles displays a higher spatial variability in -ΔQs compared to New York because of the wide range of values between the urban core and more rural-like edges of the city (Fig. 8.11b). However, like New York and Chicago, Los Angeles possesses areas of low nocturnal heat release. These are associated with the Malibu and Topanga State Parks immediately inland from the north shore of Santa Monica Bay and the Angeles National Forest and adjacent mountain areas in the northeast (Hrisko, Ramamurthy, Melecio-Vazquez, & Gonzalez, 2021b). Of the four cities, Phoenix displays the highest values for heat release from storage. Here, −ΔQs values reach 250–300 W m^{-2}, reflecting the

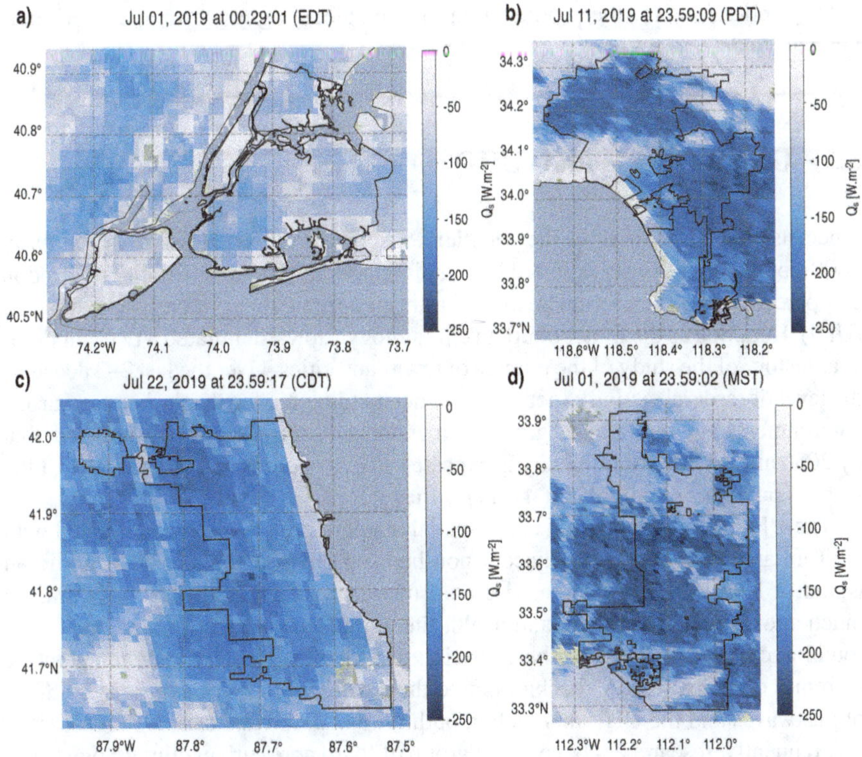

Fig. 8.11 Spatial variation of ΔQs under mostly clear nighttime skies in July 2019 across four US cities: (**a**) New York City; (**b**) Los Angeles; (**c**) Chicago; and (**d**) Phoenix. (Source: Hrisko, Ramamurthy, Melecio-Vazquez, and Gonzalez (2021b) Remote Sensing. 2021; 13(1):59. https://doi.org/10.3390/rs13010059). Used under Creative Commons License 4.0 https://creativecommons.org/licenses/by/4.0/)

nocturnal release of the significant amount heat accumulated in urban materials during the day (Fig. 8.11d).

There is no doubt that the instantaneous daily values of daytime and nighttime ΔQs presented in Figs. 8.10 and 8.11 are impressive. Similar values are apparent for several European cities. For London, ΔQ_S can swing from a maximum nighttime value of—200 W m^{-2} to a daytime value of 300 W m^{-2}. For other cities such as Basel and Heraklion, the diurnal swing can be −340 to 400 W m^{-2} and − 190 to 200 W m^{-2} respectively, with the large values for Basel driven by steep surface to air temperature gradients (Lindberg et al., 2020). Given the large values that ΔQ_S can attain, then during a prolonged heatwave event with strong shortwave radiation and temperature anomalies and assuming heat is not completely drained from storage at night, there is the clear possibility residual heat will build up in solid urban structures. Along with possible positive feedbacks related to a drying urban surface and an increasing flow of available energy (Q*) into sensible, such residual heat holds implications for the prolongation of anomalous nighttime temperatures during a heatwave event.

8.5 Urban Heat Island (UHI)

Since its first discovery by the 'gentlemen scientist' Luke Howard in the early 1800s, based on his observation of higher air temperatures across the city of London compared to its rural surrounds, much has been written about the urban heat island (UHI). Heralded as the father of urban climatology, Howard initiated a two-hundred-year history of the study of the climate of towns and cities. Over the last 2–3 decades, the level of research activity focused on understanding the feedbacks between urban environments and climate has grown tremendously, spurred on by the prospect that by 2050 almost 70% of humanity is expected to live in cities and around 10% of the world's population will inhabit 41 megacities.

The UHI phenomenon, that is the spatial temperature anomalies associated with built-up areas compared to adjacent non-built-up environments, which mimic an island of heat in a sea of otherwise cool non-urban air (Fig. 8.12), has attracted much attention in the climate and health literature. This is because the 'extra heat' found in cities is likely exacerbated during periods of regionally wide temperature extremes or heatwaves. As background to the next section on the synergistic effects of heatwaves and the UHI, this section outlines the nature and causes of the UHI.

Frequently, researchers beyond the group of 'hard core' urban climatologists use the term urban heat island quite loosely not appreciating there are in fact four types of urban heat island. These are the surface, urban canopy, boundary layer and sub-surface heat islands. A commonly used metric for describing the magnitude or intensity of each of the different heat island types is the difference between the urban temperature (T_{urban}) and a representative temperature of the rural hinterland (T_{rural}) for a specified time. In many cases temperature measurements from a single urban and rural site are used in the calculation of the heat island magnitude or intensity without much thought given to the representativeness of the associated temperature records and issues of data quality. Because few cities possess high-density urban climate station networks often a one-dimensional view of the urban heat island is only possible because measures of UHI intensity are based on a single urban to rural station comparison. For cities lucky enough to enjoy the benefits of multi-site monitoring of urban climate, urban to rural temperature differences can be used to build an idea of the morphology of the urban heat island and how this may change under different weather conditions and diurnally/seasonally. While all four urban heat island types are defined by their elevated temperature compared to their rural counterpart, the point of reference for the temperature difference, their scale and underlying processes of formation vary as described below.

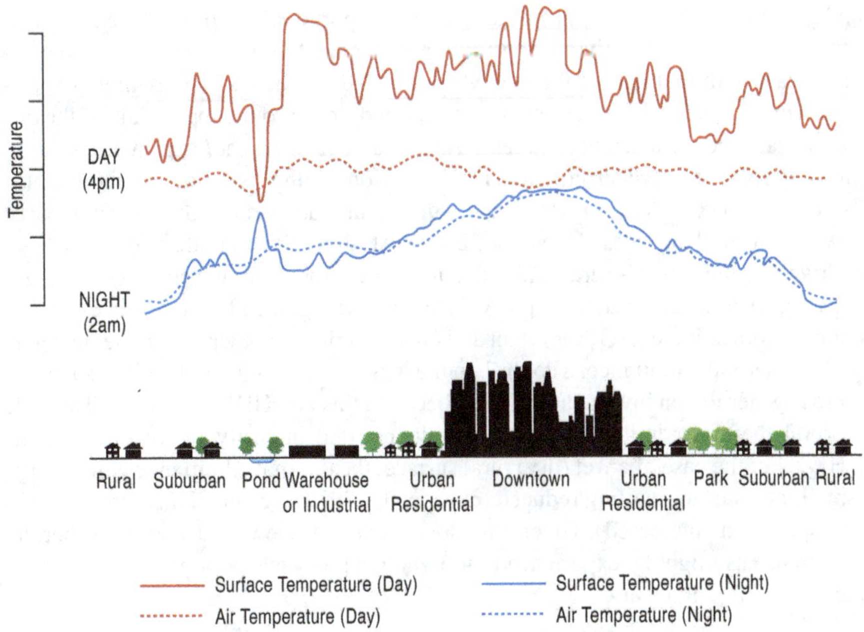

Fig. 8.12 Profile of surface and air temperature across an urban area for daytime (4 pm) and nighttime (2 am) portraying the urban canopy layer heat island. Note the temperature scale is relative and temperatures are for the urban canopy layer. Progressing from the rural to downtown area the nighttime temperature trace describes a 'cliff' at the rural-urban boundary followed by a plateau of higher temperatures interspersed within which are localised 'hills' ('valleys') of relatively high (low) temperatures before reaching the central 'peak' of high temperatures associated with the urban core. (Source: US Environmental Protection Agency. Publicly available at: https://www.usgs.gov/media/images/urban-heat-islands Accessed 05/06/2024)

8.5.1 The Surface Urban Heat Island (UHI_{surf})

The UHI_{surf} is the difference in surface temperature between urban and rural areas with surface temperature measurements derived from instruments mounted on remote platforms such as satellites, planes, helicopters, or drones. As noted previously, the urban surface is complex begging the question 'what particular surface is the reference surface in the urban case'. The same of course may apply to 'non-simple and rough' rural surfaces. Ideally, UHI_{surf} should reflect the difference in temperature between the complete urban surface and an adjacent rural surface. As one can imagine, to achieve an accurate measure of complete surface temperature poses quite a challenge given adjustments to temperature measurements are required to consider the complex urban geometry and surfaces 'not seen' by temperature sensors and the different radiative and thermal properties of the multitude of urban facets. For this reason, the seductive satellite images highlighting intra-urban and overall urban—rural contrasts in surface temperature are in many ways proxies of the actual UHI_{surf}. This is because the surface temperatures recorded are for 'visible' and generally 'hot' horizontal surfaces only. The portrayal of UHI_{surf} based on satellite images is also sensitive to the timing of the satellite overpass, which may be discordant with that part of the day when the UHI_{surf} reaches its maximum expression. This is generally for daytime clear sky conditions around the middle of the day to early afternoon, when urban areas with little vegetation and/or plan areas dominated by roofs, receive strong inputs of insolation. With this set of conditions, plan as opposed to complete surface temperatures, can reach very high levels compared to adjacent rural surfaces (Oke et al., 2017). At night, the magnitude of the UHI_{surf} is driven by the urban—rural difference in the heat lost from the surface via outgoing longwave radiation ($L\uparrow > L\downarrow = -Q^*$) and the heat gained by the surface via heat coming from storage (ΔQ_S for urban and Q_G for rural). All other factors being equal, such as thermal admittance, a control on the magnitude of $-Q^*$ is sky view, such that there is generally an inverse association between this and UHI_{surf} intensity. The state of rural thermal admittance can also influence the intensity or direction of the UHI_{surf}. For the case of a wet (dry) rural surface, the thermal admittance will be large (small) and surface cooling reduced (enhanced) with the result UHI_{surf} intensity will be suppressed (increased). Given this, low urban sky view, and low rural thermal admittance as might be experienced throughout a heatwave period, will conspire to increase UHI_{surf} intensity.

8.5.2 *The Urban Canopy Layer Urban Heat Island (UHI_{UCL})*

The UHI_{UCL} refers to the elevated temperatures found in the urban canopy (Fig. 8.5b), that is, the volume of air below roof level, compared to a rural equivalent. The reference height is usually 1–2 m above the surface. Temperature measurements from standard climate stations (e.g. thermometers in Stevenson screens), special measurement campaigns that may include high density fixed monitoring or vehicle traverses across urban areas usually provide the requisite data for UHI_{UCL} analyses (Muller et al., 2013; Chapman et al., 2015). Interestingly, several urban climatologists have begun to explore the potential of crowdsourcing data from citizen weather stations for UHI_{UCL} research (Chapman et al., 2017; Meier et al., 2017). Of the four UHI types, UHI_{UCL} has received the greatest attention in the literature. Having said that the rigour of a worrying proportion of UHI_{UCL} studies is less than ideal with issues such as the representativeness of measurement sites, inadequate metadata, small sample sizes, non-conditioning of results on the weather situation, elevation/relief, temporal factors (diurnal/season) and soil moisture status, compromising their scientific integrity (Oke et al., 2017; Stewart, 2011). These worries aside, many studies provide reliable insights into the nature of the UHI_{UCL} and its genesis.

The UHI_{UCL} is largely a nighttime phenomenon. It develops best, as manifest by a high UHI magnitude, under the ideal conditions of calm weather and clear skies when rural cooling rates are high relative to urban ones. During 'blocking episodes', which are conducive to heatwave development (Chap. 6), such conditions may persist for several days with a well-developed UHI_{UCL} 'settling in' for a number of nights and increasing in its intensity over the length of the blocking episode. Of course, at any given time, the UHI_{UCL} can be 'destroyed' by the onset of local to larger scale weather phenomena that create turbulence and/or reduce daytime heating/nighttime cooling such as sea/land and valley/mountain breezes, thunderstorms or rapidly moving frontal systems.

Climatologically, the UHI_{UCL} exhibits a clear diurnal cycle (Santamouris, 2015). It begins to develop in the late afternoon as Q* and ΔQ_S transition from positive to negative values and heat loss from the surface to the atmosphere (-Q*) commences, accompanied by heat flowing out of storage from urban materials ($-\Delta Q_S$). Following sunrise, the nocturnal UHI begins to decay. This is because in the morning hours surface heating is often less in urban compared to non-urban areas because of building shading effects, the result of which are greater rural, compared to urban, heating rates. The general UHI diurnal pattern just described is clearly demonstrated in Fig. 8.13. This shows the diurnal march of temperatures for urban and non-urban sites in London and the associated UHI magnitude. Here as the late afternoon transitions to evening and sunset, the march of urban and non-urban temperatures tends to diverge, a result of a greater rate of cooling at the non-urban compared to urban site. At the same time, the UHI begins to reveal itself, gradually growing in strength throughout the night attaining maximum values between 0000 and 0600 hours. The demise of the nocturnal UHI is quite abrupt. As the sun starts to poke its head above the horizon and daytime heating begins, non-urban temperatures start to climb rapidly. Note at around 0900 hours non-urban and urban temperatures crossover. Consequently, the value of the UHI magnitude reverses (Fig. 8.13) with a daytime 'cool island' effect apparent as daytime non-urban temperatures exceed urban ones.

For the case just described for London and acknowledging this is just one UHI event for this city, the UHI magnitude appears to peak just before 0000 hrs with a hint of a secondary peak in the hours leading up to sunrise. However, this pattern may not repeat from UHI event to UHI event or be universal to all cities. Indeed, from the large number of studies conducted to date, it is very clear the time at which the UHI_{UCL} reaches its local maximum can vary significantly from city to city (Santamouris, 2015). The cool island effect noted for London is a common feature in many cities. Occurring when $T_{rural} > T_{urban}$, it has been mainly attributed to daytime building shading effects with reduced thermal admittance in the dry urban environment probably playing a secondary role (Oke et al., 2017). For some cities, reduced daytime anthropogenic heat emissions have also been identified as important (Yang et al., 2017). Like the nocturnal UHI, the timing of the daytime cool island may vary based on local circumstances lasting for most of the day or confined to morning or afternoon hours. For example, in Adelaide, a city well known for its heatwaves, the cool island in the urban core can persist for most of the day reaching its maximum in summer on calm clear days with high temperatures and large insolation. Under such conditions, rural daytime temperatures reach significantly higher levels than in the central city where high canyon aspect ratios of inner-city streets diminish surface insolation receipts (Erell & Williamson, 2007). This contrasts with Beijing where a morning cool island may develop, lasting until around local noon. Here under calm weather conditions, high urban pollution levels in the morning

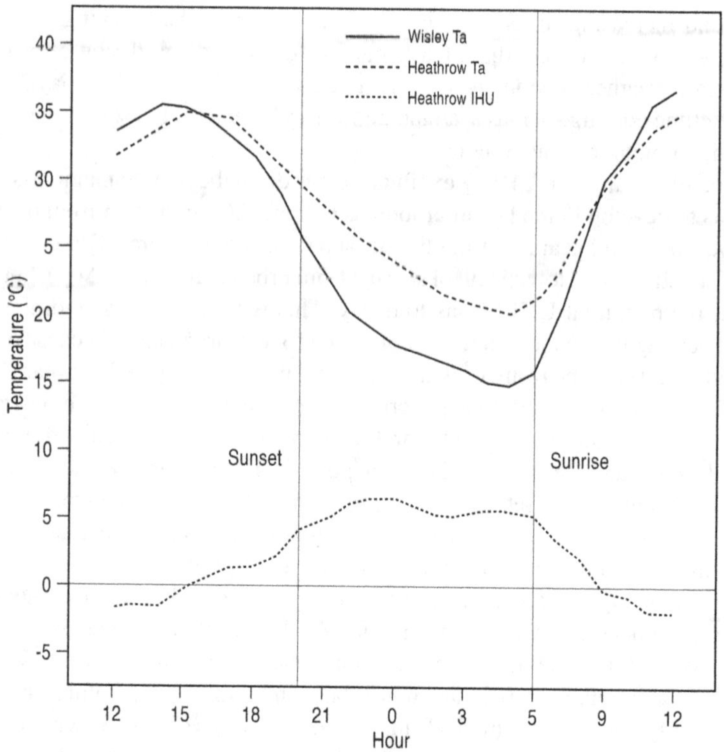

Fig. 8.13 The diurnal march of observed hourly urban (Heathrow) and non-urban (Wisley) temperatures and associated UHI for Greater London, August 9–10, 2003. Sunrise and sunset occurred at 0536 h and 2034 h respectively as indicated by the vertical lines. The horizontal at 0 °C line represents no difference between non-urban and urban temperatures with positive values indicating the development of the UHI_{UCL}. Note the vertical temperature scale for the UHI trace is read as the urban to non-urban temperature difference. This reaches a maximum of 5–6 °C in the hours leading up to midnight. (Source: Author and adaptedfrom Durham University MSc Risk Dissertation 'The Interaction Between the Urban Heat Island and Heatwaves: A Study of London 2000–2019 by Emma Beck)

lessen surface heating. In combination with urban geometry factors, this results in lower pre-noon urban temperatures in comparison to less polluted and open nearby non-urban areas (Miao et al., 2009). For Seoul it would appear afternoon cool islands are a feature of this city (Kim & Baik, 2005).

It is perhaps no surprise that the seasonal cycle of UHI_{UCL} intensity is very much dependent on seasonal variations in those conditions most conducive to the development of a well-defined UHI. All other things being equal these include strong solar irradiance, good day length and the predominance of calm clear weather conditions. The latter in many ways has an overriding impact on the seasonal pattern of UHI development. This is clearly demonstrated for Kumagaya City, Japan (Fig. 8.14). Here despite the general low values compared to much bigger cities, the greatest UHI intensity occurs in the spring months of April and May with a suggestion that high values also occur for September. During these months, the UHI intensity reaches its maximum between the hours running up to midnight and local sunrise. Interestingly, the summer months of June to August display little evidence of strong UHI development. This curious seasonal pattern of the UHI comes about due to the climatological occurrence of frequent calm, sunny conditions associated with migratory anticyclonic systems during April–May (Nakamura et al., 2018) and similarly in September. The arrival of the wet season and the Baiu front in late June/early July as well as the onset of the typhoon season and plentiful rain generally suppresses UHI development in the summer (Nakamura et al., 2018). Although there is some evidence for fleeting periods of UHI development for the early to mid-evening hours in the winter, probably associated with occasional periods of fine weather and heat inputs related to commuting and domestic activity, overall, this season lacks convincing UHI development because of the predominance of strong northwesterly flows (Nakamura et al., 2018).

The situation described in Fig. 8.14, while being unique to Kumagaya City, hints at the complex seasonal interplay between surface heating (radiative forcing), temperature and rainfall, the nature of which will vary depending on climate setting. For this reason, UHI intensity exhibits distinctive seasonal hysteretic cycles, the shape and looping direction of which varies across climatic zones (Manoli et al., 2020). Added to this is the potential role of urban geometric complexity and plan form in influencing the nature of the UHI seasonal cycle (Liang et al., 2020). These two built environment characteristics also influence the general morphology of the UHI_{UCL}.

In crude terms and for the case of a climatically featureless plain, the shape of the UHI_{UCL} will 'follow the money', that is highest canopy layer temperatures will

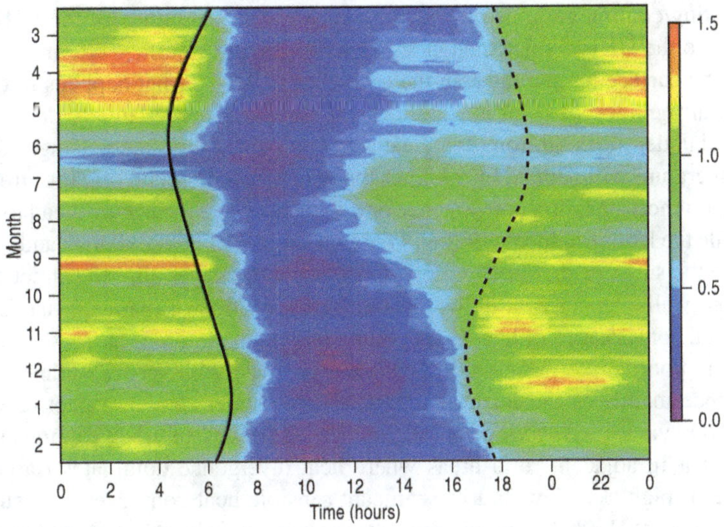

Fig. 8.14 Seasonal variation of the diurnal variation of the UHI for Kumagaya, Japan. (Source: Provided by Y Nakamura, Rissho University, Kumagaya City, Japan)

occur across the most built-up areas, often 'downtown' in the central business district. As cities take on a vast range of different shapes, determined by their topographic setting and the pattern of natural land cover, as well as their various historical legacies and functions, globally the potential range of heat island morphologies is vast. Diversity at the intra-urban scale is also a feature of the UHI_{UCL} with city thermal landscapes often pockmarked with hotspots and cool spots associated with blue and green spaces or contrasting local climate zones (Lehnert et al., 2018; Milošević et al., 2021; Sugawara et al., 2021; Top et al., 2020). Further, for any given city, the UHI_{UCL} can take on a variety of forms as determined by the direction and strength of gentle 'non-heat eroding' breezes. As these waft across a city, the centre of anomalous heat can be pushed around (Chemel & Sokhi, 2012) thus extending the influence of the UHI beyond the city, a process referred to as urban heat advection (Bassett et al., 2019).

Notwithstanding the critical role of weather conditions, a long-standing question is what key built environment, human behaviour and micro-scale climate processes conspire to produce a maximum urban canopy heat island?' Added to this, is the question, 'does city size matter in determining UHI_{UCL} intensity'? The built environment and human factors that create the potential for maximum UHI_{UCL} values are street geometry, building fabric and anthropogenic heat emissions (Oke et al., 2017). Tight narrow streets or urban canyons possessing large height:width (aspect) ratios and low sky view affect radiation exchange and ventilation and enhance heat buildup in the urban canopy; several studies have identified a positive (inverse) association between maximum UHI magnitude and canyon aspect ratio (sky view). The nature of urban materials, the building fabric, exerts a fundamental control on heat storage and release as well as whether water will be repelled or absorbed across a vast range of surfaces. Large inputs of anthropogenic heat into street canyons from vehicles and heating/cooling systems may also exacerbate the magnitude of the UHI_{UCL}. In contrast to the UHI_{surf}, for which energy exchanges at the surface are important for governing surface temperature, in the case of the UHI_{UCL} what happens in the volume of air below roof level is significant. In a typical street canyon, open at both ends and under the condition of no heat advection along the street, changes of UCL air temperature will depend on whether the volume of air occupying the street canyon experiences heat flux convergence or divergence. The former will lead to warming while the latter produces cooling. Potential sources of both radiative and sensible heat are the surfaces that bound the volume of canopy layer air—the street canyon floor and walls. Under conditions that suit maximization of the nocturnal UHI_{UCL}, horizontal convergence of heat ($-\Delta QS$) and radiation ($L\uparrow$ or $-Q^*$) from the canyon walls and floor is likely to be significant and compete with vertically directed heat divergence mainly in the form of $L\uparrow$ ($-Q^*$); high street canyon aspect ratios will curtail $-Q^*$ via a low sky view factor. Suppressed cooling of urban canopy air, relative to that in adjacent rural areas where heat divergence dominates (large $-Q^*$) because of high sky view and insignificant sensible heat convergence, results. As noted by Oke et al. (2017), cooling of urban canyon air is largely via a radiative heat loss pathway ($-Q^*$) with negligible sensible heat divergence. They suggest this as the possible 'energetic basis' underpinning the occurrence of significant UHI_{UCL}

events, as manifest by large nocturnal urban—rural canopy layer temperature differences driven by contrasts in urban-rural cooling rates.

Intuitively one might expect the magnitude of the UHI_{UCL} to be dependent on city size. For this reason, researchers have explored the association between city population and urban heat island intensity with the study of Oke (1973) on North American and European cities, setting the scene for the many similar studies that followed. Although population as a proxy of city size may indirectly capture many of the processes that determine the magnitude of UHI intensity (Roth, 2007), recent attempts at establishing the city size effect have focused on key urban variables directly relevant to climate processes. Such studies have revealed a complex interplay between city physical size, urban canyon structure, building density, urban surface fraction, and urban form in determining the level of the UHI intensity. For example, modelling studies have shown if urban area or gross building volume is held constant, the UHI_{UCL} intensity increases with building density. Further, increasing building plan area produces a more rapid increase in UHI intensity than adding vertical building height, because of the role of canyon aspect ratio and associated competing daytime shading and nighttime heat trapping effects (Li et al., 2020). Interestingly, the 'thermal playoff' due to shading and sheltering from deep urban canyons may, along with an urban induced flow of cool air into city centres, work to set an upper limit for UHI_{UCL} magnitude of around 12 degrees (Oke et al., 2017). Regarding the interaction between city size and overall urban form, as determined by measures of fractal dimension (degree of compactness) and anisometry (level of elongation), city size has the largest impact on UHI_{UCL} intensity, followed by a city's fractality and to a lesser extent anisometry. In other words, the larger, more compact (high fractal dimension), and the less elongate (low anisometry) a city is, the greater its UHI_{UCL} intensity (Zhou et al., 2017). The upshot of this of course is if avoiding high levels of UHI_{UCL} intensity is a priority for city planners, especially in relation to long-term strategies for managing heat risk during heatwave events, then urban developments should steer away from ones that are large, compact and rotund (Zhou et al., 2017). Such a potential rule of thumb is at odds with a widely held view that compact cities offer advantages in relation to energy efficiency and reduced carbon dioxide emissions. This constitutes a significant climate sensitive city design dilemma for urban planners, especially as there is evidence of a global homogenisation of urban form (Lemoine-Rodriguez et al., 2020) with the prospect that arresting unsustainable trajectories of urban development may only be achievable for small to medium-sized cities (Guneralp et al., 2020). It also raises the prospect that tradeoffs between urban efficiency and the severity of the UHI_{UCL} may be necessary in the attainment of sustainable cities and communities as encapsulated in the United Nation's Sustainable Development Goal 11.

An under researched aspect of UCL climate is humidity; there have been few attempts to establish whether a moisture equivalent of the UHI exists. In fact, the measurement and understanding of the generation of urban moisture fields is a wide-open research area, with much to be done. That aside, and climatologically speaking, urban environments are understood to be dry compared to their rural counterparts, as manifest by high Bowen ratios due to moisture limited evaporation rates and thus low levels of latent heat flux. Such urban 'dryness' is often referred to as the 'urban dry island' phenomena (Meili et al., 2022). Its intensification over time, as characterized by secular trends of reduced humidity and increased vapour pressure deficit for some locations, has been attributed to urban expansion (Lin et al., 2020; Lokoshchenko, 2017; Luo & Lau, 2019). While the idea of an urban dry island generally holds true on the climatological time scale, at the diurnal scale, and notably during summer nocturnal hours, urban moisture levels may exceed those in surrounding rural areas. Such elevated urban moisture is referred to as urban moisture excess (UME) (Holmer & Eliasson, 1999; Kuttler et al., 2007; Lee, 1991) or the urban moisture island (UMI) (Richards, 2005; Wang et al., 2021). Most likely, the nocturnal UME or UMI is generated by the tendency for evaporation to continue through the night, driven by a mildly unstable atmosphere arising from the flux of sensible heat from storage (− ΔQS) into the UCL. Evaporation can continue if the UCL maintains a surface to atmosphere moisture gradient, or in other words, a positive water vapour lapse rate (decrease of humidity with height). Given this, cities with a well-established UHI and readily available surface moisture sources, in the form of urban water bodies or irrigated parks and gardens, are likely to experience an UME/UMI, the occurrence of which may well hold implications for levels of nocturnal human thermal comfort. In addition to evaporation, other factors may also play a role in generating UME or UMI in highly urbanised cities. These include anthropogenic sources of moisture from combustion, reduced moisture sinks as a warm nighttime atmosphere will reduce the likelihood of dew fall (dew fall takes moisture out of the atmosphere and thus reduces humidity) and weak intra-urban ventilation and thus diminished removal of moisture from the UCL (Wang et al., 2021).

8.5.3 The Urban Boundary Layer Urban Heat Island (UHI_{UBL})

The UHI_{UBL} refers to the heat contained in the layer of air above the UCL, that is, the layer of air influenced by the overall urban surface and downwind of the adjacent rural boundary layer. Like the connection between the and UHI_{UCL} and UHI_{surf}, the UHI_{UBL} is partly influenced by the UHI_{UCL}. As noted earlier, the UBL can take on a dome- or plume-like shape depending on the state of the regional wind field. It is within this dome or plume where higher temperatures compared to the upwind and downwind rural boundary layers have been found using instruments usually mounted on satellites, planes or helicopters, ground-based radio acoustic sounding system (RASS) profilers, very tall towers and tethered balloons and more recently drones (Adkins et al., 2020) and even birds (Thomas et al., 2018).

The UHI_{UBL} arises due to a mixture of 'top-down' and 'bottom-up' processes (Oke et al., 2017). During the day, there is an upward flux of sensible heat into the UBL from over warm urban surfaces via the UCL and overlying roughness sublayer (RSL) (Fig. 8.5). While upward fluxes of anthropogenic heat (Q_F) from buildings, vehicles and human metabolism may also contribute to daytime UBL warming these are small relative to Q_H fluxes driven by surface to air temperature differences. The 'bubbles' of heat rising from the 'urban surface' through the RSL create turbulence and assist with the vertical growth of the UBL mixed layer. As the mixed layer grows and impinges on the capping inversion, sensible heat is sequestered from the so-called entrainment zone (Fig. 7.4) and mixed downward and stored in the UBL.

A polluted urban atmosphere is also likely to play a role in warming of the UBL via radiative flux convergence; a range of urban pollutants such as particulate matter are effective at absorbing radiation reflected (K↑) or emitted (L↑) from the urban surface and scattered in and emitted (L↓) by the urban atmosphere. At night heat coming out of storage (−ΔQS) serves as a heat source for the UBL, especially in areas of high urban density, finding its way into the UBL via the RSL. Added to this are additions of sensible and latent heat from anthropogenic sources. In the summertime in parts of the city where venting of indoor heat to the outside continues into the evening, both latent and sensible heat contributions directly into the UBL from heat exhaust systems mounted on roof tops are undoubtedly an important heat source. At night, heat contained in the daytime upwind rural inversion can be transported down into the UBL by the enhanced nocturnal mixing found over urban areas arising from a mixture of thermal and mechanical turbulence. This redistribution of heat from the upwind rural boundary layer works to raise (decrease) the temperature of the lower (higher) parts of the UBL (Oke et al., 2017).

To this point, the suggestion has been sensible heat fluxes from the surface can be an important heat source for the UBL and thus the creation of a noticeable UHI_{UBL}. However, it may not all be one-way traffic in terms of the upward flow of sensible heat from the UCL to the UBL via the RSL. There is the distinct possibility that large amounts of heat accumulated in the nocturnal urban residual layer, originating from strong upward fluxes of sensible during the previous day(s), could be mixed down to the surface, by strong daytime turbulent mixing. Such a process could potentially elevate near surface daytime temperatures through downward sensible heat flux convergence, as warm air from the UBL merges with that in the RSL and UCL. Added to this is the possible role of urban form. 'Aerodynamically smooth' cities are likely to produce less mechanically forced convection compared to cities or indeed surrounding rural areas with a higher degree of roughness and therefore have a shallower UBL. Consequently, such cities will be less efficient at the vertical dissipation of urban heat with urban warming a consequence (Zhao et al., 2014).

8.5.4 *The Subsurface Urban Heat Island (UHI_{SUB})*

Defined as the difference in below ground level temperatures between urban and rural locations, the subsurface urban heat island (UHI_{SUB}) magnitude is usually measured by comparing temperature records from groundwater monitoring stations or wells in urban and rural settings at similar depths and lithological settings. Some recent studies however have employed remote sensing technology for assessing

UHI_{SUB} characteristics (Huang et al., 2020; Zhan et al., 2014). Under normal conditions, there is a near linear increase of temperature with depth away from the surface; the earth's surface and its immediate sublayers are cooler than the deep subsurface. However, because of anthropogenic heat contributions from building basements, urban surface runoff, buried electrical and communications infrastructure, sewage systems, subsurface heating networks, and thermal wastewater, the ground beneath cities is often warm, with associated UHI_{SUB} magnitudes often exceeding five degrees Kelvin (Menberg et al., 2013; Muller et al., 2014). As might be expected, maximum heat fluxes and associated impacts on subsurface temperature usually occur beneath the most built-up parts of a city (Schweighofer et al., 2021). For example, mean fluxes from asphalt covered areas and buildings in central Osaka are significantly higher than fluxes from unpaved and grass-covered areas, with cumulative heat input to different areas of this city's groundwater system ranging from 4-60 MJ m^{-2} over the period 2003 to 2011 (Benz et al., 2018). That anthropogenic sources of heat contribute significantly to subsurface heat storage leading to long-term temperature change of the urban 'underground' in general, is apparent in several cities such as London where subsurface warming at an urban site has been shown to be 0.68 °C per decade higher than a rural equivalent (Luo & Asproudi, 2015). Similarly, as the city of Amsterdam has expanded, a systematic increase in subsurface temperature has occurred (Visser et al., 2020). Because the magnitude of the anthropogenic heat stored below cities could be enormous, there has been mounting interest in quantifying the low-temperature geothermal potential in urban ground and groundwater with the prospect that such heat once harnessed could be an important energy source for cities (Bayer et al., 2019). As yet, how the UHI_{SUB} interacts with heatwaves has not been explored. That aside, impacts are likely to be more indirect than direct. A concomitant response of the UHI_{SUB} to heatwave occurrence at the meteorological timescale is unlikely because of the typically long timescales of months to decades over which the ground at depth responds to climate forcing. Anomalously high temperatures lasting an entire summer may however add to the long-term cumulative heating of the subsurface as any associated thermal wave slowly propagates through the subsurface. On climatological timescales warmer subsurface conditions beneath cities may compromise human thermal comfort during heatwaves in underground spaces used for business or living, especially in those locations where air conditioning is not ubiquitous. Elevated UHI_{SUB} related temperatures may also challenge the operation of ventilation and air conditioning machinery installed in underground locations particularly during heatwave periods when such machinery may have to run at maximum capacity and require cooling.

8.6 Urban Heat Island—Heatwave Interactions

In the fictional world of books, movies and songs, the heat of the city during a heatwave is often used to set the scene for extraordinary events that might not otherwise have happened outside the city in cooler surrounds. Further and almost without fail, the media when reporting on heatwaves latch onto the idea that cities respond somewhat differently to non-urban environments intimating the magnitude of 'city heat' is worse than elsewhere. Perhaps unsurprisingly, the urban heat island is usually singled-out, as the 'culprit' accountable for the oppressive city heat during heatwaves. This claim has stimulated much research on the current and future synergistic effects of urban heat islands and heatwaves with investigations based on both observations and modelling (Cheval et al., 2024). At the heart of this research activity is the broad but important question 'is urban heat island intensity increased or dampened during heatwave events?' To date most of the work directed at this question has focused on the intensity of the UHI_{UCL}.

To say that the debate regarding the amplification of UHI intensity during heatwaves is not settled would be an understatement. What appears a seemingly simple question is in fact a difficult one associated with a minefield of methodological issues. These include how best to define UHI intensity and for which UHI type (see previous section), what constitutes a heatwave and to what extent might any conclusions drawn depend on whether a climatological (composite of events) or meteorological (single event) analysis framework has been applied. What's more, comparability of studies is hampered by the examination of different periods, study cities in very different climate settings and application of very different types of data (Cheval et al., 2024; Shi et al., 2024). While the intention of this section is not to pick apart studies that address the synergistic effects of the urban heat island and heatwaves, the aforementioned issues need to be borne in mind if the reader choses to dip into some of the literature mentioned here.

Studies of UHI heatwave interactions reveal a mixed picture varying from no to marked heatwave related UHI amplification. Based on just one heatwave event in 2016 Chew et al. (2021) note for Singapore while the daytime temperature for this event was 3 °C higher than previous and subsequent non-heatwave days, the UHI intensity did not increase and remained at around 2.5 °C during both heat wave and non-heat wave periods. For Lisbon, Portugal Oliveira et al. (2021) report no significant difference of the UHI intensity between a large sample of heatwave and non-heatwave days noting that wind direction/weather type is the most important influence. In a similar climatological study based on 6 years of high-density temperature observations for Dijon France, Richard et al. (2021) report strong UHIs are not restricted to only hot spells or heatwave events but can occur at any time during the summer under calm sunny conditions. They also note that maximum UHI intensities tend to precede the onset or occur in the early stages of heatwave events with UHI intensity decreasing throughout a heatwave. This is because the nocturnal urban to rural temperature gradients slacken due to rural temperatures increasing over the length of a heatwave, while urban temperatures remain relatively constant. Based on a composite approach for 54 US cities and finding that the UHI intensity decreases during heat events for 48 of the study cities, Scott et al. (2018) conclude there is no amplification of the UHI during heatwave events. They ascribe this to increases in rural temperatures during heatwave events with dampening impacts on UHI intensity greatest under warm humid weather conditions. Higher rural compared to urban temperatures during heatwaves is also identified as a reason why the surface urban heat island is not exacerbated under heatwave conditions in India (Kumar & Mishra, 2019). While these studies support a null hypothesis of no UHI intensity enhancement during heatwave events, others do not. For example, for western Sydney increases of UHI intensities of up to 8 °C have been found for heatwave compared to non-heatwave days (Khan et al., 2020; Khan et al., 2021). This is comparable to that found for a single heatwave event in Beijing. In this case the UHI intensity demonstrated a double diurnal peak, unlike the climatological pattern of UHI, due to anthropogenic heat contributions from building/dwelling cooling systems in around the beginning and end of the working day (An et al., 2020). For London, heatwaves appear to amplify the urban heat island as evidenced by a comparison of the diurnal pattern of UHI intensity for an equal number of heatwave days and non-heatwave days (random selection) and all summer days with the greatest amplification of around 1.5 °C achieved between 0200 and 0500 hrs (Fig. 8.15). Other cities, for which heatwave-related intensifications of UHI intensity have been noted include Baltimore (Li & Bou-Zeid, 2013), Phoenix and New York (Ortiz et al., 2018; Ramamurthy & Bou-Zeid, 2017; Ramamurthy et al., 2017; Tewari et al., 2019), Madison (Schatz & Kucharik, 2015), Oklahoma City (Basara et al., 2010), Athens (Founda & Santamouris, 2017), Melbourne and Adelaide (Rogers et al., 2019), Nicosia (Pyrgou et al., 2020), Seoul (Ngarambe et al., 2020), Karachi (Rizvi et al., 2019), Paris (De Ridder et al., 2017), Szeged, Hungary (Unger et al., 2020) and Berlin (Fenner et al., 2019).

That the situation regarding UHI heatwave synergies might be dependent on UHI definition, the nature of the study city and the climate type is supported by several studies. For example, for the city of Atlanta, depending on the choice of the rural reference station for defining the UHI intensity, different conclusions regarding UHI heatwave interactions can be drawn (Zhou & Shepherd, 2010). Added to this is whether a heatwave event is defined in relation to nighttime temperature trends at an urban or rural station with a decrease (increase) in UHI intensity during a heatwave apparent if the rural (urban) station is the focus (Scott et al., 2018). The choice of maximum or minimum temperature for defining UHI intensity also appears to be critical. This is because maximum and minimum temperatures are a consequence of quite different UCL climate processes and display different levels of sensitivity to daytime and nighttime weather conditions, which may assist explaining contrasts in heatwave impacts on daytime and nighttime UHI intensity or indeed conflicting results from different studies (Tewari et al., 2019). The use of non-standard climate variables for defining UHI intensity such as the wet bulb globe temperature or apparent temperature (Chap. 4) may also produce subtly different outcomes in terms of the interplay of the UHI and heatwaves (He et al., 2021). Background climate and city size also appear to influence heatwave UHI interactions as revealed by modelling studies. For example, Zhao et al. (2018) found significant daytime increases in UHI intensity, as measured by surface radiant temperature, during heatwaves for cities in temperate climate regions but not for those in dry regions, whereas in the case of nighttime UHI intensity, cities in both climate regions demonstrated UHI intensity increases. Contrasts in daytime and nighttime UHI intensity have also been found for China with Jiang et al. (2019) reporting an amplification

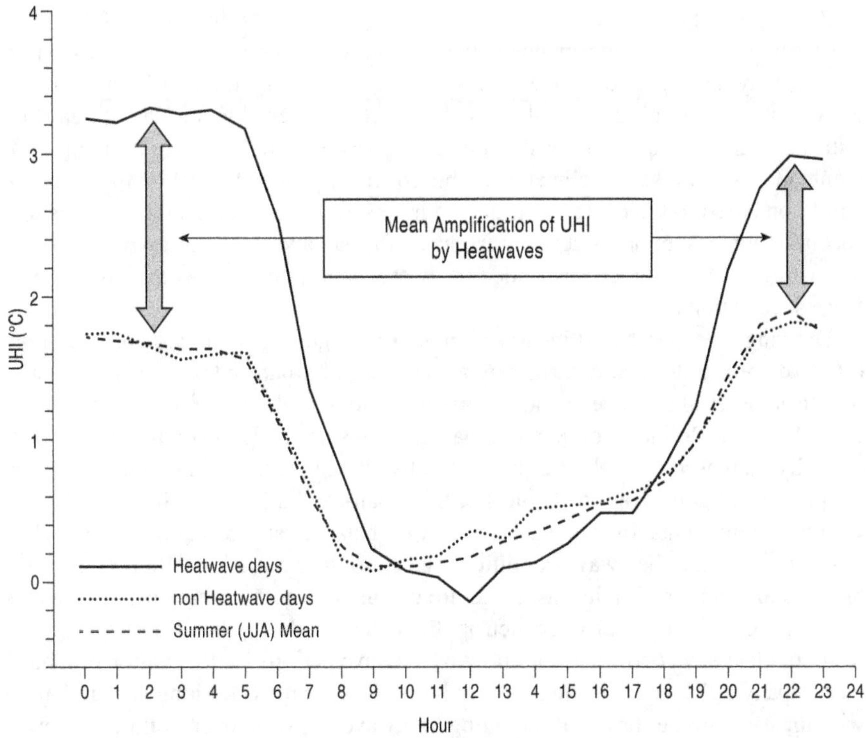

Fig. 8.15 Comparison of diurnal pattern of UHI intensity (vertical axis) for heatwave and selection of non-heatwaves and all summer days for London, UK. (Source: Author and adapted from Durham University MSc Risk Dissertation 'The Interaction Between the Urban Heat Island and Heatwaves: A Study of London 2000–2019 by Emma Beck)

of nighttime only UHI intensity for the inland cities of Beijing and Guangzhou while for coastal Shanghai it is the daytime UHI which experiences intensification because of the absence of the normal cool summertime sea breeze under heatwave conditions. In relation to city size Ramamurthy and Bou-Zeid (2017) found heatwaves generally exacerbate UHI intensities in large but not small US cities but note the response of the UHI to heatwaves in cities such as New York, Washington and Baltimore to be far from consistent with some only displaying daytime sensitivity of UHI intensity to heatwave occurrence. Using a multivariate measure of surface urban heat island magnitude Ward et al. (2016) found northern European cities with cool climates possessing a greater coverage of urban green spaces demonstrated greater amplification of the surface UHI during heatwaves when compared with warmer southern European cities. At the intra-urban scale there also appears to be a range of responses of local UHI intensities to heatwave events (He et al., 2021; Ngarambe et al., 2020; Scott et al., 2017).

As is evident from the above survey of the literature, the findings on the interplay between heatwaves and the UHI are diverse. Setting that aside, the mounting evidence and high general agreement amongst an ever-increasing number of studies lends confidence to the idea that heatwaves may amplify UHI intensities. Of course, bolstering this level of confidence is the availability of a range of physically based reasons why heatwaves enhance UHI intensities to varying degrees. Chief amongst these is the partitioning of available energy between sensible and latent heat flux with many studies speculating that decreased water availability in urban environments makes more sensible heat available for raising canopy level temperatures in cities compared to their rural surrounds during heatwaves. Fortunately, studies that specifically measure or model the response of urban and rural energy balances and the nature of UBL processes under heatwave conditions lend some credence to these speculations.

Drawing on tower-based measurements of a range of climate variables Li et al. (2015) demonstrate clear contrasts in the response of urban and rural surface energy balances for five heatwave events comprising a total of 44 days for the summers of 2009–2010 for Beijing. Compared to background summer UHI conditions, the UHI intensity increased by 2–2.5 °C during the five heatwaves, reaching a maximum in the pre-dawn hours. Composite plots of the urban-rural differences in values for the various components of the surface energy balance are revealing (Fig. 8.16). Interestingly under heatwave conditions the shortwave radiation deficit evident for background summer conditions at the urban site decreases during heatwaves. This is due to less cloud cover over metropolitan Beijing during heatwave episodes, a result of clear sky conditions common to heatwave periods for this region of China. Accompanying changes in shortwave radiative flux are those for longwave radiation with the usual higher levels of incoming longwave radiation over Beijing increased during heatwaves, because of a much warmer urban compared to the rural

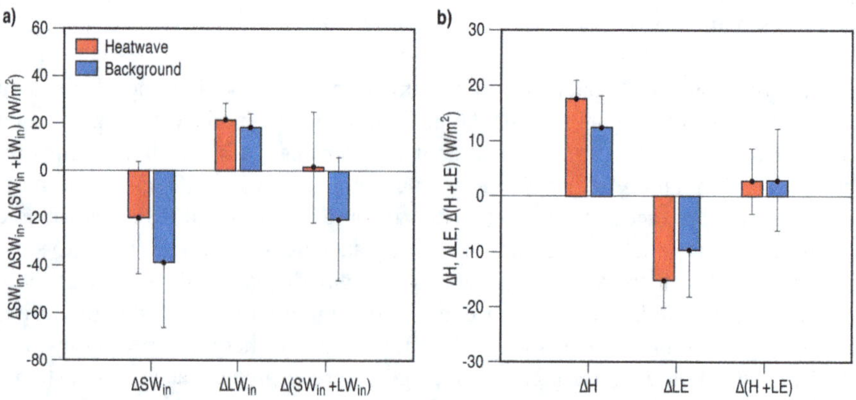

Fig. 8.16 Composite plots of the urban—rural differences in energy balance components for heatwave and background (non-heatwave) conditions for Beijing. SW and LW are shortwave and longwave radiation respectively. H and LE are sensible and latent heat respectively. Δ indicates change in associate flux (Source: Li et al. (2015). Contrasting responses of urban and rural surface energy budgets to heat waves explain synergies between urban heat islands and heat waves Environ. Res. Lett. 10 054009. https://doi.org/10.1088/1748-9326/10/5/054009. Used under the terms of the Creative Commons Attribution 3.0 Licence https://creativecommons.org/licenses/verby/3.0/)

atmosphere. Taken together urban—rural contrasts in the shortwave and longwave fluxes (K↓ and L↓) for non-heatwave and heatwave conditions reveal that the total radiative heat flux to the surface increases more at the urban than the rural site during heatwave periods—in fact a 'normal' urban deficit swings to a surplus under heatwave conditions. Just as critical, if not more so, are changes in urban—rural differences of sensible and latent heat flux. For the case of heatwaves, the sensible heat flux (Q_H) increases at the urban site. While this is partly associated with an increase in incoming radiative flux (K↓ + L↓) at the urban site, it is also driven by less of the available energy flowing into latent heat flux (Q_E) at the urban compared to the rural site under heatwave conditions—the summer background Q_E deficit is exacerbated under heatwave conditions. It is this combination of increased (decreased) Q_H (Q_E) at the urban compared to the rural site that underlies the intensification of the UHI under heatwave conditions and is thus a manifestation of the synergies of heatwaves and the UHI (Li et al., 2015). Lending strength to this contention is the negligible change in the urban—rural differences of the total energy related to Q_H and Q_E for heatwave and non-heatwave conditions. Note in Fig. 8.16b how the change in the difference is small compared to that of Q_E alone, suggesting it is alterations in the relative apportionment of energy between Q_H and Q_E that is important as opposed to changes in the total energy (Li et al., 2015).

While single city studies such as that just described for Beijing indicate the clear role of Q* apportionment between Q_H and Q_E, multi-city studies point to a possible urban background climate conditioning of the relative importance of Q_H and Q_E in UHI intensification. Using the Bowen Ratio as a descriptor of background climate, Ramamurthy and Bou-Zeid (2017) show for Washington and Baltimore a doubling of this metric and a significant increase in UHI intensity during heatwave compared to non-heatwave days. This contrasts with the larger cities of New York and Philadelphia where Bowen Ratio values and the UHI intensity only increase marginally during heatwave periods. The dissimilarity of response is attributed to the climatologically dry urban environments and thus already high Bowen Ratios of New York and Philadelphia such that the relative change in the flow of Q* into Q_H is far greater for the lower Bowen Ratio (moister background climate) cities of Washington and Baltimore where the impacts of heatwave related drying on the urban environment are significant as manifest by marked UHI amplification. Rapid swings from relatively benign to harsh (high) Bowen ratio values are most likely driven by changes in UBL structure. These include a warmer, deeper, and drier urban boundary layer aided by positive feedback processes, enhanced convection and an excessively warm nocturnal residual layer during heatwave compared to non-heatwave periods (An et al., 2020; Ramamurthy et al., 2017; Wang & Li, 2019; Zhang et al., 2020). This cocktail of processes promotes surface desiccation, the preferential flow of Q* into Q_H and UHI intensification via sensible heat enrichment of the UBL and UCL.

Sticking with the concept of the urban energy balance as a means for understanding heatwave UHI interactions, the possible roles of heat storage and anthropogenic heat flux in exacerbating UHI intensity during heatwaves is less settled than the roles of Q_H and Q_E. While partly related to the paucity of studies on urban—rural energy balance responses to heatwaves, the difficulties associated with estimating and more so measuring these fluxes accounts for the open door about their role. Turning attention back to the Beijing study mentioned above, Li et al. (2015) suggest for the heatwave events they considered, ΔQS does play a role in enhancing UHI intensity. They deduce this from the observation that while the urban—rural contrast in available energy ($Q_H + Q_E$) remains about the same, regardless of heatwave conditions, the urban incoming radiative flux ($K\downarrow + L\downarrow$) increases above background conditions during heatwaves. Hence additional radiative flux is available for absorption by urban materials with this released at night adding to the amplification of UHI intensity. This mechanism has been confirmed by numerical modeling studies for a number of large cities with a sizeable thermal mass such as London (Sun et al., 2017), Melbourne (Imran et al., 2019), and New York (Ortiz et al., 2018). Conversely for Singapore, a city with a large thermal mass, no changes in ΔQS for a single heatwave event in 2016 have been reported, a result explained by the observed increase in Q^* under heatwave conditions being partitioned to Q_H and Q_E without altering ΔQS (Chew et al., 2021).

Anthropogenic heat flux represents a possible elephant in the room in relation to the interplay between heatwaves and urban heat islands. This is because while Q_F is posited as a driver of UHI amplication, the response of this component of the urban energy balance under heatwave conditions remains under-researched. Like heat storage, Q_F is estimated from secondary data which adds to the challenge of tying down its importance in urban climate processes. Allied to the question of whether Q_F enhances the UHI, is do positive feedback processes transition this flux from 'dupe' to 'provocateur' status as a heatwave develops. As described in Sect. 8.4.2 background levels of Q_F, due to heat emissions from buildings, transport and human metabolism, can form a noticeable supplement to the net energy available in urban environments (Jin, Wang, & Wang, 2020a). Moreover at the diurnal timescale, in and around the hours of maximum temperature, Q_F values may compete with, or even exceed, those of the other energy balance fluxes, especially in high density urban settings (Cui & Chui, 2021; Dong et al., 2017; Ichinose et al., 1999; Wang et al., 2020). That there is a close association between human activity and Q_F has been aptly demonstrated by a downturn in Q_F for some cities in Japan during the COVID-19 pandemic lockdown (Nakajima et al., 2021).

There is growing evidence that Q_F contributes to the formation of the UHI through adding heat directly to the UCL and UBL (Chapman et al., 2016; Mei & Yuan, 2021; Oke et al., 2017). This effect is obvious in the cool season for mid- and high-latitude cities when Q_F may contribute substantially to the formation of the winter UHI (Varentsov, Konstantinov, et al., 2018a; Ma et al., 2017) or dampen the urban cool island effect found in some cities (Yang et al., 2017). Further, as cities have grown, the importance of Q_F as a contributor to the UHI effect has increased (Firozjaei et al., 2020) as has its influence on more extreme heat events in cities (He et al., 2020; Liu et al., 2021). Its proportion of the global net surface solar radiation (Jin, Wang, & Wang, 2020a) and thus its potential to add to the temperature forcing due to Greenhouse Gases has also grown (Lu et al., 2017). At the urban scale, a heuristic derived from modelling studies points to a 0.8–1.0 Kelvin increase in temperature per 100 W m^{-2} of Q_F input into the urban atmosphere (Kikegawa et al., 2014; Oke et al., 2017). This direct forcing effect on air temperature is central to the idea that heatwave driven increases in the use of air conditioning (AC), to maintain indoor thermal comfort, will drive up Q_F fluxes from buildings which will exacerbate already high outdoor temperatures. This in turn will elevate AC use leading to further Q_F related increases in outdoor air temperature and so on thus creating a heatwave—AC feedback on air temperature (Fig. 8.17).

While conceptually appealing, the plausibility of such a feedback mechanism rests on evidence for enhanced Q_F during heatwaves. Encouragingly, studies addressing heatwave driven Q_F trends are beginning to emerge in the literature. These build on the considerable evidence for clear increases in energy consumption related to elevated levels of cooling demand during periods of hot weather or those projected under climate change (e.g., Bird et al. (2019), Ihara et al. (2008), Lipson et al. (2019), Xu et al. (2020)). For example, Zheng and Weng (2018) revealed higher building related Q_F on hot compared to non-hot summer workdays in the city of Los Angeles. For the same city, Luo et al. (2020) have shown a clear surge in city-scale Q_F for a five-day heatwave event in 2009, with building heat emissions increasing by up to 20%. Importantly, and in relation to the feedback mechanism presented in Fig. 8.17, their study revealed heat dispersion from air-conditioning made up more than 85% of building waste heat. Further, they found significant

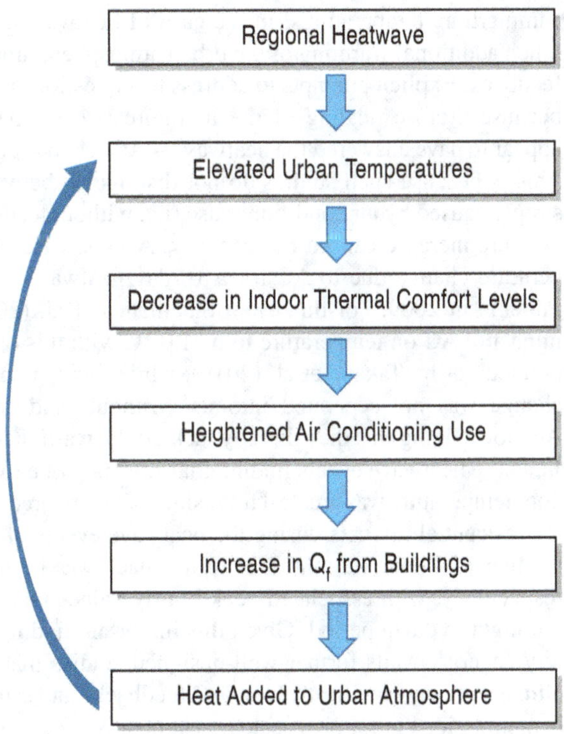

Fig. 8.17 Illustration of possible heatwave—AC feedback mechanism on air temperature via increases in heat emissions (Q_F) from AC. (Source: Author)

spatial variations in Q_F associated with AC during the heatwave event with dense urban districts located away from the coast more Q_F-sensitive than their coastal or suburban counterparts.

The extent to which AC related Q_F leads to elevated outdoor temperatures has been revealed by a number of studies. For example in the city of Wuhan, domestic air conditioners have been shown to increase the mean air temperature by 0.2 °C and 2.56 °C under neutral and inversion conditions respectively (Wen & Lian, 2009). For Paris, the UCL temperature increase due to current waste heat from AC systems has been estimated to be 0.5 °C with this reaching 2 °C if this type of Q_F was to double (de Munck et al., 2013), while for Berlin, the impact of AC is estimated to be 0.6 °C (Jin, Schubert, et al., 2020b). For Tokyo and Osaka, modelling studies have shown that waste heat from air conditioners can raise summertime air temperatures by 1.0–2.0 °C and 0.36–0.72 °C respectively (Ohashi et al., 2016; Ohashi et al., 2007). Interestingly, in an attempt to demonstrate the affect of adopting a territory wide alternative approach to building cooling, Wang et al. (2018) assessed the impact of water cooled as opposed to air cooled AC systems on urban temperatures in Hong Kong. They found for a record breaking heatwave event in 2016 that water-cooled air-conditioning systems could reduce UCL air temperatures by around 0.5–0.8 °C during the daytime, and around 1.5 °C in the early evening. This is quite the opposite effect of air-cooled AC systems because of lower Q_H fluxes to the atmosphere from water-cooled compared to air-cooled AC systems.

While the above studies demonstrate a discernible impact of AC related Q_F on summertime temperatures, in the case of heatwaves, the burning question is how much additional warming of the urban atmosphere might arise from heatwave—AC feedback. Explicit attempts to address this question are few and far between. This is because a close reading of the literature reveals that while some studies might appear to have answered the heatwave—AC feedback question they in fact have not. This is because such studies do not distinguish between the temperature increase simply caused by air conditioner use (i.e. without feedback) and the additional temperature increase caused by feedback. A good case in point is the estimated temperature change due to AC for a 10-day heatwave event that affected the city of Phoenix in 2009. For this event Salamanca et al. (2014) highlight the maximum impact of AC on temperature to be 1.0 °C with this obtained at night. However, as pointed out by Takane et al. (2019), while this is impressive the total temperature change was not seperated into its 'without' and 'with' feedback components. Acknowledging this they directly tackled the warming—AC feedback issue for two Japanese heatwave events finding that warming due to AC related feedback on outdoor temperature was greatest for residential compared to office districts. Specifically for residential districts during the heatwave events of August 2013 and July 2018, the mean 24 hour warming due to feedback was found to be 0.07 °C and 0.11 °C respectively, with associated peak hourly values reaching 0.15 and 0.25 °C in the midnight to dawn period. Given this important finding, the climatological community eagerly awaits further well-designed studies that explicitly address the urban climate outcomes of heatwave—AC feedbacks and establishing whether AC related heat emissions are a provocateur, via positive feedback on air conditioner use, of additional warming during heatwaves and thus the amplification of UHI intensity. The gravity of this research challenge is brought into sharp focus given the coalescence of three things by 2050; 68% of the worlds populaltion will be urban, the number of air conditioning units is expected to increase to 5.6 billion and for many parts of the globe the present 20-year return period heatwave will occur at least once in five years.

8.7 Conclusion

This chapter has explored the relationship between the urban surface and the overlying atmosphere, in the context of heatwave development. Urban areas are of particular interest in the case of heatwaves because of their dense populations and unique climates which makes them a focal point for heat exposure. The complexity of urban surfaces, including fabric, structure, and land cover, alters heat, moisture, and momentum fluxes, contributing to the development of the urban heat island (UHI). During heatwaves, the apportionment of net radiation between sensible, latent and subsurface heat flux changes, with sensible heat flux often dominating due to reduced surface moisture availability. Anthropogenic heat has the potential to be a significant heat source and thus contribute to the level of urban heat. While debate persists on the synergistic effects of heatwaves and UHIs, evidence suggests heatwaves enhance UHI intensity by altering energy partitioning. That said a careful reading of the UHI—heatwave interaction literature is necessary given conclusions rest on how the UHI has been defined. Though statements abound regarding air conditioning's role in exacerbating the UHI during heatwaves, empirical evidence remains limited. This is because most work directed at this problem does not distinguish between the temperature increase simply caused by air conditioner use (i.e. without feedback) and the additional temperature increase caused by feedback.

References

Adkins, K., Wambolt, P., Sescu, A., Swinford, C., & Macchiarella, N. D. (2020). Observational practices for urban microclimates using meteorologically instrumented unmanned aircraft systems. *Atmosphere, 11*(9). https://doi.org/10.3390/atmos11091008

Ado, H. Y. (1992). Numerical study of the daytime urban effect and its interaction with the sea breeze. *Journal of Applied Meteorology and Climatology, 31*(10), 1146–1164.

Allen, L., Lindberg, F., & Grimmond, C. S. B. (2011). Global to city scale urban anthropogenic heat flux: Model and variability. *International Journal of Climatology, 31*(13), 1990–2005. https://doi.org/10.1002/joc.2210

An, N., Dou, J. J., Gonzalez-Cruz, J. E., Bornstein, R. D., Miao, S. G., & Li, L. (2020). An observational case study of synergies between an intense heat wave and the urban Heat Island in Beijing. *Journal of Applied Meteorology and Climatology, 59*(4), 605–620. https://doi.org/10.1175/jamc-d-19-0125.1

Arnfield, A. J. (2003). Two decades of urban climate research: A review of turbulence, exchanges of energy and water, and the urban heat Island. *International Journal of Climatology, 23*(1), 1–26. https://doi.org/10.1002/joc.859

Barlow, J. F. (2014). Progress in observing and modelling the urban boundary layer. *Urban Climate, 10*, 216–240. https://doi.org/10.1016/j.uclim.2014.03.011

Barlow, J. F., Halios, C. H., Lane, S. E., & Wood, C. R. (2015). Observations of urban boundary layer structure during a strong urban heat Island event. *Environmental Fluid Mechanics, 15*(2), 373–398. https://doi.org/10.1007/s10652-014-9335-6

Basara, J. B., Basara, H. G., Illston, B. G., & Crawford, K. C. (2010). The impact of the urban heat Island during an intense heat wave in Oklahoma City. *Advances in Meteorology, 2010*. https://doi.org/10.1155/2010/230365

Bassett, R., Cai, X. M., Chapman, L., Heaviside, C., & Thornes, J. E. (2019). Semi-idealized urban heat advection simulations using the weather research and forecasting mesoscale model. *International Journal of Climatology, 39*(3), 1345–1358. https://doi.org/10.1002/joc.5885

Bauer, T. J. (2020). Interaction of urban heat Island effects and land-sea breezes during a New York City heat event. *Journal of Applied Meteorology and Climatology, 59*(3), 477–495. https://doi.org/10.1175/jamc-d-19-0061.1

Bayer, P., Attard, G., Blum, P., & Menberg, K. (2019). The geothermal potential of cities. *Renewable & Sustainable Energy Reviews, 106*, 17–30. https://doi.org/10.1016/j.rser.2019.02.019

Benz, S. A., Bayer, P., Blum, P., Hamamoto, H., Arimoto, H., & Taniguchi, M. (2018). Comparing anthropogenic heat input and heat accumulation in the subsurface of Osaka, Japan. *Science of the Total Environment, 643*, 1127–1136. https://doi.org/10.1016/j.scitotenv.2018.06.253

Bird, D. N., de Wit, R., Schwaiger, H. P., Andre, K., Beermann, M., & Zuvela-Aloise, M. (2019). Estimating the daily peak and annual total electricity demand for cooling in Vienna, Austria by 2050. *Urban Climate, 28*. https://doi.org/10.1016/j.uclim.2019.100452

Capel-Timms, I., Smith, S. T., Sun, T., & Grimmond, S. (2020). Dynamic anthropogenic activities impacting heat emissions (DASH v1.0): Development and evaluation. *Geoscientific Model Development, 13*(10), 4891–4924. https://doi.org/10.5194/gmd-13-4891-2020

Chapman, L., Muller, C. L., Young, D. T., Warren, E. L., Grimmond, C. S. B., Cai, X.-M., & Ferranti, E. J. (2015). The Birmingham urban climate laboratory: An open meteorological test bed and challenges of the smart city. *Bulletin of the American Meteorological Society, 96*(9), 1545–1560.

Chapman, S., Watson, J. E. M., & McAlpine, C. A. (2016). Large seasonal and diurnal anthropogenic heat flux across four Australian cities. *Journal of Southern Hemisphere Earth Systems Science, 66*(3), 342–360.

Chapman, L., Bell, C., & Bell, S. (2017). Can the crowdsourcing data paradigm take atmospheric science to a new level? A case study of the urban heat Island of London quantified using Netatmo weather stations. *International Journal of Climatology, 37*(9), 3597–3605.

Chemel, C., & Sokhi, R. S. (2012). Response of London's urban heat Island to a marine air intrusion in an easterly wind regime. *Boundary-Layer Meteorology, 144*(1), 65–81. https://doi.org/10.1007/s10546-012-9705-x

Chen, L., Ng, E., An, X., Ren, C., Lee, M., Wang, U., & He, Z. (2012). Sky view factor analysis of street canyons and its implications for daytime intra-urban air temperature differentials in high-rise, high-density urban areas of Hong Kong: A GIS-based simulation approach. *International Journal of Climatology, 32*(1), 121–136.

Cheval, S., Amihăesei, V., Chitu, Z., Dumitrescu, A., Falcescu, V., Irasoc, A., Micu, D., Mihulet, E., Ontel, I., & Paraschiv, M. G. (2024). A systematic review of urban heat Island and heat waves research (1991–2022). *Climate Risk Management*, 100603.

Chew, L. W., Liu, X., Li, X. X., & Norford, L. K. (2021). Interaction between heat wave and urban heat Island: A case study in a tropical coastal city, Singapore. *Atmospheric Research, 247*. https://doi.org/10.1016/j.atmosres.2020.105134

Chrysoulakis, N., Grimmond, S., Feigenwinter, C., Lindberg, F., Gastellu-Etchegorry, J.-P., Marconcini, M., Mitraka, Z., Stagakis, S., Crawford, B., Olofson, F., Landier, L., Morrison,

W., & Parlow, E. (2018). Urban energy exchanges monitoring from space. *Scientific Reports, 8*(1), 11498. https://doi.org/10.1038/s41598-018-29873-x

Coutts, A. M., Beringer, J., & Tapper, N. J. (2007). Impact of increasing urban density on local climate: Spatial and temporal variations in the surface energy balance in Melbourne, Australia. *Journal of Applied Meteorology and Climatology, 46*(4), 477–493. https://doi.org/10.1175/jam2462.1

Cui, W. H., & Chui, T. F. M. (2021). Measurements and simulations of energy fluxes over a high-rise and compact urban area in Hong Kong. *Science of the Total Environment, 765*. https://doi.org/10.1016/j.scitotenv.2020.142718

de Munck, C., Pigeon, G., Masson, V., Meunier, F., Bousquet, P., Tremeac, B., Merchat, M., Poeuf, P., & Marchadier, C. (2013). How much can air conditioning increase air temperatures for a city like Paris, France? *International Journal of Climatology, 33*(1), 210–227. https://doi.org/10.1002/joc.3415

De Ridder, K., Maiheu, B., Lauwaet, D., Daglis, I. A., Keramitsoglou, I., Kourtidis, K., Manunta, P., & Paganini, M. (2017). Urban heat Island intensification during hot spells-the case of Paris during the summer of 2003. *Urban Science, 1*(1). https://doi.org/10.3390/urbansci1010003

Demuzere, M., Bechtel, B., & Mills, G. (2019). Global transferability of local climate zone models. *Urban Climate, 27*, 46–63. https://doi.org/10.1016/j.uclim.2018.11.001

Dong, Y., Varquez, A. C. G., & Kanda, M. (2017). Global anthropogenic heat flux database with high spatial resolution. *Atmospheric Environment, 150*, 276–294. https://doi.org/10.1016/j.atmosenv.2016.11.040

Elhacham, E., Ben-Uri, L., Grozovski, J., Bar-On, Y. M., & Milo, R. (2020). Global human-made mass exceeds all living biomass. *Nature*. https://doi.org/10.1038/s41586-020-3010-5

Erell, E., & Williamson, T. (2007). Intra-urban differences in canopy layer air temperature at a mid-latitude city. *International Journal of Climatology, 27*(9), 1243–1255. https://doi.org/10.1002/joc.1469

Fan, Y. F., Li, Y. G., Bejan, A., Wang, Y., & Yang, X. Y. (2017). Horizontal extent of the urban heat dome flow. *Scientific Reports, 7*. https://doi.org/10.1038/s41598-017-09917-4

Fenner, D., Holtmann, A., Meier, F., Langer, I., & Scherer, D. (2019). Contrasting changes of urban heat Island intensity during hot weather episodes. *Environmental Research Letters, 14*(12). https://doi.org/10.1088/1748-9326/ab506b

Firozjaei, M. K., Weng, Q. H., Zhao, C. H., Kiavarz, M., Lu, L. L., & Alavipanah, S. K. (2020). Surface anthropogenic heat islands in six megacities: An assessment based on a triple-source surface energy balance model. *Remote Sensing of Environment, 242*. https://doi.org/10.1016/j.rse.2020.111751

Founda, D., & Santamouris, M. (2017). Synergies between urban heat Island and heat waves in Athens (Greece), during an extremely hot summer (2012). *Scientific Reports, 7*. https://doi.org/10.1038/s41598-017-11407-6

Gabey, A. M., Grimmond, C. S. B., & Capel-Timms, I. (2019). Anthropogenic heat flux: Advisable spatial resolutions when input data are scarce. *Theoretical and Applied Climatology, 135*(1–2), 791–807. https://doi.org/10.1007/s00704-018-2367-y

Grimmond, C. S. B., & Oke, T. R. (1999). Heat storage in urban areas: Local-scale observations and evaluation of a simple model. *Journal of Applied Meteorology, 38*(7), 922–940. https://doi.org/10.1175/1520-0450(1999)038<0922:hsiual>2.0.co;2

Guneralp, B., Reba, M., Hales, B. U., Wentz, E. A., & Seto, K. C. (2020). Trends in urban land expansion, density, and land transitions from 1970 to 2010: A global synthesis. *Environmental Research Letters, 15*(4). https://doi.org/10.1088/1748-9326/ab6669

He, C., Zhou, L. G., Yao, Y. R., Ma, W. C., & Kinney, P. L. (2020). Estimating spatial effects of anthropogenic heat emissions upon the urban thermal environment in an urban agglomeration area in East China. *Sustainable Cities and Society, 57*. https://doi.org/10.1016/j.scs.2020.102046

He, B. J., Wang, J. S., Liu, H. M., & Ulpiani, G. (2021). Localized synergies between heat waves and urban heat islands: Implications on human thermal comfort and urban heat management. *Environmental Research, 193*. https://doi.org/10.1016/j.envres.2020.110584

Holmer, B., & Eliasson, I. (1999). Urban-rural vapour pressure differences and their role in the development of urban heat islands. *International Journal of Climatology, 19*(9), 989–1009. https://doi.org/10.1002/(sici)1097-0088(199907)19:9<989::aid-joc410>3.3.co;2-t

Howard, E. (2003). Garden cities of tomorrow. *Organization & Environment, 16*(1), 98–107. https://doi.org/10.1177/1086026602250259

Hrisko, J., Ramamurthy, P., & Gonzalez, J. E. (2021a). Estimating heat storage in urban areas using multispectral satellite data and machine learning. *Remote Sensing of Environment, 252*. https://doi.org/10.1016/j.rse.2020.112125

Hrisko, J., Ramamurthy, P., Melecio-Vazquez, D., & Gonzalez, J. E. (2021b). Spatiotemporal variability of heat storage in major US Cities-A satellite-based analysis. *Remote Sensing, 13*(1). https://doi.org/10.3390/rs13010059

Huang, F., Zhan, W., Wang, Z.-H., Voogt, J., Hu, L., Quan, J., Liu, C., Zhang, N., & Lai, J. (2020). Satellite identification of atmospheric-surface-subsurface urban heat islands under clear sky. *Remote Sensing of Environment, 250*, 112039.

Ichinose, T., Shimodozono, K., & Hanaki, K. (1999). Impact of anthropogenic heat on urban climate in Tokyo. *Atmospheric Environment, 33*(24–25), 3897–3909. https://doi.org/10.1016/s1352-2310(99)00132-6

Ihara, T., Kikegawa, Y., Asahi, K., Genchi, Y., & Kondo, H. (2008). Changes in year-round air temperature and annual energy consumption in office building areas by urban heat-Island countermeasures and energy-saving measures. *Applied Energy, 85*(1), 12–25.

Imran, H. M., Kala, J., Ng, A. W. M., & Muthukumaran, S. (2019). Impacts of future urban expansion on urban heat Island effects during heatwave events in the city of Melbourne in Southeast Australia. *Quarterly Journal of the Royal Meteorological Society, 145*(723), 2586–2602. https://doi.org/10.1002/qj.3580

Jiang, S. J., Lee, X., Wang, J. K., & Wang, K. C. (2019). Amplified urban heat islands during heat wave periods. *Journal of Geophysical Research-Atmospheres, 124*(14), 7797–7812. https://doi.org/10.1029/2018jd030230

Jin, K., Wang, F., Chen, D. L., Liu, H. H., Ding, W. B., & Shi, S. Y. (2019). A new global gridded anthropogenic heat flux dataset with high spatial resolution and long-term time series. *Scientific Data, 6*. https://doi.org/10.1038/s41597-019-0143-1

Jin, K., Wang, F., & Wang, S. X. (2020a). Assessing the spatiotemporal variation in anthropogenic heat and its impact on the surface thermal environment over global land areas. *Sustainable Cities and Society, 63*. https://doi.org/10.1016/j.scs.2020.102488

Jin, L., Schubert, S., Hefny Salim, M., & Schneider, C. (2020b). Impact of air conditioning systems on the outdoor thermal environment during summer in Berlin, Germany. *International Journal of Environmental Research and Public Health, 17*(13), 4645.

Khan, H. S., Paolini, R., Santamouris, M., & Caccetta, P. (2020). Exploring the synergies between urban overheating and heatwaves (HWs) in western Sydney. *Energies, 13*(2), 470.

Khan, H. S., Santamouris, M., Paolini, R., Caccetta, P., & Kassomenos, P. (2021). Analyzing the local and climatic conditions affecting the urban overheating magnitude during the heatwaves (HWs) in a coastal city: A case study of the greater Sydney region. *Science of the Total Environment, 755*. https://doi.org/10.1016/j.scitotenv.2020.142515

Kikegawa, Y., Tanaka, A., Ohashi, Y., Ihara, T., & Shigeta, Y. (2014). Observed and simulated sensitivities of summertime urban surface air temperatures to anthropogenic heat in downtown areas of two Japanese major cities, Tokyo and Osaka. *Theoretical and Applied Climatology, 117*(1–2), 175–193. https://doi.org/10.1007/s00704-013-0996-8

Kim, Y. H., & Baik, J. J. (2005). Spatial and temporal structure of the urban heat Island in Seoul. *Journal of Applied Meteorology, 44*(5), 591–605. https://doi.org/10.1175/jam2226.1

Kumar, R., & Mishra, V. (2019). Decline in surface urban heat Island intensity in India during heatwaves. Environmental research. *Communications, 1*(3). https://doi.org/10.1088/2515-7620/ab121d

Kusaka, H., Kimura, F., Hirakuchi, H., & Mizutori, M. (2000). The effects of land-use alteration on the sea breeze and daytime heat Island in the Tokyo metropolitan area. *Journal of the Meteorological Society of Japan, 78*(4), 405–420. https://doi.org/10.2151/jmsj1965.78.4_405

Kuttler, W., Weber, S., Schonnefeld, J., & Hesselschwerdt, A. (2007). Urban/rural atmospheric water vapour pressure differences and urban moisture excess in Krefeld, Germany. *International Journal of Climatology, 27*(14), 2005–2015. https://doi.org/10.1002/joc.1558

Landsberg, H. (1973). Meteorologically utopian city. *Bulletin of the American Meteorological Society, 54*(2), 86–89. https://doi.org/10.1175/1520-0477(1973)054<0086:tmuc>2.0.co;2

Lee, D. O. (1991). Urban—Rural humidity differences in London. *International Journal of Climatology, 11*(5), 577–582.

Lehnert, M., Kubecek, J., Geletic, J., Jurek, M., & Frajer, J. (2018). Identifying hot and cool spots in the city Centre based on bicycle measurements: The case of Olomouc, Czech Republic. *Geographica Pannonica, 22*(4), 230–240. https://doi.org/10.5937/gp22-19750

Lemoine-Rodriguez, R., Inostroza, L., & Zepp, H. (2020). The global homogenization of urban form. An assessment of 194 cities across time. *Landscape and Urban Planning, 204*. https://doi.org/10.1016/j.landurbplan.2020.103949

Lemonsu, A., Grimmond, C. S. B., & Masson, V. (2004). Modeling the surface energy balance of the core of an old Mediterranean city: Marseille. *Journal of Applied Meteorology, 43*(2), 312–327. https://doi.org/10.1175/1520-0450(2004)043<0312:mtsebo>2.0.co;2

Li, D., & Bou-Zeid, E. (2013). Synergistic interactions between urban heat islands and heat waves: The impact in cities is larger than the sum of its parts. *Journal of Applied Meteorology and Climatology, 52*(9), 2051–2064.

Li, D., Sun, T., Liu, M. F., Yang, L., Wang, L. L., & Gao, Z. Q. (2015). Contrasting responses of urban and rural surface energy budgets to heat waves explain synergies between urban heat islands and heat waves. *Environmental Research Letters, 10*(5). https://doi.org/10.1088/1748-9326/10/5/054009

Li, Y. F., Schubert, S., Kropp, J. P., & Rybski, D. (2020). On the influence of density and morphology on the urban Heat Island intensity. *Nature Communications, 11*(1). https://doi.org/10.1038/s41467-020-16461-9

Liang, Z., Wang, Y., Huang, J., Wei, F., Wu, S., Shen, J., Sun, F., & Li, S. (2020). Seasonal and diurnal variations in the relationships between urban form and the urban heat Island effect. *Energies, 13*(22), 5909.

Lin, L. J., Chan, T. O., Ge, E. J., Wang, X. W., Zhao, Y. Q., Yang, Y. J., Ning, G. C., Zeng, Z. L., & Luo, M. (2020). Effects of urban land expansion on decreasing atmospheric moisture in Guangdong, South China. *Urban Climate, 32*. https://doi.org/10.1016/j.uclim.2020.100626

Lindberg, F., Grimmond, C. S. B., Yogeswaran, N., Kotthaus, S., & Allen, L. (2013). Impact of city changes and weather on anthropogenic heat flux in Europe 1995–2015. *Urban Climate, 4*, 1–15.

Lindberg, F., Olofson, K. F. G., Sun, T., Grimmond, C. S. B., & Feigenwinter, C. (2020). Urban storage heat flux variability explored using satellite, meteorological and geodata. *Theoretical and Applied Climatology, 141*(1–2), 271–284. https://doi.org/10.1007/s00704-020-03189-1

Lipson, M. J., Thatcher, M., Hart, M. A., & Pitman, A. (2019). Climate change impact on energy demand in building-urban-atmosphere simulations through the 21st century. *Environmental Research Letters, 14*(12). https://doi.org/10.1088/1748-9326/ab5aa5

Liu, B., Xie, Z., Qin, P., Liu, S., Li, R., Wang, L., Wang, Y., Jia, B., Chen, S., & Xie, J. (2021). Increases in anthropogenic heat release from energy consumption lead to more frequent extreme heat events in urban cities. *Advances in Atmospheric Sciences, 38*, 430–445.

Lokoshchenko, M. A. (2017). Urban heat Island and urban dry Island in Moscow and their centennial changes. *Journal of Applied Meteorology and Climatology, 56*(10), 2729–2745. https://doi.org/10.1175/jamc-d-16-0383.1

Loridan, T., & Grimmond, C. S. B. (2012). Characterization of energy flux partitioning in urban environments: Links with surface seasonal properties. *Journal of Applied Meteorology and Climatology, 51*(2), 219–241. https://doi.org/10.1175/jamc-d-11-038.1

Lu, Y., Wang, H., Qg, W., Zhang, Y., Yu, Y., & Qian, Y. (2017). Global anthropogenic heat emissions from energy consumption, 1965–2100. *Climatic Change, 145*, 459–468.

Luo, Z. W., & Asproudi, C. (2015). Subsurface urban heat Island and its effects on horizontal ground-source heat pump potential under climate change. *Applied Thermal Engineering, 90*, 530–537. https://doi.org/10.1016/j.applthermaleng.2015.07.025

Luo, M., & Lau, N. C. (2019). Urban expansion and drying climate in an urban agglomeration of East China. *Geophysical Research Letters, 46*(12), 6868–6877. https://doi.org/10.1029/2019gl082736

Luo, X., Vahmani, P., Hong, T., & Jones, A. (2020). City-scale building anthropogenic heating during heat waves. *Atmosphere, 11*(11), 1206.

Ma, S. X., Pitman, A., Hart, M., Evans, J. P., Haghdadi, N., & MacGill, I. (2017). The impact of an urban canopy and anthropogenic heat fluxes on Sydney's climate. *International Journal of Climatology, 37*, 255–270. https://doi.org/10.1002/joc.5001

Manoli, G., Fatichi, S., Bou-Zeid, E., & Katul, G. G. (2020). Seasonal hysteresis of surface urban heat islands. *Proceedings of the National Academy of Sciences of the United States of America, 117*(13), 7082–7089. https://doi.org/10.1073/pnas.1917554117

Masson, V., Grimmond, C. S. B., & Oke, T. R. (2002). Evaluation of the Town Energy Balance (TEB) scheme with direct measurements from dry districts in two cities. *Journal of Applied Meteorology, 41*(10), 1011–1026.

Masson, V., Bocher, E., Bucher, B., Chitu, Z., Christophe, S., Fortelius, C., Hamdi, R., Lemonsu, A., Perrels, A., Van Schaeybroeck, B., Schreur, B. W., Velea, L., Beddar, Y., Calvet, J. C., Delcloo, A., Druel, A., Duchene, F., Dumas, G., Dumitrescu, A., Gautier, J., Goret, M., Horttanainen, M., Kouznetsov, R., Le Bris, A., Lecorre, S., Le Roy, B., Palamarchuk, Y., Petit, G., Ruuhela, R., Saranko, O., Sofiev, M., Siljamo, P., Van de Vyverf, H., van Velthoven, P., & Votsis, A. (2020). The urban climate services URCLIM project. *Climate Services, 20*. https://doi.org/10.1016/j.cliser.2020.100194

Mei, S. J., & Yuan, C. (2021). Analytical and numerical study on transient urban street air warming induced by anthropogenic heat emission. *Energy and Buildings, 231*. https://doi.org/10.1016/j.enbuild.2020.110613

Meier, F., Fenner, D., Grassmann, T., Otto, M., & Scherer, D. (2017). Crowdsourcing air temperature from citizen weather stations for urban climate research. *Urban Climate, 19*, 170–191. https://doi.org/10.1016/j.uclim.2017.01.006

Meili, N., Paschalis, A., Manoli, G., & Fatichi, S. (2022). Diurnal and seasonal patterns of global urban dry islands. *Environmental Research Letters, 17*(5). https://doi.org/10.1088/1748-9326/ac68f8

Meir, T., Orton, P. M., Pullen, J., Holt, T., Thompson, W. T., & Arend, M. F. (2013). Forecasting the New York City urban heat Island and sea breeze during extreme heat events. *Weather and Forecasting, 28*(6), 1460–1477. https://doi.org/10.1175/waf-d-13-00012.1

Menberg, K., Bayer, P., Zosseder, K., Rumohr, S., & Blum, P. (2013). Subsurface urban heat islands in German cities. *Science of the Total Environment, 442*, 123–133. https://doi.org/10.1016/j.scitotenv.2012.10.043

Miao, S., Chen, F., LeMone, M. A., Tewari, M., Li, Q., & Wang, Y. (2009). An observational and modeling study of characteristics of urban heat Island and boundary layer structures in Beijing. *Journal of Applied Meteorology and Climatology, 48*(3), 484–501.

Milošević, D., Savić, S., Kresoja, M., Lužanin, Z., Šećerov, I., Arsenović, D., Dunjić, J., & Matzarakis, A. (2021). Analysis of air temperature dynamics in the "local climate zones" of Novi Sad (Serbia) based on long-term database from an urban meteorological network. *International Journal of Biometeorology*, 1–14.

Muller, C. L., Chapman, L., Grimmond, C. S. B., Young, D. T., & Cai, X. M. (2013). Sensors and the city: A review of urban meteorological networks. *International Journal of Climatology, 33*(7), 1585–1600. https://doi.org/10.1002/joc.3678

Muller, N., Kuttler, W., & Barlag, A. B. (2014). Analysis of the subsurface urban heat Island in Oberhausen, Germany. *Climate Research, 58*(3), 247–256. https://doi.org/10.3354/cr01195

Nakajima, K., Takane, Y., Kikegawa, Y., Furuta, Y., & Takamatsu, H. (2021). Human behaviour change and its impact on urban climate: Restrictions with the G20 Osaka summit and COVID-19 outbreak. *Urban Climate, 35*. https://doi.org/10.1016/j.uclim.2020.100728

Nakamura, Y., Shigeta, Y., & Watarai, Y. (2018). Seasonal variations of the urban heat Island in Kumagaya, Japan. *Geographical Review of Japan-Series B, 91*(2), 29–39. https://doi.org/10.4157/geogrevjapanb.91.29

Ng, E., Yuan, C., Chen, L., Ren, C., & Fung, J. C. H. (2011). Improving the wind environment in high-density cities by understanding urban morphology and surface roughness: A study in Hong Kong. *Landscape and Urban Planning, 101*(1), 59–74. https://doi.org/10.1016/j.landurbplan.2011.01.004

Ngarambe, J., Nganyiyimana, J., Kim, I., Santamouris, M., & Yun, G. Y. (2020). Synergies between urban heat Island and heat waves in Seoul: The role of wind speed and land use characteristics. *PLoS One, 15*(12). https://doi.org/10.1371/journal.pone.0243571

Offerle, B., Jonsson, P., Eliasson, I., & Grimmond, C. S. B. (2005). Urban modification of the surface energy balance in the West African Sahel: Ouagadougou, Burkina Faso. *Journal of Climate, 18*(19), 3983–3995. https://doi.org/10.1175/jcli3520.1

Ohashi, Y., Genchi, Y., Kondo, H., Kikegawa, Y., Yoshikado, H., & Hirano, Y. (2007). Influence of air-conditioning waste heat on air temperature in Tokyo during summer: Numerical experiments using an urban canopy model coupled with a building energy model. *Journal of Applied Meteorology and Climatology, 46*(1), 66–81.

Ohashi, Y., Suido, M., Kikegawa, Y., Ihara, T., Shigeta, Y., & Nabeshima, M. (2016). Impact of seasonal variations in weekday electricity use on urban air temperature observed in Osaka, Japan. *Quarterly Journal of the Royal Meteorological Society, 142*(695), 971–982. https://doi.org/10.1002/qj.2698

Oke, T. R. (1973). City size and the urban heat Island. *Atmospheric Environment, 7*(8), 769–779.

Oke, T. R. (1988). Street design and urban canopy layer climate. *Energy and Buildings, 11*(1–3), 103–113.

Oke, T. R., Mills, G., Christen, A., & Voogt, J. A. (2017). *Urban climates*. Cambridge University Press.

Oliphant, A. J., Stein, S., & Bradford, G. (2018). Micrometeorology of an ephemeral desert city, the burning man experiment. *Urban Climate, 23*, 53–70. https://doi.org/10.1016/j.uclim.2017.03.001

Oliveira, A., Lopes, A., Correia, E., Niza, S., & Soares, A. (2021). Heatwaves and summer urban heat islands: A daily cycle approach to unveil the urban thermal signal changes in Lisbon, Portugal. *Atmosphere, 12*(3), 292.

Ortiz, L. E., Gonzalez, J. E., Wu, W., Schoonen, M., Tongue, J., & Bornstein, R. (2018). New York City impacts on a regional heat wave. *Journal of Applied Meteorology and Climatology, 57*(4), 837–851.

Pal, S., Xueref-Remy, I., Ammoura, L., Chazette, P., Gibert, F., Royer, P., Dieudonne, E., Dupont, J. C., Haeffelin, M., Lac, C., Lopez, M., Morille, Y., & Ravetta, F. (2012). Spatio-temporal variability of the atmospheric boundary layer depth over the Paris agglomeration: An assessment of the impact of the urban heat Island intensity. *Atmospheric Environment, 63*, 261–275. https://doi.org/10.1016/j.atmosenv.2012.09.046

Pyrgou, A., Hadjinicolaou, P., & Santamouris, M. (2020). Urban-rural moisture contrast: Regulator of the urban heat Island and heatwaves' synergy over a mediterranean city. *Environmental Research, 182*. https://doi.org/10.1016/j.envres.2019.109102

Ramamurthy, P., & Bou-Zeid, E. (2017). Heatwaves and urban heat islands: A comparative analysis of multiple cities. *Journal of Geophysical Research-Atmospheres, 122*(1), 168–178. https://doi.org/10.1002/2016jd025357

Ramamurthy, P., Gonzalez, J., Ortiz, L., Arend, M., & Moshary, F. (2017). Impact of heatwave on a megacity: An observational analysis of New York City during July 2016. *Environmental Research Letters, 12*(5). https://doi.org/10.1088/1748-9326/aa6e59

Ribeiro, F. N. D., de Oliveira, A. P., Soares, J., de Miranda, R. M., Barlage, M., & Chen, F. (2018). Effect of sea breeze propagation on the urban boundary layer of the metropolitan region of Sao Paulo, Brazil. *Atmospheric Research, 214*, 174–188. https://doi.org/10.1016/j.atmosres.2018.07.015

Richard, Y., Pohl, B., Rega, M., Pergaud, J., Thevenin, T., Emery, J., Dudek, J., Vairet, T., Zito, S., & Chateau-Smith, C. (2021). Is urban Heat Island intensity higher during hot spells and heat waves (Dijon, France, 2014–2019)? *Urban Climate, 35*. https://doi.org/10.1016/j.uclim.2020.100747

Richards, K. (2005). Urban and rural dewfall, surface moisture, and associated canopy-level air temperature and humidity measurements for Vancouver, Canada. *Boundary-Layer Meteorology, 114*(1), 143–163. https://doi.org/10.1007/s10546-004-8947-7

Rizvi, S. H., Alam, K., & Iqbal, M. J. (2019). Spatio -temporal variations in urban heat Island and its interaction with heat wave. *Journal of Atmospheric and Solar-Terrestrial Physics, 185*, 50–57. https://doi.org/10.1016/j.jastp.2019.02.001

Roberts, S. M., Oke, T. R., Grimmond, C. S. B., & Voogt, J. A. (2006). Comparison of four methods to estimate urban heat storage. *Journal of Applied Meteorology and Climatology, 45*(12), 1766–1781. https://doi.org/10.1175/jam2432.1

Rogers, C. D. W., Gallant, A. J. E., & Tapper, N. J. (2019). Is the urban heat Island exacerbated during heatwaves in southern Australian cities? *Theoretical and Applied Climatology, 137*(1–2), 441–457. https://doi.org/10.1007/s00704-018-2599-x

Roth, M. (2007). Review of urban climate research in (sub)tropical regions. *International Journal of Climatology, 27*(14), 1859–1873. https://doi.org/10.1002/joc.1591

Sailor, D. J. (2011). A review of methods for estimating anthropogenic heat and moisture emissions in the urban environment. *International Journal of Climatology, 31*(2), 189–199. https://doi.org/10.1002/joc.2106

Sailor, D. J., Georgescu, M., Milne, J. M., & Hart, M. A. (2015). Development of a national anthropogenic heating database with an extrapolation for international cities. *Atmospheric Environment, 118*, 7–18. https://doi.org/10.1016/j.atmosenv.2015.07.016

Salamanca, F., Georgescu, M., Mahalov, A., Moustaoui, M., & Wang, M. (2014). Anthropogenic heating of the urban environment due to air conditioning. *Journal of Geophysical Research-Atmospheres, 119*(10), 5949–5965. https://doi.org/10.1002/2013jd021225

Santamouris, M. (2015). Analyzing the heat Island magnitude and characteristics in one hundred Asian and Australian cities and regions. *Science of the Total Environment, 512*, 582–598.

Schatz, J., & Kucharik, C. J. (2015). Urban climate effects on extreme temperatures in Madison, Wisconsin, USA. *Environmental Research Letters, 10*(9), 094024.

Schweighofer, J. A., Wehrl, M., Baumgärtel, S., & Rohn, J. (2021). Calculating energy and its spatial distribution for a subsurface urban heat Island using a GIS-approach. *Geosciences, 11*(1), 24.

Scott, A. A., Zaitchik, B., Waugh, D. W., & O'Meara, K. (2017). Intraurban temperature variability in Baltimore. *Journal of Applied Meteorology and Climatology, 56*(1), 159–171. https://doi.org/10.1175/jamc-d-16-0232.1

Scott, A. A., Waugh, D. W., & Zaitchik, B. (2018). Reduced urban heat Island intensity under warmer conditions. *Environmental Research Letters, 13*(6). https://doi.org/10.1088/1748-9326/aabd6c

Shen, L. X., Zhao, C. F., Ma, Z. S., Li, Z. Q., Li, J. P., & Wang, K. C. (2019). Observed decrease of summer sea-land breeze in Shanghai from 1994 to 2014 and its association with urbanization. *Atmospheric Research, 227*, 198–209. https://doi.org/10.1016/j.atmosres.2019.05.007

Shi, T., Liu, L., Wen, X., & Qi, P. (2024). Research progress on the synergies between heat waves and canopy urban heat Island and their driving factors. *Frontiers in Environmental Science, 12*, 1363837.

Stewart, I. D. (2011). A systematic review and scientific critique of methodology in modern urban heat Island literature. *International Journal of Climatology, 31*(2), 200–217. https://doi.org/10.1002/joc.2141

Stewart, I. D., & Kennedy, C. A. (2017). Metabolic heat production by human and animal populations in cities. *International Journal of Biometeorology, 61*(7), 1159–1171. https://doi.org/10.1007/s00484-016-1296-7

Stewart, I. D., & Oke, T. R. (2012). Local climate zones for urban temperature studies. *Bulletin of the American Meteorological Society, 93*(12), 1879–1900.

Stewart, I. D., Oke, T. R., & Krayenhoff, E. S. (2014). Evaluation of the 'local climate zone' scheme using temperature observations and model simulations. *International Journal of Climatology, 34*(4), 1062–1080. https://doi.org/10.1002/joc.3746

Sugawara, H., Narita, K., & Mikami, T. (2021). Vertical structure of the cool Island in a large urban park. *Urban Climate, 35*. https://doi.org/10.1016/j.uclim.2020.100744

Sun, T., Kotthaus, S., Li, D., Ward, H. C., Gao, Z. Q., Ni, G. H., & Grimmond, C. S. B. (2017). Attribution and mitigation of heat wave-induced urban heat storage change. *Environmental Research Letters, 12*(11). https://doi.org/10.1088/1748-9326/aa922a

Takane, Y., Kikegawa, Y., Hara, M., & Grimmond, C. S. B. (2019). Urban warming and future air-conditioning use in an Asian megacity: Importance of positive feedback. *npj Climate and Atmospheric Science, 2*. https://doi.org/10.1038/s41612-019-0096-2

Tejeda-Martinez, A., & Jauregui-Ostos, E. (2005). Surface energy balance measurements in the Mexico City region: A review. *Atmosfera, 18*(1), 1–23.

Tewari, M., Yang, J. C., Kusaka, H., Salamanca, F., Watson, C., & Treinish, L. (2019). Interaction of urban heat islands and heat waves under current and future climate conditions and their mitigation using green and cool roofs in New York City and Phoenix, Arizona. *Environmental Research Letters, 14*(3). https://doi.org/10.1088/1748-9326/aaf431

Thomas, R. M., MacKenzie, A. R., Reynolds, S. J., Sadler, J. P., Cropley, F., Bell, S., Dugdale, S. J., Chapman, L., Quinn, A., & Cai, X. (2018). Avian sensor packages for meteorological measurements. *Bulletin of the American Meteorological Society, 99*(3), 499–511.

Top, S., Milosevic, D., Caluwaerts, S., Hamdi, R., & Savic, S. (2020). Intra-urban differences of outdoor thermal comfort in Ghent on seasonal level and during record-breaking 2019 heat wave. *Building and Environment, 185*. https://doi.org/10.1016/j.buildenv.2020.107103

Unger, J., Skarbit, N., Kovacs, A., & Gal, T. (2020). Comparison of regional and urban outdoor thermal stress conditions in heatwave and normal summer periods: A case study. *Urban Climate, 32*. https://doi.org/10.1016/j.uclim.2020.100619

Varentsov, M., Konstantinov, P., Baklanov, A., Esau, I., Miles, V., & Davy, R. (2018a). Anthropogenic and natural drivers of a strong winter urban heat Island in a typical Arctic city. *Atmospheric Chemistry and Physics, 18*(23), 17573–17587. https://doi.org/10.5194/acp-18-17573-2018

Varentsov, M., Wouters, H., Platonov, V., & Konstantinov, P. (2018b). Megacity-induced mesoclimatic effects in the lower atmosphere: A modeling study for multiple summers over Moscow, Russia. *Atmosphere, 9*(2). https://doi.org/10.3390/atmos9020050

Varquez, A. C. G., Kiyomoto, S., Khanh, D. N., & Kanda, M. (2021). Global 1-km present and future hourly anthropogenic heat flux. *Scientific Data, 8*(1), 64.

Visser, P. W., Kooi, H., Bense, V., & Boerma, E. (2020). Impacts of progressive urban expansion on subsurface temperatures in the city of Amsterdam (The Netherlands). *Hydrogeology Journal, 28*(5), 1755–1772. https://doi.org/10.1007/s10040-020-02150-w

Wang, L., & Li, D. (2019). Modulation of the urban boundary-layer heat budget by a heatwave. *Quarterly Journal of the Royal Meteorological Society, 145*(722), 1814–1831. https://doi.org/10.1002/qj.3526

Wang, Y., Li, Y., Di Sabatino, S., Martilli, A., & Chan, P. W. (2018). Effects of anthropogenic heat due to air-conditioning systems on an extreme high temperature event in Hong Kong. *Environmental Research Letters, 13*, 3. https://doi.org/10.1088/1748-9326/aaa848

Wang, S. S., Hu, D. Y., Yu, C., Chen, S. S., & Di, Y. F. (2020). Mapping China's time -series anthropogenic heat flux with inventory method and multi -source remotely sensed data. *Science of the Total Environment, 734*. https://doi.org/10.1016/j.scitotenv.2020.139457

Wang, Z., Song, J., Chan, P. W., & Li, Y. (2021). The urban moisture Island phenomenon and its mechanisms in a high-rise high-density city. *International Journal of Climatology, 41*, E150–E170.

Ward, K., Lauf, S., Kleinschmit, B., & Endlicher, W. (2016). Heat waves and urban heat islands in Europe: A review of relevant drivers. *Science of the Total Environment, 569*, 527–539. https://doi.org/10.1016/j.scitotenv.2016.06.119

Wen, Y. G., & Lian, Z. W. (2009). Influence of air conditioners utilization on urban thermal environment. *Applied Thermal Engineering, 29*(4), 670–675. https://doi.org/10.1016/j.applthermaleng.2008.03.039

Xu, X., González, J. E., Miao, S., Shen, S., & Guo, W. (2020). District-level summertime air-conditioning electricity consumption and the sensitivity of peak cooling loads to urban weather conditions in Beijing. *Journal of Solar Energy Engineering, 142*(6), 061007.

Yamato, H., Mikami, T., & Takahashi, H. (2017). Impact of sea breeze penetration over urban areas on midsummer temperature distributions in the Tokyo Metropolitan area. *International Journal of Climatology, 37*(15), 5154–5169. https://doi.org/10.1002/joc.5152

Yang, X. Y., Li, Y. G., Luo, Z. W., & Chan, P. W. (2017). The urban cool Island phenomenon in a high-rise high-density city and its mechanisms. *International Journal of Climatology, 37*(2), 890–904. https://doi.org/10.1002/joc.4747

Yoshikado, H. (1990). Vertical structure of the sea breeze penetrating through a large urban complex. *Journal of Applied Meteorology and Climatology, 29*(9), 878–891.

You, C., Fung, J. C. H., & Tse, W. P. (2019). Response of the sea breeze to urbanization in the Pearl River Delta region. *Journal of Applied Meteorology and Climatology, 58*(7), 1449–1463. https://doi.org/10.1175/jamc-d-18-0081.1

Zhan, W., Ju, W., Hai, S., Ferguson, G., Quan, J., Tang, C., Guo, Z., & Kong, F. (2014). Satellite-derived subsurface urban heat Island. *Environmental Science & Technology, 48*(20), 12134–12140.

Zhang, Y. J., Wang, L., Santanello, J. A., Pan, Z. T., Gao, Z. Q., & Li, D. (2020). Aircraft observed diurnal variations of the planetary boundary layer under heat waves. *Atmospheric Research, 235*. https://doi.org/10.1016/j.atmosres.2019.104801

Zhao, L., Lee, X., Smith, R. B., & Oleson, K. (2014). Strong contributions of local background climate to urban heat islands. *Nature, 511*(7508), 216–219. https://doi.org/10.1038/nature13462

Zhao, L., Oppenheimer, M., Zhu, Q., Baldwin, J. W., Ebi, K. L., Bou-Zeid, E., Guan, K. Y., & Liu, X. (2018). Interactions between urban heat islands and heat waves. *Environmental Research Letters, 13*, 3. https://doi.org/10.1088/1748-9326/aa9f73

Zheng, Y. F., & Weng, Q. H. (2018). High spatial- and temporal-resolution anthropogenic heat discharge estimation in Los Angeles County, California. *Journal of Environmental Management, 206*, 1274–1286. https://doi.org/10.1016/j.jenvman.2017.07.047

Zhou, Y., & Shepherd, J. M. (2010). Atlanta's urban heat Island under extreme heat conditions and potential mitigation strategies. *Natural Hazards, 52*(3), 639–668. https://doi.org/10.1007/s11069-009-9406-z

Zhou, B., Rybski, D., & Kropp, J. P. (2017). The role of city size and urban form in the surface urban heat Island. *Scientific Reports, 7*. https://doi.org/10.1038/s41598-017-04242-2

Part III
Heatwave Consequences

Chapter 9
Conceptualising and Assessing Heatwave Impacts

Abstract This chapter discusses the multifaceted nature of heatwave impacts and proposes a methodology for their assessment. The chapter begins by defining heatwave impacts as the consequences of realised risks as influenced by the interaction of the heatwave hazard, heatwave related exposure and vulnerability and responses to heatwaves. A key focus is the differentiation between direct and indirect impacts, tangible and intangible impacts, and the propagation of these impacts through systems over time and space. The chapter underscores the need for systematic reporting and assessment of heatwave impacts to inform the development of effective risk management strategies and notes that without comprehensive heatwave impact data managing the risks associated with heatwaves will be more guesswork than evidence based. Because little is available in the literature regarding how to conduct a heatwave impact assessment, a five-step methodology for Heatwave Impact Assessment (HWIA) is introduced. This includes setting objectives and context, conceptualizing impacts, locating evidence of impacts, assessing impacts, and developing heatwave impact management strategies. This methodology is designed to provide a comprehensive framework for understanding and managing the consequences of heatwaves. In relation to the various HWIA steps, recommendations are made with regards to the type of data that can be harnessed for context setting and analysis of the heatwave hazard plus the use of various analytical techniques such as dependency and cross impact analysis matrices and event trees. A list of heatwave metrics is also suggested covering a range of sectors.

Keywords Heatwave impacts · Heatwave risk · Heatwave exposure · Vulnerability and response · Heatwave impact assessment methodology · Heatwave impact metrics

G. McGregor, *Heatwaves*, Biometeorology 6,
https://doi.org/10.1007/978-3-031-69906-1_9

9.1 Introduction

Beyond the scientific curiosity about heatwaves as pure physical events and their causative factors, we are interested in them because of how they affect us, and our surrounding world. How heatwaves manifest in terms of their consequences is a burgeoning field of enquiry largely driven by the need to understand the social and environmental outcomes of a number of recent mega-heatwaves, the threats posed by climate change and the general adage 'if we can't measure it, we can't manage it'. Accordingly, the purpose of this and the following two chapters, under the chapeau of heatwave consequences, is to provide an overview of matters related to heatwave impacts. Before addressing the factors that might moderate the scale of heatwave impacts in Chap. 10 and providing an overview of the nature of observed heatwave impacts in Chap. 11, this Chapter commences by reflecting on the meaning and nature of heatwave impacts. It then briefly considers broad approaches to heatwave impact assessment (HWIA) before proceeding to present a five-step methodology for undertaking HWIA. Chapter contents are synthesised in the last section.

9.2 Conceptualising Heatwave Impacts

What do we mean by impact? Here impact is defined as the consequence of an action, that action being in our case, the occurrence of a heatwave. More formally, the IPCC defines impacts as 'the consequences of realized risks on natural and human systems, where risks result from the interactions of climate-related hazards (including extreme weather and climate events), exposure, and vulnerability' and notes that impacts can be referred to as consequences or outcomes. Such a definition of course begs the question, what form does the consequence take and how do we establish consequence. Further when thinking of consequence or impact, it is often the negative consequences that receive attention with beneficial outcomes often overlooked, something that the impacts community appears to be predisposed to. Added to this are the semantics of impact assessment and impact analysis and the range of terms used in the literature, some of which are defined in Table 9.1. Lastly there remains the question whether impacts in general are viewed in a mechanistic fashion with the underlying assumption that an activity or system functions until an event impacts it, or impacts arise from multiple macro and micro-scale dynamic factors that condition an exposure unit to a spectrum of possible impact outcomes (Aledo-Tur & Dominguez-Gomez, 2017).

Returning to the IPCC conceptualisation of impacts being a realisation of risk as touched upon in Chap. 1, Fig. 9.1 depicts the so-called IPCC propellor diagram and portrays heatwave risk and ultimately impact as the outcome of the interaction between heatwave hazard attributes and associated hazards, heatwave exposure, heatwave related vulnerability and heatwave responses. Exposure and vulnerability as determinants of heatwave related risk and impact will be dealt with in the

Table 9.1 Common terms used in climate impact assessment relevant to heatwave impact assessment

Exposure	The presence of people, livelihoods, species or ecosystems; environmental functions, services, and resources, infrastructure or economic, social, or cultural assets in places and settings that could be adversely affected by a heatwave
Exposure unit	Activity, group, region, resource or any human or natural system exposed to a heatwave that could be adversely affected
Effect	An outcome directly produced by a heatwave acting on an exposure unit
Impact	The effect on the exposure unit or the consequence of a realised risk. The impact will possess some assigned relative or absolute value
Residual impact	Impacts that occur despite all efforts to control for them. Related to the concept of residual risk
Assessment	The scientific/systematic appraisal of effects
Sensitivity	The degree of responsiveness of an exposure unit to a given heatwave event, whether beneficial or detrimental
Vulnerability	The propensity or predisposition to be adversely affected. Vulnerability encompasses a variety of concepts and elements including sensitivity or susceptibility to harm and lack of capacity to cope and adapt. Both physical and socio-economic factors may be critical in determining vulnerability
Risk	The potential for adverse consequences/impacts for human and natural systems to occur. Can be moderated by adaptation and risk reduction strategies
Residual risk	The risk that remains despite implementation of all tenable adaptation/mitigation and risk reduction efforts. Residual risk occurs when adaptation limits have been reached and exceeded
Adaptation	The process of adjustment to actual or expected heatwave occurrence and their effects. Adjustments are made to temper harm or make the best of beneficial opportunities (see Chap. 12)
Resilience	The capacity of human and natural systems to cope with a heatwave event, with systems responding or reorganising in ways that maintain their essential function, identity and structure while also maintaining the capacity for adaptation, learning and transformation

Adapted from Carter et al. (1994) and Masson-Delmotte et al. (2022)

following chapter. As noted by Simpson et al. (2021) responses, that is well intentioned actions put in place to reduce risk, are often excluded as drivers of risk and by extension impact. This is because responses have been traditionally considered as just that, that is, risk management actions catalysed by clear evidence that a particular hazard causes harm and that harm needs to be reduced. Including responses as a risk determinant breaks with this historical view of responses and raises the possibility that some responses, broadly referred to as maladaptation, have the potential to do more harm than good thus playing a role in the possible level of impact. Responses to heatwaves, in the form of adaptation and mitigation actions designed to reduce heatwave related risk will be addressed in Chaps. 12, 13, and 14.

As described in Chap. 11, heatwave impacts may manifest in several ways across a range of realms. Although not referring to heatwaves specifically, the IPCC notes that impacts generally denote effects on lives, livelihoods, health and well-being, ecosystems and species, economic, social and cultural assets, services

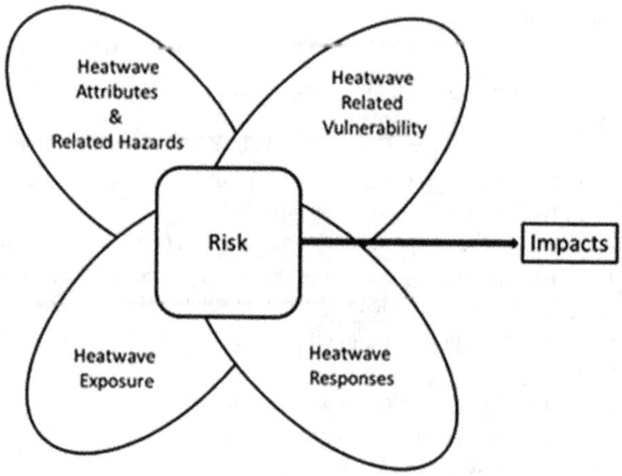

Fig. 9.1 Impacts as the realisation of risk, with risk an outcome of the interaction between the four determinants of heatwave related risk namely heatwave attributes (frequency, duration, intensity, spatial extent) and associate heatwave hazards (air pollution, wildfire, drought), heatwave related exposure and vulnerability (see Chap. 10), and responses to heatwaves (see Chaps. 12, 13, and 14). (Adapted from Simpson et al., 2021)

(including ecosystem services) and infrastructure (Allen et al., 2018) with some researchers organising these effects into general impact categories (Benzie et al., 2019; Benzie & Persson, 2019). Conceptually impacts can be categorised into tangible and intangible and direct and indirect. Tangible impacts are those which can be measured with some sort of value attached to them such as business profit or loss or the number of fatalities associated with a heatwave event. In contrast, intangible impacts cannot be readily measured, for example a diminution of personal well-being or loss of ecosystem services. Some impact analysts draw the distinction between tangible and intangible based on whether an impact can be quantified in monetary terms, but in many ways this limits tangible impacts to things for which a market value exists. A direct impact is one caused by immediate physical contact of a heatwave's physical attributes (e.g., 'the heat') with an exposure unit. Single or multiple direct impacts may arise from a heatwave. Indirect impacts on the other hand are catalysed by the direct impact. The associated chain of events that propagate through systems and unfold over space and time create so-called second to n^{th} order impacts and compound and cascading events (Carter et al., 2021; Zaidi, 2018; Zscheischler et al., 2020), with impacts sometimes manifesting at distant places via teleconnections (Challinor et al., 2017, 2018; Hedlund et al., 2018; Moser & Hart, 2015) or presenting as latent effects well after the heatwave event. As noted by Carter et al. (2021) the transmission of impacts, beyond the initial impact, can occur via a range of configurations as summarised in Table 9.2 and illustrated in Fig. 9.2.

Conventionally, most impact studies focus on specific regions, often bound by physical or political boundaries. In doing so, downstream, or remote impacts arising from a heatwave may be missed giving rise to 'blind spots' in relation to the

Table 9.2 Typology of Impact Transmission Dynamics

Impact transmission type	Characteristics
Cascade tiers	An initial impact is transmitted through a cascade of impacts on system components. At each level of the cascade, the impacts themselves are associated with a recipient risk of concern. The extent of this risk may be influenced by other features of the transmission dynamics
Escalating cascade	As impacts cascade from one system component to the next the impacts grow in magnitude—each subsequent impact is amplified compared to the previous one
Diminishing cascade	The opposite of an escalating cascade. Each subsequent impact is dampened or reduced compared to the previous one
Feedback cascade	Propagating impacts feedback to earlier links of the cascade. This adds complexity to impact transmission and its influence on the recipient risk
Compound impacts	Compound impacts may come about in a number of ways. For example impacts at two different locations may propagate through separate systems, but with these systems possessing common end points in terms of an exposed human or natural asset. Eventual convergence between the two systems when they interact affects the common exposure unit or asset

Based on Carter et al. (2021)

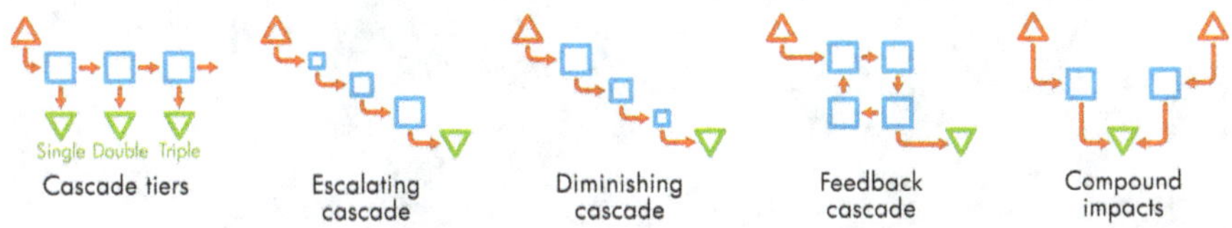

Fig. 9.2 Types of impact transmission dynamics. See Table 9.2 for characteristics. Δ is the initial impact, ∇ is recipient risk, □ is system component. Arrows represent transmission of impact. (Adapted from: Carter et al. (2021). Used under the terms of the Creative Commons CC-BY license 4.0 https://creativecommons.org/licenses/by/4.0/)

interactions and amplification of impacts, beyond a narrowly defined impact study area (Brimicombe et al., 2021; Challinor et al., 2017, 2018). Moser and Hart (2015) refer to the processes, structures and substances that might connect non-physical social impacts between distance locations as 'societal teleconnections' while Challinor et al. (2018) outline some of the domestic and international factors that might influence trans-boundary and trans-sector climate impact transmission. That climate impacts might 'travel', has catalysed a burgeoning discussion on trans-national/trans-boundary climate impact exposure (Benzie et al., 2019; Gaupp, 2020; Hedlund et al., 2018). Recognising this, Carter et al. (2021) have outlined a framework for considering the transmission of cross-border climate impacts and responses (Fig. 9.3). In relation to the subject matter of this book, the chain of events leading to cross-border impacts begins with a heatwave, the climate trigger that produces a short-term climate shock. This generates an initial impact in the region of heatwave occurrence. What follows is a downstream set of consequences that can propagate

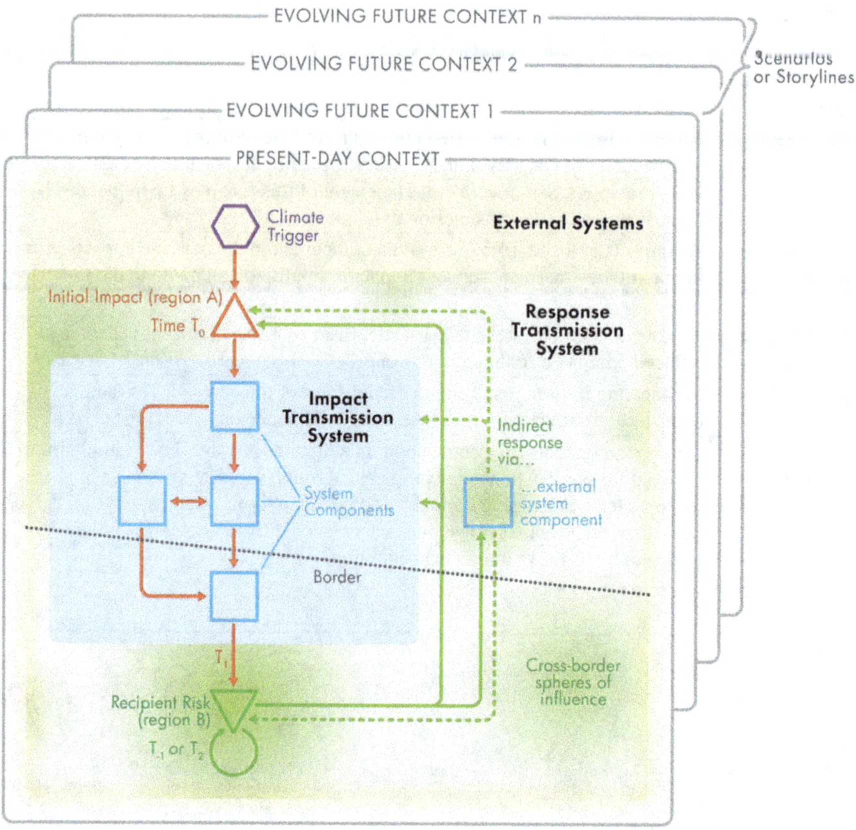

Fig. 9.3 Conceptual framework for consideration of cross-border climate impacts for present and future time domains. Δ is the initial impact, ∇ is recipient risk, □ is system component. Arrows represent transmission of impact or response. T refers to the response time. Both the impact and response transmissions systems interact with wider natural and social processes comprising so-called external systems. (Adapted from: Carter et al. (2021). Used under the terms of the Creative Commons CC-BY license 4.0 https://creativecommons.org/licenses/by/4.0/)

across space, including proximal and distant single and multiple borders, and over time with delayed effects likely. The propagation of impacts makes up what Carter et al. (2021) call an impact transmission system with the net impact transmitted into a region by flows of various kinds (e.g., raw, and processed materials, information, natural processes and species), referred to as the recipient risk. Countering the impact transmission system is the so-called response transmission system (Carter et al., 2021), such that responses may be anticipatory or reactionary comprising a mixture of response strategies (see Chaps. 12, 13, and 14). Importantly the conceptual model put forward by Carter et al. (2021) acknowledges that non-climatic triggers (e.g., economic, geophysical, health-related or geopolitical shocks), or in other words vulnerabilities (Chap. 11) could precondition or exacerbate impacts.

Although a possible hostage to fortune, it is worth distinguishing between impact assessment and impact analysis in the case of heatwaves. Heatwave impact assessment (HWIA) can be defined as the process of identifying the consequences of heatwave occurrence. In contrast, impact analysis is the investigation of how observed impacts may have come about by understanding the associated causal chain of events and any influencing factors. Essentially impact analysis refers to breaking down the complexity of an impact into its multiple component drivers, that is, understanding its pathology. Seen in this way, HWIA as dealt with in this chapter is more ex-poste (retrospective) than ex-ante (prospective) in nature and focuses on the 'after the fact' known or observed impacts of heatwaves.

A question that comes to mind when contemplating HWIA is are there any parallels with other forms of impact assessment, such as environmental impact assessment (EIA), social impact assessment (SIA) and health impact assessment (HIA). On the face of it they would appear the same but, there are differences. This is because in the case of HWIA as conceived here, the physical event of heatwave occurrence is the 'provocateur'. This contrasts with EIA and SIA. Evolving in response to the development of the United States National Environment Policy Act of 1969 (Esteves et al., 2012; Vanclay, 2006) and setting aside the debate of whether environment refers to realms beyond the biophysical (e.g. socio-economic and cultural), EIA is broadly defined as "a process of identifying and predicting the potential environmental impacts…of proposed actions, policies, programmes and projects, and communicating this information to decision makers before they make their decisions on the proposed actions" (Harvey, 1998). Similarly, SIA which has emerged from its initial subaltern status to EIA and in some ways broadened into socio-economic impact assessment. It is generally defined as analysing, monitoring and managing the social consequences of development (Esteves et al., 2012; Vanclay, 2020). From these definitions it is clear in the case of EIA and SIA the 'forcing function' that gives rise to an impact or consequence is human related and mainly associated with 'development' whether resource extraction (e.g., mining or forestry), the construction of new infrastructure (e.g., a dam, transport links or housing) or any other planned alteration of the qualities and characteristics of locations, places, or areas. Viewed as lying at the intersection of EIA and SIA (Vanclay, 2006) and with its conceptual roots in EIA and policy appraisal and the promotion of health public policy (Kemm et al., 2004), HIA has been defined as any combination of procedures or methods by which a proposed policy or program may be judged as to the effects it may have on the health of the population (Ratner et al., 1997). Allied definitions point to the health of the population and the prediction of the future consequences of health-related decisions yet to be made as the central focus of HIA, with HIA practiced in a variety of forms globally (Winkler et al., 2020). As for 'environment', the definition of health can be all encompassing going beyond the physical health of people and extending to the wider aspects of mental and social well-being including subjective feelings, a sense of well-being and quality of life with positive aspects of health also encapsulated by HWIA (Bednar-Friedl et al., 2022; Kemm et al., 2004; Lock, 2000).

While intuitively HWIA is most likely understood as studying the impact of heatwaves on environment and society, including health and economy, the subtleties of grammar make it different to EIA, SIA and HIA. This is because in the case of EIA, SIA and HIA, environment, society and health are the object of the assessment (that is, having something done to it) with 'development' or 'policy' implicit as the subject (the thing doing something to the object). In the case of heatwave impact assessment, the physical heatwave event is the subject, implicitly environment and society are the object. Considered in this way, heatwave impact assessment aligns with the overarching goals of climate impact assessment, that is working out both in an ex-poste and ex-ante sense, using a variety of methods, how climate affects a range of human and non-human domains. More exactly and building on general definitions of climate impact assessment (Allen et al., 2018; Carter et al., 1994; Kates et al., 1985), the primary purpose of HWIA is to systematically identify, analyse and evaluate the impacts of heatwaves, measured in monetary or non-monetary terms, on human and natural systems including animate and inanimate objects and estimate the uncertainties enveloping the identified impacts. Of course, HWIA shouldn't be for its own sake. Rather in the context of what constitutes actionable knowledge (Alexander & Dessai, 2019; Asrar et al., 2012; Jacobs & Street, 2020; Kandlikar et al., 2011; Keller et al., 2021; McGregor, 2015; Weichselgartner & Arheimer, 2019), HWIA's end game should be one focused on providing knowledge to inform the development of strategies to build multi-scale societal heatwave resilience and where possible, manage the effects of heatwave events on natural systems.

9.3 Approaches to Heatwave Impact Assessment (HWIA)

The systematic identification, analysis and evaluation of the impacts of heatwaves can be approached in one of three ways as is the case for climate impact assessment in general. These include impact, interaction and integrated approaches (Kates et al., 1985). Over time integrated approaches have become the most favoured by the climate impacts community, especially since the need for assessments of the impacts of climate change on society that incorporate physical, socio-economic, adaptation and mitigation dimensions has become obvious if a good sense of the implications of anthropogenic climate change for human and non-human systems is to be gauged.

A direct cause to effect pathway typifies the impact approach. In this case a heatwave stresses an exposure unit with the resultant strain manifesting as a particular type of impact. As outlined by Carter et al. (1994) this approach is adequate if simple dose-response relationships need to be established for individual activities or sectors and the relationship between the climate event and the exposure unit is not fettered by non-climate confounding factors. However, because of the underlying assumption of a simple deterministic relationship between a climate event, such as a heatwave and the resulting impact, plus the non-consideration of feedbacks, the impact approach is not normally applied in impact assessment (Carter et al., 1994).

The interactive approach offers an improvement over the impact approach. This acknowledges that factors in addition to the climate event itself might condition the type of impact experienced. Take for example the case of a spatially homogenous heatwave event covering several countries, with some affected countries, or even cities within a particular country, possessing well-developed heatwave warning systems (see Chap. 13) whiles others do not. In this case, the impact in places that have been forewarned about the heatwave and have implemented a well-honed heat action plan may well experience a lower level of impact (e.g., heat-related mortality and/or morbidity) compared to places where this type of heatwave risk management strategy is not in place. Consequently, the nature of the impact is contingent on the level and efficacy of heatwave risk management strategies in place. A rapidly growing city offers a good example of how feedback may modify heatwave impacts over time assuming no change in heatwave risk management strategies. As cities grow and, in many cases densify, the volume of human made material increases, altering the scale of exchanges and thus feedbacks of energy between the urban surface and the overlying atmosphere. Under such a scenario the intensity of the urban heat island and thus levels of urban heat stress are likely to increase with time. At the heatwave event timescale city size may also be important with regard to surface-atmosphere feedbacks and thus the nature of heatwave impacts. As seen in the previous chapter, heatwaves may amplify the urban heat island intensity largely as a response to contrasts in the nature of urban and rural energy balances attributable to the role of urban heat storage and the maintenance of nocturnal urban temperatures above those of their rural counterparts. Given this, two nearby cities that differ in size might experience contrasting heatwave impacts due to not only the impact of urban volume on the level of heat storage and release and the elevation of nocturnal temperatures but also the duration of anomalously high nocturnal temperatures as surface-atmosphere feedbacks prolong heatwave effects.

Ideally any HWIA conducted in an ex-poste frame should address all aspects of a heatwave's impact including downstream indirect or secondary and beyond effects as well as direct consequences. To achieve a holistic end-to-end analysis of any heatwave event, the physical drivers at multiple spatial and temporal scales (Chaps. 5, 6, 7, and 8) and the myriad of possible non-climate factors that may condition the response of the exposure unit of interest (Chap. 10) should preferably accompany an assessment of heatwave consequences. This type of approach that incorporates considerations of multiple outcomes and drivers and aspires to deliver a comprehensive treatment of the interactions and feedbacks underlying the interplay of a climate event and human or non-human systems is often referred to as integrated. Rotmans and VanAsselt (1996) capture nicely the ideals of an integrated approach by considering this as an interdisciplinary and participatory process of combining, interpreting and communicating knowledge from diverse scientific disciplines to allow a better understanding of complex phenomena. Further they suggest such an approach possesses two main characteristics, namely the value should be greater than that derived from pure disciplinary research and the emergent information should be useful to decision-makers and society in general (Rotmans & VanAsselt, 1996). Notwithstanding the aspirations of the integrated approach, data limitations often prejudice the attainment of what might be considered a complete integrated assessment with current knowledge on the understanding of the physical drivers and outcomes of heatwaves exceeding that of the non-climatic dimensions. With regards to this, the analysis of the 1995 Chicago heatwave (Klinenberg, 1999) contains many of the elements of what might comprise an integrated assessment of non-climate drivers and consequences in its exploration of the social and political complexities that gave rise to the substantial impacts associated with this event.

9.4 Procedures for Undertaking a Heatwave Impact Assessment (HWIA)

In this section some generic steps for undertaking a HWIA will be presented. The emphasis is on ex-poste assessments, that is, establishing impacts after the event. Figure 9.4 presents a general framework for undertaking an ex-post assessment of heatwave impacts soon after the occurrence of an event. Each of the steps is elaborated below. The framework is part informed by evolving methodologies for natural hazard and social impact assessment approaches to disaster analysis (Aznar-Crespo et al., 2021; Corvo et al., 2021; Dominguez et al., 2021; Lee et al., 2021; Mendoza

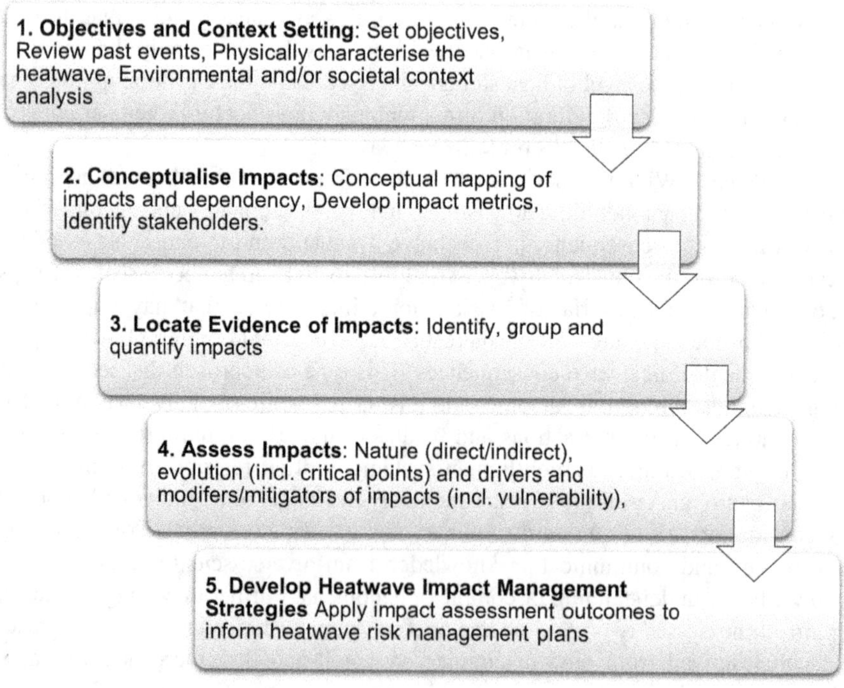

Fig. 9.4 Framework for undertaking an ex-poste heatwave impact assessment. (Source: Author)

& Schwarze, 2019; Sword-Daniels et al., 2015) such as FORIN (Forensic Investigation of Disasters) (Ferreira et al., 2023; Oliver-Smith et al., 2017), PERC (Post Event Review Capability) (Keating et al., 2016), the Centre for Disaster Risk Management and Risk Technology's Disaster Analysis in Near Real Time (Köhler et al., 2006) and methods of rapid assessment following disasters (Arii, 2013; Goode et al., 2021; Kryvasheyeu et al., 2016).

9.4.1 Step 1: Objectives and Context Setting

At the outset it is important to define the objectives for the assessment. Typically for human activity systems this would entail clarifying the target exposure units. This might comprise a particular economic sector (e.g., transport) and associated subsectors (e.g., rail), a well-defined social group or community or indeed an all-encompassing assessment for a region or city. In the case of natural systems, the exposure unit might comprise a distinct terrestrial, freshwater or marine ecosystem or a certain plant or animal community for a specified area. In many ways defining the exposure unit of interest will determine the resolution of the assessment in terms of its spatial (e.g., local to regional) and temporal (e.g., short to long term) coverage. Context setting facilitates situating the event of interest in a wider historical, physical and societal context. It could usefully comprise assembling information about and reviewing past heatwave events, the physical characterisation of the heatwave and horizon scanning in relation to the state of the environment and the general social, economic, and political climate at the time of the heatwave event. Reviewing past events, if any, will assist establishing the type of exposure units affected by heatwaves and associated impacts. A review will also reveal any past actions taken to mitigate heatwave consequences. As the purpose of such a review is to establish a reference for the heatwave event for which an assessment is to be undertaken, its outcomes should be seen as complementary, with direct extrapolations to the event being investigated to be avoided (Aznar-Crespo et al., 2021).

Likely sources of information for a review of past events include published studies in academic journals, the 'grey' literature, media reports and social media activity. Although disaster data bases such as EMDAT (Emergency Events Database: The Centre for Research on the Epidemiology of Disasters (CRED)), NATCAT (The Natural Hazards Assessment Network—NATHAN) and proprietary NatCatSERVICE (Munich Re) and DesInventar (United Nations Centre for Disaster Risk Reduction—UNDRR) are possible sources of information about historical events, very few contain statistics about past heatwaves. This also applies to the indicators developed as part of the Sendai Framework for Disaster Risk Reduction (Zaidi, 2018). Further, several important biases permeate global and national databases that monitor losses from natural hazards. These work against the attainment of a reliable picture of the spatial and temporal characteristics of heatwave impacts. Such biases include hazard bias, temporal bias, threshold bias, accounting bias, geographic bias, and systemic bias (Gall et al., 2009).

To physically characterise present and any past heatwave events will necessitate access to the requisite meteorological data usually recorded at weather stations. Typically, daily or hourly temperature data are used to describe attributes such as magnitude/intensity, duration and spatial coverage and thus physically characterise heatwaves (Chap. 3). However, there is always the possibility that some heatwaves will be complex with the synergistic effects of all the constituents of heat (temperature, wind/ventilation, humidity and radiative load) playing a role in the scale and type of impact. Due consideration therefore needs to be accorded to whether the assessment is dealing with a meteorologically simple or complex event. Spatially complexity may also be an issue with the challenge of defining the core and peripheral areas of a heatwave, especially as heatwaves may evolve spatially over their duration with cascading spatial impacts the result, with some of these occurring across borders. Associated with characterisation is the thorny issue of defining the heatwave event as different heatwave definitions (see Chap. 2) may well result in heatwave attributes taking on a range of possible values depending on how the heatwave is demarcated both in time and space. This holds implications for any exercise designed to assess the association between heatwave attributes and scale of impact as might take place in step 4 of the proposed framework. In the situation where standard meteorological observations are not available for event characterisation, as might be the case in some regions where observational networks are scant, assessors will need to explore alternative data sources. For large regions where meteorological observations are completely absent, such as across vast tracts of Africa and some parts of South America and Eurasia, reanalysis products (Chap. 3) may assist with heatwave characterisation, bearing in mind spatial resolution limits and in some cases the faithfulness of estimates of heat variables to actual conditions (Di Napoli et al., 2023; Kruger & Di Napoli, 2022). At the intra-urban scale or indeed the regional scale, observations from citizen science networks may assist with heatwave characterisation assuming the requisite checks have been undertaken concerning reliability of observations. Once the meteorological variables have been assembled their values can be considered in a longer-term climatological context by applying Extreme Value Theory approaches as described in Chap. 3, assuming long meteorological records for weather stations that manage to record the heatwave event are available. Of course, reanalysis data, some of which extends well over 50 years, can be used for establishing the probability/return period characteristics for the heatwave event.

Horizon scanning of the broader environmental and societal climate can play a powerful role in an assessment of heatwave impacts. This is because many of the impacts may be conditioned on the general state of the physical and/or social environment at the time of the heatwave event. For example, antecedent land surface conditions, as discussed in Chaps. 7 and 8, may prime intense, prolonged and spatially extensive heatwaves. For coastal areas, pre-existing sea surface temperature conditions in combination with certain atmospheric circulation configurations may serve to increase the likelihood of an extreme heatwave event as was partly the case for the record breaking 2021 Pacific Northwest heatwave event.

The burgeoning literature on the social determinants and/or the social construction of disaster risk (Birkmann et al., 2013; Oliver-Smith et al., 2017) and heat risk (Kovats & Hajat, 2008), has highlighted the role of place and situation as a modifier of the human consequences arising from extreme events such as heatwaves. For this reason, an analysis of the social, economic, political, and cultural forces at play may assist with understanding the nature of the impacts and any differential responses across exposure units arising from a heatwave event. Some of these factors will be expanded on in the following chapter. For natural systems or human production systems based on natural processes, such as agriculture, the general health of such systems is likely to influence their response to heatwaves. For example, a crop suffering disease or pest infestation is likely to be far less resilient to the impacts of a heatwave compared to a healthy crop.

9.4.2 Step 2: Conceptualise Impacts

Based on a review of past events for the area of interest a suite of known impacts may exist and form the basis of a list of potential impacts. Consultation of the broader literature may also reveal a range of impacts not immediately obvious based on the history of heatwaves for the assessment area or elsewhere. Beyond a mere listing of possible impacts much more useful is a conceptualisation of interconnections between impacts. Interconnections may take the form of primary and higher order impacts, the sequence of which might reveal the nature of possible cascading impacts distributed over time and space. Added to an assessment of interconnections and depending on existing knowledge and data availability, is an evaluation of the potential dependency of elements within (e.g., health system) and between (e.g., health and transport) systems, or in other words system intra-dependency and system interdependency respectively (Dominguez et al., 2021), sometimes referred to as a system of systems analysis (Cavallo & Ireland, 2014). The justification for this relates to the idea that intra-dependency and inter-dependency may generate unidentified risks by initiating new pathways for the propagation of impacts within and across systems, bearing implications for system fragility and thus resilience to 'shocks' (Cuartas et al., 2021; Pescaroli & Alexander, 2016, 2018; Vespignani, 2010). A range of tools exist to aid such an exercise including relational diagrams and dependency matrices. For example, causal loop diagrams (Bala et al., 2017) have been applied to an analysis of the observed cascading effects and dependencies associated with the European 2003 heatwave event (Berariu et al., 2015). At a more conceptual level, Zaidi (2018) has presented an example of a heatwave causal loop analysis (Fig. 9.5). While such analyses possess a high degree of abstraction, they provide a useful basis for contemplating the potential direct (primary) and indirect (secondary, tertiary etc.) impacts of heatwaves, the identification of nodes of vulnerability and system interdependency and the development of impact metrics (Zaidi, 2018).

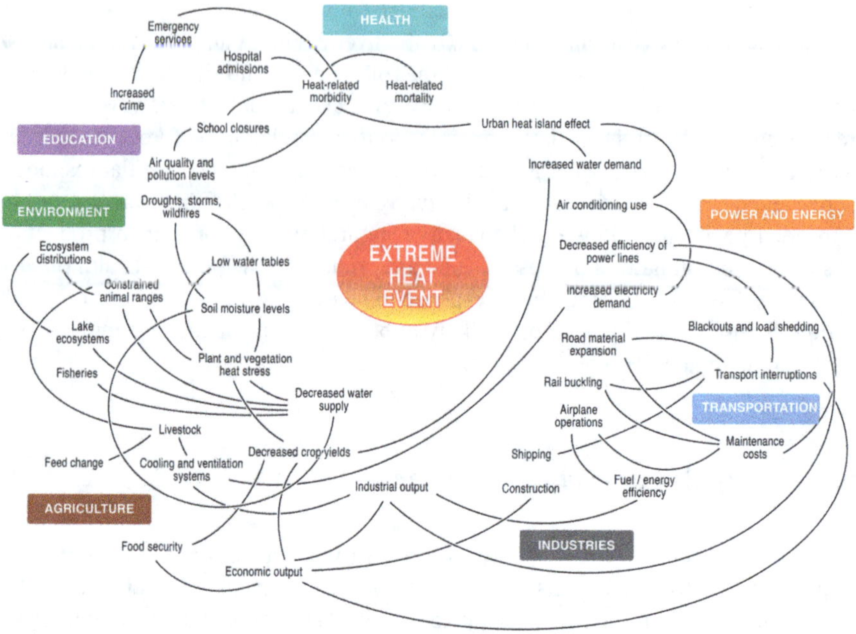

Fig. 9.5 Causal loop diagram for heatwave impacts. (Adapted from original provided by R Zehra Zaidi, Ca' Foscari, University of Venice)

Complementing relational diagrams are devices such as dependency and cross impact analysis matrices. Dependency matrices provide an insight into possible system intra- and inter-dependencies especially in relation to critical infrastructure, services and facilities. For a hypothetical heatwave and location Table 9.3 shows a completed dependency matrix based on a mock review of historical evidence of heatwave impacts. This reveals the dependency between several elements of infrastructure, emergency services and critical facilities. For each matrix cell a score can be allocated to indicate the level of dependency between the interacting elements. This could vary from no dependency for totally independent systems through to high dependency with the latter state signifying interacting elements are totally dependent on each other and are highly likely unable to operate if one is compromised. Intervening categories of low and medium dependency point to only partial relevance in terms of interaction and discernible relevance such that operations may continue albeit not at full capacity respectively. Based on the hypothetical situation outlined in Table 9.3 it is evident for the location of study that schools and care homes are highly dependent on power, such that power blackouts or brown-outs may cause schools to shut. This level of dependency on power is higher than that for hospitals, which in the case of this hypothetical location have their own power generation plant as a backup in case of power outages. Sticking with the hypothetical example, road operations show less dependency on power than rail as for this location, large parts of the rail network are electrified. Rail displays a medium level of

Table 9.3 Example of hypothetical heatwave related dependency matrix with two-colour and four-graded shading scheme to indicate level of dependency (none to high) and whether an intra-dependency (gold) or inter-dependency (blue)[a]

	Infrastructure								Emergency Services		Facilities		
	Power	Water	Waste Water & Sewage	Waste Disposal	Road	Rail	Airport & Air Traffic	Telecommunications	Fire	Ambulance	Hospitals	Care Homes	Schools
Power	0.33	--	0	--	0	0	0	0	0	0	0	0	0
Water	0.33	--	--	--	0	0	0	0	--	--	--	--	--
Waste Water & Sewage	0.33	0.33	--	--	--	--	--	0	--	--	--	--	--
Waste Disposal	0.33	--	--	0.33	0.33	0.33	0	0	--	--	--	--	--
Road	0	0.33	--	0	0.33	--	--	--	0	0	0	0	0
Rail	0.33	--	--	0	--	0.66	0	0.66	0	0	0	0	0
Airport & Air Traffic	0.33	--	0.33	0.33	--	--	0.33	0.66	0.33	0.66	0	0	0
Telecommunications	0.66	0	0	0	--	--	--	0.33	0	0	0	0	0
Fire	0.33	0.66	0	--	0.33	--	--	0.66	0.66	0.66	--	--	--
Ambulance	0.33	0	0	0	0.33	0	--	0.66	0.66	0.66	0.33	--	--
Hospitals	0.66	0.66	0.66	0.66	0.66	0	--	--	--		0.66	0	0
Care Home	1.0	0.66	0.66	0.66	0.66	0	--				--	0	--
Schools	1.0	0.66	0.66	0.66	0.33	0.33	0	0.33	0.33	0.33	--	0	0.33

Source: Author

[a]The matrix is read with elements in the rows depending on the elements in the column. Cells can be scored using a uniform distribution scale to indicate level of dependency. For example, following Dominguez et al. (2021), 0 = no dependency; 0.33 = low dependency; 0.66 = medium dependency and 1 = high dependency with an associated graded shading scheme from light to dark. Further cells can be differentiated using two colours based on whether they signify intra-dependencies (e.g., between the same infrastructure elements such as 'roads' and 'roads') or inter-dependencies (e.g., between elements of infrastructure and facilities such as power and hospitals). -- means no evidence of an association found in the review stage

intra-dependency as the review of historical evidence has revealed that rail buckling under extreme temperatures associated with heatwaves has knock-on effects for the rail system beyond the immediate location of interest. Similarly, there are intra-dependencies for power. These arise from the necessity to reduce generating capacity from nuclear power plants because of reactor cooling challenges during heatwaves. Given this and assuming there is capacity to do so, other sources of energy such as solar and hydropower must be turned to until conditions are once again conducive for full generation capacity to be reinstated at the nuclear power plant. Intra-dependencies within the power system also relate to some instances when very calm conditions associated with heatwaves have meant that wind turbines operate well below capacity requiring the downturn in energy supply from this renewable energy source to be counter balanced by supply to the power grid from other sources which might include gas and coal, although not shown in the dependency table. The blank squares in the dependency matrix indicate interactions/dependencies for which there is no historical evidence, but that is not to say future heatwaves and a unique set of circumstances will not give rise to dependencies. Notwithstanding this and acknowledging that lack of evidence may introduce some bias into the conclusions drawn about relative dependencies, the scores for the matrix columns and rows can be summed to indicate the elements that are the most crucial and the most dependent respectively. For the hypothetical study location this reveals power and water being the most critical during heatwaves with care homes the least critical. In contrast hospitals, schools and care homes are the most dependent on the various elements of infrastructure and so on, while power and water are the least dependent (Table 9.4).

Cross impact analysis for heatwaves facilitates an appraisal of the sensitivity of a range of activities or sectors to a suite of driving forces related to heatwave attributes and the four main elements of heat. As for dependency tables, the cross

Table 9.4 Criticality and dependency scores and raw ranks for elements of infrastructure, services and facilities for a set of hypothetical heatwave events as revealed by historical evidence

Element	Criticality score	Rank	Dependency score	Rank
Power	5.96	1	0.33	9
Water	3.3	2	0.33	9
Waste water & sewage	2.31	5	0.66	8
Waste disposal	2.64	4	1.32	6
Road	2.97	3	0.66	8
Rail	1.32	7	1.32	6
Airport & air traffic	0.33	9	2.64	5
Telecommunications	3.3	2	0.99	7
Fire	1.98	6	3.3	3
Ambulance	2.31	5	2.97	4
Hospitals	0.99	8	3.96	1
Care home	0	10	3.64	2
Schools	0.33	9	3.96	1

Source: Author

Table 9.5 Example of cross impact analysis matrix for electricity system

		Generation				Transmission	Demand				Grid operation	
		Nuclear	Solar	Hydro	Wind	High voltage cables	Domestic	Commercial	Industrial	Transport	Demand/ Supply balance	Driving force
	HW attributes											
Heatwave characteristics	Intensity	2	3	1	2	3	3	2	1	3	2	22
	Duration	3	2	3	3	2	3	3	2	3	3	27
	Onset: rapid	2	2	0	2	2	2	1	1	2	2	16
	Onset: slow	0	1	2	1	0	1	1	1	0	1	8
	Cessation: rapid	0	0	0	0	0	1	1	2	0	0	4
	Cessation: slow	1	2	2	1	1	2	2	1	2	2	16
	Sensitivity	8	10	8	9	8	12	10	8	10	10	
	Dimensions of heat											
	Tmax	3	3	2	1	3	3	3	2	3	3	26
	Tmin	1	0	1	0	1	2	1	1	1	1	9
	Tmean	2	2	1	0	2	2	1	1	1	2	14
	Tmean: water	3	0	1	0	0	1	1	2	0	0	8
	Humidity: high	0	2	0	0	1	2	2	1	1	1	10
	Humidity: low	0	2	2	0	0	2	2	1	1	0	10
	Rad load: high	1	3	2	1	3	3	1	1	3	2	20

(continued)

Table 9.5 (continued)

		Generation				Transmission	Demand				Grid operation	
		Nuclear	Solar	Hydro	Wind	High voltage cables	Domestic	Commercial	Industrial	Transport	Demand/ Supply balance	Driving force
	Rad load: low	0	2	0	0	0	0	0	0	0	0	2
	Wind 0–1[a]	1	2	1	3	1	2	2	1	2	1	18
	Wind 2–3	1	1	1	1	1	1	1	1	1	1	10
	Wind 4–5	1	1	2	2	2	1	1	1	1	1	13
	Wind >5	2	1	2	3	3	1	1	1	1	1	15
	Sensitivity	14	19	15	11	17	20	17	13	15	13	

Source: Author

[a]Wind is expressed as wind force such that 0–1 = 0-1 m/s; 2–3 = 2-5 m/s; 4–5 = 6-11 m/s and > 5 = > 11 m/s

impacts matrix can be developed based on historical evidence of heatwave impacts with the associated information contained within and across the matrix serving as a pointer to where to look for evidence of impacts in an ex-poste assessment. For a hypothetical case, Table 9.5 presents a cross impact analysis for an electricity system comprised of non fossil fuel generation, transmission, demand and grid management components—transport is included as for the hypothetical location the rail system is electrified. As for the dependency matrix, cells can be populated with numerical scores indicating the relative response or sensitivity of the various components (exposure units) of the electricity system to heatwave driving forces. For this example, weighted scores of 0, 1 2 and 3 are used to indicate no, low, medium, and high sensitivity respectively. Summing the weighed scores in the columns and rows provides a sense of the relative sensitivity of each of the exposure units under examination and the degree of forcing associated with the individual heatwave characteristics respectively. In the example presented in Table 9.5, sensitivity scores for the various components of the electricity system are presented separately for heatwave attributes and dimensions of heat to avoid possible double counting, for example, there is a close association between intensity and maximum temperature. Bearing in mind this is just an example and the allocated weightings are purely hypothetical, the cross-impact analysis matrix demonstrates that domestic demand is the most sensitive with regards to heatwave attributes followed by commercial demand, solar generation, and grid operation. With regards to the dimensions of heat, domestic demand remains the most sensitive with some slight reordering of the sensitivity for other components of the electricity system. Of the heatwave attribute forcings, heatwave duration emerges as the most powerful, strongly influencing several components of the electricity system. This is followed closely by heatwave intensity with drawn out heatwave cessation also signalled as important. Maximum temperature and high radiation load prove to be the dominant forcings in relation to the dimensions of heat, with near calm wind conditions also notable. Whilst this example has been presented for a hypothetical electricity system, cross-impact analysis can be applied at the general level for broad socio-economic sectors such as agriculture, forestry, tourism, energy, water, health, and transport or indeed their sub-sectors as well as natural systems that might include components of terrestrial, freshwater and marine ecosystems.

Designing impact metrics is an important part of the conceptualising impacts stage. Essentially this entails settling on a set of indicators that facilitate the measurement of impact. By definition, these are different to metrics that measure the physical characteristics of heatwaves as established in stage 1. While the preference will be for quantitative metrics, the importance of qualitative metrics of impact should not be dismissed. A non-exhaustive list of possible heatwave impact metrics organised by broad activity area is provided in Table 9.6. Although units of measurement are not given for the metrics listed, these are likely to fall across the nominal, ordinal and ratio scales of measurement. For example, a human behaviour indicator such as 'staying indoors' during a heatwave could be registered simply as a yes or no response on the nominal scale of measurement. In contrast, a measure of psychological well-being for a heatwave period may use a seven-point ordinal scale,

Table 9.6 Indicative listing of heatwave impact metrics—units of measurement not specified

Health	**Insurance**	**Retail**	**Agriculture**
Crude mortality rate Cause-specific mortality rate (especially heat) Age/gender-specific mortality rate Proportionate mortality Death-to-case ratio All cause hospital admissions Heat-related hospital admissions Heat-related injuries GP visits Visits to mental health services Psychological Well-being Drownings Suicides Calls to health service helpline **Emergency services** Ambulance call-outs Fire brigade call-outs Life-saver rescues **Labour** Worker productivity Work hours Heat related absences **Water demand/ consumption** All sectors Evaporative loss from reservoirs, ponds, recreational pools etc.	Heat-related claims **Transport** Rail journey interruptions Road transport interruptions Aviation operations Traffic/cycle accidents **Infrastructure** Rail buckling Road surface integrity Bridge expansion stress Building superstructure stress Insulated glass unit stress Building coating material effects **Energy** Electricity/gas consumption Timing and magnitude of peak energy demand Transmission efficiency/load shedding Number and extent of blackouts/brownouts Renewable energy efficiency (solar/ wind) Fuel efficiency	Food. drink, clothing sales Electrical appliance sales Pharmaceutical sales **Tourism** Visits to national parks Theme park visits Visits to beach side attractions **Societal** School closures/ restricted times Crime (domestic violence/road rage) Social unrest Human behaviour change (e.g. staying indoors) Public heat warnings (level and duration) TV, radio and print media activity Social media activity (social sensing) **Construction** Work hours lost (see labour) Heat related building material damage Water usage Machine operation efficiency/ maintenance Worker injuries (see health)	Crop phenology Crop yield Domestic animal health/ mortality (e.g. poultry, sheep, cattle, horses) Veterinary service visits/ callouts Milk production Wine production Fruit ripening Animal feed availability Stock sales Farm management practices (e.g. delayed tilling) Labour (see column 1) Deployment of heat adaptation measures (e.g. water misting facilities and provision of cool water for workers) **Environmental** Fire index Air quality (e.g. ozone/$PM_{2.5}$) Water quality/algal blooms Soil moisture/ground water levels Streamflow Vegetation health/mortality Animal mortality/morbidity Animal behaviour Ecosystem processes and rates (terrestrial, freshwater, marine)

Source: Author

in line with the Likert scale approach (Likert et al., 1934), ranging from strongly disagree to strongly agree in response to a question such as 'would you agree that your mental well-being was negatively affected by the recent heatwave?'. A metric that typifies measurement on the ratio scale is daily standardised mortality in the form of deaths per 100,000 population for each day of a heatwave event, or the number of school closures during a heatwave event.

Associated with settling on any one or more of a range of possible indicators is the assumption background levels for the metric are known. This is because impact is normally established in terms of the degree of departure from the long-term average or in climate speak the 'climatology' or normal pattern of behaviour. In many cases securing long term records for a particular metric can prove problematic. This leaves the analyst with the challenge of deciding how anomalous the observed impact is. Some of the suggested metrics listed in Table 9.5 are widely used in heatwave impact assessments. For example, human mortality has been the go-to metric because the human health consequences of heatwaves are considered the most obvious as is apparent from daily mortality records usually kept by national or regional statistical agencies. That said, many heatwave—health studies use the exceedance of all-cause or all non-accidental deaths above the long term average for the period of analysis as a proxy of heatwave related deaths. This is because heat is often not diagnosed as a specific cause of death. Insurance is also another good example of this type of data challenge as any impact analysis in relation to this sector will need to 'triage' non-heat from heat-related claims. Given this, heatwave impact analysts when working with data that might form the basis of an impact metric need to be aware of what that data represents and not blindly accept a time series of values or other information at face value. An associated issue is the extent to which the available data used to form an impact metric will reveal clear impact signatures. This will not only depend on the direct link between what is measured and the impact, that is, the relationship between the supposed driving force and the consequence, but the temporal scale of the data. Many metrics will call for at least daily data, so as to capture immediate acute heatwave impacts such as heat related mortality or hospital admissions. Other metrics which measure the cumulative and delayed impacts may be based on measures over timescales beyond the daily, such as weekly, monthly, or seasonal and for periods well after a heatwave. For example, the impact of an intense period of heat on viticulture yield will not be apparent until grape harvest. This is because intervening conditions are also likely to affect yield and confound the association between heatwave occurrence and grape yield and thus complicate attempts to directly attribute a non-concurrent impact to a given heatwave event.

Stakeholders are those directly or indirectly affected by an event. Stakeholders can range across several social, economic and political scales from individuals to distinct communities, agencies and non-governmental and governmental organisations. Aznar-Crespo et al. (2021) categorise stakeholders into three broad groups namely decision makers who possess some if not all responsibility for the management of impactful events, exposure units (human, environmental and material) which experience the consequences of an event and third parties or other social actors that participate indirectly in the management of an event. Highly relevant to the case of heatwaves, this categorisation can assist with two matters. Firstly, either based on a review of existing information for the location of interest or theoretical considerations, a list of stakeholders can be drawn up. After this an assessment of each stakeholder's degree of influence in the governance of heatwaves as well as their vulnerability and exposure can be undertaken. Such a triumvirate of influence, vulnerability and exposure assists with identifying what Aznar-Crespo et al. (2021) refer to as the structural position of each stakeholder to face the impacts of an event and participate in the risk management process. To achieve the stakeholder assessment a simple scoring system related to influence, vulnerability and exposure could be established with each stakeholder categorised according to one of the three stakeholder types. Undertaking such a stakeholder analysis as part of the conceptualisation stage of HWIA will assist with signposting the analyst in the direction of those stakeholders, whether they be decision makers, exposure units or other actors, who are likely to have been impacted by a heatwave event. In a 'heads-up' sense this could prove invaluable as a component of an ex-post impact assessment.

9.4.3 Step 3: Locate Evidence of Impact

Equipped with information from the conceptualisation of impacts stage, in particular the outcomes of the dependency and cross impact analysis matrices exercise, the impact analyst should be in a good place to proceed to the step of locating evidence of impacts arising from the heatwave event under investigation. That said, heatwaves unlike many climate-related natural hazards pose the challenge of low visibility as their direct impacts will not be as obvious as those associated with more spectacular events such as hurricanes, tornadoes, and floods for example. Added to this are the spatial and temporal dimensions of heatwave related impacts as these are likely to emerge slowly over space and time, especially with regards to geographical teleconnections and delayed secondary and higher order impacts. A vast range of technocratic (top-down) or participative (bottom-up) approaches exist for gathering information on human and non-human systems. Not specific to heatwaves these include case studies, social and biophysical field survey research, in-depth interviews, qualitative and quantitative content analysis, community-based participatory research, statistical analysis of primary and secondary social, economic, technical or environmental data, focus groups, direct observation, and geospatial analysis including remote sensing (Peek et al., 2020). In undertaking information gathering involving direct contact with human subjects the need for ethical considerations (Louis-Charles et al., 2020; Van Brown, 2020) and cultural competence (Wu et al., 2022) has been emphasised. Furthermore, awareness of the two broad technocratic and constructivist paradigms associated with social impact analysis has been encouraged given these contrasting paradigms have fundamental influences on the assumptions, motivations, and values that inform social impact assessment research practices (Aledo-Tur & Dominguez-Gomez, 2017).

The complex nature of heatwave impacts is likely to require an integrated identification strategy that draws on a range of complementary approaches for information gathering and this seems to have been recognised by impact assessment analysts. For example, a recent survey of people involved in natural hazard impact analysis revealed most investigators use a combination of three to five methodological approaches to assemble a description of event impacts (Peek et al., 2020). In the case of social impacts arising from natural hazards, Aznar-Crespo et al. (2021) have suggested a staged approach. Applicable in the case of heatwaves, this begins with a primary data review to identify the most immediate impacts. Interviews of those with direct experience of an event (the exposure units) would then follow, ensued by interviews with agencies/institutions responsible for risk management (the decision makers) with impacts divided into broad categories related to loss of life and property, loss of livelihood, migratory processes (short/long-term), psychosocial effects, barriers to economic growth and development and political consequences.

9.4.4 Step 4: Assess Impacts

The individual impacts identified in the previous stage are the focus of the impact assessment stage. Of specific concern is the nature of the impacts, their evolutionary history including an assessment of critical points and a consideration of the impact drivers and modifiers.

In the context of understanding genesis, impacts can be classified into direct and indirect. Direct impacts are those arising unambiguously from a heatwave such as heat related mortality or morbidity, buckled rail lines or the melting of tarmac road surfaces (see Chap. 11). These types of impacts are often referred to as first order or primary impacts because any resulting 'damage' to human or natural systems can be attributed unequivocally to the heatwave event whether it be solely the extreme temperatures or these in combination with any one or more of the elements of heat such as radiative load, atmospheric moisture, or wind. For example, sunburn browning and sunburn necrosis in grapes, usually associated with heatwaves and related clear sky conditions, arises from the synergistic effects of excessive photosynthetically active radiation, UV radiation, temperature, and low atmospheric humidity (Gambetta et al., 2021). A characteristic of direct impacts is their immediacy in time and space. For heatwaves, the spatial extent of extreme heat anomalies and their duration is likely to determine the geographical range and degree of protraction of direct heat effects.

Unlike the tightly coupled link between the heatwave trigger and primary or direct impacts, indirect, secondary, and higher order impacts are the outcome of the propagation of the effects of direct damages to other parts of a system or indeed across systems. Generally, second order impacts can be thought of in terms of a loss of functionality arising from the 'damage' associated with the primary/first order impact. In this way third order impacts are those arising from functionality loss. In the context of human systems higher order impacts can be viewed as impacts experienced in individual social, economic, political, and cultural sectors. Often referred to as cascading effects, the chain of events emanating from the direct impact to cause increasingly remote second to higher order impacts in both space and time can be visualised using analytical tools such as Causal Loops (Fig. 9.5), Fishbone, Fault Tree, Event Tree and Bow Tie diagrams (Dominguez et al., 2021). The extent of the ripple effect inducing secondary to higher order impacts will be determined by the degree of dependency of nested activities within hierarchical systems. Given this, the levels of dependencies exposed by dependency matrices may offer insight into the potential ordering of impacts with primary or secondary impacts spilling over into third and higher order social and economic impacts. For example, a heatwave may result in an increase in river water temperature which affects water availability at the requisite temperature to achieve cooling related processes in nuclear power plants. This impacts the generating capacity of nuclear power plants which may comprise an important energy source for the regional or national power grid. Such a loss of functionality and an associated downturn in power generation leads to brown or even worse blackouts, which holds implications for a range of sectors dependent on a reliable energy supply. For natural systems, constructs such as food webs and the trophic relationships they visualise offer useful tools for tracking second to higher order impacts as might arise for example from coral bleaching induced by a marine heatwave or heatwave related reductions in the primary productivity of a deciduous forest. Such nested configurations of functions and activities forms a bounded system of systems with increasing levels of complexity at greater distance from the initial impacted system or system component. Understandably this can pose challenges in terms of higher order impact detection and establishing their evolutionary history.

To assist with understanding the evolutionary history and root causes of an impact, cascading event trees can be helpful. Figure 9.6a presents an example of how a range of second order scenarios could play out in relation to the influence of extreme temperatures on the diminished transmission efficiency or de-rating of power lines. For example, the presence or otherwise of a de-rating early warning system (EWS) and an associated rapid detection of de-rating with a requisite response strategy may determine the difference between no, and severe (e.g widespread and sustained) power supply disruptions. The bifurcations in the event tree represent critical points at which escalation of impacts may occur. For example, if de-rating only occurs on a part of the transmission system, that is a non-critical component of a geographically wider distribution grid, this holds implications for the scale of the power supply disruption as measured by geographical extent and the number of people exposed to the disruption.

Power disruptions have the potential to generate a range of downstream effects, or third order and higher impacts in sectors that are heavily dependent on a reliable power supply. The possible impact of a heatwave related power disruption on the so-called 2–8 °C vaccine cold chain and whether vaccine loss/wastage occurs is illustrated in Fig. 9.6b. This starts with the interruption of power supply to a pharmaceutical distribution company where vaccines received from a vaccine producer are stored within the critical range of 2-8 °C in large refrigeration units before

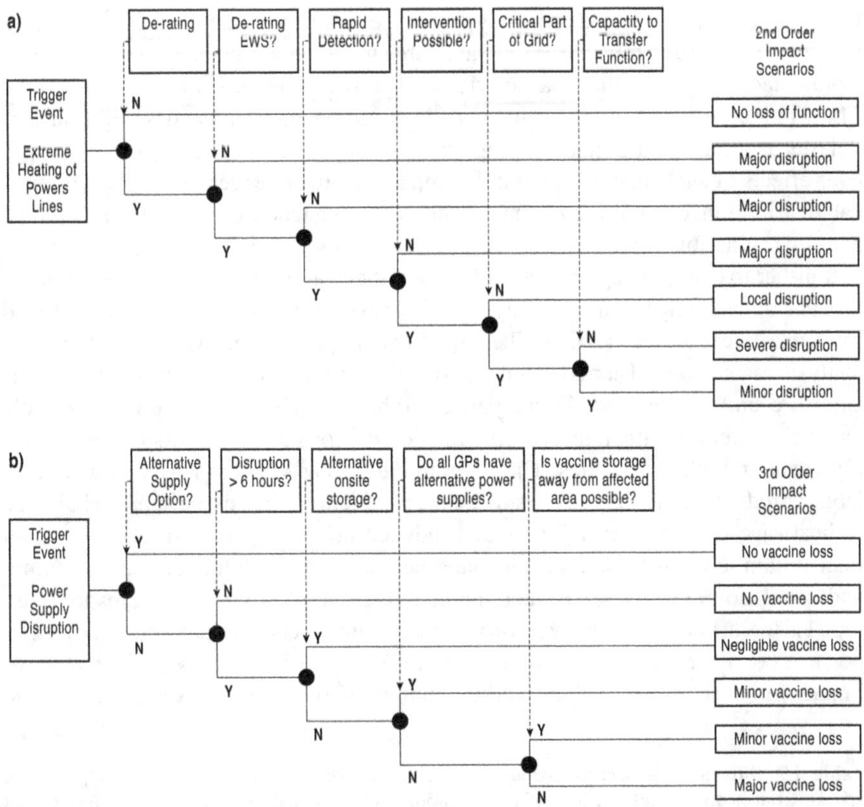

Fig. 9.6 (**a**) Event tree demonstrating second order impact scenarios arising from loss of transmission capacity due to extreme heating of overhead power cables and (**b**) possible third order impacts in the form of vaccine wastage if the vaccine cold chain cannot be maintained. Y is yes and N is no. Black dots represent nodes or critical points. (Source: Author)

distribution to pharmacies and general practitioners (GPs). Assuming the distribution company has an alternative power supply provided by onsite generators with sufficient capacity to maintain the functioning of all refrigeration units for the duration of the heatwave event, then no vaccine loss may occur. In the case alternative onsite power sources are not available, the critical determinant is the duration of the disruption, as the refrigeration units are only able to maintain temperatures within the critical limit of 2-8 °C for up to 6 h. This represents a critical point in the vaccine cold chain. For a longer lasting disruption alternative vaccine storage will need to be sought. If this exists in the form of refrigerated trucks or containers that comprise part of the company's distribution infrastructure, then negligible vaccine loss might occur. However, if the capacity of the alternative onsite storage is less than required vaccines may need to be transported to pharmacies or GPs assuming they have alternative power supplies and storage capacity so they can take receipt of vaccines with little spoilage. In the case that not all GPs and pharmacies are able to absorb the vaccines in stock at the distribution company, access to alternative storage away from the heatwave affected area may offset the risk of vaccine loss. This of course assumes refrigerated transporters are available to ship the vaccines to the alternative storage facility outside the area experiencing power supply disruption in good time. The worst-case scenario of vaccine loss due to temperature breaches will transpire if failure or lack of alternative refrigerated storage occurs. The higher order spill over effects of such an outcome will be impacts on the finances of the pharmaceutical distribution company and the seamless maintenance of any ongoing public health vaccination programme.

Further to categorising impacts in terms of their order, the properties and characteristics of each impact can be analysed using several assessment criteria as outlined by Aznar-Crespo et al. (2021) albeit for flood impacts. This type of multicriteria analysis can be applied across a range of stakeholders whether they be policy makers, exposure units or interested/third parties. Table 9.7 presents a fictitious example using six criteria for the impact 'economic losses' for stakeholders in the sugar supply chain and allied activities. The economic losses arise because of a downturn in labour productivity mainly in the sugar cane production stage of the supply chain due to heatwave related worker illness and reduced hours of work. An outcome is an interruption to the provision of raw materials for upstream refining, manufacturing, and retail components of the commodity chain. Following Aznar-Crespo et al. (2021), the criteria are defined as follows. Nature refers to whether the impact is positive (+), negative (−) or neutral (n). In the case of sugar supply chain, a positive impact for stakeholders such as workers' union, labour policy maker and economic

Table 9.7 Illustrative impact assessment matrix for economic losses along sugar supply chain arising from heatwave related loss of labour productivity due to sugar cane worker illness and reduced hours of work. Scores are 1 = not at all important, 2 = slightly Important, 3 = important, 4 = fairly Important, and 5 = very Important. (Source: Author)

	Impact criteria					
Stakeholder	Nature	Intensity	Duration	Governance	Reversibility	Manageability
Sugar cane workers	−	5	5	1	1	3
Plantation owner	−	4	4	5	4	5
Workers' union	−	4	4	4	4	3
Sugar cane mill operator	−	3	4	1	4	4
Refinery operator	−	3	3	1	4	4
Sugar producer	−	2	2	1	5	4
Retailer	−	2	2	1	5	4
Labour policy maker	+	3	3	3	3	2
Economic planner	+	3	4	3	4	4

planner, is taken to mean an increase in related stakeholder activity. Intensity is the magnitude of the economic loss as measured by a suitable impact metric whereas duration is the length of time from the impact's appearance to disappearance or mitigation. Governance refers to the level of responsibility a particular stakeholder has for managing the impact. In part this is tied to the degree of connection of the stakeholder to the origin of economic loss which in this example is the heat related downturn in worker productivity. Impact reversibility is the degree to which the loss will be dissipated with the pre-heatwave event status quo restored. Manageability describes the prospect for managing and mitigating the impact and/or the feasibility to implement loss reduction strategies. Using a Likert scoring system, the impact assessment matrix can be populated with values indicating the level of importance associated with each of the impact characteristics by stakeholder. Turning to the illustrative example in Table 9.7 and focusing on sugar cane workers paid on an hours-worked only basis, the importance of the intensity of economic loss due to not being able to work is high (score of 5), with the same applying to impact duration especially if a worker is absent for a prolonged period due to heat-related illness and there is no facility for sickness pay. In relation to governance, the individual sugar cane worker has little influence over the level of responsibility for managing the economic loss apart from how they might manage their own finances if off work. The same applies to reversibility hence the low score of 1 indicating no or little importance. Lastly the sugar cane worker may possess some ability to put in place personal heat avoidance strategies to manage the risk of heat related illness and loss of wages. While working under heatwave conditions if compelled, these might include maintaining good hydration status, controlling radiative heat absorption via the type of clothing worn and indeed where possible, keeping that clothing moist to encourage evaporative heat loss. In the best-case situation such actions at the individual level may generate a manageability score of 3. Note for the stakeholder 'plantation owner', high scores are accorded to governance and manageability. This is because plantation owners possess the ability to put in place measures to reduce heatwave related risk amongst workers and thus avoid unwanted high levels of worker heat-related illness and associated economic loss. Acknowledging the complexity of the suite of possible impact criteria it is recommended that the scores for nature and intensity be estimated by the stakeholders themselves, whereas the scoring for the remaining criteria are undertaken by the impact analyst (Aznar-Crespo et al., 2021).

Armed with the information related to impact properties and characteristics, their order and any quantitative or qualitative measures of impact, ranking of impacts may be possible. Such a ranking of course will be influenced by the axiological values underlying any impact assessment and thus the measure thought best to describe the scale of impact (Aledo-Tur & Dominguez-Gomez, 2017). Although the Analytic Hierarchy Process (Forman & Gass, 2001) has been largely applied to hazards other than heatwaves (Alvarez et al., 2013; Dano, 2020; Dung et al., 2022; Malmasi et al., 2010), it would appear that it may offer some potential for assisting the development of an impact rank list based on importance, notwithstanding the different views of what constitutes importance.

An integral part of the assessment of impacts arising from a specific heatwave event is an evaluation of potential heatwave impact modifiers. These generally fall within the broad categories of exposure and vulnerability. As these risk elements are dynamic in both time and space their variation often manifests in heterogeneous patterns of heatwave impacts as might be observed via spatial variations in heat related mortality across a large city or the degree of grape sunburn within a vineyard. The general importance of exposure and vulnerability as impact modifiers will be expanded on in the next chapter.

9.4.5 *Step 5: Develop Heatwave Impact Management Strategies*

This last step in many ways encapsulates the reason for undertaking heatwave impact assessment that is the generation of information to assist the active management of heatwave related risk and thus impacts in human and natural systems. Given the vast range of management options this subject is treated separately in Chaps. 12, 13, and 14 under the chapeau of responses to heatwaves.

9.5 Conclusion

In this chapter, heatwave impacts have been viewed as the realisation of risk, with heatwave related risk a product of the synergistic interaction between four risk determinants namely heatwave attributes, heatwave related vulnerability, heatwave exposure and responses to heatwaves. Heatwave impacts can be tangible, intangible, direct or indirect and propagate through systems unfolding over space and time. Without knowledge about the nature of heatwave impacts, developing impact management solutions will remain a challenge. This is certainly the situation presently, as the systematic reporting of heatwaves and their impacts either locally, nationally, or globally has not yet been achieved. Added to this is the general lack of a methodology for assessing the impacts arising from heatwaves. To address this issue, some generic steps for undertaking a systematic heatwave impact assessment (HWIA) have been presented. These include settling on the purpose of the HWIA and establishing context, undertaking a conceptual mapping of likely impacts and dependencies, developing impact metrics and identifying stakeholders, locating evidence of impacts and assessing the impacts in terms of their nature, evolution, drivers and modifiers and finally, developing heatwave risk management strategies.

References

Aledo-Tur, A., & Dominguez-Gomez, J. A. (2017). Social impact assessment (SIA) from a multidimensional paradigmatic perspective: Challenges and opportunities. *Journal of Environmental Management, 195*, 56–61. https://doi.org/10.1016/j.jenvman.2016.10.060

Alexander, M., & Dessai, S. (2019). What can climate services learn from the broader services literature? *Climatic Change, 157*(1), 133–149. https://doi.org/10.1007/s10584-019-02388-8

Allen, M., Dube, O., Solecki, W., Aragón-Durand, F., Cramer, W., Humphreys, S., Kainuma, M., Kala, J., Mahowald, N., & Mulugetta, Y. (2018) *Global warming of 1.5 °C* (An IPCC special report on the impacts of global warming of 1.5 C above pre-industrial levels and related global greenhouse gas emission pathways, in the context of strengthening the global response to the threat of climate change, sustainable development, and efforts to eradicate poverty). Sustainable Development, and Efforts to Eradicate Poverty.

Alvarez, M., Moreno, A., & Mataix, C. (2013). The analytic hierarchy process to support decision-making processes in infrastructure projects with social impact. *Total Quality Management & Business Excellence, 24*(5–6), 596–606. https://doi.org/10.1080/14783363.2012.669561

Arii, M. (2013). Rapid assessment in disasters. *Journal of the Medical Association of Japan, 56*(1), 19–24.

Asrar, G. R., Ryabinin, V., & Detemmerman, V. (2012). Climate science and services: Providing climate information for adaptation, sustainable development and risk management. *Current Opinion in Environmental Sustainability, 4*(1), 88–100. https://doi.org/10.1016/j.cosust.2012.01.003

Aznar-Crespo, P., Aledo, A., Melgarejo-Moreno, J., & Vallejos-Romero, A. (2021). Adapting social impact assessment to flood risk management. *Sustainability, 13*(6). https://doi.org/10.3390/su13063410

Bala, B. K., Arshad, F. M., & Noh, K. M. (2017). Causal loop diagrams. *System Dynamics: Modelling and Simulation:37–51*. https://doi.org/10.1007/978-981-10-2045-2_3

Bednar-Friedl, B., Biesbroek, R., & Schmidt, D. N. (2022). *IPCC sixth assessment report (AR6): Climate change 2022-impacts, adaptation and vulnerability*. Regional Factsheet Europe.

Benzie, M., & Persson, A. (2019). Governing borderless climate risks: Moving beyond the territorial framing of adaptation. *International Environmental Agreements-Politics Law and Economics, 19*(4–5), 369–393. https://doi.org/10.1007/s10784-019-09441-y

Benzie, M., Carter, T. R., Carlsen, H., & Taylor, R. (2019). Cross-border climate change impacts: Implications for the European Union. *Regional Environmental Change, 19*(3), 763–776. https://doi.org/10.1007/s10113-018-1436-1

Berariu, R., Fikar, C., Gronalt, M., & Hirsch, P. (2015). Understanding the impact of cascade effects of natural disasters on disaster relief operations. *International Journal of Disaster Risk Reduction, 12*, 350–356. https://doi.org/10.1016/j.ijdrr.2015.03.005

Birkmann, J., Cardona, O. D., Carreno, M. L., Barbat, A. H., Pelling, M., Schneiderbauer, S., Kienberger, S., Keiler, M., Alexander, D., Zeil, P., & Welle, T. (2013). Framing vulnerability, risk and societal responses: The MOVE framework. *Natural Hazards, 67*(2), 193–211. https://doi.org/10.1007/s11069-013-0558-5

Brimicombe, C., Di Napoli, C., Cornforth, R., Pappenberger, F., Petty, C., & Cloke, H. L. (2021). Borderless heat hazards with bordered impacts. Earths. *Futures, 9*(9). https://doi.org/10.1029/2021ef002064

Carter, T. R., Intergovernmental Panel on Climate Change, Working G, II, Carter, T. R., Kokuritsu Kankyō Kenkyūjo, Chikyū Kankyō Kenkyū S, University College LDoG. (1994). *IPCC Technical guidelines for assessing climate change impacts and adaptations: Part of the IPCC special report to the first session of the Conference of the parties to the UN Framework Convention on Climate Change. Climate change impacts & adaptions*. Department of Geography, University College London Center for Global Environmental Research.

Carter, T. R., Benzie, M., Campiglio, E., Carlsen, H., Fronzek, S., Hilden, M., Reyer, C. P. O., & West, C. (2021). A conceptual framework for cross-border impacts of climate change. *Global*

Environmental Change-Human and Policy Dimensions, 69, 102307. https://doi.org/10.1016/j.gloenvcha.2021.102307

Cavallo, A., & Ireland, V. (2014). Preparing for complex interdependent risks: A system of systems approach to building disaster resilience. *International Journal of Disaster Risk Reduction, 9*, 181–193. https://doi.org/10.1016/j.ijdrr.2014.05.001

Challinor, A. J., Adger, W. N., & Benton, T. G. (2017). Climate risks across borders and scales. *Nature Climate Change, 7*(9), 621–623. https://doi.org/10.1038/nclimate3380

Challinor, A. J., Adger, W. N., Benton, T. G., Conway, D., Joshi, M., & Frame, D. (2018). Transmission of climate risks across sectors and borders. *Philosophical Transactions of the Royal Society a-Mathematical Physical and Engineering Sciences, 376*(2121), 20170301. https://doi.org/10.1098/rsta.2017.0301

Corvo, L., Pastore, L., Manti, A., & Iannaci, D. (2021). Mapping social impact assessment models: A literature overview for a future research agenda. *Sustainability, 13*(9), 4750.

Cuartas, J. B., Frazier, T., & Wood, E. (2021). The application of cascading consequences for emergency management operations. *Natural Hazards, 108*(3), 2919–2938. https://doi.org/10.1007/s11069-021-04807-5

Dano, U. L. (2020). Flash flood impact assessment in Jeddah City: An analytic hierarchy process approach. *Hydrology, 7*(1). https://doi.org/10.3390/hydrology7010010

Di Napoli, C., Allen, T., Mendez-Lazaro, P. A., & Pappenberger, F. (2023). Heat stress in the Caribbean: Climatology, drivers, and trends of human biometeorology indices. *International Journal of Climatology, 43*(1), 405–425. https://doi.org/10.1002/joc.7774

Dominguez, L., Bonadonna, C., Frischknecht, C., Menoni, S., & Garcia, A. (2021). Integrative post-event impact assessment framework for volcanic eruptions: A disaster forensic investigation of the 2011–2012 eruption of the Cordón Caulle volcano (Chile). *Frontiers in Earth Science, 9*, 645945.

Dung, N. B., Long, N. Q., Goyal, R., An, D. T., & Minh, D. T. (2022). The role of factors affecting flood hazard zoning using analytical hierarchy process: A review. *Earth Systems and Environment., 6*, 697–713. https://doi.org/10.1007/s41748-021-00235-4

Esteves, A. M., Franks, D., & Vanclay, F. (2012). Social impact assessment: The state of the art. *Impact Assessment and Project Appraisal, 30*(1), 34–42. https://doi.org/10.1080/14615517.2012.660356

Ferreira, A. M., Marchezini, V., Mendes, T. S. G., Trejo-Rangel, M. A., & Iwama, A. Y. (2023). A systematic review of forensic approaches to disasters: Gaps and challenges. *International Journal of Disaster Risk Science, 14*(5), 722–735. https://doi.org/10.1007/s13753-023-00515-9

Forman, E. H., & Gass, S. I. (2001). The analytic hierarchy process – An exposition. *Operations Research, 49*(4), 469–486. https://doi.org/10.1287/opre.49.4.469.11231

Gall, M., Borden, K. A., & Cutter, S. L. (2009). When do losses count? Six fallacies of natural hazards loss data. *Bulletin of the American Meteorological Society, 90*(6), 799–810. https://doi.org/10.1175/2008bams2721.1

Gambetta, J. M., Holzapfel, B. P., Stoll, M., & Friedel, M. (2021). Sunburn in grapes: A review. *Frontiers in Plant Science, 11*. https://doi.org/10.3389/fpls.2020.604691

Gaupp, F. (2020). Extreme events in a globalized food system. *One Earth, 2*(6), 518–521. https://doi.org/10.1016/j.oneear.2020.06.001

Goode, E. J., Thomas, E., Landeg, O., Duarte-Davidson, R., Hall, L., Roelofs, J., Schulpen, S., De Bruin, A., Wigenstam, E., Liljedahl, B., Waleij, A., Simonsson, L., & Nyberg, A. G. (2021). Development of a rapid risk and impact assessment tool to enhance response to environmental emergencies in the early stages of a disaster: A tool developed by the European multiple environmental threats emergency NETwork (EMETNET) project. *International Journal of Disaster Risk Science, 12*(4), 528–539. https://doi.org/10.1007/s13753-021-00352-8

Harvey, N. (1998). *Environmental impact assessment: Procedures, practice and prospects*. Oxford University Press.

Hedlund, J., Fick, S., Carlsen, H., & Benzie, M. (2018). Quantifying transnational climate impact exposure: New perspectives on the global distribution of climate risk. *Global Environmental Change-Human and Policy Dimensions, 52*, 75–85. https://doi.org/10.1016/j.gloenvcha.2018.04.006

Jacobs, K. L., & Street, R. B. (2020). The next generation of climate services. *Climate Services, 20*, 100199. https://doi.org/10.1016/j.cliser.2020.100199

Kandlikar, M., Zerriffi, H., & Lem, C. H. (2011). Science, decision-making and development: Managing the risks of climate variation in less-industrialized countries. *Wiley Interdisciplinary Reviews-Climate Change, 2*(2), 201–219. https://doi.org/10.1002/wcc.98

Kates, R. W., Ausubel, J., Berberian, M., & Ausubel, J. H. (1985). *Climate impact assessment: Studies of the interaction of climate and society* (SCOPE) (Vol. 27). Wiley/Chichester West Sussex.

Keating, A., Venkateswaran, K., Szoenyi, M., MacClune, K., & Mechler, R. (2016). From event analysis to global lessons: Disaster forensics for building resilience. *Natural Hazards and Earth System Sciences, 16*(7), 1603–1616. https://doi.org/10.5194/nhess-16-1603-2016

Keller, K., Helgeson, C., & Srikrishnan, V. (2021). Climate risk management. *Annual Review of Earth and Planetary Sciences, 49*, 95–116. https://doi.org/10.1146/annurev-earth-080320-055847

Kemm, J. R., Parry, J., Palmer, S., & Oxford Scholarship Online Public Health and E. (2004). *Health impact assessment : Concepts, theory, techniques, and applications*. Oxford Medical Publications/Oxford University Press.

Klinenberg, E. (1999). Denaturalizing disaster: A social autopsy of the 1995 Chicago heat wave. *Theory and Society, 28*(2), 239–295. https://doi.org/10.1023/a:1006995507723

Köhler, P., Müller, M., Sanders, M., & Wächter, J. (2006). Data management and GIS in the Center for Disaster Management and Risk Reduction Technology (CEDIM): From integrated spatial data to the mapping of risk. *Natural Hazards and Earth System Sciences, 6*(4), 621–628.

Kovats, R. S., & Hajat, S. (2008). Heat stress and public health: A critical review. *Annual Review of Public Health, 29*, 41–55. https://doi.org/10.1146/annurev.publhealth.29.020907.090843

Kruger, E. L., & Di Napoli, C. (2022). Feasibility of climate reanalysis data as a proxy for onsite weather measurements in outdoor thermal comfort surveys. *Theoretical and Applied Climatology, 149*(3–4), 1645–1658. https://doi.org/10.1007/s00704-022-04129-x

Kryvasheyeu, Y., Chen, H. H., Obradovich, N., Moro, E., Van Hentenryck, P., Fowler, J., & Cebrian, M. (2016). Rapid assessment of disaster damage using social media activity. *Science Advances, 2*(3). https://doi.org/10.1126/sciadv.1500779

Lee, C.-C., Huang, K.-C., Kuo, S.-Y., Cheng, C.-K., Tung, C.-P., & Liu, T.-M. (2021). Development of a social impact assessment for the water environment: A professional perspective. *Water, 13*(23), 3355.

Likert, R., Roslow, S., & Murphy, G. (1934). A simple and reliable method of scoring the thurstone attitude scales. *Journal of Social Psychology, 5*(2), 228–238. https://doi.org/10.1080/00224545.1934.9919450

Lock, K. (2000). Health impact assessment. *British Medical Journal, 320*(7246), 1395–1398. https://doi.org/10.1136/bmj.320.7246.1395

Louis-Charles, H. M., Howard, R., Remy, L., Nibbs, F., & Turner, G. (2020). Ethical considerations for postdisaster fieldwork and data collection in the Caribbean. *American Behavioral Scientist, 64*(8), 1129–1144.

Malmasi, S., Jozi, S. A., Monavari, S. M., & Jafarian, M. E. (2010). Ecological impact analysis on Mahshahr petrochemical industries using analytic hierarchy process method. *International Journal of Environmental Research, 4*(4), 725–734.

Masson-Delmotte, V., Zhai, P., Pörtner, H.-O., Roberts, D., Skea, J., & Shukla, P. R. (2022). *Global warming of 1.5 C: IPCC special report on impacts of global warming of 1.5 C above pre-industrial levels in context of strengthening response to climate change, sustainable development, and efforts to eradicate poverty*. Cambridge University Press.

McGregor, G. (2015). Climatology in support of climate risk management: A progress report. *Progress in Physical Geography, 39*(4), 536–553. https://doi.org/10.1177/0309133315578941

Mendoza, M. T., & Schwarze, R. (2019). Sequential disaster forensics: A case study on direct and socio-economic impacts. *Sustainability, 11*(21), 5898.

Moser, S. C., & Hart, J. A. F. (2015). The long arm of climate change: Societal teleconnections and the future of climate change impacts studies. *Climatic Change, 129*(1–2), 13–26. https://doi.org/10.1007/s10584-015-1328-z

Oliver-Smith, A., Alcantara-Ayala, I., Burton, I., & Lavell, A. (2017). The social construction of disaster risk: Seeking root causes. *International Journal of Disaster Risk Reduction, 22*, 469–474. https://doi.org/10.1016/j.ijdrr.2016.10.006

Peek, L., Champeau, H., Austin, J., Mathews, M., & Wu, H. (2020). What methods do social scientists use to study disasters? An analysis of the social science extreme events research network. *American Behavioral Scientist, 64*(8), 1066–1094.

Pescaroli, G., & Alexander, D. (2016). Critical infrastructure, panarchies and the vulnerability paths of cascading disasters. *Natural Hazards, 82*(1), 175–192. https://doi.org/10.1007/s11069-016-2186-3

Pescaroli, G., & Alexander, D. (2018). Understanding compound, interconnected, interacting, and cascading risks: A holistic framework. *Risk Analysis, 38*(11), 2245–2257. https://doi.org/10.1111/risa.13128

Ratner, P. A., Green, L. W., Frankish, C. J., Chomik, T., & Larsen, C. (1997). Setting the stage for health impact assessment. *Journal of Public Health Policy, 18*(1), 67–79. https://doi.org/10.2307/3343358

Rotmans, J., & VanAsselt, M. (1996). Integrated assessment: A growing child on its way to maturity – An editorial essay. *Climatic Change, 34*(3–4), 327–336. https://doi.org/10.1007/bf00139296

Simpson, N. P., Mach, K. J., Constable, A., Hess, J., Hogarth, R., Howden, M., Lawrence, J., Lempert, R. J., Muccione, V., Mackey, B., New, M. G., O'Neill, B., Otto, F., Portner, H. O., Reisinger, A., Roberts, D., Schmidt, D. N., Seneviratne, S., Strongin, S., van Aalst, M., Totin, E., & Trisos, C. H. (2021). A framework for complex climate change risk assessment. *One Earth, 4*(4), 489–501. https://doi.org/10.1016/j.oneear.2021.03.005

Sword-Daniels, V. L., Twigg, J., & Loughlin, S. C. (2015). Time for change? Applying an inductive timeline tool for a retrospective study of disaster recovery in Montserrat, West Indies. *International Journal of Disaster Risk Reduction, 12*, 125–133. https://doi.org/10.1016/j.ijdrr.2014.12.006

Van Brown, B. L. (2020). Disaster research "Methics": Ethical and methodological considerations of researching disaster-affected populations. *American Behavioral Scientist, 64*(8), 1050–1065.

Vanclay, F. (2006). The concepts, process and methods of social impact assessment. *Geographical Research, 44*(3), 329–330.

Vanclay, F. (2020). Reflections on social impact assessment in the 21st century. *Impact Assessment and Project Appraisal, 38*(2), 126–131.

Vespignani, A. (2010). Complex networks: The fragility of interdependency. *Nature, 464*(7291), 984–985. https://doi.org/10.1038/464984a

Weichselgartner, J., & Arheimer, B. (2019). Evolving climate services into knowledge-action systems. *Weather Climate and Society, 11*(2), 385–399. https://doi.org/10.1175/wcas-d-18-0087.1

Winkler, M. S., Furu, P., Viliani, F., Cave, B., Divall, M., Ramesh, G., Harris-Roxas, B., & Knoblauch, A. M. (2020). Current global health impact assessment practice. *International Journal of Environmental Research and Public Health, 17*(9). https://doi.org/10.3390/ijerph17092988

Wu, H., Peek, L., Mathews, M. C., & Mattson, N. (2022). Cultural competence for hazards and disaster researchers: Framework and training module. *Natural Hazards Review, 23*(1), 06021005.

Zaidi, R. Z. (2018). Beyond the Sendai indicators: Application of a cascading risk lens for the improvement of loss data indicators for slow-onset hazards and small-scale disasters. *International Journal of Disaster Risk Reduction, 30*, 306–314. https://doi.org/10.1016/j.ijdrr.2018.03.022

Zscheischler, J., Martius, O., Westra, S., Bevacqua, E., Raymond, C., Horton, R. M., van den Hurk, B., AghaKouchak, A., Jezequel, A., Mahecha, M. D., Maraun, D., Ramos, A. M., Ridder, N. N., Thiery, W., & Vignotto, E. (2020). A typology of compound weather and climate events. *Nature Reviews Earth & Environment, 1*(7), 333–347. https://doi.org/10.1038/s43017-020-0060-z

Chapter 10
Exposure and Vulnerability as Modifiers of Heatwave Impacts

Abstract This chapter considers how exposure and vulnerability work to shape the impacts of heatwaves across various scales. It begins by defining exposure as the extent to which an entity encounters heat, emphasizing the critical dimensions of frequency, duration, and intensity. To lend a geographical perspective to heat exposure the concept of heatwave climates is introduced, identifying regions where significant heatwave-related human mortality has been recorded, using the Köppen-Geiger climate classification as a framework. The chapter then explores vulnerability, defined as the propensity to be adversely affected, and outlines factors that determine vulnerability for human populations, including demographic, health, physical, socioeconomic, behavioural, and institutional factors. How socioeconomic factors, such as poverty and social isolation, may exacerbate heatwave related vulnerability is emphasised. Because smoothly functioning infrastructure is critical to society and economy and heatwaves can compromise the integrity of transportation, energy, and water systems, heatwave related vulnerability of infrastructure is also outlined with the issue of how failures in one system can cascade into others, exacerbating the overall impact of heatwaves highlighted. In addition to human populations, the chapter examines the vulnerability of natural systems. It discusses some of the factors that determine exposure and make terrestrial, marine, and aquatic ecosystems heatwave vulnerable, noting that the health status of these systems significantly affects their ability to withstand heat stress. The chapter concludes by emphasizing the need for further research on the determinants of heatwave exposure and vulnerability especially in the case of natural systems and infrastructure.

Keywords Exposure unit · Heatwave exposure · Heatwave vulnerability · Determinants · Heatwave climates · Human and animal populations · Infrastructure · Marine ecosystems · Terrestrial and aquatic ecosystems

G. McGregor, *Heatwaves*, Biometeorology 6,
https://doi.org/10.1007/978-3-031-69906-1_10

10.1 Introduction

In the previous chapter impacts were conceptualised as the consequences arising from the realisation of risk, with risk a product of the synergistic effects of hazard, exposure, vulnerability, and response. Framed in this way heatwave risk and therefore impact can be dynamic, depending not only on the physical attributes of the heatwave event, but also the degree of exposure and vulnerability of the object experiencing the heatwave. Given this, the impacts arising from a specific heatwave can be context specific with the context largely determined by the exposure unit's degree of exposure and vulnerability. For this reason, how exposure and vulnerability might modify heatwave related risk and thus influence the level of impact across a range of scales from the individual to the societal or ecosystem level is explored in this chapter. In doing so, the material presented here lays the groundwork for understanding many of the impacts of heatwaves on human and natural systems described in the following chapter. Apart from the closing section that offers a synthesis of contents, this chapter is comprised of two major sections. The section on exposure explores the nature of exposure, its controlling factors, and the broad approaches to establishing heatwave related exposure. It also introduces the idea of heatwave climates in relation to understanding a climate regime-based understanding of exposure. Vulnerability as a concept is then explored within the general framing that vulnerability modifies the relationship between exposure and impact. An overview of the factors that determine vulnerability, based on a consideration of the heatwave vulnerability literature, for human populations, economy and society, infrastructure, animal populations and marine, terrestrial and aquatic ecosystems follows.

10.2 Exposure

Exposure is largely a product of the degree to which an exposure unit, whether animate or inanimate, encounters heat. The critical dimensions of exposure are how often, for how long and how much an object is subject to critically damaging heat or in other words the frequency, duration and intensity of the exposure. That said, what constitutes a critical level plays a role in defining what type of exposure has occurred. This issue is at the heart of the debate related to exposure levels based purely on physical parameters of heat irrespective of any impact outcome. For example, in human population level studies, the 95th percentile value of maximum temperature is often adopted as a proxy for exposure intensity—in essence it is a statistical construct of criticality. Given this, an increasing number of researchers are turning their attention to ascertaining impact-sensitive levels of heat, that is levels of heat beyond which discernible impacts occur. Understandably, the range of critical impact-sensitive heat values is potentially vast as these will be tied to the heat tolerance of the object or exposure unit under consideration, with levels of heat tolerance dependent on an exposure unit's level of acclimatisation or 'heat-hardening'. Added to the complexity of exposure unit—to—exposure unit variability in tolerance to heat exposure are temporal and spatial dimensions. In the case of human populations this can be clearly seen in the geographical variation of the so-called minimum mortality temperature with a combination of physiological and societal adaptations to heat often moderating the response to heat exposure and thus the level of heatwave related health impacts. Impact sensitive levels of exposure may also alter over the duration of the heat season especially in living organisms as levels of intra-seasonal heat adaptation evolve. Again, taking the example of human populations, the health effects of early heat-season heatwaves can be greater than those later in the season. This is because of the low level of early heat-season acclimatisation, especially amongst vulnerable populations such as the elderly. Accordingly, what might be considered as a 'lethal' exposure will vary throughout the heat-season as heat acclimatisation builds thus moderating the level of health impact for a given exposure. There is also the possibility that tolerance to exposure may alter over time because of adaptation to heat stress as evidenced from trends in the minimum mortality temperature for humans (Folkerts et al., 2020; Yin et al., 2019). For some animal and plant organisms, there is evidence that phenotypic plasticity, that is the ability of an organism to alter its form, developmental processes, biochemical and physiological processes as well as behaviour, is invoked as a response to coping with heat exposure over a range of timescales (Mottola et al., 2022; Rodrigues & Beldade, 2020).

For a single heatwave event, the frequency dimension of exposure may be irrelevant if heat is maintained above a level critical for the duration of the heatwave. On the other hand, exposure frequency may be important in the case of episodic exposure, as might arise with the diurnal cycle of radiative load as one component of heat, with the duration and intensity of those exposures critical over the lifetime of the heatwave event. Therefore, at the intra-event timescale the cumulative effects of heat associated with the heatwave become important as determined by the breaks between any episodic exposure and any associated respite or the length of uninterrupted exposure. Indeed, for heat exposure to be damaging, prolonged exposure to moderate levels of heat may not be necessary during a heatwave. For example, human mental health effects may emerge among some members of the population because of exposure to levels of heat above average but below those considered critical for other health related manifestations of heat illness. Further, the notions of acute and chronic exposure may be relevant with the former related to exposure during a single event while the later refers to heatwave exposure over a longer duration as measured by the number of heatwaves experienced over a given period such as a season or multiple years making the frequency dimension additionally important.

At the macro-level and assuming a heatwave is defined in relation to the degree of excursion of heat from a climatological value for a given reference period, then exposure to the effects of heatwaves is not geographically bound—heatwave exposure could occur anywhere. However, in the context of limits, for example heat-related physiological limits for humans and animal and plant species, heatwave exposure will be confined to those areas where such limits have a reasonable chance of being exceeded. Again, at the macro-level this will be broadly set by the climate type whether it be for terrestrial or marine environments. In this respect the idea of heatwave climates, as defined by those climates where discernible heatwave impacts have been recorded might serve as a useful construct for signposting where impactful heatwave exposure is most likely. To this end and for impacts on humans, Fig. 10.1, by adopting a bottom-up approach (impact used to identify causative physical conditions), identifies the Köppen-Geiger climate types for which significant heatwave related human mortality has been recorded in the EM-DAT data base and established via a review of the heatwave literature. The Köppen-Geiger climate types and their associated subtypes are set out in Table 10.1.

Of the tropical climate types, the wet/dry or savannah climate type (Aw) appears important and mainly so for the Asian region, especially southern Asia and India where regions such as Odisha, Andhra Pradesh and Tamil Nadu have recorded high heatwave related death tolls. Interestingly, the Aw climate is one which has a wet summer and dry winter. This may seem contradictory as a wet summer, an outcome of the arrival of the summer monsoon rains, might be expected to reduce summer heat extremes, and therefore reduce the risk of heatwaves. While this is true, a closer look at the timing of deadly heatwaves in regions with Aw climate reveals that these occur in the hot dry period immediately preceding the arrival of monsoon rains. The severe heatwaves experienced over India in the 2022 pre-monsoon period with

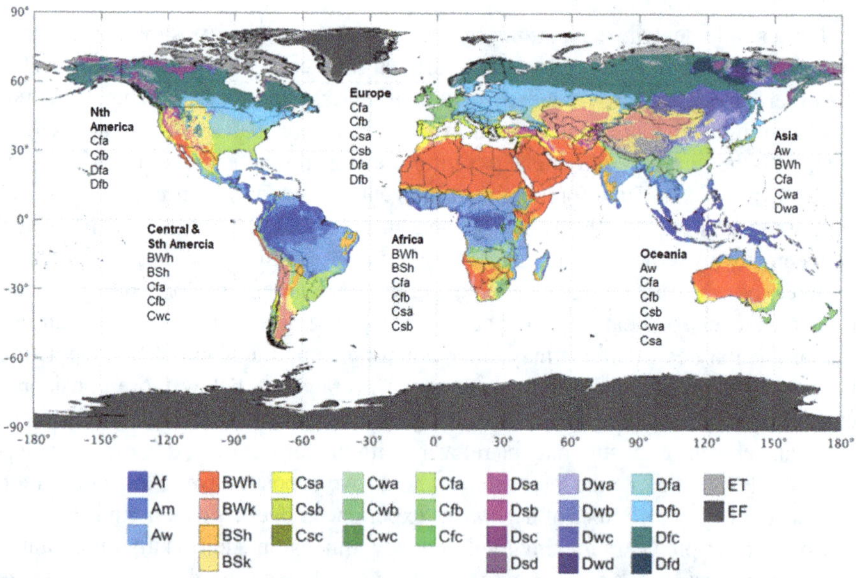

Fig. 10.1 Köppen-Geiger climate classification with climate types by broad geographical region for which heatwave related human mortality has been recorded in the EM-DAT data base or in the literature. Listing of climate types by major regions done by author. See Table 10.1 for legend description. Köppen-Geiger climate classification map from Beck et al. (2018) Present and future Köppen-Geiger climate classification maps at 1-km resolution. Sci Data 5, 180214. https://doi.org/10.1038/sdata.2018.214 and used under Creative Commons Attribution 4.0 International License http://creativecommons.org/licenses/by/4.0/

Table 10.1 Köppen-Geiger climate types

Climate Type and Subtype	Description
A	Tropical climates
Af	Tropical rainforest climate
Am	Tropical monsoon climate
Aw	Tropical savanna climate with dry-winter characteristics
As	Tropical savanna climate with dry-summer characteristics
B	Dry (desert and semi-arid) climates
BWh	Arid hot desert
BWc	Arid cold desert
BSh	Semi-arid hot
BSc	Semi-arid cold
C	Temperate climates
Csa	Mediterranean hot summer
Csb	Mediterranean warm/cool summer
Csc	Mediterranean cold summer
Cfa	Humid subtropical
Cfb	Oceanic climate
Cfc	Subpolar oceanic climate
Cwa	Dry-winter humid subtropical climate
Cwb	Dry-winter subtropical highland climate
Cwc	Dry-winter cold subtropical highland climate
Group D	Continental climates
Dfa	Hot-summer humid continental climate
Dfb	Warm-summer humid continental climate
Dfc	Subarctic climate
Dfd	Extremely cold subarctic climate
Dwa	Monsoon-influenced hot-summer humid continental
Dwb	Monsoon-influenced warm-summer humid continental
Dwc	Monsoon-influenced subarctic climate
Dwd	Monsoon-influenced extremely cold subarctic climate
Dsa	Mediterranean-influenced hot-summer humid continental
Dsb	Mediterranean-influenced warm-summer humid continental climate
Dsc	Mediterranean-influenced subarctic climate
Dsd	Mediterranean influenced extremely cold subarctic climate
E	Polar climates
ET	Tundra climate
EF	Ice cap climate

Based on Beck et al. (2018)

record breaking temperatures is a good example of this. For Africa, the hot desert (BWh) and hot semi-arid desert (BSh) climates associate with recorded heat related mortalities. As yet, locations falling within the Aw climate regions of Africa appear to have not recorded significant heatwave related mortality. This of course could be a result of a lack of reporting as opposed to no incidences of heatwave related impacts. Arid climates are also important for parts of the Americas and northern India, Western Asia, and the Middle East. There are few temperate climate types for which heatwave associated mortality has not been recorded. The humid subtropical climate type (Cfa) is where mega-cities such as Shanghai, Tokyo, New York and Sao Paulo are found, all of which have experienced noteworthy heatwave related human impacts often associated with tropical maritime airmasses. Other major Cfa cities where discernible heatwave impacts have been recorded include Sydney and Buenos Aires. Located poleward of the Cfa climate type, the temperate oceanic climate type (Cfb), which has a cooler summer climate than Cfa locations, typifies cities where significant heatwave impacts have occurred on several occasions. These include Melbourne, London and Paris where the regular passage of high and low pressure systems, and occasional perturbations of the upper westerly wind pattern can lead to periods of heatwave conducive blocking (see Chap. 6). A pertinent example of this is the unprecedented Pacific Northwest heatwave of June 2021 and its associated heat dome that arose from a period of intense blocking. This rendered the relatively benign Cfb climates of Seattle and Vancouver as ones akin to a desert like BSh climate for approximately 2 weeks. These conditions and the associated high human and animal death tolls were in many ways a reminder that locations with summer climates that are often perceived as mild can be exposed to severe heat conditions. Taking its name from the Mediterranean region, harsh summers typify the hot dry Mediterranean climate type (Csa). As summer temperature extremes are a characteristic of this climate type, unsurprisingly the likelihood of heatwaves in regions typified by Csa conditions is high. Accordingly, Mediterranean cities such as Rome, Marseilles, Seville, Athens and Casablanca have reported deaths associated with heatwaves. Outside the Mediterranean region, Csa climates are found across the mid to southern regions of California and the southwestern parts of South Africa and Australia where heatwaves with clear societal impacts have occurred in cities such as Los Angeles, Cape Town and Perth respectively. Frequently heatwaves associated with the Csa climate type are connected to the advection of hot continental airmasses from upwind semi-arid to arid regions. The movement of such hot dry air over major urban areas can expose large numbers of the population to extreme heat with advected hot air possessing the potential to amplify the urban heat island effect. Amongst the continental climate types, notable heatwave related impacts appear to be associated with the hot summer humid (Dfa), warm summer humid (Dfb) and monsoon-influenced hot-summer humid (Dwa) continental types. Example cities representing these climate types and where significant levels of heat related health effects have been recorded include Chicago (Dfa), Montreal, Stockholm and Moscow (Dfb), and Seoul and Beijing (Dwa). Unlike the ‘A’, ‘B’ and ‘C’ climate types (Table 10.1), the continental ‘D’ climate types are characterised by stark contrasts in their summer and winter temperatures with both heat and cold extremes typical. Generally, the populations that reside in locations possessing this climate type are better adapted to cold as opposed to heat extremes which may well play a role in determining exposure levels to extreme heat and thus impacts.

An alternative method for establishing population heatwave exposure is to adopt a top-down approach. This involves three broad steps. Firstly, a heat index is chosen with the identification of an associated threshold index value as well as the specification of the minimum number of days that threshold must be exceeded to be considered a heatwave event. As indicated earlier (Chap. 4), the heat index and its threshold are often selected without direct regard to impacts. Next the number of heatwave days is calculated for the period of analysis such as an individual year.

The third step involves finding the product of the number of heatwaves days and the total number of people in the population or a specific population group for the area and period of interest (Coffel et al., 2018; Tuholske et al., 2021). The result is the number of person days of heatwave exposure. An example of this type of analysis for changes in urban extreme heat exposure with a duration of 2 days or more—the study's definition of a heatwave—for the period 1983 to 2016 is presented in Fig. 10.2. This clearly reveals the regions where exposure has changed most drastically as measured by the rate of increase of urban population exposure to extreme heat (Fig. 10.2a), the rate of increase in the total number of days per year that a threshold value of a heat index is exceeded (Fig. 10.2b) and the degree of contribution of observed warming or population growth to changes in heat exposure (Fig. 10.2c). Turning back to the idea of heatwave climates, the contiguous regions for which large changes in the number of days exceeding a WBGT maximum of 30 °C have been recorded and for which warming accounts for most of the exposure change, align with the broad Tropical and Arid climate categories, especially Am/Aw and Bwh/Bsh respectively.

Shifting from the macro-scale to smaller scales, several factors can influence the duration and intensity dimensions of exposure to heat during a heatwave. For humans and animals, direct exposure to the radiative heat load from the sun can be avoided by shade-seeking behaviour as in shaded environments not only is the radiative load less but also the air temperature. Such behaviour is displayed for example by savannah elephants during extreme periods of heat (Mole et al., 2016). Active adjustment to levels of heat exposure and the potential to moderate the level of heat-related impacts through mobility is not afforded to sessile living systems such as terrestrial, marine, and freshwater plant communities and some marine and freshwater animal communities. Such communities during prolonged heatwave events must endure high levels of heat exposure whether in the form of radiative load and/or air or water temperature. That said, research over the last decade or so has revealed sessile organisms in non-arid and non-semi-arid climates have developed heat adaptation mechanisms such as heat stress memory and plasticity to temporarily cope with short term heat stress as experienced during atmospheric and marine heatwaves and thus reduce the possibility of severe heat related impacts (Ahrens et al., 2021; Clemente et al., 2023; Bäurle, 2016; Hackerott et al., 2021; Nguyen et al., 2020). Such adaptations differ from the evolutionary ones evident for example for sessile plant communities found in arid and semi-arid environments including root structure, leaf waxing, night blooming, reproducing without seeds and general desiccation resistance.

While the general importance of heat exposure at the individual organism level has been understood for several decades as a moderator of impact, the last 10–15 years has witnessed a significant uptick in research activity specifically directed at the controls on and impacts of individual heat exposure (Gaughan et al., 2020; Kuras et al., 2017). In the case of livestock, this has been spurred on by the impacts of a changing frequency of heatwaves and ever-increasing levels of heat stress on animal health and productivity (Lees et al., 2019). The significant surge in activity related to individual human heat exposure emanates from the fields of

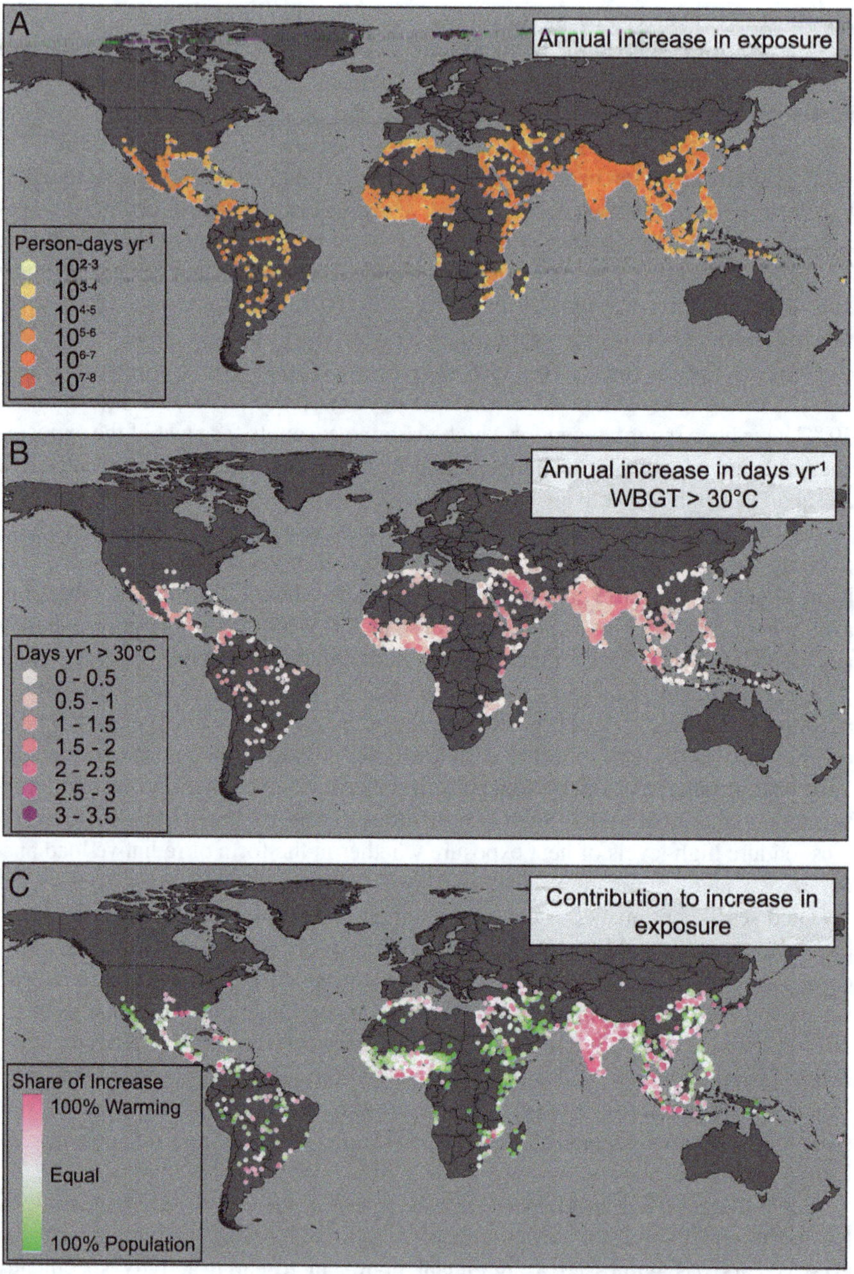

Fig. 10.2 Changes in heat exposure 1983–2016. (**a**) urban-level rate of change of population heat exposure, (**b**) the rate of increase in the total number of days per year when maximum value of WBGT >30 °C, and (**c**) the degree to which exposure can be attributed to warming or population growth. (Source: Tuholske et al. (2021). Used under Creative Commons Attribution Non-Commercial No-Derivatives License 4.0 (CC BY-NC-ND), https://creativecommons.org/licenses/by-nc-nd/4.0/)

occupational health, sports performance and urban climate science and urban planning. Of pressing interest is the extent to which the degree of heat exposure amongst people involved in outdoor work, recreational or competitive activities during heatwaves influences a range of health impacts (Alele et al., 2020a; Varghese et al., 2020; Hosokawa et al., 2019; Morabito et al., 2021; Vanos et al., 2017; Venugopal et al., 2021; Xiang et al., 2014). At the street canyon to city scale, how heatwaves may amplify urban heat and increase levels of heat exposure has attracted much interest amongst the urban climate science and urban planning communities, especially in the context of the increasing emergence of megacities in climate regions conducive to heatwaves and the ideal of planning healthy cities (Ren & McGregor, 2021). As discussed later, where a person lives or works in a large urban area or indeed a building may influence their level of heat exposure. For humans there are a range of personal heat exposure modifying factors that may well moderate heat related impacts on health and well-being. These include behavioural patterns, variations in metabolic rate and physiological strain as determined by activity level, and psychological determinants (Kuras et al., 2015; Kuras et al., 2017; Nazarian & Lee, 2021). In addition to shade seeking behaviour, adjusting body orientation is a way that terrestrial animals, and humans for that matter, can control heat load to minimise exposure and any associated impacts (Lees et al., 2019).

10.3 Vulnerability

Table 9.1 defines vulnerability as the propensity or predisposition to be adversely affected. The term, which is equally applicable to natural and human systems, encompasses a variety of concepts and elements including sensitivity or susceptibility to harm and lack of capacity to cope and adapt. In the case of heatwaves, the idea of vulnerability has been largely applied to human based activity systems, in particular individuals or specific societal groups, with little attention given to natural systems and inanimate objects such as built infrastructure. Essentially vulnerability traits modify the relationship between exposure and impact. This section will present on overview of the heatwave related vulnerability factors for human and animal populations, with a focus on health, and attempt to outline in general terms some of the determinants of vulnerability for economy and society in general as well as infrastructure and ecosystems.

10.3.1 Human Populations

Although the concept of vulnerability when applied to human populations carries with it a range of interpretations and criticisms (Carraro et al., 2021; Eriksen, 2022; Mikulewicz, 2019) it provides a useful framework to consider the interconnected factors that may moderate heatwave related health risk and thus impacts. The search for the broad socioeconomic, political and cultural drivers of heatwave related vulnerability cuts across the social determinants of health risk, political ecology and environmental justice paradigms (Braveman & Gottlieb, 2014; Robbins, 2019; Schlosberg, 2007). It is also tied in with the concept of 'hazard of place vulnerability' as espoused by Cutter et al. (2003) given that the balance of vulnerability drivers may vary from place to place. This section will primarily focus on those vulnerability factors important for population health.

There is a burgeoning literature on the risk factors that underlie heatwave related vulnerability (Benmarhnia et al., 2015; Bezgrebelna et al., 2021; Ellena et al., 2020; Mayrhuber et al., 2018; Sera et al., 2019; Shih & Mabon, 2021). In general heatwave risk factors can be divided into the categories of demographic, health, physical, socioeconomic, behavioural/cultural and institutional (Table 10.2). How some of these factors might interact to elevate the chances of heatwave related health risk are illustrated in Fig. 10.3. As well as being interconnected, many of the individual heat-risk factors are ever-changing such that vulnerability is never static (McGregor et al., 2017). Further the synergistic effects of heat-risk factors manifest as the level

Table 10.2 Determinants of heatwave related vulnerability: health

Category	Factor
Demographic	Age (especially the elderly and young), gender, race and/or ethnicity
Health	Physical (e.g. cardiovascular, chronic obstructive pulmonary and renal diseases, diabetes and fluid/electrolyte disorders), neurological (e.g. epilepsy, learning disabilities, neuromuscular, autism, attention deficit disorder, brain tumours, cerebral palsy) and mental health disorders (e.g. psychosis). Medication dependency. Obesity. Physical disability. Bed confinement
Physical	Buildings including private dwellings, residential care homes, hospitals, schools and prisons with low thermal mass, poor insulation, meagre shading and no climate control (air conditioning) Inadequate shelter arising from displacement or homelessness Place of abode: urban versus suburban versus non-urban. Meagre green space
Socioeconomic	Socially isolated/lack of social capital, deprivation, energy precarity, lack of access to social, health and insurance, emergency, communication and heat information services, poorly educated, living in rented accommodation. Welfare dependency. Certain employment types such as agricultural and construction workers and those working indoors exposed to direct heat from industrial/manufacturing processes and wearers of personal protective equipment
Institutional	Understanding gravity of heat as a hazard. Heat governance across multiple scales, building codes, heat mitigation embedded in urban design and planning principles, presence of heat-health warning systems and provision of heatwave specific community heat avoidance strategies (e.g. cooling centres, suspending power disconnections arising from unpaid energy bills)
Behavioural & cultural	Linguistic isolation. Drug and alcohol dependency, impaired cognitive ability, mass gatherings (e.g. religious and sports events) and celebration of the heat through swimming and outdoor events. Fear of opening windows and accessing external environments due to security concerns. Reluctance to use cooling devices because of energy costs (energy precarity as above)

Adapted from: McGregor et al. (2017)

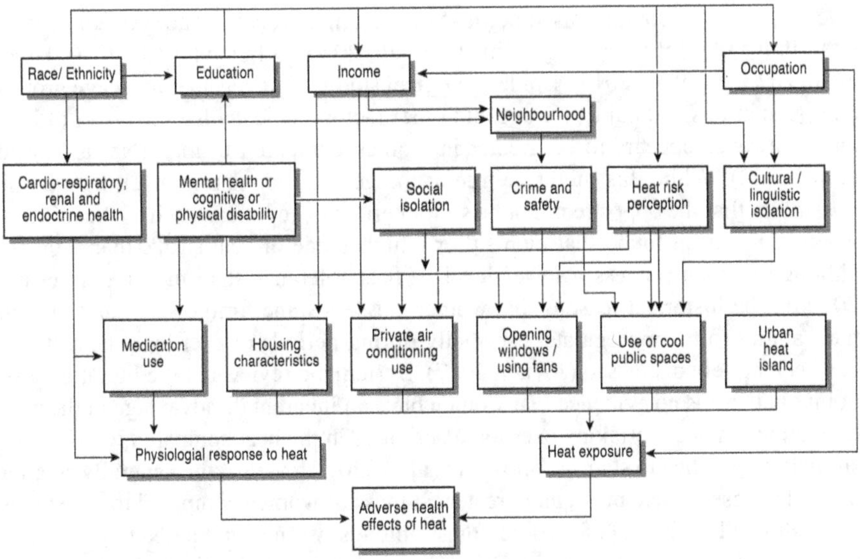

Fig. 10.3 An illustration of how adverse heat related health effects may arise from the interaction of a range of interconnected vulnerability factors. (Source: Gronlund (2014). Use granted under Springer License Number 5803140124495)

of population heatwave adaptive capacity and resilience, which can be fundamental determinants of heatwave related impacts. Given the dynamic nature of vulnerability this system characteristic provides an important context for understanding societal heatwave impacts across space and time.

Given the plethora of heat risk factors that can influence vulnerability at the individual to community scale, the question regarding relative importance comes to mind. Notwithstanding the idea of 'hazard of place vulnerability' mentioned above and the fact that some occupations make people more vulnerable to the effects of heatwaves than others (Chapman et al., 2021; Morris et al., 2020, 2021), the factors that generally predispose individuals to the negative effects of heatwaves include, the presence of pre-existing disease, being elderly or very young, being female, living alone or being institutionalised, having low economic status and low income and living in urban areas with little access to green space. Such factors can operate alone or, in the worst case, conspire to create the perfect cocktail of heat vulnerability, for example, being elderly, living alone, suffering from cardiovascular disease and reliant on social welfare benefits.

People with chronic health conditions such as cardiovascular disease are more vulnerable during heatwaves compared to healthy individuals because such diseases place strain on the heart which needs to work much harder to pump heat to the skin surface via blood flow so that it can be dissipated through sensible heat loss to the atmosphere via convection or through sweating and evaporative heat loss (see Chap. 4). Being elderly or very young increases heat vulnerability because the ability to control heat dissipation using physiological mechanisms of convective heat loss and sweating is either compromised because of the aging process or not yet fully developed in the case of the young (Millyard et al., 2020; O'Lenick et al., 2020; Uibel et al., 2022; Xu et al., 2014). Epidemiological studies that consider most age groups and populations and adjust for age and other factors indicate that the effect of heat and heatwaves appears to be greater in women compared to men (Navas-Martín et al., 2022). While this might be true at the general population level there is the possibility that the opposite might be so for certain occupations such as the military where it has been found that men suffer a higher rate of heat stroke than women, although women possess a lower level of heat tolerance than men (Alele et al., 2020b). The historical reasons that women possess some form of "impairment" in their ability to thermoregulate, especially during periods of exertional heat stress, has been largely dismissed (Joyner, 2017). Systematic reviews of the literature also point out there is no evidence that women have an inherent disadvantage in thermoregulation when undertaking exertional activities in the heat compared with men of similar age and health status (Yanovich et al., 2020). However, the generally smaller body size, less muscle mass, and greater fat mass of women compared to men tends to result in a lower level of aerobic fitness amongst women compared to men at the population level. This may constitute an important point of difference as increases in body temperature and/or heart rate are inversely related to aerobic fitness. In addition to possible thermophysiological factors, the complexity of which are not fully understood, a range of socio-economic factors may be powerful determinants of increased heat vulnerability amongst women (Morrissey-Basler et al., 2023). For example, in many non-industrialised countries women spend far more time undertaking domestic tasks such as travelling considerable distances to gather fuel and water, activities which bring with them greater exposure to heat. This along with male absences from households due to them working elsewhere for prolonged periods to earn money places extra-burden on female heads of household which in addition to female's limited access to resources may result in their increased vulnerability to heat risks and constrained adaptive capacity (Lee et al., 2021). Further as important income earners women may have no other option than to work in the heat (Chanana-Nag & Aggarwal, 2020; Trahan et al., 2023). Being pregnant also increases heat vulnerability (Spencer et al., 2022). Evidence is also emerging from refugee camps that women may be at greater risk to heat exposure during extreme heat events because of segregation in non-breathable tents and often restricted access to food, water, and medical care (Rashid & Adnan, 2009). Added to this is the possibility that female workers in some workplaces may be reluctant to hydrate and do not use sanitation facilities either because of their unhygienic state or total absence (Venugopal et al., 2016; Schmidt et al., 2020). Unfortunately, many of these female-specific factors reach their maximum expression in regions typified by Tropical and Arid climates, especially the Am/Aw and Bwh/Bsh types respectively, both of which are prone to heat extremes.

Living alone is an important heatwave vulnerability factor (Arthurson & Baum, 2015; Bouchama et al., 2007; Kafeety et al., 2020; Orlando et al., 2021). This is because the lack of social connections has the potential to impair executive functioning, sleep, and mental and physical well-being (Cacioppo & Cacioppo, 2014) all of which increase the likelihood of negative health effects during heatwaves. For example, impaired executive functioning will work against taking decisive steps to mitigate the effects of heat such as maintaining good rates of hydration or simply closing the blinds to reduce solar radiation load indoors. These factors are especially important for socially isolated and lonely older adults who generally demonstrate higher rates of morbidity and mortality even in the absence of heat (Kafeety et al., 2020). In fact, the Chicago 1995 and Paris 2003 heatwaves provide good evidence of social isolation as a key risk factor for heat-related mortality (Klinenberg, 1999; Poumadère et al., 2005) with the likelihood that this factor was also an important driver of some of the heat related deaths during the 2021 Pacific Northwest heatwave (Henderson et al., 2022).

Being institutionalised is often cited as a heatwave risk factor. While there are a range of institutional settings, residential care or nursing homes are frequently identified as places where heatwave related death rates are higher than the general population. While such rates may reflect the high proportion of people living with heat risk factors, such as being elderly and possessing a chronic health condition or constrained perceptions of heat (Malmquist et al., 2022), many residential care homes lack air conditioning and are prone to over-heating during heatwave events, which increases the chances of heat exposure (Brown & Walker, 2008; Gupta et al., 2021). Hospitalisation may also constitute a heatwave vulnerability factor because in some hospital buildings overheating occurs during periods of elevated outdoor temperatures thus exposing patients to conditions of thermal discomfort. In England it has been found that wards in hospitals built in Victorian times with great thermal mass have greater resilience to overheating compared to thermally lightweight, well insulated, naturally ventilated hospital wards (Gough et al., 2019; Iddon et al., 2015). In some cases, the overheating problem in hospital wards and nursing stations can reach levels requiring the installation of temporary air conditioning to protect patients and staff from heat-related health effects (Fifield et al., 2018). In some tropical settings hospital wards are built without air conditioning to avoid high construction and utility costs, increase infection control and reduce airborne infections all of which may raise the risk of overheating during anomalously warm conditions (Abd Rahmam et al., 2022). Prisons have been largely overlooked as places where individuals may be susceptible to the health effects of heatwaves (Motanya & Valera, 2016). Taking the United States as an example, which has nearly 25 percent of the world's incarcerated population (Golembeski et al., 2021), compared to the non-institutionalized population, there are a high proportion of individuals with heat risk factors such as cardiovascular disease, diabetes and dependency on prescription medications (Skarha et al., 2020). Further, because of the physical nature of prisons and their associated accommodation, indoor temperatures exceeding 38 °C during heat events often occur with plaintiffs in relation to unbearable heat stress citing broken or lack of air conditioning, sealed windows, faulty ventilation systems and limited cooling resources such as water and access to fans as causes of personal heat stress and overheating in general (Skarha et al., 2020, 2023). In a study that assessed the risk of dangerous heat exposure across 4078 carceral facilities in the continental United States, Tuholske et al., 2024 revealed a rise in the annual count of hot days over the period 1982 to 2020 for 1739 facilities, predominantly situated in the southern regions of the country. Notably, state-operated carceral facilities in Texas and Florida contributed to 52 percent of the total heat exposure, despite housing only 12 percent of the incarcerated population. As noted by Tuholske et al. (2024) these findings underscore the pressing need for improved infrastructure, healthcare interventions, and support systems to reduce heat exposure amongst incarcerated individuals.

Because people spend a lot of our time indoors, especially in their homes, there is the possibility that domestic building type and its construction may make some members of the population more vulnerable to the potential impacts of heatwaves than others (Gabbe et al., 2022). This is because there is tendency for some dwelling types to experience overheating during anomalously warm periods including heatwaves. Such overheating has the potential to increase an occupant's exposure to heat stress. For this reason, the number of assessments of indoor overheating have increased over the past 10–15 years. For example, surveys of UK housing have revealed that overheating was greater in social housing, with low incomes or with members aged over the state pension age (Mavrogianni et al., 2015; Lomas et al., 2021). Studies elsewhere have also shown overheating in warm summer conditions in houses typified by high solar gain, poor roof-space ventilation and low thermal mass, and in the case of multi-story apartment blocks overheating in upper floor accommodation (Birchmore et al., 2017; Schunemann et al., 2022; Sharifi et al., 2019; Kuczynski et al., 2021; Wilson & Black, 2022). A housing type, or perhaps more accurately a shelter type, that has not received much attention until relatively recently is that found in informal settlements (e.g., slums, shantytowns, favelas) (Ramsay et al., 2021, 2024). These are often made from construction materials (e.g. wrought iron), or in a way (e.g. few windows because of security concerns), that exacerbates the effects of heatwaves, which manifest as debilitating life-threatening temperatures (Naicker et al., 2017; Pasquini et al., 2020; Pohl et al., 2021; Scott et al., 2017; Teare et al., 2020; Wilby et al., 2021) (Fig. 10.4). For example, in a survey of twelve informal settlements in Makassar, Indonesia, Ramsay et al. (2021) found indoor conditions in informal housing reached chronic heat stress levels with wet bulb temperatures approaching the uppermost limits of human survivability, conditions underestimated by nearby weather stations (Fig. 10.5).

Poverty is a major driver of increased vulnerability to heat and other climate-sensitive hazards (Ebi et al., 2021; Hallegatte & Walsh, 2021). Deprivation can lead to greater exposure and vulnerability to unfavourable social, economic, and environmental circumstances including high heat exposure and poor air quality (Rauh et al., 2008). These often intersect with gender, ethnicity and disability with decimating health consequences during periods of extreme heat (Jay et al., 2021). For example, strong associations between neighbourhood racial composition and heat exposure have been found for the US (Renteria et al., 2022) with studies revealing that people of colour and those living below the poverty line generally reside in census tracts with a higher urban heat island intensity (Hsu et al., 2021; Mitchell et al., 2021; Voelkel et al., 2018). The co-occurrence of temperature extremes with impoverished areas within cities with low adaptative capacity has been referred to

Fig. 10.4 Informal settlement at Imizamo Yethu in the greater Hout Bay Valley area near Capetown, South Africa. The settlement, also referred to as Mandela Park, covers around 18 hectares. It is home to around 34,000 people most of whom live in low quality housing comprising materials and built in a way which increases the likelihood indoor overheating during heatwaves and elevates the heat vulnerability of occupants. (Image courtesy of Professor Jennifer Fitchett, University of Witwatersrand, Johannesburg, South Africa)

as heat or thermal inequity and considered as an environmental justice issue (Dialesandro et al., 2021; Mitchell et al., 2021; Saverino et al., 2021). In the US such heat inequity manifests strongly in former 'redlined' districts, redlining being the discriminatory practice implemented in the 1930s of preventing the provision of financial services to residents of certain areas based on their race or ethnicity (Li et al., 2023). That such social engineering has left a legacy of increased heat exposure and thus heat vulnerability is supported by observations that land surface temperatures as much as 7 °C greater can be attained in former redlined districts compared to their non-redlined neighbours (Hoffman et al., 2020) with the poor provision of trees and parks possibly contributing to heat inequity (Schinasi et al., 2022). Worryingly evidence is emerging that areas within cities with low levels of green infrastructure experience high levels of heat related health outcomes compared to districts that benefit from the cooling effects of green and blue infrastructure (Burkart et al., 2016; Graham et al., 2016; Venter et al., 2020). That said, reduced vulnerability to heat achieved through urban green spaces may not be the experience everywhere (Song et al., 2022). Tied up with income status and impoverishment in general is the notion of energy insecurity (Jessel et al., 2019). This is

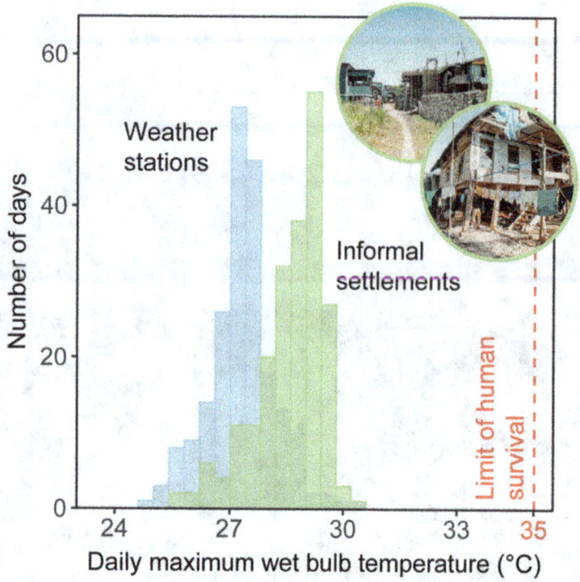

Fig. 10.5 Daily frequency distribution of wet bulb temperature (°C) recorded at weather stations and inside housing in 12 informal settlements in Makassar, Indonesia. The difference in wet bulb temperature distributions demonstrates that conventional climate information recorded at weather stations gives little indication of the extreme conditions of thermal discomfort experienced inside dwellings in the informal settlements surveyed. (Source: Ramsay et al. (2021). Used under open access CC BY license: http://creativecommons.org/licenses/by/4.0/)

an important determinant of whether air conditioning is used as a solution for mitigating the thermal discomfort and health effects associated with heatwaves (Madureira et al., 2021). Although in many parts of the world air conditioning is the go-to solution for heat risk management and thus the reduction of heat vulnerability (Jay et al., 2021; Sera et al., 2020; Turek-Hankins et al., 2021), it is neither affordable nor practical for many. This is concerning as there is evidence that heat inequity arising from no air conditioning can result in increased heat vulnerability and ultimately higher levels of heat-related health impacts compared to areas where air conditioning is close to ubiquitous (Cardoza et al., 2020; Guirguis et al., 2018; O'Neill et al., 2005). The inability to access cooled public spaces or heat refuges may also play a role in elevating heat vulnerability during heatwave events (Eisenman et al., 2016; Fraser et al., 2018; Kim et al., 2021; Quick et al., 2022). Such socio-economic factors that create heat inequity constitute serious environmental injustice issues (McIntyre et al., 2022). That such issues were potentially critical in elevating vulnerability during the COVID-19 pandemic is slowly becoming evident (Knox-Hayes et al., 2023; Ortiz et al., 2022).

As the understanding of the multiplicity of heat vulnerability factors has grown and data directly related to these, or proxies of, have become available, the mapping of heat vulnerability has grown. In many ways this activity has been inspired by the formulation of the Social Vulnerability Index by Susan Cutter and her group of researchers at the University of South Carolina (Cutter et al., 2003). The idea of the social vulnerability index is to shed light on the spatial variation of social vulnerability to a range of hazards, such as heatwaves, for a predefined area. The index is dimensionless and usually formulated by applying the statistical technique of principal components analysis (PCA) to large data sets comprising values for various variables that represent heat vulnerability factors at the intra-urban scale. In the case of heat and heatwaves, the resultant heat vulnerability index is an amalgam of scores for individual heat vulnerability variables. Typical variables that serve as input to the creation of a heat vulnerability index fall into the general categories of demographic characteristics, socio-economic conditions, nature of the built environment, underlying health conditions and level of heat exposure (Cheng et al., 2021; Conlon et al., 2020; Niu et al., 2021). Since the first attempts at creating and mapping a heat vulnerability index (Johnson et al., 2012; Reid et al., 2009; Wolf & McGregor, 2013) there has been a proliferation of heat vulnerability index mappings at a range of spatial scales from the city, county to national levels. Notwithstanding the attractiveness of the idea of mapping heat vulnerability to reveal spatial variations in this dimension of heat risk and thus apply such mappings to communicate policy recommendations, raise awareness and induce institutional networking and learning (Wolf et al., 2015), they come with some 'baggage'. Chiefly this relates to the quality and timeliness of the input data including its spatial resolution, whether variables are weighted or not, the dependency of mapping outcomes on the methodology adopted such as so-called supervised versus unsupervised approaches to index development and whether the indices are validated against actual heat-related health outcome data such as heat related mortality or morbidity in the form of hospitalisations and/or ambulance call-out. Added to these issues is whether indices of heat vulnerability have been developed in a pure mechanistic top-down as opposed to a more user-informed bottom-up approach involving stakeholder input.

Figure 10.6 presents an example of a heat vulnerability index (HVI) map for London's 4765 census districts. The developers of the London HVI (Wolf & McGregor, 2013) followed the general methodology used by Cutter et al. (2003) in the development of the Social Vulnerability Index. Based on a review of the heat-risk factor literature and the type of variables available in the UK census, Wolf and McGregor (2013) identified nine heat-related risk variables for input into a principal component analysis (PCA). Emerging from the PCA were four principal components that accounted for the majority of heat risk factor variance across the 4765 census districts. The associated PC scores were then weighted and aggregated to produce a combined PC score. This formed the HVI. Subsequently, the 4765 HVI values were grouped into ten ordinal vulnerability classes based on deciles. These were then mapped, As can be seen in Fig. 10.6 the mapping revealed a rather heterogeneous pattern of heat vulnerability across London. Noteworthy features are the generally higher heat vulnerability in central London, in particular areas north of the Thames, and single pockets of high vulnerability across greater London. There is also evidence of rapid changes in heat vulnerability over short distances. Of course, the ultimate test of such a theoretical index is whether any association exists between

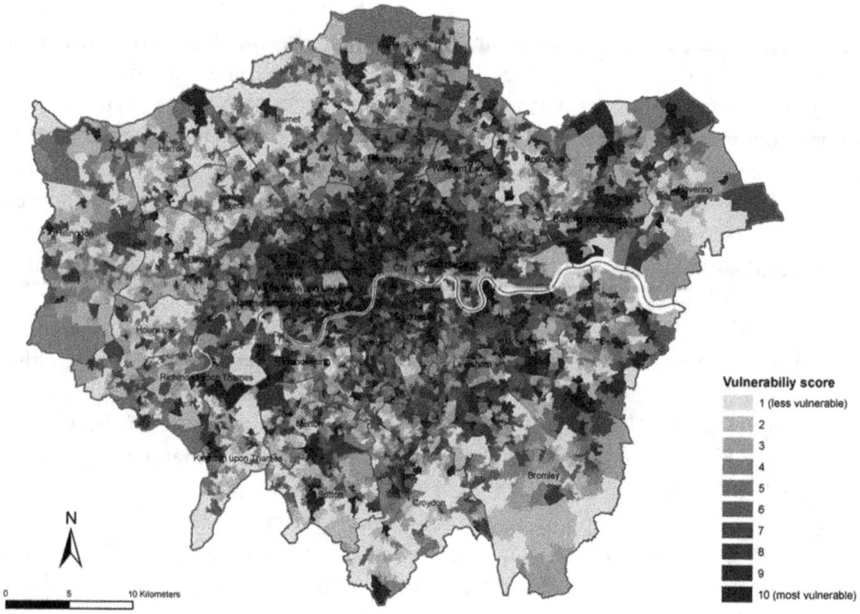

Fig. 10.6 Spatial distribution of heat vulnerability across London as indicated by ten heat vulnerability classes. (Source: Wolf et al. (2014) © American Meteorological Society. Used with permission)

the HVI categories and a heatwave related health outcome. To this end Wolf et al. (2014) tested the efficacy of the London HVI by considering the change in relative risk of mortality and ambulance call-out for a set of 133 heatwave days over the period 1990–2006. They found that as the HVI category increases from a low to high level of heat vulnerability so does the relative risk of mortality and ambulance call-out indicating that the mapping of a HVI has the potential to signpost areas at the intra-urban scale where the relative risk of heat-related health responses is expected to be greatest during heatwave events. Encouragingly, subsequent studies on other cities have demonstrated the utility of a HVI for managing heat-related risk during heatwaves.

10.3.2 Animal Populations

Animal populations can be divided into two broad categories namely endotherms which comprise the mammals and birds and ectotherms which includes animal groups such as the fishes, reptiles, and insects. The former possesses the ability to control their body temperature via a range of thermophysiological mechanisms while that latter relies mainly on the external environment and behavioural adaptation to regulate their body temperature. Endotherms are commonly referred to as 'warm-blooded' and ectotherms as 'cold-blooded'.

Like humans, the health status of animals, regardless of whether endotherms or ectotherms, is likely to be an important heat vulnerability factor as this will impact the extent to which animals are able to deal with heat load conditions. For example, domestic stock that have suffered from cardiovascular or respiratory problems have been found to have a higher risk of heat-related mortality compared to their healthy counterparts (Brown-Brandl et al., 2006). Similarly, animals plagued by disease caused by vectors or parasites are also likely to be vulnerable to heatwaves as sickness will affect the ability to physiologically control body temperature in the case of endotherms and in both endotherms and ectotherms have an impact on mobility and the exercise of shade seeking behaviour. Lack of access to water will compromise the hydration status of animals and thus their ability to achieve evaporative cooling making them vulnerable to heat effects especially during heatwaves that are embedded in longer periods of dryness. For some animals, body size is an important heat vulnerability factor (Buckley, 2021; Fuller et al., 2021; Lindmark et al., 2022; Peralta-Maraver & Rezende, 2021). For example, smaller insects are more susceptible to dehydration and overheating during heatwaves with such conditions disproportionately affecting adults and influencing reproductive potential (Ma et al., 2021). In experimental situations smaller crabs standing on a heated sand surface, akin to inter-tidal flats, have been found to succumb to heat effects sooner than larger crabs (Levinton, 2020).

10.3.3 Economy and Society

To date most studies that consider the heat sensitivity of the economy and its constituent sectors do so at temporal scales beyond that of heatwaves with much activity focused on the economic impacts of climate change. As a range of economic sectors are heat sensitive in general terms, by association these same sectors, because of their intrinsic characteristics, are theoretically vulnerable to short term heat shocks in the form of heatwaves. A prime example is the agricultural sector. Livestock-based agricultural systems are highly vulnerable to heatwaves, especially those organised around feedlots where animals are housed in close quarters. As outlined in the previous section, animals can succumb to the effects of extreme heat. This can affect animal health and livestock productivity. In the worst case, overheating or hyperthermia in farm animals will result in death and economic loss. The heatwave vulnerability of crop-based agricultural systems is likely to vary by crop, cultivar, and development stage with implications for crop production and quality (Parker et al., 2020). For example, some grape cultivars are especially heatwave vulnerable at the veraison (ripening) stage (Greer & Weston, 2010). As outlined in the following section on infrastructure, electricity generation and supply systems are heatwave vulnerable. This holds huge implications for the heatwave vulnerability of any economic or societal sector heavily dependent on reliable power supplies with the spawning of cascading vulnerabilities and failures along non-linear paths likely (Pescaroli & Alexander, 2016) if heatwave related brown or blackouts occur.

Some communities may be heatwave vulnerable because of their situation. The general lack of services related to water, sanitation, and hygiene as well as the type housing and its construction materials are the chief drivers of heatwave vulnerability in informal settlements (Teare et al., 2020). Immigrants and newcomer populations are also heatwave vulnerable as factors such as language, culture, and religion intersect to increase heatwave related risk (Varickanickal & Newbold, 2021; Wiesbock et al., 2016). For example, not being familiar with the common language of their new place of settlement will influence the access to heatwave related information for immigrant and newcomer communities (Hansen et al., 2014). Low levels of social capital, the connections among individuals including networks and norms and reciprocity (Putnam, 1995; Pelling & High, 2005), also influence a communities heatwave vulnerability (Guardaro et al., 2022; Klinenberg, 2015).

10.3.4 Infrastructure

Hard infrastructure, or in other words the basic engineered structures and facilities that serve a region including transportation, energy, water, communication, and building systems are vital for supporting economic development and quality of life. Many facets of these systems, such as roads and rail are at risk to extreme heat events. While there are plenty of examples of the impacts of heatwaves on infrastructure, as described in the next chapter, material on what makes engineered structures vulnerable to the effects of heat is meagre in comparison. By vulnerable it is meant the innate characteristics of a structure that predisposes it to sub-optimal performance or indeed failure under conditions of extreme environmental heat. In many cases exposure will be a function of the level of direct radiative load on a structure with sensible heat transfer from a warm atmosphere to a cooler surface or conduction from another infrastructural element playing a secondary role in determining the level of heat exposure.

An important infrastructure property that may influence heat related vulnerability or sensitivity is the thermal characteristics of construction materials. Chapter 7 touched upon the thermal property of specific heat capacity and thermal conductivity, noting that some materials are thermally more conservative than others because of their high specific heat capacity and low thermal conductivity. The same applies to construction materials (Table 10.3). A further important property is described by a material's linear thermal expansion coefficient (Table 10.3). This is an indicator of the extent to which an object's length changes when heated or cooled which is proportional to the original object length and the change in temperature. Related coefficients are area and volume thermal expansion coefficients. Thermal expansion is an especially important determinant of buckling in structural components made of metal materials, particularly in the case of insufficient provision of expansion joints. For example, comparing a steel and aluminium beam, each with a length of 10 m

Table 10.3 Thermal properties of a range of materials typical of infrastructure systems

Material	Specific Heat (c)—J kg^{-1} K^{-1} x 10^{3}	Thermal Conductivity (k)—W m^{-1} K^{-1}	Linear Expansion Coefficient (m/m °C)[a]
Copper	0.38	400	0.000017
Steel	0.42	16–80	0.000012
Iron	0.45	52	0.000012
Brick	0.84	0.5–1.0	0.000005–0.00009
Concrete	0.88	1.4	0.000013
Marble	0.88	2.0–3.0	0.000005–0.000014
Aluminium	0.90	237	0.000023
Timber	1.200	0.14–0.17	0.000005

Compiled by Author

[a]The unit of measurement for thermal expansion coefficients is the inverse of temperature, °C^{-1} or K^{-1}

and subject to a temperature change from 20 °C to 50 °C, the expansion would be 0.0036 and 0.0069 m respectively. Under conditions of longitudinal restriction this indicates aluminium beams to be more susceptible to buckling compared to steel beams because their expansion is almost twice that of steel beams. Extending the idea that different materials have a range of linear expansion coefficients, structures made up of a mix of construction materials will be vulnerable to the effects of extreme heat if the materials are not designed to expand at the same rate. Take for example the steel rods that reinforce the concrete used to build bridges and buildings. If the steel rods were to expand at a different rate to that of the concrete, the concrete would most likely crack. With time and in the worst case, the concrete could crumble and the structure fail.

For bricks and mortar buildings, time and the nature of the mortar is an important determinant of the level of heat vulnerability and the extent to which structural integrity might be compromised under conditions of thermal expansion. While bricks and mortar are quite resistant to compressive forces that arise under thermal expansion, they are less resilient to tensile forces as experienced during the cooling phase of the diurnal heating/cooling cycle. Repeated expansion and contraction under strong heating and cooling cycles, as might occur during heatwave events, may result in the growth of cracks in masonry. With time this may result in serious damage. For the UK this is potentially more of a problem in new, especially twenty-first century, compared to old or 'period' builds as soft lime mortars, which have greater tolerance to thermal expansion and contraction, were used widely in the construction of buildings prior to the 1920s–1930s. This contrasts with the 'hard' less flexible cement generally used in modern buildings.

As highlighted in previous chapters, albedo is a critical determinant of the amount of energy absorbed by surface materials and thus surface temperature. Albedo is therefore an important material property that may make some infrastructure or parts of infrastructure systems more vulnerable to heat related effects during heatwaves. This is clear in the case of pavement coverings with very high surface temperatures achieved on dark (low albedo) compared to relatively light (high albedo) surfaces (see Chap. 14) which may affect the level of tarmac melting/asphalt deformation, rutting or loss of firmness on roads or airport runways, or road surface/pavement heave as road surface segments, often in the form of concrete slabs, butt up against each other when expanding. Added to this is the heat vulnerability of roads/runways composed of layers of different materials with the lower layers heating at slower rates compared to the surface layers with surface cracking the result. Notwithstanding the susceptibility to softening under extreme heat, roads and runways constructed from asphalt/tarmac are less prone to buckling as these materials are viscoelastic with old concrete slab-based roads/runways the most likely to experience heat-related heave.

Like roads, rail tracks are susceptible to the effects of exceptional heat during heatwaves, mainly in the form of buckling. Because rail tracks, including their supporting structures, are complex entities, there are a range of engineering properties that may work independently or in unison to cause rail tracks to buckle. These include rail size, peak lateral resistance, level of tie/ballast friction, torsional and longitudinal resistance, track curvature, misalignment amplitude and wavelength, track foundation modulus, axle load, and track centre spacing (Mulholland & Feyen, 2021). Beyond these, other determinants of heatwave related risk include whether maintenance regimes are in place to ensure pre-stressing of rail tracks and ballast strength is maintained. These are important as the former enables safe longitudinal track expansion while the latter restricts lateral movements (Palin et al., 2013). Evidence also points to slab track being less vulnerable to buckling compared to ballasted track (Palin et al., 2021). Track albedo, through its impact on the amount of solar heat gain, is also important. Electronic signalling equipment is also prone to overheating in hot weather, especially if it is not installed in well-ventilated housing. For rail networks that are electrified, both overhead lines and conductor rails are used to transfer energy to a train's operating equipment. During heatwave events, overhead lines can sag as is the case for cable conductors in electricity distribution networks. Especially vulnerable to the effects of heatwaves are old overhead lines with fixed tension connectors as these, unlike their modern equivalents, do not have auto-tension systems that adjust overhead line length according to both line and air temperature. This makes old overhead line systems far more vulnerable to failure during heatwaves compared to their modern counterparts. Similarly, the stress-free temperature for rail tracks in good condition is much higher for those in poor condition. Added to this the stress-free temperature may vary by country with countries for which heatwaves have not been historically an issue for the rail network possessing lower temperatures. For example, the stress-free temperature is around 27 °C for the UK, whereas for the US it is in the range of 35 °C to 43 °C (Dobney et al., 2009).

Bridges, as critical elements of road and rail transport infrastructure, have begun to receive some attention in relation to what makes them heat vulnerable. This is especially so in the case of bridges with joints given the tendency for debris to build up in these spaces designed to accommodate movement with debris preventing the expansion and contraction of bridge components during temperature extremes. The main outcomes of this are local buckling of girder flanges, spalling of concrete on the abutments, and cracking and crushing of the roadway deck (Fig. 10.7). Further

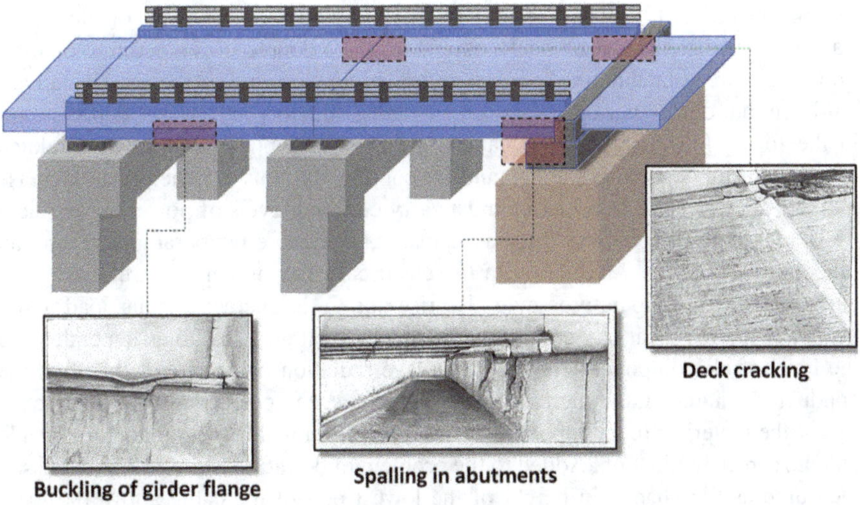

Fig. 10.7 Bridge damages arising from a combination of clogged joints preventing free movement of bridge structural elements during periods of heat-related thermal expansion. (Source: Palu and Mahmoud (2019). This is an open access article distributed under the terms of the Creative Commons Attribution License 4: https://creativecommons.org/licenses/by/4.0/)

to the issue of joint-clogging, joint deterioration allows infiltration of water, salts, and other materials beneath a bridge deck with accumulation in and around areas where the bridge bearings are located. This can promote corrosion and ultimately seizure of these vital elements which transfer load from the bridge superstructure to substructure. Consequently, during heatwave events compromised bridge bearings will be ineffective in their role as nodes of load transfer thus preventing the bridge from accommodating heat related movements of major structural elements such as bridge girders (Palu & Mahmoud, 2019), with the result that significant buckling/twisting occurs. Research has also shown that uneven heating across bridge decks in conditions akin to heatwaves produces large nonlinear temperature gradients in bridge girder sections which bears implications for the heat vulnerability of bridges (Hagedorn et al., 2019).

For electric power infrastructure systems there are several factors which make these vulnerable to heatwaves. Electricity generation at nuclear, coal-fired, and natural gas power stations that use water-cooled turbine generators are at risk to increases in intake-water temperature during heatwaves, the result of which is reduced electricity output if influent water temperatures are too high. Added to this are temperature limits for not only the effluent water discharged into terrestrial water courses or coastal waters but threshold temperatures for the receiving waters which must be maintained below safe environmental limits for ecosystem protection. Dry power generators such as dry steam power plants that use geothermal fluids are also prone to reduced output during heatwaves as such plants function less efficiently during periods of high ambient temperatures typical of heatwaves

(Burillo et al., 2017; Burillo et al., 2019). Individual components of electricity transmission systems can be heat-vulnerable. For example, if temperature calibration checks are not maintained for power flow protection devices these may inadvertently cut out during periods of extreme heat. Conductors in transmission systems, in the form of overhead power cables, are also susceptible to heatwave related effects. During heatwaves the demand for cooling and thus electricity can increase markedly. As a result, conductors must carry elevated levels of power across electricity distribution networks. Increased load causes cable temperatures to rise, an outcome of which is a reduction in the electrical carrying capacity referred to as de-rating. Cable temperatures may also rise due to the increased solar load under heatwave related clear sky conditions with solar heat gain to cables dependent on the level of all components of the shortwave radiation balance (K↓ and K↑), the conductor's diameter, the thermal properties (e.g. specific heat capacity, conductivity) of the materials (e.g., copper or aluminium) from which the conductor is made and the cables inclination. Added to the problem of de-rating is the risk of cable sag measured as the change in height of the lowest part of the cable above the local ground level. Sag happens because as the temperature of the cable increases expansion/elongation of the cable occurs. This increases the slack between transmission line structures such as pylons. In the worst case, irreversible deformation of overhead power cables may result from extreme heat conditions. An important factor that influences the degree of sag is the composition of the cables as this determines the extent of longitudinal stretching when heated (see Table 10.3 for expansion coefficients). For example, over a temperature range of below freezing to around 40 °C, the change in length of a copper cable is around 1.5 m for a 300 m length. In comparison, the elongation for an aluminium cable is greater at around 2 m for a 300 m long cable making conductors composed of aluminium more susceptible to heat-related sag compared to copper ones.

The main parts of water supply systems that are vulnerable to heatwaves are the pipes that distribute water, some of which may be above or below ground. During heatwave events, there is usually a rise in water use. This can exert added water pressure in under- or over-ground pipes causing pipe bursts where pre-existing weaknesses in the pipework exist. Especially vulnerable are weak pipe connections, particularly those in over-ground distribution networks which may expand due to solar heat load. Water pipes located within drought-prone soils may become vulnerable to breakage during heatwaves. This is because enhanced soil desiccation during heatwaves, especially if heatwaves are embedded in a prolonged period of dryness, may bring about differential settling or movement of water and sewerage pipes. Such motion will place additional stress on pipe bends, connectors and joints. The state of a reservoir may also be an important determinant of how vulnerable it may be to increased evaporation rates and thus loss of water to the atmosphere during heatwave events. Take for example a reservoir that has experienced drastic drawdown in its water level due to high domestic and/or irrigation demand. In such a case and assuming equal levels of mixing, a shallow water body's temperature will be higher compared to a deeper water body. Given that the latent heat of vapourisation decreases with an increase in water temperature, or in other words less energy is required to evaporate water from a 'warm' compared to 'cool' water body, then a reservoir with a shallow cover of water is likely to be highly susceptible to evaporative loss under heatwave conditions. This susceptibility will increase with the increase in water temperature.

To this point the focus has been on the heat vulnerability of single infrastructure systems such as energy or transport. Infrastructure systems possess, however, a high degree of interdependency (Lu et al., 2018). Such interdependency may amplify the effects of heatwaves because of spiralling feedbacks between infrastructure systems. Take for example the vulnerability of nuclear power generators to the elevated temperature of influent waters for turbine cooling during heatwaves with the result that generation capacity is severely affected. In the worst-case scenario, and in the absence of electricity generation from other sources, brownouts or blackouts may occur rendering electrified transportation networks inoperable or indoor temperatures unbearable because of no air conditioning. The latter may result in elevated heat illness requiring increased activity of ambulance services with their ability to attend emergency situations eventually compromised if refuelling is dependent on fuel from electric gas pumps or recharging stations in the case of an electrical fleet of ambulances. The falloff in electricity generation capacity will also bear implications for water, especially for those activities dependent on electric pumping stations for water supply. Acknowledging the complexity of relationships between urban infrastructure systems and their linked vulnerabilities, Clark et al. (2019) imagine the calamity that might arise from a 'Katrina of heat' in the US city of Phoenix, given that this tumultuous hurricane precipitated a cascade of failures attributable to the complex interdependency of energy, water, transport and communication systems.

10.4 Ecosystems

Compared to human systems, the understanding of factors that determine heatwave vulnerability in natural systems is undeveloped. In attempt to address this knowledge gap this section will consider what makes marine, terrestrial and aquatic ecosystems vulnerable to heatwaves.

10.4.1 Marine Ecosystems

In a global assessment of the threats posed by heatwaves on marine ecosystems, the central west Atlantic, the northeast Atlantic and the northwest Pacific have been identified as particularly vulnerable because anthropogenic stressors such as high levels of overfishing and pollution, have the potential to exacerbate marine heatwave impacts. Further, multiple regions in the Pacific, Atlantic and Indian Oceans are particularly vulnerable to the heat stress exerted by marine heatwaves because many species in these regions are near their thermal limits (Smale et al., 2019). Further, for those regions where successive heatwave events occur, ecosystem recovery may be greatly compromised with increased fragility resulting in highly heat-sensitive marine ecosystems. Of a range of marine ecosystems, coral reefs and kelp forests have received considerable attention in relation to what makes them vulnerable to marine heatwaves.

Most coral ecosystems can be found between 22 °N and 22 °S, a latitudinal band characterised by damped seasonal variations in water temperature of around 4–5 °C with maximum temperatures averaging around 30 °C. That said, coral reefs can also be found in some extreme thermal environments such as in the Persian/Arabian Gulf region where the seasonal water temperature range can be 5–6 times that of the large equatorial/tropical oceans (Schoepf et al., 2015). For the iconic Great Barrier Reef of Australia, the thermal thresholds for reefs dominated by Acropora coral communities are in the range of 29–31 °C. However, these limits reduce drastically if water temperatures exceed such limits for multiple days making this coral type and others like it more vulnerable during marine heatwave events (Berkelmans, 2002). The extent to which corals are heat vulnerable also depends on a range of morphological and physiological factors. Branching corals are generally more susceptible to the effects of marine heatwaves than massive corals. High levels of stored energy reserves and thick tissue also appear to enhance resistance to heat stress while coral starvation and death during extreme heat events can be reduced if heterotrophic feeding is increased (Schoepf et al., 2015). The physiological and cellular makeup of corals may also determine whether they are bleaching susceptible or not under marine heatwave conditions (Innis et al., 2021).

Local factors can interact synergistically with spatially extensive marine heatwaves to increase the likelihood of coral death. For example, a high abundance of macroalgae or urchins has been found to magnify coral loss following marine heatwave events (Donovan et al., 2021). In the case of macroalgae, which can induce a range of biochemical impacts on coral health, it is likely that plentiful nutrients often from land-based pollution sources (Burkepile et al., 2020; Fabricius, 2005) and reduced herbivory arising from overfishing of herbivorous fish, drive up their abundance (Donovan et al., 2021). Sea urchins impact coral through direct predation and bio-erosion (Birkeland, 1989) with their populations often exploding because of overfishing of their predators (McClanahan & Muthiga, 1988). Consequently, coral reefs already under considerable stress from urchin-related predation and bioerosion are prone to elevated levels of die-off from heatwave-related coral bleaching events (Donovan et al., 2021). The local wave climate also appears to be important with greater coral loss on reefs exposed to higher physical disturbance compared to sheltered reefs with a more benign wave climate. Along with less physical disturbance, sheltered coral reefs or back reefs may be less heat-vulnerable because of coral acclimation to higher temperatures and generally better heat-tolerance because of the greater swings in temperature experienced here compared to exposed or fore reefs where water is generally well mixed over greater depths (DeCarlo et al., 2019; Palumbi et al., 2014; Safaie et al., 2018; Schubert et al., 2021).

Because kelp is essentially a cool-water species it responds rapidly to changing temperature conditions typical of marine heatwaves. In addition to an individual species' intrinsic thermal tolerance and its relative position within the broad thermal range of kelp species (Smale, 2020), other stressors including overgrazing from sea urchins (Ling et al., 2015) and declining water quality in coastal waters near major urban developments (Connell et al., 2008) may make kelp ecosystems more vulnerable to marine heatwaves compared to their counterparts not subject to these non-climatic stressors. Added to this is the possibility that variations in local thermal adaptation in kelps and intra-species variability in thermal tolerance may cause some mid-range kelp populations to be susceptible to the effects of marine heatwaves (Smale, 2020). Poor nutrient supply, as might arise from decreases in oceanic upwelling along coastal margins during MHW, and associated impacts on ecosystem health (Edwards, 2019) are also likely to render kelp ecosystems vulnerable to marine heatwave effects and other physical disturbances.

10.4.2 Terrestrial Ecosystems

Terrestrial ecosystems can be generally grouped into six types, namely taiga, tundra, deciduous forest, grasslands, tropical rain forests, and deserts. As for all living organisms, a fundamental determinant of the vulnerability of terrestrial ecosystems to heat is the level of intrinsic heat tolerance for the individual species that make up an ecosystem and whether thermal limits of a range of vegetation types will be exceeded during heatwave periods (Teskey et al., 2015). The same of course applies to animals that inhabit terrestrial ecosystems, whether they be ectotherms or endotherms (González-Tokman et al., 2020; Ma et al., 2021; McKechnie & Wolf, 2019; Pollock et al., 2021). Generally, the optimal temperature range for boreal, temperate and tropical angiosperm and gymnosperm trees lies between 20–30 °C with some species capable of tolerating temperatures up to 35–40 °C. Some tropical tree species however are adapted to much higher temperatures being able to maintain essential functions at over 45 °C. While on the surface it might appear knowing the thermal tolerance of individual plant types or species might be enough to establish vulnerability, the fact is that this information is very limited with cross-species studies of the thermal sensitivity of tree/plant species still rare (Kunert & Hajek, 2022). Much therefore remains to be discovered about the critical bounding conditions of plant vulnerability to heatwaves (Breshears et al., 2021).

Perhaps unsurprisingly, a critical determinant of the vulnerability of terrestrial ecosystems to heatwaves is their general health status which affects their ability to function normally during periods of extreme heat; the same of course applies to marine and freshwater ecosystems and human populations especially at the individual level. Broadly speaking, terrestrial ecosystem health is determined by three main things, the ability of vegetation to uptake carbon and its efficiency in using carbon, and water (Migliavacca et al., 2021). Anything that affects these three critical processes will have an impact on vegetation health and thus its vulnerability to the effects of heat. For example, amongst a range of factors, carbon uptake is negatively influenced by a lack of water availability and high temperatures, as occur in heatwave periods. An associated response will be a decline in tree health and compromised self-defence mechanisms, such as a fall-off of tree sap production. Consequently, the susceptibility to other stressors such as insect attack is likely to increase (Hartmann et al., 2022). Added to this, is the general deterioration of health over the lifetime of vegetation because of reduced carbon uptake which makes aged stands of trees particularly vulnerable to the effects of heat and heat-related stressors.

Trees already stressed by drought may be highly susceptible to heatwave effects. This is because paradoxically, some tree species when under water stress open their stomata in heatwaves to prevent leaves from critically overheating thus accelerating water loss. The outcome is that dangerous tipping points in terms of tree survival are arrived at much more quickly than otherwise might occur in the absence of heat extremes (Marchin et al., 2022). Leaf morphology factors can also be critical in determining leaf temperature and thus influence heat vulnerability. For example, large dark leaves will absorb more heat via direct solar radiation load compared to vegetation types with small leaf area and higher albedo characteristics thus influencing leaf temperature and the susceptibility to leaf burn/leaf scorch (Leigh et al., 2017). Similarly thick and hardy leaves possess low thermal sensitivity making associated plants less prone to wilting and more resistant to the deleterious effects related to periods of heat stress (Münchinger et al., 2023). At the vegetation stand level those plants/trees located near the edges will be far more vulnerable to heatwaves. This is because mutual shading effects are reduced making whole of organism solar radiation loads higher compared to central stand locations. Added to this is the exposure to heat advected from elsewhere and the direct battering of trees/plants by hot dry winds and the operation of the so-called clothesline effect that results in high evapotranspiration rates and soil moisture depletion at stand edges.

10.4.3 Aquatic Ecosystems

Aquatic ecosystems comprise wetlands, rivers, lakes, and coastal estuaries. Despite their critical role in ecosystem service provision, much remains to be learned about the factors that contribute to their heatwave vulnerability. Much of the literature that attends to the climate related vulnerability of aquatic systems is set within a climate change context as opposed to an assessment of heatwave risk factors. Given this what follows is more speculative than based on evidence arising from systematic research.

As for other ecosystems, heat vulnerability can be considered at the whole system or individual component (physical, chemical, and biological) or species scale. For organisms that inhabit aquatic systems the factors that increase the likelihood of elevated water temperatures, which will interact with organism scale vulnerability factors, are important. A likely determinant of heat vulnerability in aquatic systems is the hydrodynamics of water with sluggish shallow water far more vulnerable to overheating than deep fast flowing water because of contrasts in mixing and advection (Folegot et al., 2018). In the case of streams, channel morphology may exert an influence on heat vulnerability of stream ecosystems. For example, stream width influences the water surface area over which surface energy exchange occurs such that small narrow streams may benefit from bank and riparian vegetation related shading, whereas more open channels will experience greater exposure to atmospheric heat inputs and be more vulnerable to heating from high air temperatures and solar radiation loads. Channel orientation may also influence the vulnerability of stream waters to heating via the length of exposure to incoming solar radiation. Added to this is the nature of the stream bed material, especially in terms of its albedo and thermal conductivity characteristics, as well as whether a stream bed might be covered in dark heat absorbing organic matter.

The moisture status of wetlands and peatlands will be a critical determinant of their vulnerability to heatwaves. This will be especially so if an intense long duration heatwave occurs within a period of extended dryness and low moisture status. As a result, evapotranspiration rates from wetlands and peatlands will be enhanced exacerbating drying and in in the worst case, physical breakdown of the substrate that supports vegetation, often referred to aligatoring in the case of peatland (Evans & Warburton, 2011) . The heat vulnerability of coastal estuaries will depend very much on the co-occurrence of low tide conditions and the timing of the diurnal distribution as well as duration of heat. The worst-case scenario is likely to comprise the co-occurrence of low tide, when large areas of an estuary will be exposed directly to atmospheric conditions, and the period of maximum diurnal heating. This will determine whether organism specific lethal heat conditions will be obtained (Dominguez et al., 2021; Miranda et al., 2019). This applies especially to sessile organisms which are not fossorial nor can physically move off the exposed 'hot' estuary surface. Notwithstanding the evaporative cooling effects across a generally wet estuary surface between tides, estuary microtopography and its control on short term inter-tidal heat exposure and drying patterns will undoubtedly exert a strong influence on heat vulnerability at the sub-estuary scale (He et al., 2022; Hesketh & Harley, 2023; Zhou et al., 2022). Added to this, the physiological plasticity and thus the ability of a species to adapt to recurrent exposure under heatwave conditions will be an important survival-determining factor.

Compared to streams and estuaries, lakes are land bound with only limited throughflow of water. Factors that are likely to influence the vulnerability of lake waters to heating are overall lake depth, the depth of the surface mixed layer and turbidity. In very shallow lakes the possibility exists for heating to occur right throughout the water column in severe heat conditions. This will also diminish lake levels via evaporation and reduce the depth of solar attenuation making shallow lakes vulnerable to heating. In lakes, how well mixed the surface layer is will influence the thermal state of near surface waters. Under very calm conditions associated with heatwaves, mixing may shut down because of lack of wind driven turbulence with lake surface skin temperatures attaining anomalously high levels. This will bear implications for the survival of surface/near-surface dwelling plants and animals. Turbidity affects water temperature through the way in which suspended particles absorb and scatter incoming solar radiation. Generally, there is a positive association between turbidity and water temperature such that the less the clarity of water the higher the temperature. This is because suspended solids decrease water column albedo and increase the absorption of incident solar radiation and limit the so-called extinction of solar radiation to the upper parts of the turbid layer. Turbidity's role in elevating the heat vulnerability of aquatic ecosystems will reach maximum potential in very shallow turbid water bodies, or in other words, shallow turbid water is much more vulnerable to the heating effects of heatwaves compared to shallow clear water. Although based on artificially constructed water pools, Paaijmans et al. (2008) have demonstrated that near surface layers in very turbid water pools can attain temperatures 2-3 °C higher than their clear low turbidity equivalents. As well as suspended inorganic solids increasing water turbidity, algae are also important. During warm weather and reduced mixing of surface waters, the growth and abundance of algae can increase substantially. This raises the possibility that with rapidly increasing algal growth the turbidity-albedo effect will work through positive feedback to further increase surface water temperatures which promotes further algal growth and so on. This kind of process will make water bodies with significant algal growth susceptible to the effects of enhanced heating during heatwaves. This contention of course assumes that both the temperature and ultraviolet radiation tolerance limits of the 'offending' algal population are not exceeded.

As for all organisms, the heat vulnerability of those that occupy aquatic systems will be largely dependent on the level of intrinsic thermal tolerance, whether thermal microclimate refugia can be accessed and the ability to implement behavioural thermoregulation. In contrast to the literature devoted to sorting out the array of physical and chemical factors that make aquatic systems heatwave sensitive, that on heatwave related tolerance of a range of vertebrate and invertebrate animal groups is more developed and continues to grow (Pottier et al., 2022). Fish are very sensitive to changes in water temperature being able to detect temperature changes of less than a tenth of a degree. Based on a wide-ranging comparison of the thermal tolerance of freshwater tropical and temperate fish species, Nati et al. (2021) found that the intra-specific variation in thermal tolerance for tropical fish species is less than their temperate counterparts. This raises the possibly of increased vulnerability to thermal stress amongst tropical fish species and thus stream or lake heatwaves. Souchon and Tissot (2012) also note that fish in the larval stage are far more vulnerable to heat stress than adults because of their undeveloped ability to regulate membrane fluidity and undergo metabolic acclimation or behavioural thermoregulation. While there is general concern about the vulnerability of freshwater fish to heatwaves, laboratory-based work has revealed the remarkable resilience of some fish species to heatwave-like conditions. For example, the fathead minnow, an important prey species across North America, has been shown to be resilient to simulated heatwaves typical of agricultural streams through short-term acclimation (Dai et al., 2022).

The Chironomids (Diptera: Chironomidae) occupy an important place in all aquatic environments making up an significant source of food for higher order animals in the trophic chain including other invertebrates, fish, birds and amphibians.

Of all the macroinvertebrates they are probably the most abundant in terms of number of species and individuals occurring across all continents making up about a quarter of all aquatic insects. It is estimated there could be as many as 10,000 species of Chironomids with each individual species having adapted to its thermal environment whether it be a cold or warm water one (Rossaro, 1991). Of the many species of Chironomid, it is likely the stenothermal species, the ones that can only survive over a narrow thermal range, will be the most vulnerable to high water temperature extremes that occur during heatwaves. For example, many cold stenothermal species have an upper thermal tolerance of around 15 °C so variations of water temperature beyond this will have fundamental impacts on survivability. Other species that are eurythermal and able to tolerate a wide range of temperatures may be better adapted to temperature extremes. That said the abundance of many eurythermal species drops dramatically at water temperatures in the range of 26–30 °C indicating even these species may be susceptible to short term increases in water temperature associated with heatwaves. While the narrow temperature tolerance of some Chironomid species may make them vulnerable to increases in water temperature associated with heatwaves there is the possibility that as a multivoltine organism some Chironomids may adjust to rapidly changing temperature conditions associated with prolonged heatwaves via genetic adaptation over 2–3 generations at the intra-seasonal timescale (Foucault et al., 2018). As well as water temperature, recent research has revealed that Chironomids are vulnerable to high air temperatures associated with heatwaves as this influences their flight activity and ability to reach new sites suitable for breeding (Vebrová et al., 2018).

10.5 Conclusion

From the material presented in this chapter it is clear exposure and vulnerability are dynamic factors that shape heatwave risk and its consequences across time and space. The critical dimensions of exposure encompass frequency, duration, and intensity. Along with heat tolerance, these influence an entity's resilience to heat stress. Simplistic reliance on fixed thresholds, as has been the tendency with many studies focused on exposure, has the potential to overlook the cumulative impact of sub-lethal heat exposure—there is an inherent danger in thinking impacts will only arise with the exceedance of a predefined threshold. Human population vulnerability to heatwaves has been extensively explored with the understanding of how this risk determinant modifies the relationship between heatwave exposure and heatwave related impact slowly increasing. That said, most studies have been conducted in high income countries with a dearth of analyses for low and middle-income country situations. This disparity in understanding of heatwave vulnerability needs to be addressed. There is a pressing need for further investigation of the underlying drivers of heatwave vulnerability within marine, terrestrial, and aquatic ecosystems. Similarly, for infrastructure, there is a considerable knowledge gap relating to the intricate interplay between various infrastructure components (such as energy, transport, and health systems) and how these interdependencies may exacerbate heatwave-related vulnerability. Understanding these complex interactions is crucial, as in the event of so-called mega-heatwaves, they could precipitate unprecedented impacts through feedback loops and non-linear effects.

References

Abd Rahmam, N. M., Haw, L. C., Fazlizan, A., Hussin, A., & Imran, M. S. (2022). Thermal comfort assessment of naturally ventilated public hospital wards in the tropics. *Building and Environment, 207*. https://doi.org/10.1016/j.buildenv.2021.108480

Ahrens, C. W., Challis, A., Byrne, M., Leigh, A., Nicotra, A. B., Tissue, D., & Rymer, P. (2021). Repeated extreme heatwaves result in higher leaf thermal tolerances and greater safety margins. *New Phytologist, 232*(3), 1212–1225. https://doi.org/10.1111/nph.17640

Alele, F., Malau-Aduli, B., Malau-Aduli, A., & Crowe, M. (2020a). Systematic review of gender differences in the epidemiology and risk factors of exertional heat illness and heat tolerance in the armed forces. *BMJ Open, 10*(4). https://doi.org/10.1136/bmjopen-2019-031825

Alele, F. O., Malau-Aduli, B. S., Malau-Aduli, A. E. O., & Crowe, M. J. (2020b). Epidemiology of exertional heat illness in the military: A systematic review of observational studies. *International Journal of Environmental Research and Public Health, 17*(19). https://doi.org/10.3390/ijerph17197037

Arthurson, K., & Baum, S. (2015). Making space for social inclusion in conceptualising climate change vulnerability. *Local Environment, 20*(1), 1–17. https://doi.org/10.1080/13549839.2013.818951

Bäurle, I. (2016). Plant heat adaptation: Priming in response to heat stress. *F1000Research, 5*.

Beck, H. E., Zimmermann, N. E., McVicar, T. R., Vergopolan, N., Berg, A., & Wood, E. F. (2018). Present and future Koppen-Geiger climate classification maps at 1-km resolution. *Scientific Data, 5*. https://doi.org/10.1038/sdata.2018.214

Benmarhnia, T., Deguen, S., Kaufman, J. S., & Smargiassi, A. (2015). Vulnerability to heat-related mortality: A systematic review, meta-analysis, and meta-regression analysis. *Epidemiology, 26*(6), 781–793.

Berkelmans, R. (2002). Time-integrated thermal bleaching thresholds of reefs and their variation on the great barrier reef. *Marine Ecology Progress Series, 229*, 73–82.

Bezgrebelna, M., McKenzie, K., Wells, S., Ravindran, A., Kral, M., Christensen, J., Stergiopoulos, V., Gaetz, S., & Kidd, S. A. (2021). Climate change, weather, housing precarity, and homelessness: A systematic review of reviews. *International Journal of Environmental Research and Public Health, 18*(11). https://doi.org/10.3390/ijerph18115812

Birchmore, R., Davies, K., Etherington, P., Tait, R., & Pivac, A. (2017). Overheating in Auckland homes: Testing and interventions in full-scale and simulated houses. *Building Research and Information, 45*(1–2), 157–175. https://doi.org/10.1080/09613218.2017.1232857

Birkeland, C. (1989). The influence of echinoderms on coral-reef communities. *Echinoderm Studies, 3*(1).

Bouchama, A., Dehbi, M., Mohamed, G., Matthies, F., Shoukri, M., & Menne, B. (2007). Prognostic factors in heat wave-related deaths – A meta-analysis. *Archives of Internal Medicine, 167*(20), 2170–2176. https://doi.org/10.1001/archinte.167.20.ira70009

Braveman, P., & Gottlieb, L. (2014). The social determinants of health: it's time to consider the causes of the causes. *Public Health Reports, 129*(1_Suppl_2), 19–31.

Breshears, D. D., Fontaine, J. B., Ruthrof, K. X., Field, J. P., Feng, X., Burger, J. R., Law, D. J., Kala, J., & Hardy, G. E. S. (2021). Underappreciated plant vulnerabilities to heat waves. *New Phytologist, 231*(1), 32–39. https://doi.org/10.1111/nph.17348

Brown, S., & Walker, G. (2008). Understanding heat wave vulnerability in nursing and residential homes. *Building Research and Information, 36*(4), 363–372. https://doi.org/10.1080/09613210802076427

Brown-Brandl, T. M., Eigenberg, R. A., & Nienaber, J. A. (2006). Heat stress risk factors of feedlot heifers. *Livestock Science, 105*(1–3), 57–68. https://doi.org/10.1016/j.livsci.2006.04.025

Buckley, L. B. (2021). Body size shapes thermal stress. *Nature Climate Change, 11*(1), 5–6. https://doi.org/10.1038/s41558-020-00948-w

Burillo, D., Chester, M. V., Ruddell, B., & Johnson, N. (2017). Electricity demand planning forecasts should consider climate non-stationarity to maintain reserve margins during heat waves. *Applied Energy, 206*, 267–277. https://doi.org/10.1016/j.apenergy.2017.08.141

Burillo, D., Chester, M. V., Pincetl, S., & Fournier, E. (2019). Electricity infrastructure vulnerabilities due to long-term growth and extreme heat from climate change in Los Angeles County. *Energy Policy, 128*, 943–953. https://doi.org/10.1016/j.enpol.2018.12.053

Burkart, K., Meier, F., Schneider, A., Breitner, S., Canario, P., Alcoforado, M. J., Scherer, D., & Endlicher, W. (2016). Modification of heat-related mortality in an elderly urban population by vegetation (urban green) and proximity to water (urban blue): Evidence from Lisbon, Portugal. *Environmental Health Perspectives, 124*(7), 927–934. https://doi.org/10.1289/ehp.1409529

Burkepile, D. E., Shantz, A. A., Adam, T. C., Munsterman, K. S., Speare, K. E., Ladd, M. C., Rice, M. M., Ezzat, L., McIlroy, S., & Wong, J. C. (2020). Nitrogen identity drives differential impacts of nutrients on coral bleaching and mortality. *Ecosystems, 23*, 798–811.

Cacioppo, J. T., & Cacioppo, S. (2014). Social relationships and health: The toxic effects of perceived social isolation. *Social and Personality Psychology Compass, 8*(2), 58–72.

Cardoza, J. E., Gronlund, C. J., Schott, J., Ziegler, T., Stone, B., & O'Neill, M. S. (2020). Heat-related illness is associated with lack of air conditioning and pre-existing health problems in Detroit, Michigan, USA: A community-based participatory co-analysis of survey data. *International Journal of Environmental Research and Public Health, 17*(16), 5704.

Carraro, V., Visconti, C., & Inzunza, S. (2021). Neoliberal urbanism and disaster vulnerability on the Chilean central coast. *Geoforum, 121*, 83–92. https://doi.org/10.1016/j.geoforum.2021.02.023

Chanana-Nag, N., & Aggarwal, P. K. (2020). Woman in agriculture, and climate risks: Hotspots for development. *Climatic Change, 158*(1), 13–27.

Chapman, C. L., Hess, H. W., Lucas, R. A. I., Glaser, J., Saran, R., Bragg-Gresham, J., Wegman, D. H., Hansson, E., Minson, C. T., & Schlader, Z. J. (2021). Occupational heat exposure and the risk of chronic kidney disease of nontraditional origin in the United States. *American Journal of Physiology-Regulatory Integrative and Comparative Physiology, 321*(2), R141–R151. https://doi.org/10.1152/ajpregu.00103.2021

Cheng, W. W., Li, D. Y., Liu, Z. X., & Brown, R. D. (2021). Approaches for identifying heat-vulnerable populations and locations: A systematic review. *Science of the Total Environment, 799*. https://doi.org/10.1016/j.scitotenv.2021.149417

Clark, S. S., Chester, M. V., Seager, T. P., & Eisenberg, D. A. (2019). The vulnerability of interdependent urban infrastructure systems to climate change: Could Phoenix experience a Katrina of extreme heat? *Sustainable and Resilient Infrastructure, 4*(1), 21–35. https://doi.org/10.1080/23789689.2018.1448668

Clemente, K. J. E., Thomsen, M. S., & Zimmerman, R. C. (2023). The vulnerability and resilience of seagrass ecosystems to marine heatwaves in New Zealand: A remote sensing analysis of seascape metrics using PlanetScope imagery. *Remote Sensing in Ecology and Conservation.* https://doi.org/10.1002/rse2.343

Coffel, E. D., Horton, R. M., & de Sherbinin, A. (2018). Temperature and humidity based projections of a rapid rise in global heat stress exposure during the 21st century. *Environmental Research Letters, 13*(1). https://doi.org/10.1088/1748-9326/aaa00e

Conlon, K. C., Mallen, E., Gronlund, C. J., Berrocal, V. J., Larsen, L., & O'Neill, M. S. (2020). Mapping human vulnerability to extreme heat: A criticalaAssessment of heat vulnerability indices created using principal components analysis. *Environmental Health Perspectives, 128*(9). https://doi.org/10.1289/ehp4030

Connell, S. D., Russell, B. D., Turner, D. J., Shepherd, S. A., Kildea, T., Miller, D., Airoldi, L., & Cheshire, A. (2008). Recovering a lost baseline: Missing kelp forests from a metropolitan coast. *Marine Ecology Progress Series, 360*, 63–72.

Cutter, S. L., Boruff, B. J., & Shirley, W. L. (2003). Social vulnerability to environmental hazards. *Social Science Quarterly, 84*(2), 242–261. https://doi.org/10.1111/1540-6237.8402002

Dai, Q. H., Hostert, L. E., Rondon, J. K., Cao, Y., & Suski, C. D. (2022). Thermal tolerance of fish to heatwaves in agricultural streams: What does not kill you makes you stronger? *Freshwater Biology, 67*(10), 1801–1814. https://doi.org/10.1111/fwb.13976

DeCarlo, T. M., Harrison, H. B., Gajdzik, L., Alaguarda, D., Rodolfo-Metalpa, R., D'Olivo, J., Liu, G., Patalwala, D., & McCulloch, M. T. (2019, 1898). Acclimatization of massive reef-building corals to consecutive heatwaves. *Proceedings of the Royal Society B-Biological Sciences, 286*. https://doi.org/10.1098/rspb.2019.0235

Dialesandro, J., Brazil, N., Wheeler, S., & Abunnasr, Y. (2021). Dimensions of thermal inequity: Neighborhood social demographics and urban heat in the southwestern US. *International Journal of Environmental Research and Public Health, 18*(3). https://doi.org/10.3390/ijerph18030941

Dobney, K., Baker, C. J., Quinn, A. D., & Chapman, L. (2009). Quantifying the effects of high summer temperatures due to climate change on buckling and rail related delays in south-East United Kingdom. *Meteorological Applications, 16*(2), 245–251. https://doi.org/10.1002/met.114

Dominguez, R., Olabarria, C., Woodin, S. A., Wethey, D. S., Peteiro, L. G., Macho, G., & Vazquez, E. (2021). Contrasting responsiveness of four ecologically and economically important bivalves to simulated heat waves. *Marine Environmental Research, 164*. https://doi.org/10.1016/j.marenvres.2020.105229

Donovan, M. K., Burkepile, D. E., Kratochwill, C., Shlesinger, T., Sully, S., Oliver, T. A., Hodgson, G., Freiwald, J., & van Woesik, R. (2021). Local conditions magnify coral loss after marine heatwaves. *Science, 372*(6545), 977–980.

Ebi, K. L., Capon, A., Berry, P., Broderick, C., de Dear, R., Havenith, G., Honda, Y., Kovats, R. S., Ma, W., Malik, A., Morris, N. B., Nybo, L., Seneviratne, S. I., Vanos, J., & Jay, O. (2021). Hot weather and heat extremes: Health risks. *Lancet, 398*(10301), 698–708.

Edwards, M. S. (2019). Comparing the impacts of four ENSO events on giant kelp (Macrocystis pyrifera) in the Northeast Pacific Ocean. *Algae, 34*(2), 141–151. https://doi.org/10.4490/algae.2019.34.5.4

Eisenman, D. P., Wilhalme, H., Tseng, C.-H., Chester, M., English, P., Pincetl, S., Fraser, A., Vangala, S., & Dhaliwal, S. K. (2016). Heat death associations with the built environment, social vulnerability and their interactions with rising temperature. *Health & Place, 41*, 89–99.

Ellena, M., Breil, M., & Soriani, S. (2020). The heat-health nexus in the urban context: A systematic literature review exploring the socio-economic vulnerabilities and built environment characteristics. *Urban Climate, 34*. https://doi.org/10.1016/j.uclim.2020.100676

Eriksen, S. H. (2022). *Is my vulnerability so different from yours?* Progress in Human Geography. https://doi.org/10.1177/03091325221083221

Evans, M., & Warburton, J. (2011). *Geomorphology of upland peat: Erosion, form and landscape change*. Wiley.

Fabricius, K. E. (2005). Effects of terrestrial runoff on the ecology of corals and coral reefs: Review and synthesis. *Marine Pollution Bulletin, 50*(2), 125–146.

Fifield, L. J., Lomas, K. J., Giridharan, R., & Allinson, D. (2018). Hospital wards and modular construction: Summertime overheating and energy efficiency. *Building and Environment, 141*, 28–44. https://doi.org/10.1016/j.buildenv.2018.05.041

Folegot, S., Hannah, D. M., Dugdale, S. J., Kurz, M. J., Drummond, J. D., Klaarr, M. J., Lee-Cullin, J., Keller, T., Marti, E., Zarnetske, J. P., Ward, A. S., & Krause, S. (2018). Low flow controls on stream thermal dynamics. *Limnologica, 68*, 157–167. https://doi.org/10.1016/j.limno.2017.08.003

Folkerts, M. A., Brode, P., Botzen, W. J. W., Martinius, M. L., Gerrett, N., Harmsen, C. N., & Daanen, H. A. M. (2020). Long term adaptation to heat stress: Shifts in the minimum mortality temperature in the Netherlands. *Frontiers in Physiology, 11*. https://doi.org/10.3389/fphys.2020.00225

Foucault, Q., Wieser, A., Waldvogel, A. M., Feldmeyer, B., & Pfenninger, M. (2018). Rapid adaptation to high temperatures in Chironomus riparius. *Ecology and Evolution, 8*(24), 12780–12789. https://doi.org/10.1002/ece3.4706

Fraser, A. M., Chester, M. V., & Eisenman, D. (2018). Strategic locating of refuges for extreme heat events (or heat waves). *Urban Climate, 25*, 109–119.

Fuller, A., Mitchell, D., Maloney, S. K., Hetem, R. S., Fonseca, V. F. C., Meyer, L. C. R., van de Ven, T., & Snelling, E. P. (2021). How dryland mammals will respond to climate change: The effects of body size, heat load and a lack of food and water. *Journal of Experimental Biology, 224*. https://doi.org/10.1242/jeb.238113

Gabbe, C. J., Mallen, E., & Varni, A. (2022). Housing and urban heat: Assessing risk disparities. *Housing Policy Debate*. https://doi.org/10.1080/10511482.2022.2093938

Gaughan, J. B., Lees, A. M., & Sejian, V. (2020). Sixty years of animal biometeorology. *International Journal of Biometeorology, 64*(1), 157–163. https://doi.org/10.1007/s00484-017-1459-1

Golembeski, C. A., Dong, K. R., & Irfan, A. (2021). Carceral and climate crises and health inequities: A call for greater transparency, accountability, and human rights protections. *World Medical & Health Policy, 13*(1), 69–96. https://doi.org/10.1002/wmh3.382

González-Tokman, D., Córdoba-Aguilar, A., Dáttilo, W., Lira-Noriega, A., Sánchez-Guillén, R. A., & Villalobos, F. (2020). Insect responses to heat: Physiological mechanisms, evolution and ecological implications in a warming world. *Biological Reviews, 95*(3), 802–821. https://doi.org/10.1111/brv.12588

Gough, H., Faulknall-Mills, S., King, M. F., & Luo, Z. W. (2019). Assessment of overheating risk in gynaecology scanning rooms during near-heatwave conditions: A case study of the Royal Berkshire Hospital in the UK. *International Journal of Environmental Research and Public Health, 16*(18). https://doi.org/10.3390/ijerph16183347

Graham, D. A., Vanos, J. K., Kenny, N. A., & Brown, R. D. (2016). The relationship between neighbourhood tree canopy cover and heat-related ambulance calls during extreme heat events in Toronto, Canada. *Urban Forestry & Urban Greening, 20*, 180–186. https://doi.org/10.1016/j.ufug.2016.08.005

Greer, D. H., & Weston, C. (2010). Heat stress affects flowering, berry growth, sugar accumulation and photosynthesis of Vitis vinifera cv. Semillon grapevines grown in a controlled environment. *Functional Plant Biology, 37*(3), 206–214. https://doi.org/10.1071/fp09209

Gronlund, C. J. (2014). Racial and socioeconomic disparities in heat-related health effects and their mechanisms: A review. *Current Epidemiology Reports, 1*, 165–173.

Guardaro, M., Hondula, D. M., & Redman, C. L. (2022). Social capital: Improving community capacity to respond to urban heat. *Local Environment, 27*(9), 1133–1150. https://doi.org/10.1080/13549839.2022.2103654

Guirguis, K., Basu, R., Al-Delaimy, W. K., Benmarhnia, T., Clemesha, R. E. S., Corcos, I., Guzman-Morales, J., Hailey, B., Small, I., Tardy, A., Vashishtha, D., Zivin, J. G., & Gershunov, A. (2018). Heat, disparities, and health outcomes in San Diego County's diverse climate zones. *Geohealth, 2*(7), 212–223. https://doi.org/10.1029/2017gh000127

Gupta, R., Howard, A., Davies, M., Mavrogianni, A., Tsoulou, I., Oikonomou, E., & Wilkinson, P. (2021). Examining the magnitude and perception of summertime overheating in London care homes. *Building Services Engineering Research & Technology, 42*(6), 653–675. https://doi.org/10.1177/01436244211013645

Hackerott, S., Martell, H. A., & Eirin-Lopez, J. M. (2021). Coral environmental memory: Causes, mechanisms, and consequences for future reefs. *Trends in Ecology & Evolution, 36*(11), 1011–1023. https://doi.org/10.1016/j.tree.2021.06.014

Hagedorn, R., Marti-Vargas, J. R., Dang, C. N., Hale, W. M., & Floyd, R. W. (2019). Temperature gradients in bridge concrete I-girders under heat wave. *Journal of Bridge Engineering, 24*(8). https://doi.org/10.1061/(asce)be.1943-5592.0001454

Hallegatte, S., & Walsh, B. (2021). Natural disasters, poverty and inequality: New metrics for fairer policies. In *The Routledge handbook of the political economy of the environment* (pp. 111–131). Routledge.

Hansen, A., Nitschke, M., Saniotis, A., Benson, J., Tan, Y., Smyth, V., Wilson, L., Han, G. S., Mwanri, L., & Bi, P. (2014). Extreme heat and cultural and linguistic minorities in Australia: Perceptions of stakeholders. *BMC Public Health, 14*. https://doi.org/10.1186/1471-2458-14-550

Hartmann, H., Bastos, A., Das, A. J., Esquivel-Muelbert, A., Hammond, W. M., Martinez-Vilalta, J., McDowell, N. G., Powers, J. S., Pugh, T. A. M., Ruthrof, K. X., & Allen, C. D. (2022). Climate change risks to global forest health: Emergence of unexpected events of elevated tree mortality worldwide. *Annual Review Plant Biology, 73*, 673–702. https://doi.org/10.1146/annurev-arplant-102820-012804

He, G. X., Zou, J., Liu, X. L., Liang, F. L., Liang, J., Yang, K., Masanja, F., Xu, Y., Zheng, Z., Deng, Y. W., & Zhao, L. Q. (2022). Assessing the impact of atmospheric heatwaves on intertidal clams. *Science of the Total Environment, 841*. https://doi.org/10.1016/j.scitotenv.2022.156744

Henderson, S. B., McLean, K. E., Lee, M. J., & Kosatsky, T. (2022). Analysis of community deaths during the catastrophic 2021 heat dome: Early evidence to inform the public health response during subsequent events in greater Vancouver, Canada. Environmental. *Epidemiology, 6*(1). https://doi.org/10.1097/ee9.0000000000000189

Hesketh, A. V., & Harley, C. D. G. (2023). Extreme heatwave drives topography-dependent patterns of mortality in a bed-forming intertidal barnacle, with implications for associated community structure. *Global Change Biology, 29*(1), 165–178. https://doi.org/10.1111/gcb.16390

Hoffman, J. S., Shandas, V., & Pendleton, N. (2020). The effects of historical housing policies on resident exposure to intra-urban heat: A study of 108 US urban areas. *Climate, 8*(1). https://doi.org/10.3390/cli8010012

Hosokawa, Y., Casa, D. J., Trtanj, J. M., Belval, L. N., Deuster, P. A., Giltz, S. M., Grundstein, A. J., Hawkins, M. D., Huggins, R. A., Jacklitsch, B., Jardine, J. F., Jones, H., Kazman, J. B., Reynolds, M. E., Stearns, R. L., Vanos, J. K., Williams, A. L., & Williams, W. J. (2019). Activity modification in heat: Critical assessment of guidelines across athletic, occupational, and military settings in the USA. *International Journal of Biometeorology, 63*(3), 405–427. https://doi.org/10.1007/s00484-019-01673-6

Hsu, A., Sheriff, G., Chakraborty, T., & Manya, D. (2021). Disproportionate exposure to urban heat Island intensity across major US cities. Nature. *Communications, 12*(1). https://doi.org/10.1038/s41467-021-22799-5

Iddon, C. R., Mills, T. C., Giridharan, R., & Lomas, K. J. (2015). The influence of hospital ward design on resilience to heat waves: An exploration using distributed lag models. *Energy and Buildings, 86*, 573–588. https://doi.org/10.1016/j.enbuild.2014.09.053

Innis, T., Allen-Waller, L., Brown, K. T., Sparagon, W., Carlson, C., Kruse, E., Huffmyer, A. S., Nelson, C. E., Putnam, H. M., & Barott, K. L. (2021). Marine heatwaves depress metabolic activity and impair cellular acid-base homeostasis in reef-building corals regardless of bleaching susceptibility. *Global Change Biology, 27*(12), 2728–2743. https://doi.org/10.1111/gcb.15622

Jay, O., Capon, A., Berry, P., Broderick, C., de Dear, R., Havenith, G., Honda, Y., Kovats, R. S., Ma, W., Malik, A., Morris, N. B., Nybo, L., Seneviratne, S. I., Vanos, J., & Ebi, K. L. (2021). Reducing the health effects of hot weather and heat extremes: From personal cooling strategies to green cities. *Lancet, 398*(10301), 709–724.

Jessel, S., Sawyer, S., & Hernandez, D. (2019). Energy, poverty, and health in climate change: A comprehensive review of an emerging literature. *Frontiers in Public Health, 7*. https://doi.org/10.3389/fpubh.2019.00357

Johnson, D. P., Stanforth, A., Lulla, V., & Luber, G. (2012). Developing an applied extreme heat vulnerability index utilizing socioeconomic and environmental data. *Applied Geography, 35*(1–2), 23–31.

Joyner, M. J. (2017). Physiological limits to endurance exercise performance: Influence of sex. *Journal of Physiology-London, 595*(9), 2949–2954. https://doi.org/10.1113/jp272268

Kafeety, A., Henderson, S. B., Lubik, A., Kancir, J., Kosatsky, T., & Schwandt, M. (2020). Social connection as a public health adaptation to extreme heat events. *Canadian Journal of Public Health-Revue Canadienne De Sante Publique, 111*(6), 876–879. https://doi.org/10.17269/s41997-020-00309-2

Kim, K., Jung, J., Schollaert, C., & Spector, J. T. (2021). A comparative assessment of cooling center preparedness across twenty-five US cities. *International Journal of Environmental Research and Public Health, 18*(9). https://doi.org/10.3390/ijerph18094801

Klinenberg, E. (1999). Denaturalizing disaster: A social autopsy of the 1995 Chicago heat wave. *Theory and Society, 28*(2), 239–295. https://doi.org/10.1023/a:1006995507723

Klinenberg, E. (2015). *Heat wave: A social autopsy of disaster in Chicago*. University of Chicago press.

Knox-Hayes, J., Osorio, J. C., Stamler, N., Dombrov, M., Winer, R., Smith, M. H., Blake, R. A., & Rosenzweig, C. (2023). The compound risk of heat and COVID-19 in new York City: Riskscapes, physical and social factors, and interventions. *Local Environment, 28*(6), 699–727. https://doi.org/10.1080/13549839.2023.2187362

Kuczynski, T., Staszczuk, A., Ziembicki, P., & Paluszak, A. (2021). The effect of the thermal mass of the building envelope on summer overheating of dwellings in a temperate climate. *Energies, 14*(14). https://doi.org/10.3390/en14144117

Kunert, N., & Hajek, P. (2022). Shade-tolerant temperate broad-leaved trees are more sensitive to thermal stress than light-demanding species during a moderate heatwave. *Trees Forests and People, 9*. https://doi.org/10.1016/j.tfp.2022.100282

Kuras, E., Hondula, D., & Brown-Saracino, J. (2015). Heterogeneity in individually experienced temperatures (IETs) within an urban neighborhood: Insights from a new approach to measuring heat exposure. *International Journal of Biometeorology, 59*, 1363–1372.

Kuras, E. R., Richardson, M. B., Calkins, M. M., Ebi, K. L., Hess, J. J., Kintziger, K. W., Jagger, M. A., Middel, A., Scott, A. A., & Spector, J. T. (2017). Opportunities and challenges for personal heat exposure research. *Environmental Health Perspectives, 125*(8), 085001.

Lee, Y., Haile, B., Seymour, G., & Azzarri, C. (2021). The heat never bothered me anyway: Gender-specific response of agricultural labor to climatic shocks in Tanzania. *Applied Economic Perspectives and Policy, 43*(2), 732–749. https://doi.org/10.1002/aepp.13153

Lees, A. M., Sejian, V., Wallage, A. L., Steel, C. C., Mader, T. L., Lees, J. C., & Gaughan, J. B. (2019). The impact of heat load on cattle. *Animals, 9*(6). https://doi.org/10.3390/ani9060322

Leigh, A., Sevanto, S., Close, J., & Nicotra, A. (2017). The influence of leaf size and shape on leaf thermal dynamics: Does theory hold up under natural conditions? *Plant, Cell & Environment, 40*(2), 237–248.

Levinton, J. (2020). Thermal stress: The role of body size and the giant major claw in survival and heat transfer of a fiddler crab (Leptuca pugilator). *Journal of Experimental Marine Biology and Ecology, 530*. https://doi.org/10.1016/j.jembe.2020.151428

Li, X. J., Ma, X. L., Lyu, F., & Song, Y. (2023). Examine the environmental inequity impact of urban heat mitigation on redlining legacy: Case study of Charlotte's retrofitting, 2001-2020. *Frontiers in Environmental Science, 11*. https://doi.org/10.3389/fenvs.2023.1218819

Lindmark, M., Ohlberger, J., & Gardmark, A. (2022). Optimum growth temperature declines with body size within fish species. *Global Change Biology, 28*(7), 2259–2271. https://doi.org/10.1111/gcb.16067

Ling, S., Scheibling, R., Rassweiler, A., Johnson, C., Shears, N., Connell, S., Salomon, A., Norderhaug, K., Pérez-Matus, A., & Hernández, J. (2015). Global regime shift dynamics of catastrophic sea urchin overgrazing. *Philosophical Transactions of the Royal Society B: Biological Sciences, 370*(1659), 20130269.

Lomas, K. J., Watson, S., Allinson, D., Fateh, A., Beaumont, A., Allen, J., Foster, H., & Garrett, H. (2021). Dwelling and household characteristics' influence on reported and measured summertime overheating: A glimpse of a mild climate in the 2050's. *Building and Environment, 201*. https://doi.org/10.1016/j.buildenv.2021.107986

Lu, L. Q., Wang, X., Ouyang, Y. F., Roningen, J., Myers, N., & Calfas, G. (2018). Vulnerability of interdependent urban infrastructure networks: Equilibrium after failure propagation and cascading impacts. *Computer-Aided Civil and Infrastructure Engineering, 33*(4), 300–315. https://doi.org/10.1111/mice.12347

Ma, C. S., Ma, G., & Pincebourde, S. (2021). Survive a warming climate: Insect responses to extreme high temperatures. *Annual Review of Entomology, 66*(66), 163–184. https://doi.org/10.1146/annurev-ento-041520-074454

Madureira, H., Pacheco, M., Sousa, C., Monteiro, A., De-Donato, F., & De-Sario, M. (2021). Evidences on adaptive mechanisms for cardiorespiratory diseases regarding extreme temperatures and air pollution: A comparative systematic review. *Geography and Sustainability, 34.* https://doi.org/10.1016/j.geosus.2021.08.001

Malmquist, A., Hjerpe, M., Glaas, E., Karlsson, H., & Lassi, T. (2022). Elderly people's perceptions of heat stress and adaptation to h eat: An interview study. *International Journal of Environmental Research and Public Health, 19*(7). https://doi.org/10.3390/ijerph19073775

Marchin, R. M., Backes, D., Ossola, A., Leishman, M. R., Tjoelker, M. G., & Ellsworth, D. S. (2022). Extreme heat increases stomatal conductance and drought-induced mortality risk in vulnerable plant species. *Global Change Biology, 28*(3), 1133–1146. https://doi.org/10.1111/gcb.15976

Mavrogianni, A., Taylor, J., Davies, M., Thoua, C., & Kolm-Murray, J. (2015). Urban social housing resilience to excess summer heat. *Building Research and Information, 43*(3), 316–333. https://doi.org/10.1080/09613218.2015.991515

Mayrhuber, E. A. S., Duckers, M. L. A., Wallner, P., Arnberger, A., Allex, B., Wiesbock, L., Wanka, A., Kolland, F., Eder, R., Hutter, H. P., & Kutalek, R. (2018). Vulnerability to heatwaves and implications for public health interventions – A scoping review. *Environmental Research, 166*, 42–54. https://doi.org/10.1016/j.envres.2018.05.021

McClanahan, T. R., & Muthiga, N. A. (1988). Changes in Kenyan coral reef community structure and function due to exploitation. *Hydrobiologia, 166*, 269–276.

McGregor, G. R., Bone, A., & Pappenberger, F. (2017). Meteorological risk: Extreme temperatures. In K. Poljanšek (Ed.), *Science for disaster risk management 2017: Knowing better and losing less* (Vol. 1, pp. 257–270). European Union. https://doi.org/10.2788/842809

McIntyre, A. M., Scammell, M. K., Martinez, M. P. B., Heidari, L., Negassa, A., Bongiovanni, R., & Fabian, M. P. (2022). Facilitators and barriers for keeping cool in an urban heat Island: Perspectives from residents of an environmental justice community. *Environmental Justice.* https://doi.org/10.1089/env.2022.0019

McKechnie, A. E., & Wolf, B. O. (2019). The physiology of heat tolerance in small endotherms. *Physiology, 34*(5), 302–313. https://doi.org/10.1152/physiol.00011.2019

Migliavacca, M., Musavi, T., Mahecha, M. D., Nelson, J. A., Knauer, J., Baldocchi, D. D., Perez-Priego, O., Christiansen, R., Peters, J., Anderson, K., Bahn, M., Black, T. A., Blanken, P. D., Bonal, D., Buchmann, N., Caldararu, S., Carrara, A., Carvalhais, N., Cescatti, A., Chen, J. Q., Cleverly, J., Cremonese, E., Desai, A. R., El-Madany, T. S., Farella, M. M., Fernandez-Martinez, M., Filippa, G., Forkel, M., Galvagno, M., Gomarasca, U., Gough, C. M., Gockede, M., Ibrom, A., Ikawa, H., Janssens, I. A., Jung, M., Kattge, J., Keenan, T. F., Knohl, A., Kobayashi, H., Kraemer, G., Law, B. E., Liddell, M. J., Ma, X. L., Mammarella, I., Martini, D., Macfarlane, C., Matteucci, G., Montagnani, L., Pabon-Moreno, D. E., Panigada, C., Papale, D., Pendall, E., Penuelas, J., Phillips, R. P., Reich, P. B., Rossini, M., Rotenberg, E., Scott, R. L., Stahl, C., Weber, U., Wohlfahrt, G., Wolf, S., Wright, I. J., Yakir, D., Zaehle, S., & Reichstein, M. (2021). The three major axes of terrestrial ecosystem function. *Nature, 598*(7881), 468. https://doi.org/10.1038/s41586-021-03939-9

Mikulewicz, M. (2019). Thwarting adaptation's potential? A critique of resilience and climate-resilient development. *Geoforum, 104*, 267–282. https://doi.org/10.1016/j.geoforum.2019.05.010

Millyard, A., Layden, J. D., Pyne, D. B., Edwards, A. M., & Bloxham, S. R. (2020). Impairments to thermoregulation in the elderly during heat exposure events. *Gerontology and Geriatric Medicine, 6*. https://doi.org/10.1177/2333721420932432

Miranda, N. A. F., Peer, N., Ishak, M. Z. B., & Marshall, D. J. (2019). Heat-wave tolerance in tropical intertidal animals: Accounting for thermal and desiccation tolerances. *Ecological Indicators, 107*. https://doi.org/10.1016/j.ecolind.2019.105561

Mitchell, B. C., Chakraborty, J., & Basu, P. (2021). Social inequities in urban heat and greenspace: Analyzing climate justice in Delhi, India. *International Journal of Environmental Research and Public Health, 18*(9). https://doi.org/10.3390/ijerph18094800

Mole, M. A., Daraujo, S. R., Van Aarde, R. J., Mitchell, D., & Fuller, A. (2016). Coping with heat: Behavioural and physiological responses of savanna elephants in their natural habitat. *Conservation Physiology, 4*. https://doi.org/10.1093/conphys/cow044

Morabito, M., Messeri, A., Crisci, A., Bao, J. Z., Ma, R., Orlandini, S., Huang, C. R., & Kjellstrom, T. (2021). Heat-related productivity loss: Benefits derived by working in the shade or work-time shifting. *International Journal of Productivity and Performance Management, 70*(3), 507–525. https://doi.org/10.1108/ijppm-10-2019-0500

Morris, N. B., Jay, O., Flouris, A. D., Casanueva, A., Gao, C., Foster, J., Havenith, G., & Nybo, L. (2020). Sustainable solutions to mitigate occupational heat strain – An umbrella review of physiological effects and global health perspectives. *Environmental Health, 19*(1). https://doi.org/10.1186/s12940-020-00641-7

Morris, N. B., Piil, J. F., Morabito, M., Messeri, A., Levi, M., Ioannou, L. G., Ciuha, U., Pogacar, T., Bogataj, L. K., Kingma, B., Casanueva, A., Kotlarski, S., Spirig, C., Foster, J., Havenith, G., Mayor, T. S., Flouris, A. D., & Nybo, L. (2021). The HEAT-SHIELD project – Perspectives from an inter-sectoral approach to occupational heat stress. *Journal of Science and Medicine in Sport, 24*(8), 747–755. https://doi.org/10.1016/j.jsams.2021.03.001

Morrissey-Basler, M. C., Eason, C. M., Clines, S. H., Kaufman, C. E., & Casa, D. J. (2023). Perceived challenges and barriers for females working in the heat. *Journal of Occupational and Environmental Hygiene*. https://doi.org/10.1080/15459624.2023.2268725

Motanya, N. C., & Valera, P. (2016). Climate change and its impact on the incarcerated population: A descriptive review. *Social Work in Public Health, 31*(5), 348–357. https://doi.org/10.1080/19371918.2015.1137513

Mottola, G., Lopez, M. E., Vasemagi, A., Nikinmaa, M., & Anttila, K. (2022). Are you ready for the heat? Phenotypic plasticity versus adaptation of heat tolerance in three-spined stickleback. *Ecosphere, 13*(4). https://doi.org/10.1002/ecs2.4015

Mulholland, E., & Feyen, L. (2021). Increased risk of extreme heat to European roads and railways with global warming. *Climate Risk Management, 34*. https://doi.org/10.1016/j.crm.2021.100365

Münchinger, I. K., Hajek, P., Akdogan, B., Caicoya, A. T., & Kunert, N. (2023). Leaf thermal tolerance and sensitivity of temperate tree species are correlated with leaf physiological and functional drought resistance traits. *Journal of Forestry Research, 34*(1), 63–76.

Naicker, N., Teare, J., Balakrishna, Y., Wright, C. Y., & Mathee, A. (2017). Indoor temperatures in low cost housing in Johannesburg, South Africa. *International Journal of Environmental Research and Public Health, 14*(11), 1410.

Nati, J., Svendsen, M., Marras, S., Killen, S., Steffensen, J., Mckenzie, D. J., & Domenici, P. (2021). Intraspecific variation in thermal tolerance differs between tropical and temperate fishes. *Scientific Reports, 11*(1), 21272.

Navas-Martín, M. A., López-Bueno, J. A., Ascaso-Sánchez, M. S., Sarmiento-Suárez, R., Follos, F., Vellón, J. M., Mirón, I. J., Luna, M. Y., Sánchez-Martínez, G., Culqui, D., Linares, C., & Díaz, J. (2022). Gender differences in adaptation to heat in Spain (1983–2018). *Environmental Research, 215*. https://doi.org/10.1016/j.envres.2022.113986

Nazarian, N., & Lee, J. K. W. (2021). Personal assessment of urban heat exposure: A systematic review. *Environmental Research Letters, 16*, 3. https://doi.org/10.1088/1748-9326/abd350

Nguyen, H. M., Kim, M., Ralph, P. J., Marín-Guirao, L., Pernice, M., & Procaccini, G. (2020). Stress memory in seagrasses: First insight into the effects of thermal priming and the role of epigenetic modifications. *Frontiers in Plant Science, 11*, 533085.

Niu, Y. L., Li, Z. C., Gao, Y., Liu, X. B., Xu, L., Vardoulakis, S., Yue, Y. J., Wang, J., & Liu, Q. Y. (2021). A systematic review of the development and validation of the heat vulnerability index: Major factors, methods, and spatial units. *Current Climate Change Reports, 7*(3), 87–97. https://doi.org/10.1007/s40641-021-00173-3

O'Lenick, C. R., Baniassadi, A., Michael, R., Monaghan, A., Boehnert, J., Yu, X., Hayden, M. H., Wiedinmyer, C., Zhang, K., Crank, P. J., Heusinger, J., Hoel, P., Sailor, D. J., & Wilhelmi, O. (2020). A case-crossover analysis of indoor heat exposure on mortality and hospitalizations among the elderly in Houston, Texas. *Environmental Health Perspectives, 128*(12). https://doi.org/10.1289/ehp6340

O'Neill, M. S., Zanobetti, A., & Schwartz, J. (2005). Disparities by race in heat-related mortality in four US cities: The role of air conditioning prevalence. *Journal of Urban Health, 82*, 191–197.

Orlando, S., Mosconi, C., De Santo, C., Gialloreti, L. E., Inzerilli, M. C., Madaro, O., Mancinelli, S., Ciccacci, F., Marazzi, M. C., Palombi, L., & Liotta, G. (2021). The effectiveness of intervening on social isolation to reduce mortality during heat waves in aged population: A retrospective ecological study. *International Journal of Environmental Research and Public Health, 18*(21). https://doi.org/10.3390/ijerph182111587

Ortiz, L., Mustafa, A., Cantis, P. H., & McPhearson, T. (2022). Overlapping heat and COVID-19 risk in new York City. *Urban Climate, 41*. https://doi.org/10.1016/j.uclim.2021.101081

Paaijmans, K., Takken, W., Githeko, A., & Jacobs, A. (2008). The effect of water turbidity on the near-surface water temperature of larval habitats of the malaria mosquito Anopheles gambiae. *International Journal of Biometeorology, 52*, 747–753.

Palin, E. J., Thornton, H. E., Mathison, C. T., McCarthy, R. E., Clark, R. T., & Dora, J. (2013). Future projections of temperature-related climate change impacts on the railway network of Great Britain. *Climatic Change, 120*(1–2), 71–93. https://doi.org/10.1007/s10584-013-0810-8

Palin, E. J., Oslakovic, I. S., Gavin, K., & Quinn, A. (2021). Implications of climate change for railway infrastructure. *Wiley Interdisciplinary Reviews-Climate Change, 12*(5). https://doi.org/10.1002/wcc.728

Palu, S., & Mahmoud, H. (2019). Impact of climate change on the integrity of the superstructure of deteriorated US bridges. *PLoS One, 14*(10). https://doi.org/10.1371/journal.pone.0223307

Palumbi, S. R., Barshis, D. J., Traylor-Knowles, N., & Bay, R. A. (2014). Mechanisms of reef coral resistance to future climate change. *Science, 344*(6186), 895–898.

Parker, L. E., McElrone, A. J., Ostoja, S. M., & Forrestel, E. J. (2020). Extreme heat effects on perennial crops and strategies for sustaining future production. *Plant Science, 295*. https://doi.org/10.1016/j.plantsci.2019.110397

Pasquini, L., van Aardenne, L., Godsmark, C. N., Lee, J., & Jack, C. (2020). Emerging climate change-related public health challenges in Africa: A case study of the heat-health vulnerability of informal settlement residents in Dar es Salaam. *Tanzania. Science of the Total Environment, 747*, 141355.

Pelling, M., & High, C. (2005). Understanding adaptation: What can social capital offer assessments of adaptive capacity? *Global Environmental Change-Human and Policy Dimensions, 15*(4), 308–319. https://doi.org/10.1016/j.gloenvcha.2005.02.001

Peralta-Maraver, I., & Rezende, E. L. (2021). Heat tolerance in ectotherms scales predictably with body size. Nature. *Climate Change, 11 (1):58-+*. https://doi.org/10.1038/s41558-020-00938-y

Pescaroli, G., & Alexander, D. (2016). Critical infrastructure, panarchies and the vulnerability paths of cascading disasters. *Natural Hazards, 82*(1), 175–192. https://doi.org/10.1007/s11069-016-2186-3

Pohl, B., Dos Santos, S., Bai, G. M., Compaoré, Y., Dianou, K., Diallo-Dudek, J., Soura, A., & Janicot, S. (2021). Indoor temperature variability in the Sahel: A pilot study in Ouagadougou. *Burkina Faso. Theoretical and Applied Climatology, 146*(3–4), 1403–1420. https://doi.org/10.1007/s00704-021-03800-z

Pollock, H. S., Brawn, J. D., & Cheviron, Z. A. (2021). Heat tolerances of temperate and tropical birds and their implications for susceptibility to climate warming. *Functional Ecology, 35*(1), 93–104. https://doi.org/10.1111/1365-2435.13693

Pottier, P., Lin, H. Y., Oh, R. R. Y., Pollo, P., Rivera-Villanueva, A. N., Valdebenito, J. O., Yang, Y. F., Amano, T., Burke, S., Drobniak, S. M., & Nakagawa, S. (2022). A comprehensive database of amphibian heat tolerance. *Scientific Data, 9*(1). https://doi.org/10.1038/s41597-022-01704-9

Poumadère M, Mays C, Le Mer S, Blong R (2005) The 2003 heat wave in France: Dangerous climate change here and now. Risk Analysis 25 (6):1483–1494. doi:https://doi.org/10.1111/j.1539-6924.2005.00694.x.

Putnam, R. D. (1995). Tuning in, tuning out – The strange disappearance of social capital in America. *Ps-Political Science & Politics, 28*(4), 664–683. https://doi.org/10.2307/420517

Quick, M., Christidis, T., Olaniyan, T., Newstead, N., & Pinault, L. (2022). *Exploring the associations between cooling Centre accessibility and marginalization in Montreal, Toronto, and Vancouver*. Canadian Geographer-Geographe Canadien. https://doi.org/10.1111/cag.12805

Ramsay, E. E., Fleming, G. M., Faber, P. A., Barker, S. F., Sweeney, R., Taruc, R. R., Chown, S. L., & Duffy, G. A. (2021). Chronic heat stress in tropical urban informal settlements. *Iscience, 24*(11). https://doi.org/10.1016/j.isci.2021.103248

Ramsay, E. E., Hamel, P., Chown, S. L., & Duffy, G. A. (2024). Humid heat stress overlooked for one billion people in urban informal settlements. *One Earth, 7*(1), 2–5. https://doi.org/10.1016/j.oneear.2023.12.005

Rashid, A., & Adnan, M. N. (2009). Pakistan's refugees face uncertain future. *Lancet, 374*(9683), 13–14. https://doi.org/10.1016/s0140-6736(09)61218-6

Rauh, V. A., Landrigan, P. J., & Claudio, L. (2008). Housing and health – Intersection of poverty and environmental exposures. *Reducing the Impact of Poverty on Health and Human Development: Scientific Approaches, 1136*, 276–288. https://doi.org/10.1196/annals.1425.032

Reid, C. E., O'neill, M. S., Gronlund, C. J., Brines, S. J., Brown, D. G., Diez-Roux, A. V., & Schwartz, J. (2009). Mapping community determinants of heat vulnerability. *Environmental Health Perspectives, 117*(11), 1730–1736.

Ren, C., & McGregor, G. (2021). *Urban climate science for planning healthy cities*. Springer.

Renteria, R., Grineski, S., Collins, T., Flores, A., & Trego, S. (2022). Social disparities in neighborhood heat in the Northeast United States. *Environmental Research, 203*. https://doi.org/10.1016/j.envres.2021.111805

Robbins, P. (2019). *Political ecology: A critical introduction. Critical introductions to geography* (3rd ed.). Wiley Blackwell.

Rodrigues, Y. K., & Beldade, P. (2020). Thermal plasticity in insects' response to climate change and to multifactorial environments. *Frontiers in Ecology and Evolution, 8*. https://doi.org/10.3389/fevo.2020.00271

Rossaro, B. (1991). CHIRONOMIDS and water temperature. *Aquatic Insects, 13*(2), 87–98. https://doi.org/10.1080/01650429109361428

Safaie, A., Silbiger, N. J., McClanahan, T. R., Pawlak, G., Barshis, D. J., Hench, J. L., Rogers, J. S., Williams, G. J., & Davis, K. A. (2018). High frequency temperature variability reduces the risk of coral bleaching. *Nature Communications, 9*(1), 1–12.

Saverino, K. C., Routman, E., Lookingbill, T. R., Eanes, A. M., Hoffman, J. S., & Bao, R. (2021). Thermal inequity in Richmond, VA: The effect of an unjust evolution of the urban landscape on urban heat islands. *Sustainability, 13*(3). https://doi.org/10.3390/su13031511

Schinasi, L. H., Kanungo, C., Christman, Z., Barber, S., Tabb, L., & Headen, I. (2022). Associations between historical redlining and present-day heat vulnerability housing and land cover characteristics in Philadelphia, PA. *Journal of Urban Health-Bulletin of the New York Academy of Medicine, 99*(1), 134–145. https://doi.org/10.1007/s11524-021-00602-6

Schlosberg, D. (2007). *Defining environmental justice: Theories, movements, and nature*. Oxford University Press.

Schmidt, W.-P., Chauhan, K., Bhavsar, P., Yasobant, S., Patwardhan, V., Aunger, R., Mavalankar, D., Saxena, D., & Curtis, V. (2020). Cluster-randomised trial to test the effect of a behaviour change intervention on toilet use in rural India: Results and methodological considerations. *BMC Public Health, 20*, 1–16.

Schoepf, V., Stat, M., Falter, J. L., & McCulloch, M. T. (2015). Limits to the thermal tolerance of corals adapted to a highly fluctuating, naturally extreme temperature environment. *Scientific Reports, 5*(1), 17639.

Schubert, N., Santos, R., & Silva, J. (2021). Living in a fluctuating environment increases tolerance to marine heatwaves in the free-living coralline Alga Phymatolithon lusitanicum. *Frontiers in Marine Science, 8*. https://doi.org/10.3389/fmars.2021.791422

Schunemann, C., Son, S., & Ortlepp, R. (2022). Heat resilience of apartment buildings in Korea and Germany: Comparison of building design and climate. *International Journal of Energy and Environmental Engineering*. https://doi.org/10.1007/s40095-022-00476-7

Scott, A. A., Misiani, H., Okoth, T., Jordan, A., Gohlke, J., Ouma, G., Arrighi, J., Zaitchik, B. F., Jjemba, E., Verjee, S., & Waugh, D. W. (2017). Temperature and heat in informal settlements in Nairobi. *PLoS One, 12*(11). https://doi.org/10.1371/journal.pone.0187300

Sera, F., Armstrong, B., Tobias, A., Vicedo-Cabrera, A. M., Astrom, C., Bell, M. L., Chen, B. Y., Coelho, M., Correa, P. M., Cruz, J. C., Dang, T. N., Hurtado-Diaz, M., Van, D. D., Forsberg, B., Guo, Y. L., Guo, Y. M., Hashizume, M., Honda, Y., Iniguez, C., Jaakkola, J. J. K., Kan, H. D., Kim, H., Lavigne, E., Michelozzi, P., Ortega, N. V., Osorio, S., Pascal, M., Ragettli, M. S., Ryti, N. R., Saldiva, P. H. N., Schwartz, J., Scortichini, M., Seposo, X., Tong, S. L., Zanobetti, A., & Gasparrini, A. (2019). How urban characteristics affect vulnerability to heat and cold: A multi-country analysis. *International Journal of Epidemiology, 48*(4), 1101–1112. https://doi.org/10.1093/ije/dyz008

Sera, F., Hashizume, M., Honda, Y., Lavigne, E., Schwartz, J., Zanobetti, A., Tobias, A., Iniguez, C., Vicedo-Cabrera, A. M., Blangiardo, M., Armstrong, B., & Gasparrini, A. (2020). Air conditioning and heat-related mortality a multi-country longitudinal study. *Epidemiology, 31*(6), 779–787. https://doi.org/10.1097/ede.0000000000001241

Sharifi, S., Saman, W., & Alemu, A. (2019). Identification of overheating in the top floors of energy-efficient multilevel dwellings. *Energy and Buildings, 204*. https://doi.org/10.1016/j.enbuild.2019.109452

Shih, W. Y., & Mabon, L. (2021). Understanding heat vulnerability in the subtropics: Insights from expert judgements. *International Journal of Disaster Risk Reduction, 63*. https://doi.org/10.1016/j.ijdrr.2021.102463

Skarha, J., Peterson, M., Rich, J. D., & Dosa, D. (2020). An overlooked crisis: Extreme temperature exposures in incarceration settings. *American Journal of Public Health, 110*, S41–S42. https://doi.org/10.2105/ajph.2019.305453

Skarha, J., Spangler, K., Dosa, D., Rich, J. D., Savitz, D. A., & Zanobetti, A. (2023). Heat-related mortality in US state and private prisons: A case-crossover analysis. *PLoS One, 18*, 3. https://doi.org/10.1371/journal.pone.0281389

Smale, D. A. (2020). Impacts of ocean warming on kelp forest ecosystems. *New Phytologist, 225*(4), 1447–1454. https://doi.org/10.1111/nph.16107

Smale, D. A., Wernberg, T., Oliver, E. C. J., Thomsen, M., Harvey, B. P., Straub, S. C., Burrows, M. T., Alexander, L. V., Benthuysen, J. A., Donat, M. G., Feng, M., Hobday, A. J., Holbrook, N. J., Perkins-Kirkpatrick, S. E., Scannell, H. A., Sen Gupta, A., Payne, B. L., & Moore, P. J. (2019). Marine heatwaves threaten global biodiversity and the provision of ecosystem services. *Nature Climate Change, 9 (4):306*-+. https://doi.org/10.1038/s41558-019-0412-1

Song, J. L., Lu, Y., Zhao, Q., Zhang, Y. Q., Yang, X. C., Chen, Q., Guo, Y. M., & Hu, K. J. (2022). Effect modifications of green space and blue space on heat-mortality association in Hong Kong, 2008–2017. *Science of the Total Environment, 838*. https://doi.org/10.1016/j.scitotenv.2022.156127

Souchon, Y., & Tissot, L. (2012). Synthesis of thermal tolerances of the common freshwater fish species in large Western Europe rivers. *Knowledge and Management of Aquatic Ecosystems, 405*. https://doi.org/10.1051/kmae/2012008

Spencer, S., Samateh, T., Wabnitz, K., Mayhew, S., Allen, H., & Bonell, A. (2022). The challenges of working in the heat whilst pregnant: Insights from gambian women farmers in the face of climate change. *Frontiers in Public Health, 10*. https://doi.org/10.3389/fpubh.2022.785254

Teare, J., Mathee, A., Naicker, N., Swanepoel, C., Kapwata, T., Balakrishna, Y., du Preez, D. J., Millar, D. A., & Wright, C. Y. (2020). Dwelling characteristics influence indoor temperature and may pose health threats in LMICs. *Annals of Globalization and Health, 86*(1). https://doi.org/10.5334/aogh.2938

Teskey, R., Wertin, T., Bauweraerts, I., Ameye, M., McGuire, M. A., & Steppe, K. (2015). Responses of tree species to heat waves and extreme heat events. *Plant Cell and Environment, 38*(9), 1699–1712. https://doi.org/10.1111/pce.12417

Trahan, A., Walshe, R., & Mehta, V. (2023). Extreme heat, gender, and access to preparedness measures: An analysis of the heatwave early warning system in Ahmedabad, India. *International Journal of Disaster Risk Reduction, 99*. https://doi.org/10.1016/j.ijdrr.2023.104080

Tuholske, C., Caylor, K., Funk, C., Verdin, A., Sweeney, S., Grace, K., Peterson, P., & Evans, T. (2021). Global urban population exposure to extreme heat. *Proceedings of the National Academy of Sciences of the United States of America, 118*(41). https://doi.org/10.1073/pnas.2024792118

Tuholske, C., Lynch, V. D., Spriggs, R., Ahn, Y., Raymond, C., Nigra, A. E., & Parks, R. M. (2024). Hazardous heat exposure among incarcerated people in the United States. *Nature Sustainability, 7*(4), 394–398.

Turek-Hankins LL, Coughlan de Perez E, Scarpa G, Ruiz-Diaz R, Nayna Schwerdtle P, Joe ET, Galappaththi EK, French EM, Austin SE, Singh C, Siña M, Siders AR, van Aalst MK, Templeman S, Nunbogu AM, Berrang-Ford L, Agrawal TA, Team TGAMI, Mach KJ (2021) Climate change adaptation to extreme heat: A global systematic review of implemented action. Oxford Open Climate Change 1 (1):13. doi:https://doi.org/https://doi.org/10.1093/oxfclm/kgab005.

Uibel, D., Sharma, R., Piontkowski, D., Sheffield, P. E., & Clougherty, J. E. (2022). Association of ambient extreme heat with pediatric morbidity: A scoping review. *International Journal of Biometeorology*. https://doi.org/10.1007/s00484-022-02310-5

Vanos, J. K., McKercher, G. R., Naughton, K., & Lochbaum, M. (2017). Schoolyard shade and sun exposure: Assessment of personal monitoring during children's physical activity. *Photochemistry and Photobiology, 93*(4), 1123–1132.

Varghese, B. M., Hansen, A. L., Williams, S., Bi, P., Hanson-Easey, S., Barnett, A. G., Heyworth, J. S., Sim, M. R., Rowett, S., Nitschke, M., Di Corleto, R., & Pisaniello, D. L. (2020). Heat-related injuries in Australian workplaces: Perspectives from health and safety representatives. *Safety Science, 126*. https://doi.org/10.1016/j.ssci.2020.104651

Varickanickal J, Newbold KB (2021) Extreme heat events and health vulnerabilities among immigrant and newcomer populations. Environmental Health Review 64 (2):28–34. doi:https://doi.org/https://doi.org/10.5864/d2021-011.

Vebrová, L., van Nieuwenhuijzen, A., Kolár, V., & Boukal, D. S. (2018). Seasonality and weather conditions jointly drive flight activity patterns of aquatic and terrestrial chironomids. *BMC Ecology, 18*. https://doi.org/10.1186/s12898-018-0175-y

Venter, Z. S., Krog, N. H., & Barton, D. N. (2020). Linking green infrastructure to urban heat and human health risk mitigation in Oslo, Norway. *Science of the Total Environment, 709*. https://doi.org/10.1016/j.scitotenv.2019.136193

Venugopal, V., Rekha, S., Manikandan, K., Latha, P. K., Vennila, V., Ganesan, N., Kumaravel, P., & Chinnadurai, S. J. (2016). Heat stress and inadequate sanitary facilities at workplaces – An occupational health concern for women? *Global Health Action, 9*. https://doi.org/10.3402/gha.v9.31945

Venugopal, V., Latha, P. K., Shanmugam, R., Krishnamoorthy, M., Omprashanth, R., Lennqvist, R., & Johnson, P. (2021). Epidemiological evidence from south Indian working population-the heat exposures and health linkage. *Journal of Exposure Science and Environmental Epidemiology, 31*(1), 177–186. https://doi.org/10.1038/s41370-020-00261-w

Voelkel, J., Hellman, D., Sakuma, R., & Shandas, V. (2018). Assessing vulnerability to urban heat: A study of disproportionate heat exposure and access to refuge by socio-demographic status in

Portland, Oregon. *International Journal of Environmental Research and Public Health, 15*(4). https://doi.org/10.3390/ijerph15040640

Wiesbock, L., Wanka, A., Mayrhuber, E. A. S., Allex, B., Kolland, F., Hutter, H. P., Wallner, P., Arnberger, A., Eder, R., & Kutalek, R. (2016). Heat vulnerability, poverty and health inequalities in urban migrant communities: A pilot study from Vienna. In *Climate change and health: Improving resilience and reducing risks* (pp. 389–401). https://doi.org/10.1007/978-3-319-24660-4_22

Wilby, R. L., Kasei, R., Gough, K. V., Amankwaa, E. F., Abarike, M., Anderson, N. J., Codjoe, S. N. A., Griffiths, P., Kaba, C., Abdullah, K., Kayaga, S., Matthews, T., Mensah, P., Murphy, C., & Yankson, P. W. K. (2021). Monitoring and moderating extreme indoor temperatures in low-income urban communities. *Environmental Research Letters, 16*(2). https://doi.org/10.1088/1748-9326/abdbf2

Wilson, L. A., & Black, D. A. (2022). The intersection between heatwaves, high-rise living and the aged: A narrative review of the literature. *Atmosphere, 13*(9). https://doi.org/10.3390/atmos13091461

Wolf, T., & McGregor, G. (2013). The development of a heat wave vulnerability index for London, United Kingdom. *Weather and Climate Extremes, 1*, 59–68. https://doi.org/10.1016/j.wace.2013.07.004

Wolf, T., McGregor, G., & Analitis, A. (2014). Performance assessment of a heat wave vulnerability index for greater London, United Kingdom. *Weather Climate and Society, 6*(1), 32–46. https://doi.org/10.1175/wcas-d-13-00014.1

Wolf, T., Chuang, W. C., & McGregor, G. (2015). On the science-policy bridge: Do spatial heat vulnerability assessment studies influence policy? *International Journal of Environmental Research and Public Health, 12*(10), 13321–13349. https://doi.org/10.3390/ijerph121013321

Xiang, J. J., Bi, P., Pisaniello, D., & Hansen, A. (2014). Health impacts of workplace heat exposure: An epidemiological review. *Industrial Health, 52*(2), 91–101. https://doi.org/10.2486/indhealth.2012-0145

Xu, Z. W., Sheffield, P. E., Su, H., Wang, X. Y., Bi, Y., & Tong, S. L. (2014). The impact of heat waves on children's health: A systematic review. *International Journal of Biometeorology, 58*(2), 239–247. https://doi.org/10.1007/s00484-013-0655-x

Yanovich, R., Ketko, I., & Charkoudian, N. (2020). Sex differences in human thermoregulation: Relevance for 2020 and beyond. *Physiology, 35*(3), 177–184. https://doi.org/10.1152/physiol.00035.2019

Yin, Q., Wang, J. F., Ren, Z. P., Li, J., & Guo, Y. M. (2019). Mapping the increased minimum mortality temperatures in the context of global climate change. *Nature Communications, 10*. https://doi.org/10.1038/s41467-019-12663-y

Zhou, Z. Q., Bouma, T. J., Fivash, G. S., Ysebaert, T., van Ijzerloo, L., van Dalen, J., van Dam, B., & Walles, B. (2022). Thermal stress affects bioturbators' burrowing behavior: A mesocosm experiment on common cockles (Cerastoderma edule). *Science of the Total Environment, 824*. https://doi.org/10.1016/j.scitotenv.2022.153621

Chapter 11
Observed Impacts of Heatwaves

Abstract This chapter provides an overview of the observed impacts of heatwaves, focusing on human populations, the economy, society, infrastructure, domesticated animals, terrestrial wildlife, and marine environments. For human populations heat-related mortality is a major concern, with heatwaves causing more deaths than any other natural hazard in some countries. Vulnerable groups, such as the elderly and children, are disproportionately affected, with significant increases in mortality rates during heatwave events. The health impacts of heatwaves can be exacerbated by chronic diseases, pregnancy, and behavioural changes. Economically, heatwaves impose substantial costs on infrastructure, labour productivity, and healthcare systems. The agricultural sector is also affected, with heatwaves causing significant reductions in crop yields and livestock productivity. Increased aggression, mental health issues, and changes in social behaviour as social impacts arising from heatwaves are also touched upon. Because infrastructure systems, especially energy and transport, are highly vulnerable to heatwaves, impacts often manifest in the form increased cooling demand during heatwaves leading to grid failures and brownouts and blackouts and damage to rail and road networks via road heave and rail buckling, resulting in delays and increased maintenance costs. Domesticated animals and wildlife are not spared from the impacts of heatwaves. High mortality rates among cattle, poultry, and other farm animals are common during heatwave events. Wildlife, including some iconic species, also suffer significant mortality due to heatwave related heat stress. Marine environments are equally affected, with marine heatwaves causing widespread mortality among some marine species, disrupting food webs, and altering ecosystem structures. The chapter underscores the urgent need for systematic data collection on heatwave impacts to develop effective risk management strategies. It calls for integrated global systems to accurately record heatwave-related deaths and economic losses. The chapter concludes by emphasizing the importance of understanding the full scale of heatwave impacts to inform policy and adaptation and mitigation measures.

Keywords Heatwave impacts · Human and animal populations · Economic and infrastructure impacts · Terrestrial and marine ecosystems · Mortality and morbidity · Productivity loss · Disruption to energy and transport systems · Alterations to community structure and food webs

G. McGregor, *Heatwaves*, Biometeorology 6,
https://doi.org/10.1007/978-3-031-69906-1_11

11.1 Introduction

To achieve evidence-based heatwave risk management, a solid understanding of the impacts precipitated by heatwaves is essential. Without this knowledge, risk managers and decision makers would simply be stumbling around in the dark when developing heatwave related risk management strategies. In the case of heatwaves, the necessity to develop policy based on more than feeling or intuition has reached a critical juncture given this often-overlooked natural hazard is becoming increasingly palpable as a driver of loss and damage at a range of scales in both human and natural systems. Accordingly, the purpose of this chapter is to provide an overview of observed heatwave impacts, that is impacts that have been recorded. Accordingly, the chapter avoids speculation about possible impacts. In assembling the material for this chapter, material beyond the traditional academic literature has also been used. Structured similarly to the previous chapter, impacts on humans, economy and society and infrastructure systems along with animal populations and a range of ecosystems will be outlined. For impacts on life forms, the primary focus will be at the population to ecosystem level with eco-physiological and individual level impacts, as manifest through thermoregulatory and physiological modifications and changes in reproduction and growth, given less attention.

11.2 Human Populations

Heat-related mortality is the most widely written about impact of heatwaves on human populations. However, many such studies originate from mid-latitude, high-income countries of low- to medium-population density and neglect regions where heatwave risk is the most concerning (Campbell et al., 2018; Klingelhöfer et al., 2023; Sapari et al., 2023). For some countries, such as the United States, Australia and the United Kingdom and a range of European countries, it is estimated that heatwaves kill more people than any other type of natural hazard (Vecellio & Vanos, 2024).

Notwithstanding some of the criticisms levelled at the Centre for Research on the Epidemiology of Disasters (CRED) Emergency Events Database (EM-DAT), including variability in reporting and recording of heatwave related death tolls, the United Nations Office for Disaster Risk Reduction (UNISDR) in an analysis of heatwave deaths recorded in EM-DAT estimate that 166,000 people died from heatwaves between 1998 and 2017 of which 70,000 can be attributed to the 2003 European heatwave. Querying the same database for the period 2018–2022 reveals a further 10,667 heatwave deaths making the total of 176,667 for the period 1998–2022. While these are impressive numbers that speak to the scale of the impact of heatwaves on human populations it is highly likely the actual numbers of heatwave deaths are severely underestimated. Setting aside the challenges of when a heatwave is not a heatwave—the definition problem—as well as how heatwave deaths are measured—the estimation problem—until there is a fully integrated global scale system with a robust protocol for recording heatwave deaths, the true scale of these may never be known ((Brimicombe et al., 2021; Henderson et al., 2022). Humans succumb to the effects of heat during heatwaves because the body's core temperature starts to drift upwards from the normal resting temperature of around 37 °C (McGregor & Vanos, 2018; Ebi et al., 2021) resulting in a suite of heat-related illnesses. These can be subdivided into four categories, namely acute heat illnesses, chronic disease exacerbations, pregnancy complications, and behavioural and other heat-related emergencies (Hess et al., 2022).

Acute heat illness manifests in several ways. Generally, the gradation of severity is from heat cramps and edema (swelling), to heat syncope (dizziness) through to heat exhaustion which can progress to heat stroke, the most dangerous and life threating of the heat illnesses. While heat stroke is characterised by many of the symptoms associated with heat exhaustion, such as weakness, intense thirst, gastrointestinal symptoms and dehydration, its defining trait is hyperthermia which occurs as the body core temperature starts to exceed 40 °C with executive function compromised and seizure and coma highly likely (Hess et al., 2022). With hyperthermia, the body struggles to manage its heat balance with heat loss mechanisms such as evaporative and convective cooling shutting down (McGregor & Vanos, 2018). At this point, critical proteins begin to breakdown and a sepsis-like inflammatory response follows with multi-organ failure and death occurring if untreated (Hess et al., 2022). In all, 27 different pathways to death arising from the excessive heat load associated with heatwaves have been identified (Mora et al., 2017) with classic heatstroke found to contribute to 9–37% of heat-related fatalities during heatwaves (Bouchama et al., 2022). As the elderly are especially heatwave vulnerable, deaths amongst this population group are usually disproportionately higher than other age groups (Bunker et al., 2016). For example, during the London August 2003 heatwave, the mortality rate amongst the elderly increased by 59% (Johnson et al., 2005). This is similar in magnitude to the percentage increases in mortality observed for the elderly across England and Wales for the most intense of a series of heatwave events in 2022 (Fig. 11.1). At the other end of the age spectrum there is clear evidence of the impact of heatwaves on children and young people (Niu et al., 2023; Wilk et al., 2021; Xu et al., 2014) with the most horrific outcomes being fatal heat stroke in children left in parked vehicles (Adato et al., 2016; Booth et al., 2010; Breitfeld, 2020), a manifestation of the so-called 'forgotten baby syndrome' (Anselmi et al., 2020). Deaths from heat stroke during periods of extreme heat, largely brought on by exertional heat stress, are also evident for a range of activities including agriculture, construction, some forms of sport and military exercises (Alele et al., 2020; Gamage et al., 2020; Grundstein et al., 2017; Kerr et al., 2020; Lee et al., 2022; McMahon et al., 2021). For example, in the United States between 1992 and 2016, construction workers, who composed 6% of the total workforce at that time, accounted for 36% of all occupational heat-related deaths (Dong et al., 2019). In some instances, heat-related deaths amongst construction workers may be under-reported (Pradhan et al., 2019). Since the COVID-19 pandemic there has been a growing interest in the health impacts of the co-occurrence of heatwaves and

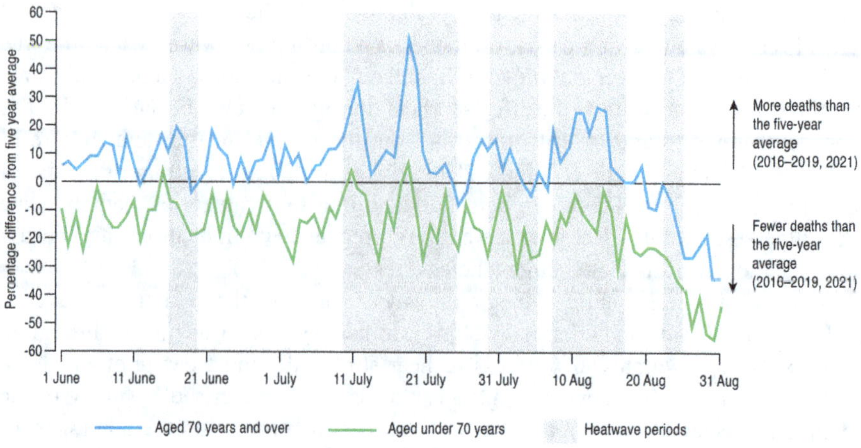

Fig. 11.1 Percentage increase in daily mortality compared to the 2016–2019 and 2021 (five year average) for the over 70 and under 70 year old age groups across England and Wales for the period June 1 to August 31, 2022. Numbers above (below) 0% indicate an increase (decrease) in daily mortality compared to the 5-year average. The shaded periods indicate when heatwaves occurred based on the UK Met Office definition of heatwaves. Noteworthy are the high percentage increases in daily mortality amongst the over 70-year-olds for the heatwave period in the middle of July. (Adapted from: Office for National Statistics (ONS) and UK Health Security Agency (UKHSA) (2022). Excess mortality during heat-periods: 1 June to 31 August 2022. Available at: https://www.ons.gov.uk/peoplepopulationandcommunity/birthsdeathsandmarriages/deaths/articles/excessmortalityduringheatperiods/englandandwales1juneto31august2022 (Accessed 28/03/2024)

COVID-19. To date various studies have revealed a range of outcomes of the compounding of extreme heat and COVID-19 (Hatakeyama et al., 2021; Hatakeyama & Seposo, 2022; Lane et al., 2023; Park et al., 2023; Seposo et al., 2021).

Chronic diseases such as cardiovascular, respiratory, and renal disease can intensify during heatwaves, with plenty of evidence that hospital admissions for such diseases increase accordingly (Anderson et al., 2013; Borg et al., 2017; Chan et al., 2014; Cheng et al., 2019; Desai et al., 2023; McTavish et al., 2018; Schwartz et al., 2004). The number of people presenting at hospital accident and emergency departments with fluid disorders, urinary tract infections, septicaemia and unintentional injuries also increases during heatwave events (Goggins & Chan, 2017; Phung et al., 2017). This was especially obvious during the 2021 Pacific Northwest heatwave (Nori-Sarma et al., 2022a; Wettstein et al., 2024). With regards to occupational injuries there appears to be some evidence that for heatwaves the relative risk is greatest in regions that possess an oceanic or humid subtropical climate (Fatima et al., 2021). Although some people affected by heatwaves can get to the hospital without using the emergency services, some can't. Accordingly increased pressure is exerted on ambulance services during heatwave events as manifest by an upturn in ambulance callouts (Campbell et al., 2021; Davis et al., 2020; Mason et al., 2022; Nakamura et al., 2022; Papadakis et al., 2018; Romani et al., 2020; Xu et al., 2022). Worryingly, evidence has emerged that heatwaves bear implications for the health of pregnant people, embryos and foetuses, with a number of adverse foetal and neonatal outcomes identified. These include preterm delivery, stillbirth, low birth weight, hypertensive disorders of pregnancy, gestational diabetes, and pregnancy-related emergency hospital admissions (Hess et al., 2022; McElroy et al., 2022; Nyadanu et al., 2024; Syed et al., 2022). The effect of heat on adverse pregnancy outcomes is especially evident for countries where extreme heat and heatwaves have become endemic with socio-economic status often an exacerbating factor (Andriano, 2023; Davenport et al., 2020; Le & Nguyen, 2021; Part et al., 2022; Rekha et al., 2024; Scorgie et al., 2023; Shankar et al., 2023).

A noticeable behavioural change during heatwaves is people seeking to cool off in terrestrial water bodies or the sea. An outcome of this behaviour change, as increasingly highlighted in the academic literature (Hills et al., 2024; Hu et al., 2023; Huang et al., 2023; Peden et al., 2024; Sindall et al., 2022) and widely reported in the press for the 2022 heatwave in England and the same in 2021 in the Pacific Northwest, is an increase in drownings (Anon, 2021; Gregory, 2022; Videtich, 2021). This is especially so amongst children and less able swimmers (Chauvin et al., 2020; Fralick et al., 2013). In some circumstances the consumption of alcohol during hot weather adds to the risk of drowning (Parks et al., 2023). Increases in aggression during extreme heat events is another example of behaviour change (Anderson, 2001).

As can be inferred from the above, mortality and morbidity impacts from heatwaves are often the focus of heatwave—population physical health studies. An area which remains undeveloped is research on heatwave—mental health relationships. While evidence is slowly emerging on the association between heatwaves and an elevation of suicide rates (Burke et al., 2018; Kim et al., 2019; Ngu et al., 2021), poor mental health outcomes in general (Cloud et al., 2023; Liu et al., 2021; Schwarz et al., 2022; Zhang et al., 2023), increases in mental health related visits to hospital emergence departments (Chang et al., 2023; Crank et al., 2023; Hansen et al., 2008; Nori-Sarma et al., 2022b; Niu et al., 2023; Parks et al., 2023) and possible physiological mechanisms linking mental health with extreme heat (Lohmus, 2018), significant knowledge gaps exist on the impact of heatwaves on many common mental health disorders (Liu et al., 2021; Thompson et al., 2018; Thompson et al., 2023). Framed within the broader issue of the impacts of climate change on mental health ((Clayton, 2021; Vafeiadou et al., 2023; Walinski et al., 2023), much remains to be discovered about the acute and chronic mental health effects of heatwaves.

11.3 Economy and Society

While it is generally understood that the economy of a region or nation is sensitive to weather (Lazo et al., 2011; Moon et al., 2018; Parnaudeau & Bertrand, 2018; Rose & Dolega, 2022), systematic analyses of the economic impacts of specific heatwaves or heatwaves in general are thin on the ground. That said, the limited number of estimates reveal the significant scale of these. Although almost forgotten, the prolonged heatwave of 1954 covering eleven states across the mid-western US with temperatures reaching a peak of 47.2 °C in mid-July, incurred substantial infrastructure repair costs from buckled streets and warped railroad ties, strained energy, and water resources. This heatwave also exacted heavy costs arising from governmental interventions to assist the agricultural community (Westcott, 2011). Based on the analysis of several heatwave events and economic modelling, the state of Victoria in Australia, where the large city of Melbourne is found, estimates that severe heatwaves, which can be expected once every 2 years, come with an economic cost of (AUS) $131 million with the agricultural and construction sectors most affected. For extreme and very extreme heatwaves (recurrence intervals of 25 and 110 years) the economic costs are valued at (AUS) $291 million and (AUS) $1000 million respectively (State of Victoria Environment LW, Planning, 2019). For the extreme heatwave of January/February 2009 across southeast Australia, which resulted in 500 fatalities in Adelaide and Melbourne, the financial losses mainly due to power outages, transport service disruptions and response costs have been estimated at (AUS) $800 million (QUT, 2010). For four European heatwaves (2003, 2010, 2015, and 2018) the economic damages due to reduced labour productivity have been estimated at 0.3–0.5% of European gross domestic product (GDP). However, the impacts were found to be heterogeneous with the more vulnerable warm summer regions of Europe experiencing damages of 1.5–2% of the national GDP (Fig. 11.2). Generally, the greatest impacts were felt in the services and construction sectors and for the European 2003 heatwave event overall (Garcia-Leon et al., 2021). Electricity spot markets in Europe also appear to be affected by the type of extreme temperatures associated with heatwaves (Pechan & Eisenack, 2014).

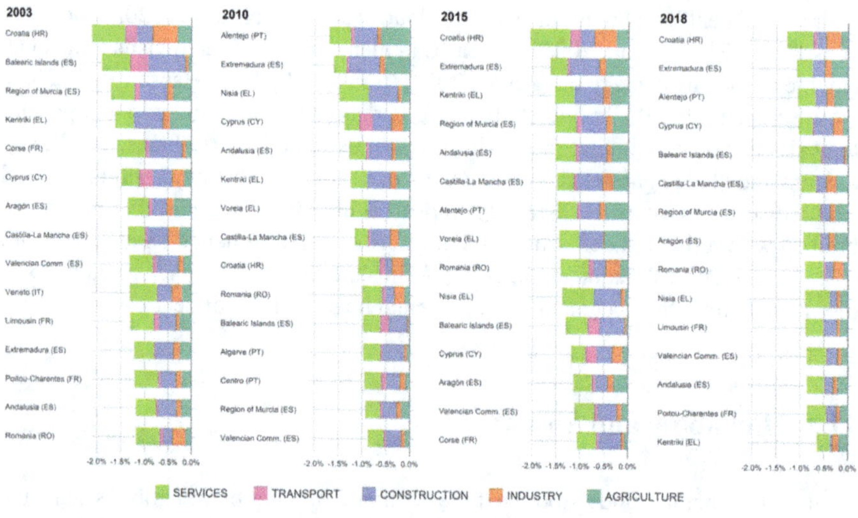

Fig. 11.2 Regional economic damages associated with reduced labour productivity for four European heatwave events across five sectors expressed as percentage downturn in gross domestic product (GDP). (Source: Garcia-Leon et al. (2021). Used under the Creative Commons Attribution 4.0 International License http://creativecommons.org/licenses/by/4.0/)

Outside Europe, heatwaves have also been found to have a large economic impact. For a 14-day heatwave that affected the Chinese city of Nanjing in 2013, the total economic cost due to productivity loss was 27.49 billion Yuan (US$4.5 billion). At the time, this was equivalent to 3.4% of the city's gross value of production with the manufacturing sector sustaining around 63% of the total economic loss (Xia et al., 2018). Using a different approach to estimating the economic cost of heatwaves, namely value of a statistical life, and knowing the total number of heatwave related deaths, Yan et al. (2022) estimated the death-related economic burden of the exceptional heatwaves across China in 2017 at 61.3 billion Yuan (US $9.5 billion). Such estimates of the economic losses attributed specifically to heatwaves corroborate the generally held view of the effects of heat on economy via downturns in labour productivity arising from heat-related occupational health concerns (Flouris et al., 2018; Morris et al., 2021). For example, based on a systematic review of the literature, Borg et al. (2021) estimate the global costs from heat-related lost work time for 1995 and 2010 to be US $280 billion and US $311 billion respectively, with the greatest losses as a proportion of gross domestic product (GDP) experienced in low- and middle-income countries and countries with warmer climates. Complementing these estimates are those of Callahan and Mankin (2022) who estimate cumulative losses attributed to heatwaves from 1992 to 2013 to range globally from US $5 trillion to US $29.3 trillion with the most pronounced negative impact on economic productivity in impoverished tropical areas. Estimated losses equate to 6.7% of GDP per capita annually for regions in the lowest income decile, while regions in the highest income decile experience a comparatively lower impact at 1.5%, a clear indication of the globally unequal effect of extreme heat on economic growth (Callahan & Mankin, 2022).

A general unknown for heatwaves is the economic cost to health services with much remaining to be learned for both developed and developing country situations (Wondmagegn et al., 2019). To date the limited number of studies on this topic point to females, the elderly, low-income families, and ethnic minorities possessing the highest healthcare costs for a range of health services (Wondmagegn et al., 2019). An examination of the 2006 California heatwave revealed a notable surge in heat related outcomes with 16,000 additional emergency room visits and 1100 excess hospitalizations across the state. The impact was particularly pronounced among young children and the elderly, particularly along the Central Coast. The emergency room visits demonstrated significant spikes in cases related to acute renal failure, cardiovascular diseases, diabetes, electrolyte imbalance, and nephritis (Knowlton et al., 2009). The healthcare costs attributed to this event are estimated in US dollars to range from approximately $120 million to $175 million statewide. The societal cost, considering over 650 premature deaths, has been estimated to surpass $5 billion (Srinavasan, 2008). For China the annual cost to the healthcare system attributable to extreme heat has been estimated at 274 million Chinese Yuan ($US 44.91 million) (Wang et al., 2021), while for France a significant heat-related economic burden on the health system has been identified (Adelaide et al., 2022). For the heatwave prone city of Adelaide in South Australia, the cost of 1161 heatwave-attributable all-cause emergency department visits over the period of 2014–2017 has been estimated at AU$1.02 million with heat-related illness accounting for most of this total ((Wondmagegn et al., 2021). Further for the same city, the projected expense for implementing the city's heat-health warning system (HHWS) for a duration of one week has been estimated at AU$593,000 with an associated cost-benefit ratio of 2.0–3.3 related to the potential expenses avoided due to HHWS-related decreases in hospital admissions and ambulance callouts (Williams et al., 2022). Borg et al. (2023) also note significant costs associated with heatwave related occupational illness and injuries in Australia.

Specifically for the European 2003 heatwave event, Munich Re, a provider of reinsurance, primary insurance and insurance-related risk solutions, estimated the economic impact of this event to be (US)$10 billion (Trigo et al., 2009). For France, which experienced an exceptionally high death toll during the 2003 heatwave, the economic impacts took a variety of forms including increases in compensation claims for disrupted journeys by rail estimated at US$1.25 to 3.75 million, a downturn in clothing sales by 8.9%, an upturn in the sales of bottled water and ice cream by 18 and 14% respectively with tourism in northern France benefitting while that in the south of France was quite the opposite (Létard et al., 2004). Agriculture in France also suffered with the maize grain and fresh fruit harvests down by 30 and 25% respectively compared to the previous year (Ciais et al., 2005). For Switzerland, where water withdrawals for agriculture were banned, the agricultural deficit exceeded approximately (US) $230 million. Across Europe, cereal production fell by more than 23 million tonnes compared to the previous year with this needing to be topped up from imports and carry-over stocks. In Italy, durum wheat yields were severely impacted during the 2003 heatwave ((Fontana et al., 2015). The impact of the heatwave on green fodder production and thus animal feed was also severe with countries such as Germany, Austria, Spain, and Italy experiencing fodder deficits of 30–40% with these reaching 60% in France, necessitating the import of fodder from eastern European countries, most notably Ukraine. The economic impacts of heatwaves also extend to international aid agencies. For example, the International Federation of the Red Cross' Disaster Response Emergency Fund received its first appeal for a heatwave in 2018, from North Korea. The next came in 2021 from Vietnam, then from Kyrgyzstan and Tajikistan in 2022. This was followed by three appeals in 2023 (Greece, Bangladesh and Kyrgyzstan). In total the amount provided from the fund to heatwave appeals over the period 2018–2023 was equivalent to ($US) 2.34 m.

Beyond the academic literature, there is evidence in online blogs from companies specialising in retail intelligence gathering of the immediate impact of heatwaves on the economy. For example, Rigby (2022) chronicles the effects of the July 2022 UK heatwave on retail and shopping activity noting that shoppers heeded the heatwave warnings issued by the UK Met Office and rushed to buy items in advance of the arrival of the heatwave with shoppers subsequently delaying spending until the peak temperatures of 40 °C on July 18 and 19 had waned. Ahead of the heatwave strong sales of fans, air conditioning units, paddling pools and general hot weather items were reported by retailers. With the arrival of the heatwave spending fell by 47.8%, week-on-week, with total web revenues depressed by 30.7% compared to the same period a week earlier. Remarkably, once the heatwave had passed online revenues bounced back achieving an upturn of 51.2% on the previous week (Rigby, 2022). In addition to shop and online sales, footfall in shops was also impacted. According to a shopper traffic analyst company, footfall fell by 11% across the UK over the 2 days of the heatwave peak with shops across London and Nottingham reporting a decrease in shoppers by 9.2 and 24% respectively (Rigby, 2022). This set of events clearly demonstrates the acute effects of an intense heatwave on retail activity and raises the question whether in the absence of daily retail and shopper activity data would these effects have been detected if longer timescale data (e.g., monthly) was only available for analysis. Although not specific to heatwaves events, it would appear hot weather affects trading activity on the French stock market with a fall in trading volume between 4 and 10% when maximum daily temperatures exceed 30 °C (Peillex et al., 2021). Whether such an effect exists for other stock markets remains to be explored.

Besides the known impacts on human physical and mental health as described in Sect. 11.2, societal impacts of heatwaves manifest in several ways. These include impacts on sleep patterns and associated cognitive performance (Buguet et al., 2023), the decline in learning amongst school children (Park et al., 2020), increases in school absenteeism (Pule et al., 2021), increases in climate anxiety (Bratu et al., 2022), an upturn in intentional violence and self-harm and the possibility of social unrest (Auliciems & Dibartolo, 1995; Kubo et al., 2021; Otrachshenko et al., 2021; Sanz-Barbero et al., 2018; Yeeles, 2015) and crime (Hou et al., 2023) and changes in the reporting behaviour of the press with regards to human induced climate change (Pianta & Sisco, 2020). Changes in social media activity amongst the public are also evident during heatwaves (Dzyuban et al., 2022). For example, a social sensing study of heatwaves in the UK, US and Australia unveiled a more pronounced response to heatwaves from users in the UK compared to those in Australia, even though the UK experiences comparatively milder conditions. Analysis of language patterns also indicated that the UK user population tends to express concern about individual well-being and discomfort, while Australian and US users typically direct their attention toward the environmental consequences (Young et al., 2021). Some research has also suggested that high temperatures associated with heatwaves influence judicial decision-making processes (Heyes & Saberian, 2019), while experimental approaches to understanding the impacts of heatwaves on workers highlight how human capacity can be undermined via an increase in the number of mistakes made and time spent on unplanned breaks (Ioannou et al., 2021).

Although most of the estimates of the economic and societal impacts for heatwaves are for atmospheric heatwaves, the economic and social costs of marine heatwaves are beginning to receive attention. For example, the reported economic costs of individual MHW arising from direct, and indirect losses due to harm to ecosystem services, has been estimated to exceed US$800 million and US$3.1 billion respectively (Smith et al., 2021). Further for the US, marine heatwaves have been identified as one of the leading causes of fishery disasters (Bellquist et al., 2021). There is also emerging evidence that marine heatwaves redistribute pelagic fishing fleets (Farchadi et al., 2024). At the fishing community scale or for specific fisheries, the socioeconomic consequences and cascading effects of marine heatwaves and associated extreme ecological events like harmful algal blooms, can be significant and non-uniform (Fisher et al., 2021; Holbrook et al., 2022; Jardine et al., 2020; Kourantidou et al., 2022; Ritzman et al., 2018).

11.4 Infrastructure

Critical to the economy and society in general is a functioning infrastructure. As established in the previous Chapter, infrastructure is heatwave vulnerable for a range of reasons. This is especially so with regards to transport and energy infrastructure.

During heatwaves the use of electricity surges due to increased cooling demand with 'heat disasters' likely because of electric grid failures (Sailor et al., 2019). How the impacts of heatwaves on energy infrastructure play out in the form of brown outs (downturn in electricity output) or black outs (complete failure of electricity supply) is clear for several locations (Aivalioti, 2015). The European 2003 heatwave exacted a clear impact on the energy sector in France. Here around 80% of the country's electricity supply is from nineteen nuclear power plants. As noted earlier, such plants are heavily reliant on water for reactor cooling with the restriction if water intake or discharge temperatures are too high, reactor cooling will be affected with environmental impacts likely to occur from the discharge of warm water into natural water bodies. During the August 2003 heatwave, river water temperatures across France reached record levels. This resulted in the need to decrease the magnitude of reactor cooling with a reduction in electricity output a consequence. In fact, seventeen nuclear reactors had to shut down or lower their production due to water temperature intake and discharge restrictions. To manage the downturn in electricity production and protect against local blackouts during peak hours, Electricité de France, France's major energy producer, reduced its export of electricity to Switzerland, Britain, Italy, Belgium, and Spain by half (Guardian, 2003). Interestingly this situation reversed during an intense heatwave in July 2006 when France imported electricity to avoid shortages and possible brown outs or black outs. At the same time, Spain was forced to shut down the Santa Maria de Garona nuclear power station on the River Ebro because intake waters exceeded the threshold for effective reactor cooling while in Germany similar reasons necessitated output reduction from several nuclear reactors. Special permission was also granted to several nuclear power plants in France and Germany to discharge reactor cooling water to the environment that exceeded the regulatory temperature thresholds to prevent interruptions to power supply (Jowit & Espinoza, 2006). Waivers for discharging hot water to rivers from the Bugey, Saint Alban, Tricastin, Blayais and Golfech power plants in France were also implemented during the 2022 heatwaves (Reuters, 2022).

Heatwaves across California in the United States have had profound impacts on electricity supply with blackouts lasting from several hours to days across parts of the state (Burillo et al., 2019; Vine, 2012), with the city of Los Angeles especially affected (Antczak, 2011). For example, the July 2006 heatwave across California resulted in transformers overheating. This tripped circuit breakers with fuses breaking and insulation burning, causing short circuits within transformers. Consequently, 80,000 people in Los Angeles experienced a blackout for several days due to around 860 transformers either malfunctioning or completely failing. At the same time in northern California 1.2 million Pacific Gas and Electric Company customers experienced power shortages (Bernstein, 2006; Jurgens, 2006). Heatwaves in 2010 and 2014 had similar impacts on the electricity distribution network. Again, in September 2022, California's electric grid was almost brought to its knees, with rolling blackouts a possibility, due to an exceptional record breaking heatwave (Toohey et al., 2022). On the opposite side of the country heatwave events have clearly demonstrated that New York City's power grid is not immune to the effects of extreme temperatures. For a compound Derecho and heatwave event across the eastern US in June 2012, four million people lost power with an increase in heat-related fatalities a result (Stone et al., 2021; Stone et al., 2023). Similarly, a two-day heatwave in August 2003 across New York City led to blackouts and a 28% increase in all-cause mortality partly attributed to the lack of access to air conditioning for many of the heat-vulnerable population (Anderson & Bell, 2011). Brown outs were a feature of New York heatwaves in 2002, 2006, 2010, 2011 and 2013 (Aivalioti, 2015), with blackouts occurring in the heatwaves of July 2022, notably in many areas of the city where heat-vulnerability is considered high (Rajamani, 2022). In the fashion of a compound event, power outages arising from the July 2022 heatwave event were exacerbated by storms that rolled into New York City during the heatwave with the result that trees toppled by strong winds fell across powerlines further disrupting power supply, adding to the misery of the extreme heat (Grymes & Westbrook, 2022). The June 2021 heatwave event across the Pacific Northwest of the United States stretching into Western Canada also serves as a good example of how record-breaking temperatures and the demand for electricity result in blackouts (Geranios & Selsky, 2021). This is aptly demonstrated by power outage data reported by the Puget Sound Energy (PSE) company for the period June 26—July 6, 2021 (Table 11.1) which recorded 951 outages with an average duration of 3.22 hours across eight counties including the one in which Seattle is located. That 2021 was exceptional with 951 outages, is clearly obvious when the period of the heatwave is compared with the same period in the previous years of 2018, 2019 and 2020 with outages for all categories totalling 421, 425 and 406 respectively (PSE, 2021).

As discovered in previous sections of this book, Australia is no stranger to heatwaves, these being a distinct feature of its climate. Amongst several significant heatwaves over the period of 2014–2017, those in January/February 2009 and December 2019 in Melbourne standout as exceptional, with both exacting notable impacts on the power grid. Temperatures 12-15 °C above normal for 5 days during the 2009 event exerted immense pressure on power supply systems as Melbourne struggled to stay cool in the intense heat. Those who had access to it turned to air conditioning systems for relief driving electricity demand 7% higher than previous loading records. In and around the peak of the heatwave on January 29–30 several

Table 11.1 Puget Sound Energy outage data for the period June 26—July 6, 2021. (Source: PSE (2021))

County	Total number of outages	Total customer minutes interrupted	Total customers interrupted	Average outage duration per customer interruption (hr:min)
Whatcom	83	1,591,572	8902	2:58
Skagit	67	418,357	4871	1:25
Island	33	219,668	777	4:42
King	518	14,258,556	62,489	3:48
Kittitas	20	850,051	6879	2:03
Pierce	66	1,709,686	7220	3:56
Thurston	92	594,360	2732	3:37
Kitsap	72	3,494,629	20,446	2:50
Total	**951**	**23,137,879**	**114,316**	**3:22**

incidences conspired to produce rolling blackouts. These included the shutdown of the Basslink interconnector cable that carries electricity from Tasmania to the mainland as land based low to high voltage converters reached their maximum operating temperatures in the intense heat, the difficulty of power plants to meet the extra electricity demand, the explosion of one instrumentation transformer and faults in up to 50 local voltage transformers all of which necessitated load-shedding. Across central and western Melbourne, this situation precipitated blackouts with a duration of a few of hours to two days affecting 500,000 residents (QUT, 2010). The late January 2019 heatwave occurred at a time when the energy supply system was already under strain because three electricity generation units at coal-fired power plants were not functioning due to unexpected outages and planned maintenance ((Harrison, 2019). To meet the high cooling demand for air conditioning and manage the electricity supply deficit across Victoria, the Australian Energy Market Operator (AEMO) drew on emergency energy reserves, imported electricity from neighbouring states, paid major energy consumers such as aluminium smelters to power down and encouraged people not to use domestic appliances unnecessarily. However, these actions were insufficient given the level of electricity demand with the AEMO ultimately turning to load shedding to avoid blackouts for all users. Consequently, around 200,000 domestic customers across Victoria were without power during the heatwave peak (Harrison, 2019).

The fragility of electricity supply systems during heatwaves is clear for many locations beyond those mentioned above. For example, between 2006 and 2013, twenty heatwave related blackouts occurred in Buenos Aires (Santagata et al., 2017). Although historically heatwaves have not been a major issue for the power grid in Russia, these have emerged as a threat as Russia experiences an increasing number of heatwaves and associated demands for cooling during exceptionally warm summers (Klimenko et al., 2020). During the intense heatwaves across Japan in late June 2022, the power grid system was pushed to its limits as power demand for air conditioning spiralled, while in the southern China province of Guangdong, the power grid was pushed beyond its limits during a heatwave in late July 2002 with the provincial capital of Guangzhou experiencing extensive blackouts (Davidson, 2022). The pre-monsoon heatwave that affected India in May 2022 caused acute power shortages and extensive blackouts as electricity consumption surged to record levels. Interestingly, although the peak demands were within the total capacity of the power system, coal delivery problems via rail to many of India's coal-fired power stations appear to be the reason why power supplies were compromised. To ensure enough supplies, many passenger services on the Indian rail network were cancelled to prioritise the lines for coal related rail freight a totally unforeseen consequence of the heatwave (Pathi & Ghosa, 2022). In nearby Pakistan during the same heatwave, similar power supply problems meant load shedding was applied extensively with cities such as Turbat barely receiving any electricity to power air conditioners and fridges with daily blackouts of nine hours (Ellis-Petersen & Baloch, 2022).

The transport sector including rail, road and air are all heat-vulnerable with this vulnerability manifesting in a variety of ways. In the case of rail, the impacts of heatwaves are demonstrated by ample examples of rail buckling or 'sun kinks' along with problems with overhead lines and signalling equipment. For example, during the August 2003 heatwave in the UK, the number of rail buckles reached 137 amounting to 3–4 times the annual average (Dobney et al., 2009). Such failures can lead to delays to rail journeys or in the worst case, complete cancellation which through the disruptive effects on commuters, travellers and freighters, can produce heavy costs for rail network operating companies. For the UK and the period of 2006 to 2016, it has been estimated that the company that operates the rail network paid £20 million in compensation payments to train operating companies because of heatwave related delays arising from a range of heat-incidences (NR, 2017). In this regard, the July 2015 heatwave event across England, Scotland and Wales exemplifies how breakdown and impairment of assets, along with speed restrictions, can result in a significant increase in 'delay minutes' beyond what might be expected. During this heatwave delays doubled, totalling 220,000 minutes with damage to the national economy estimated at £16 million (Ferranti et al., 2018). For England, the rail network in the southeast is particularly affected by heat. For this region, which climatologically experiences the greatest frequency of heatwaves across the UK, it has been estimated that signalling failures account for around 53% of heat-related incident costs and 51% of delays with 'failure harvesting'—the occurrence of failures early in the warm season—often causing an increase in heat-related incidences by 80% (Ferranti et al., 2016). As well as overground rail networks being affected, underground systems such as the iconic London Underground can be severely impacted by extreme heat as described for the heatwaves of July 2013 and 2015 and September 2016 (Greenham et al., 2020). The record heatwave of July 2022 across England, with ambient temperatures more than 40 °C and rail track temperatures reaching 62 °C in some parts of the country plus heat damage to overhead wires and trackside fires, also wreaked havoc on the rail network (Topham, 2022), as did the heatwave in July 2019, which until the July 2022 heatwave, recorded the highest temperature for England.

Beyond the UK heat also presents a significant issue for rail networks. In the United States the occurrence of heat related delays has increased as the number of heatwave events has risen (Bruzek et al., 2013). In India safety teams are dispatched during heatwaves to monitor the state of rail tracks and apply heat mitigation techniques when necessary, including covering tracks with leaves (Kulkarni, 2018). As for England, the extreme 2022 heatwave caused problems on the French and Spanish rail networks with not only delays due to speed restrictions as rail tracks responded to the extreme heat, but challenges for passenger thermal comfort as train air conditioning systems broke down (Geerts, 2022; Slavicek, 2022). In some cases, the uncontrolled effects of extreme thermal discomfort on trains in the absence of air conditioning can lead to passengers suffering heat illness and even death as recorded during the severe 2019 heatwave across India when temperatures approached 48 °C ((Baynes, 2019).

The January–February 2009 heatwave event across southeast and south Australia serves well to demonstrate the heat-related impacts on rail networks in this country. With temperatures well over 40 °C this compound event, with two extreme heat peaks separated by a few days with temperatures of 30–35 °C, has been one of Australia's harshest heatwave events. During the heatwave, buckling of tracks occurred across Melbourne's tram and rail networks (Fig. 11.3). Buckling was attributed to new tracks with continuous welded joints which have no expansion gap. This contrasts with former tracks connected by 'fish plates' which permitted expansion.

Fig. 11.3 Buckled tram lines, Melbourne, January 2009. (Photo credit: Gary Tillotson, Yarra Trams, 2009)

For Melbourne trams, most buckling incidences were confined to the first day of the exceptional temperatures (January 28) and places with elevated (or ballast) tram tracks. Consequently, only a few tram services were cancelled with delays the major impact because of 10 km/hr. speed restrictions on sections of tracks with buckling (QUT, 2010). In contrast, the scale of the impact on the rail network was significant as most of the rail stock runs on elevated train tracks which are extremely heat susceptible; 29 instances of buckling were reported for the Melbourne rail network (McEvoy et al., 2012). As heat built over the first three days of this compound event the number of train cancellations grew. By the third day of the heatwave (January 30) train services had dropped to 64% with more than 750 services out of 2400 cancelled (Fig. 11.4). As well as buckling, service cancellations were complicated by a power failure. On the evening of January 30 electricity to the overhead powerlines that power the trains was cut because of an explosion at a transmission station. Consequently, thousands of commuters were stranded adding to the chaos of train disruptions. Over the following 6–7 days there was some respite from the heat with temperatures in the low 30 s with train cancellations dropping significantly. However, with a return of extreme temperatures to just over 46 °C on February 7, train cancellations once again rose dramatically (Fig. 11.4). Added to the misery of train disruptions many incidences of air conditioning failure on older style trains were recorded, resulting in unbearable levels of thermal discomfort in carriages and possible train cancellations (McEvoy et al., 2012).

Heatwaves can increase thermal loading on road surfaces causing expansion (buckling), bleeding and rutting. During an exceptional heatwave in July 2006 when temperatures approached 35–38 °C in England the rural highways of Leicestershire were severely heat-damaged. A significant factor contributing to the damage of the roads was their history. Being former cart tracks, many dating back to Roman times, the modern roads have been built up by the application of consecutive layers of thin

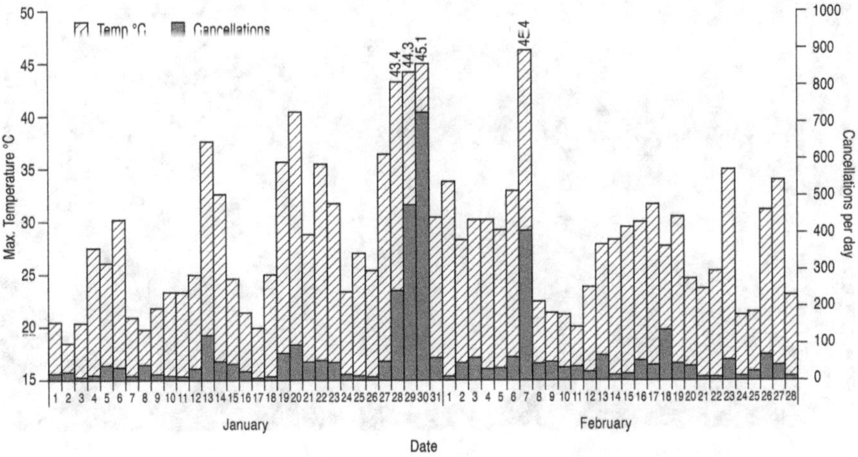

Fig. 11.4 Train cancellations and daily maximum temperature for Melbourne during the January—February 2009 heatwave event. (Adapted from QUT, 2010)

bitumen surfacing. Lacking full depth construction, the roads are more vulnerable to heat compared to their counterparts that form the principal roads across the county (Willway et al., 2008). An upshot of the severe temperatures was damage to around 80 km of the county's rural roads with heat-softened surfaces leading to rapid formation of pothole clusters, wheel track rutting and skid resistance reduction due to loss of texture depth. At the time, the cost to repair the heatwave related damage was estimated at £2 million (Willway et al., 2008). Roads in England also suffered during the July 2022 heatwave with buckling mostly occurring in the central and southeastern regions of England where record temperatures were recorded. Spectacular examples of road pavement buckling are evident for the June 2021 Pacific Northwest heatwave across the US states of Oregon and Washington and the Canadian province of British Columbia. The state of Washington reported that State Route 544 failed when temperatures hit 40 °C, with failure attributed to the nature of highway construction, namely asphalt laid over concrete. Under the extreme heat the slabs of concrete forming the base of the highway expanded compressing the inter-slab buffers with buckling ensuing (Fig. 11.5). Added to this, the bitumen mix in the top layer of asphalt simultaneously softened, causing rutting and cracking with eventual road closure so repairs could be undertaken (Marshall, 2021).

Fig. 11.5 Buckled road pavement during the June 2021 Pacific Northwest 'heat dome' event. (Source: Washington State Department of Transportation North Official Twitter account. Photo posted at @wsdot_north 4:55 PM June 28, 2021)

As well as roads, bridges were also affected by the Pacific Northwest heatwave event requiring steel drawbridges to be doused with water to temper the heat extremes and avoid structural damage (Marshall, 2021). For the US in general, Palu and Mahmoud (2019) have reported on the general impacts of heat on the country's aging bridge infrastructure. The 133 year old cast iron Hammersmith Bridge in west London England demonstrates how aged infrastructure can be severely impacted by heatwaves. During a routine engineering inspection in 2019 micro-fractures were revealed in the cast iron casing surrounding the bridge pedestals. In the following year these expanded significantly during a heatwave in August when daily maximum temperatures in London reached the low- to mid-30 s for six consecutive days. Vain attempts by contractors to douse the cracks with cold water to retard expansion resulted, with the bridge soon deemed unsafe with closure following This situation was exacerbated by the July 2022 heatwave when temperatures in London hovered around 39-40 °C for 2 days. To mitigate the effects of heat on the bridge and associated structural damage a cooling system was installed along with wrapping its four pedestal chains in energy reflecting foil to keep this critical part of the bridge at temperatures below 18 °C (Fig. 11.6), the temperature threshold beyond which the bridge will become unsafe due to expansion of its structural elements.

The aviation sector can be impacted in several ways by heatwaves given planes require stable surfaces in the form of well-maintained runways for take-off and landing (Burbidge, 2018). Further, air temperature, through air density effects, affects aircraft lifting power and thus aircraft take-off weight (Coffel et al., 2017; McRae et al., 2021). In July 2018 with local temperatures over 36 °C in Hanover, Germany, the north runway at Hanover Airport had to be closed due to heat-related heaving of parts of the runway. This resulted in the cancellation of 41 flights and the

Fig. 11.6 Hammersmith Bridge west London wrapped in silver insulation foil as a protective heat measure during the London July 2022 heatwave. (Source: @masato_jones via Twitter)

severe delay of 44 arrivals (De Vivo et al., 2022). Similarly, runways were affected by the 2022 heatwave across England necessitating Brize Norton a military airport, and Luton Airport a civilian airport, to delay flights because of heat-related defects on the runways. De Vivo et al. (2022) also note technical problems with air traffic control radars at Gatwick and Heathrow airports during the July 2019 across England. For the US, Borsky and Unterberger (2019) report on a tendency for higher average aircraft departure delays when temperatures exceed 45 °C, although they note that a limited sample size for high temperature events prevents drawing statistically robust conclusions about this tendency. That said, as an example of the impact of extreme heat on aviation, they cite what occurred in Phoenix on June 20, 2017, when temperatures climbed to 49 °C resulting in the cancellation of 40 regional flights because air temperatures exceeded the safe operating temperatures for the type of aircraft usually used on regional routes. Carpenter (2019) also highlights this event as a good example of how heatwaves impact aviation operations, especially aircraft passenger and freight carrying capacity.

Although there is a burgeoning literature on the potential impact of extreme heat and hence heatwaves on buildings, as touched upon in the previous chapter, there appears to be few examples of observed impacts on the structural elements of buildings. Although mainly a drought impact, the subsidence of buildings and thus cracking in foundations and walls can be exacerbated during heatwaves. For example, following the July 2022 heatwave event across England a number of reports appeared in the press about a surge in the volume of insurance claims related to building subsidence (Smith, 2022). If the heatwave did result in building subsidence, it is more likely than not that antecedent soil moisture conditions, related to a prolonged period of dryness, were the precursor for any heatwave-related subsidence, with the heatwave essentially acting as a tipping point. In contrast to the evidence for structural damage to buildings there are ample examples of heatwave-related overheating in buildings. For the same 2022 heatwave as above, residents living in high-rise apartments in some parts of England reported they couldn't cope and were struggling in the heat (Gerrard, 2022). For Italy, Pyrgou et al. (2017) note how during extreme heat episodes, well insulated buildings require far more cooling—up to three times—than traditional non-insulated buildings. School buildings are equally affected by overheating during heatwaves with the result that school closures may be necessary as occurred in the city of Detroit during a heatwave in May 2022 (Bakull, 2022). In Delhi India, during a heatwave in May 2016, the government took the decision to close schools due to fears of overheating in school buildings. This caused havoc with the planned schedule of end of year exams (Doshi, 2016). For similar reasons, primary schools closed in Islamabad during the extreme May 2022 heatwave across Pakistan (Anon, 2022).

As for buildings the number of theoretical studies of the impacts of heatwaves on other types of infrastructure such as telecommunication and water systems outweigh the observations of direct impacts. Interestingly for the Melbourne 2009 heatwave, McEvoy et al. (2012) found the impacts on telecommunication and water systems were far less than those on transport and electric power infrastructure. In contrast, the Christmas 2021 heatwave that hit Perth Australia had a discernible impact on telecommunication systems as the extreme heat pushed data centre cooling systems to their limit with the consequence that many people were unable to access IT services (Crozier, 2021). Heatwaves also push up water consumption rates placing extra burden on already strained water supply systems, especially during long dry spells. This leads to water restrictions as seen during the 2022 heatwave across southern England with hose pipe bans the result. For Adelaide in Australia, almost 20% of this cities peak water demand occurs during heatwaves with notable anomalies in daily water supply observable when temperatures exceed 30 °C (Hatvani-Kovacs et al., 2016). That heatwave driven water scarcity can cause political tensions is evident from northern India where during the 2016 heatwave various political jurisdictions battled over the control of rivers, with armed guards deployed at dams to protect dwindling water supplies (Niazi, 2016).

11.5 Domesticated Animals and Terrestrial Wildlife

Notwithstanding animal populations are integral components of natural and human-constructed ecosystems (e.g., farming ecosystems), this section focuses on the impacts of heatwaves on specific animal populations. The impacts of heatwaves on ecosystems are dealt with in later sections.

Heatwaves have rendered serious impacts on domesticated animal populations. Evidence comes from not only verified reports in the press and other media but also from systematic studies of the association of measures of heat stress and animal death during heatwaves. As reported widely in the press, the extended heatwave experienced across Kansas and nearby states in June 2022 resulted in cattle deaths more than 2000 (Asma-Sadeque, 2022). Such high numbers are typical amongst feedlot cattle which are held in large numbers in close quarters, conditions that can create high levels of heat stress during heatwave periods. For example, the July 1995 heatwave which had devastating consequences for Chicago resulted in around 4000 cattle deaths across the mid-central US with associated production losses estimated at $28 million (Hahn, 1999). Lees et al. (2019) in an assessment of the impact of heat load on cattle cite, amongst others, examples of the devastating impact of heatwaves on feedlot cattle including 4000 deaths in Queensland in 1991 during a period of high temperatures and humidity with little ventilation and the death of 4000–6000 dairy cows during a heatwave across the Californian counties of Fresno, Kings and Tulare in 2017. Using systematically collected weather and mortality data, Morignat et al. (2014) estimate a relative excess mortality of 24% amongst cattle for the 2003 European heatwave across France. Using similar data for Southern Ontario, Bishop-Williams et al. (2015) observed that cattle deaths in this part of Canada increase by 1.27 times during heatwave periods. Others also note the clear impacts of heatwaves on cattle either in terms of mortality or downturns in production such as milk yield from cows (Manica et al., 2022; Vitali et al., 2015). Heatwaves also impact other domesticated farm animals such as sheep and horses (Al-Dawood, 2017; Brownlow et al., 2016; Takahashi & Takahashi, 2020). As the world turns to sources of cheap protein there has been a rapid growth in poultry production. However, the intense way in which poultry is farmed makes this type of animal production system highly vulnerable to the detrimental effects of heatwaves (Biswal et al., 2022; Vandana et al., 2021) the result of which can be livestock mortality and downturns in production (Vale et al., 2010). For example, around 500,000 hens perished in a heatwave that gripped the Bastos region in the state of Sao Paulo, Brazil in 2012 (Lamarca et al., 2018). Although not corroborated by the poultry industry, accounts of large numbers of deaths amongst chickens arising from heatwave events occasionally appear in the press with 'whistle blowers' and animal rights groups often the source of these statements. 'Man's best friend', the dog, is also impacted by periods of extreme heat with canine deaths often reported during heatwaves as well as increases in heat illness for a range of pets (Hall et al., 2020; Hall et al., 2022). Stray animals, many of them dogs, also suffer during heatwaves as reported widely for the extreme heat conditions experienced across India in 2022 (Majumdar, 2022).

There is burgeoning evidence of the impacts of heatwaves on wildlife (non-domesticated animals). For example, heatwaves impact iconic species such as the Koala in Australia with these animals often unable to recover from periods of extreme temperature exposure with death an outcome (Narayan & Williams, 2016). Severe heatwaves have also resulted in high bat/flying fox death counts (Welbergen et al., 2008) et al. 2008). The impacts can be especially severe amongst orphaned pups, because of abandonment by adults who seek shelter from the heat away from the roost (Mo et al., 2022). Interestingly, the large number of heatwave related bat deaths in the Australian heatwaves of 2019–2020 revealed a potential public health risk associated with the clean-up of dead bats in urban areas (Merone et al., 2020). This represents a good example of heatwave-related human-wildlife conflict as also recorded for the marine environment during marine heatwaves (Samhouri et al., 2021) and possibly evidenced by press reports of polar bears venturing into populated areas because of local sea ice loss and diminution of hunting grounds during heatwave events.

Almost unnoticed, the impacts of heatwaves on insects can be significant. At the individual level impacts manifest in the form of reduced survival via negative impacts on physiological homeosis, development, reproduction, and fitness while at the population level there are downturns in abundance, changes in phenology and the number of generations in a season with these outcomes having knock on effects on community structure, diversity, and species interaction strength (Harvey et al., 2020; Ma et al., 2021). Although the field evidence for the impact of heatwaves on insect populations is limited (Gols et al., 2021), laboratory-based experimental programmes are revealing interesting details about the nature of insect-related heatwave impacts such as compromised learning behaviour in bumble bees (Gerard et al., 2022), altered host-parasitoid interactions (Moore et al., 2022) and the importance of small-scale habitat heterogeneity for buffering the effects of heatwave related temperature extremes (Terlau et al., 2023).

Heatwaves can have devasting impacts on bird populations (Swanson et al., 2022). Examples of mass mortality events include that related to a short duration heatwave across Santa Barbara Island, California in June 1979 when catastrophic chick mortality occurred within the population of Western Gulls (Salzman, 1982), the death of 264 adult and 90 juvenile Magellanic Penguins during a heatwave in January 2019 at Punta Tombo, Argentina (Holt & Boersma, 2022), a similar event in December 2016 at Punta Leon, Patagonia when 86% of Imperial Cormorant Leucocarbo atriceps chicks perished in historically high temperatures (Quintana et al., 2022), the mortality of nesting Bushtits during the unprecedented June 2021 heatwave across the US Northwest Pacific coast (Sloane et al., 2022) and widespread bird mortality during a heatwave event in eastern South Africa which also affected bats (McKechnie et al., 2021). Analyses of the impact of the 2003 European heatwave on avian populations have also revealed that amongst 71 avian species the most significant decreases in population growth rate occurred in those with small thermal tolerance ranges in regions with the greatest temperature anomalies (Jiguet et al., 2006). Although unsubstantiated by systematic research, the 2022 heatwave events across Europe and India appear to have had devastating impacts on bird populations with it widely reported in the press that 'birds were falling from the sky' due to the extreme temperatures. In addition to these observed impacts in nature, some rather novel laboratory-based experimental evidence indicates that cognitive and motor performance of land-based birds, and even song production and salience, as well as body mass is affected by temperature conditions akin to heatwaves (Blackburn et al., 2022; Coomes & Derryberry, 2021; Danner et al., 2021; Sharpe et al., 2019). Notably, the impact of heatwaves on cognition of animals in general remains an open question (Soravia et al., 2021).

11.6 Marine Environments

Marine heatwaves (MHW) have emerged as a notable aspect of the climate of oceans. Because they are typified by temperature departures often beyond the thermal limits of many marine organisms with ocean biophysical processes also impacted, they can have disastrous consequences for specific marine fauna and flora and marine ecosystems in general (Frolicher & Laufkotter, 2018; Holbrook et al., 2022; Smale et al., 2019; Smith et al., 2023; Straub et al., 2019).

Seabirds, apart than nesting on land, spend most of their existence living on the oceans. Impacts of MHW on these ocean dwellers is mainly through a downturn in food supply because of changes in near surface water temperatures and an associated decrease in the upwelling of cool nutrient rich waters which form the basis of marine food chains, with starvation amongst some bird populations the result (Jones et al., 2018; Piatt et al., 2020). Associated with poor food supplies, seabirds increase the distances over which they forage during MHW (Osborne et al., 2020). The impacts of MHW on specific animal populations other than seabirds is evident from a growing number of reports that consider population dynamics during and following these extreme events. Examples include the survival and reproductive success of the humpback whale (Gabriele et al., 2022), and similarly some dolphin species

(Mann et al., 2021; Wild et al., 2019), the development, diet, and foraging activity of the Guadalupe fur seal (Amador-Capitanachi et al., 2020; Galvez et al., 2020) and the settlement rates of spotted sand bass larvae in an estuarine nursery habitat in southern California (Basilio et al., 2017). Beyond the response of specific species, broad readjustments of biota have been recorded for MHW. For example, anomalous water temperatures in the north-western Mediterranean associated with the notable 2003 European heatwave resulted in mass mortality amongst at least 25 rocky benthic macro-invertebrate species (Garrabou et al., 2009; Garrabou et al., 2022). The impacts on macro-invertebrates have also been reported for the exceptional 2010–2011 MHW event off the coast of Western Australia (Smale et al., 2017). For the intense Tasman Sea MHW of 2015 to 2016, when temperatures approached or exceeded the upper thermal limits of numerous native species, Brown et al. (2024) noted a significant transformation in the composition of pelagic microbial communities. This shift not only disrupted the usual seasonal variations in diversity but also reshaped environmental conditions to resemble those found over 1000 kilometres towards the equator.

The 2014–2016 northeast Pacific MHW, known colloquially as the 'Blob', has attracted much attention in relation to its short- and longer-term impact on marine life (Robinson et al., 2022; Thompson et al., 2022). For northern California, widespread shifts in coastal biota occurred during the 'Blob' including substantial changes in the spatial distribution and/or abundance of 67 southern species, the poleward extension of 37 species and increases in the recruitment of some limpet and barnacle species (Sanford et al., 2019). Drawing on data from the Gulf Watch Alaska Long-Term Monitoring programme, established following the Exxon Valdez oil spill disaster in 1989, Suryan et al. (2021) note the abrupt changes in the Gulf of Alaska across all trophic levels because of the 'Blob', with many of these changes persisting up to five years. The 'Blob' also affected the nutritional value of some key forage fish in the Prince William Sound in the Gulf of Alaska resulting in the disruption of energy flow through some pelagic food webs. This downturn in energy flow most likely contributed to population declines and/or breeding failures amongst several predators in this oceanic region (von Biela et al., 2019). That the impacts of the 'Blob' extended to inter-tidal environments is also apparent from data collected as part of the Gulf Watch Alaska programme with a decline of macroalgal foundation species observed because of this event. Specifically, monitoring sites revealed a region-wide shift from an autotroph-macroalgal dominated rocky intertidal to a heterotroph-filter-feeder dominated state, with these effects likely extending to rocky intertidal communities stretching along the entire Gulf of Alaska coastline (Weitzman et al., 2021). Clear effects of the 'Blob' on the marine ecosystem off central Oregon also occurred. Here impacts were especially notable for lower trophic levels with the occurrence of an unusual assemblage of plankton species, the abundance of which pointed to oligotrophic (low nutrient) ocean conditions. Added to this, the 'Blob' precipitated a toxic diatom bloom throughout the California Current. Associated changes in the nature of the food web produced knock-on effects for several commercially and ecologically important species (Peterson et al., 2017; Pena et al., 2019). That MHW cause harmful algal blooms (HAB) is also evident for other regions such as Japan. Korea and Australia. In mid-to-late September 2021, following an intense and extensive MHW an unprecedented large-scale HAB occurred off the coast of Hokkaido, Japan ((Kuroda et al., 2021; Kuroda & Setou, 2021; Takagi et al., 2022). The weather patterns at the time favoured pushing toxic offshore waters towards the coast of east Hokkaido with the result that some fishing ports were badly affected causing mass mortality of several aquacultured fish species and echinoderms (Hasegawa et al., 2022). Similarly, a large HAB event occurred in Korean coastal waters during a MHW event in July 2018 when water temperatures increased from 24.1 °C C to 29.2 °C for two weeks with associated ecological impacts (Lim et al., 2021). For Australia, an extensive MHW driven, HAB-related fish kill event occurred in 2013. This covered 2900 km of the south Australian coastline affecting 29 fish species as well as abalone (Roberts et al., 2019). Interestingly for coastal waters, there appears to be some evidence that riverine discharge might mediate phytoplankton related HAB such that low river discharge, through extension of saline habitat area and longer residence times, contributes to the development of HAB in estuaries (Dodrill et al., 2023).

There is a growing interest in the impacts of MHW on seaweed ecosystems because of their importance as habitats for a range of marine organisms (Smale, 2020). Although based on a limited number of observational studies it appears turf-forming seaweeds, especially invasive species, increase in abundance following MHW events. In contrast, MHW force declines in the abundance of native canopy-forming kelps and fucoids (brown algae) and result in general changes in community structure (Arafeh-Dalmau et al., 2019; McPherson et al., 2021; Straub et al., 2019). Massive and cryptic loss of genetic diversity in kelp forests also appears to be the outcome of some severe MHW (Gurgel et al., 2020). Worryingly, there is evidence of extinctions of kelp and fucoids following some MHW events. For example, the loss of a local low intertidal bull kelp occurred during a MHW affecting the coast of New Zealand during the summer of 2017–2018 with an invasive kelp species replacing the native one (Thomsen et al., 2019). For this same MHW event, the loss of giant kelp in some locations along the New Zealand coast has also been noted (Tait et al., 2021). The massive 2010–2011 MHW off the coast of the state of Western Australia precipitated the extinction of two important canopy-forming seaweeds at the edges of its southern range with a 100 km northward contraction culminating in the loss of 43% of marine forests along 800 km of the West Australian coast and a regime shift to turf-dominated reefs (Wernberg et al., 2013; Wernberg, 2021). Localized die-off of giant kelp along Tasmania's east coast also occurred during the 2015–2016 Tasman Sea MHW event (Straub et al., 2019). While these examples point to the extensive spatial scale of the impact MHW on kelp, subtle variations in microclimate and substrate can result in a patchy pattern of kelp forest extinction especially in near coastal waters ((Starko et al., 2022; Verdura et al., 2021). Extinction of course represents the worst-case scenario response for seaweed species to MHW events with a spectrum of impacts having been observed (Fig. 11.7).

Like seaweeds, coral reefs harbour a wide range of marine species. Frequently referred to as the rainforests of the oceans, coral reef ecosystems account for nearly

Response Group	Response	Examples	MHW
1 Resistance	Not detected	*Ecklonia radiata*[10;12], CCA[12], turf forming algae[12] *Macrocystis pyrifera*[16]	2011 Ningaloo Niño, WA (Hamelin Bay) 2013/14 the blob, Southern California
2 Health	Bleaching	*Egregia menziesii* [3] *Porphyra perforata*[4], *Iridaea cordata*[4], *Laurencia spectabilis*[4], *Prionitis lanceolata*[4], *Gastroclonium subarticulatum*[4], *Gigartina canaliculata* [4]	1982/83 ENSO California
3 Ecological performance	Increased recruitment potential	*Pelvetia fastigiata*[3], *Sargassum muticum* [3]	1982/83 ENSO California
	Increased herbivory	Range-expansion of sub-tropical and tropical herbivores [5, 10]	1997/98 ENSO 2011 Ningaloo Niño; WA
4 Abundance change to altered biodiversity indices	Increased abundance natives	*Ulva spp.*[2], *Enteromorpha spp.*[2], *Iridaea spp.* [2] Geniculate coralline algae [8; 9] CCA[4] and turf-algae cover [7-10; 13;15] *Chorda filum*[14]	1982/83 ENSO, Chile 2009/10 ENSO, California 2011 Ningaloo Niño; WA
	Increased abundance invasives	*Giffordia mitchelliae* [1] *Agardhiella tenera*[6], *Amphiroa misakiensis*[6], *Caulerpa sertularoides*[6], *Padina durvillaei*[6], *Jania capillacea*[6], *Jania mexicana*[6] *Undaria pinnatifida*[17]	1982/83 ENSO; South America 1997/98 ENSO, Mexico 2017/18 Tasman Sea MHW
	Declined abundance natives	*Sargassum sp.*[1], *Blossevillea galapagensis*[1], *Ulva spp*[1], *Spermothamnio spp.*[1], *Centroceras spp.*[1] *Lessonia nigrescens*[2], *Macrocystis integrifolia*[2] *Egregia menziesii*[3], *Eisenia arborea* [3] *Mastocarpus papillatus*[4] *Lessonia trabeculata*[5], *Macrocystis pyrifera*[5] *Ecklonia cava*[7] CCA [14] *Saccharina latissima*[15]	1982/83 ENSO; South America 1982/83 ENSO; Chile 1982/83 ENSO; California 1997/98 ENSO; Chile 2011 Ningaloo Niño; WA 1997, 2002, 2006 hot summers, Norway
5 Extinction	Range shift, local to regional extinction	*Ecklonia radiata*[10, 11], *Scytothalia dorycarpa* [10, 11] *Macrocystis pyrifera* [c] *Durvillaea poha* [17]	2011 Ningaloo Niño; WA 2015/16 Tasmanian MHW 2017/18 Tasman Sea MHW
	Global extinction	Not yet observed	

Severity

Fig. 11.7 Range of observed impacts on seaweeds, grouped by severity, to recorded MHW events. Responses range from no observed effect (e.g., resistance) to very severe impacts (e.g., local to global extinctions). (Source: Straub et al. (2019). Used under the terms of the Creative Commons Attribution License (CC BY) https://creativecommons.org/licenses/by/4.0/)

25% of total marine biodiversity (Wagner et al., 2020). Accordingly, they have attracted much attention from scientists and environmental protection groups given threats to their survival because of increases in ocean heat content under climate change and episodic MHW events superimposed on gradual ocean warming. That coral reefs are impacted by MHW is abundantly evident from an increasing number of reports of coral bleaching and alterations to community structure. The rapid accumulation of heat over a matter of several days to a couple of weeks typical of MHW, can have quite different ecological outcomes for corals compared to bleaching events caused by the slow accumulation of heat and the development of anomalous ocean temperatures over several months driven by modes of atmospheric and ocean variability such as El Nino (Fordyce et al., 2019). For intense MHW, extensive bleaching and mortality typically occurs in corals that otherwise would be thermally tolerant during moderate warming events (Fordyce et al., 2019). For example, Hoogenboom et al. (2017) report on how the 2016 MHW event in and around Lizard Island on Australia's Great Barrier Reef wiped out some coral species that had exhibited thermal tolerance during previous bleaching events. Rapid and non-selective bleaching of thermally tolerant coral populations has also been reported for the Arabian Gulf, where corals with some of the highest temperature related bleaching thresholds globally can be found (Burt et al., 2019). Along with other rapid MHW onset events and associated coral bleaching, such as for Dongsha Atoll in the north of the South China Sea (DeCarlo et al., 2017), Iriomote Island south-western Japan (Baird et al., 2018) and the Red Sea (Anton et al., 2020; Genevier et al., 2019), this lends weight to the idea that rapidly developing high intensity MHW events can overwhelm even heat tolerant coral species (Fordyce et al., 2019) with almost immediate heat-induced mortality followed by coral skeletal dissolution accelerated by microbial excretions, and the eventual loss of the three-dimensional reef structure (Leggat et al., 2019). Intense MHW events may also have greater impacts on larger members of coral reef communities. For example, the intense MHW that occurred in the waters of Moorea Island, French Polynesia in 2019 killed a disproportionately high number of the largest individuals of the coral genera, *Pocillopora* and *Acropora*. A decrease in the overall fecundity of these spatially dominant corals by over 60% followed, reducing the recovery capacity at the whole reef scale (Speare et al., 2022). As well as direct impacts of MHW events on coral ecosystems a range of indirect impacts arise such as changes in fish populations (Brown et al., 2021; Olsen et al., 2022) and microbiomes (Corinaldesi et al., 2022). While the literature tends to purvey the view that coral reefs stand little chance against the overwhelming effects of MHW, there is evidence that some corals develop resilience via acclimatizing because of the exposure to multiple MHW events (DeCarlo et al., 2019; Fox et al., 2021). That said, Brown and Barott (2022) have emphasised the importance of understanding the acclimatization versus sensitization trajectories of various coral species as ultimately this may determine the makeup of coral reefs under severe MHW conditions.

Because seagrass ecosystems not only provide habitats for a vast range of marine organisms and are important stores of carbon (Fourqurean et al., 2012), the impacts of MHW on their ecological structure and function and the ecosystem services they provide is of interest (Serrano et al., 2021). Perhaps one of the most replete accounts of MHW impacts on seagrass meadows is that for the event that occurred in Shark Bay, Western Australia during the austral summer of 2010–2011. Here the MHW induced collapse of seagrass meadows and associated canopy loss over an area of around 1300 km^2 resulted in the remineralisation of large carbon stocks tied up in seagrass and its substrate and impacts on marine food webs in general. Because of their dependence on seagrass meadows, individual species such as dugongs, blue-nosed dolphins, pied cormorants and green turtles, as well as several species of sea snakes and fish communities and invertebrates, were severely impacted (Serrano et al., 2021) while other marine animals were not due to their life cycle being less susceptible to downturns in seagrass community health (Smith et al., 2023). Other accounts of impacted seagrass ecosystems include those for Chesapeake Bay in the US when a MHW in 2005 resulted in more than 50% loss of one seagrass species with indirect affects manifesting as a downturn in three seagrass-dependent commercially important fishery species ((Kendrick et al., 2019), the Mediterranean where significant negative impacts on seagrass shoot mortality and flowering induction occurred during MHW events in 2003 and 2006 (Diaz-Almela et al., 2007; Marba & Duarte, 2010) plus evidence of MHW induced supplanting of some native seagrass species by non-native ones (Pergent et al., 2014), as found for seagrass meadows along the Californian coast as result of the MHW associated with the 1997–1998 El Niño event (Johnson et al., 2003). Overall, it would seem generalist and opportunistic marine organisms are far less affected by MHW driven alterations to seagrass structure and function than seagrass-associated specialists (Smith et al., 2023).

Although occupying the margins of the marine environment, intertidal spaces and the organisms that live there are susceptible to the effects of atmospheric heatwaves as outlined in the section on vulnerability. In experimental studies intertidal bivalves have been shown to display a variety of responses to heatwave exposure with heatwave duration and bivalve recovery time critical in terms of physiological adjustments and survival (Amorim et al., 2020; He et al., 2022a, b). That atmospheric heatwaves can have massive impacts on the intertidal zone is amply demonstrated by the effects of the 2021 Pacific Northwest 'heat dome'. This event resulted in widespread mortality of intertidal marine organisms, including barnacles, mussels, clams, and oysters (Raymond et al., 2022). Interestingly, associated with this event was a rise in the reported cases of vibriosis in the Pacific Northwest. This illness, often associated with eating raw or undercooked shellfish infected with the Vibrio bacteria or exposing a wound to seawater containing Vibrio, has also been reported in parts of Northern Europe following heatwave events (Amato et al., 2022; Baker-Austin et al., 2016; Brehm et al., 2021). This type of impact forms a good example of the indirect impacts of coastal/marine heatwaves on human health via temperature related increases in harmful water-borne bacteria.

Further to direct field evidence, researchers have turned to laboratory-based experimental approaches for establishing the impacts of MHW events on marine life. Such approaches commonly involve exposing the flora or fauna of interest to conditions akin to those of actual MHW. Acknowledging the increasing frequency, duration, and magnitude of compound MHW and deoxygenation events, and by developing experiments to establish the biological impacts of these, Lucey et al. (2022) found that short duration compound MHW events with hypoxic conditions proved catastrophic for tropical water dwelling sea urchins. Kellet's Whelk, a potentially important wild shellfishery along the Californian coast, has also been found to be prone to simulated MHW events with various impacts including mortality and formation abnormalities uncovered for two of its larval stages (Clare et al., 2022). Similarly, Balogh and Byrne (2020) have shown how simulated heatwave conditions can impact the life stages of an intertidal starfish in south-eastern Australia. For the Pacific Crown of Thorns starfish, Lang et al. (2021) have demonstrated while generally sensitive to increasing water temperatures, as evidenced through changes in metabolic pathways and rates, the adults, and especially sub-adults of this macroinvertebrate, may have some resilience to short-term MHW.

Corroborating field-based studies, Atkinson et al. (2020) found the replacement of native seaweed species with invasive ones is likely under MHW conditions. Also focusing on seaweed, Nauer et al. (2022) established that tropical and transition specimens used in their experiments suffered irreversibility under a range of simulated MHW scenarios, while warm-temperate specimens were able to recover, highlighting the threat of extinction of tropical and transition seaweed species if the frequency and magnitude of MHW continue to increase. Resonating with these findings are those of Diehl et al. (2021). These researchers measured the physiological and biochemical traits of the broadly distributed European kelp *Saccharina latissimi* along a latitudinal gradient under a range of MHW scenarios. They concluded that the southernmost specimens were most affected by heat stress pointing to greater MHW impacts amongst populations of this kelp living near their thermal limits or in other words rear-edge populations. For the same kelp, Filbee-Dexter et al. (2020) experimentally determined that the strong and severe North Atlantic MHW events of 2018 exacted a heavy toll on kelp forests along the coasts of Norway and the eastern US with kelp mortality patterns mapping onto areas with water temperature anomalies that exceeded deadly thermal thresholds. Adding to the body of knowledge for this geographically important kelp, Niedzwiedz et al. (2022) have provided experimental evidence that the thermal tolerance of *Saccharina latissimi* to summer MHW is dependent on its environmental history, especially general thermal conditions preceding heatwave events.

Complementing the experimental work on seaweed communities is that for coral ecosystems. For a range of species in the Coral Sea Marine Park, Marzonie et al. (2023), by analysing the outcomes of coral heat stress experiments, suggest that MHW select for heat tolerant individuals and populations because of acclimatization associated with a modest thermal history. That said they caution study results also indicate that this adaptation response could have been compromised by the exposure to recent MHW events. Focusing on 'leather coral'—a soft coral—Travesso et al. (2023), via experimental manipulations mimicking MHW, established that this coral type is especially sensitive to the synergistic effects of heat and high light stress with the chief impacts a decline in photosynthetic efficiency plus heat stress induced bleaching and pigment damage under high light stress. Experimental work on seagrass responses to MHW related heat stress has also been undertaken revealing impacts such as a reduction in seagrass nutritional value and palatability (Stipcich et al., 2023), spatial variability in the decline of seagrass meadow related carbon stocks in sediments (Aoki et al., 2021) and a decrease in seagrass internal carbon reserves (Vivanco-Bercovich et al., 2022). Lastly, some interesting experimental work on the effects of MHW on seafloor ecosystem functioning in the Baltic Sea region, as established through monitoring macrofauna bioturbation and solute fluxes in sediment cores, has shown that the nature of changes to biogeochemical and biological processes and biogeochemical cycling depend on MHW intensity and duration (Kauppi et al., 2023; Kauppi & Villnas, 2022).

11.7 Terrestrial Ecosystems

The response of different terrestrial ecosystems to heatwave events appears to be quite diverse (Wang et al., 2023). That said, most observation-based heatwave impact studies focus on trees as important components of terrestrial ecosystems. For trees, Teskey et al. (2015) have outlined the main physiological responses during extreme heat events. At the leaf level a reduction in photosynthesis, an increase in photooxidative stress, leaf abscission and a decline in the growth rate of any remaining leaves are common. At the whole tree level, growth is impaired and biomass allocation shifts with tree mortality likely if drought stress accompanies heat waves (Teskey et al., 2015). All that said, these effects are not ubiquitous as some tree and plant species demonstrate remarkable tolerance to heatwave events (Andrew et al., 2023; Drake et al., 2018) with so-called photosynthesis—transpiration decoupling, or in other words leaf or canopy cooling, often implemented to avoid lethal leaf temperatures (De Kauwe et al., 2019; Krich et al., 2022).

A challenge to establishing heatwave impacts on trees and plants in general is distinguishing between the effects of drought and short-term heatwave events, acknowledging that some heatwave events are embedded within a period of drought while others may occur independent of drought. Apart from acute wilting, a widely applied tangible indicator of the impact of heat stress on forests is leaf (foliar) browning. However, foliar browning may arise for a variety of reasons. If the result of drought induced hydraulic failure, browning may take several weeks if not months to manifest (Hammond et al., 2019). This contrasts with the timescale typical of heatwaves. Given this what is pertinent in establishing heatwave impacts, if applying foliar browning as a metric, is this response occurs during or immediately following a heatwave, typically a matter of days (Still et al., 2023). Further, at this timescale it is most likely foliar scorching, like the scorching experienced as the result of nearby fire, is relevant. That is, browning is predominantly the result of atmospheric heat and solar loading, with hydraulic failure playing a minor role. The Pacific Northwest (US) heat dome of June 2021 provides a very good example of how heatwaves impact vegetation via scorching. For this event widespread foliar scorching occurred in coastal forest stands that were only under drought conditions considered as abnormally dry, the lowest drought category of the US Drought Monitor's drought categorisation. In contrast forests away from the coast experiencing exceptional drought were little affected (Sang & Hamann, 2023; Still et al., 2023). Further for completely solar-exposed tree canopies, whole of canopy scorching was a noticeable feature. This contrasts with the patchy pattern of foliar browning that often typifies the impacts of hydraulic failure. Local exposure and strong aspect control on scorching patterns also occurred. Trees on forest edges and sunlit exposed south to west facing slopes experienced the highest levels of foliar scorch. In contrast, foliage on the relatively cool and shaded backside of trees and hillsides showed few to no signs of scorching (Still et al., 2023), (Figure 11.8a and b).

Other examples of the impact of heatwaves on trees include rapid forest canopy collapse during an extreme drought-associated heatwave in south-western Australia

Fig. 11.8 Foliar scorching in Oregon due to the Pacific Northwest June 2021 'heat dome' event as observed for trees on (**a**) west facing slopes and (**b**) forest edges. (Images courtesy of Adam Sibley, Research Associate, Oregon State University, College of Forestry and Daniel DePinte, US Forest Service, Pacific Northwest Region)

(Matusick et al., 2018), 30–50% canopy loss amongst 'urban' London Plane trees during the 2019 heatwave in Melbourne when temperatures exceeded 43 °C (Sanusi & Livesley, 2020), greater leaf damage to non-native tree species during heatwaves in Western Sydney in 2019 and 2020 (Esperon-Rodriguez et al., 2021), variable stem growth and stem shrinkage across 21 tree species during the 2018 European heatwave (Salomon et al., 2022), plus severe impacts on Scots Pine across Central Europe during the same heatwave event (Haberstroh et al., 2022). As well as canopy cover, heatwaves affect seed set (Enright et al., 2015). Observed secondary impacts of heatwaves on terrestrial ecosystems include the loss of food sources for birds (Steel et al., 2019), alteration of reptile assemblages (Dundas et al., 2021), an increase in fuel for wildfires (Ruthrof et al., 2016), outbreaks of vegetation damaging insects (Seaton et al., 2015) and alteration of community structures in the rhizosphere beneath trees (Hopkins et al., 2018). Because there is a pressing need to establish how heatwaves impact vegetation there has been an uptick in experimental impact studies, albeit with more of a focus on agricultural plants compared to natural terrestrial ecosystems (Breshears et al., 2021).

11.8 Aquatic Ecosystems

In general, much remains to be learned about the observed impacts of heatwaves on aquatic systems. The speculations regarding the heat-vulnerability of aquatic ecosystems put forward in the previous chapter and those contained in some overview papers (Leigh et al., 2015) are partly borne out by a limited number of field-based observations and experimental work. Although primarily focusing on the combined effects of heatwaves and micropollutants on freshwater ecosystems, Polazzo et al.

(2022) have effectively summarized many of the observed impacts of heatwaves on aquatic food webs arising from field- and experiment-based observations (Fig. 11.9). In relation to their primary purpose however, Polazzo et al. (2022) found no consistent interactive heatwave—micropollutant impacts at any level of biological organization, noting this to be more a consequence of the meagre number of studies than a total absence of effects.

For lakes there is clear evidence of the impacts of heatwaves on their physical properties and biota. For example, during the extreme European 2003 heatwave, eutrophication and deoxygenation of some lakes in Central Europe occurred (Jankowski et al., 2006). Similarly, reduced mixing and elevated temperatures of surface lake waters induce stronger stratification in lakes (Chou et al., 2021). Depending on a range of biotic and abiotic factors, higher levels of cyanobacteria biomass have been observed in lakes during heatwave events (Bartosiewicz et al., 2019; Huber et al., 2012; Jeppesen et al., 2021a, b; Johnk et al., 2008; Li et al., 2023). Heatwave related changes in the volume of chlorophyll-a in lakes have also

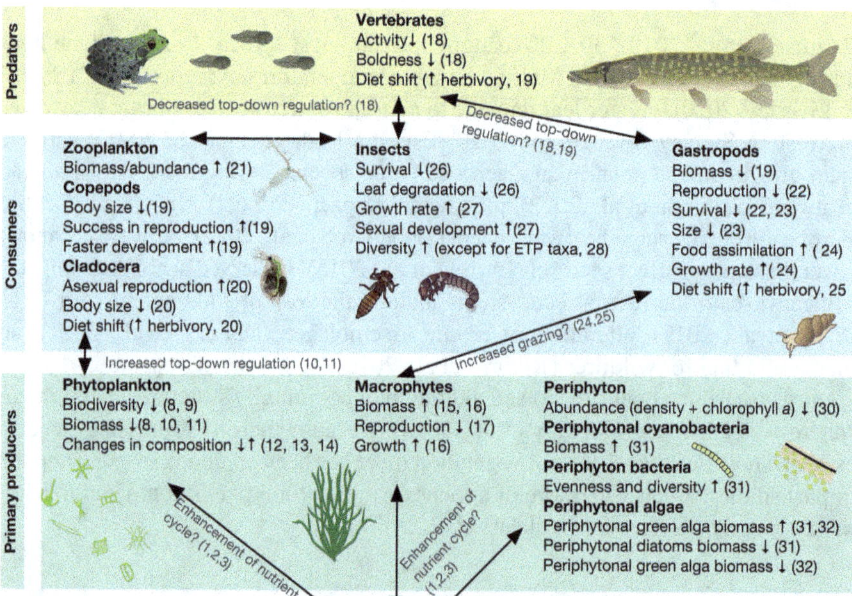

Fig. 11.9 Effects of heatwaves on aquatic food-webs. Trophic and/or indirect effects are indicated by arrows connecting different organisms. An increase or decrease of an ecological parameter is indicated by an upward or downward arrow respectively. (Adapted from: Polazzo et al. (2022). Used under conditions of Creative Commons Attribution Licence BY-NC/4.0: https://creativecommons.org/licenses/by-nc/4.0/)

been reported through observation and modelling (Chen et al., 2020; Free et al., 2022; Woolway et al., 2021) along with assemblages of other phytoplankton and zooplankton (Huber (Huber et al., 2010; Maazouzi et al., 2008) as indicated in Fig. 11.9. The degradation of lake water quality is also another observed impact of heatwaves (Bartosiewicz et al., 2016; Blagrave et al., 2022). Impacts of heatwaves on the death of freshwater organisms such as mayflies, molluscs and fish have also been reported (Durham et al., 2006; Macaulay et al., 2021; Matthews et al., 1982; Mouthon & Daufresne, 2006; Mundahl, 1990). For estuarine environments there is emerging evidence of the impacts of the synergistic effects of elevated water and air temperatures on the dissolved oxygen and pH status of estuarine waters, although the indirect effects of these heatwave-related water quality changes on organisms remains unknown (Tassone et al., 2022). Complementing field- and indoor laboratory-based studies are those using mesocosms—outdoor experimental systems. Results from these have revealed a range phytoplankton, zooplankton, metabolism and submerged macrophyte community responses to heatwaves (Filiz et al., 2020; Iskin et al., 2020; Jeppesen et al., 2021a; Li et al., 2017; Sun & Arnott, 2022; Zhang et al., 2022) as well as how heatwaves influence the level of Greenhouse Gas emissions from shallow lakes (Audet et al., 2017).

11.9 Conclusion

From the observational evidence presented in this chapter it is clear heatwaves have far-reaching consequences for human-related and natural systems. However, in the absence of systematic data collection programmes on heatwave occurrence and impacts, the true scale of the loss and damage associated with this natural hazard will remain unknown. Further and worryingly, without recourse to heatwave impact data for both human and natural systems, progress towards achieving well-honed heatwave risk management strategies will be compromised. Besides the impacts of heatwaves on human and domesticated animal health, economy and society and infrastructure, heatwaves seriously affect natural systems and ecosystem services. Nowhere is this more obvious than in marine and aquatic ecosystems where anomalously warm water can influence physical and biological processes and negatively affect organisms across a range of trophic levels. While a variety of heatwave impacts have been chronicled for land-based natural systems much remains unknown about the deleterious effects of heatwaves on terrestrial flora and fauna. To augment the 'observed' evidence base for heatwave impacts in natural systems, researchers are increasingly turning to laboratory-based experimental approaches whereby particular animal and plant species are subject to simulated heatwaves with their responses from the cellular to whole of organism scale monitored.

References

Adato, B., Dubnov-Raz, G., Gips, H., Heled, Y., & Epstein, Y. (2016). Fatal heat stroke in children found in parked cars: Autopsy findings. *European Journal of Pediatrics, 175*, 1249–1252.

Adelaide, L., Chanel, O., & Pascal, M. (2022). Health effects from heat waves in France: An economic evaluation. *European Journal of Health Economics, 23*(1), 119–131. https://doi.org/10.1007/s10198-021-01357-2

Aivalioti, S. (2015). *Electricity sector adaptation to heat waves*. Sabin Center for Climate Change Law.

Al-Dawood, A. (2017). Towards heat stress management in small ruminants – A review. *Annals of Animal Science, 17*(1), 59–88. https://doi.org/10.1515/aoas-2016-0068

Alele, F. O., Malau-Aduli, B. S., Malau-Aduli, A. E. O., & Crowe, M. J. (2020). Epidemiology of exertional heat illness in the military: A systematic review of observational studies. *International Journal of Environmental Research and Public Health, 17*(19). https://doi.org/10.3390/ijerph17197037

Amador-Capitanachi, M. J., Moreno-Sanchez, X. G., Ventura-Dominguez, P. D., Juarez-Ruiz, A., Gonzalez-Rodriguez, E., Galvez, C., Norris, T., & Elorriaga-Verplancken, F. R. (2020). Ecological implications of unprecedented warm water anomalies on interannual prey preferences and foraging areas of Guadalupe fur seals. *Marine Mammal Science, 36*(4), 1254–1270. https://doi.org/10.1111/mms.12718

Amato, E., Riess, M., Thomas-Lopez, D., Linkevicius, M., Pitkanen, T., Wolkowicz, T., Rjabinina, J., Jernberg, C., Hjertqvist, M., MacDonald, E., Antony-Samy, J. K., Bjerre, K. D., Salmenlinna, S., Fuursted, K., Hansen, A., & Naseer, U. (2022). Epidemiological and microbiological investigation of a large increase in vibriosis, northern Europe, 2018. *Eurosurveillance, 27*(28). https://doi.org/10.2807/1560-7917.es.2022.27.28.2101088

Amorim, V. E., Goncalves, O., Capela, R., Fernandez-Boo, S., Oliveira, M., Dolbeth, M., Arenas, F., & Cardoso, P. G. (2020). Immunological and oxidative stress responses of the bivalve Scrobicularia plana to distinct patterns of heatwaves. *Fish & Shellfish Immunology, 106*, 1067–1077. https://doi.org/10.1016/j.fsi.2020.09.024

Anderson, C. A. (2001). Heat and violence. *Current Directions in Psychological Science, 10*(1), 33–38. https://doi.org/10.1111/1467-8721.00109

Anderson, G. B., & Bell, M. L. (2011). Heat waves in the United States: Mortality risk during heat waves and effect modification by heat wave characteristics in 43 U.S. communities. *Environmental Health Perspectives, 119*(2), 210–218. https://doi.org/10.1289/ehp.1002313

Anderson, G. B., Dominici, F., Wang, Y., McCormack, M. C., Bell, M. L., & Peng, R. D. (2013). Heat-related emergency hospitalizations for respiratory diseases in the medicare population. *American Journal of Respiratory and Critical Care Medicine, 187*(10), 1098–1103. https://doi.org/10.1164/rccm.201211-1969OC

Andrew, S. C., Arnold, P. A., Simonsen, A. K., & Briceno, V. F. (2023). Consistently high heat tolerance acclimation in response to a simulated heatwave across species from the broadly distributed Acacia genus. *Functional Plant Biology, 50*(1), 71–83. https://doi.org/10.1071/fp22173

Andriano, L. (2023). On the health impacts of climatic shocks: How heatwaves reduce birthweight in sub-Saharan Africa. *Population and Development Review, 49*(4), 737–769. https://doi.org/10.1111/padr.12583

Anon. (2021). *Several people die in lakes, rivers, during heat wave*. US News and World Health Report.

Anon. (2022). Sweltering heatwave: Islamabad's primary schools observe early summer vacation. *The News International*.

Anselmi, N., Montaldo, S., Pomilla, A., & Maraone, A. (2020). Forgotten baby syndrome: Dimensions of the phenomenon and new research perspectives. *Rivista di Psichiatria, 55*(2), 112–118.

Antczak, J. (2011). *Los Angeles heat wave bakes at record 113 degrees*. Huffington Post.

Anton, A., Randle, J. L., Garcia, F. C., Rossbach, S., Ellis, J. I., Weinzierl, M., & Duarte, C. M. (2020). Differential thermal tolerance between algae and corals may trigger the proliferation of algae in coral reefs. *Global Change Biology, 26*(8), 4316–4327. https://doi.org/10.1111/gcb.15141

Aoki, L. R., McGlathery, K. J., Wiberg, P. L., Oreska, M. P. J., Berger, A. C., Berg, P., & Orth, R. J. (2021). Seagrass recovery following marine heat wave influences sediment carbon stocks. *Frontiers in Marine Science, 7.* https://doi.org/10.3389/fmars.2020.576784

Arafeh-Dalmau, N., Montano-Moctezuma, G., Martinez, J. A., Beas-Luna, R., Schoeman, D. S., & Torres-Moye, G. (2019). Extreme marine heatwaves alter kelp forest community near its equatorward distribution limit. *Frontiers in Marine Science, 6.* https://doi.org/10.3389/fmars.2019.00499

Asma-Sadeque, S. (2022). *Thousands of cattle dead due to heatwave in Kansas*. The Guardian.

Atkinson, J., King, N. G., Wilmes, S. B., & Moore, P. J. (2020). Summer and Winter Marine heatwaves favor an invasive over native seaweeds. *Journal of Phycology, 56*(6), 1591–1600. https://doi.org/10.1111/jpy.13051

Audet, J., Neif, E. M., Cao, Y., Hoffmann, C. C., Lauridsen, T. L., Larsen, S. E., Sondergaard, M., Jeppesen, E., & Davidson, T. A. (2017). Heat-wave effects on greenhouse gas emissions from shallow lake mesocosms. *Freshwater Biology, 62*(7), 1130–1142. https://doi.org/10.1111/fwb.12930

Auliciems, A., & Dibartolo, L. (1995). Domestic violence in a subtropical environment – Police calls and weather in BRISBANE. *International Journal of Biometeorology, 39*(1), 34–39. https://doi.org/10.1007/bf01320891

Baird, A. H., Keith, S. A., Woolsey, E., Yoshida, R., & Naruse, T. (2018). Rapid coral mortality following unusually calm and hot conditions on Iriomote, Japan. *F1000Research, 6*(1728). https://doi.org/10.12688/f1000research.12660.2

Baker-Austin, C., Trinanes, J. A., Salmenlinna, S., Lofdahl, M., Siitonen, A., Taylor, N. G. H., & Martinez-Urtaza, J. (2016). Heat wave-associated Vibriosis, Sweden and Finland, 2014. *Emerging Infectious Diseases, 22*(7), 1216–1220. https://doi.org/10.32032/eid2207.151996

Bakull, E. (2022). *Detroit's half day for heat spotlights district's air conditioning problem.* Chalk Beat.

Balogh, R., & Byrne, M. (2020). Developing in a warming intertidal, negative carry over effects of heatwave conditions in development to the pentameral starfish in Parvulastra exigua. *Marine Environmental Research, 162.* https://doi.org/10.1016/j.marenvres.2020.105083

Bartosiewicz, M., Laurion, I., Clayer, F., & Maranger, R. (2016). Heat-wave effects on oxygen, nutrients, and phytoplankton can alter global warming potential of gases emitted from a small shallow lake. *Environmental Science & Technology, 50*(12), 6267–6275. https://doi.org/10.1021/acs.est.5b06312

Bartosiewicz, M., Przytulska, A., Deshpande, B. N., Antoniades, D., Cortes, A., MacIntyre, S., Lehmann, M. F., & Laurion, I. (2019). Effects of climate change and episodic heat events on cyanobacteria in a eutrophic polymictic lake. *Science of the Total Environment, 693*, 133414. https://doi.org/10.1016/j.scitotenv.2019.07.220

Basilio, A., Searcy, S., & Thompson, A. R. (2017). Effects of the Blob on settlement of spotted sand bass, *Paralabrax maculatofasciatus*, to Mission Bay, San Diego, CA. *Plos One, 12*(11). https://doi.org/10.1371/journal.pone.0188449

Baynes, C. (2019). India heatwave: Four people die in train carriage without air conditioning as temperatures hit 48C. *Independent.*

Bellquist, L., Saccomanno, V., Semmens, B. X., Gleason, M., & Wilson, J. (2021). The rise in climate change-induced federal fishery disasters in the United States. *PeerJ, 9.* https://doi.org/10.7717/peerj.11186

Bernstein, S. (2006). DWP scrambles to gauge Power needs before the next heat wave. *Los Angeles Times.*

Bishop-Williams, K. E., Berke, O., Pearl, D. L., Hand, K., & Kelton, D. F. (2015). Heat stress related dairy cow mortality during heat waves and control periods in rural southern Ontario from 2010-2012. *BMC Veterinary Research, 11*. https://doi.org/10.1186/s12917-015-0607-2

Biswal, J., Vijayalakshmy, K., Bhattacharya, T. K., & Rahman, H. (2022). Impact of heat stress on poultry production. *Worlds Poultry Science Journal, 78*(1), 179–196. https://doi.org/10.1080/00439339.2022.2003168

Blackburn, G., Broom, E., Ashton, B. J., Thornton, A., & Ridley, A. R. (2022). Heat stress inhibits cognitive performance in wild Western Australian magpies, Cracticus tibicen dorsalis. *Animal Behaviour, 188*, 1–11. https://doi.org/10.1016/j.anbehav.2022.03.016

Blagrave, K., Moslenko, L., Khan, U. T., Benoit, N., Howell, T., & Sharma, S. (2022). Heatwaves and storms contribute to degraded water quality conditions in the nearshore of Lake Ontario. *Journal of Great Lakes Research, 48*(4), 903–913. https://doi.org/10.1016/j.jglr.2022.04.008

Booth, J. N., Davis, G. G., Waterbor, J., & McGwin, G. (2010). Hyperthermia deaths among children in parked vehicles: An analysis of 231 fatalities in the United States, 1999–2007. *Forensic Science, Medicine, and Pathology, 6*, 99–105.

Borg, M., Bi, P., Nitschke, M., Williams, S., & McDonald, S. (2017). The impact of daily temperature on renal disease incidence: An ecological study. *Environmental Health, 16*. https://doi.org/10.1186/s12940-017-0331-4

Borg, M. A., Xiang, J. J., Anikeeva, O., Pisaniello, D., Hansen, A., Zander, K., Dear, K., Sim, M. R., & Bi, P. (2021). Occupational heat stress and economic burden: A review of global evidence. *Environmental Research, 195*. https://doi.org/10.1016/j.envres.2021.110781

Borg, M. A., Xiang, J. J., Anikeeva, O., Ostendorf, B., Varghese, B., Dear, K., Pisaniello, D., Hansen, A., Zander, K., Sim, M. R., & Bi, P. (2023). Current and projected heatwave-attributable occupational injuries, illnesses, and associated economic burden in Australia. *Environmental Research, 236*. https://doi.org/10.1016/j.envres.2023.116852

Borsky, S., & Unterberger, C. (2019). Bad weather and flight delays: The impact of sudden and slow onset weather events. *Economics of Transportation, 18*, 10–26. https://doi.org/10.1016/j.ecotra.2019.02.002

Bouchama, A., Abuyassin, B., Lehe, C., Laitano, O., Jay, O., O'Connor, F. G., & Leon, L. R. (2022). Classic and exertional heatstroke. *Nature Reviews Disease Primers, 8*(1). https://doi.org/10.1038/s41572-021-00334-6

Bratu, A., Card, K. G., Closson, K., Aran, N., Marshall, C., Clayton, S., Gislason, M. K., Samji, H., Martin, G., Lem, M., Logie, C. H., Takaro, T. K., & Hogg, R. S. (2022). The 2021 Western north American heat dome increased climate change anxiety among British Columbians: Results from a natural experiment. *Journal of Climate Change and Health, 6*, 100116. https://doi.org/10.1016/j.joclim.2022.100116

Brehm, T. T., Berneking, L., Martins, M. S., Dupke, S., Jacob, D., Drechsel, O., Bohnert, J., Becker, K., Kramer, A., Christner, M., Aepfelbacher, M., Schmiedel, S., & Rohde, H. (2021). Heatwave-associated vibrio infections in Germany, 2018 and 2019. *Eurosurveillance, 26*(41). https://doi.org/10.2807/1560-7917.es.2021.26.41.2002041

Breitfeld, E. (2020). Hot-car deaths and forgotten-baby syndrome: A case against prosecution. *Berkeley J Crim L, 25*, 72.

Breshears, D. D., Fontaine, J. B., Ruthrof, K. X., Field, J. P., Feng, X., Burger, J. R., Law, D. J., Kala, J., & Hardy, G. E. S. (2021). Underappreciated plant vulnerabilities to heat waves. *New Phytologist, 231*(1), 32–39. https://doi.org/10.1111/nph.17348

Brimicombe, C., Di Napoli, C., Cornforth, R., Pappenberger, F., Petty, C., & Cloke, H. L. (2021). Borderless heat hazards with bordered impacts. Earths. *Futures, 9*(9). https://doi.org/10.1029/2021ef002064

Brown, K. T., & Barott, K. L. (2022). The costs and benefits of environmental memory for reef-building corals coping with recurring marine heatwaves. *Integrative and Comparative Biology, 62*(6), 1748–1755. https://doi.org/10.1093/icb/icac074

Brown, C. J., Mellin, C., Edgar, G. J., Campbell, M. D., & Stuart-Smith, R. D. (2021). Direct and indirect effects of heatwaves on a coral reef fishery. *Global Change Biology, 27*(6), 1214–1225. https://doi.org/10.1111/gcb.15472

Brown, M. V., Ostrowski, M., Messer, L. F., Bramucci, A., van de Kamp, J., Smith, M. C., Bissett, A., Seymour, J., Hobday, A. J., & Bodrossy, L. (2024). A marine heatwave drives significant shifts in pelagic microbiology. *Communications Biology, 7*(1). https://doi.org/10.1038/s42003-023-05702-4

Brownlow, M. A., Dart, A. J., & Jeffcott, L. B. (2016). Exertional heat illness: A review of the syndrome affecting racing thoroughbreds in hot and humid climates. *Australian Veterinary Journal, 94*(7), 240–247. https://doi.org/10.1111/avj.12454

Bruzek R, Biess L, Al-Nazer L, Asme (2013) development of rail temperature predictions to minimize risk of track buckle derailments. Proceeding of the Asme Joint Rail Conference, 2013.

Buguet, A., Radomski, M. W., Reis, J., & Spencer, P. S. (2023). Heatwaves and human sleep: Stress response versus adaptation. *Journal of the Neurological Sciences, 454*. https://doi.org/10.1016/j.jns.2023.120862

Bunker, A., Wildenhain, J., Vandenbergh, A., Henschke, N., Rocklöv, J., Hajat, S., & Sauerborn, R. (2016). Effects of air temperature on climate-sensitive mortality and morbidity outcomes in the elderly; a systematic review and meta-analysis of epidemiological evidence. *eBioMedicine, 6*, 258–268.

Burbidge, R. (2018). Adapting aviation to a changing climate: Key priorities for action. *Journal of Air Transport Management, 71*, 167–174. https://doi.org/10.1016/j.jairtraman.2018.04.004

Burillo, D., Chester, M. V., Pincetl, S., & Fournier, E. (2019). Electricity infrastructure vulnerabilities due to long-term growth and extreme heat from climate change in Los Angeles County. *Energy Policy, 128*, 943–953. https://doi.org/10.1016/j.enpol.2018.12.053

Burke, M., Gonzalez, F., Bayliss, P., Heft-Neal, S., Baysan, C., Basu, S., & Hsiang, S. (2018). Higher temperatures increase suicide rates in the United States and Mexico. *Nature Climate Change, 8 (8):723-+*. https://doi.org/10.1038/s41558-018-0222-x

Burt, J. A., Paparella, F., Al-Mansoori, N., Al-Mansoori, A., & Al-Jailani, H. (2019). Causes and consequences of the 2017 coral bleaching event in the southern Persian/Arabian gulf. *Coral Reefs, 38*(4), 567–589. https://doi.org/10.1007/s00338-019-01767-y

Callahan, C. W., & Mankin, J. S. (2022). Globally unequal effect of extreme heat on economic growth. Science. *Advances, 8*(43). https://doi.org/10.1126/sciadv.add3726

Campbell, S., Remenyi, T. A., White, C. J., & Johnston, F. H. (2018). Heatwave and health impact research: A global review. *Health & Place, 53*, 210–218. https://doi.org/10.1016/j.healthplace.2018.08.017

Campbell, S. L., Remenyi, T., Williamson, G. J., Rollins, D., White, C. J., & Johnston, F. H. (2021). Ambulance dispatches and heatwaves in Tasmania, Australia: A case-crossover analysis. *Environmental Research, 202*. https://doi.org/10.1016/j.envres.2021.111655

Carpenter, B. (2019). An overview and analysis of the impacts of extreme heat on the aviation industry. *Pursuit – The Journal of Undergraduate Research at The University of Tennessee, 9*(1).

Chan, F., Francis, O., Dodd, L., Mahdi, Z., & Koblar, S. A. (2014). A hot topic – Heat waves and stroke. *International Journal of Stroke, 9*(7), 858–859. https://doi.org/10.1111/ijs.12319

Chang, H. H., Zhang, H. S., Latimore, A. D., Murray, B. P., D'Souza, R. R., Scovronick, N., Gribble, M. O., & Ebelt, S. T. (2023). Associations between short-term ambient temperature exposure and emergency department visits for amphetamine, cocaine, and opioid use in California from 2005 to 2019. *Environment International, 181*. https://doi.org/10.1016/j.envint.2023.108233

Chauvin, M., Kosatsky, T., Bilodeau-Bertrand, M., Gamache, P., Smargiassi, A., & Auger, N. (2020). Hot weather and risk of drowning in children: Opportunity for prevention. *Preventive Medicine, 130*. https://doi.org/10.1016/j.ypmed.2019.105885

Chen, W. Y., Nielsen, A., Andersen, T. K., Hu, F. J., Chou, Q. C., Sondergaard, M., Jeppesen, E., & Trolle, D. (2020). Modeling the ecological response of a temporarily summer-stratified Lake to extreme heatwaves. *Water, 12*(1). https://doi.org/10.3390/w12010094

Cheng, J., Xu, Z. W., Bambrick, H., Su, H., Tong, S. L., & Hu, W. B. (2019). Impacts of exposure to ambient temperature on burden of disease: A systematic review of epidemiological evidence. *International Journal of Biometeorology, 63*(8), 1099–1115. https://doi.org/10.1007/s00484-019-01716-y

Chou, Q. C., Nielsen, A., Andersen, T. K., Hu, F. J., Chen, W. Y., Cao, T., Ni, L. Y., Sondergaard, M., Johansson, L. S., Jeppesen, E., & Trolle, D. (2021). The impacts of extreme climate on summer-stratified temperate lakes: Lake Soholm, Denmark, as an example. *Hydrobiologia, 848*(15), 3521–3537. https://doi.org/10.1007/s10750-021-04607-9

Ciais, P., Reichstein, M., Viovy, N., Granier, A., Ogée, J., Allard, V., Aubinet, M., Buchmann, N., Bernhofer, C., Carrara, A., Chevallier, F., De Noblet, N., Friend, A. D., Friedlingstein, P., Grünwald, T., Heinesch, B., Keronen, P., Knohl, A., Krinner, G., Loustau, D., Manca, G., Matteucci, G., Miglietta, F., Ourcival, J. M., Papale, D., Pilegaard, K., Rambal, S., Seufert, G., Soussana, J. F., Sanz, M. J., Schulze, E. D., Vesala, T., & Valentini, R. (2005). Europe-wide reduction in primary productivity caused by the heat and drought in 2003. *Nature, 437*(7058), 529–533. https://doi.org/10.1038/nature03972

Clare, X. S., Kui, L., & Hofmann, G. E. (2022). Larval thermal tolerance of kellet's whelk (Kelletia kelletii) as a window into the resilience of a wild shellfishery to marine heatwaves. *Journal of Shellfish Research, 41*(2), 283–290. https://doi.org/10.2983/035.041.0214

Clayton, S. (2021). Climate change and mental health. *Current Environmental Health Reports, 8*(1), 1–6. https://doi.org/10.1007/s40572-020-00303-3

Cloud, D. H., Williams, B., Haardörfer, R., Brinkley-Rubinstein, L., & Cooper, H. L. F. (2023). Extreme heat and suicide watch incidents among incarcerated men. *JAMA Network Open, 6*(8). https://doi.org/10.1001/jamanetworkopen.2023.28380

Coffel, E. D., Thompson, T. R., & Horton, R. M. (2017). The impacts of rising temperatures on aircraft takeoff performance. *Climatic Change, 144*(2), 381–388. https://doi.org/10.1007/s10584-017-2018-9

Coomes, C. M., & Derryberry, E. P. (2021). High temperatures reduce song production and alter signal salience in songbirds. *Animal Behaviour, 180*, 13–22. https://doi.org/10.1016/j.anbehav.2021.07.020

Corinaldesi, C., Varrella, S., Tangherlini, M., Dell'Anno, A., Canensi, S., Cerrano, C., & Danovaro, R. (2022). Changes in coral forest microbiomes predict the impact of marine heatwaves on habitat-forming species down to mesophotic depths. *Science of the Total Environment, 823*. https://doi.org/10.1016/j.scitotenv.2022.153701

Crank, P. J., Hondula, D. M., & Sailor, D. J. (2023). Mental health and air temperature: Attributable risk analysis for schizophrenia hospital admissions in arid urban climates. *Science of the Total Environment, 862*. https://doi.org/10.1016/j.scitotenv.2022.160599

Crozier, R. (2021). iiNet says heatwave conditions behind data Centre outage. itNews,

Danner, R. M., Coomes, C. M., & Derryberry, E. P. (2021). Simulated heat waves reduce cognitive and motor performance of an endotherm. *Ecology and Evolution, 11*(5), 2261–2272. https://doi.org/10.1002/ece3.7194

Davenport, F., Dorélien, A., & Grace, K. (2020). Investigating the linkages between pregnancy outcomes and climate in sub-Saharan Africa. *Population and Environment, 41*(4), 397–421. https://doi.org/10.1007/s11111-020-00342-w

Davidson, H. (2022). *Blackouts in China as heatwave pushes electricity usage to record levels.* The Guardian.

Davis, R. E., Houck, M., Markle, E., Windoloski, S., Enfield, K. B., Kang, H., Balling, R. C., Kuehl, D. R., Burton, J. H., Farthing, W., Rubio, E. R., & Novicoff, W. M. (2020). The impact of heat waves on emergency department visits in Roanoke. *Virginia. Academic Emergency Medicine, 27*(7), 614–617. https://doi.org/10.1111/acem.13919

De Kauwe, M. G., Medlyn, B. E., Pitman, A. J., Drake, J. E., Ukkola, A., Griebel, A., Pendall, E., Prober, S., & Roderick, M. (2019). Examining the evidence for decoupling between photosynthesis and transpiration during heat extremes. *Biogeosciences, 16*(4), 903–916. https://doi.org/10.5194/bg-16-903-2019

De Vivo C, Ellena M, Capozzi V, Budillon G, Mercogliano P (2022) Risk assessment framework for Mediterranean airports: A focus on extreme temperatures and precipitations and sea level rise. Natural Hazards 111 (1):547–566. doi:https://doi.org/10.1007/s11069-021-05066-0.

DeCarlo, T. M., Cohen, A. L., Wong, G. T. F., Davis, K. A., Lohmann, P., & Soong, K. (2017). Mass coral mortality under local amplification of 2 °C Ocean warming. *Scientific Reports, 7*. https://doi.org/10.1038/srep44586

DeCarlo TM, Harrison HB, Gajdzik L, Alaguarda D, Rodolfo-Metalpa R, D'Olivo J, Liu G, Patalwala D, Mcculloch MT (2019) acclimatization of masssive reef-building corals to consecutive heatwaves. Proceedings of the Royal Society B-Biological Sciences 286 (1898). doi:https://doi.org/10.1098/rspb.2019.0235.

Desai, Y., Khraishah, H., & Alahmad, B. (2023). Heat and the heart. *Yale Journal of Biology and Medicine, 96*(2), 197–203.

Diaz-Almela, E., Marba, N., & Duarte, C. M. (2007). Consequences of Mediterranean warming events in seagrass (Posidonia oceanica) flowering records. *Global Change Biology, 13*(1), 224–235. https://doi.org/10.1111/j.1365-2486.2006.01260.x

Diehl, N., Roleda, M. Y., Bartsch, I., Karsten, U., & Bischof, K. (2021). Summer heatwave impacts on the European kelp Saccharina latissima across its latitudinal distribution gradient. *Frontiers in Marine Science, 8*. https://doi.org/10.3389/fmars.2021.695821

Dobney, K., Baker, C. J., Quinn, A. D., & Chapman, L. (2009). Quantifying the effects of high summer temperatures due to climate change on buckling and rail related delays in south-East United Kingdom. *Meteorological Applications, 16*(2), 245–251. https://doi.org/10.1002/met.114

Dodrill, T. N., Pan, Y. D., & Peterson, T. D. (2023). Riverdischarge mediates extent of phytoplankton and harmful algal bloom habitat in the Columbia River estuary (USA) during North Pacific marine heat waves. *Estuaries and Coasts, 46*(1), 166–181. https://doi.org/10.1007/s12237-022-01129-y

Dong, X., West, G. H., Holloway-Beth, A., Wang, X. W., & Sokas, R. K. (2019). Heat-related deaths among construction workers in the United States. *American Journal of Industrial Medicine, 62*(12), 1047–1057. https://doi.org/10.1002/ajim.23024

Doshi, V. (2016). *Delhi schools ordered to close early for summer as temperatures soar*. The Guardian.

Drake, J. E., Tjoelker, M. G., Vårhammar, A., Medlyn, B. E., Reich, P. B., Leigh, A., Pfautsch, S., Blackman, C. J., López, R., Aspinwall, M. J., Crous, K. Y., Duursma, R. A., Kumarathunge, D., De Kauwe, M. G., Jiang, M., Nicotra, A. B., Tissue, D. T., Choat, B., Atkin, O. K., & Barton, C. V. M. (2018). Trees tolerate an extreme heatwave via sustained transpirational cooling and increased leaf thermal tolerance. *Global Change Biology, 24*(6), 2390–2402. https://doi.org/10.1111/gcb.14037

Dundas, S. J., Ruthrof, K. X., Hardy, G. E. S., & Fleming, P. A. (2021). Some like it hot: Drought-induced forest die-off influences reptile assemblages. *Acta Oecologica-International Journal of Ecology, 111*. https://doi.org/10.1016/j.actao.2021.103714

Durham, B. W., Wilde, G. R., & Pope, K. L. (2006). Temperature-caused fish kill in a flowing Great Plains river. *Southwestern Naturalist, 51*(3), 397–401. https://doi.org/10.1894/0038-4909(2006)51[397:tfkiaf]2.0.co;2

Dzyuban, Y., Ching, G. N. Y., Yik, S. K., Tan, A. J., Crank, P. J., Banerjee, S., Pek, R. X. Y., & Chow, W. T. L. (2022). Sentiment analysis of weather-related tweets from cities within hot climates. *Weather Climate and Society, 14*(4), 1133–1145. https://doi.org/10.1175/wcas-d-21-0159.1

Ebi, K. L., Capon, A., Berry, P., Broderick, C., de Dear, R., Havenith, G., Honda, Y., Kovats, R. S., Ma, W., Malik, A., Morris, N. B., Nybo, L., Seneviratne, S. I., Vanos, J., & Jay, O. (2021). Hot weather and heat extremes: Health risks. *Lancet, 398*(10301), 698–708.

Ellis-Petersen, H., & Baloch, S. (2022). *We are living in hell: Pakistan and India suffer extreme spring heatwaves*. The Guardian.

Enright NJ, Fontaine JB, Bowman DM, Bradstock RA, Williams RJ (2015) Interval squeeze: Altered fire regimes and demographic responses interact to threaten woody species persistence as climate changes. Frontiers in Ecology and the Environment 13 (5):265–272. doi:https://doi.org/https://doi.org/10.1890/140231.

Esperon-Rodriguez, M., Power, S. A., Tjoelker, M. G., Marchin, R. M., & Rymer, P. D. (2021). Contrasting heat tolerance of urban trees to extreme temperatures during heatwaves. *Urban Forestry & Urban Greening, 66*. https://doi.org/10.1016/j.ufug.2021.127387

Farchadi, N., Welch, H., Braun, C. D., Allyn, A. J., Bograd, S. J., Brodie, S., Hazen, E. L., Kerney, A., Lezama-Ochoa, N., Mills, K. E., Pugh, D., Young-Morse, R., & Lewison, R. L. (2024). Marine heatwaves redistribute pelagic fishing fleets. *Fish and Fisheries*. https://doi.org/10.1111/faf.12828

Fatima, S. H., Rothmore, P., Giles, L. C., Varghese, B. M., & Bi, P. (2021). Extreme heat and occupational injuries in different climate zones: A systematic review and meta-analysis of epidemiological evidence. *Environment International, 148*. https://doi.org/10.1016/j.envint.2021.106384

Ferranti, E., Chapman, L., Lowe, C., McCulloch, S., Jaroszweski, D., & Quinn, A. (2016). Heat-related failures on Southeast England's railway network: Insights and implications for heat risk management. *Weather Climate and Society, 8*(2), 177–191. https://doi.org/10.1175/wcas-d-15-0068.1

Ferranti, E., Chapman, L., Lee, S., Jaroszweski, D., Lowe, C., McCulloch, S., & Quinn, A. (2018). The hottest July day on the railway network: Insights and thoughts for the future. *Meteorological Applications, 25*(2), 195–208. https://doi.org/10.1002/met.1681

Filbee-Dexter, K., Wernberg, T., Grace, S. P., Thormar, J., Fredriksen, S., Narvaez, C. N., Feehan, C. J., & Norderhaug, K. M. (2020). Marine heatwaves and the collapse of marginal North Atlantic kelp forests. *Scientific Reports, 10*(1). https://doi.org/10.1038/s41598-020-70273-x

Filiz, N., Iskin, U., Beklioglu, M., Oglu, B., Cao, Y., Davidson, T. A., Sondergaard, M., Lauridsen, T. L., & Jeppesen, E. (2020). Phytoplankton community response to nutrients, temperatures, and a heat wave in Shallow Lakes: An experimental approach. *Water, 12*(12). https://doi.org/10.3390/w12123394

Fisher, M. C., Moore, S. K., Jardine, S. L., Watson, J. R., & Samhouri, J. F. (2021). Climate shock effects and mediation in fisheries. *Proceedings of the National Academy of Sciences of the United States of America, 118*(2). https://doi.org/10.1073/pnas.2014379117

Flouris, A. D., Dinas, P. C., Ioannou, L. G., Nybo, L., Havenith, G., Kenny, G. P., & Kjellstrom, T. (2018). Workers' health and productivity under occupational heat strain: A systematic review and meta-analysis. *Lancet Planetary Health, 2*(12), E521–E531. https://doi.org/10.1016/s2542-5196(18)30237-7

Fontana, G., Toreti, A., Ceglar, A., & De Sanctis, G. (2015). Early heat waves over Italy and their impacts on durum wheat yields. *Natural Hazards and Earth System Sciences, 15*(7), 1631–1637. https://doi.org/10.5194/nhess-15-1631-2015

Fordyce, A. J., Ainsworth, T. A., Heron, S. F., & Leggat, W. (2019). Marine heatwave hotspots in coral reef environments: Physical drivers, ecophysiological outcomes, and impact upon structural complexity. *Frontiers in Marine Science, 6*. https://doi.org/10.3389/fmars.2019.00498

Fourqurean, J. W., Duarte, C. M., Kennedy, H., Marba, N., Holmer, M., Mateo, M. A., Apostolaki, E. T., Kendrick, G. A., Krause-Jensen, D., McGlathery, K. J., & Serrano, O. (2012). Seagrass ecosystems as a globally significant carbon stock. *Nature Geoscience, 5*(7), 505–509. https://doi.org/10.1038/ngeo1477

Fox, M. D., Cohen, A. L., Rotjan, R. D., Mangubhai, S., Sandin, S. A., Smith, J. E., Thorrold, S. R., Dissly, L., Mollica, N. R., & Obura, D. (2021). Increasing coral reef resilience through successive marine heatwaves. *Geophysical Research Letters, 48*(17). https://doi.org/10.1029/2021gl094128

Fralick, M., Denny, C. J., & Redelmeier, D. A. (2013). Drowning and the influence of hot weather. *PLoS One, 8*(8), e71689.

Free, G., Bresciani, M., Pinardi, M., Simis, S., Liu, X. H., Albergel, C., & Giardino, C. (2022). Investigating lake chlorophyll-a responses to the 2019 European double heatwave using satellite remote sensing. *Ecological Indicators, 142*. https://doi.org/10.1016/j.ecolind.2022.109217

Frolicher, T. L., & Laufkotter, C. (2018). Emerging risks from marine heat waves. *Nature Communications, 9*. https://doi.org/10.1038/s41467-018-03163-6

Gabriele, C. M., Amundson, C. L., Neilson, J. L., Straley, J. M., Baker, C. S., & Danielson, S. L. (2022). Sharp decline in humpback whale (Megaptera novaeangliae) survival and reproductive success in southeastern Alaska during and after the 2014-2016 Northeast Pacific marine heatwave. *Mammalian Biology*. https://doi.org/10.1007/s42991-021-00187-2

Galvez, C., Pardo, M. A., & Elorriaga-Verplancken, F. R. (2020). Impacts of extreme ocean warming on the early development of a marine top predator: The Guadalupe fur seal. *Progress in Oceanography, 180*. https://doi.org/10.1016/j.pocean.2019.102220

Gamage, P. J., Fortington, L. V., & Finch, C. F. (2020). Epidemiology of exertional heat illnesses in organised sports: A systematic review. *Journal of Science and Medicine in Sport, 23*(8), 701–709. https://doi.org/10.1016/j.jsams.2020.02.008

Garcia-Leon, D., Casanueva, A., Standardi, G., Burgstall, A., Flouris, A. D., & Nybo, L. (2021). Current and projected regional economic impacts of heatwaves in Europe. Nature. *Communications, 12*(1). https://doi.org/10.1038/s41467-021-26050-z

Garrabou, J., Coma, R., Bensoussan, N., Bally, M., Chevaldonne, P., Cigliano, M., Diaz, D., Harmelin, J. G., Gambi, M. C., Kersting, D. K., Ledoux, J. B., Lejeusne, C., Linares, C., Marschal, C., Perez, T., Ribes, M., Romano, J. C., Serrano, E., Teixido, N., Torrents, O., Zabala, M., Zuberer, F., & Cerrano, C. (2009). Mass mortality in northwestern Mediterranean rocky benthic communities: Effects of the 2003 heat wave. *Global Change Biology, 15*(5), 1090–1103. https://doi.org/10.1111/j.1365-2486.2008.01823.x

Garrabou, J., Gomez-Gras, D., Medrano, A., Cerrano, C., Ponti, M., Schlegel, R., Bensoussan, N., Turicchia, E., Sini, M., Gerovasileiou, V., Teixido, N., Mirasole, A., Tamburello, L., Cebrian, E., Rilov, G., Ledoux, J. B., Ben Souissi, J., Khamassi, F., Ghanem, R., Benabdi, M., Grimes, S., Ocana, O., Bazairi, H., Hereu, B., Linares, C., Kersting, D. K., la Rovira, G., Ortega, J., Casals, D., Pages-Escola, M., Margarit, N., Capdevila, P., Verdura, J., Ramos, A., Izquierdo, A., Barbera, C., Rubio-Portillo, E., Anton, I., Lopez-Sendino, P., Diaz, D., Vazquez-Luis, M., Duarte, C., Marba, N., Aspillaga, E., Espinosa, F., Grech, D., Guala, I., Azzurro, E., Farina, S., Gambi, M. C., Chimienti, G., Montefalcone, M., Azzola, A., Mantas, T. P., Fraschetti, S., Ceccherelli, G., Kipson, S., Bakran-Petricioli, T., Petricioli, D., Jimenez, C., Katsanevakis, S., Kizilkaya, I. T., Kizilkaya, Z., Sartoretto, S., Elodie, R., Ruitton, S., Comeau, S., Gattuso, J. P., & Harmelin, J. G. (2022). Marine heatwaves drive recurrent mass mortalities in the Mediterranean Sea. *Global Change Biology, 28*(19), 5708–5725. https://doi.org/10.1111/gcb.16301

Geerts, E. (2022). *Are railways prepared for more frequent heat waves?* RailTech.

Genevier, L. G. C., Jamil, T., Raitsos, D. E., Krokos, G., & Hoteit, I. (2019). Marine heatwaves reveal coral reef zones susceptible to bleaching in the Red Sea. *Global Change Biology, 25*(7), 2338–2351. https://doi.org/10.1111/gcb.14652

Geranios N, Selsky A (2021) Blackouts in US northwest due to heat wave, deaths reported. AP News,

Gerard, M., Amiri, A., Cariou, B., & Baird, E. (2022). Short-term exposure to heatwave-like temperatures affects learning and memory in bumblebees. *Global Change Biology, 28*(14), 4251–4259. https://doi.org/10.1111/gcb.16196

Gerrard, I. (2022). *How people stuck in Hull's high rise flats suffer the most in record-breaking heatwave*. Hull Daily Mail.

Goggins, W. B., & Chan, E. Y. Y. (2017). A study of the short-term associations between hospital admissions and mortality from heart failure and meteorological variables in Hong Kong weather and heart failure in Hong Kong. *International Journal of Cardiology, 228*, 537–542. https://doi.org/10.1016/j.ijcard.2016.11.106

Gols, R., Ojeda-Prieto, L. M., Li, K., Van der Putten, W. H., & Harvey, J. A. (2021). Within-patch and edge microclimates vary over a growing season and are amplified during a heatwave: Consequences for ectothermic insects. *Journal of Thermal Biology, 99*. https://doi.org/10.1016/j.jtherbio.2021.103006

Greenham, S., Ferranti, E., Quinn, A., & Drayson, K. (2020). The impact of high temperatures and extreme heat to delays on the London underground rail network: An empirical study. *Meteorological Applications, 27*, 3. https://doi.org/10.1002/met.1910

Gregory A (2022) At least 10 people die in open water during UK's searing heatwave. Independent.

Grundstein, A., Knox, J. A., Vanos, J., Cooper, E. R., & Casa, D. J. (2017). American football and fatal exertional heat stroke: A case study of Korey stringer. *International Journal of Biometeorology, 61*(8), 1471–1480. https://doi.org/10.1007/s00484-017-1324-2

Grymes, A., & Westbrook, E. (2022). *NYC heat wave: Utility companies working to keep up with demand and keep grid online New York.* CBS News.

Guardian, T. (2003). *Heatwave hits French power production*. The Guardian.

Gurgel, C. F. D., Camacho, O., Minne, A. J. P., Wernberg, T., & Coleman, M. A. (2020). Marine heatwave drives cryptic loss of genetic diversity in underwater forests. *Current Biology, 30*(7), 1199. https://doi.org/10.1016/j.cub.2020.01.051

Haberstroh, S., Werner, C., Grun, M., Kreuzwieser, J., Seifert, T., Schindler, D., & Christen, A. (2022). Central European 2018 hot drought shifts scots pine forest to its tipping point. *Plant Biology, 24*(7), 1186–1197. https://doi.org/10.1111/plb.13455

Hahn, G. L. (1999). Dynamic responses of cattle to thermal heat loads. *Journal of Animal Science, 77*, 10–20.

Hall, E. J., Carter, A. J., & O'Neill, D. G. (2020). Dogs don't die just in hot cars-exertional heat-related illness (heatstroke) is a greater threat to UK dogs. *Animals, 10*(8). https://doi.org/10.3390/ani10081324

Hall, E., Radford, A., & Carter, A. (2022). Surveillance of heat-related illness in small animals presenting to veterinary practices in the UK between 2013 and 2018. *Open Veterinary Journal, 12*(1), 5–16. https://doi.org/10.5455/OVJ.2022.v12.i1.2

Hammond, W. M., Yu, K. L., Wilson, L. A., Will, R. E., Anderegg, W. R. L., & Adams, H. D. (2019). Dead or dying? Quantifying the point of no return from hydraulic failure in drought-induced tree mortality. *New Phytologist, 223*(4), 1834–1843. https://doi.org/10.1111/nph.15922

Hansen, A., Bi, P., Nitschke, M., Ryan, P., Pisaniello, D., & Tucker, G. (2008). The effect of heat waves on mental health in a temperate Australian city. *Environmental Health Perspectives, 116*(10), 1369–1375. https://doi.org/10.1289/ehp.11339

Harrison, D. (2019). *What caused the blackouts in Melbourne, and do Victorians need to get used to power cuts?*

Harvey, J. A., Heinen, R., Gols, R., & Thakur, M. P. (2020). Climate change-mediated temperature extremes and insects: From outbreaks to breakdowns. *Global Change Biology, 26*(12), 6685–6701. https://doi.org/10.1111/gcb.15377

Hasegawa, N., Watanabe, T., Unuma, T., Yokota, T., Izumida, D., Nakagawa, T., Kurokawa, T., Takagi, S., Azumaya, T., Taniuchi, Y., Kuroda, H., Kitatsuji, S., & Abe, K. (2022). Repeated reaching of the harmful algal bloom of Karenia spp. around the Pacific shoreline of Kushiro, eastern Hokkaido, Japan, during autumn 2021. *Fisheries Science, 88*(6), 787–803. https://doi.org/10.1007/s12562-022-01642-w

Hatakeyama, K., & Seposo, X. (2022). Heatstroke-related ambulance dispatch risk before and during COVID-19 pandemic: Subgroup analysis by age, severity, and incident place. *Science of the Total Environment, 821*. https://doi.org/10.1016/j.scitotenv.2022.153310

Hatakeyama, K., Ota, J., Takahashi, Y., Kawamitsu, S., & Seposo, X. (2021). Effect of the COVID-19 pandemic on heatstroke-related ambulance dispatch in the 47 prefectures of Japan. *Science of the Total Environment, 768*. https://doi.org/10.1016/j.scitotenv.2021.145176

Hatvani-Kovacs, G., Belusko, M., Pockett, J., & Boland, J. (2016). Assessment of heatwave impacts. *Fourth International Conference on Countermeasures to Urban Heat Island, (Uhi 2016), 169*, 316–323. https://doi.org/10.1016/j.proeng.2016.10.039

He, G. X., Peng, Y. L., Liu, X. L., Liu, Y., Liang, J., Xu, X., Yang, K., Masanja, F., Xu, Y., Deng, Y. W., & Zhao, L. Q. (2022a). Post-responses of intertidal bivalves to recurrent heatwaves. *Marine Pollution Bulletin, 184*. https://doi.org/10.1016/j.marpolbul.2022.114223

He, G. X., Zou, J., Liu, X. L., Liang, F. L., Liang, J., Yang, K., Masanja, F., Xu, Y., Zheng, Z., Deng, Y. W., & Zhao, L. Q. (2022b). Assessing the impact of atmospheric heatwaves on intertidal clams. *Science of the Total Environment, 841*. https://doi.org/10.1016/j.scitotenv.2022.156744

Henderson, S. B., McLean, K. E., Lee, M. J., & Kosatsky, T. (2022). Analysis of community deaths during the catastrophic 2021 heat dome: Early evidence to inform the public health response during subsequent events in greater Vancouver, Canada. *Environmental Epidemiology, 6*(1). https://doi.org/10.1097/ee9.0000000000000189

Hess, J. J., Errett, N. A., McGregor, G., Isaksen, T. B., Wettstein, Z. S., Wheat, S. K., & Ebi, K. L. (2022). Public health preparedness for extreme heat events. *Annual Review of Public Health, 44*, 301–321. https://doi.org/10.1146/annurev-publhealth-071421-025508

Heyes, A., & Saberian, S. (2019). Temperature and decisions: Evidence from 207,000 court Cases. *American Economic Journal-Applied Economics, 11*(2), 238–265. https://doi.org/10.1257/app.20170223

Hills, S. P., Hobbs, M., Brown, P., Tipton, M., & Barwood, M. (2024). Association between air temperature and unintentional drowning risk in the United Kingdom 2012-2019: A nationwide case-crossover study. *Preventive Medicine, 179*. https://doi.org/10.1016/j.ypmed.2023.107832

Holbrook, N. J., Hernaman, V., Koshiba, S., Lako, J., Kajtar, J. B., Amosa, P., & Singh, A. (2022). Impacts of marine heatwaves on tropical western and Central Pacific Island nations and their communities. *Global and Planetary Change, 208*. https://doi.org/10.1016/j.gloplacha.2021.103680

Holt, K. A., & Boersma, P. D. (2022). Unprecedented heat mortality of Magellanic penguins. *Ornithological Applications, 124*(1). https://doi.org/10.1093/ornithapp/duab052

Hoogenboom, M. O., Frank, G. E., Chase, T. J., Jurriaans, S., Alvarez-Noriega, M., Peterson, K., Crítchell, K., Berry, K. L. E., Nicolet, K. J., Ramsby, B., & Paley, A. S. (2017). Environmental drivers of variation in bleaching severity of Acropora species during an extreme thermal anomaly. *Frontiers in Marine Science, 4*. https://doi.org/10.3389/fmars.2017.00376

Hopkins, A. J. M., Ruthrof, K. X., Fontaine, J. B., Matusick, G., Dundas, S. J., & Hardy, G. E. S. (2018). Forest die-off following global-change-type drought alters rhizosphere fungal communities. *Environmental Research Letters, 13*(9). https://doi.org/10.1088/1748-9326/aadc19

Hou, K., Zhang, L. Q., Xu, X., Yang, F., Chen, B. Z., Hu, W., & Shu, R. (2023). High ambient temperatures are associated with urban crime risk in Chicago. *Science of the Total Environment, 856*. https://doi.org/10.1016/j.scitotenv.2022.158846

Hu, J., He, G., Meng, R., Gong, W., Ren, Z., Shi, H., Lin, Z., Liu, T., Zeng, F., Yin, P., Bai, G., Qin, M., Hou, Z., Dong, X., Zhou, C., Pingcuo, Z., Xiao, Y., Yu, M., Huang, B., Xu, X., Lin, L., Xiao, J., Zhong, J., Jin, D., Zhao, Q., Li, Y., Gama, C., Xu, Y., Lv, L., Zeng, W., Li, X., Luo, L., Zhou, M., Huang, C., & Ma, W. (2023). Temperature-related mortality in China from specific injury. Nature. *Communications, 14*(1). https://doi.org/10.1038/s41467-022-35462-4

Huang, Z., Li, Z., Hu, J., Zhu, S., Gong, W., Zhou, C., Meng, R., Dong, X., Yu, M., Xu, X., Lin, L., Xiao, J., Zhong, J., Jin, D., Xu, Y., Liu, T., Lin, Z., He, G., & Ma, W. (2023). The association of heatwave with drowning mortality in five provinces of China. *Science of the Total Environment, 903*. https://doi.org/10.1016/j.scitotenv.2023.166321

Huber, V., Adrian, R., & Gerten, D. (2010). A matter of timing: Heat wave impact on crustacean zooplankton. *Freshwater Biology, 55*(8), 1769–1779. https://doi.org/10.1111/j.1365-2427.2010.02411.x

Huber, V., Wagner, C., Gerten, D., & Adrian, R. (2012). To bloom or not to bloom: Contrasting responses of cyanobacteria to recent heat waves explained by critical thresholds of abiotic drivers. *Oecologia, 169*(1), 245–256. https://doi.org/10.1007/s00442-011-2186-7

Ioannou, L. G., Mantzios, K., Tsoutsoubi, L., Panagiotaki, Z., Kapnia, A. K., Ciuha, U., Nybo, L., Flouris, A. D., & Mekjavic, I. B. (2021). Effect of a simulated heat wave on physiological strain and labour productivity. *International Journal of Environmental Research and Public Health, 18*(6). https://doi.org/10.3390/ijerph18063011

Iskin, U., Filiz, N., Cao, Y., Neif, E. M., Oglu, B., Lauridsen, T. L., Davidson, T. A., Sondergaard, M., Tavsanoglu, U. N., Beklioglu, M., & Jeppesen, E. (2020). Impact of nutrients, temperatures, and a heat wave on zooplankton community structure: An experimental approach. *Water, 12*(12). https://doi.org/10.3390/w12123416

Jankowski, T., Livingstone, D. M., Buhrer, H., Forster, R., & Niederhauser, P. (2006). Consequences of the 2003 European heat wave for lake temperature profiles, thermal stability, and hypolimnetic oxygen depletion: Implications for a warmer world. *Limnology and Oceanography, 51*(2), 815–819. https://doi.org/10.4319/lo.2006.51.2.0815

Jardine, S. L., Fisher, M. C., Moore, S. K., & Samhouri, J. F. (2020). Inequality in the economic impacts from climate shocks in fisheries: The case of harmful algal blooms. *Ecological Economics, 176*. https://doi.org/10.1016/j.ecolecon.2020.106691

Jeppesen, E., Audet, J., Davidson, T. A., Neif, E. M., Cao, Y., Filiz, N., Lauridsen, T. L., Larsen, S. E., Beklioglu, M., Sh, T., & Sondergaard, M. (2021a). Nutrient loading, temperature and heat wave effects on nutrients, oxygen and metabolism in shallow lake mesocosms pre-adapted for 11 years. *Water, 13*(2). https://doi.org/10.3390/w13020127

Jeppesen, E., Pierson, D., & Jennings, E. (2021b). Effect of extreme climate events on lake ecosystems. *Water, 13*(3). https://doi.org/10.3390/w13030282

Jiguet, F., Julliard, R., Thomas, C. D., Dehorter, O., Newson, S. E., & Couvet, D. (2006). Thermal range predicts bird population resilience to extreme high temperatures. *Ecology Letters, 9*(12), 1321–1330. https://doi.org/10.1111/j.1461-0248.2006.00986.x

Johnk, K. D., Huisman, J., Sharples, J., Sommeijer, B., Visser, P. M., & Stroom, J. M. (2008). Summer heatwaves promote blooms of harmful cyanobacteria. *Global Change Biology, 14*(3), 495–512. https://doi.org/10.1111/j.1365-2486.2007.01510.x

Johnson, M. R., Williams, S. L., Lieberman, C. H., & Solbak, A. (2003). Changes in the abundance of the seagrasses Zostera marina L. (eelgrass) and Ruppia maritima L. (widgeongrass) in San Diego, California, following an El Nino event. *Estuaries, 26*(1), 106–115. https://doi.org/10.1007/bf02691698

Johnson, H., Kovats, R. S., McGregor, G., Stedman, J., Gibbs, M., & Walton, H. (2005). The impact of the 2003 heat wave on daily mortality in England and Wales and the use of rapid weekly mortality estimates. *Euro surveillance : Bulletin Europeen Sur les maladies transmissibles = European Communicable Disease Bulletin, 10*(7), 168–171.

Jones, T., Parrish, J. K., Peterson, W. T., Bjorkstedt, E. P., Bond, N. A., Ballance, L. T., Bowes, V., Hipfner, J. M., Burgess, H. K., Dolliver, J. E., Lindquist, K., Lindsey, J., Nevins, H. M., Robertson, R. R., Roletto, J., Wilson, L., Joyce, T., & Harvey, J. (2018). Massive mortality of a planktivorous seabird in response to a marine heatwave. *Geophysical Research Letters, 45*(7), 3193–3202. https://doi.org/10.1002/2017gl076164

Jowit, J., & Espinoza, J. (2006). *Heatwave shuts down nuclear power plants*. The Guardian.

Jurgens, R. (2006). Outages identify PG&E's limits: After heat wave caused 1.2 million customers to lose power, experts assess utility's vulnerabilities. *Contra Costa Times* (Walnut Creek, CA),

Kauppi, L., & Villnas, A. (2022, 1986). Marine heatwaves of differing intensities lead to distinct patterns in seafloor functioning. *Proceedings of the Royal Society B-Biological Sciences, 289*. https://doi.org/10.1098/rspb.2022.1159

Kauppi, L., Gobeler, N., Norkko, J., Norkko, A., Romero-Ramirez, A., & Bernard, G. (2023). Changes in macrofauna bioturbation during repeated heatwaves mediate changes in biogeochemical cycling of nutrients. *Frontiers in Marine Science, 9*. https://doi.org/10.3389/fmars.2022.1070377

Kendrick, G. A., Nowicki, R. J., Olsen, Y. S., Strydom, S., Fraser, M. W., Sinclair, E. A., Statton, J., Hovey, R. K., Thomson, J. A., Burkholder, D. A., McMahon, K. M., Kilminster, K., Hetzel, Y., Fourqurean, J. W., Heithaus, M. R., & Orth, R. J. (2019). A systematic review of how multiple stressors from an extreme event drove ecosystem-wide loss of resilience in an iconic seagrass community. *Frontiers in Marine Science, 6*. https://doi.org/10.3389/fmars.2019.00455

Kerr, Z. Y., Yeargin, S. W., Hosokawa, Y., Hirschhorn, R. M., Pierpoint, L. A., & Casa, D. J. (2020). The epidemiology and management of exertional heat illnesses in high school sports during the 2012/2013–2016/2017 academic years. *Journal of Sport Rehabilitation, 29*(3), 332–338. https://doi.org/10.1123/jsr.2018-0364

Kim, Y., Kim, H., Gasparrini, A., Armstrong, B., Honda, Y., Chung, Y., Ng, C. F. S., Tobias, A., Iniguez, C., Lavigne, E., Sera, F., Vicedo-Cabrera, A. M., Ragettli, M. S., Scovronick, N., Acquaotta, F., Chen, B. Y., Guo, Y. L. L., Seposo, X., Dang, T. N., Coelho, M., Saldiva, P. H. N., Kosheleva, A., Zanobetti, A., Schwartz, J., Bell, M. L., & Hashizume, M. (2019). Suicide and ambient temperature: A multi-country multi-city study. *Environmental Health Perspectives, 127*(11). https://doi.org/10.1289/ehp4898

Klimenko, V. V., Ginzburg, A. S., Fedotova, E. V., & Tereshin, A. G. (2020). Heat waves: A new danger for the Russian power system. *Doklady Physics, 65*(9), 349–354. https://doi.org/10.1134/s1028335820090050

Klingelhöfer, D., Braun, M., Brüggmann, D., & Groneberg, D. A. (2023). Heatwaves: Does global research reflect the growing threat in the light of climate change? *Globalization and Health, 19*(1), 56.

Knowlton, K., Rotkin-Ellman, M., King, G., Margolis, H. G., Smith, D., Solomon, G., Trent, R., & English, P. (2009). The 2006 California heat wave: Impacts on hospitalizations and emergency department visits. *Environmental Health Perspectives, 117*(1), 61–67. https://doi.org/10.1289/ehp.11594

Kourantidou, M., Jin, D., & Schumacker, E. J. (2022). Socioeconomic disruptions of harmful algal blooms in indigenous communities: The case of Quinault Indian nation. *Harmful Algae, 118*. https://doi.org/10.1016/j.hal.2022.102316

Krich, C., Mahecha, M. D., Migliavacca, M., De Kauwe, M. G., Griebel, A., Runge, J., & Miralles, D. G. (2022). Decoupling between ecosystem photosynthesis and transpiration: A last resort against overheating. *Environmental Research Letters, 17*(4). https://doi.org/10.1088/1748-9326/ac583e

Kubo, R., Ueda, K., Seposo, X., Honda, A., & Takano, H. (2021). Association between ambient temperature and intentional injuries: A case-crossover analysis using ambulance transport records in Japan. *Science of the Total Environment, 774*. https://doi.org/10.1016/j.scitotenv.2021.145511

Kulkarni N (2018) Mumbai: Rail tracks bend due to heat, leaves used to cool them down. The Indian Express,

Kuroda, H., & Setou, T. (2021). Extensive marine heatwaves at the sea surface in the northwestern Pacific Ocean in summer 2021. *Remote Sensing, 13*(19). https://doi.org/10.3390/rs13193989

Kuroda, H., Azumaya, T., Setou, T., & Hasegawa, N. (2021). Unprecedented outbreak of harmful algae in Pacific coastal waters off Southeast Hokkaido, Japan, during late summer 2021 after record-breaking marine heatwaves. *Journal of Marine Science and Engineering, 9*(12). https://doi.org/10.3390/jmse9121335

Lamarca, D. S. F., Pereira, D. F., Magalhaes, M. M., & Salgado, D. D. (2018). Climate change in layer poultry farming: Impact of heat waves in region of Bastos. *Brazil. Brazilian Journal of Poultry Science, 20*(4), 657–664. https://doi.org/10.1590/1806-9061-2018-0750

Lane, K., Smalls-Mantey, L., Hernández, D., Watson, S., Jessel, S., Jack, D., Spaulding, L., & Olson, C. (2023). Extreme heat and COVID-19 in new York City: An evaluation of a large air conditioner distribution program to address compounded public health risks in summer 2020. *Journal of Urban Health-Bulletin of the New York Academy of Medicine, 100*(2), 290–302. https://doi.org/10.1007/s11524-022-00704-9

Lang, B. T. J., Donelson, J. M., Caballes, C. F., Doll, P. C., & Pratchett, M. S. (2021). Metabolic responses of Pacific crown-of-Thorns Sea stars (Acanthaster sp.) to acute warming. *Biological Bulletin, 241*(3), 347–358. https://doi.org/10.1086/717049

Lazo, J. K., Lawson, M., Larsen, P. H., & Waldman, D. M. (2011). US economic sensitivity to weather variability. *Bulletin of the American Meteorological Society, 92*(6), 709–720. https://doi.org/10.1175/2011bams2928.1

Le, K., & Nguyen, M. (2021). The mpacts of temperature shocks on birth weight in Vietnam. *Population and Development Review, 47*(4), 1025–1047. https://doi.org/10.1111/padr.12428

Lee, J., Lee, Y. H., Choi, W. J., Ham, S., Kang, S. K., Yoon, J. H., Yoon, M. J., Kang, M. Y., & Lee, W. (2022). Heat exposure and workers' health: A systematic review. *Reviews on Environmental Health, 37*(1), 45–59. https://doi.org/10.1515/reveh-2020-0158

Lees, A. M., Sejian, V., Wallage, A. L., Steel, C. C., Mader, T. L., Lees, J. C., & Gaughan, J. B. (2019). The impact of heat load on cattle. *Animals, 9*(6). https://doi.org/10.3390/ani9060322

Leggat, W. P., Camp, E. F., Suggett, D. J., Heron, S. F., Fordyce, A. J., Gardner, S., Deakin, L., Turner, M., Beeching, L. J., Kuzhiumparambil, U., Eakin, C. M., & Ainsworth, T. D. (2019). Rapid coral decay is associated with marine heatwave mortality events on reefs. *Current Biology, 29 (16):2723-+*. https://doi.org/10.1016/j.cub.2019.06.077

Leigh, C., Bush, A., Harrison, E. T., Ho, S. S., Luke, L., Rolls, R. J., & Ledger, M. E. (2015). Ecological effects of extreme climatic events on riverine ecosystems: Insights from Australia. *Freshwater Biology, 60*(12), 2620–2638. https://doi.org/10.1111/fwb.12515

Létard V, Flandre H, Lepeltier S (2004) La France et les Français face à la canicule: les leçons d'une crise. Report No. 195 (2003-2004) to the Sénat, Government of France.

Li, Z. Q., He, L., Zhang, H., Urrutia-Cordero, P., Ekvall, M. K., Hollander, J., & Hansson, L. A. (2017). Climate warming and heat waves affect reproductive strategies and interactions between submerged macrophytes. *Global Change Biology, 23*(1), 108–116. https://doi.org/10.1111/gcb.13405

Li, N., Zhang, Y. L., Zhang, Y. B., Shi, K., Qian, H. M., Yang, H. Y., Niu, Y. K., Qin, B. Q., Zhu, G. W., Jeppesen, E., & Woolway, R. I. (2023). The unprecedented 2022 extreme summer heatwaves increased harmful cyanobacteria blooms. *Science of the Total Environment, 896*. https://doi.org/10.1016/j.scitotenv.2023.165312

Lim, Y. K., Park, B. S., Kim, J. H., Baek, S. S., & Baek, S. H. (2021). Effect of marine heatwaves on bloom formation of the harmful dinoflagellate Cochlodinium polykrikoides: Two sides of the same coin? *Harmful Algae, 104*. https://doi.org/10.1016/j.hal.2021.102029

Liu, J., Varghese, B., Hansen, A., & Bi, P. (2021). Effects of high temperatures on poor mental health outcomes: A systematic review and meta-analysis. *European Journal of Public Health, 31*.

Lohmus, M. (2018). Possible biological mechanisms linking mental health and heat-a contemplative review. *International Journal of Environmental Research and Public Health, 15*(7). https://doi.org/10.3390/ijerph15071515

Lucey, N., Aube, C., Herwig, A., & Collin, R. (2022). Compound extreme events induce rapid mortality in a Tropical Sea urchin. *The Biological Bulletin, 243*(2), 239–254. https://doi.org/10.1086/722283

Ma, C. S., Ma, G., & Pincebourde, S. (2021). Survive a warming climate: Insect responses to extreme high temperatures. *Annual Review of Entomology, 66*(66), 163–184. https://doi.org/10.1146/annurev-ento-041520-074454

Maazouzi, C., Masson, G., Izquierdo, M. S., & Pihan, J. C. (2008). Midsummer heat wave effects on lacustrine plankton: Variation of assemblage structure and fatty acid composition. *Journal of Thermal Biology, 33*(5), 287–296. https://doi.org/10.1016/j.jtherbio.2008.03.002

Macaulay, S. J., Hageman, K. J., Piggott, J. J., & Matthaei, C. D. (2021). Time-cumulative effects of neonicotinoid exposure, heatwaves and food limitation on stream mayfly nymphs: A multiple-stressor experiment. *Science of the Total Environment, 754*. https://doi.org/10.1016/j.scitotenv.2020.141941

Majumdar, R. (2022). *India's stray animals suffer under record heat.* DW In Focus.

Manica, E., Coltri, P. P., Pacheco, V. M., & Martello, L. S. (2022). Changes in the pattern of heat waves and the impacts on Holstein cows in a subtropical region. *International Journal of Biometeorology*. https://doi.org/10.1007/s00484-022-02374-3

Mann, J., Foroughirad, V., McEntee, M. H. F., Miketa, M. L., Evans, T. C., Karniski, C., Krzyszczyk, E., Patterson, E. M., Strohman, J. C., & Wallen, M. M. (2021). Elevated calf mortality and long-term responses of wild bottlenose dolphins to extreme climate events: Impacts of foraging specialization and provisioning. *Frontiers in Marine Science, 8*. https://doi.org/10.3389/fmars.2021.617550

Marba, N., & Duarte, C. M. (2010). Mediterranean warming triggers seagrass (Posidonia oceanica) shoot mortality. *Global Change Biology, 16*(8), 2366–2375. https://doi.org/10.1111/j.1365-2486.2009.02130.x

Marshall, A. (2021). *Oregon's buckled roads and melted cables are warning signs*. WIRED.

Marzonie, M. R., Bay, L. K., Bourne, D. G., Hoey, A. S., Matthews, S., Nielsen, J. J. V., & Harrison, H. B. (2023). The effects of marine heatwaves on acute heat tolerance in corals. *Global Change Biology, 29*(2), 404–416. https://doi.org/10.1111/gcb.16473

Mason, H., King, J. C., Peden, A. E., & Franklin, R. C. (2022). Systematic review of the impact of heatwaves on health service demand in Australia. *BMC Health Services Research, 22*(1). https://doi.org/10.1186/s12913-022-08341-3

Matthews, W. J., Surat, E., & Hill, L. G. (1982). Heat death of the ORANGETHROAT darter ETHEOSTOMA-SPECTABILE (PERCIDAE) in a natural-environment. *Southwestern Naturalist, 27*(2), 216–217. https://doi.org/10.2307/3671147

Matusick, G., Ruthrof, K. X., Kala, J., Brouwers, N. C., Breshears, D. D., & Hardy, G. E. S. (2018). Chronic historical drought legacy exacerbates tree mortality and crown dieback during acute heatwave-compounded drought. *Environmental Research Letters, 13*(9). https://doi.org/10.1088/1748-9326/aad8cb

McElroy, S., Ilango, S., Dimitrova, A., Gershunov, A., & Benmarhnia, T. (2022). Extreme heat, preterm birth, and stillbirth: A global analysis across 14 lower-middle income countries. *Environment International, 158*. https://doi.org/10.1016/j.envint.2021.106902

McEvoy, D., Ahmed, I., & Mullett, J. (2012). The impact of the 2009 heat wave on Melbourne's critical infrastructure. *Local Environment, 17*(8), 783–796. https://doi.org/10.1080/13549839.2012.678320

McGregor, G. R., & Vanos, J. K. (2018). Heat: A primer for public health researchers. *Public Health, 161*, 138–146. https://doi.org/10.1016/j.puhe.2017.11.005

McKechnie, A. E., Rushworth, I. A., Myburgh, F., & Cunningham, S. J. (2021). Mortality among birds and bats during an extreme heat event in eastern South Africa. *Austral Ecology, 46*(4), 687–691. https://doi.org/10.1111/aec.13025

McMahon, S., Gamage, P., & Fortington, L. V. (2021). Sports related heat injury in Victoria, Australia: An analysis of 11 years of hospital admission and emergency department data. *Journal of Science and Medicine in Sport, 24*(3), 224–228. https://doi.org/10.1016/j.jsams.2020.08.011

McPherson, M. L., Finger, D. J. I., Houskeeper, H. F., Bell, T. W., Carr, M. H., Rogers-Bennett, L., & Kudela, R. M. (2021). Large-scale shift in the structure of a kelp forest ecosystem co-occurs with an epizootic and marine heatwave. *Communications Biology, 4*(1). https://doi.org/10.1038/s42003-021-01827-6

McRae, M., Lee, R. A., Steinschneider, S., & Galgano, F. (2021). Assessing aircraft performance in a warming climate. *Weather Climate and Society, 13*(1), 39–55. https://doi.org/10.1175/wcas-d-20-0098.1

McTavish, R. K., Richard, L., McArthur, E., Shariff, S. Z., Acedillo, R., Parikh, C. R., Wald, R., Wilk, P., & Garg, A. X. (2018). Association between high environmental heat and risk of acute kidney injury among older adults in a northern climate: A matched case-control study. *American Journal of Kidney Diseases, 71*(2), 200–208. https://doi.org/10.1053/j.ajkd.2017.07.011

Merone, L., Thirlwell, C., Esmonde, J., & Gair, R. (2020). A mass mortality event in bats caused by extreme heat: Surprising public health challenges. *Public Health Research & Practice, 30*(4). https://doi.org/10.17061/phrp3042032

Mo, M., Roache, M., Davies, J., Hopper, J., Pitty, H., Foster, N., Guy, S., Parry-Jones, K., Francis, G., Koosmen, A., Colefax, L., Costello, C., Stokes, J., Curran, S., Smith, M., Daly, G., Simmons, C. M., Hansen, R., Prophet, D., Judge, S., Major, F., Hogarth, T., McGarry, C. A., Pope, L., Brend, S., Coxon, D., Baker, K., Kaye, K., Collins, L., Wallis, M., Brown, R., Roberts, L., Taylor, S., Pearson, T., Bishop, T., Dunne, P., Coutts-McClelland, K., Oliver, L., Dawe, C.,

& Welbergen, J. A. (2022). Estimating flying-fox mortality associated with abandonments of pups and extreme heat events during the austral summer of 2019-20. *Pacific Conservation Biology, 28*(2), 124–139. https://doi.org/10.1071/pc21003

Moon, S., Kwon, J., Jung, S. U., & Bae, Y. H. (2018). The impact of individual differences in weather sensitivity on weather-related purchase intentions. *International Journal of Market Research, 60*(1), 104–117. https://doi.org/10.1177/1470785317744855

Moore, M. E., Hill, C. A., & Kingsolver, J. G. (2022). Developmental timing of extreme temperature events (heat waves) disrupts host-parasitoid interactions. *Ecology and Evolution, 12*, 3. https://doi.org/10.1002/ece3.8618

Mora, C., Counsell, C. W. W., Bielecki, C. R., & Louis, L. V. (2017). Twenty-seven ways a heat wave can kill you: Deadly heat in the era of climate change. *Circulation Cardiovascular Quality and Outcomes, 10*(11). https://doi.org/10.1161/circoutcomes.117.004233

Morignat, E., Perrin, J. B., Gay, E., Vinard, J. L., Calavas, D., & Henaux, V. (2014). Assessment of the impact of the 2003 and 2006 heat waves on cattle mortality in France. *PLoS One, 9*(3). https://doi.org/10.1371/journal.pone.0093176

Morris, N. B., Piil, J. F., Morabito, M., Messeri, A., Levi, M., Ioannou, L. G., Ciuha, U., Pogacar, T., Bogataj, L. K., Kingma, B., Casanueva, A., Kotlarski, S., Spirig, C., Foster, J., Havenith, G., Mayor, T. S., Flouris, A. D., & Nybo, L. (2021). The HEAT-SHIELD project – Perspectives from an inter-sectoral approach to occupational heat stress. *Journal of Science and Medicine in Sport, 24*(8), 747–755. https://doi.org/10.1016/j.jsams.2021.03.001

Mouthon, J., & Daufresne, M. (2006). Effects of the 2003 heatwave and climatic warming on mollusc communities of the Saône: A large lowland river and of its two main tributaries (France). *Global Change Biology, 12*.

Mundahl, N. D. (1990). Heat death of fish in shrinking stream pools. *American Midland Naturalist, 123*(1), 40–46. https://doi.org/10.2307/2425758

Nakamura, D., Kinoshita, H., Asada, K., Arimitsu, T., Yasumatsu, M., & Ishiwata, T. (2022). Trends in ambulance dispatches related to heat illness from 2010 to 2019: An ecological study. *PLoS One, 17*(11), e0275641.

Narayan, E. J., & Williams, M. (2016). Understanding the dynamics of physiological impacts of environmental stressors on Australian marsupials, focus on the koala (Phascolarctos cinereus). *BMC Zoology, 1*. https://doi.org/10.1186/s40850-016-0004-8

Nauer, F., Oliveira, M. C., Plastino, E. M., Yokoya, N. S., & Fujii, M. T. (2022). Coping with heatwaves: How a key species of seaweed responds to heat stress along its latitudinal gradient. *Marine Environmental Research, 177*. https://doi.org/10.1016/j.marenvres.2022.105620

Ngu, F. F., Kelman, I., Chambers, J., & Ayeb-Karlsson, S. (2021). Correlating heatwaves and relative humidity with suicide (fatal intentional self-harm). *Scientific Reports, 11*(1). https://doi.org/10.1038/s41598-021-01448-3

Niazi, S. (2016). *Armed guards protect last water in drought-parched Indian city*. Thomson Reuters Foundation News.

Niedzwiedz, S., Diehl, N., Fischer, P., & Bischof, K. (2022). Seasonal and inter-annual variability in the heatwave tolerance of the kelp Saccharina latissima (Laminariales, Phaeophyceae). *Phycological Research, 70*(4), 212–222. https://doi.org/10.1111/pre.12501

Niu, L., Girma, B., Liu, B., Schinasi, L. H., Clougherty, J. E., & Sheffield, P. (2023). Temperature and mental health-related emergency department and hospital encounters among children, adolescents and young adults. *Epidemiology and Psychiatric Sciences, 32*. https://doi.org/10.1017/s2045796023000161

Nori-Sarma, A., Milando, C., Weinberger, K. R., Hess, J. J., Errett, N. A., & Wellenius, G. A. (2022a). Association between the 2021 heat wave in Portland, Oregon, and Seattle, Washington, and emergency department visits. *Jama-Journal of the American Medical Association, 328*(23), 2360–2362. https://doi.org/10.1001/jama.2022.20665

Nori-Sarma, A., Sun, S. Z., Sun, Y. T., Spangler, K. R., Oblath, R., Galea, S., Gradus, J. L., & Wellenius, G. A. (2022b). Association between ambient heat and risk of emergency department visits for mental health among US adults, 2010 to 2019. *JAMA Psychiatry, 79*(4), 341–349. https://doi.org/10.1001/jamapsychiatry.2021.4369

NR. (2017). *Weather resilience and climate change adaptation strategy 2017–2019*. Network Rail.

Nyadanu, S. D., Dunne, J., Tessema, G. A., Mullins, B., Kumi-Boateng, B., Bell, M. L., Duko, B., & Pereira, G. (2024). Maternal exposure to ambient air temperature and adverse birth outcomes: An umbrella review of systematic reviews and meta-analyses. *Science of the Total Environment, 917*. https://doi.org/10.1016/j.scitotenv.2024.170236

Olsen, A. Y., Larson, S., Padilla-Gamino, J. L., & Klinger, T. (2022). Changes in fish assemblages after marine heatwave events in West Hawai'i Island. *Marine Ecology Progress Series, 698*, 95–109. https://doi.org/10.3354/meps14156

Osborne, O. E., O'Hara, P. D., Whelan, S., Zandbergen, P., Hatch, S. A., & Elliott, K. H. (2020). Breeding seabirds increase foraging range in response to an extreme marine heatwave. *Marine Ecology Progress Series, 646*, 161–173. https://doi.org/10.3354/meps13392

Otrachshenko, V., Popova, O., & Tavares, J. (2021). Extreme temperature and extreme violence: Evidence from Russia. *Economic Inquiry, 59*(1), 243–262. https://doi.org/10.1111/ecin.12936

Palu, S., & Mahmoud, H. (2019). Impact of climate change on the integrity of the superstructure of deteriorated US bridges. *PLoS One, 14*(10). https://doi.org/10.1371/journal.pone.0223307

Papadakis, G., Chalabi, Z., & Thornes, J. E. (2018). Ambulance service resource planning for extreme temperatures: Analysis of ambulance 999 calls during episodes of extreme temperature in London, UK. *Atmosphere, 9*(5). https://doi.org/10.3390/atmos9050182

Park, R. J., Goodman, J., Hurwitz, M., & Smith, J. (2020). Heat and learning. *American Economic Journal-Economic Policy, 12*(2), 306–339. https://doi.org/10.1257/pol.20180612

Park, H., Lee, S. M., Kim, W. J., & Chae, Y. (2023). Analysis of compound health impacts of heatwave and COVID-19 in Korea from May to September in 2020. *Scientific Reports, 13*(1). https://doi.org/10.1038/s41598-023-41880-1

Parks, R. M., Rowland, S. T., Do, V., Boehme, A. K., Dominici, F., Hart, C. L., & Kioumourtzoglou, M. A. (2023). The association between temperature and alcohol- and substance-related disorder hospital visits in New York state. *Communications Medicine, 3*(1). https://doi.org/10.1038/s43856-023-00346-1

Parnaudeau, M., & Bertrand, J. L. (2018). The contribution of weather variability to economic sectors. *Applied Economics, 50*(43), 4632–4649. https://doi.org/10.1080/00036846.2018.1458200

Part, C., le Roux, J., Chersich, M., Sawry, S., Filippi, V., Roos, N., Fairlie, L., Nakstad, B., de Bont, J., Ljungman, P., Stafoggia, M., Kovats, S., Luchters, S., & Hajat, S. (2022). Ambient temperature during pregnancy and risk of maternal hypertensive disorders: A time-to-event study in Johannesburg. *South Africa. Environmental Research, 212*. https://doi.org/10.1016/j.envres.2022.113596

Pathi, K., & Ghosa, A. (2022). *Heat wave sparks blackouts, questions on India's coal usage*. AP News.

Pechan, A., & Eisenack, K. (2014). The impact of heat waves on electricity spot markets. *Energy Economics, 43*, 63–71. https://doi.org/10.1016/j.eneco.2014.02.006

Peden, A. E., Mason, H. M., King, J. C., & Franklin, R. C. (2024). Examining the relationship between heatwaves and fatal drowning: A case study from Queensland, Australia. *Injury Prevention, 30*(1), 7–13. https://doi.org/10.1136/ip-2023-044938

Peillex, J., El Ouadghiri, I., Gomes, M., & Jaballah, J. (2021). Extreme heat and stock market activity. *Ecological Economics, 179*. https://doi.org/10.1016/j.ecolecon.2020.106810

Pena, M. A., Nemcek, N., & Robert, M. (2019). Phytoplankton responses to the 2014-2016 warming anomaly in the northeast subarctic Pacific Ocean. *Limnology and Oceanography, 64*(2), 515–525. https://doi.org/10.1002/lno.11056

Pergent, G., Bazairi, H., Bianchi, C. N., Boudouresque, C. F., Buia, M. C., Calvo, S., Clabaut, P., Harmelin-Vivien, M., Mateo, M. A., Montefalcone, M., Morri, C., Orfanidis, S., Pergent-Martini, C., Semroud, R., Serrano, O., Thibaut, T., Tomasello, A., & Verlaque, M. (2014). Climate change and Mediterranean seagrass meadows: A synopsis for environmental managers. *Mediterranean Marine Science, 15*(2), 462–473. https://doi.org/10.12681/mms.621

Peterson, W. T., Fisher, J. L., Strub, P. T., Du, X. N., Risien, C., Peterson, J., & Shaw, C. T. (2017). The pelagic ecosystem in the northern California current off Oregon during the 2014-2016 warm anomalies within the context of the past 20 years. *Journal of Geophysical Research-Oceans, 122*(9), 7267–7290. https://doi.org/10.1002/2017jc012952

Phung, D., Chu, C., Rutherford, S., Nguyen, H. L. T., Do, C. M., & Huang, C. R. (2017). Heatwave and risk of hospitalization: A multi-province study in Vietnam. *Environmental Pollution, 220*, 597–607. https://doi.org/10.1016/j.envpol.2016.10.008

Pianta, S., & Sisco, M. R. (2020). A hot topic in hot times: How media coverage of climate change is affected by temperature abnormalities. *Environmental Research Letters, 15*(11). https://doi.org/10.1088/1748-9326/abb732

Piatt, J. F., Parrish, J. K., Renner, H. M., Schoen, S. K., Jones, T. T., Arimitsu, M. L., Kuletz, K. J., Bodenstein, B., Garcia-Reyes, M., Duerr, R. S., Corcoran, R. M., Kaler, R. S. A., McChesney, G. J., Golightly, R. T., Coletti, H. A., Suryan, R. M., Burgess, H. K., Lindsey, J., Lindquist, K., Warzybok, P. M., Jahncke, J., Roletto, J., & Sydeman, W. J. (2020). Extreme mortality and reproductive failure of common murres resulting from the Northeast Pacific marine heatwave of 2014-2016. *PLoS One, 15*(1). https://doi.org/10.1371/journal.pone.0226087

Polazzo, F., Roth, S. K., Hermann, M., Mangold-Doring, A., Rico, A., Sobek, A., van den Brink, P. J., & Jackson, M. C. (2022). Combined effects of heatwaves and micropollutants on freshwater ecosystems: Towards an integrated assessment of extreme events in multiple stressors research. *Global Change Biology, 28*(4), 1248–1267. https://doi.org/10.1111/gcb.15971

Pradhan, B., Kjellstrom, T., Atar, D., Sharma, P., Kayastha, B., Bhandari, G., & Pradhan, P. K. (2019). Heat stress impacts on cardiac mortality in Nepali migrant workers in Qatar. *Cardiology, 143*(1–2), 37–48. https://doi.org/10.1159/000500853

PSE. (2021). *How Puget Sound energy handled the historic 2021 heat dome*. Puget Sound Energy, Puget Sound.

Pule, V., Mathee, A., Melariri, P., Kapwata, T., Abdelatif, N., Balakrishna, Y., Kunene, Z., Mogotsi, M., Wernecke, B., & Wright, C. Y. (2021). Classroom temperature and learner absenteeism in public primary schools in the eastern cape, South Africa. *International Journal of Environmental Research and Public Health, 18*(20). https://doi.org/10.3390/ijerph182010700

Pyrgou, A., Castaldo, V. L., Pisello, A. L., Cotana, F., & Santamouris, M. (2017). On the effect of summer heatwaves and urban overheating on building thermal-energy performance in Central Italy. *Sustainable Cities and Society, 28*, 187–200. https://doi.org/10.1016/j.scs.2016.09.012

Quintana, F., Uhart, M. M., Gallo, L., Mattera, M. B., Rimondi, A., & Gomez-Laich, A. (2022). Heat-related massive chick mortality in an Imperial cormorant Leucocarbo atriceps colony from Patagonia, Argentina. *Polar Biology, 45*(2), 275–284. https://doi.org/10.1007/s00300-021-02982-6

QUT. (2010). *Impacts and adaptation response of infrastructure and communities to heatwaves: The southern Australian experience of 2009* (Vol. 13/10). National Climate Change Adaptation Research Facility (NCCAR), Queensland University of Technology.

Rajamani M (2022) Thousands of new Yorkers lose power amid heat wave. Spectrum News NY1.

Raymond, W. W., Barber, J. S., Dethier, M. N., Hayford, H. A., Harley, C. D. G., King, T. L., Paul, B., Speck, C. A., Tobin, E. D., Raymond, A. E. T., & McDonald, P. S. (2022). Assessment of the impacts of an unprecedented heatwave on intertidal shellfish of the Salish Sea. *Ecology, 103*(10). https://doi.org/10.1002/ecy.3798

Rekha, S., Nalini, S. J., Bhuvana, S., Kanmani, S., Hirst, J. E., & Venugopal, V. (2024). Heat stress and adverse pregnancy outcome: Prospective cohort study. *Bjog-an International Journal of Obstetrics and Gynaecology, 131*(5), 612–622. https://doi.org/10.1111/1471-0528.17680

Reuters. (2022). *France tweaks rules to keep nuclear plants running during heatwave*. Reuters.

Rigby, C. (2022). *How shoppers bought in a heatwave*.

Ritzman, J., Brodbeck, A., Brostrom, S., McGrew, S., Dreyer, S., Klinger, T., & Moore, S. K. (2018). Economic and sociocultural impacts of fisheries closures in two fishing-dependent communities following the massive 2015 US West coast harmful algal bloom. *Harmful Algae, 80*, 35–45. https://doi.org/10.1016/j.hal.2018.09.002

Roberts, S. D., Van Ruth, P. D., Wilkinson, C., Bastianello, S. S., & Bansemer, M. S. (2019). Marine heatwave, harmful algae blooms and an extensive fish kill event during 2013 in South Australia. *Frontiers in Marine Science, 6*. https://doi.org/10.3389/fmars.2019.00610

Robinson, C. L. K., Yakimishyn, J., & Evans, R. (2022). Minimal effects of the 2014-16 marine heatwave on fish assemblages found in eelgrass meadows on the southwestern coast of Vancouver Island, British Columbia, Canada. *Frontiers in Marine Science, 9*. https://doi.org/10.3389/fmars.2022.980703

Romani, S. G., Roye, D., Santos, L. S., & Figueiras, A. (2020). Impact of extreme temperatures on ambulance dispatches due to cardiovascular causes in north-West Spain. *International Journal of Environmental Research and Public Health, 17*(23). https://doi.org/10.3390/ijerph17239001

Rose, N., & Dolega, L. (2022). It's the weather: Quantifying the impact of weather on retail sales. *Applied Spatial Analysis and Policy, 15*(1), 189–214. https://doi.org/10.1007/s12061-021-09397-0

Ruthrof, K. X., Fontaine, J. B., Matusick, G., Breshears, D. D., Law, D. J., Powell, S., & Hardy, G. (2016). How drought-induced forest die-off alters microclimate and increases fuel loadings and fire potentials. *International Journal of Wildland Fire, 25*(8), 819–830.

Sailor, D. J., Baniassadi, A., O'Lenick, C. R., & Wilhelmi, O. V. (2019). The growing threat of heat disasters. *Environmental Research Letters, 14*(5). https://doi.org/10.1088/1748-9326/ab0bb9

Salomon, R. L., Peters, R. L., Zweifel, R., Sass-Klaassen, U. G. W., Stegehuis, A. I., Smiljanic, M., Poyatos, R., Babst, F., Cienciala, E., Fonti, P., Lerink, B. J. W., Lindner, M., Martinez-Vilalta, J., Mencuccini, M., Nabuurs, G. J., van der Maaten, E., von Arx, G., Bar, A., Akhmetzyanov, L., Balanzategui, D., Bellan, M., Bendix, J., Berveiller, D., Blazenec, M., Cada, V., Carraro, V., Cecchini, S., Chan, T., Conedera, M., Delpierre, N., Delzon, S., Ditmarova, L., Dolezal, J., Dufrene, E., Edvardsson, J., Ehekircher, S., Forner, A., Frouz, J., Ganthaler, A., Gryc, V., Guney, A., Heinrich, I., Hentschel, R., Janda, P., Jezik, M., Kahle, H. P., Knusel, S., Krejza, J., Kuberski, L., Kucera, J., Lebourgeois, F., Mikolas, M., Matula, R., Mayr, S., Oberhuber, W., Obojes, N., Osborne, B., Paljakka, T., Plichta, R., Rabbel, I., Rathgeber, C. B. K., Salmon, Y., Saunders, M., Scharnweber, T., Sitkova, Z., Stangler, D. F., Sterenczak, K., Stojanovic, M., Strelcova, K., Svetlik, J., Svoboda, M., Tobin, B., Trotsiuk, V., Urban, J., Valladares, F., Vavrcik, H., Vejpustkova, M., Walthert, L., Wilmking, M., Zin, E., Zou, J. L., & Steppe, K. (2022). The 2018 European heatwave led to stem dehydration but not to consistent growth reductions in forests. *Nature Communications, 13*(1). https://doi.org/10.1038/s41467-021-27579-9

Salzman, A. G. (1982). The selective importance of heat-stress in gull nest location. *Ecology, 63*(3), 742–751. https://doi.org/10.2307/1936795

Samhouri, J. F., Feist, B. E., Fisher, M. C., Liu, O., Woodman, S. M., Abrahms, B., Forney, K. A., Hazen, E. L., Lawson, D., Redfern, J., & Saez, L. E. (2021). Marine heatwave challenges solutions to human-wildlife conflict. *Proceedings of the Royal Society B-Biological Sciences, 288*(1964). https://doi.org/10.1098/rspb.2021.1607

Sanford, E., Sones, J. L., Garcia-Reyes, M., Goddard, J. H. R., & Largier, J. L. (2019). Widespread shifts in the coastal biota of northern California during the 2014-2016 marine heatwaves. *Scientific Reports, 9*. https://doi.org/10.1038/s41598-019-40784-3

Sang, Z. H. H., & Hamann, A. (2023). A record-setting 2021 heat wave in western Canada had a significant temporary impact on greenness of the world's largest protected temperate rainforest. *Remote Sensing, 15*(8). https://doi.org/10.3390/rs15082162

Santagata, D. M., Castesana, P., Rossler, C. E., & Gomez, D. R. (2017). Extreme temperature events affecting the electricity distribution system of the metropolitan area of Buenos Aires (1971-2013). *Energy Policy, 106*, 404–414. https://doi.org/10.1016/j.enpol.2017.04.006

Sanusi, R., & Livesley, S. J. (2020). London plane trees (Platanus x acerifolia) before, during and after a heatwave: Losing leaves means less cooling benefit. *Urban Forestry & Urban Greening, 54*. https://doi.org/10.1016/j.ufug.2020.126746

Sanz-Barbero, B., Linares, C., Vives-Cases, C., Gonzalez, J. L., Lopez-Ossorio, J. J., & Diaz, J. (2018). Heat wave and the risk of intimate partner violence. *Science of the Total Environment, 644*, 413–419. https://doi.org/10.1016/j.scitotenv.2018.06.368

Sapari, H., Selamat, M. I., Isa, M. R., Ismail, R., & Mahiyuddin, W. R. W. (2023). The impact of heat waves on health care services in low- or middle-income countries: Protocol for a systematic review. *Jmir Research Protocols, 12*. https://doi.org/10.2196/44702

Schwartz, J., Samet, J. M., & Patz, J. A. (2004). Hospital admissions for heart disease – The effects of temperature and humidity. *Epidemiology, 15*(6), 755–761. https://doi.org/10.1097/01.ede.0000134875.15919.0f

Schwarz, L., Castillo, E. M., Chan, T. C., Brennan, J. J., Sbiroli, E. S., Carrasco-Escobar, G., Nguyen, A., Clemesha, R. E. S., Gershunov, A., & Benmarhnia, T. (2022). Heat waves and emergency department visits among the homeless, San Diego, 2012–2019. *American Journal of Public Health, 112*(1), 98–106. https://doi.org/10.2105/ajph.2021.306557

Scorgie, F., Lusambili, A., Luchters, S., Khaemba, P., Filippi, V., Nakstad, B., Hess, J., Birch, C., Kovats, S., & Chersich, M. (2023). "Mothers get really exhausted!" the lived experience of pregnancy in extreme heat: Qualitative findings from Kilifi. *Kenya. Social Science & Medicine, 335*, 116223.

Seaton, S., Matusick, G., Ruthrof, K. X., & Hardy, G. E. S. J. (2015). Outbreak of Phoracantha semipunctata in response to severe drought in a Mediterranean eucalyptus forest. *Forests, 6*(11), 3868–3881.

Seposo, X., Madaniyazi, L., Ng, C. F. S., Hashizume, M., & Honda, Y. (2021). COVID-19 pandemic modifies temperature and heat-related illness ambulance transport association in Japan: A nationwide observational study. *Environmental Health, 20*(1). https://doi.org/10.1186/s12940-021-00808-w

Serrano, O., Arias-Ortiz, A., Duarte, C. M., Kendrick, G. A., & Lavery, P. S. (2021). Impact of marine heatwaves on seagrass ecosystems. In R. B. Jackson & J. G. Canadell (Eds.), *Ecosystem collapse and climate change* (Vol. 241, pp. 345–364). Ecological Studies-Analysis and Synthesis. https://doi.org/10.1007/978-3-030-71330-0_13

Shankar, K., Hwang, K., Westcott, J. L., Saleem, S., Ali, S. A., Jessani, S., Patel, A., Kavi, A., Somannavar, M. S., Goudar, S. S., Hibberd, P. L., Derman, R. J., Hoffman, M., Wylie, B. J., Goldenberg, R. L., Thorsten, V. R., McClure, E. M., & Krebs, N. F. (2023). Associations between ambient temperature and pregnancy outcomes from three south Asian sites of the global network maternal newborn health registry: A retrospective cohort study. *Bjog-an International Journal of Obstetrics and Gynaecology*. https://doi.org/10.1111/1471-0528.17616

Sharpe, L., Cale, B., & Gardner, J. L. (2019). Weighing the cost: The impact of serial heatwaves on body mass in a small Australian passerine. *Journal of Avian Biology, 50*(11). https://doi.org/10.1111/jav.02355

Sindall, R., Mecrow, T., Queiroga, A. C., Boyer, C., Koon, W., & Peden, A. E. (2022). Drowning risk and climate change: A state-of-the-art review. *Injury Prevention, 28*(2), 185–191. https://doi.org/10.1136/injuryprev-2021-044486

Slavicek, M. (2022). *How the heat wave is damaging French rail infrastructure and forcing trains to run at low speed*. Le Monde,

Sloane, S. A., Gordon, A., & Connelly, I. D. (2022). Bushtit (Psaltriparus minimus) nestling mortality associated with unprecedented June 2021 heatwave in Portland, Oregon. *Wilson Journal of Ornithology, 134*(1), 155–162. https://doi.org/10.1676/21-00080

Smale, D. A. (2020). Impacts of ocean warming on kelp forest ecosystems. *New Phytologist, 225*(4), 1447–1454. https://doi.org/10.1111/nph.16107

Smale, D. A., Wernberg, T., & Vanderklift, M. A. (2017). Regional-scale variability in the response of benthic macroinvertebrate assemblages to a marine heatwave. *Marine Ecology Progress Series, 568*, 17–30. https://doi.org/10.3354/meps12080

Smale, D. A., Wernberg, T., Oliver, E. C. J., Thomsen, M., Harvey, B. P., Straub, S. C., Burrows, M. T., Alexander, L. V., Benthuysen, J. A., Donat, M. G., Feng, M., Hobday, A. J., Holbrook, N. J., Perkins-Kirkpatrick, S. E., Scannell, H. A., Sen Gupta, A., Payne, B. L., & Moore, P. J. (2019). Marine heatwaves threaten global biodiversity and the provision of ecosystem services. Nature. *Climate Change, 9*(4), 306. https://doi.org/10.1038/s41558-019-0412-1

Smith, I. (2022). Heatwave leads to surge in housing subsidence claims across southern England. *Financial Times*, FT Group,

Smith, K. E., Burrows, M. T., Hobday, A. J., Sen Gupta, A., Moore, P. J., Thomsen, M., Wernberg, T., & Smale, D. A. (2021). Socioeconomic impacts of marine heatwaves: Global issues and opportunities. *Science, 374*(6566), 419. https://doi.org/10.1126/science.abj3593

Smith, K. E., Burrows, M. T., Hobday, A. J., King, N. G., Moore, P. J., Sen Gupta, A., Thomsen, M. S., Wernberg, T., & Smale, D. A. (2023). Biological impacts of marine heatwaves. *Annual Review of Marine Science, 15*, 119–145. https://doi.org/10.1146/annurev-marine-032122-121437

Soravia, C., Ashton, B. J., Thornton, A., & Ridley, A. R. (2021). The impacts of heat stress on animal cognition: Implications for adaptation to a changing climate. Wiley interdisciplinary reviews-climate. *Change, 12*(4). https://doi.org/10.1002/wcc.713

Speare, K. E., Adam, T. C., Winslow, E. M., Lenihan, H. S., & Burkepile, D. E. (2022). Size-dependent mortality of corals during marine heatwave erodes recovery capacity of a coral reef. *Global Change Biology, 28*(4), 1342–1358. https://doi.org/10.1111/gcb.16000

Srinavasan, T. (2008). *Cost of excess hospitalizations and emergency department visits for the 2006 California heat wave*.

Starko, S., Neufeld, C. J., Gendall, L., Timmer, B., Campbell, L., Yakimishyn, J., Druehl, L., & Baum, J. K. (2022). Microclimate predicts kelp forest extinction in the face of direct and indirect marine heatwave effects. *Ecological Applications, 32*(7). https://doi.org/10.1002/eap.2673

State of Victoria Environment LW, Planning. (2019). *The economic impact of heatwaves on the state of Victoria*. The State of Victoria Environment, Land, Water and Planning.

Steel, E. J., Fontaine, J. B., Ruthrof, K. X., Burgess, T. I., & Hardy, G. E. S. (2019). Changes in structure of over- and midstory tree species in a Mediterranean-type forest after an extreme drought-associated heatwave. *Austral Ecology, 44*(8), 1438–1450. https://doi.org/10.1111/aec.12818

Still, C. J., Sibley, A., DePinte, D., Busby, P. E., Harrington, C. A., Schulze, M., Shaw, D. R., Woodruff, D., Rupp, D. E., Daly, C., Hammond, W. M., & Page, G. F. M. (2023). Causes of widespread foliar damage from the June 2021 Pacific northwest heat dome: More heat than drought. *Tree Physiology, 43*(2), 203–209. https://doi.org/10.1093/treephys/tpac143

Stipcich, P., Beca-Carretero, P., Alvarez-Salgado, X. A., Apostolaki, E. T., Chartosia, N., Efthymiadis, P. T., Jimenez, C. E., La Manna, G., Pansini, A., Principato, E., Resaikos, V., Stengel, D. B., & Ceccherelli, G. (2023). Effects of high temperature and marine heat waves on seagrasses: Is warming affecting the nutritional value of Posidonia oceanica? *Marine Environmental Research, 184*. https://doi.org/10.1016/j.marenvres.2022.105854

Stone, B., Mallen, E., Rajput, M., Broadbent, A., Krayenhoff, E. S., Augenbroe, G., & Georgescu, M. (2021). Climate change and infrastructure risk: Indoor heat exposure during a concurrent heat wave and blackout event in Phoenix, Arizona. *Urban Climate, 36*. https://doi.org/10.1016/j.uclim.2021.100787

Stone, B., Gronlund, C. J., Mallen, E., Hondula, D., O'Neill, M. S., Rajput, M., Grijalva, S., Lanza, K. V., Harlan, S., Larsen, L., Augenbroe, G., Krayenhoff, E. S., Broadbent, A., & Georgescu, M. (2023). How blackouts during heat waves amplify mortality and morbidity risk. *Environmental Science & Technology, 57*(22), 8245–8255. https://doi.org/10.1021/acs.est.2c09588

Straub, S. C., Wernberg, T., Thomsen, M. S., Moore, P. J., Burrows, M. T., Harvey, B., & Smale, D. A. (2019). Resistance, extinction, and everything in between – The diverse responses of seaweeds to marine heatwaves. *Frontiers in Marine Science, 6*. https://doi.org/10.3389/fmars.2019.00763

Sun, X. Y., & Arnott, S. E. (2022). Interactive effects of increased salinity and heatwaves on freshwater zooplankton communities in simultaneous and sequential treatments. *Freshwater Biology, 67*(9), 1604–1617. https://doi.org/10.1111/fwb.13964

Suryan, R. M., Arimitsu, M. L., Coletti, H. A., Hopcroft, R. R., Lindeberg, M. R., Barbeaux, S. J., Batten, S. D., Burt, W. J., Bishop, M. A., Bodkin, J. L., Brenner, R., Campbell, R. W., Cushing, D. A., Danielson, S. L., Dorn, M. W., Drummond, B., Esler, D., Gelatt, T., Hanselman, D. H.,

Hatch, S. A., Haught, S., Holderied, K., Iken, K., Irons, D. B., Kettle, A. B., Kimmel, D. G., Konar, B., Kuletz, K. J., Laurel, B. J., Maniscalco, J. M., Matkin, C., McKinstry, C. A. E., Monson, D. H., Moran, J. R., Olsen, D., Palsson, W. A., Pegau, W. S., Piatt, J. F., Rogers, L. A., Rojek, N. A., Schaefer, A., Spies, I. B., Straley, J. M., Strom, S. L., Sweeney, K. L., Szymkowiak, M., Weitzman, B. P., Yasumiishi, E. M., & Zador, S. G. (2021). Ecosystem response persists after a prolonged marine heatwave. *Scientific Reports, 11*(1). https://doi.org/10.1038/s41598-021-83818-5

Swanson, D. L., Vezina, F., McKechnie, A. E., & Nord, A. (2022). Editorial: Avian behavioral and physiological responses to challenging thermal environments and extreme weather events. *Frontiers in Ecology and Evolution, 10*. https://doi.org/10.3389/fevo.2022.1034659

Syed, S., O'Sullivan, T. L., & Phillips, K. P. (2022). Extreme heat and pregnancy outcomes: A scoping review of the epidemiological evidence. *International Journal of Environmental Research and Public Health, 19*(4). https://doi.org/10.3390/ijerph19042412

Tait, L. W., Thoral, F., Pinkerton, M. H., Thomsen, M. S., & Schiel, D. R. (2021). Loss of Giant kelp, Macrocystis pyrifera, driven by marine heatwaves and exacerbated by poor water clarity in New Zealand. *Frontiers in Marine Science, 8*. https://doi.org/10.3389/fmars.2021.721087

Takagi, S., Kuroda, H., Hasegawa, N., Watanabe, T., Unuma, T., Taniuchi, Y., Yokota, T., Izumida, D., Nakagawa, T., Kurokawa, T., & Azumaya, T. (2022). Controlling factors of large-scale harmful algal blooms with <i>Karenia selliformis</i> after record-breaking marine heatwaves. *Frontiers in Marine Science, 9*. https://doi.org/10.3389/fmars.2022.939393

Takahashi, Y., & Takahashi, T. (2020). Risk factors for exertional heat illness in thoroughbred racehorses in flat races in Japan (2005-2016). *Equine Veterinary Journal, 52*(3), 364–368. https://doi.org/10.1111/evj.13179

Tassone, S. J., Besterman, A. F., Buelo, C. D., Walter, J. A., & Pace, M. L. (2022). Co-occurrence of aquatic heatwaves with atmospheric heatwaves, low dissolved oxygen, and low pH events in estuarine ecosystems. *Estuaries and Coasts, 45*(3), 707–720. https://doi.org/10.1007/s12237-021-01009-x

Terlau, J. F., Brose, U., Eisenhauer, N., Amyntas, A., Boy, T., Dyer, A., Gebler, A., Hof, C., Liu, T., Scherber, C., Schlägel, U. E., Schmidt, A., & Hirt, M. R. (2023). Microhabitat conditions remedy heat stress effects on insect activity. *Global Change Biology, 29*(13), 3747–3758. https://doi.org/10.1111/gcb.16712

Teskey, R., Wertin, T., Bauweraerts, I., Ameye, M., McGuire, M. A., & Steppe, K. (2015). Responses of tree species to heat waves and extreme heat events. *Plant Cell and Environment, 38*(9), 1699–1712. https://doi.org/10.1111/pce.12417

Thompson, R., Hornigold, R., Page, L., & Waite, T. (2018). Associations between high ambient temperatures and heat waves with mental health outcomes: A systematic review. *Public Health, 161*, 171–191. https://doi.org/10.1016/j.puhe.2018.06.008

Thompson, A. R., Ben-Aderet, N. J., Bowlin, N. M., Kacev, D., Swalethorp, R., & Watson, W. (2022). Putting the Pacific marine heatwave into perspective: The response of larval fish off southern California to unprecedented warming in 2014–2016 relative to the previous 65 years. *Global Change Biology, 28*(5), 1766–1785. https://doi.org/10.1111/gcb.16010

Thompson, R., Lawrance, E. L., Roberts, L. F., Grailey, K., Ashrafian, H., Maheswaran, H., Toledano, M. B., & Darzi, A. (2023). Ambient temperature and mental health: A systematic review and meta-analysis. *Lancet Planetary Health, 7*(7).

Thomsen, M. S., Mondardini, L., Alestra, T., Gerrity, S., Tait, L., South, P. M., Lilley, S. A., & Schiel, D. R. (2019). Local extinction of bull kelp (Durvillaea spp.) due to a marine heatwave. Frontiers in marine. *Science, 6*. https://doi.org/10.3389/fmars.2019.00084

Toohey, G., Petri, A. E., & Yee, G. (2022). California pushed to the limit by a relentless heat wave that broke the mold. *The Anchorage Times*,

Topham, G. (2022). Britain's rail services severely disrupted as heatwave damage is repaired. *The Guardian*,

Travesso, M., Missionario, M., Cruz, S., Calado, R., & Madeira, D. (2023). Combined effect of marine heatwaves and light intensity on the cellular stress response and photophysiology of the leather coral Sarcophyton cf. glaucum. *Science of the Total Environment, 861*. https://doi.org/10.1016/j.scitotenv.2022.160460

Trigo, R. M., Ramos, A. M., Noqueira, P. J., Santos, F. D., Garcia-Herrera, R., Gouveia, C., & Santo, F. E. (2009). Evaluating the impact of extreme temperature based indices in the 2003 heatwave excessive mortality in Portugal. *Environmental Science & Policy, 12*(7), 844–854. https://doi.org/10.1016/j.envsci.2009.07.007

Vafeiadou, A., Banissy, M. J., Banissy, J. F. M., Higgins, J. P. T., & Howard, G. (2023). The influence of climate change on mental health in populations of the western Pacific region: An umbrella scoping review. *Heliyon, 9*(11). https://doi.org/10.1016/j.heliyon.2023.e21457

Vale, M. M., Moura, D. J., Naas, I. A., & Pereira, D. F. (2010). Characterization of heat waves affecting mortality rates of broilers between 29 days and market age. *Brazilian Journal of Poultry Science, 12*(4), 279–285.

Vandana, G. D., Sejian, V., Lees, A. M., Pragna, P., Silpa, M. V., & Maloney, S. K. (2021). Heat stress and poultry production: Impact and amelioration. *International Journal of Biometeorology, 65*(2), 163–179. https://doi.org/10.1007/s00484-020-02023-7

Vecellio, D. J., & Vanos, J. K. (2024). Aligning thermal physiology and biometeorological research for heat adaptation and resilience in a changing climate. *Journal of Applied Physiology*.

Verdura, J., Santamaria, J., Ballesteros, E., Smale, D. A., Cefali, M. E., Golo, R., de Caralt, S., Verges, A., & Cebrian, E. (2021). Local-scale climatic refugia offer sanctuary for a habitat-forming species during a marine heatwave. *Journal of Ecology, 109*(4), 1758–1773. https://doi.org/10.1111/1365-2745.13599

Videtich, A. (2021). *Spring waters run cold: Preventing needless drownings in King County*.

Vine E (2012) Adaptation of California's electricity sector to climate change. Climatic Change 111 (1):75–99. doi:https://doi.org/10.1007/s10584-011-0242-2.

Vitali, A., Felici, A., Esposito, S., Bernabucci, U., Bertocchi, L., Maresca, C., Nardone, A., & Lacetera, N. (2015). The effect of heat waves on dairy cow mortality. *Journal of Dairy Science, 98*(7), 4572–4579. https://doi.org/10.3168/jds.2015-9331

Vivanco-Bercovich, M., Belando-Torrentes, M. D., Figueroa-Burgos, M. F., Ferreira-Arrieta, A., Macias-Carranza, V., Garcia-Pantoja, J. A., Cabello-Pasini, A., Samperio-Ramos, G., Cruz-Lopez, R., & Sandoval-Gil, J. M. (2022). Combined effects of marine heatwaves and reduced light on the physiology and growth of the surfgrass Phyllospadix torreyi from Baja California, Mexico. *Aquatic Botany, 178*. https://doi.org/10.1016/j.aquabot.2021.103488

von Biela, V. R., Arimitsu, M. L., Piatt, J. F., Heflin, B., Schoen, S. K., Trowbridge, J. L., & Clawson, C. M. (2019). Extreme reduction in nutritional value of a key forage fish during the Pacific marine heatwave of 2014–2016. *Marine Ecology Progress Series, 613*, 171–182. https://doi.org/10.3354/meps12891

Wagner, D., Friedlander, A. M., Pyle, R. L., Brooks, C. M., Gjerde, K. M., & Wilhelm, T. A. (2020). Coral reefs of the high seas: Hidden biodiversity hotspots in need of protection. *Frontiers in Marine Science, 7*. https://doi.org/10.3389/fmars.2020.567428

Walinski, A., Sander, J., Gerlinger, G., Clemens, V., Meyer-Lindenberg, A., & Heinz, A. (2023). The effects of climate change on mental health. Deutsches Arzteblatt. *International, 120*(8), 117. https://doi.org/10.3238/arztebl.m2022.0403

Wang, Y., Liu, Y., Ye, D. X., Li, N., Bi, P., Tong, S. L., Cheng, Y. B., Li, Y. H., & Yao, X. Y. (2021). Temperatures and health costs of emergency department visits: A multisite time series study in China. *Environmental Research, 197*. https://doi.org/10.1016/j.envres.2021.111023

Wang, L. H., Zhang, Y. Q., Ma, N., Song, P. L., Tian, J., Zhang, X. Z., & Xu, Z. W. (2023). Diverse responses of canopy conductance to heatwaves. *Agricultural and Forest Meteorology, 335*. https://doi.org/10.1016/j.agrformet.2023.109453

Weitzman, B., Konar, B., Iken, K., Coletti, H., Monson, D., Suryan, R., Dean, T., Hondolero, D., & Lindeberg, M. (2021). Changes in rocky intertidal community structure during a marine heatwave in the northern Gulf of Alaska. *Frontiers in Marine Science, 8*. https://doi.org/10.3389/fmars.2021.556820

Welbergen, J. A., Klose, S. M., Markus, N., & Eby, P. (2008). Climate change and the effects of temperature extremes on Australian flying-foxes. *Proceedings of the Royal Society B-Biological Sciences, 275*(1633), 419–425. https://doi.org/10.1098/rspb.2007.1385

Wernberg, T. (2021). Marine heatwave drives collapse of kelp forests in Western Australia. *Ecosystem Collapse and Climate Change, 241*, 325–343. https://doi.org/10.1007/978-3-030-71330-0_12

Wernberg, T., Smale, D. A., Tuya, F., Thomsen, M. S., Langlois, T. J., de Bettignies, T., Bennett, S., & Rousseaux, C. S. (2013). An extreme climatic event alters marine ecosystem structure in a global biodiversity hotspot. *Nature Climate Change, 3*(1), 78–82. https://doi.org/10.1038/nclimate1627

Westcott, N. E. (2011). The prolonged 1954 Midwestern U.S. heat wave: Impacts and responses. *Weather Climate and Society, 3*(3), 165–176. https://doi.org/10.1175/wcas-d-10-05002.1

Wettstein, Z. S., Hall, J., Buck, C., Mitchell, S. H., & Hess, J. J. (2024). Impacts of the 2021 heat dome on emergency department visits, hospitalizations, and health system operations in three hospitals in Seattle, Washington. *Journal of the American College of Emergency Physicians Open, 5*(1). https://doi.org/10.1002/emp2.13098

Wild, S., Krutzen, M., Rankin, R. W., Hoppitt, W. J. E., Gerber, L., & Allen, S. J. (2019). Long-term decline in survival and reproduction of dolphins following a marine heatwave. *Current Biology, 29*(7), R239–R240. https://doi.org/10.1016/j.cub.2019.02.047

Wilk, P., Gunz, A., Maltby, A., Ravichakaravarthy, T., Clemens, K. K., Lavigne, É., Lim, R., & Vicedo-Cabrera, A. M. (2021). Extreme heat and paediatric emergency department visits in southwestern Ontario. *Paediatrics & Child Health, 26*(5), 305–309.

Williams, S., Nitschke, M., Wondmagegn, B. Y., Tong, M. C., Xiang, J. J., Hansen, A., Nairn, J., Karnon, J., & Bi, P. (2022). Evaluating cost benefits from a heat health warning system in Adelaide, South Australia. *Australian and New Zealand Journal of Public Health, 46*(2), 149–154. https://doi.org/10.1111/1753-6405.13194

Willway, T., Baldachin, L., Reeves, S., Harding, M., McHale, M., & Nunn, M. (2008). *The effects of climate change on highway pavements and how to minimise them*. TRL Limited.

Wondmagegn, B. Y., Xiang, J. J., Williams, S., Pisaniello, D., & Bi, P. (2019). What do we know about the healthcare costs of extreme heat exposure? A comprehensive literature review. *Science of the Total Environment, 657*, 608–618. https://doi.org/10.1016/j.scitotenv.2018.11.479

Wondmagegn, B. Y., Xiang, J. J., Dear, K., Williams, S., Hansen, A., Pisaniello, D., Nitschke, M., Nairn, J., Scalley, B., Varghese, B. M., Xiao, A., Jian, L., Tong, M., Bambrick, H., Karnon, J., & Bi, P. (2021). Impact of heatwave intensity using excess heat factor on emergency department presentations and related healthcare costs in Adelaide, South Australia. *Science of the Total Environment, 781*. https://doi.org/10.1016/j.scitotenv.2021.146815

Woolway, R. I., Kraemer, B. M., Zscheischler, J., & Albergel, C. (2021). Compound hot temperature and high chlorophyll extreme events in global lakes. *Environmental Research Letters, 16*(12). https://doi.org/10.1088/1748-9326/ac3d5a

Xia, Y., Li, Y., Guan, D. B., Tinoco, D. M., Xia, J. J., Yan, Z. W., Yang, J., Liu, Q. Y., & Huo, H. (2018). Assessment of the economic impacts of heat waves: A case study of Nanjing, China. *Journal of Cleaner Production, 171*, 811–819. https://doi.org/10.1016/j.jclepro.2017.10.069

Xu, Z. W., Sheffield, P. E., Su, H., Wang, X. Y., Bi, Y., & Tong, S. L. (2014). The impact of heat waves on children's health: A systematic review. *International Journal of Biometeorology, 58*(2), 239–247. https://doi.org/10.1007/s00484-013-0655-x

Xu, E., Li, Y. N., Li, T. T., & Li, Q. (2022). Association between ambient temperature and ambulance dispatch: A systematic review and meta-analysis. *Environmental Science and Pollution Research, 29*(44), 66335–66347. https://doi.org/10.1007/s11356-022-20508-3

Yan, M. L., Xie, Y., Zhu, H. H., Ban, J., Gong, J. C., & Li, T. T. (2022). The exceptional heatwaves of 2017 and all-cause mortality: An assessment of nationwide health and economic impacts in China. *Science of the Total Environment, 812*. https://doi.org/10.1016/j.scitotenv.2021.152371

Yeeles, A. (2015). Weathering unrest: The ecology of urban social disturbances in Africa and Asia. *Journal of Peace Research, 52*(2), 158–170. https://doi.org/10.1177/0022343314557508

Young, J. C., Arthur, R., Spruce, M., & Williams, H. T. P. (2021). Social sensing of heatwaves. *Sensors, 21*(11). https://doi.org/10.3390/s21113717

Zhang, P. Y., Wang, T., Zhang, H., Wang, H., Hilt, S., Shi, P. L., Cheng, H. W., Feng, M. J., Pan, M., Guo, Y. L., Wang, K., Xu, X. Q., Chen, J. L., Zhao, K. S., He, Y. H., Zhang, M., & Xu, J. (2022). Heat waves rather than continuous warming exacerbate impacts of nutrient loading and herbicides on aquatic ecosystems. *Environment International, 168*. https://doi.org/10.1016/j.envint.2022.107478

Zhang, X., Chen, F. L., & Chen, Z. F. (2023). Heatwave and mental health. *Journal of Environmental Management, 332*. https://doi.org/10.1016/j.jenvman.2023.117385

Part IV
Heatwave Responses

Chapter 12
Heatwave Responses: Adaptation

Abstract The chapter 'Heatwave Responses: Adaptation' explores adaptation to heatwaves. It begins by defining adaptation as intentional, sustained actions to reduce heatwave-related risks and impacts, distinguishing it from mitigation. The historical context of adaptation is explored, highlighting its evolution from traditional methods to systematic approaches enunciated in international frameworks. The chapter outlines generic adaptation strategies, categorized into anticipatory vs. reactive, autonomous vs. planned, and incremental vs. transformational, emphasizing the importance of context-specific responses. Rather than describing the vast range of heatwave adaptation options the Chapter focuses on heatwave risk governance, heatwave insurance, the establishment of Chief Heat Officers, and the development of knowledge networks as emerging topics in the wider field of heatwave adaptation. Heatwave risk governance is presented as a critical component, involving multi-level coordination and the integration of various stakeholders to effectively manage heatwave risks. The chapter underscores the nascent state of heatwave insurance, exploring the potential of parametric insurance products as a risk transfer mechanism. The role of Chief Heat Officers in promoting heatwave awareness, coordinating responses, and implementing urban cooling projects is assessed. Also examined is the role of knowledge networks, such as the Global Heat Health Information Network, in facilitating information exchange and promoting heat resilience noting that such knowledge networks may be crucial in the absence of strong government leadership related to heatwave adaptation. In addressing what catalyses heatwave responses at the municipal to government level, the chapter explores various theories of policy change, including the role of extreme events as focusing events that create windows of opportunity for policy shifts. For potentially understanding what drives heatwave responses at the individual to small group level several theories related to autonomous decision making are presented. Complementing this, the Cultural Theory of Risk is also introduced as a possible framework for analysing heatwave responses, whether they be adaptation or mitigation oriented, at the individual to government level.

Keywords Concept of adaptation · Heatwave adaptation options · Heatwave governance · Heatwave insurance · Chief heat officers · Heatwave risk reduction knowledge networks · Catalysts of heatwave policy change · Disaster-reform

G. McGregor, *Heatwaves*, Biometeorology 6,
https://doi.org/10.1007/978-3-031-69906-1_12

hypothesis · Focusing events · Autonomous decision making · Cultural theory of risk

12.1 Introduction

Interpreted in its purest sense, a response is an action evoked by a stimulus. In the context of natural hazards, responses have often been defined as the impacts arising from an extreme event, for example, the type of heatwave impacts recorded for human and natural systems as outlined in the previous chapter. In the realm of disaster risk reduction, response typically refers to the measures taken just before, during, or immediately after a disaster occurs. These actions aim to safeguard public safety, protect lives, mitigate health risks, and fulfil the basic survival requirements of affected individuals. Rather than these interpretations, this chapter and the following two, considers responses to mean intentional short to long-term and sustained actions and measures to reduce heatwave related risk and thus impact. As noted earlier, alongside responses, hazard, exposure, and vulnerability make up the other elements of the determinants of heatwave risk. In discussing heatwave responses, it is assumed effective responses can influence the determinants of risk over time. For example, successful response efforts that reduce general or specific heatwave impacts may lead to changes in heatwave related exposure and vulnerability. The antithesis of this of course is some well-intentioned actions may have inadvertent negative effects on exposure and vulnerability and exacerbate heatwave related risk.

Focusing specifically on human responses to heatwaves, this chapter outlines actions that fall under the broad category of adaptation. Those related to mitigation will be dealt with in Chap. 14. Often defined as two distinct overall actions, it is important to remember that adaptation and mitigation are not mutually exclusive. Instead, they are complementary. In tandem, both can reduce climate-related risk (Broto, 2017; Spencer et al., 2017). Although it may seem so, adaptation is not a new term as it has been used to portray an array of biological, psychological, and social processes since the nineteenth century. However, with the realisation that human-induced climate change is likely to have wide ranging impacts, adaptation has taken on a new meaning within the context of climate risk planning and response (Orlove, 2022). It is in this context that adaptation, as an action in response to heatwave risk and impact, is defined here as adjustments in human systems (e.g., social, economic, cultural, political, technological, institutional) in anticipation of heatwaves and their effects. Often linked with the notion of adaptation is resilience. Rather than a distinct response to heatwave risk, resilience is treated here as pertaining to the inherent qualities of individuals, locations, or entities, including infrastructure, to bear the challenges posed by heatwaves. While sharing common ground with adaptation, by encompassing the capacity to engage in adaptation efforts, resilience distinguishes itself as a trait rather than an active process. For this reason, resilience is not considered in this chapter. That said it is acknowledged that acting on adaptation has implications for resilience, either bolstering it or in the worst case diminishing it.

This chapter is organised in the following way. It begins with a section on the concept of adaptation. Material on heatwave adaptation options then follows. Rather than systematically considering the vast range of heatwave adaptation options only a limited number are focused on here. These include heatwave risk governance, heatwave insurance, the establishment of Chief Heat Officers and the development of heatwave related knowledge networks dedicated to promoting heatwave risk reduction. These have been chosen for attention as they constitute rapidly emerging topics in the broader field of heatwave adaptation. Because of the enormity of the subject, early warning systems as an adaptation option are considered in the next chapter along with the related topics of heatwave prediction and heatwave naming and categorisation. Before closing with a synthesis of chapter contents, an attempt is made to address the infrequently considered topic of what might catalyse heatwave risk reduction actions. The focus of this section is not exclusively related to adaptation responses, however. Rather it is more generic in nature. Therefore, it should be read with not only adaptation actions in mind but also those related to heatwave mitigation.

12.2 Concept of Adaptation

While adaptation to extreme heat and heatwaves has gained prominence in recent decades, the roots of the concept of adaptation can be traced back to early human civilizations. Essentially the history of climate change adaptation is marked by an evolution from traditional, often implicit, methods of coping with climate variability to a more systematic and intentional approach. Historically, as humanity's understanding of weather and climate extremes and their profound impacts on societies has evolved, human communities have adapted to the challenges of heat and heatwaves through a combination of cultural practices and technological innovations. The construction of infrastructure like aqueducts and reservoirs in ancient Rome, for instance, was a form of adaptation to water scarcity during the sweltering months of July and August, and the so-called 'dog days' of summer especially throughout the Roman Climatic Optimum, so that water supply could be maintained for public baths and cooling gardens which were often refuges from the heat of the day. Similarly, the development of drought and heat-resistant crop varieties and advanced farming techniques in various cultures can be seen as early instances of adaptation to periods of drought and extreme heat.

With the arrival of the latter half of the twentieth century and mounting evidence of the link between human activities, climate change and climate extremes, efforts to moderate the impacts of heatwaves via a range of adaptation options has gained momentum. This has been assisted by the formal recognition of the general concept of adaptation with the adoption of the United Nations Framework Convention on Climate Change (UNFCCC) in 1992, which followed on from the formation of the

Intergovernmental Panel on Climate Change (IPCC) in 1988. Alongside mitigation and technology transfer, adaptation was identified as a key pillar of global climate action in the convention. However, despite its apparent importance, the status of adaptation as measured by the amount of material on this subject in early IPCC reports was minimal. In fact, it wasn't until the publication IPCC's Working Group II report in 2007 (the Fourth Assessment Report—4AR) that a dedicated chapter on adaptation appeared. This marked a clear turning point for the status of adaptation, albeit with a somewhat subaltern status compared to mitigation (Pielke et al., 2007; Khan & Roberts, 2013). Recognising the criticality of adaptation as an action for managing the inherent risks associated with an increasing frequency of climate extremes, the IPCC expanded the material on adaptation to four chapters in the 2014 WGII Fifth Assessment Report (5AR) with this topic comprehensively mainstreamed throughout the IPCC 2022 WGII Sixth Assessment Report (6AR). In parallel with the IPCC process, the conferences of the parties (COPs) to the UNFCCC have also emphasized the importance of adaptation, eventually leading to the establishment of the Adaptation Fund in 2001 and the Green Climate Fund in 2010. An important moment, complementing the IPCC's 2007 4AR, was the development of a new negotiating process on climate change in form of the Bali Action Plan. Formulated at the 2007 COP13 in Bali this ensured the status of adaptation as a priority issue. Subsequent COPs especially COP16 (2010), COP17 (2011) and COP21 (2015) further elevated the visibility of adaptation. Respectively, these COPs saw the establishment of the Cancun Adaptation Framework and an associated Adaptation Committee as well as the Green Climate Fund, a mechanism for the development of National Adaptation Plans (NAPs) and the Paris Agreement and its associated Article 7 focused specifically on adaptation, including a request for parties (countries) to track climate adaptation practice (Orlove, 2022).

The general process of adaptation has been referred to as iterative risk management with adaptation viewed as involving the general stages of awareness, assessment, planning, implementation, and monitoring and evaluation (Ara Begum et al., 2022). Together these stages comprise the 'adaptation cycle' with three archetypal modes of the adaptation cycle generally recognisable (Stafford-Smith et al., 2022). A range of approaches to adaptation can be distinguished. These include anticipatory versus reactive, autonomous versus planned and incremental versus transformational adaptation (Table 12.1). In practice, adaptation strategies often involve a combination of these couplets depending on the context and the nature of the natural hazard-related challenges individuals, societal and economic sectors, or governments face.

Adaptation can occur across a range of spatial and temporal scales. Small scale adaptation is usually typified by autonomous responses by individuals usually in climate affected communities which draws on locally available knowledge and resources (Rahman et al., 2021). Much of the evidence for autonomous adaptation to heatwaves derives from the global south Turek-Hankins et al. (2021) and comprises community-based and community-led adaptation (McNamara & Buggy, 2017; Piggott-McKellar et al., 2019). In the case of heatwaves autonomous adaptation might involve staying hydrated, remaining indoors out of the direct heat, using

Table 12.1 Definition of different generic approaches to adaptation

Type of adaptation approach	Definition
Anticipatory vs. reactive adaptation	**Anticipatory adaptation**: This involves proactively preparing for future changes or challenges based on forecasts, predictions, or an understanding of potential risks. It aims to reduce vulnerabilities and enhance resilience before the actual impact occurs. This is an ex-ante action **Reactive adaptation**: This occurs in response to an immediate threat or a realized change. Reactive adaptation is often a result of the need to address unexpected events or disruptions that have already taken place. It is usually an ex-poste action
Autonomous vs. planned adaptation	**Autonomous adaptation**: This type of adaptation happens spontaneously or without central coordination. Individuals or small groups independently adjust in response to changing conditions, often based on their own knowledge, experience, or intuition **Planned adaptation**: This involves a deliberate and organized effort to prepare for and respond to changes. It often requires coordination, strategic planning, and the implementation of specific measures or policies to enhance adaptive capacity
Incremental vs. transformational adaptation	**Incremental adaptation**: This approach involves making gradual and incremental adjustments to existing systems, processes, or strategies. It focuses on building on current practices and improving them over time **Transformational adaptation**: This is a more radical and fundamental change in response to a significant shift in circumstances. It often requires rethinking and restructuring existing systems, sometimes leading to a complete transformation of the way things are done

Partly based on McCarthy (2001)

a personal fan, seeking out shade, adjusting agricultural practices or making a temporary move away from a heatwave affected area (Hlahla & Hill, 2018; Kemen et al., 2021). Large-scale adaptation usually takes the form of top-down planned conscious intervention. Normally government (public) led, it contrasts with autonomous spontaneous adjustments (Teebken et al., 2023). Large-scale adaptation might include a national plan to address heatwaves as a threat to society and economy. The cumulative actions of individuals across a large geographical area in response to a national heatwave plan could be considered as an example of large-scale adaptation. Notably, there has been a growing recognition in recent years of the need for localized and context-specific heat and heatwave adaptation strategies (Richmond & Hill, 2023), with the possibilities what autonomous adaptation might offer as part of a seamless approach to changing climate conditions often overlooked, especially in the global north (Teebken et al., 2023). This shift is reflected in the emphasis on climate-resilient development and the integration of adaptation into broader sustainability goals given incessant heatwaves may prejudice the achievement of sustainable development at a range of spatial scales (Debnath et al., 2023; Nunes, 2023). Theoretically—there are few documented examples—adaptation speed can vary between fast and slow. Fast responses are generally those occurring within a year. These are often autonomous actions at the small scale. Slow responses in contrast necessitate an extended period of several years for planning and implementation. Usually, slow responses are a result of governmental initiatives and occur at the equivalent of the urban scale and beyond.

Allied with the concept of adaptation are the constructs of adaptation limits and maladaptation. Adaptation limits refer to the constraints or boundaries beyond which the ability of a system to adapt to changing conditions is compromised. This applies to both human and natural systems. In the case of human systems, linked to the notion of adaptation limits is the term loss and damage. Although there is no universally agreed definition of the term, it has been at the centre of ongoing discussions at successive COPs (Serdeczny, 2023). Loss and damage, which has also been linked to the idea of residual risk, generally refers to the risks and impacts, positioned beyond what adaptation can achieve (Orlove, 2022), or in other words, the risks and impacts lying beyond the so-called solutions space (Haasnoot et al., 2020). In the context of extreme heat and heatwaves, understanding adaptation limits is crucial for developing effective strategies to address the challenges posed by changes in heatwave climatology whether it be associated with climate variability or climate change. Adaptation limits can be categorized as either hard or soft. Soft limits denote conditions where no current adaptation options are viable, but future possibilities may emerge. On the other hand, hard limits indicate situations where existing adaptation strategies become ineffective, and no further alternatives are feasible. The likelihood of encountering hard limits for heatwave adaptation will no doubt rise with higher levels of global warming, especially for regions that presently experience a high frequency of heatwaves and are projected to become heatwave hotspots (Thompson et al., 2023) with heatwave related heat possibly exceeding the limits of human adaptability (Pal & Eltahir, 2016; Sherwood & Huber, 2010; Roberts et al., 2023; Vecellio et al., 2023). The constraints that influence adaptation limits are determined by factors that may or may not be surmountable through adaptive actions with the pace at which heatwave impacts manifest also likely to play a pivotal role in shaping limits. In some geographical areas and social and economic sectors, the primary factors contributing to soft limits are predominantly rooted in financial, governance, institutional, and policy constraints. The capacity of stakeholders to navigate and overcome these socioeconomic limitations significantly shapes the feasibility of implementing further adaptation measures, thereby preventing transformational change. For example, vulnerable communities experiencing heat inequity may face challenges in accessing information, resources, and support systems necessary for successful heatwave adaptation.

Maladaptation refers to actions, policies, or strategies that, despite being implemented with the intention of adapting to a changing environment or addressing a particular challenge, result in unintended and adverse consequences (Schipper, 2020, 2022). Maladaptation is an example of 'response risk' with this contrasting starkly with successful adaptation. Perhaps the worst outcome of maladaptation is the so-called lock-in effect in which improperly designed and costly adaptation strategies diminish the ability of future generations to overturn these (Teebken et al., 2023). Essentially the construct of maladaptation acknowledges that not all responses related to adaptation necessarily mitigate risks and may fundamentally effect resilience in a negative way. In general, maladaptation can arise from various impediments, encompassing administrative, human, financial, and technical resource limitations (Ara Begum et al., 2022). Factors contributing to maladaptation include deficiencies in transparency and governance capacity, unreliable climate impact information, and the absence of essential policy guidelines. Entrenched institutional, legal, and technical hurdles, along with low literacy levels, including environmental and scientific literacy contribute to the maladaptive landscape (Ara Begum et al., 2022). In the context of heatwaves, maladaptation involves instances where the efforts to cope with changes in heatwave characteristics inadvertently exacerbate other heatwave-related problems or create new heatwave vulnerabilities. For this reason, understanding maladaptation to heatwaves helps to identify and rectify strategies that may hinder rather than enhance resilience to heatwaves (Bertana et al., 2022; Schipper, 2022). An often-cited maladaptation is the heavy reliance of individuals and communities on air conditioning in response to heatwaves. While air conditioning provides immediate relief, widespread dependence can lead to unintended consequences such as increased energy consumption and associated anthropogenic heat flux (see Chap. 8), strain on power grids (Viguie et al., 2020, 2021) and higher greenhouse gas emissions, all of which contribute to local and larger scale feedback processes that exacerbate the heat hazard.

12.3 Heatwave Adaptation Options

As one can imagine there is a plethora of potential adaptation actions/options in response to climate extremes and it is probably only 'human' to want to categorise such actions and discern which are most effective and the underlying reasons for their success. Despite the vast and rapidly expanding body of literature on climate related adaptation, which spans diverse academic communities, documented syntheses on adaptation actions are few. Those that exist usually are non-specific in terms of the hazard, sector, or actor. Perhaps the most widely recognised categorisation of generic adaptation options in relation to changing climate conditions is that developed by the IPCC in its WGII 5AR. This identifies structural/physical, social, and institutional as the three main categories of options.

In general, structural/physical options involve the development and implementation of various types of infrastructure, engineering works and elements of the built environment to reduce vulnerability and enhance resilience. These adaptations are often tangible and involve constructed or natural objects or the modification of existing ones. In Chap. 14 a case is made for these to be considered as mitigation options. Social options for adaptation refer to strategies and initiatives that involve communities, individuals, and social networks in addressing and coping with the climate-related impacts. These options recognize the importance of social structures, cultural practices, and community resilience in enhancing heatwave adaptive capacity. Institutional options refer to various organizational and governance structures that can be implemented to address and respond to heatwaves as a risk management problem. Potentially institutions can play a crucial role in coordinating and facilitating heat and heatwave adaptation efforts at different levels, from local to national and even international. Of course, these options should not be viewed as operating in isolation. Rather integrating social with institutional and structural/physical options creates the potential for a more holistic and effective adaptation strategy.

Complementing the IPCC's categorisation of adaptation options is that of the Global Adaptation Mapping Initiative (GAMI) which has grouped climate related adaptation responses under behavioural, technical/infrastructural, ecosystem or nature-based and institutional (Berrang-Ford et al., 2021). Other typologies related to adaptation consider approaches by which climate change adaptation can be tracked (Ford et al., 2013), the type of actions funded through the Global Environment Facility (Biagini et al., 2014), levels of adaptation related transformation (Hurlimann et al., 2021), perspectives on loss and damage (Boyd et al., 2017), community-based responses (Paterson & Charles, 2019), discourses about adaptation (Cotton & Stevens, 2019) and adaptation measures for specific locations (Kalbarczyk & Kalbarczyk, 2020).

As yet, few typologies of heat and heatwave related adaptation responses have emerged. Although based on a small sample size of 57 papers, eleven areas related to adaptation to heatwaves have been identified by Kiarsi et al. (2023). These include education and awareness raising, adaptation of critical infrastructure, government measures, health-related measures, early warning systems, workplace-based protective behaviours, physical condition, individual adaptive behaviours, building design and architecture, green infrastructure, and urban design. Using a subsample of 301 heatwave-related papers covering 98 countries from the GAMI project, Turek-Hankins et al. (2021) have identified four broad groupings of heatwave adaptation responses namely, behavioural and cultural, infrastructural and technological, institutional, and ecosystem-based, across seven broad human activity categories. These are presented in Table 12.2 which reveals a rich assortment of heatwave adaptation responses. However, these are not applied evenly across broad geographical regions. Beyond Northern America, Europe, and Eastern Asia, the prevalent responses are predominantly behavioural or cultural in low- and middle-income countries. In Southern, Northern, and Western Europe, infrastructural or technological (28%), behavioural or cultural (26%), and institutional (43%) responses are evenly distributed across countries. In North America, institutional responses take precedence, constituting the majority at 55%. Regardless of region, ecosystem- or nature-based responses comprised less than 15% of adaptations (Turek-Hankins et al., 2021). According to this research four types of actors appear to be involved in the implementation of heatwave adaptation actions namely, communities and/or individuals, the private sector, technical professionals and/or industry, academics, non-profits and/or civil societies, and government. Regardless of region, communities and/or individuals are the main actors in heat adaptation applying mainly autonomous behavioural and cultural responses. However, when broken down by region individual/community-based actions are mainly found in Africa, Eastern, Southern and Southeast Asia, Latin America and the Caribbean. In contrast, government is the

Table 12.2 Heatwave response types and examples for a range of activity categories

Type	Category	Example
BC	L&AC	Changing livestock breed or animal type; changing livestock or animal management practice; supplementing or changing livestock food or water types or quantities; providing medicinal plants to livestock; relocating livestock to cool areas
	FFS	Diversify cropping system; changing planting and harvesting schedules; land fallowing or not planting; changing land use decisions; diversifying crops; combining varieties and pasture diversification; switching to monoculture or commercial crops; shade management and shelter for crops or aquatic environments; seed saving programs, banks, or kits; multi-cropping, mixed cropping, intercropping, mixed grazing and/or integrating animals with crops; purchasing water or fodder; reducing herd size, farm size, or food expenditure; switching from crops to livestock or animal-based production; changing crop management, preparation, planting, or harvesting practices; creating community, home, or urban gardens; using weather or climate information; consulting experts; marketing methods or processing crops pre-sale; changing water management or allocation decisions; adopting, building or changing irrigation practices; water rationing, harvesting and/or storage
	ES	Diversification of income; migration; remittances, selling or renting assets; reduce expenses; diversification of industry
	P&O	Ritual ceremonies or prayer; lobbying or advocating for policies; community members teaching traditional skills
	OH&S	Ceasing or rescheduling work; taking more frequent breaks and hydrating; changing employment or livelihood based on the seasons; applying sun or heat protection; wearing light or wet clothes; taking traditional medicines
	BE&I	Government providing energy efficiency incentives; urban water conservation; reducing energy consumption (BC);
	PC&W-B	Changing clothes or wearing protective gear; drinking more water; eating cold foods and/or changing diet; endure discomfort to acclimatize; limiting physical or outdoor activity; mobilizing community networks or asking for help from others; relocating to cool or shaded places; sleeping or spending time outside; opening windows or creating airflow; using wet sheets or towels; turning off or avoiding heat generating appliances; avoiding excessive alcohol; bathing frequently or swimming; altering mobility or transportation preferences; accessing and using weather information; visiting homes and checking on others; taking traditional medicines
IT	L&AC	Improving shade, ventilation, access to water bodies or cooling systems for livestock; exchanging tidal water; increasing pond depth and taking measures to reduce water salinity; improved methods to store fish
	FFS	New technical equipment or forms of farm organization; changing or using an improved crop species or type; use of pesticides and fertilizer; community storage facilities and cooperatives; manure and/or other chemical aids; switching to organic fertilizers, pesticides or production; creating water storage or management infrastructure; creating irrigation infrastructure; diversifying; changing; or creating new water sources
	P&O	Inspecting homes; community organizing and developing social networks and capacity; creating health surveillance systems; land use and urban planning

(continued)

Table 12.2 (continued)

Type	Category	Example
	OH&S	Shading or actively cooling workspaces
	BE&I	Building structural retrofits and weatherization; greening outdoor spaces, creating green corridors or using green or blue infrastructure; greening walls or roofs; improved shading of structures or buildings; increasing structure or surface albedo; installing solar panels; pavement surface watering or improving evapotranspiration; resurfacing and improving roads; improving or providing additional natural ventilation in and between buildings; installing; updating or improving air conditioning (AC) unit; installing shutters, reflective curtains or blinds; using energy efficient appliances or switching off heat producing units
	PC&W-B	Using AC, fans or cooling units
IS	L&AC	Using livestock loans or insurance
	FFS	Using crop loans or insurance; water governance, policy reform and education (IS);
	ES	Government aid or social assistance programs
	P&O	Communicating or educating the public about risks; coordinating government and non-government stakeholders; creating communication networks; creating a network of professionals to support vulnerable populations; creating cooling centres or distributing water, fans or AC; creating or updating a climate action plan; developing data monitoring and evaluation systems; identifying vulnerable communities or conducting risk assessments; issuing extreme heat alerts or weather related information; implementing or updating a heat action plan; relocating people to cool spaces; targeting resources to vulnerable communities; improving emergency response capacity and coordination; heat emergency or help hotlines
	OH&S	Changing work pace or rate; heat stress training; developing or recommending policies for workers; rotating staff through roles
	BE&I	Changing building occupant practices; improving or creating building codes
	PC&W-B	
EB	FFS	Using nature-based solutions; agroforestry, afforestation, reforestation and/or planting trees; forest protection, natural regeneration and/or biological conservation; using water and soil moisture conservation practices or technologies
	BE&I	Restoring natural spaces

Adapted from Turek-Hankins et al. (2021)

For the types, BC is behavioural and cultural, IT is infrastructural and technological, IS is institutional and EB is ecosystem-based. For the categories, L&AC is livestock & animal care, FFS is farming, forestry and subsistence, ES is economic stability, P&O is planning and organisation, OH&S is occupational heath and safety, BE&I is built environment and infrastructure, PC&W-B is personal care and well-being. The absence of a category for a type indicates no heatwave response actions were recorded in the literature analysed

major actor in Northern America, across Europe and Western and Central Asia (Turek-Hankins et al., 2021).

Rather than delving into the detail of the vast range of adaptation options identified by Turek-Hankins et al. (2021) and others (Kiarsi et al., 2023; Lanza et al., 2023), the rest of this chapter focuses on the subject matter of heatwave risk governance, heatwave insurance, Chief Heat Officers and the development of knowledge networks and portals related to managing the risk arising from heatwaves. Heatwave risk governance has become an ever-increasing topic of discussion with experience to date indicating that without effective governance little progress is able to be made with implementing and overseeing heatwave risk reduction strategies. Unlike many other natural hazards such as floods and earthquakes, insurance as an adaptation option remains very much in its nascency in the context of heatwaves and therefore deserves some attention. Two relatively new heatwave responses are the establishment of Chief Heat Officer positions in several cities and the rise of several knowledge networks that direct their efforts to heatwave risk reduction. As noted in the introduction to this chapter, heatwave early warning systems as an adaptation option are dealt with in the next chapter.

12.3.1 Heatwave Risk Governance

Broadly speaking governance refers to the processes, structures, and mechanisms through which governments, organizations, societies, or communities make decisions, implement policies, and ensure accountability. It encompasses the way in which authority, power, and resources are distributed and exercised. Governance is also linked to the notion of responsibility (Pellizzoni, 2004). Although the concept of governance continues to be a subject of debate, there is general agreement that governance encapsulates the dynamics of guiding collective efforts toward a common objective (Lange et al., 2013). In essence it can be viewed as a steering tool for achieving a needed outcome (van Zeijl-Rozema et al., 2008). In relation to the subject matter of this book, that outcome would be in the form of heatwave risk reduction and thus a lessening of heatwave impact with collective efforts focused on the design and implementation of heatwave related adaptation (and mitigation) options.

Much discussion in the literature gravitates towards governance modes and the type best suited to achieve a particular outcome. For example, would a particular combination of actors (e.g., public authorities, private entities, and civil society), power or institutional structures be best suited to achieving a prescribed goal? To understand governance modes various frameworks have been suggested (Arnouts et al., 2012; Bednar & Henstra, 2018; Driessen et al., 2012; Hysing, 2009; Lange et al., 2013), some of which focus on understanding the interaction of political processes (politics), institutional structures (polity) and policy content (policy) (Lange et al., 2013). Some viewpoints posit that governance signifies a departure from traditional state-centric, 'top-down' decision-making, transitioning toward more intricate networks of actors and non-hierarchical processes (Pahl-Wostl, 2009; Walker et al., 2014). Often this is portrayed as a shift from 'government to governance', or in other words a general swing from state-centric to society-centric governance (Hamman, 2020; Swyngedouw, 2005). Occasionally, and perhaps rather optimistically (Mukherji et al., 2021), this transition is referred to as new governance (Howlett et al., 2017). However, this view of a unidirectional evolution of governance has been contested with the observation that hybrid governance forms appear to coexist, residing along a continuum amid the dichotomy of state intervention and societal autonomy (Jordan et al., 2005; Lange et al., 2013; Shiroyama et al., 2012) with the state visible in governance structures more often than not (Bell & Hindmoor, 2009; Hochrainer-Stigler et al., 2023). An issue related to the effectiveness of governance is the dimension and tiers at which governance processes unfold. There is a growing body of evidence that monocentric governance, focused on structural reforms and defining responsibilities across various administrative levels, struggles to deliver on solutions and challenges for which the governance was designed, especially in relation to complex problems. Accordingly, this has given rise to concepts such as multi-level governance, polycentricity and adaptive governance as alternatives to monocentric governance (Alexander et al., 2016).

Heatwave risk governance is defined here as the institutional structures, systems, policies, practices, actors and multi-level coordination mechanisms in place to manage and mitigate the risks associated with this extreme event. Because of the ever-increasing toll on society and economy of more frequent and intense heatwaves, the imperative for well-formed heatwave risk governance has become strong (Klinke & Renn, 2021). However, the gap between heatwave risk governance and that for other natural hazards is significant. This is less than ideal because well-functioning and credible heatwave risk governance arrangements are essential for the successful implementation of a range of heatwave risk management strategies. This is a point made clearly in Priority 2 of the Sendai Framework for Disaster Risk Reduction which highlights the importance of strong risk governance to enable the development of resilience against natural hazards in general. Moreover, the lack of, or ill-configured heatwave-related risk governance, may give rise to what has been coined institutionally configured risk, a result of the non-inclusion of some key players in heatwave risk reduction (Zaidi & Pelling, 2015). Table 12.3 sets out some of the key aspects of governance. In many ways these could be viewed as a set of normative goals for the effective development and functioning of heatwave risk management strategies.

Building upon the assertion the elements outlined in Table 12.3 embody a range of potential normative goals and drawing inspiration from broader research on natural hazard risk governance (Walker et al., 2014) and more specifically, flood risk governance (Alexander et al., 2016), a profiling framework is introduced in Fig. 12.1 to stimulate conversations about the evaluation of heatwave risk governance characteristics. The framework is illustrated using examples for two hypothetical countries where heatwaves represent a significant risk to society and economy. The governance profile in Fig. 12.1a demonstrates while the national policy framework covering heatwave risk management and associated resourcing is very good and the decision-making authority and the process itself is well-articulated, with the level of risk communication also good, other aspects of heatwave risk governance are viewed as less favourable only scoring fair to poor. Such a profile may well describe the case of a mainly top-down government-led approach to heatwave risk management in which there is significant stakeholder/public disenfranchisement, or low legitimacy (Alexander et al., 2016), little mainstreaming of the principles of

Table 12.3 Aspects of governance

Aspect of governance	Type of governance activity
Resourcing	Organisations involved in managing the heatwave hazard are well resourced. This will determine whether organisations are able to undertake their role effectively. For example, are there dedicated heatwave risk management personnel and is there a sustained programme of heatwave risk research
Decision-making	Governance involves the formulation and implementation of policies, strategies, and decisions. In the context of heatwaves, it will define who has the authority to make decisions and how those decisions are made. For example, is the role of local authorities or municipalities in decision making clearly set out and what is their part in specifying their own local heatwave risk management strategies and/or implementing nation/regional level guidelines? Are there clearly defined linkages between all administrative levels associated with heatwave risk management?
Accountability	Governance establishes mechanisms for individuals and entities to be answerable for their actions and decisions. This often involves transparency, reporting, and oversight. In the case of heatwave risk management, accountability will reside with those empowered with making decisions whether they be operational actions in the case of a heatwave action plan or financial in relation to releasing funds for the development and implementation of heatwave adaptation and mitigation options
National Policy and legal framework	Good governance is often associated with a clear articulation via national level policy of the objectives associated with the natural hazard (e.g., heatwaves) related risk reduction. Adherence to the rule of law, ensuring that decisions and actions are in accordance with established laws and regulations is also a critical factor in good governance. For example, in the United Kingdom the legal responsibility for the heatwave related heat-health alert system covering England and Wales resides with the UK Health Security Agency (UKHSA) as part of its Adverse Weather and Health Plan. In turn the UKHSA, reports directly to the Secretary of State for Health and Social Care whose responsibilities are set out in the parliamentary bill entitled the Health and Care Act, a bill to make provision about health and social care
Participation and representation	In democratic governance, the active participation of citizens, stakeholders and non-governmental organisations in decision-making processes is essential. Participation and representation ensures that diverse interests are considered. For example, in the design and development of a heatwave management plan, a range of actors (multi-stakeholders) should be involved, with the principles associated with co-design and/or co-production mainstreamed throughout the development process

(continued)

Table 12.3 (continued)

Aspect of governance	Type of governance activity
Transparency	Openness and transparency are crucial for effective governance. Information about decisions, processes, and outcomes should be readily available to those affected. This might also include a visible process via which interested parties may contest decisions. An example of transparency might include regular reports related to heatwave events including details related to warning processes and an evaluation of outcomes
Responsiveness	Governance systems need to be responsive to the needs and concerns of the people they serve. This involves not only making decisions but also adapting to changing circumstances. Responsiveness may involve, for example, moving from a hazard-based (heatwave event only) to an impact-based (e.g., likelihood of a certain level of heat-related mortality being exceeded) warning system in response to requests from stakeholders who want to manage risk in line with an expected impact
Equity and inclusiveness	Good governance promotes fairness and inclusivity, ensuring that the benefits and burdens of decisions are distributed fairly among different segments of the population. Ensuring that all sectors of the population making up a community receive heatwave related information represents an example of inclusiveness. That is, regardless of a person's or group's status there should be equal access to heatwave risk related information, whether it be in the form of educational material or warnings
Risk communication	Effective, frequent and substantive communication with major heatwave risk stakeholders and the public also forms a critical element of effective governance

Source: Author

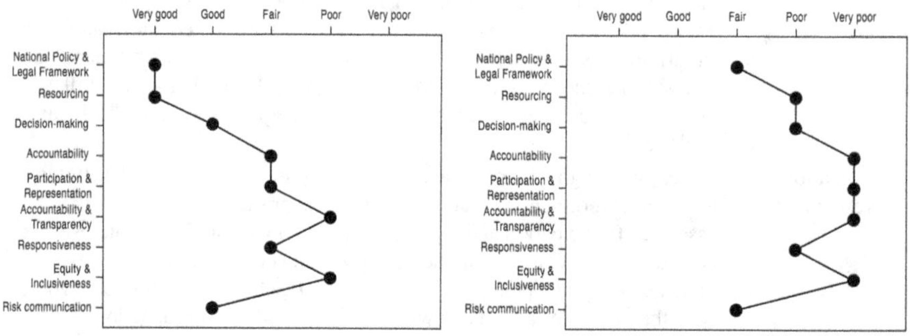

Fig. 12.1 Example of heatwave risk governance profiles for two contrasting situations. The qualitative scale of very good to very poor describes the status of each of the heatwave risk governance facets. (Source: Author)

co-design/co-production into the development of heatwave risk management strategies and responsiveness by the government to feedback is not overly enthusiastic or possibly manufactured. The profile portrayed in Fig. 12.1b represents a rather woeful situation. Despite a significant level of heatwave risk, the government response in terms of a national policy as well as communication of the level of risk to its citizens is no better than fair, or perhaps fair in name only. This may portray a situation in which the governing authority is in a possible state of denial about the threat posed by heatwaves or where the capacity to respond to heatwave risk is constrained by spending on other priorities, for example defence spending or other distractions. Predictably, the remaining aspects of governance are nowhere near achieving any degree of maturity, raising concerns about the potential emergence of a self-sustaining cycle of dysfunction. Consequently, in this case, there is a looming risk that some, if not all, aspects of governance currently rated as fair or poor, may further deteriorate towards a classification of very poor. Of course, this worsening situation may be ameliorated through communities self-organising in response to injustices, inequality, or political dissatisfaction by acting autonomously and achieving heatwave risk reduction via drawing on what limited local resources are available in combination with applying local (indigenous) knowledge. This process of self-determination in response to a pressing need related to an environmental threat such as heatwaves has often been referred to as a socio-ecological system of governance (Seng, 2012).

Admittedly the governance aspects listed in Table 12.3 and included in the profiling framework (Fig. 12.1) are not exhaustive. For example, additional criteria such as the degree of involvement of regional and local/municipal organisations, the extent to which those at risk bear the responsibility for protecting themselves (Walker et al., 2014), as well as resource efficiency (Alexander et al., 2016), could be included. That aside, the two examples demonstrate how profiling could usefully reveal contrasts in governance characteristics between different jurisdictions, an outcome of which might be critical reflection regarding the present state of governance and what might be needed to exact any improvements and who may assist with this. Comparisons could also be made between the governance profiles created by different individuals or groups involved in heatwave risk governance for the same location. As noted by Walker et al. (2014), the qualitative aspect of such a risk profiling tool can prove highly beneficial in the local context, allowing each user to generate their own unique version with discussions related to similar or divergent perspectives to follow, hopefully leading to collaborative efforts in settling on a shared profile. This becomes especially valuable in situations where there are evident differences of opinion that require deeper understanding or external pressures for change that are disrupting existing governance structures (Walker et al., 2014). Moreover, considering the historically subordinate status of heatwaves in comparison to other natural threats, one could leverage a comparative analysis of the risk governance profiles for heatwaves and another hazard, such as floods. This comparison could serve as a persuasive tool for urging policymakers and decision-makers responsible for developing natural hazard risk management policies to address and rectify any inadequacies related to heatwave risk governance and thus achieve 'levelling-up' with other natural hazards.

So, what is the empirical evidence for heatwave risk governance? This is a question more easily asked than answered. This is because despite the burgeoning literature addressing heatwave risk management, heat risk governance remains a relatively nascent subject. Consequently, there is an increasing number of calls for systematic programmes of research directed towards establishing the nature of heat and heatwave risk governance (Keith et al., 2023; Meerow & Keith, 2022). One of the standout issues related to heatwave risk governance is which organisations or sectors are best placed to lead in the deployment and coordination of heatwave risk related interventions whether they be adaptation or mitigation oriented (Lass et al., 2011). Or alternatively, is it the responsibility of individual citizens to protect themselves against heatwave impacts? Added to this is the question, might the way extreme heat is problematised influence its governance (Oppermann et al., 2017)? For example, in India the way in which heat is framed, either as a disaster or as a local issue, has a profound impact on heat (thermal) governance (Khandekar et al., 2023).

Guyer et al. (2019) report from a workshop, diverse perspectives on heat and heatwave governance for Phoenix Arizona. Suggestions included empowering communities, non-profits, faith-based organizations, and grassroots groups to play a substantial role in ensuring heatwave risk reduction of vulnerable individuals because of their community-based knowledge and trust, as opposed to government agencies or private corporations. Conversely, another viewpoint emphasized the need for leadership from private utility companies, given their crucial role in providing heat refuge. The need for the involvement of state and local emergency management services in preparing communities for heatwave-related challenges was also articulated. However, participants from this sector expressed uncertainty regarding how heat preparedness aligns with their current responsibilities, with some professionals perceiving heat as outside the scope of their work (Guyer et al., 2019). Other examples that illustrate heat and heatwave governance uncertainty include those from The Netherlands. Here, disagreement as to who bears responsibility for the implementation of heatwave related protection measures and the degree to which stakeholders in health and social services are aware of associated plans or their role in them, has been noted (Mees et al., 2015; Loenhout et al., 2016). Further, some local authorities in The Netherlands have asserted a lack of comprehension regarding the significance of heatwave impacts, hindering them from taking the essential measures for implementing practical actions to mitigate the risks linked to an escalating frequency of heatwave events (Klok & Kluck, 2018). Specifically for the pre-school education sector in Sweden, Malmquist et al. (2022) report little organizational readiness resulting in uncoordinated responses to heatwaves which bear implications for increased heatwave related vulnerability amongst not only preschoolers but the staff as well. For California it has been found that cities often take an ad hoc approach to heat governance with a lack of a central authority providing coordination, technical assistance and strategic funding to address extreme heat (Gabbe et al., 2021; Schmeltz et al., 2023). More broadly across the US, there appears to be two emerging systems of heat-related governance. Heatwaves/extreme heat events are often associated with institutional actions focused on hazard planning, whereas heat in the context of the urban heat island is more frequently linked with interventions involving green and physical infrastructure as part of overall city planning (Turner et al., 2022). Further, although aspects of heat planning occur across a variety of US city municipal plans, only a small number of strategies appear to be explicitly framed in terms of heat and heatwaves, suggesting an opportunity to better connect this hazard with other policy goals (Keith et al., 2023). Randazza et al. (2023) also note for the US, the majority of heat and heatwave related plans are led by the local Office of Emergency Management (81%) with public health departments in the minority (28%) as far as taking a leading governance role. For Scotland, there appears to be a clear view emerging that the public health service sector in the form of Public Health Scotland should take the lead on any heat-health strategy (Wan et al., 2023). This is the situation presently in place in England where the heat-health alert system is overseen by the UK Health Security Agency (previously Public Health England), an executive agency sponsored by the Department of Health and Social Care. Interestingly, the view that public health agencies should be involved in heat and heatwave governance appears to be a trend for several countries (Sheehan et al., 2022). This chimes with a wider call for progress beyond merely reacting to extreme heat emergencies through shifting from an emergency-centric approach to adopting a conventional public health governance structure focused on tackling the underlying structural inequities and root causes of heatwave related health risks as this would foster the development of enduring policies, facilitating the transition of public health care systems towards enhancing resilience (Guardaro, 2023). Notwithstanding a 'one-size-fits-all' model of heatwave governance is unlikely, because of between and within country variations in political structures, the apparent preference in several cases for the public health agency to lead heatwave-health risk reduction initiatives is a reassuring trend. This is because one of the necessary starting points for putting strong governance structures in place is the early identification of a lead agency. This approach in many ways underlies the success of the establishment of heat-health governance arrangements for Adelaide, south Australia (Akompab et al., 2013) and the development of a harmonised Heat Warning and Information System (HWIS) for Ontario ((Henderson et al., 2020).

Of course, once a lead agency has been identified, a necessary corollary is securement of the necessary capital and human resources as well as sorting out the associated legal, political and general governance architecture (e.g., major stakeholders) issues including accounting for distributional as well as procedural and recognitional equity (Meerow et al., 2019) and rupturing the silos in which heat governance may have already settled, so that heatwave risk management actions can be delivered with maximum effect (Sovacool & Martiskainen, 2020). Perhaps it is no surprise that these ideals are often not achieved with some distance between theory and practice. This has been laid bare in an honest and forthright assessment of heatwave risk governance for the city of Karachi, Pakistan where researchers have concluded that the governance of heatwave related extreme heat is inherently set up for failure. Specifically the lack of reliable, long-term empirical data and thorough analysis, combined with a restricted comprehension of the city's inequalities at the policy level results in the inadequacy of planning for heat governance, with this deficiency not only rooted in the inflexible perceptions of heatwave risk held by planners and policymakers but also the limited administrative capabilities and resources available at the local level to effectively address and mitigate heatwave-related risks (Anwar et al., 2022). Similarly for Europe, although most heatwave related action plans prioritize governance as a key goal, a review of governance arrangements has revealed the need for a more deliberate and structured approach to governance (Lass et al., 2011; Vanderplanken et al., 2021). This is especially with regards to adopting a cross-sectoral approach to identifying key stakeholders which can lay the basis for comprehensive heatwave preparedness and response with clear definition and assignment of roles and responsibilities among stakeholders to minimize confusion and enhance effective implementation. In this regard, Vanderplanken et al. (2021) outline seven key roles and their commonly associated stakeholders as a basis for heatwave risk governance (Table 12.4). Notwithstanding their derivation from an analysis of the European situation, these roles appear to have general applicability in the context of heatwave risk governance arrangements for heatwave related action plans.

To conclude this section, it's worth noting that while governance has been discussed here as a response (an adaptation action), there's likely an argument, although not explored at all here, for considering governance as an enabler of responses, whether they take the form of adaptation or mitigation actions. Given this, governance could be viewed as anticipatory and facilitative rather than merely reactive.

12.3.2 Heatwave Insurance

As a response to increased climate variability and the prospect of climate change there has been a rapid growth in what can be broadly referred to as climate risk insurance (Bernards, 2022; Johnson, 2021b; Kousky, 2019; Kousky & Light, 2019; Lin et al., 2023) and the unfurling of opportunities for climatologists to work with the insurance industry (Changnon et al., 1999; Changnon, 2007; Hoeppe, 2016; Randalls, 2010). Broadly climate risk insurance serves as a fundamental risk transfer mechanism, providing individuals and businesses with a means to mitigate the financial impact of extreme weather and climate events. At its core, climate risk insurance involves the transfer of risk from policyholders to insurers. In exchange for a premium, the insured party receives a contractual promise from the insurer to compensate for covered weather and climate related losses or damages or if a specified weather or climate related event exceeding some predefined threshold occurs. In essence climate risk insurance as a risk transfer mechanism provides a safety net, allowing individuals and organizations to assign the uncertainties of the future to a specialized entity equipped to manage and absorb climate related risks. As a result, insurance not only safeguards against unexpected losses but also fosters resilience and confidence in the face of unpredictable climate related events (Botzen et al., 2019; Jarzabkowski et al., 2019; Lamm et al., 2020). Traditionally viewed as a compensatory measure to strengthen economic and social recovery post-event, climate risk transfer can also be employed to play a role in both proactive and remedial risk management strategies (Sarmiento & Torres-Muñoz, 2020).

Generally, risk transfer mechanisms fall into the broad categories of indemnity insurance, bond financing instruments (e.g. catastrophe bonds, environmental impact bonds, forest resilience bonds, resilience bonds), captive insurance, and

Table 12.4 Key stakeholder roles for heatwave action plan governance

Role	Definition, responsibilities, skills, resources
Author	Stakeholder responsible for overseeing heatwave plan development and writing in a continuous, iterative process. Relevant national and subnational stakeholders from health and other sectors should be involved. Strongly recommended that one stakeholder (e.g., public health agency) takes the lead
Activator	Stakeholder responsible for activating the plan and/or warning system Responsibilities include decision-making, communicating the activation to internal and external stakeholders as well as the public, activating plan implementation and collaborating with the monitor(s), coordinator(s) and informer(s). Essential to this role is authority and communication lines with other stakeholders
Coordinator	Stakeholder responsible for inter-organizational coordination of plan implementation and collaboration. Performing this role requires authority, an inter-organizational network, coordinating skills and sufficient human and financial resources
Evaluator	Stakeholder responsible for evaluating the heatwave related action plan, the evolving heatwave situation and plan implementation during heatwave occurrence. Assessment of the effectiveness and relevance of implemented measures, the accuracy and impact of parameters and any triggers, identifying difficulties and lessons learned and updating the heatwave action plan are key to this role. Further a systematic approach and clear definition of evaluation indicators is critical
Implementer	Stakeholder responsible for implementing the measures described in the heatwave action plan. Performing this role requires a clear definition of actions and how to realize them
Informer	Stakeholder responsible for diffusing information to internal and/or external stakeholders and/or the public, preferably in a bi-directional communication flow. Performing this role requires access to information material, identification of target audiences; connections to internal and/or external stakeholders and communication skills
Monitor	Stakeholder responsible for monitoring parameter(s). Responsibilities include anticipating when critical values may be reached, informing relevant stakeholders and providing advice on the definition of parameters and thresholds. This may very well involve collaborations between national/local meteorological and public health agencies

Adapted from Vanderplanken et al. (2021)

parametric insurance (Jarzabkowski et al., 2019; Keucheyan, 2018; Lamm et al., 2020; Vajjhala & Rhodes, 2019). Although there is significant convergence in the risk transfer mechanisms devised by insurers and governments to manage risks associated with hurricanes, storms, tornadoes, floods, wildfires and droughts, the application of climate risk transfer methods to heatwaves and extreme heat is still in its infancy with theoretical considerations pointing to parametric insurance as perhaps the most practical in the case of heatwave risk transfer (Larsson, 2023), especially at the individual or small business level in low latitude to subtropical countries where heatwaves have become endemic. As a form of risk transfer mechanism parametric insurance has been present since the late 1990s having developed from the practice of catastrophe bond issuance (commonly referred to as CAT bonds). In the

Table 12.5 Basic differences between weather-indexed insurance and weather derivatives[a]

Characteristic	Weather indexed insurance	Weather derivatives
Structure	Weather-index insurance is based on predefined trigger points in the form of weather indices or parameters with a fixed value. These indices are often based on objective weather data from specific locations. Weather-indexed insurance usually covers low probability events. Policies are often simple and straightforward. There is generally no tiered structure in the case of weather-indexed insurance—The payout does not increase with the severity of the event	Weather derivatives are financial contracts whose value is linked to weather conditions represented by an objective weather index. Weather derivative contracts are traded on financial markets (e.g., heating degree day (HDD), and cooling degree day (CDD), futures traded by the CME Group—https://www.cmegroup.com/trading/weather/), with their value determined by changes in specified weather indices. Weather derivatives usually cover higher probability events than those covered by weather-indexed insurance such as a drier than expected summer. Weather derivatives often use a tiered structure, where the payout increases in line with the severity of the event—there may be different trigger levels for different levels of severity
Purpose	The primary purpose is to compensate policyholders for losses or expenses incurred due to adverse weather conditions. For example, a farmer might purchase weather-indexed insurance to protect against crop losses caused by a period of extremely high temperatures during the growing season	The primary purpose of weather derivatives is to provide financial hedging against the impact of weather-related risks. Market participants, such as businesses in weather-sensitive industries (e.g., energy, agriculture), use weather derivatives to manage their exposure to weather variability and possible decline in demand for their services resulting from unusual weather conditions such as a downturn in the sales of fans during a cool summer
Customisation	While weather-indexed insurance can involve customisation, the scope is often straightforward. The terms are frequently designed to align with specific risks faced by the insured, with customization options not as extensive as those for weather derivatives	Weather derivatives offer a high degree of customization. The parties involved can tailor the contract to cover specific weather parameters, trigger levels, and settlement mechanisms based on their unique needs. This flexibility increases the complexity as it requires careful consideration and negotiation of contract terms

(continued)

Table 12.5 (continued)

Characteristic	Weather-indexed insurance	Weather derivatives
Payout determination	Payouts are typically determined based on the deviation of the observed weather indices from predetermined thresholds. If the weather conditions trigger the agreed-upon parameters, the policyholder receives a payout. This makes the payout structure typically straightforward, directly tied to the occurrence of predefined weather conditions. In short the payout structure for weather-indexed insurance is less complex than that for weather derivatives	The payout for weather derivatives is not compensatory in the same way as insurance. Instead, the value of the derivative is settled based on the changes in weather indices, and it is often used as a financial tool for risk management. Generally, the payout for weather derivatives increases as the severity of adverse weather conditions worsens. For instance, there might be a base payout for moderate deviations, an additional payout for more extreme deviations, or payouts scaled to how far conditions have deviated from a specified level. The payout structure for weather derivatives can vary widely, and settlement mechanisms may involve financial instruments such as options or futures contracts. The complexity increases when designing sophisticated financial instruments to meet the specific risk management needs of the parties involved

[a]Based on Alaton et al. (2002), Bems and Aydin (2022), Benth and Benth (2013), Larsson (2023)

case of weather-related hazards, parametric insurance often takes the form of weather-indexed insurance. In a strict sense this type of insurance should not be confused with weather derivatives which, although often viewed as a form of parametric insurance (Alaton et al., 2002; Bems & Aydin, 2022; Benth & Benth, 2013; Larsson, 2023), are relatively complex in comparison (Table 12.5).

At the heart of weather-indexed insurance is the use of specific, quantifiable indices, often derived from meteorological measurements. These indices (parameters hence parametric) serve as triggers, indicating the occurrence of a predefined event with an automatic insurance payout triggered. This contrasts with traditional indemnity insurance which often involves lengthy claims investigations and loss assessments that lead to delays in receiving funds crucial for financial recovery. Their relatively low cost and ability to provide rapid financial assistance to those affected by extreme weather events makes weather-indexed insurance policies particularly suited to individuals and small businesses with limited financial resources and whose income is weather-dependent. For this reason, weather-indexed insurance as a type of micro-insurance—small, simplified insurance policies targeting the poorest—has emerged as a potential risk transfer mechanism in several low- and middle-income countries especially in the agricultural sector (Adetoro et al., 2022; Collier et al., 2021; Singh & Agrawal, 2019). However, despite the apparent attractiveness of this risk transfer mechanism, uptake of weather-indexed insurance in the agricultural sector has not been as extensive as imagined with factors such as lack of understanding of insurance as a concept, heterogeneous preferences between men and women for insurance, product cost and quality, limited trust in the insurer, willingness to pay and liquidity constraints all forming barriers (Akter et al., 2016; Bernards, 2022; Bogale, 2015; Hill et al., 2013; Maganga et al., 2021; Murphy & Ichinkhorloo, 2023; Platteau et al., 2017; Tadesse et al., 2017). Added to this there is some suggestion that parametric based insurance instruments do not deliver what they promise (Johnson, 2021a). Moreover, even when in place, although insurance may improve recovery outcomes, its impact on mitigating risks through preparatory adaptation measures may be somewhat limited (Kousky, 2019).

While concerning in their scope, the limiting factors identified for the adoption of micro-insurance in the agricultural sector could provide valuable insights for the development of micro-insurance products tailored to address heatwave-related concerns. Furthermore, assessing the insurability of heatwaves as a risk will require consideration if this form of insurance is to be taken forward as a credible adaptation option. This will necessarily involve examining the extent to which heatwaves as a physical hazard and a threat to society and economy are fully understood and whether they will continue to be genuinely random occurrences under the influence of anthropogenic climate change, as well as ensuring that heatwave insurance doesn't induce moral hazard, and confirmation of the feasibility of risk pooling (Biener & Eling, 2012; Stranks, 2023). In developing indices as a basis for heatwave-indexed insurance it will also be necessary to acknowledge that the simplicity that makes parametric insurance attractive can also be a drawback, as certain nuances of loss may not be fully captured by predefined parameters. For example, heatwave effects may manifest below a predefined trigger temperature and duration with loss incurred with no compensation via policy payout. Additionally, accurately defining the weather index parameters to ensure they truly reflect the heatwave hazard for the area of interest (Leppert et al., 2021) and losses incurred may prove a challenge because of meteorological and impact data constraints and ill-defined heatwave-impact relationships. There is also the issue of whether the triggering mechanism for payout should be a fixed index value in time that serves as a single trigger point, or be composed of multiple trigger points that reflect diurnal variations in critical health or economic sector relevant levels of heat (e.g., maximum and minimum temperature thresholds) or indeed points throughout a heatwave event for example based on the rate of change of a critical heatwave related variable or thermal index. Further a one size fits all model, may prove insensitive to the precarity of vulnerable communities or sectors of the working population in the case of heatwaves making a differential risk transfer parametric approach preferable (Lopera et al., 2022).

Setting aside the above mainly theoretical considerations, what empirical evidence is there for heatwave related insurance? Historically, the insurance industry has largely dismissed the notion that heatwaves, as a meteorological phenomenon, constitute a significant peril. Consequently, insurance products specifically designed to address heatwaves have been notably absent from the array of offerings in the portfolios of insurance companies. This holds particularly true in affluent nations, where potential impacts of heatwaves on infrastructure are typically addressed through comprehensive indemnity insurance policies. Comprehensive personal health insurance also renders specialized heat illness coverage redundant. Further in work environments, robust safeguards against heat, as well as paid sick leave and unemployment benefits, are often established under employment laws. Additionally, employers in high income countries frequently carry workers' compensation insurance, encompassing on-the-job injuries, including those associated with heat. That said, there are signs of an emerging market for heatwave insurance (Stranks, 2023) with the appearance of several heatwave related insurance products on the market. Those for which reasonable detail exists include extreme heat income insurance for women workers in India (Dabrowski, 2023; Ro, 2023), heat stroke insurance in Japan (Kohyama, 2023), parametric heat-stress insurance for the UK dairy industry and that in the state of Kerala India (Anon, 2023) and a climate insurance-linked resilient infrastructure financing instrument (Richmond & Ortega-Pastor, 2022). Added to this some cities are at the early stages of investigating the scope and potential use of heatwave insurance (Lamm et al., 2020).

Recognizing the challenges that many Indian women experience working in prolonged dangerously hot conditions in occupations where they are exposed to heat with very high levels of heat-related health risk, the Self-Employed Women's Association, a prominent Indian women's trade union, an insurance technology firm, and a global non-profit worked collaboratively to develop one of the world's first weather(heat)-indexed insurance products to assist informal-sector women workers to recover lost income because of the inability to work during heatwaves. With the overarching goal to prevent female workers from facing the dilemma of choosing between safeguarding their health and ensuring their family's financial security, the experimental Extreme Heat Income Insurance scheme was launched in the state of Gujarat in February 2023 and implemented at the beginning of the heat season in May of the same year. This parametric insurance product uses an algorithm that ingests 3 days of satellite-derived temperature data, occurring within six 10-day cycles and tailored to unique local conditions to trigger payouts to the Self-Employed Women's Association with payments then distributed to those registered for the income protection scheme. Crafted to facilitate multiple payouts within a single heat season, this micro-insurance product aims to substitute income of around (US)$3 per day whenever the prescribed temperature conditions are met. Taking Ahmedabad as an example, a payout would be triggered when the minimum cumulative three-day temperature reaches 134 °C, equivalent to a daily average of 44.7 °C (Dabrowski, 2023; Ro, 2023). Importantly and in the spirit of co-production via focus group discussions, members of the women's trade union worked with the producers of this microinsurance product to identify the temperature trigger point based on their views that work becomes challenging at temperatures around 45 °C. Focus groups also revealed that members of the trade union would be willing to contribute an insurance premium ranging from 250 to 300 rupees (US$ 3.02–3.63), particularly if the option to make payments throughout the heat season rather than in a lump sum was available. Interestingly in the trial phase of the Extreme Heat Income Insurance the cost of the premium has been subsidised by the global non-profit organisation involved in the development of the product. As the premium at approximately US$10 before taxes is notably higher than that indicated as bearable by the trade union members, the long-term sustainability of this parametric insurance scheme will very much depend on whether funds can be raised from sources other than its members and the initiating non-profit organisation (Dabrowski, 2023; Ro, 2023). In the long term for the scheme to be effective from both worker health and safety and insurer profitability perspectives there will need to be a balance between the number of false positives (payments triggered unnecessarily) and the number of false negatives (payments should have been triggered) which will depend on the robustness of the satellite-based air temperature prediction algorithm and whether this can accurately estimate the heat conditions experienced by those for whom the insurance product was designed. This is an important issue, especially as the accuracy of satellite-based air temperature estimates will be heavily reliant on robust transfer functions that faithfully describe the association between surface temperature and minimum air temperature (the trigger metric), something that is quite challenging in the case of morphologically complex urban environments such as the ones where the Extreme Heat Income Insurance has been trialled. Given the growing accuracy of short-range temperature forecasts and the improving predictability at the medium-range, and assuming the Extreme Heat Income Insurance scheme evolves into an enduring adaptation option, it will be interesting to see if the insurance scheme eventually transitions from a reactive one—payments are made once the triggering conditions are surpassed—to a forecast based one so that both workers and the insurer can manage their respective risks proactively.

As noted in Chap. 11 heatwaves have a clear impact on the agricultural sector with mounting evidence that extreme heat results in a downturn in milk production from dairy cows. In response to an increasing number of impactful heatwave events and a concern amongst dairy farmers that heat stress poses a serious threat to cattle and thus milk production related earnings, two parametric insurance products have appeared on the market, one in the UK and the other in India. Although developed completely independently, the two schemes possess the common aim to alleviate the risks and financial losses linked to heat stress in cattle with insurance coverage functioning based on a temperature index-driven parametric insurance model, providing payouts to dairy farmers based on specified trigger events. Monitoring and recording of independent satellite (UK and India) and weather-station data (UK only) are applied to the assessment of whether temperature thresholds outlined in the respective policies have been exceeded. Although the details have not been made publicly available, the UK insurance product provides a selection of four coverage levels, enabling policyholders to customize their policies according to their risk tolerance, financial constraints, and herd size (Mukhopadhyay, 2023). In the case of the scheme for Indian state of Kerala, the prescribed temperature limits that trigger payouts vary across the state being 37 °C in Palakkad and Wayanad, 34.5 °C in Kannur and Kasaragod, 33.5 °C in Malappuram and 33 °C in Kozhikode. The scheme is also tiered such that farmers will receive 140 rupees (~US$1.70) in compensation if the prescribed temperature is exceeded continuously for six days, 440 rupees (~ US$5.30) for 8 days, 900 rupees (US$11.90) for 10 days, and 2000 rupees (US$24.00) for 25 days or more (Anon, 2023).

Perhaps somewhat of a surprise, and no doubt as a reaction to the extremely hot summer of 2022 in Japan, during which temperatures in Tokyo remained above 35 °C for nine consecutive days over the period June 26—July 3, two major Japanese insurance companies broke new ground in 2022 by offering heatstroke insurance. This was designed to cover medical expenses arising from heatstroke including outpatient treatments such as intravenous drips, longer term hospitalisations and even death (Kohyama, 2023). Unlike the weather-indexed parametric insurance products developed in India for informal workers and for the dairy industry in the UK and India, the heatstroke insurance took the form of indemnity insurance with customers offered the option to purchase insurance daily at a rate of 115 Yen (~US$0.79), if purchased before a morning cut-off time, and around 230 Yen (US $1.57) for 1 month's cover (Kohyama, 2023). Initially designed to only cover children, the heatstroke insurance was promptly made available to all ages driven by demand as heatwave conditions descended on the country during the 2022 summer. Most coverage was purchased by people over 65 years old as anxiousness amongst this heatwave vulnerable group, which comprised around 29% of the total population in 2022, grew. The success of this indemnity product can be attributed, in part, to its development within a well-established insurance market (Kohyama, 2023). Further being a high-income country potential customers could readily afford the premiums with the value of insurance well understood. Likewise, the heatstroke insurance premiums could be kept low because Japan has a well-funded health care system supported by the taxpayer, necessitating only relatively small heatstroke insurance co-payments for heat illness related treatment (Kohyama, 2023; Stranks, 2023; Scherer, 2020b).

Championed by the United Nations Capital Development Fund (UNCDF) and the public and private sectors, the Climate Insurance Linked Resilient Infrastructure Financing (CILRIF) initiative has been suggested as a lasting insurance solution designed to encourage municipalities to allocate resources toward climate resilient infrastructure to de-risk potentially crippling heatwave and flood events. The fundamental idea underlying this scheme is to facilitate the access of cities in Least Developed Countries and Small Island Developing States to affordable, long-term climate insurance with predetermined premiums, contingent upon evidence that the city possesses the requisite administrative, legal, regulatory and technical capacities, is moderately politically stable, has access to finance, heatwave risk assessment has been undertaken and appropriate assurances exist that the city will invest in resiliency measures to de-risk heatwaves. Although heatwaves are not included as a peril, other insurance instruments broadly comparable to the CILRIF with similar aims and risk transfer strategies include the African Risk Capacity (Iyahen & Syroka, 2018; Scherer, 2020a), Philippine City Disaster Insurance Pool (ADB, 2018) and the Caribbean Catastrophe Risk Insurance Facility (Grove, 2012; Hochrainer-Stigler et al., 2023; Scherer, 2020b).

Still at the conceptual stage, the initial plan for the CILRIF involves establishing a centralized financial structure (CILRIF Private Facility) within which an insurance facility and a financing facility will reside (Fig. 12.2). The insurance facility will be designed to provide city governments with a climate (heatwave) insurance product of 10–15 years in duration with clearly articulated coverage terms and fixed premiums, the level of which will depend on the city's dedication to investing in climate adaptation projects aimed at bolstering the resilience of essential urban infrastructure. The expectation is the insurance facility will adopt a hybrid approach, comprising a parametric product for providing initial post-heatwave event liquidity and an indemnity product to support longer-term post-heatwave disaster reconstruction efforts such as repairing heatwave related road heave and rail buckling and rectifying any issues related to the integrity of other heatwave affected infrastructure such as electricity grids and water reticulation systems. A sense of the type and scale of the impacts the indemnity product would be applied to are akin to those outlined for an imagined 'Katrina of heat' for the US city of Phoenix (Clark et al., 2019). Complementing the insurance facility, the infrastructure finance facility (Fig. 12.2) is envisioned as a commercial investment entity with a specific emphasis on funding robust municipal infrastructure. A city's eligibility for financial support through this facility would hinge on procuring a CILRIF insurance policy. The aspiration is to achieve a global fund equivalent to US$1 billion supported by

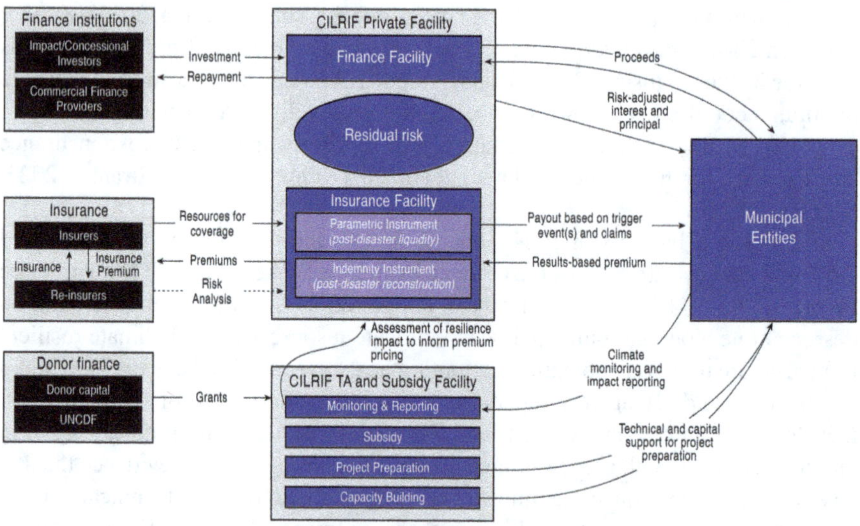

Fig. 12.2 Overall mechanics of the proposed CILRIF instrument The CILRIF approach is a result of four years of consultative dialogue of a voluntary group of individuals representing different stakeholders listed at www.uncdf.org/cilrif CILRIF has been created and led by Abhisheik Dhawan at UNCDF in an attempt to find a blended finance solution to a complex problem specially in least developed countries and other low income countries with the support of the Global Innovation Lab for Climate Finance (https://www.climatepolicyinitiative.org/the-programs/the-global-innovation-lab-for-climate-finance/) at the Climate Policy Initiative (https://www.climatepolicyinitiative.org/). Figure provided by Abhisheik Dhawan (UNCDF)

development finance institutions and donors (Fig. 12.2) capable of making direct investments in resilience initiatives outlined in the insurance contract. Accompanying the insurance and finance facilities a CILRIF Technical Assistance (TA) and Subsidy Facility would provide technical assistance to cities in low to middle income countries through grants to aid them in obtaining funding for proposed heatwave resilience initiatives, complementing the financial assistance provided by the financing facility.

It is anticipated the CILRIF will apply to two broad city types, one being a medium-sized city with moderate heatwave risk and the other a large city with high heatwave risk. For swift payouts a parametric trigger structured into three layers—10 °C, 12 °C and 15 °C above the average temperature in its hottest month of the year for 3 days or more—is imagined with the probability of achieving the trigger temperatures higher in the large high heatwave risk city compared to the medium sized moderate heatwave risk city. The types of heatwave interventions anticipated that would lead to reductions in insurance premiums include the development of an extreme heat early detection system, cooling centres, a district cooling network, and green corridors. At the time of writing UNCDF plans to pilot the CILRIF mechanism for extreme heat in Freetown, Sierra Leone, a city that experiences challenging heat conditions in relation to both high temperature and humidity levels, a pronounced urban heat island, significant changes in land surface temperatures as a result of land surface change as well as facing a number of planning issues related to rapid urban development and the proliferation of informal settlements (Kaloop et al., 2022; Mustafa et al., 2020; Tarawally et al., 2018, 2019).

Understandably for the UNCDF to achieve its CILRIF related aspirations several barriers and challenges will need to be surmounted and addressed. Potential obstacles may arise from the lack of widespread availability of long-term fixed-price climate insurance coverage, primarily due to underdeveloped insurance markets compounded by cities often experiencing financial challenges in affording existing climate insurance, particularly municipalities with sub-investment grade ratings, as their access to capital markets is constrained. The lack of incentives for cities to integrate resiliency measures into both existing and new infrastructure may also pose a barrier (Richmond & Ortega-Pastor, 2022). The challenges openly recognized by the UNCDF include issues related to the affordability of premiums, estimated to range between US$20–30 million, accurately pricing the impact of implemented resilience strategies, demonstrating the value proposition to cities and limited investment in cities from development finance institutions. Furthermore, loss of trust in the parametric insurance product could arise due to the financial loss experienced by a city exceeding the payout received, resulting in so-called basis risk. Additionally, there may be too few cities with similar climate (heatwave) risk profiles that map onto the offered insurance product to constitute a feasible pool required to spread and diversify the financial impact of heatwave risks and promote stability, affordability, and fair distribution of financial burdens (Richmond & Ortega-Pastor, 2022).

Overcoming these barriers and challenges will be important for achieving the UNCDF's ambition to create a new multi-city heatwave related risk transfer mechanism. Ideally in implementing this initiative lessons learned from other large scale extreme event risk transfer schemes will need to be heeded. In the case of the Africa Risk Capacity these include heterogeneous benefits and mixed efficiencies across risk pools and apparent contradictions of the climate-development-finance nexus (Awondo, 2019; Chamberlain & Bernards, 2024), and for the Caribbean Catastrophe Risk Insurance Facility, despite its advantages, (Hochrainer-Stigler et al., 2023), and concerns it has led to the financialization of disaster management and has become a peripheral tool for many establishments (Grove, 2012; Joyette et al., 2015). As an upshot, some critics have gone as far as to advise recipient countries to maintain a sense of scepticism regarding the supposed advantages of insurance-based mechanisms noting that numerous debt resolutions, including those related to disasters, are influenced by the same economic principles and geopolitical entities that initially led to many low and middle income countries' debilitating debt challenges in the first place (Gahman et al., 2021; Sheller, 2018).

12.3.3 Chief Heat Officers

In the face of escalating heatwave risk and the ever-increasing issue of heat as a chronic problem in cities, an emerging adaptation response is the appointment of Chief Heat Officers (CHO) within several city administrations. At the time of writing ten city-based Chief Heat Officer positions had been established as well as the appointment of a Global Chief Heat Officer. Eight of the city-based positions have been created and financially supported as part of a collaboration between the Arsht-Rock Foundation and the Extreme Heat Resilience Alliance and the respective city governments, the first being for the US city of Miami in April 2021. This was followed by Chief Heat Officer appointments in the Athens, Greece (July 2021), Freetown Sierra Leone (October 2021), Santiago, Chile (November 2021), Monterrey, Mexico (April 2022), Melbourne, Australia (two appointed October 2022) and North Dhaka, Bangladesh (May 2023). Complementing these is the Global Heat Officer post which arose from an agreement reached at the World Urban Forum in June 2022 between UN-Habitat and the Arsht-Rock Foundation to establish a partnership to address the impact of extreme urban heat by mainstreaming heat as a pressing issue across UN-Habitat's various areas of work and other UN agencies (Witze, 2023). The remaining two Chief Heat Officer positions are for the US cities of Los Angeles and Phoenix which have been funded from their respective municipal government budgets. In 2021, the city of Los Angeles established a new publicly funded Office of Climate Emergency Mobilization to specifically concentrate on the nexus of climate change, social justice, community engagement, and advocacy, aiming to bring about systemic change. Notably, the Director of this office also serves concurrently as the Chief Heat Officer for Los Angeles. In the same year, the city of Phoenix in Arizona established the United States' first publicly funded Office of Heat Response and Mitigation with its Director widely referred to as the city's Chief Heat Officer. More recently and following an unprecedented summer of extreme heat in 2023, the Arizona Governor in March 2024 appointed the US's first state-wide Chief Heat Officer.

The overarching aim of Chief Heat Officers is to work towards implementing solutions to counteract the fragmentation of authority and control often associated with heat and heatwave risk management, while actively contributing to the formulation and execution of strategies aimed at addressing heat and heatwave-related challenges, ensuring the protection of public health, and fortifying the overall heat resilience of urban areas. More specifically, some of the primary duties associated with Chief Heat Officer positions include promoting awareness of the risks associated with extreme heat and proposing heatwave risk reduction solutions to both the public and fellow professionals as well as identifying communities and neighbourhoods that face heightened vulnerability to extreme heat, enhancing planning and responses to heatwaves, orchestrating collaboration among stakeholders, and overseeing the execution of enduring heatwave adaptation and urban heat mitigation (urban cooling) projects aimed at reducing chronic heat and acute heatwave related risks. As well as facilitating the UN-Habitat Arsht-Rock partnership and advancing heat response and resilience measures in cities around the globe, one of the main tasks associated with the Global Heat Officer position is to put in place processes to establish an Urban Heat Task Force at the global level to complement the many such task forces that have sprung up in cities worldwide irrespective of whether they have an 'official' Chief Heat Officer or not.

As an example of how some of the Chief Heat Officers are working to achieve their aims, Freetown's Chief Heat Officer has focused on reducing heat exposure in informal settlements and markets with attention given to ways in which surface albedo across the city can be increased to reduce solar radiation absorption by the urban fabric. Collaborations with the Freetown City Council have led to the implementation of market shade covers, safeguarding many street vendors from solar heat load. Crafted from heat-reflective plexiglass, these covers offer protection against the intense sun and heavy rainfall. Beyond providing shade, the covers incorporate solar panels to capture energy throughout the day. This stored energy is then utilized to power lights during the night, enabling store holders, the majority of whom are women, to prolong their opening hours. In a similar fashion, the Freetown Chief Heat Officer has worked with the non-profit organization Mirrors for Earth's Energy Rebalancing and the local council on a pilot 'cool roofs' programme in one of Freetown's informal settlements where the roofs of 55 households, covering a combined area of 1200 square meters have been coated with a reflective mirror-like film to lower indoor temperatures. In Santiago, the Chief Heat Officer is championing the planting of trees to increase urban shade and the implementation of green roofs on public facilities, including schools and hospitals. For example, the retrofitting of the Hospital de Maipú located in the commune of Maipú within the wider Santiago Metropolitan Region, via the installation 1000 square metres of vegetated rooftop to enhance the building's cooling capabilities. Working within the wider framework of the city's heat action plan, Miami's Chief Heat Office is focused on achieving three goals related to extreme heat and heatwave risk reduction organised under the themes of 'inform, prepare, and protect people', 'cool our homes and emergency facilities' and 'cool our neighbourhoods'.

While the appointment of Chief Heat Officers reflects a possible paradigm shift in how cities approach extreme heat and heatwave related risk and is to be applauded, Chief Heat Officers potentially face a spectrum of challenges that span technical, social, and political dimensions given the complexity of extreme heat and heatwaves as a risk management problem. These are likely to include resourcing, data gaps and accuracy related to intra-urban variability in heat exposure, vulnerability and impact, intra-governmental coordination and policy integration, urban planning and infrastructure constraints arising from inflexible urban morphologies, catalysing public awareness and behaviour change, social equity issues related to unequal access to risk reduction measures, long-term planning versus immediate pressures, climate change scepticism and global scale climate factors. Amongst these resourcing and intra-governmental co-ordination and policy integration are of particular concern. Resourcing will undoubtedly pose a challenge in low- and middle-income cities whose budgets might not stretch to covering extreme heat and heatwave risk reduction strategies or the scaling up of pilot projects. A pertinent case in point is that of Freetown where many of the heat risk reduction initiatives facilitated by the local Chief Heat Officer rely on the availability of international funding creating uncertainty about the long-term sustainability of what is an ambitious and wholly worthwhile heat risk reduction programme. This local issue chimes with the general challenge of securing climate adaptation finance in the wider Global South. As noted in the section on governance, the range of players involved in the extreme heat and heatwave risk management problem is potentially large. Imperative for the successful development and implementation of heat mitigation strategies will be garnering close collaboration across a diverse group of municipal government departments, agencies, and policy makers as well as civil society interest groups. Coordinating such a diverse array of players as well as policy strategies represents a genuine governance challenge for Chief Heat Officers (Kearl & Vogel, 2023). While some of the challenges touched upon here may be surmountable with time, some factors contributing to urban heat, such as global climate patterns, are likely to be beyond the control of Chief Heat Officers. In brief addressing heat at the local level may be insufficient if broader climate change issues are not tackled on a global scale. One way to overcome this would be to establish an influential lobby group of Chief Heat Officers that could form a unified front and influence global climate change negotiations and policy, especially in the areas of sustainable climate sensitive urban planning, green infrastructure, and equitable climate action. For example, encouraging members of the C40 group of cities (see Sect. 12.3.4) to appoint Chief Heat Officers with an initial focus, for heat inequity reasons, on cities in low- and middle-income countries at risk to heatwaves might be a good starting point—presently only 9 of the C40 cities have such positions. This action could well form one of the remits for the Global Chief Heat Officer. That said, some have observed that international networks such as C40 are still some distance from gaining a seat at the table of multilateral negotiations and national policymaking related to climate risk (Smeds & Acuto, 2018).

12.3.4 Knowledge Networks and Portals

Reflecting the need to put in place strategies for heatwave risk reduction, or perhaps more sceptically an inadequacy of national and state governments in recognizing the imperative for a coordinated approach to address the risk arising from heatwaves, there has been a rapid growth in heatwave related knowledge networks and portals that focus on extreme heat and the promotion of heat and heatwave related resilience over the last one—two decades. From a risk reduction practice perspective, such networks serve as conduits for the exchange of information, enabling countries, organizations, and individuals to tap into a vast reservoir of heatwave related information and expertise. In the realm of research and innovation, knowledge networks catalyse progress by connecting researchers and professionals. Given the boom in heatwave related knowledge networks this section provides a brief overview of the general nature of knowledge networks before considering some of the more notable not- for-profit ones that focus on extreme heat and heatwave related matters.

Generally, three types of outcomes arise from knowledge networks. Knowledge creation involves the generation of novel insights, commonly manifested as ideas, practices, research papers, technical innovations, or products. Knowledge transfer encompasses the endeavours of the knowledge network (provider) to communicate information and knowledge to a recipient, with the recipient actively seeking to acquire and assimilate (i.e., learn) the shared knowledge. Knowledge adoption pertains to the choice and capability to utilize or put into practice a distinct element of knowledge, frequently taking the form of a product, procedure, or document (Phelps et al., 2012). Knowledge networks are comprised of nodes. These function as repositories of knowledge, and agents that actively seek, adopt, transmit, and generate knowledge. Nodes play a dual role as both providers and receivers of information and knowledge (Phelps et al., 2012). Nodes can take the form of individuals or collectives, including teams, organizational units, entire organizations, cities or even nation-states. The inherent characteristics, traits, and available resources of these network nodes collectively constitute what is known as network composition (Phelps et al., 2012). Additionally, knowledge networks encompass social relationships nodes, serving as avenues through which nodes seek information and knowledge. These relationships act as channels for the diffusion and flow of information and knowledge, while also serving as a framework through which nodes assess and evaluate each other (Phelps et al., 2012; Podolny, 2001).

The Global Heat Health Information Network (GHHIN) is an information-sharing initiative spearheaded by the World Health Organization and World Meteorological Organization Joint Office for Climate and Health based in Geneva, and the United States National Oceanic and Atmospheric Administration (NOAA). As well as being supported by these agencies it is partly funded by the Wellcome Trust. The network was initiated in June 2016 by experts from several founding institutions and officially launched with a call to action in December 2018 at the First Global Forum for Heat and Health. GHHIN stands as an autonomous, voluntary, and member-oriented platform comprising scientists, practitioners, and policy makers dedicated to enhancing the capability to safeguard populations from the preventable health hazards associated with extreme heat and heatwaves. Its inception aimed to swiftly amplify initiatives addressing the intricate human-health threats posed by escalating ambient heat and heatwaves, fostering coordination and enhancement of information exchange among communities of health professionals, decision makers, and scientists committed to tackling this challenge. To deliver its vision, GHHIN is focused on the five shared objectives to (i) enhance global awareness of the urgent threat that rising extreme heat poses to human health, well-being, and productivity on a worldwide scale, (ii) foster and maintain interdisciplinary collaborations and mutual learning between researchers and practitioners across relevant sectors, including government, academia, the private sector, and civil society, (iii) synthesize and advance the available science and technology to support decision-making and heat risk reduction across various sectors and time frames, (iv) improve accessibility to expert resources and opportunities for learning, exchange, and engagement and (v) identify and expedite actions to address critical gaps in research, knowledge, and practice. GHHIN has been active in supporting the development of regional nodes that capitalise on heat risk expertise with one of the first established for the Southeast Asia region at the Heat Resilience and Performance Centre at the National University of Singapore Yong Loo Lin School of Medicine.

The Red Cross Climate Centre, established in 2002 as a specialized unit within the International Federation of Red Cross and Red Crescent Societies (IFRC), is dedicated to enhancing the resilience of communities to climate related hazards worldwide. Its establishment marked a strategic move by the Red Cross movement to integrate climate adaptation into its humanitarian work. The Climate Centre operates under the mandate of the IFRC, a global network comprising national Red Cross and Red Crescent Societies, and is headquartered in The Hague, Netherlands. Extreme heat and heatwaves are special focus areas within the Climate Centre's resilient cities programme of activities. In response to the growing threat of heatwaves, the Climate Centre has developed a comprehensive Heatwave Guide for Cities. This resource provides practical information and tools for urban planners, policymakers, and communities to mitigate the impact of extreme heat events and has served as the foundation for crafting a City Heatwave Guide for Red Cross Red Crescent Branches. This guide is customized to offer practical guidance on actions that Red Cross Red Crescent branches can initiate to prepare for and respond to heatwaves in towns and cities. Added to this, the IFRC's Global Disaster Preparedness Centre and Climate Centre have jointly produced a Heat Toolkit. A noticeable initiative of the IFRC Climate Centre is so-called Forecast-based-Financing a mechanism designed for the release of funds in advance of a forecast extreme climate event. Initially imagined for flood and drought events, it will be intriguing to see if this financing mechanism will be applied in the future to heatwaves given the rapid development in intra-seasonal to seasonal forecasting technology and improvements in heatwave predictability at these timescales (see Chap. 13). The IFRC through its Climate Centre is also partnering with the C40 Cities network to call on city officials, urban planners, and city residents in every region of the world to prepare for more dangerous and deadly heatwaves.

Founded in 2005, the C40 Cities Climate Leadership Group, commonly known as C40, is a global network that includes 94 of the world's largest cities, covering 650 million people, an estimated 25 per cent of the global GDP and covering a wide range of low to high-income countries (Davidson et al., 2019a). Although committed to addressing the challenges of climate change and fostering sustainable urban development (Barthold, 2019; Sancino et al., 2022) a close analysis of its history indicates an ever-evolving agenda and a high fluidity of people that shape its agenda (Acuto & Ghojeh, 2019; Davidson et al., 2019b). Generally, C40 focuses on a range of critical areas, including sustainable transportation, energy efficiency, waste management, and climate resilience with the issue of extreme heat and the impact of heatwaves on cities a key area of focus within its portfolio of climate resilience activities. An especially notable initiative of the C40 alliance has been the development of an Urban Cooling Toolbox which outlines options to lower urban temperatures and reduce the impact of the urban heat effect and a Heat Resilient Cities Benefit Tool designed to help city planners and decision-makers quantify the health, economic and environmental benefits of adaptation actions. While on the face of it such tools provide interesting outcomes presented in a graphically captivating way, it must be remembered they are based on a range of assumptions and ideal nature-based solution scenarios which may not necessarily hold true or be achievable for a range of constraining reasons. Further some analyses of how transformative proposed C40 solutions are, have revealed that most proposed climate actions focus on infrastructure and technology with only a few transformational climate measures envisaged, for example, as modelled by some of the C40 'tools' (Heikkinen et al., 2019). Added to this, Smeds and Acuto (2018) posit that the climate initiatives pursued by cities within the C40 network lead more to scaling out of experiments (repeating climate initiatives within similar contexts—(Luederitz et al., 2017)), as opposed to the scaling up of these (integrating and applying innovative actions at higher system levels—(Luederitz et al., 2017)), beyond the urban level of governance.

Co-existing with the C40 Cities network are the more traditional Local Governments for Sustainability (ICLEI) and the United Cities and Local Government (UCLG) networks (Acuto (Acuto & Ghojeh, 2019). Active in over 125 countries, ICLEI is a global network working with more than 2500 local and regional governments committed to sustainable urban development. Founded in 1990 as the 'International Council for Local Environmental Initiatives,' ICLEI has its headquarters in Bonn, Germany with regional offices worldwide. ICLEI facilitates collaboration among cities, towns, and regions to address environmental challenges, combat climate change, and promote resilience. As part of its remit which includes climate change mitigation and adaptation the ICLEI, often working with the Climate and Development Knowledge Network, has undertaken several heatwave assessments and developed several resources for heatwave risk reduction such as a City Heat Resilience Toolkit for Indian Cities. United Cities and Local Governments (UCLG) is a global network that serves as an advocate and platform for local and regional governments. Established in 2004, UCLG aims to promote the role of local governance in addressing global challenges, fostering sustainable urban development, and enhancing local democracy. With its headquarters in Barcelona, Spain, UCLG represents a diverse membership, bringing together over 240,000 cities, regions, and associations worldwide. Amongst a range of issues UCLG provides a forum for local leaders to exchange knowledge, share best practices, and collaborate on urbanization and climate change with extreme heat included as an issue of concern in the wider context of the achievement of the Sustainable Development Goals. Heatwaves receive attention in UCLG resources such as 'peer learning notes' associated with climate resilience and urban development. Of note UCLG Africa will work with the Global Centre on Adaptation to scale up locally led adaptation in Africa including that related to heatwaves.

The National Integrated Heat Health Information System (NIHHIS) is a comprehensive initiative between the National Oceanic and Atmospheric Administration and the Centre for Disease Control and Prevention designed to address the challenges posed by extreme heat events/heatwaves and their impacts on public health in the United States. Established to enhance the nation's preparedness, response, and resilience to heat-related threats, NIHHIS was launched in June 2015 with its inception driven by the need for a coordinated and integrated system that could provide timely and accurate information to mitigate the health risks associated with extreme heat. The key components of NIHHIS are data Integration, early warning systems, health surveillance, community engagement and research and innovation. By consolidating data, research, and communication strategies, NIHHIS aims to build a more robust framework for heat/heatwave-health management. NIHHIS operates through collaborations between federal, state, and local agencies, as well as non-governmental organizations and academic institutions. At the local level, health departments, emergency management agencies, and community organizations actively participate in the implementation of NIHHIS strategies tailored to their specific needs. In July 2022, under the auspices of the Biden's administration's Climate Task Force and its Interagency Working Group on Extreme Heat, NIHHIS launched 'Heat.gov', its new web-based portal designed to furnish the US public and decision-makers with accessible, timely, and science-backed information, aiding in the comprehension and mitigation of health risks associated with extreme heat and heatwaves. Contained within Heat.gov are a range of tools related to climate and health outlooks, probabilistic extreme temperature forecasts, mapping of heatwave risk as well as extreme heat vulnerability, tracking heat and health outcomes, as well as information on the evolution of extreme heat over time, the health consequences associated with extreme heat, and visual depictions showcasing how the Low Income Home Energy Assistance Program (LIHEAP) can mitigate the stress of extreme heat for vulnerable populations. As yet, no formal evaluation of the effectiveness of NIHHIS activities has been undertaken as the programme is still in its early days.

The Extreme Heat Resilience Alliance (EHRA) was created in 2020 by the Adrienne Arsht-Rockefeller Foundation Resilience Centre (Arsht-Rock) at the Atlantic Council. Its aim is to deepen, expand, and accelerate action, aiming to deliver heat resilient solutions to 500 million people worldwide by 2030. Like GHHIN, the EHRA is comprised of experts from a wide range of disciplines but distinguishes itself from GHHIN in having city leaders as part of its alliance. EHRA's priority areas include educating decision makers linked to vulnerable communities about the dangers of extreme heat, co-developing heat risk reduction policies, creating affordable capital and risk transfer products for heat resilience (see above section on heat insurance) and supporting on the ground implementation of heatwave risk reduction strategies. Notably the EHRA in conjunction with Arsht-Rock have promoted the idea of heatwave naming (see Chap. 13) and the establishment of Chief Heat Officer positions.

The Resilient Cities Network (RCN) builds upon the legacy of the 100 Resilient Cities (100RC) programme initiated by the Rockefeller Foundation in 2013 which was wrapped up in 2019 when the Rockefeller Foundation withdrew funding for the 100RC programme. Launched in the wake of the 100RC and leveraging on the success of and lessons learned from the 100RC programme, the RCN operates as a collaborative platform that brings together cities to share knowledge, best practices, and resources to address not only the immediate impacts of climate change but also the interconnected social, economic, and environmental challenges that cities face. One key aspect of 100RC was the development of a Chief Resilience Officer (CRO) position in each participating city. This initiative has been carried forward and expanded on in support of the RCN's aims and objectives which broadly cover the promotion of knowledge sharing and collaboration, capacity building, inclusive resilience, innovation and technology, policy advocacy and global collaboration to achieve climate resilient cities in the face of climate change. While RCN acknowledges the multifaceted challenges associated with rising global temperatures, heat-related considerations in its climate resilience initiatives has a relatively low profile as far as can be established by a review of materials on the RCN's website. Given this, there is considerable potential for extreme heat as an urban resilience issue to be integrated into RCN's priorities with possible projects directed at urban heat island mitigation, heatwave adaptation strategies (e.g., heat action plans), infrastructure resilience to heat (e.g., transport, water and energy), community engagement on heat resilience, data and technology solutions for heat monitoring and inclusive heat resilience.

From this brief review of some of the knowledge networks and portals operating in the extreme heat and heatwaves space it is clear the potential benefits for knowledge exchange and progressing understanding of heatwave risk reduction strategies are substantial. That said, sustaining knowledge networks whether they be national or international poses various challenges (Estrada, 2017). Perhaps one of the greatest is resourcing as funding and resource disparities among participating organizations may impede sustainability with unequal access to financial support possibly leading to varying levels of engagement and commitment. Indeed, many knowledge networks are heavily dependent on securing funds from major international businesses and philanthropic actors, something which has been referred to as 'philanthrocapitalism' (McGoey, 2012; Haydon et al., 2021), the dependency-dangers of which are aptly demonstrated by the Rockefeller Foundation's decision to suspend its (US)$200 m support of the 100RC initiative (Davidson et al., 2019a, b; Webber et al., 2020). Other challenges include ensuring transparent governance and consensus decision-making while balancing inclusivity and efficiency, ensuring diversity of participants, effective communication and coordination, balancing perspectives at a range of geographical scales (e.g., global vs local), volatility and turnover of organisations and individuals (the network nodes) and their long-term commitment and engagement. Lastly, but importantly, measuring the impact and effectiveness of collaborative efforts will require the application of comprehensive evaluation methods. As a heatwave adaptation response, knowledge networks have been an intriguing development inviting further investigation beyond what is often the hubris conveyed via some of their websites. Borrowing from Davidson et al. (2019a) in relation to city networks, a worthy research agenda related to extreme heat and heatwave knowledge networks might include establishing who has power and influence to shape the direction and coordinating capacity of such knowledge networks, the nature of the politics of decision-making in extreme heat/heatwave networks and the implications such networks hold for traditional institutions that have in the past shaped heatwave risk reduction strategies.

12.4 What Catalyses Heatwave Responses?

Responses to heatwaves, either in the form of adaptation or mitigation potentially occur at a range of scales from the individual, interpersonal and community levels, extending to the institutional, regional, national, and global levels. At scales beyond the community level responses are generally in the form of policy change often led by a government agency or ministry, while at the individual and interpersonal levels responses usually take the form of a range of autonomous actions that are often independent of a policy making environment. This of course begs the question 'what precipitates responses at these two broad levels of response'.

For explaining policy change, three diverse approaches have been adopted spanning prescriptive, descriptive, and explanatory perspectives (Travis, 2014; Nohrstedt, 2022). The prescriptive viewpoint underscores the urgent requirement for transformation to reduce disaster risks. Consequently, communities must implement measures to diminish vulnerabilities, address root causes of risk, and enhance their preparedness, response, and recovery capabilities in the face of disruptive extreme events. This perspective begins with the recognition that disasters arise when natural hazards intersect with societal vulnerability, a vulnerability that is exacerbated by poverty and inequality (Schipper, 2020). In the realm of international policy discourse, the prescriptive perspective has instigated a shift in emphasis, moving from risk management approaches primarily responding to external risks towards sustained endeavours to tackle the fundamental risk factors (Nohrstedt, 2022; Nohrstedt et al., 2022). From the descriptive standpoint, change is viewed as a policy choice that may not inherently be superior or inferior to other options, which encompasses maintaining the current state or making incremental adjustments

(Nalau & Handmer, 2015; Nohrstedt et al., 2022). Therefore, within this perspective, transformation is one of a range of policy approaches aimed at decreasing the risk arising from extreme events (Nohrstedt, 2022). The explanatory perspective places emphasis on the role of extreme events as catalysts for policy change. In relation to this perspective is the often-cited notion of tipping points, which alludes to instances of regime shifts, thresholds, and critical transitions in engendering policy change (Novalia & Malekpour, 2020; Giordono et al., 2021; Nohrstedt, 2022).

Associated with these three perspectives are a range of theories of how and why policy change may come about. An external shock or catalyst, a combination of external pressures, leading to large leaps with the status quo disrupted, or a window of opportunity, represent the three classic theories for significant policy change (Penning-Rowsell et al., 2014). Of these the shock/catalyst and windows of opportunity theories generally fall within the explanatory perspective of approaches to understanding the drivers of policy change, while the idea of mounting external pressures aligns with the prescriptive perspective. Other theories on how policy change happens include prospect theory, political elite theory and grass-roots theory (Stachowiak, 2009). Prospect theory advocates for the role of how policy options are framed and presented in actioning change. Political elite theory makes the case for the role of powerful and highly influential individuals in precipitating policy change while grass-roots theory is the antithesis of this, emphasising the formidable agency of community organizations which through collective action can bring about change.

Although some researchers have suggested it is still early days to unequivocally attribute policy change to extreme weather related shocks with more research on the extreme event—policy change nexus required (Kopp et al., 2016) and others have raised empirically and theoretically grounded doubts about whether disasters genuinely serve as a significant catalyst for policy change (Nohrstedt et al., 2022; Rowan, 2023; Yeganeh et al., 2020; Zhang & Maroulis, 2021), there is a growing body of literature that adopts an explanatory perspective and suggests that extreme events that cause significant loss do catalyse or accelerate actions related to risk reduction (Arnell et al., 1984; Cabral et al., 2023; Nguyen et al., 2022; Page & Dilling, 2020; Penning-Rowsell et al., 2006; Rouhaud & Vanderlinden, 2022; Smith et al., 2017; Travis, 2014; Wanner, 2022). Such catalytic events have been referred to as focusing events. Because of their suddenness, rarity, level of harm and harbinger effects, focusing events are thought to precipitate simultaneous conversations amongst policy makers and the public about whether something needs to be done to reduce harms arising from similar or even worse events in the future (Birkland, 2006). In this way extreme events as focusing events, create windows of opportunity acting as alarms regarding the vulnerabilities and exposures associated with a highly impactful event. They may also accentuate an underlying trend in an extreme and impel purposeful adaptation tied to the trend (Travis et al. 2014). Rather than being seen as alarms engendering rapid policy change, extreme events may serve as adaptive ratchets such that responses to a series of extreme events accrue to achieve net adaptation that increases coping capacity (Travis, 2014), 2014). In this way extreme events are viewed as pacemakers of policy change with incremental adaptation as a continuous cyclical process in the form of adaptation action cycles or learning loops (Park et al., 2012; Travis, 2014).

Given the evidence of discernible links between extreme events such as floods and policy actions (Penning-Rowsell et al., 2006; Smith et al., 2017), it is intuitively appealing to suggest that significant heatwave events have bolstered actions related to reducing heatwave-related risk. For example, following the impactful 1995 Chicago heatwave event, which resulted in many heat-related fatalities (Whitman et al., 1997), the Mayor's Office, City Council, and the Chicago Health Department agreed to develop a "heat warning plan" (Changnon et al., 1996). By autumn 1995 the city's Extreme Weather Operations Plan had been established. This included a set of actions for the city to take during heatwaves to reduce heatwave related impacts. The occurrence of a significant heatwave event in 1999 further consolidated the city's response in terms of heatwave related policies. Although these policy uplifts constituted actions more akin to disaster response plans, as opposed to purposeful and enduring heatwave adaptation strategies, it would appear that the 1995 and 1999 heatwave events may have formed part of an adaptive ratchet that eventually led to the development of the 2008 Chicago Climate Action Plan an all-encompassing plan that set out a range of climate change related adaptation and mitigation actions with management options specific to heatwaves occupying a central place. Similarly, the August 2003 European heatwave event appears to have played a game changing role in relation to a range of heatwave related adaptation actions, the most obvious of which were the development of heatwave-related warning systems in countries such as England and France as manifest by the Heatwave Plan for England and France's Plan National Canicule respectively. In the case of France, it is highly likely that the political fallout from the 2003 heatwave hastened reflection on the need for policy adjustments. In India, the severe heatwave experienced in the city of Ahmedabad in 2010, which resulted in a 30% increase in excess mortality (Azhar et al., 2014), appears to have not only galvanized the city's public health community and fuelled concern about the public health threats of heatwaves, but acted as a catalyst for the development of heatwave related adaptive actions, eventually leading to the launch of south Asia's first heat-health action plan (Knowlton et al., 2014). This also seems to be true for the city of Karachi in Pakistan which experienced a severe heatwave in June 2015 with 1181 fatalities officially reported. To that point most of the city's hazard planning as laid out in its Multi-hazard Contingency Plan was focused on urban flooding. However, the 2015 heatwave seems to have trigger extensive discussions in the media and amongst officials about heatwaves as a serious threat to life and economy. As a result, the Karachi Heatwave Management Plan was drafted (Anwar et al., 2022). As we have seen in previous chapters, extreme heat and heatwaves are a distinct feature of Australia's climate. Although historically this country has experienced several major heatwaves it was probably not until the devasting 2009 heatwave that wreaked havoc across southeast and southern Australia (see Chap. 11) that the requirements for a fully-fledged adaptative response to heatwaves landed heavily with policy makers. Consequently, a full-scale heatwave plan for Victoria, focused on protecting health and reducing harm from heatwaves was put into operation in 2011. Acknowledging that heatwaves are a national issue and with support from the national government, the Bureau of Meteorology eventually launched a national level heatwave service in 2014.

Admittedly the above examples of heatwaves acting as shocks, focusing events, adaptive ratchets, or alarms, and thus instigators of policy change, are possibly speculative. That said, in a global stocktake of evidence on human adaptation to climate change, Berrang-Ford et al. (2021) note that extreme heat appears to play a role in motivating policy responses across most regions. Given this the heatwave examples provided here, as well as many others not cited here, do invite cross disciplinary conversations about the various processes associated with the heatwave—policy response nexus. This chimes with wider calls for an increased understanding of how policy-led adaptation is implemented in response to climate risks with the associated testing of the so-called 'disaster-reform hypothesis' which suggests that disasters are important external drivers that enable major policy changes (Nohrstedt, 2022). In relation to heatwaves, perhaps a furtive place to begin would be considering the relevance of the scenarios put forward by Nohrstedt (2022) that describe possible pathways to policy change following extreme events as presented in Fig. 12.3. The bouncing forward scenario (Fig. 12.3a) represents the reference case. This is the situation an extreme event triggers speedy policy action leading to significant change, culminating in the establishment of a new order as manifest by a radical and irreversible regime shift (Nohrstedt, 2022), This resonates with the focusing event ideas of Birkland (2006) and falls within the explanatory set of approaches to understanding policy change. Perhaps this scenario is the one most aligned with what happened following the 2003 heatwave in England and France as alluded to above and indeed Italy starting with the city of Rome. The interest decline scenario (Fig. 12.3b) illustrates a period of intense policy initiatives. What follows however is a decline in public attention and support for reform. Such a recession may well be related to the non-occurrence of extreme events over an extended period. Of course, a period of stasis could be overturned by a massive event with deep societal impacts thus catalysing something like a bounce forward scenario. The tyranny of the urgent scenario (Fig. 12.3c) sets out the possibility that a series of extreme events, following the initial attention raising one, could interfere with the move to new or improved policy. This is because attention is diverted towards short-term response and recovery akin to disaster response, putting at risk the achievement of longer-term changes aimed at reducing risk and vulnerability. Lastly, the lagged effects scenario (Fig. 12.3d) suggests that the cumulative experiences from one or more extreme events gradually engender an uplift in policy action aimed at transformation to a new policy state. In simple terms this is a lingering form of the bouncing forward scenario as although new policy is eventually obtained a considerable amount of time passes before a regime shift is fully realised. This is analogous to the pacemaker effect espoused by Travis (2014).

Theories, generally rooted in the psychological and behavioural sciences, abound in relation to what motivates individuals and like-minded groups of individuals to undertake change in relation to a new set of circumstances such as being directly or indirectly affected by heatwaves. As yet, research on how individuals or small

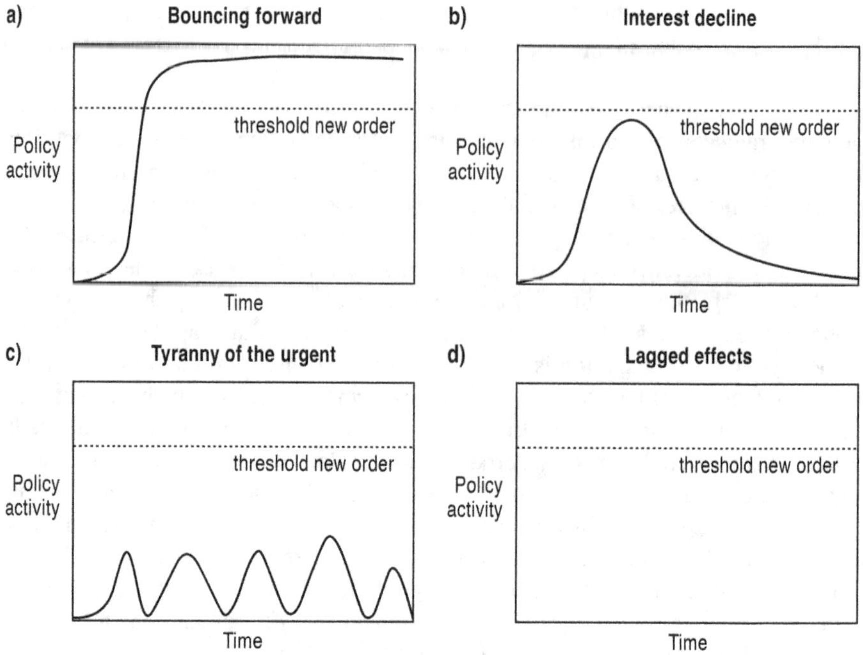

Fig. 12.3 Simplified theoretical scenarios of policy activity and change following disaster causing extreme events. (**a**) bouncing forward, (**b**) interest decline, (**c**) tyranny of the urgent, (**d**) lagged effects. (Source: Nohrstedt (2022). Used with permission under Creative Commons Licence CC BY 4.0 DEED Attribution 4.0 International. https://creativecommons.org/licenses/by/4.0/)

groups of people consciously or unconsciously reach a decision related to short or long-term heatwave risk reduction actions is undeveloped. A possible rectification of this situation might fruitfully entail the design of systematic programmes of research focused on testing one or more of a range of theories related to autonomous decision-making applicable to individuals and groups of concurring individuals (Table 12.6). Such programmes of research could yield insightful information of benefit to advocates for heatwave related adaptation and mitigation actions and those involved in heatwave risk communication.

From Table 12.6 It is clear there is a host of possibilities for exploring the generative processes underlying autonomous decision making related to heatwave risk reduction at the sub-community scale. Although all theories/models listed in Table 12.6 are conceivable candidates for offering explanations about why action related decisions may be arrived at, some leap out as distinct prospects in the case of heatwaves. Given that many risk reduction actions taken by individuals/small groups will probably have health protection as a driving motivation, the health oriented Transtheoretical, Health Belief and Health Action Process Approach models may be highly relevant in the context of heatwave related heat illness. For example, the Health Belief model may be a suitable explanatory device in the case of an elderly person choosing to maintain hydration, keep cool and reduce activity

Table 12.6 Theories related to individual to small group autonomous decision making of possible relevance to understanding actions taken in the context of heatwave risk reduction

Theory	Description
Self-Determination Theory (SDT)	SDT (Deci & Ryan, 1980), emphasizes the role of autonomy, competence, and relatedness in human motivation. According to SDT, individuals are more likely to engage in autonomous decision-making when their basic psychological needs for autonomy are satisfied. Autonomy is considered a fundamental motivator for behaviour
Theory of Planned Behaviour (TPB)	The TPB (Ajzen, 1991), suggests that behaviour is determined by intentions, which are influenced by attitudes toward the behaviour, subjective norms (social influences), and perceived behavioural control (self-efficacy). Autonomous decision-making can be driven by an intention to act in a way that aligns with their beliefs and perceived control
Dual Systems Theory (DST)	DST (Shulman et al., 2016) posits that decision-making involves two systems: a reflective, controlled system, and an impulsive, automatic system. Originally focused on teenager behaviour (Casey et al., 2008), autonomous decisions are said to arise from the reflective system, which considers long-term goals, values, and reasoning, rather than impulsive reactions
Expectancy-Value Theory (EVT)	EVT suggests that individuals make decisions based on their expectations of success and the subjective value they place on the outcomes (Atkinson, 1950). Autonomous decision-making is influenced by a belief in the ability to succeed (expectancy) and the perceived value of the goal or outcome
Cognitive Evaluation Theory (CET)	Building on SDT, CET proposes that external factors such as rewards and feedback can either enhance or diminish intrinsic motivation and autonomous decision-makin (Rummel & Feinberg, 1988). Autonomy support, in the form of choice and involvement in decision-making, is considered important for maintaining intrinsic motivation
Social Cognitive Theory (SCT)	Social Cognitive Theory (Bandura, 1989) emphasizes observational learning, modelling, and self-efficacy in shaping behaviour. Autonomous decision-making is influenced by an individual's observation of others, their beliefs in their own capabilities, and the outcomes of their actions
Sensemaking Theory	Sensemaking Theory (Weick, 1988) explores how individuals make sense of complex and ambiguous situations. Autonomous decision-making is driven by the efforts to create meaning and understanding in their environment, often in response to uncertainty.
Ecological Rationality	Ecological rationality (Gigerenzer, 2000), suggests decision-making should be adapted to the specific environment or context in which it occurs. Autonomous decisions are driven by heuristics and cognitive shortcuts that are well-suited to the demands of the particular situation

(continued)

Table 12.6 (continued)

Theory	Description
Control Balance Theory (CBT)	CBT (Tittle, 2004) although somewhat contested as a theory, suggests that individuals seek to maintain a balance between the control they exert over their own lives and the control exercised by external influences. Autonomous decision-making is influenced by the individual's perception of control and the desire to maintain a sense of agency
Prospect Theory	Prospect Theory (Kahneman & Tversky, 1979) describes how individuals evaluate potential outcomes and make decisions under risk and uncertainty. Autonomous decision-making may involve a preference for choices that are perceived as gains rather than losses, and decisions may be influenced by the framing of options
Transtheorectical	The health-oriented Transtheoretical model (Prochaska & Velicer, 1997) sometimes referred to as the stages of change model, proposes that individuals and consensual groups follow a number of iterative steps in achieving autonomous decision making namely, precontemplation, contemplation, preparation, action, maintenance, and termination
Health Belief Model (HBM)	The HBM (Rosenstock, 1974) suggests that a person's belief in a personal health threat, combined with a belief in the effectiveness of a recommended health behaviour, will predict the likelihood of the person adopting the behaviour. It considers perceived susceptibility, perceived severity, perceived benefits, perceived barriers, cues to action, and self-efficacy
Health Action Process Approach (HAPA):	HAPA integrates concepts from several theories, including the health belief model and the theory of planned behaviour. It distinguishes between a pre-intentional motivational phase and a post-intentional volitional phase, recognising the importance of both motivational and volitional processes in behaviour change (Schwarzer, 2016)
Triggers and Habit Formation Theories	Some theories emphasise the role of triggers or cues in initiating behaviour change. Habit formation theories, such as the habit loop—cue, routine, reward—suggest that habits are formed through repeated associations between cues and behaviours (Duhigg, 2013)
Positive Deviance	This approach focuses on identifying individuals or small groups of people within a community who, despite facing the same challenges as others, exhibit positive behaviours (Herington & van de Fliert, 2018). Positive deviance theory seeks to understand what enables individuals/groups to deviate positively from the norm and how those behaviours can be replicated

Compiled by Author

throughout a heatwave, if they believe heat illness constitutes a substantial personal health threat, combined with a belief in the effectiveness of the adaptation options applied (Akompab et al., 2013). Trigger and habit-forming theories such as the habit loop may be informative when studying individual and small group responses to cues such as heatwave messaging, either in the context of heatwave warnings or longer-term heatwave education programmes. One can imagine the Ecological Rationality model being highly applicable in the case of farmers who will no doubt arrive at decisions based on the necessity to change farming practices in one way or another because of challenges posed by an increasing frequency and intensity of heatwaves. Further, the Positive Deviance model may be pertinent in the case of studying heatwave risk reduction diffusion processes with 'positive deviants' acting as seed points and catalysing change amongst their neighbours with this change slowly spreading throughout a community. In this regard there may be a good deal of synergism between the Ecological Rationality and Positive Deviance models in explaining behaviour change related to heatwave risk reduction actions in small communities. Such synergism of course applies across the gamut of theories presented in Table 12.6 as these should not be viewed as mutually exclusive with many complementing each other with such complementarity potentially enhancing explanations of why certain heatwave related risk interventions are enacted on by individual and like-minded groups.

Complementing the theories outlined above related to what might influence governmental or municipal policy change and autonomous decision making is the Cultural Theory of Risk framework (Douglas & Wildavsky, 1983). At the heart of this concept is the idea that the interpretation of and response to risks is in accordance with the cultural worldviews possessed by those making decisions about a particular risk whether they be governments, cities, small groups or individuals (McNeeley & Lazrus, 2014). The Cultural Theory of Risk proposes that four idealized cultural biases—hierarchist, individualist, egalitarian, and fatalist—influence orientations towards risk and related response. Accordingly, this theory provides a further lens through which to explore the way in which heatwaves are perceived and responded to at a variety of scales as a particular government's, group's or individual's orientation and associated response in relation to heatwave risk reduction may influence their level of heatwave vulnerability and exposure. Typically portrayed as a grid, the four positions can be used to situate societies, small groups or individuals along a two-dimensional continuum related to the ability to self-determine and the degree of social cohesion (Fig. 12.4). While such a portrayal may imply that cultural biases are fixed this is not the case as these can evolve over time in response to social, economic, and environmental changes.

In relation to heatwave responses, the hierarchist cultural bias (high grid, high group) refers to the situation of centralized governance structures for responding to heatwaves. With a tendency for these to be top-down, government agencies and scientific experts take the lead in developing policy related to heatwave preparedness plans and implementing interventions with social cohesion maintained at all levels by adherence to traditional norms and hierarchical structures. In this case, individuals possess little self-determination and follow official guidance and protocols related to heatwave risk reduction. Conversely, in individualist cultural contexts (low grid, high group) the emphasis is on personal responsibility and individual preparedness and adaptation to heatwaves. In this case people may rely on market-based solutions such as purchasing air conditioning units, seeking out cooling centres or purchasing heatwave-related insurance. Influencing this position is the underlying wider political philosophy that individuals should have the autonomy to make their own choices regarding heatwave risk management, with less or little

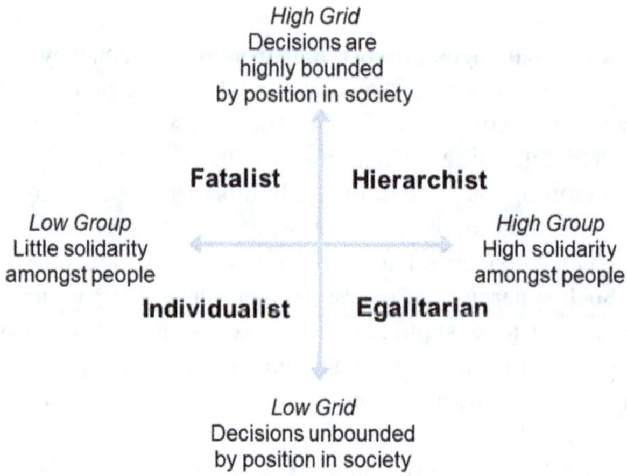

Fig. 12.4 Cultural Theory of Risk framework. Portrayed as a grid this represents the possible position of societies, groups or individuals in a two-dimensional continuum related to self-determination and societal cohesion. (Adapted from McNeeley and Lazrus (2014))

reliance on government intervention. This approach however may exacerbate inequalities, because not everyone has equal access to resources leading to heat inequity. The egalitarian dimension (low Grid, low Group) orients towards a focus on social justice and community-based responses to heatwaves. An egalitarian approach will lean towards efforts to ensure that vulnerable populations, such as the elderly, low-income individuals, and marginalized communities, are adequately supported related to heatwave risk reduction. Egalitarians may advocate for policies that address underlying social inequalities and injustices that contribute to differential heatwave impacts with collective action and structural changes central for the protection of vulnerable populations. Under a fatalist cultural bias position (high grid, low group), heatwaves will be viewed as an inescapable force of nature that society and individuals have little control over with a resigned acceptance of heatwave risks as part of life. In such a situation individuals will generally feel powerless to change their circumstances leading to a lack of proactive planning and preparedness for heatwaves. Fatalism may also contribute to a sense of apathy or fatalism regarding broader environmental issues further undermining efforts to address heatwave risks.

12.5 Conclusion

This chapter has explored responses to heatwave-related risks, emphasizing adaptation as behavioural adjustments in human systems to anticipate and reduce heatwave impacts. It concludes that a distinction between adaptation and mitigation is necessary and has planted the seeds for making a clear division between these concepts, the boundaries of which have become increasingly blurred notwithstanding their complementarity in reducing heatwave risk. In compiling this chapter, a conscious decision was taken to avoid running through all the possible heatwave adaptation options. Instead, a limited number of emerging areas have been focused on including heatwave risk governance, heatwave insurance, the appointment of Chief Heat Officers, and the development of heatwave risk reduction knowledge networks, and platforms. In relation to heatwave risk governance this chapter concludes that getting this 'right' is imperative for successful heatwave adaptation. Whether heatwave insurance is successful or not amongst those that could benefit from it most remains to be seen, especially in the absence of heavily subsidised micro-insurance schemes. While the appointment of Chief Heat Officers and the development of knowledge networks are worthy adaptation options, the long-term sustainability and thus effectiveness of these is heavily dependent on a range of factors not least financing. In asking the question what precipitates policy action relating to adaptation and indeed mitigation, this chapter has provided an overview of a range of theories for understanding the impetus for and determinants of heatwave responses at a range of societal levels. In the case of the development of governmental and municipal policies specific to heatwaves, the chapter concludes there is probably a strong case to be made that impactful heatwave events like those in Europe 2003, Chicago 1995, and Ahmedabad 2010 can be deemed as catalysing or focusing events that precipitated significant policy change.

References

Acuto, M., & Ghojeh, M. (2019). C40 cities inside out. *Global Policy, 10*(4), 709–711. https://doi.org/10.1111/1758-5899.12760

ADB. (2018). *Philippine City disaster insurance Pool: Rationale and design*. Asian Development Bank. doi:https://doi.org/10.22617/TIM189799-2

Adetoro, A. A., Ngidi, M. S. C., Ojo TO, Danso-Abbeam, G., Ogundeji, A. A., & Orimoloye, I. R. (2022). Weather-index insurance as an adaptation strategy to climate change: A global insight. *Climate Research, 88*, 73–85. https://doi.org/10.3354/cr01697

Ajzen, I. (1991). The theory of planned behaviour. *Organizational Behavior and Human Decision Processes, 50*(2), 179–211. https://doi.org/10.1016/0749-5978(91)90020-t

Akompab, D. A., Bi, P., Williams, S., Grant, J., Walker, I. A., & Augoustinos, M. (2013). Heat waves and climate change: Applying the health belief model to identify predictors of risk perception and adaptive behaviours in Adelaide, Australia. *International Journal of Environmental Research and Public Health, 10*(6), 2164–2184. https://doi.org/10.3390/ijerph10062164

Akter, S., Krupnik, T. J., Rossi, F., & Khanam, F. (2016). The influence of gender and product design on farmers' preferences for weather-indexed crop insurance. *Global Environmental Change-Human and Policy Dimensions, 38*, 217–229. https://doi.org/10.1016/j.gloenvcha.2016.03.010

Alaton, P., Djehiche, B., & Stillberger, D. (2002). On modelling and pricing weather derivatives. *Applied Mathematical Finance, 9*(1), 1–20. https://doi.org/10.1080/13504860210132897

Alexander, M., Priest, S., & Mees, H. (2016). A framework for evaluating flood risk governance. *Environmental Science & Policy, 64*, 38–47. https://doi.org/10.1016/j.envsci.2016.06.004

Anon. (2023). Heat stress impacting milk yield: Milma launches insurance scheme for dairy farmers. *Mathrubhami News*,

Anwar, N. H., Khan, H. F., Abdullah, A., Macktoom, S., & Fatima, A. (2022). Designed to fail? Heat governance in urban South Asia: The case of Karachi, a scoping study. *University of Edinburgh*. https://doi.org/10.7488/era/2180

Ara Begum, R., Lempert, R., Elham, A., Benjaminsen, T. A., Bernauer, T., Cramer, W., Cui, X., Mach, K., Nagy, G., & Stenseth, N. C. (2022). Point of departure and key concepts. In H.-O. Pörtner, D. C. Roberts, M. Tignor, et al. (Eds.), *Climate change 2022: Impacts, adaptation and vulnerability. Contribution of working group II to the sixth assessment report of the intergovernmental panel on climate change* (pp. 121–196). Cambridge University Press Cambridge. https://doi.org/10.1017/9781009325844.003

Arnell, N. W., Clark, M. J., & Gurnell, A. M. (1984). Flood insurance and extreme events – The role of crisis in prompting changes in British institutional response to flood hazard. *Applied Geography, 4*(2), 167–181. https://doi.org/10.1016/0143-6228(84)90020-1

Arnouts, R., van der Zouwen, M., & Arts, B. (2012). Analysing governance modes and shifts – Governance arrangements in Dutch nature policy. *Forest Policy and Economics, 16*, 43–50. https://doi.org/10.1016/j.forpol.2011.04.001

Atkinson, J. W. (1950). *Studies in projective measurement of achievement motivation*. University of Michigan.

Awondo, S. N. (2019). Efficiency of region-wide catastrophic weather risk pools: Implications for African risk capacity insurance program. *Journal of Development Economics, 136*, 111–118. https://doi.org/10.1016/j.jdeveco.2018.10.004

Azhar, G. S., Mavalankar, D., Nori-Sarma, A., Rajiva, A., Dutta, P., Jaiswal, A., Sheffield, P., Knowlton, K., Hess, J. J., & Ahmedabad HeatClimate Study G. (2014). Heat-related mortality in India: Excess all-cause mortality associated with the 2010 Ahmedabad heat wave. *PLoS One, 9*(3). https://doi.org/10.1371/journal.pone.0091831

Bandura, A. (1989). Human agency in social cognitive theory. *American Psychologist, 44*(9), 1175–1184. https://doi.org/10.1037/0003-066x.44.9.1175

Barthold, S. (2019). *Greening the global city the role of C40 cities as actors in global environmental governance*.

Bednar, D., & Henstra, D. (2018). Applying a typology of governance modes to climate change adaptation. *Politics and Governance, 6*(3), 147–158. https://doi.org/10.17645/pag.v6i3.1432

Bell, S., & Hindmoor, A. (2009). Rethinking governance: The centrality of the state in modern society. *Rethinking Governance: the Centrality of the State in Modern Society*, 1–234. https://doi.org/10.1017/cbo9780511814617

Bems, J., & Aydin, C. (2022). Introduction to weather derivatives. *Wiley Interdisciplinary Reviews-Energy and Environment, 11*, 3. https://doi.org/10.1002/wene.426

Benth, F. E., & Benth, J. S. (2013). Modeling and pricing in financial markets for weather derivatives. *Modeling and Pricing in Financial Markets for Weather Derivatives, 17*, 1–242.

Bernards, N. (2022). Waiting for the market? Microinsurance and development as anticipatory marketization. *Environment and Planning a-Economy and Space, 54*(5), 949–965. https://doi.org/10.1177/0308518x221073986

Berrang-Ford, L., Siders, A. R., Lesnikowski, A., Fischer, A. P., Callaghan, M. W., Haddaway, N. R., Mach, K. J., Araos, M., Shah, M. A. R., Wannewitz, M., Doshi, D., Leiter, T., Matavel, C., Musah-Surugu, J. I., Wong-Parodi, G., Antwi-Agyei, P., Ajibade, I., Chauhan, N., Kakenmaster, W., Grady, C., Chalastani, V. I., Jagannathan, K., Galappaththi, E. K., Sitati, A., Scarpa, G., Totin, E., Davis, K., Hamilton, N. C., Kirchhoff, C. J., Kumar, P., Pentz, B., Simpson, N. P., Theokritoff, E., Deryng, D., Reckien, D., Zavaleta-Cortijo, C., Ulibarri, N., Segnon, A. C., Khavhagali, V., Shang, Y. Y., Zvobgo, L., Zommers, Z., Xu, J. R., Williams, P. A., Canosa, I. V., van Maanen, N., van Bavel, B., van Aalst, M., Turek-Hankins, L. L., Trivedi, H., Trisos, C. H., Thomas, A., Thakur, S., Templeman, S., Stringer, L. C., Sotnik, G., Sjostrom, K. D., Singh, C., Siña, M. Z., Shukla, R., Sardans, J., Salubi, E. A., Chalkasra, L. S. S., Ruiz-Díaz, R., Richards, C., Pokharel, P., Petzold, J., Penuelas, J., Avila, J. P., Murillo, J. B. P., Ouni, S., Niemann, J., Nielsen, M., New, M., Schwerdtle, P. N., Alverio, G. N., Mullin, C. A., Mullenite, J., Mosurska, A., Morecroft, M. D., Minx, J. C., Maskell, G., Nunbogu, A. M., Magnan, A. K.,

Lwasa, S., Lukas-Sithole, M., Lissner, T., Lilford, O., Koller, S. F., Jurjonas, M., Joe, E. T., Huynh, L. T. M., Hill, A., Hernandez, R. R., Hegde, G., Hawxwell, T., Harper, S., Harden, A., Haasnoot, M., Gilmore, E. A., Gichuki, L., Gatt, A., Garschagen, M., Ford, J. D., Forbes, A., Farrell, A. D., Enquist, C. A. F., Elliott, S., Duncan, E., de Perez, E. C., Coggins, S., Chen, T., Campbell, D., Browne, K. E., Bowen, K. J., Biesbroek, R., Bhatt, I. D., Kerr, R. B., Barr, S. L., Baker, E., Austin, S. E., Arotoma-Rojas, I., Anderson, C., Ajaz, W., Agrawal, T., & Abu, T. Z. (2021). A systematic global stocktake of evidence on human adaptation to climate change. Nature. *Climate Change, 11*, 989–1000. https://doi.org/10.1038/s41558-021-01170-y

Bertana, A., Clark, B., Benney, T. M., & Quackenbush, C. (2022). Beyond maladaptation: Structural barriers to successful adaptation. *Environmental Sociology, 8*(4), 448–458. https://doi.org/10.1080/23251042.2022.2068224

Biagini, B., Bierbaum, R., Stults, M., Dobardzic, S., & McNeeley, S. M. (2014). A typology of adaptation actions: A global look at climate adaptation actions financed through the global environment facility. *Global Environmental Change-Human and Policy Dimensions, 25*, 97–108. https://doi.org/10.1016/j.gloenvcha.2014.01.003

Biener, C., & Eling, M. (2012). Insurability in microinsurance markets: An analysis of problems and potential solutions. *Geneva Papers on Risk and Insurance-Issues and Practice, 37*(1), 77–107. https://doi.org/10.1057/gpp.2011.29

Birkland, T. A. (2006). *Lessons of disaster: Policy change after catastrophic events*. American Governance and Public Policy/Georgetown University Press.

Bogale, A. (2015). Weather-indexed insurance: An elusive or achievable adaptation strategy to climate variability and change for smallholder farmers in Ethiopia. *Climate and Development, 7*(3), 246–256. https://doi.org/10.1080/17565529.2014.934769

Botzen, W. J. W., Kunreuther, H., & Michel-Kerjan, E. (2019). Protecting against disaster risks: Why insurance and prevention may be complements. *Journal of Risk and Uncertainty, 59*(2), 151–169. https://doi.org/10.1007/s11166-019-09312-6

Boyd, E., James, R. A., Jones, R. G., Young, H. R., & Otto, F. E. L. (2017). A typology of loss and damage perspectives. Nature. *Climate Change, 7*, 723–729. https://doi.org/10.1038/nclimate3389

Broto, V. C. (2017). Urban governance and the politics of climate change. *World Development, 93*, 1–15. https://doi.org/10.1016/j.worlddev.2016.12.031

Cabral, L. C. D., Pot, W., van Oel, P., Kchouk, S., Neto, G. R., & Dewulf, A. (2023). From creeping crisis to policy change: The adoption of drought preparedness policy in Brazil. *Water Policy, 25*(10), 949–965. https://doi.org/10.2166/wp.2023.073

Casey, B. J., Getz, S., & Galvan, A. (2008). The Adolescent Brain. *Developmental Review, 28*(1), 62–77.

Chamberlain, J., & Bernards, N. (2024). Insurance and the contradictions of the climate-development-finance nexus: The case of the African risk capacity. *Competition & Change*. https://doi.org/10.1177/10245294241226985

Changnon, S. A. (2007). New risk assessment products for dealing with financial exposure to weather hazards. *Natural Hazards, 43*(3), 295–301. https://doi.org/10.1007/s11069-007-9114-5

Changnon, S. A., Kunkel, K. E., & Reinke, B. C. (1996). Impacts and responses to the 1995 heat wave: A call to action. *Bulletin of the American Meteorological Society, 77*(7), 1497–1506. https://doi.org/10.1175/1520-0477(1996)077<1497:iartth>2.0.co;2

Changnon, S. A., Fosse, E. R., & Lecomte, E. L. (1999). Interactions between the atmospheric sciences and insurers in the United States. *Climatic Change, 42*(1), 51–67. https://doi.org/10.1023/a:1005408131413

Clark, S. S., Chester, M. V., Seager, T. P., & Eisenberg, D. A. (2019). The vulnerability of interdependent urban infrastructure systems to climate change: Could Phoenix experience a Katrina of extreme heat? *Sustainable and Resilient Infrastructure, 4*(1), 21–35. https://doi.org/10.1080/23789689.2018.1448668

Collier, S. J., Elliott, R., & Lehtonen, T. K. (2021). Climate change and insurance. *Economy and Society, 50*(2), 158–172. https://doi.org/10.1080/03085147.2021.1903771

Cotton, M., & Stevens, E. (2019). Mapping discourses of climate change adaptation in the United Kingdom. *Weather Climate and Society, 11*(1), 17–32. https://doi.org/10.1175/wcas-d-18-0024.1

Dabrowski, J. (2023). *First-of-its-kind extreme heat microinsurance launched in India to protect women workers*. Arsht-Rockefeller Foundation.

Davidson, K., Coenen, L., Acuto, M., & Gleeson, B. (2019a). Reconfiguring urban governance in an age of rising city networks: A research agenda. *Urban Studies, 56*(16), 3540–3555. https://doi.org/10.1177/0042098018816010

Davidson, K., Coenen, L., & Gleeson, B. (2019b). A decade of C40: Research insights and agendas for city networks. *Global Policy, 10*(4), 697–708. https://doi.org/10.1111/1758-5899.12740

Debnath, R., Bardhan, R., & Bell, M. L. (2023). Lethal heatwaves are challenging India's sustainable development. *PLOS Climate, 2*(4), e0000156.

Deci, E. L., & Ryan, R. M. (1980). Self-determination theory – The iteration of psychophysiology and motivation. *Psychophysiology, 17*(3), 321–321.

Douglas, M., & Wildavsky, A. (1983). *Risk and culture: An essay on the selection of technological and environmental dangers*. University of California Press.

Driessen, P. P. J., Dieperink, C., van Laerhoven, F., Runhaar, H. A. C., & Vermeulen, W. J. V. (2012). Towards a conceptual framework for the study of shifts in modes of environmental governance – Experiences from The Netherlands. *Environmental Policy and Governance, 22*(3), 143–160. https://doi.org/10.1002/eet.1580

Duhigg, C. (2013). *The power of habit: Why we do what we do and how to change*. Random House.

Estrada, M. S. (2017). Exploring tensions in knowledge networks: Convergences and divergences from social capital, actor-network theory and sociologies of the south. *Current Sociology, 65*(6), 886–908. https://doi.org/10.1177/0011392116676298

Ford, J. D., Berrang-Ford, L., Lesnikowski, A., Barrera, M., & Heymann, S. J. (2013). How to track adaptation to climate change: A typology of approaches for national-level application. *Ecology and Society, 18*(3). https://doi.org/10.5751/es-05732-180340

Gabbe, C. J., Pierce, G., Petermann, E., & Marecek, A. (2021). Why and how do cities plan for extreme heat? *Journal of Planning Education and Research., 44*, 1316–1330. https://doi.org/10.1177/0739456x211053654

Gahman L, Thongs G, Greenidge A (2021) Disaster, debt, and 'underdevelopment': The cunning of colonial-capitalism in the Caribbean. Development (Society for International Development) 64 (1–2):112–118. doi:https://doi.org/10.1057/s41301-021-00282-4, 112.

Gigerenzer, G. (2000). Ecological rationality. *International Journal of Psychology, 35*(3–4), 284–284.

Giordono, L., Gard-Murray, A., & Boudet, H. (2021). From peril to promise? Local mitigation and adaptation policy decisions after extreme weather. *Current Opinion in Environmental Sustainability, 52*, 118–124. https://doi.org/10.1016/j.cosust.2021.10.002

Grove, K. (2012). Preempting the next disaster: Catastrophe insurance and the financialization of disaster management. *Security Dialogue, 43*(2), 139–155. https://doi.org/10.1177/0967010612438434

Guardaro, M. (2023). Strengthening heat action plans in the United States. *American Journal of Public Health, 113*(5), 465–467. https://doi.org/10.2105/ajph.2023.307260

Guyer, H. E., Putnam, H. F., Roach, M., Iñiguez, P., & Hondula, D. M. (2019). Cross-sector management of extreme heat risks in Arizona. *Bulletin of the American Meteorological Society, 100*, ES101–ES104. https://doi.org/10.1175/bams-d-18-0183.1

Haasnoot, M., Biesbroek, R., Lawrence, J., Muccione, V., Lempert, R., & Glavovic, B. (2020). Defining the solution space to accelerate climate change adaptation. *Regional Environmental Change, 20*(2). https://doi.org/10.1007/s10113-020-01623-8

Hamman, P. (2020). Rethinking hierarchy in sustainability governance a literature review. *Sustainability Governance and Hierarchy*, 3–59.

Haydon, S., Jung, T., & Russell, S. (2021). 'You've been Framed': A critical review of academic discourse on philanthrocapitalism. *International Journal of Management Reviews, 23*(3), 353–375. https://doi.org/10.1111/ijmr.12255

Heikkinen, M., Ylä-Anttila, T., & Juhola, S. (2019). Incremental, reformistic or transformational: What kind of change do C40 cities advocate to deal with climate change? *Journal of Environmental Policy & Planning, 21*(1), 90–103. https://doi.org/10.1080/1523908x.2018.1473151

Henderson, D., Aubin, L., Behan, K., Chen, H., Doyle, H., Gower, S., MacDonald, M., Mee, C., Richardson, G. R. A., Rochon, G., Shnabel, M., Storfer, J., Yagouti, A., & Yusa, A. (2020). Developing a harmonized heat warning and information system for Ontario: A case study in collaboration. *Canadian Journal of Public Health-Revue Canadienne De Sante Publique, 111*(3), 426–432. https://doi.org/10.17269/s41997-020-00337-y

Herington, M. J., & van de Fliert, E. (2018). Positive deviance in theory and practice: A conceptual review. *Deviant Behavior, 39*(5), 664–678. https://doi.org/10.1080/01639625.2017.1286194

Hill, R. V., Hoddinott, J., & Kumar, N. (2013). Adoption of weather-index insurance: Learning from willingness to pay among a panel of households in rural Ethiopia. *Agricultural Economics, 44*(4–5), 385–398. https://doi.org/10.1111/agec.12023

Hlahla, S., & Hill, T. R. (2018). Responses to climate variability in urban poor communities in Pietermaritzburg, KwaZulu-Natal, South Africa. *SAGE Open, 8*(3), 215824401880091. https://doi.org/10.1177/2158244018800914

Hochrainer-Stigler, S., Zhu, Q., Ciullo, A., Peisker, J., & Van den Hurk, B. (2023). Differential fiscal performances of plausible disaster events: AsStoryline approach for the Caribbean and central American governments under CCRIF. *Economics of Disasters and Climate Change, 7*, 209–229. https://doi.org/10.1007/s41885-023-00126-0

Hoeppe, P. (2016). Trends in weather related disasters – Consequences for insurers and society. *Weather and Climate Extremes, 11*, 70–79. https://doi.org/10.1016/j.wace.2015.10.002

Howlett, M., Kekez, A., & Poocharoen, O. O. (2017). Understanding co-production as a policy tool: Integrating new public governance and comparative policy theory. *Journal of Comparative Policy Analysis, 19*(5), 487–501. https://doi.org/10.1080/13876988.2017.1287445

Hurlimann, A. C., Moosavi, S., & Browne, G. R. (2021). Climate change transformation: A definition and typology to guide decision making in urban environments. *Sustainable Cities and Society, 70*, 102890. https://doi.org/10.1016/j.scs.2021.102890

Hysing, E. (2009). From government to governance? A comparison of environmental governing in Swedish forestry and transport. *Governance-an International Journal of Policy Administration and Institutions, 22*(4), 647–672. https://doi.org/10.1111/j.1468-0491.2009.01457.x

Iyahen, E., & Syroka, J. (2018). *Managing risks from climate change on the African continent: The African risk capacity (ARC) as an innovative risk financing mechanism.*

Jarzabkowski, P., Chalkias, K., Clarke, D., Iyahen, E., & Stadtmueller, D. (2019). *Insurance for climate adaptation: Opportunities and limitations*. University of London.

Johnson, L. (2021a). Paying ex gratia: Parametric insurance after calculative devices fail. *Geoforum, 125*, 120–131. https://doi.org/10.1016/j.geoforum.2021.06.018

Johnson, L. (2021b). Rescaling index insurance for climate and development in Africa. *Economy and Society, 50*(2), 248–274. https://doi.org/10.1080/03085147.2020.1853364

Jordan, A., Wurzel, R. K. W., & Zito, A. (2005). The rise of 'new' policy instruments in comparative perspective: Has governance eclipsed government? *Political Studies, 53*(3), 477–496. https://doi.org/10.1111/j.1467-9248.2005.00540.x

Joyette, A. R. T., Nurse, L. A., & Pulwarty, R. S. (2015). Disaster risk insurance and catastrophe models in risk-prone small Caribbean islands. *Disasters, 39*(3), 467–492. https://doi.org/10.1111/disa.12118

Kahneman, D., & Tversky, A. (1979). Prospect theory – Analysis of decision under risk. *Econometrica, 47*(2), 263–291. https://doi.org/10.2307/1914185

Kalbarczyk, E., & Kalbarczyk, R. (2020). Typology of climate change adaptation measures in polish cities up to 2030. *Land, 9*(10). https://doi.org/10.3390/land9100351

Kaloop, M. R., Iqbal, M., Elnabwy, M. T., Mustafa, E. K., & Hu, J. W. (2022). A novel AI approach for modeling land surface temperature of Freetown, Sierra Leone, based on land-cover changes. *International Journal of Digital Earth, 15*(1), 1236–1258. https://doi.org/10.1080/17538947.2022.2099022

Kearl, Z., & Vogel, J. (2023). Urban extreme heat, climate change, and saving lives: Lessons from Washington state. *Urban Climate, 47*, 101392. https://doi.org/10.1016/j.uclim.2022.101392

Keith, L., Gabbe, C. J., & Schmidt, E. (2023). Urban heat governance: Examining the role of urban planning. *Journal of Environmental Policy & Planning., 25*, 642–662. https://doi.org/10.1080/1523908x.2023.2244446

Kemen, J., Schäffer-Gemein, S., Grünewald, J., & Kistemann, T. (2021). Heat perception and coping strategies: A structured interview-based study of elderly people in Cologne, Germany. *International Journal of Environmental Research and Public Health, 18*(14). https://doi.org/10.3390/ijerph18147495

Keucheyan, R. (2018). Insuring climate change: New risks and the financialization of nature. *Development and Change, 49*(2), 484–501. https://doi.org/10.1111/dech.12367

Khan, M. R., & Roberts, J. T. (2013). Adaptation and international climate policy. *Wiley Interdisciplinary Reviews-Climate Change, 4*(3), 171–189. https://doi.org/10.1002/wcc.212

Khandekar, A., Cross, J., & Maringanti, A. (2023). Scale and modularity in thermal governance: The replication of India's heat action plans. *Urban Studies*. https://doi.org/10.1177/00420980231195193

Kiarsi, M., Amiresmaili, M., Mahmoodi, M. R., Farahmandnia, H., Nakhaee, N., Zareiyan, A., & Aghababaeian, H. (2023). Heat waves and adaptation: A global systematic review. *Journal of Thermal Biology, 116*, 103588. https://doi.org/10.1016/j.jtherbio.2023.103588

Klinke, A., & Renn, O. (2021). The coming of age of risk governance. *Risk Analysis, 41*(3), 544–557. https://doi.org/10.1111/risa.13383

Klok, E. J., & Kluck, J. (2018). Reasons to adapt to urban heat (in The Netherlands). *Urban Climate, 23*, 342–351. https://doi.org/10.1016/j.uclim.2016.10.005

Knowlton, K., Kulkarni, S. P., Azhar, G. S., Mavalankar, D., Jaiswal, A., Connolly, M., Nori-Sarma, A., Rajiva, A., Dutta, P., Deol, B., Sanchez, L., Khosla, R., Webster, P. J., Toma, V. E., Sheffield, P., Hess, J. J., & Ahmedabad Heat Climate Study G. (2014). Development and implementation of South Asia's first heat-health action plan in Ahmedabad (Gujarat, India). *International Journal of Environmental Research and Public Health, 11*(4), 3473–3492. https://doi.org/10.3390/ijerph110403473

Kohyama, K. (2023). Iapanese insurers sell heatstroke coverage as temperatures soar to record levels. *Insurance journal*.

Kopp, R. E., Shwom, R., Wagner, G., & Yuan, J. (2016). Tipping elements and climate-economic shocks: Pathways toward integrated assessment. *Rutgers University Community Repository*. https://doi.org/10.7282/t3h997bn

Kousky, C. (2019). The role of natural disaster insurance in recovery and risk reduction. *Annual Review of Resource Economics, 11*(11), 399–418. https://doi.org/10.1146/annurev-resource-100518-094028

Kousky, C., & Light, S. E. (2019). Insuring nature. *Duke Law Journal, 69*(2), 323–376.

Lamm, T., Blumberg, L., & Elkind, E. N. (2020). *Insuring extreme heat risks: Policy report*. Los Angeles.

Lange, P., Driessen, P. P. J., Sauer, A., Bornemann, B., & Burger, P. (2013). Governing towards sustainability-conceptualizing modes of governance. *Journal of Environmental Policy & Planning, 15*(3), 403–425. https://doi.org/10.1080/1523908x.2013.769414

Lanza, K. V., Jones, J., Acuna, F., Coudert, M., Bixler, R. P., Kamath, H., & Niyogi, D. (2023). Heat vulnerability of Latino and Black residents in a low-income community and their recommended adaptation strategies: A qualitative study. *Urban Climate, 51*, 101656. https://doi.org/10.1016/j.uclim.2023.101656

Larsson, K. (2023). Parametric heat wave insurance. *Journal of Commodity Markets, 31*, 100345. https://doi.org/10.1016/j.jcomm.2023.100345

Lass, W., Haas, A., Hinkel, J., & Jaeger, C. (2011). Avoiding the avoidable: Towards a European heat waves risk governance. *International Journal of Disaster Risk Science, 2*(1), 1–14. https://doi.org/10.1007/s13753-011-0001-z

Leppert, D., Dalhaus, T., & Lagerkvist, C. J. (2021). Accounting for geographic basis risk in heat index insurance: How spatial interpolation can reduce the cost of risk. *Weather Climate and Society, 13*(2), 273–286. https://doi.org/10.1175/wcas-d-20-0070.1

Lin, Y.-H., Wang, L.-J., Shi, X.-Y., & Chen, M.-P. (2023). Evolution of research on climate risk insurance: A bibliometric analysis from 1975 to 2022. *Advances in Climate Change Research, 14*(4), 592–604.

Loenhout, J. A. F., Rodriguez-Llanes, J. M., & Guha-Sapir, D. (2016). Stakeholders' perception on national heatwave plans and their local implementation in Belgium and The Netherlands. *International Journal of Environmental Research and Public Health, 13*(11). https://doi.org/10.3390/ijerph13111120

Lopera, C. C. F., Mendes, J. M., & Barata, E. J. (2022). The differential risk transfer: A new approach for reducing vulnerability to climate-related hazards. *Disaster Prevention and Management, 31*(5), 550–564. https://doi.org/10.1108/dpm-05-2021-0185

Luederitz, C., Schaepke, N., Wiek, A., Lang, D. J., Bergmann, M., Bos, J. J., Burch, S., Davies, A., Evans, J., Koenig, A., Farrelly, M. A., Forrest, N., Frantzeskaki, N., Gibson, R. B., Kay, B., Loorbach, D., McCormick, K., Parodi, O., Rauschmayer, F., Schneidewind, U., Stauffacher, M., Stelzer, F., Trencher, G., Venjakob, J., Vergragt, P. J., von Wehrden, H., & Westley, F. R. (2017). Learning through evaluation – A tentative evaluative scheme for sustainability transition experiments. *Journal of Cleaner Production, 169*, 61–76. https://doi.org/10.1016/j.jclepro.2016.09.005

Maganga, A. M., Chiwaula, L. S., & Kambewa, P. (2021). Parametric and non-parametric estimates of willingness to pay for weather index insurance in Malawi. *International Journal of Disaster Risk Reduction, 62*, 102406. https://doi.org/10.1016/j.ijdrr.2021.102406

Malmquist, A., Hjerpe, M., Glaas, E., Karlsson, H., & Lassi, T. (2022). Elderly people's perceptions of heat stress and adaptation to h eat: An interview study. *International Journal of Environmental Research and Public Health, 19*(7). https://doi.org/10.3390/ijerph19073775

McCarthy, J. J. (2001). *Climate change 2001: Impacts, adaptation, and vulnerability: Contribution of working group II to the third assessment report of the intergovernmental panel on climate change* (Vol. 2). Cambridge University Press.

McGoey, L. (2012). Philanthrocapitalism and its critics. *Poetics, 40*(2), 185–199. https://doi.org/10.1016/j.poetic.2012.02.006

McNamara, K. E., & Buggy, L. (2017). Community-based climate change adaptation: A review of academic literature. *Local Environment, 22*(4), 443–460. https://doi.org/10.1080/13549839.2016.1216954

McNeeley, S. M., & Lazrus, H. (2014). The cultural theory of risk for climate change adaptation. *Weather, Climate, and Society, 6*(4), 506–519.

Meerow, S., & Keith, L. (2022). Planning for extreme heat: A national survey of US planners. *Journal of the American Planning Association, 88*(3), 319–334. https://doi.org/10.1080/01944363.2021.1977682

Meerow, S., Pajouhesh, P., & Miller, T. R. (2019). Social equity in urban resilience planning. *Local Environment, 24*(9), 793–808. https://doi.org/10.1080/13549839.2019.1645103

Mees, H. L. P., Driessen, P. P. J., & Runhaar, H. A. C. (2015). "Cool" governance of a "hot" climate issue: Public and private responsibilities for the protection of vulnerable citizens against extreme heat. *Regional Environmental Change, 15*(6), 1065–1079. https://doi.org/10.1007/s10113-014-0681-1

Mukherji, A., Ganapati, N. E., & Manandhar, B. (2021). Panacea or problem: New governance structures for disaster recovery. *International Journal of Disaster Risk Reduction, 52*, 101960. https://doi.org/10.1016/j.ijdrr.2020.101960

Mukhopadhyay, A. (2023). NFU mutual launches innovative parametric heat-stress insurance for dairy farmers. *Reinsurance News*.

Murphy, D. J., & Ichinkhorloo, B. (2023). Index insurance and the moral economy of pastoral risk management in Mongolia. *Journal of Peasant Studies., 51*, 1185–1207. https://doi.org/10.1080/03066150.2023.2284972

Mustafa EK, Liu GX, Hassan A, Damos MA, Tarawally M (2020) Predicting of land surface temperature distribution in Freetown City, Sierra Leone by using polynomial curve fitting model. Journal of Geographic Information System 12. doi:https://doi.org/https://doi.org/10.4236/jgis.2020.125031, 531, 544.

Nalau, J., & Handmer, J. (2015). When is transformation a viable policy alternative? *Environmental Science & Policy, 54*, 349–356. https://doi.org/10.1016/j.envsci.2015.07.022

Nguyen, C. T., Ngo, T. Q., & Nguyen, Q. H. (2022). Weather-induced shocks, coping strategies and farmer food consumption in the Vietnamese Mekong Delta: A case in long an province. *Journal of Agribusiness in Developing and Emerging Economies., 14*, 251–291. https://doi.org/10.1108/jadee-01-2022-0002

Nohrstedt, D. (2022). When do disasters spark transformative policy change and why? *Policy and Politics, 50*(3), 425–441. https://doi.org/10.1332/030557321x16508834302815

Nohrstedt, D., Hileman, J., Mazzoleni, M., Di Baldassarre, G., & Parker, C. F. (2022). Exploring disaster impacts on adaptation actions in 549 cities worldwide. *Nature Communications, 13*(1). https://doi.org/10.1038/s41467-022-31059-z

Novalia, W., & Malekpour, S. (2020). Theorising the role of crisis for transformative adaptation. *Environmental Science & Policy, 112*, 361–370. https://doi.org/10.1016/j.envsci.2020.07.009

Nunes, A. R. (2023). Mapping interactions between sustainable development and heatwave resilience. *Environment Development and Sustainability, 25*(11), 12707–12733. https://doi.org/10.1007/s10668-022-02587-x

Oppermann, E., Brearley, M., Law, L., Smith, J. A., Clough, A., & Zander, K. (2017). Heat, health, and humidity in Australia's monsoon tropics: A critical review of the problematization of 'heat' in a changing climate. Wiley interdisciplinary reviews-climate. *Change, 8*(4). https://doi.org/10.1002/wcc.468

Orlove, B. (2022). The concept of adaptation. *Annual Review of Environment and Resources, 47*, 535–581. https://doi.org/10.1146/annurev-environ-112320-095719

Page, R., & Dilling, L. (2020). How experiences of climate extremes motivate adaptation among water managers. *Climatic Change, 161*(3), 499–516. https://doi.org/10.1007/s10584-020-02712-7

Pahl-Wostl, C. (2009). A conceptual framework for analysing adaptive capacity and multi-level learning processes in resource governance regimes. *Global Environmental Change-Human and Policy Dimensions, 19*(3), 354–365. https://doi.org/10.1016/j.gloenvcha.2009.06.001

Pal, J. S., & Eltahir, E. A. B. (2016). Future temperature in Southwest Asia projected to exceed a threshold for human adaptability. *Nature Climate Change, 6*(2), 197–200. https://doi.org/10.1038/nclimate2833

Park, S. E., Marshall, N. A., Jakku, E., Dowd, A. M., Howden, S. M., Mendham, E., & Fleming, A. (2012). Informing adaptation responses to climate change through theories of transformation. *Global Environmental Change-Human and Policy Dimensions, 22*(1), 115–126. https://doi.org/10.1016/j.gloenvcha.2011.10.003

Paterson, B., & Charles, A. (2019). Community-based responses to climate hazards: Typology and global analysis. *Climatic Change, 152*(3–4), 327–343. https://doi.org/10.1007/s10584-018-2345-5

Pellizzoni, L. (2004). Responsibility and environmental governance. *Environmental Politics, 13*(3), 541–565. https://doi.org/10.1080/0964401042000229034

Penning-Rowsell, E., Johnson, C., & Tunstall, S. (2006). 'Signals' from pre-crisis discourse: Lessons from UK flooding for global environmental policy change? *Global Environmental Change-Human and Policy Dimensions, 16*(4), 323–339. https://doi.org/10.1016/j.gloenvcha.2006.01.006

Penning-Rowsell, E. C., Priest, S., & Johnson, C. (2014). The evolution of UK flood insurance: Incremental change over six decades. *International Journal of Water Resources Development, 30*(4), 694–713. https://doi.org/10.1080/07900627.2014.903166

Phelps, C., Heidl, R., & Wadhwa, A. (2012). Knowledge, networks, and knowledge networks: A review and research agenda. *Journal of Management, 38*(4), 1115–1166. https://doi.org/10.1177/0149206311432640

Pielke, R., Prins, G., Rayner, S., & Sarewitz, D. (2007). Lifting the taboo on adaptation. *Nature, 445*(7128), 597–598. https://doi.org/10.1038/445597a

Piggott-McKellar, A. E., McNamara, K. E., Nunn, P. D., & Watson, J. E. M. (2019). What are the barriers to successful community-based climate change adaptation? A review of grey literature. *Local Environment, 24*(4), 374–390. https://doi.org/10.1080/13549839.2019.1580688

Platteau, J. P., De Bock, O., & Gelade, W. (2017). The demand for microinsurance: A literature review. *World Development, 94*, 139–156. https://doi.org/10.1016/j.worlddev.2017.01.010

Podolny, J. M. (2001). Networks as the pipes and prisms of the market. *American Journal of Sociology, 107*(1), 33–60. https://doi.org/10.1086/323038

Prochaska, J. O., & Velicer, W. F. (1997). The transtheoretical model – Introduction. *American Journal of Health Promotion, 12*(1), 6–7. https://doi.org/10.4278/0890-1171-12.1.6

Rahman, H. M. T., Albizua, A., Soubry, B., & Tourangeau, W. (2021). A framework for using autonomous adaptation as a leverage point in sustainable climate adaptation. *Climate Risk Management, 34*, 100376. https://doi.org/10.1016/j.crm.2021.100376

Randalls, S. (2010). Weather profits: Weather derivatives and the commercialization of meteorology. *Social Studies of Science, 40*(5), 705–730. https://doi.org/10.1177/0306312710378448

Randazza, J. M., Hess, J., Bostrom, A., Hartwell, C., Adams, Q. H., Nori-Sarma, A., Spangler, K. R., Sun, Y. T., Weinberger, K. R., Wellenius, G. A., & Errett, N. A. (2023). Planning to reduce the health impacts of extreme heat: A content analysis of heat action plans in local United States jurisdictions. *American Journal of Public Health, 113*(5), 559–567. https://doi.org/10.2105/ajph.2022.307217

Richmond, J. G., & Hill, R. (2023). Rethinking local resilience for extreme heat events. *Public Health, 218*, 146–148. https://doi.org/10.1016/j.puhe.2023.03.005

Richmond, M., & Ortega-Pastor, A. (2022). *Climate insurance-linked resilient infrastructure financing instrument analysis*. Climate Policy Initiative.

Ro, C. (2023). *The new type of insurance that protects Indian women during extreme heat.* FORBES INNOVATION SCIENCE.

Roberts, M., Deuskar, C., Jones, N., & Park, J. (2023). *Unlivable: What the urban heat Island effect means for East Asia's cities*. World Bank.

Rosenstock, I. M. (1974). Health belief model and preventive health behavior. *Health Education Monographs, 2*(4), 354–386. https://doi.org/10.1177/109019817400200405

Rouhaud, E., & Vanderlinden, J. P. (2022). 10 years and going strong? Coastal flood risk management in the wake of a major coastal event (the 2010 Xynthia storm, Charente maritime, France). *Climate Risk Management, 35*, 100413. https://doi.org/10.1016/j.crm.2022.100413

Rowan, S. (2023). Extreme weather and climate policy. *Environmental Politics, 32*(4), 684–707. https://doi.org/10.1080/09644016.2022.2127478

Rummel, A., & Feinberg, R. (1988). Cognitive evaluation theory – A meta-analytic review of the literature. *Social Behavior and Personality, 16*(2), 147–164. https://doi.org/10.2224/sbp.1988.16.2.147

Sancino, A., Stafford, M., Braga, A., & Budd, L. (2022). What can city leaders do for climate change? Insights from the C40 cities climate leadership group network. *Regional Studies, 56*(7), 1224–1233. https://doi.org/10.1080/00343404.2021.2005244

Sarmiento, J. P., & Torres-Muñoz, A. M. (2020). Risk transfer for populations in precarious urban environments. *International Journal of Disaster Risk Science, 11*(1), 74–86. https://doi.org/10.1007/s13753-020-00252-3

Scherer, N. (2020a). The African risk capacity (ARC). In *Insuring against climate change: The emergence of regional catastrophe risk pools* (pp. 130–181). Routledge.

Scherer, N. (2020b). The Caribbean catastrophe risk insurance facility (CCRIF). In *Insuring against climate change: The emergence of regional catastrophe risk pools* (pp. 69–129). Routledge.

Schipper, E. L. F. (2020). Maladaptation: When adaptation to climate change goes very wrong. *One Earth, 3*(4), 409–414. https://doi.org/10.1016/j.oneear.2020.09.014

Schipper, E. L. F. (2022). Catching maladaptation before it happens. *Nature Climate Change, 12*(7), 617–618. https://doi.org/10.1038/s41558-022-01409-2

Schmeltz, M. T., Smith, J. A., Olmos, I., & Quintero, E. (2023). Extreme heat governance: A critical analysis of heat action plans in California. *American Journal of Public Health, 113*(1), 15–19. https://doi.org/10.2105/ajph.2022.307117

Schwarzer, R. (2016). Health action process approach (HAPA) as a theoretical framework to understand behavior change. *Actualidades en Psicología, 30*(121), 119–130.

Seng, D. S. C. (2012). Improving the governance context and framework conditions of natural hazard early warning systems. *Integrated Disaster Risk Management, 2*(1), 1–25.

Serdeczny, O. (2023). The effects of political knowledge use by developing country negotiators in loss and damage negotiations. *Global Environmental Politics, 23*(3), 12–31. https://doi.org/10.1162/glep_a_00727

Sheehan MC, Khudairi F, Swaich GS, Hines W, Mehta S, Fox MA (2022) Urban climate-health governance: Charting the role of public health in large global city adaptation plans. PLOS Climate 1 (3). doi:https://doi.org/https://doi.org/10.1371/journal.pclm.0000012.

Sheller, M. (2018). Caribbean futures in the offshore Anthropocene: Debt, disaster, and duration. *Environment and Planning D-Society & Space, 36*(6), 971–986. https://doi.org/10.1177/0263775818800849

Sherwood, S. C., & Huber, M. (2010). An adaptability limit to climate change due to heat stress. *Proceedings of the National Academy of Sciences of the United States of America, 107*(21), 9552–9555. https://doi.org/10.1073/pnas.0913352107

Shiroyama, H., Yarime, M., Matsuo, M., Schroeder, H., Scholz, R., & Ulrich, A. E. (2012). Governance for sustainability: Knowledge integration and multi-actor dimensions in risk management. *Sustainability Science, 7*, 45–55. https://doi.org/10.1007/s11625-011-0155-z

Shulman, E. P., Smith, A. R., Silva, K., Icenogle, G., Duell, N., Chein, J., & Steinberg, L. (2016). The dual systems model: Review, reappraisal, and reaffirmation. *Developmental Cognitive Neuroscience, 17*, 103–117.

Singh, P., & Agrawal, G. (2019). Efficacy of weather index insurance for mitigation of weather risks in agriculture an integrative review. *International Journal of Ethics and Systems, 35*(4), 584–616. https://doi.org/10.1108/ijoes-09-2018-0132

Smeds, E., & Acuto, M. (2018). Networking cities after Paris: Weighing the ambition of urban climate change experimentation. *Global Policy, 9*(4), 549–559. https://doi.org/10.1111/1758-5899.12587

Smith, A., Porter, J. J., & Upham, P. (2017). "We cannot let this happen again": Reversing UK flood policy in response to the Somerset levels floods, 2014. *Journal of Environmental Planning and Management, 60*(2), 351–369. https://doi.org/10.1080/09640568.2016.1157458

Sovacool, B. K., & Martiskainen, M. (2020). Hot transformations: Governing rapid and deep household heating transitions in China, Denmark, Finland and the United Kingdom. *Energy Policy, 139*, 111330. https://doi.org/10.1016/j.enpol.2020.111330

Spencer, B., Lawler, J., Lowe, C., Thompson, L., Hinckley, T., Kim, S. H., Bolton, S., Meschke, S., Olden, J. D., & Voss, J. (2017). Case studies in co-benefits approaches to climate change mitigation and adaptation. *Journal of Environmental Planning and Management, 60*(4), 647–667. https://doi.org/10.1080/09640568.2016.1168287

Stachowiak, S. (2009). *Pathways for change: 6 theories about how policy change happens*. Organisational Research Services.

Stafford-Smith, M., Rissik, D., Street, R., Lin, B., Doerr, V., Webb, R., Andrew, L., & Wise, R. M. (2022). Climate change adaptation guidance: Clarifying three modes of planning and implementation. *Climate Risk Management, 35*, 100392. https://doi.org/10.1016/j.crm.2021.100392

Stranks, P. (2023). *Assessing the feasibility of risk transfer for extreme heat in developing, Urban Environments*. Masters of Risk Dissertation. Durham University, Durham, UK.

Swyngedouw, E. (2005). Governance innovation and the citizen: The Janus face of governance-beyond-the-state. *Urban Studies, 42*(11), 1991–2006. https://doi.org/10.1080/00420980500279869

Tadesse, M. A., Alfnes, F., Erenstein, O., & Holden, S. T. (2017). Demand for a labor-based drought insurance scheme in Ethiopia: A stated choice experiment approach. *Agricultural Economics, 48*(4), 501–511. https://doi.org/10.1111/agec.12351

Tarawally, M., Xu, W. B., Hou, W. M., & Mushore, T. D. (2018). Comparative analysis of responses of land surface temperature to long-term land use/cover changes between a coastal and inland city: A case of Freetown and Bo town in Sierra Leone. *Remote Sensing, 10*(1). https://doi.org/10.3390/rs10010112

Tarawally, M., Xu, W. B., Hou, W. M., Mushore, T. D., & Kursah, M. B. (2019). Land use/land cover change evaluation using land change modeller: A comparative analysis between two main cities in Sierra Leone. *Remote Sensing Applications-Society and Environment, 16*, 100262. https://doi.org/10.1016/j.rsase.2019.100262

Teebken, J., Mitchell, N., Jacob, K., & Heimann, T. (2023). Classifying social adaptation practices to heat stress-learning from autonomous adaptations in two small towns in Germany. *Weather Climate and Society, 15*(1), 95–108. https://doi.org/10.1175/wcas-d-22-0003.1

Thompson, V., Mitchell, D., Hegerl, G. C., Collins, M., Leach, N. J., & Slingo, J. M. (2023). The most at-risk regions in the world for high-impact heatwaves. Nature. *Communications, 14*(1). https://doi.org/10.1038/s41467-023-37554-1

Tittle, C. R. (2004). Refining control balance theory. *Theoretical Criminology, 8*(4), 395–428. https://doi.org/10.1177/1362480604046657

Travis, W. R. (2014). Weather and climate extremes: Pacemakers of adaptation? *Weather and Climate Extremes, 5-6*, 29–39. https://doi.org/10.1016/j.wace.2014.08.001

Turek-Hankins LL, Coughlan de Perez E, Scarpa G, Ruiz-Diaz R, Nayna Schwerdtle P, Joe ET, Galappaththi EK, French EM, Austin SE, Singh C, Siña M, Siders AR, van Aalst MK, Templeman S, Nunbogu AM, Berrang-Ford L, Agrawal Ta, team TGAMI, Mach KJ (2021) Climate change adaptation to extreme heat: A global systematic review of implemented action. Oxford Open Climate Change 1 (1):13. https://doi.org/10.1093/oxfclm/kgab005.

Turner, V. K., French, E. M., Dialesandro, J., Middel, A., Hondula, D. M., Weiss, G. B., & Abdellati, H. (2022). How are cities planning for heat? Analysis of United States municipal plans. *Environmental Research Letters, 17*(6). https://doi.org/10.1088/1748-9326/ac73a9

Vajjhala S, Rhodes J (2019) Insurance innovation and community-based adaptation finance, *Federal Reserve Bank of San Francisco Community Development Innovation Review, 14*. San Francisco.

van Zeijl-Rozema, A., Cörvers, R., Kemp, R., & Martens, P. (2008). Governance for sustainable development: A framework. *Sustainable Development, 16*(6), 410–421. https://doi.org/10.1002/sd.367

Vanderplanken, K., van den Hazel, P., Marx, M., Shams, A. Z., Guha-Sapir, D., & van Loenhout, J. A. F. (2021). Governing heatwaves in Europe: Comparing health policy and practices to better understand roles, responsibilities and collaboration. *Health Research Policy and Systems, 19*(1), 20. https://doi.org/10.1186/s12961-020-00645-2

Vecellio, D. J., Kong, Q. Q., Kenney, W. L., & Huber, M. (2023). Greatly enhanced risk to humans as a consequence of empirically determined lower moist heat stress tolerance. *Proceedings of the National Academy of Sciences of the United States of America, 120*(42), e2305427120. https://doi.org/10.1073/pnas.2305427120

Viguie, V., Lemonsu, A., Hallegatte, S., Beaulant, A. L., Marchadier, C., Masson, V., Pigeon, G., & Salagnac, J. L. (2020). Early adaptation to heat waves and future reduction of air-conditioning energy use in Paris. *Environmental Research Letters, 15*(7). https://doi.org/10.1088/1748-9326/ab6a24

Viguié, V., Juhel, S., Ben-Ari, T., Colombert, M., Ford, J. D., Giraudet, L. G., & Reckien, D. (2021). When adaptation increases energy demand: A systematic map of the literature. *Environmental Research Letters, 16*, 3. https://doi.org/10.1088/1748-9326/abc044

Walker, G., Tweed, F., & Whittle, R. (2014). A framework for profiling the characteristics of risk governance in natural hazard contexts. *Natural Hazards and Earth System Sciences, 14*(1), 155–164. https://doi.org/10.5194/nhess-14-155-2014

Wan, K., Lane, M., & Feng, Z. Q. (2023). Heat-health governance in a cool nation: A case study of Scotland. *Environmental Science & Policy, 147*, 57–66. https://doi.org/10.1016/j.envsci.2023.05.019

Wanner, M. S. T. (2022). Change in policy regimes for disaster risk reduction in Fiji and Nepal. *International Journal of Disaster Risk Reduction, 77*, 103030. https://doi.org/10.1016/j.ijdrr.2022.103030

Webber, S., Leitner, H., & Sheppard, E. (2020). Wheeling out urban resilience: Philanthrocapitalism, marketization, and local practice. *Annals of the American Association of Geographers, 111*(2), 343–363. https://doi.org/10.1080/24694452.2020.1774349

Weick, K. E. (1988). Enacted sensemaking in crisis situations. *Journal of Management Studies, 25*(4), 305–317. https://doi.org/10.1111/j.1467-6486.1988.tb00039.x

Whitman, S., Good, G., Donoghue, E. R., Benbow, N., Shou, W. Y., & Mou, S. X. (1997). Mortality in Chicago attributed to the July 1995 heat wave. *American Journal of Public Health, 87*(9), 1515–1518. https://doi.org/10.2105/ajph.87.9.1515

Witze, A. (2023). How Earth's first global heat officer is tackling climate change. *Nature, 624*(7992), 502–502. https://doi.org/10.1038/d41586-023-03924-4

Yeganeh, A. J., McCoy, A. P., & Schenk, T. (2020). Determinants of climate change policy adoption: A meta-analysis. *Urban Climate, 31*, 100547. https://doi.org/10.1016/j.uclim.2019.100547

Zaidi, R. Z., & Pelling, M. (2015). Institutionally configured risk: Assessing urban resilience and disaster risk reduction to heat wave risk in London. *Urban Studies, 52*(7), 1218–1233. https://doi.org/10.1177/0042098013510957

Zhang, F., & Maroulis, S. (2021). Experience is not enough: A dynamic explanation of the limited adaptation to extreme weather events in public organizations. *Global Environmental Change-Human and Policy Dimensions, 70*, 102358. https://doi.org/10.1016/j.gloenvcha.2021.102358

Chapter 13
Heatwave Responses: Early Warning Systems

Abstract This chapter provides an overview of the development, implementation, and evaluation of early warning systems designed to reduce the risks associated with heatwaves. It begins by discussing the historical context of early warning systems (EWS) as an adaptation option and underscores the importance of EWS in disaster risk reduction in general. The chapter details the four essential elements of EWS: risk knowledge, detection and monitoring, warning dissemination and communication, and preparedness and response capabilities. It emphasizes the significance of these elements in creating effective heatwave early warning systems (HEWS) and Heat-Health Warning Systems (HHWS). Various case studies from around the world, including the United States, Europe, Australia, and China, are presented to illustrate different approaches to HEWS and HHWS. These highlight the diversity in methodologies, from hazard-based systems to impact-based systems, and the importance of tailoring these systems to local climatic and demographic conditions. The chapter also discusses the challenges and limitations of current HEWS, such as the need for accurate and timely predictions, effective communication strategies, and the integration of local knowledge and practices. The importance of rigorous process and outcome evaluations to assess the efficacy of HEWS and HHWS and to ensure their continuous improvement in mitigating the impacts of heatwaves is emphasised. In relation to this, the findings from some HEWS and HHWS process and outcome evaluations are reviewed with the conclusion drawn there is mixed evidence regarding the efficacy of these EWS. Because accurate and timely predictions of heatwave occurrence are one of necessities for effective HEWS and HHWS, the technical aspects of heatwave prediction, covering short to medium-range forecasts and the potential for extended-range and seasonal predictions of heatwaves are explored. Finally, the chapter addresses the controversial topic of heatwave naming and categorisation, examining its potential benefits and drawbacks. It highlights the need for systematic research to determine the effectiveness of heatwave naming in enhancing risk communication and public awareness.

Keywords Early warning systems · Early warning systems for all (EWS4All) · Disaster risk reduction · Heatwave early warning systems · Heat-health warning systems · Warning system process and outcome evaluations · Heatwave prediction · Heatwave naming and categorisation

G. McGregor, *Heatwaves*, Biometeorology 6,
https://doi.org/10.1007/978-3-031-69906-1_13

13.1 Introduction

It is probably safe to say that for most humans staying out of harm's way is a basic goal on a day-to-day and longer-term basis. To achieve this, humans naturally look out for warning signs and once detected set into action a range of responses if possible. In the broad realm of dealing with the threats posed by weather and climate hazards humans in the past have often applied 'weather lore' or implemented traditional (indigenous) weather and climate knowledge. While weather lore and indigenous weather and climate knowledge are distinct concepts, they share similar objectives. Weather lore typically refers to a broader range of traditional sayings, proverbs, and observations related to weather and climate. In contrast, indigenous knowledge is deeply embedded within the cultural and ecological context of specific indigenous communities, adopting a holistic approach that integrates meteorological observations with ecological, cultural, and spiritual aspects. It often includes sustainable practices and adaptive strategies developed over time in response to local weather patterns. Despite their differences, the goals are essentially the same, that is the reduction of weather and climate related risk (Balehegn et al., 2019; Filho et al., 2022; Hadlos et al., 2022; McKemey et al., 2021; Poudel et al., 2022). Notwithstanding the value of indigenous weather and climate knowledge, the emergence of systematic weather observation and forecasting in the nineteenth century and its further development in the twentieth century marked a significant advancement in early warnings. Meteorological observations and the understanding and modelling of atmospheric patterns allowed for the prediction of weather systems and extreme events facilitating timely alerts for a range of weather sensitive sectors. Added to this, advancements in intra-seasonal to seasonal climate forecasting in the twenty-first century offer the tantalising prospect of advanced climate risk related information to be used in medium to long-range early warning systems (EWS) thus opening longer windows of opportunity to achieve actions related to climate risk reduction (de Perez et al., 2022). Although only one component of the arsenal of available heatwave adaptation options, the opening decades of the twenty-first century have witnessed the widespread development and implementation of heatwave early warning systems (HEWS) and associated heat-health warning systems (HHWS). The purpose of this chapter therefore is to first present an overview of EWS and then focus specifically on the nature and evaluation of HEWS and HHWS. Because a necessary basis for effective HEWS and HHWS is a reliable system for predicting heatwaves and periods of extreme heat, the processes involved with, and the advances made in heatwave prediction at a range of timescales follow. In the context of the communication and dissemination dimension of EWS, the topic of heatwave naming is introduced in the latter parts of this chapter.

13.2 Nature and Emergence of Early Warning Systems for Natural Hazards

Although there is no universally agreed definition of early warning systems, because such systems are context, scale and hazard dependent (Kelman & Glantz, 2014), they are commonly understood to be an integrated system of hazard monitoring, forecasting and prediction, disaster risk assessment, communication and preparedness activities, and systems and processes that enables individuals, communities, governments, businesses and others to reduce disaster risks in advance of hazardous events (UNDRR, 2016). As a rule, EWS are composed of four inter-related and complementary elements, those being risk knowledge, observation, detection and monitoring, warning dissemination and communication, and preparedness and response capability (Table 13.1).

In the latter parts of the twentieth century there was somewhat of an epiphany regarding EWS as it dawned on many that much stood to be gained by preventing and preparing for extreme weather and climate events as opposed to responding to and recovering from them once they had occurred. In this respect, the 1994 World Conference on Natural Disaster Reduction held in Yokohama, Japan was a defining moment. At this meeting a *Plan of Action for a Safer World* with EWS at the centre

Table 13.1 Four elements of EWS (adapted from WMO, 2022b) early warnings for all: the UN global early warning initiative for the implementation of climate adaptation executive action plan 2023–2027 (Adapted from WMO, 2022b)

Element 1: Risk knowledge Systematic collection of risk data and perform risk assessments Establish whether hazards, exposures and vulnerabilities are well known Establish whether there are patterns and trends in hazards, exposure and vulnerability Undertake hazard and risk mapping where possible	**Element 2: Detection, observation, monitoring, analysis and Hazard forecasting** Develop hazard monitoring and early warning systems Establish whether the correct parameters for the hazard are being monitored and whether there is a sound scientific basis for making hazard forecasts Determine if accurate and timely warnings can be generated
Element 4: Preparedness and response capabilities Build response capabilities at a range of scales—community to national levels Determine if response plans exist and if up to date Ascertain if local capabilities and knowledge (e.g., indigenous) are being utilised Establish the general level of preparedness to react to warnings amongst the population	**Element 3: Warning dissemination and communication** Communicate risk information and early warnings Establish whether warnings reach all who are at risk Assess whether the risks and warnings are understood and if warning information is clear and can be out into action

of this plan was hatched. Essentially the so-called *Yokohama Strategy* highlighted the necessity of creating or enhancing EWS and urged support in developing such systems for countries that are particularly susceptible to natural hazards (IDNDR, 1994). Despite the international level hubris related to achieving a major leap forward in the development and implementation of EWS, little progress was made until a global footprint for disaster risk reduction in the form of the *Hyogo Framework for Action* 2002–2015 was agreed (Tozier de la Poterie & Baudoin, 2015; Golding, 2022). Like the *Yokohama Strategy*, central to achieving major advances in disaster risk reduction were EWS.

Following on from the *Hyogo Framework*, the next major global agreement focused on disaster risk reduction took the form of the *Sendai Framework for Disaster Risk Reduction 2015–2030*. Unlike its predecessor, the *Sendai Framework* elevated EWS to the status of one of its global targets (Golding, 2022). Specifically, Target 'G' sets a timeline of 2030 for countries to significantly enhance public accessibility to multi-hazard early warning systems (MHEWS) and disaster risk information (UNDRR 2015). Associated with Target 'G' are five indicators, designed to measure global progress in achieving the goals associated with this target (Table 13.2). These indicators formed the basis of the mid-term review of the progress towards achieving Target G. Drawing on data from the *Sendai Framework Monitor* and information collected by the WMO, the Target G mid-term report, which presents an overview and a baseline of the status of MHEWS coverage around the world, was published in October 2022. Despite some indications supporting the efficacy of MHEWS in reducing disaster-related mortality and the number of affected individuals, worryingly the mid-term report revealed that only half of the countries worldwide reported having MHEWS coverage. Additionally, regions exhibited varying levels of progress and commitment in establishing MHEWS, highlighting the need for specific support, especially in the case of least

Table 13.2 Progress indicators for sendai framework target G

Target G Sub-targets	Indicator
G-1(compound G2-G5)	Number of countries that have multi-hazard early warning systems
G-2	Number of countries that have multi-hazard monitoring and forecasting systems
G-3	Number of people per 100,000 that are covered by early warning information through local governments or through national dissemination mechanisms
G-4	Percentage of local governments having a plan to act on early warnings
G-5	Number of countries that have accessible, understandable, usable and relevant disaster risk information and assessment available to the people at the national and local levels
G-6	Percentage of population exposed to or at risk from disasters protected through pre-emptive evacuation following early warning. *Member states in a position to do so are encouraged to provide information on the number of evacuated people*

Source: UNDRR (2015)

developed countries, small island developing states, and African nations. The report also emphasized the imperative for increased investments across the entire MHEWS value cycle. Perhaps in a move to temper concerns about the mounting evidence of a lack of tangible progress with implementing the *Sendai Framework's* Target G, and recognising the integral role that MHEWS potentially play in achieving the overall aims of the *Sendai Framework*, the Secretary-General of the United Nations, in advance of the release of the October 2022 mid-term review of Target G progress, announced the UN backed Early Warnings for All (EW4All) initiative on March 23, 2022. Officially launched at COP27 in Egypt in November 2022, the EW4All Action Plan outlines how universal protection of human populations from the impacts of natural hazards through the implementation of early warning systems by the year 2027 is to be achieved by making these systems accessible and effective for all communities, with a special focus on those most vulnerable to natural hazards. Operating through a collaborative effort involving the WMO, the UNDRR, the International Telecommunications Union (ITU), and the International Federation of Red Cross and Red Crescent Societies (IFRC), EW4All encompasses the four essential elements of EWS and is designed to support the achievement of Target G goals and those of the *Sendai Framework* more widely.

Although at the time the concept of MHEWS was not new, the *Sendai Framework* was perhaps the first time that a multi-hazard approach and early warnings were linked (Golding, 2022). While the *Sendai Framework* has been recognised as a well-intentioned aspiration, a range of shortcomings have been aired in the academic literature. These include a failure to deal with root causes of disaster (Wisner, 2020), challenges with disaster loss reporting (Wilkins et al., 2021; Khoza et al., 2022), missed opportunities for addressing fundamental gender-based issues and problems with embedding health resilience in disaster risk reduction (Wright et al., 2020; Zaidi & Fordham, 2021) and the preponderance of governance systems dominated by a top-down, technocentric, centralized management which are not attuned to the needs of peripheral and vulnerable populations (Zia & Wagner, 2015).

Beyond global agreements for disaster risk reduction, EWS are a visible part of the Sustainable Development Goals (SDGs), especially in relation to SDGs 1 (no poverty), 11 (sustainable cities and communities) and 13 (climate action). This is because EWS are considered pivotal to reducing vulnerability and enhancing resilience at a range of scales. EWS also receive mentions in Articles 7 and 8 of the United Nations Framework Convention on Climate Change Paris Agreement with the establishment of the Climate Risk Early Warning Systems (CREWS) initiative during COP21, a clear illustration of the interconnectedness of risk reduction and climate action at the global level (Golding, 2022). The importance of EWS is also recognised in many countries' climate change related National Adaptation Plans.

Returning to the *Hyogo Framework*, a defining characteristic of this global blueprint for risk reduction was a swing towards a so-called 'last mile' as opposed to a 'first mile' approach to EWS. Essentially the 'last mile' approach focuses on the crucial final stages of the early warning process, where the information reaches the at-risk individuals and communities. This involves effective communication strategies and community engagement to ensure that people understand the warnings, know how to respond, and can take appropriate actions to protect themselves. Key aspects of the last mile approach include community level participation in the development of the EWS (Tan et al., 2022) and associated localized and culturally sensitive communication methods, community education and awareness programmes, and the integration of local knowledge and practices into the warning dissemination process. This approach recognizes the importance of addressing the unique needs and vulnerabilities of specific communities and has been referred to as people-centred and/or bottom-up (Basher, 2006; Garcia & Fearnley, 2016), often manifesting in the form of community EWS, community-centric EWS, community-based EWS or participatory EWS (Baudoin et al., 2016; Macherera & Chimbari, 2016; Marchezini et al., 2018). While notionally attractive, reviews of the progress on converting the rhetoric of people-centred approaches into reality for natural hazards in general point to several issues. These include varying definitions of what people-centred and participation means, inadequate resources at the community level, a reluctance among the at-risk public to collaborate with authorities in shouldering the responsibility for disaster risk management, conflicts between public and private interests, inadequate community engagement across the four elements of EWS often with a focus on the risk knowledge element, lack of understanding of community members' viewpoints and expectations regarding responsibility, and poor communication of authorities with communities (Clegg et al., 2022; Scolobig et al., 2015; Sufri et al., 2020).

In clear contrast to the 'last mile' approach and increasingly less favoured, the 'first mile' approach focuses on the initial stages of the early warning process, where information is generated and transmitted from authoritative sources (e.g., meteorological agencies, disaster management organizations, health agencies) to communities at risk. This involves the collection and analysis of data, detection of potential hazards, and the initiation of warnings. The goal is to ensure timely and accurate information flow from the source to the local level. Key aspects of the first mile approach include the use of advanced technologies for hazard monitoring, the establishment of efficient communication channels between meteorological agencies and local authorities, and the development of clear protocols for issuing warnings. Essentially a top-down process, operated by regional or national agencies and dominated by technocentric and centralized management, the first mile approach is typically hazard focused with such systems often referred to as hazard-based EWS.

13.3 Heatwave and Heat-Health Early Warning Systems (HEWS and HHWS)

As established elsewhere in this volume, heatwaves pose a significant threat to society, economy and natural systems. In response to the escalating challenges posed by heatwaves, there has been a growing demand for the establishment of heatwave early warning systems (HEWS) with some tailored to heat-health related issues in the form of heat-health warning systems (HHWS) (Linares et al., 2020; McGregor et al., 2015).

Building on the general definition of EWS, the objective of HEWS and HHWS is to provide timely and pertinent weather and climate information regarding upcoming heatwave events. This information is utilized to prompt decisions and actions aimed at achieving tangible reductions in heatwave related risks in the case of general HEWS or heatwave-related health risks in the case of HHWS. In essence, HEWS and HHWS function as tools in preparation for heatwave-related challenges, serving as adaptive measures to mitigate associated risks. Because the general outcome of HEWS is a forecast of the likelihood of heatwave occurrence and in some cases heatwave intensity, they are often referred to as hazard based EWS, generally falling within the 'first mile' approach to early warning. In contrast if HHWS are focused on producing estimates of the level of impact arising from a heatwave, for example the expected levels of hospital admissions for a prescribed area, they are often referred to as impact based EWS. While some impact based HHWS may adopt a 'last mile' approach in their development others may not. Although HEWS and HHWS may exist as standalone systems they generally reside within a wider heat action plan (HAP) or Heat-Health Action Plan (HHAP) or form part of a MHEWS. Typically, effective HEWS and HHWS require dependable meteorological forecasts tailored to the population or region under consideration, a comprehensive understanding of the cause-and-effect relationships between the heatwave hazard and societal impacts, especially health in the case of HHWS, the identification of a system trigger usually based on an inflection point in an identified hazard and impact relationship, a system of graded alerts/warnings, the implementation of response measures within the lead-time provided by the alert/warning with responses often tailored to the alert/warning level, and good governance as determined by active engagement from institutions and civil society equipped with ample resources, capacity, knowledge, and political will to execute the required response measures. Overall, the efficacy of HEWS and HHWS hinges on proactively and systematically addressing issues related to awareness, preparedness, organizational aspects, and actor networks (Bernard & McGeehin, 2004; Lass et al., 2011; McGregor et al., 2015).

13.3.1 Development of HEWS and HHWS

Although temperature forecasts have been part of daily national weather service activities since the inauguration of public weather forecasts in August 1861 in England, it was probably not until the late 1970s when specific forecasts related to heat began. In 1979, NOAA's NWS adopted the heat index (see Chap. 4) into its operations. By 1984, the heat index parameter and guidance on its usage had been introduced in an NWS policy document with this index subsequently used for warning the US public about potentially dangerous heatwave conditions (Hawkins et al., 2017). However, by the early to mid-1990s research was beginning to question the effectiveness of the HI and its associated heat products as a basis for heatwave related warnings. Although the HI is still widely used in the US, this critique continues today with recent studies highlighting health responses occurring at levels different to the longstanding NWS HI alert ranges (Hondula et al., 2022; Vaidyanathan et al., 2019), the inability of the HI to model heat stress under very high temperature and humidity conditions (Lu & Romps, 2023; Romps & Lu, 2022) and its possible displacement by new heatwave products such as 'HeatRisk' (Kimutis et al., 2024).

Based on the tenet that humans respond to not only temperature and humidity as modelled by the HI, but the synergistic effects of all heat-relevant variables acting simultaneously, plus consecutive days of oppressive heat, in the mid-1990s members of the University of Delaware school of synoptic climatology advocated for an airmass-based approach to developing HHWS using a temporal synoptic index (TSI) as an indicator of dangerous heatwave conditions. Undoubtedly spurred on by the impactful heatwaves of 1993 and 1994, the airmass-based Philadelphia Hot Weather-Health Watch/Warning System (PWWS) was inaugurated in 1995 for the purpose of alerting the city's population about impending heatwave conditions and associated health risks (Kalkstein et al., 1996). Although launched with a different phrasing this was probably the first ever impact based HHWS in the world. System mechanics involved using a suite of weather forecast variables and the TSI to establish whether known health-impactful airmasses were likely to occur. For Philadelphia the most oppressive airmass was identified as a maritime tropical plus airmass typified by high temperature and humidity levels, moderate cloud cover and gentle southerly winds off the Atlantic. Depending on the temporal immediacy of this airmass a heat-health watch (next 2 days) or alert (next day) was issued. For the situation of an alert, a maritime tropical airmass specific algorithm for predicting mortality was then run to determine whether elevated mortality might be expected given the evolving airmass conditions. In the case of predicted elevated mortality, a level 1, 2 or 3 heat-health warning was issued depending on percentage departure from the 'normal' level of mortality (Fig. 13.1).

While synoptic based systems proved popular, becoming established in several US cities as well as in cities in Italy, China and Canada over the period 1995–2004 (Sheridan & Kalkstein, 2004), only a few presently exist. For example, although not central to the current HHWS for Philadelphia, outputs from the original air mass-based system for here are still consulted. The same applies for other US cities such as Kansas City and Des Moines. For Italy, where the air mass-based approach was widely adopted in its early days of developing HHWS, synoptic/air mass models are presently used in conjunction with simple apparent temperature-based threshold models in large cities such as Rome and Naples whereas only apparent temperature models are used in smaller cities. Interestingly the synoptic approach is the basis of an experimental heatwave categorisation and naming system for the Spanish city Seville (see Sect. 13.5).

That the synoptic approach has not achieved 'world domination' may well relate to its complexity compared to less complicated systems based on the exceedance of values of single meteorological or thermal index values and much tighter and more

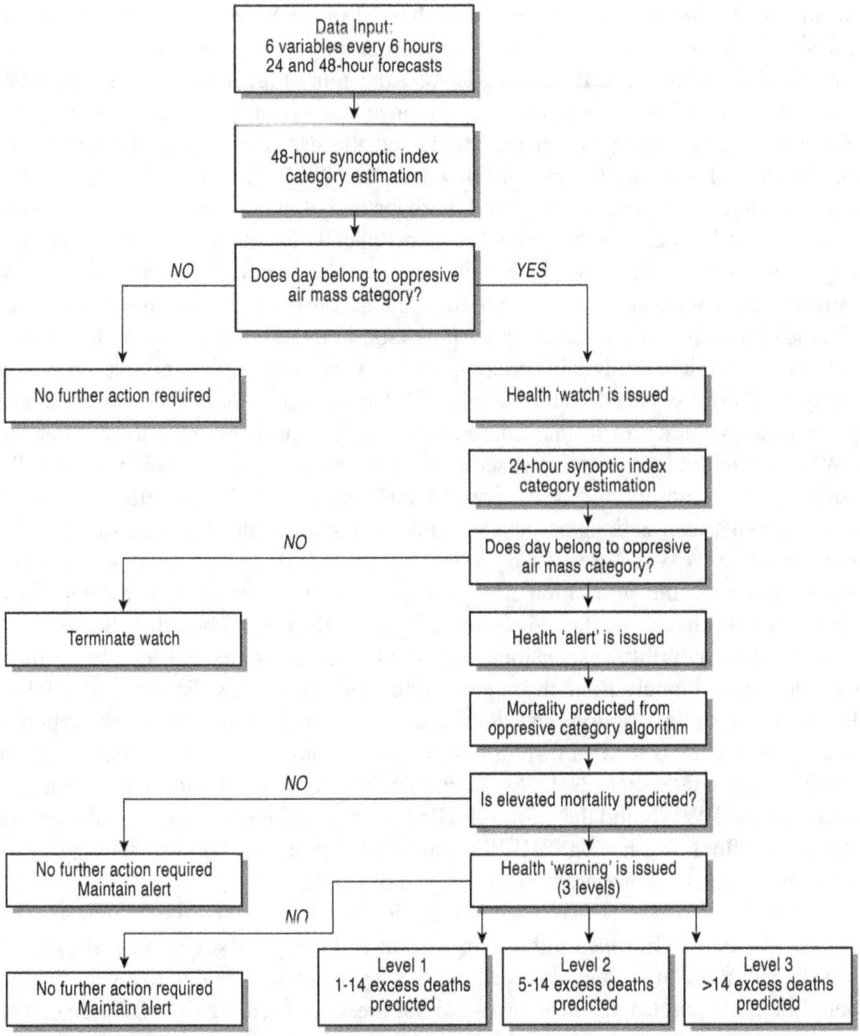

Fig. 13.1 Structure of the Philadelphia Hot Weather-Health Watch/Warning System (PWWS). (Source: Kalkstein et al. (1996). © American Meteorological Society, used with permission)

transparent statistical associations between heat and health. In comparison with other approaches to identifying the link between heat and health, the airmass-based approach does not clearly out-perform other methods (Hajat et al., 2010; Urban & Kysely, 2018). Plus, it does not provide a continuous estimate of excess death and risk warnings as these are only generated when a dangerous or oppressive airmass is identified. The TSI on which the airmass-based approach was originally based is also not transportable and thus requires redevelopment for every location for which a system is proposed. This is not a simple task. To solve the issue related to its location dependency a spatial version of the TSI in the form of the spatial synoptic classification (SSC), which could be applied simultaneously across large regions, was developed (Davis & Kalkstein, 1990; Kalkstein et al., 1996; Sheridan, 2002). Although the SSC has stimulated research on airmass health associations (Fonseca-Rodríguez et al., 2020; Lee et al., 2018) and the use of synoptic climatology in environmental analysis in general (Bower et al., 2007; Hondula et al., 2014), its application in HHWS remains limited. Perhaps one of the greatest hurdles related to airmass based systems is the accuracy of forecasted values for a range of meteorological variables required to determine, using the TSI or SSC algorithms, if a day falls into an airmass category deemed to be dangerously oppressive in terms of heat. This is especially so for variables such as cloud cover, wind and humidity which have relatively low predictability compared to temperature, the variable on which many non-airmass based HHWS are based. Criticisms aside concerning the complexity and accuracy of airmass based systems, it is probably safe to say that the PWWS as the first ever impact based HHWS broke new ground. This is especially with regards to presenting a framework for HEWS and HHWS and setting out many of the downstream actions associated with the issuance of a heat-health warning. For example, PWWS response actions included dedicated media heatwave announcements, the promotion of a 'buddy system', activation of a 'Heatline', home visits by health workers, nursing and personal care and boarding home interventions, halting utility suspensions, increased emergency medical service staffing, outreach to the homeless and the opening of cooling centres (Kalkstein et al., 1996). In many ways these actions provided a general blueprint for heatwave response plans globally as well as an insight into the type of governance structures required for the implementation of HHWS. Given this, it is perhaps difficult to deny that the legacy of the PWWS and the various HHWS it spawned has left a lasting impact on the field, influencing not only HEWS and HHWS practices but also shaping heatwave awareness and preparedness at a range of levels.

While HEWS and HHWS continued to develop across the US from the mid-1990s, across Europe, with the exception of Portugal (Nogueira, 2005), there was a general malaise about the need for such systems, largely due to a widely held belief in policy circles that heatwaves did not present a credible risk. Of course, that all changed with the occurrence of the 2003 European heatwave event with its wide-ranging impacts. As a result there was a rapid expansion of European HEWS and HHWS from virtually none in 2003 to 26 countries in 2006 and 28 by 2009 (Casanueva et al., 2019; Lass et al., 2011; Martinez et al., 2022) as well as the funding of a number of major research projects such as CASHh (WHO, 2004), EuroHeat (Matthies & Menne, 2009; D'Ippoliti et al., 2010), SCORCH (Vanderplanken et al., 2019; Vanderplanken et al., 2021) and HEAT-SHIELD (Morabito et al., 2019) all of which possessed significant HEWS and HHWS elements. Perhaps the most notable and immediate responses to the European 2003 heatwave event in terms of heatwave related EWS were the HEWS/HHWS established for England and Wales, France and Germany.

The French heat warning system, instituted in 2004, is jointly operated by the French public health agency Santé Publique France and the French weather agency

Météo-France. The system focuses on issuing warnings when sustained periods of unusual heat, identified as the 99.5th percentile of the 1981–2010 temperature distribution, occur. Heatwaves are defined as periods of time during which the average minimum and maximum temperatures over a three-day period surpass specific local thresholds (Pascal et al., 2013) (Fig. 13.2), At the 'departmental' level heatwaves are described based on their daily severity. This is computed as the difference between the observed temperatures and locally specific thresholds with duration defined as the number of days when the minimum and maximum thresholds are exceeded simultaneously (Pascal et al., 2021). Based on values of the daily severity, overall heatwave severity is calculated as the sum of the daily severities with heatwave intensity defined as the maximum of the daily severities during the heatwave. A heatwave warning is issued based on the probability that the forecasted 3-day averaged temperatures exceeds the identified temperature thresholds, taking into account a critical analysis of the forecast uncertainties, possible aggravating factors and when available, health surveillance data (Pascal et al., 2012a). Warnings are graded on a four-colour scale; green (no risk), yellow (moderate risk, e.g., forecasts slightly below thresholds), orange (high risk, forecasts above thresholds) and red (extreme risk, forecasts above thresholds associated with an exceptional intensity). Because of the changing nature of heatwaves, the thresholds which trigger warning issuance are periodically reviewed (Pascal et al., 2021).

Whereas the French system has essentially remained a hazard-based system, the one for England and Wales, initially hatched in 2004 as part of the inaugural Heatwave Plan for England, has evolved from a hazard-based to an impact-based system. Launched in June 2023, the new United Kingdom Health Security Agency (UKHSA) Heat-Health Alerting (HHA) system, run in partnership with the UK Met

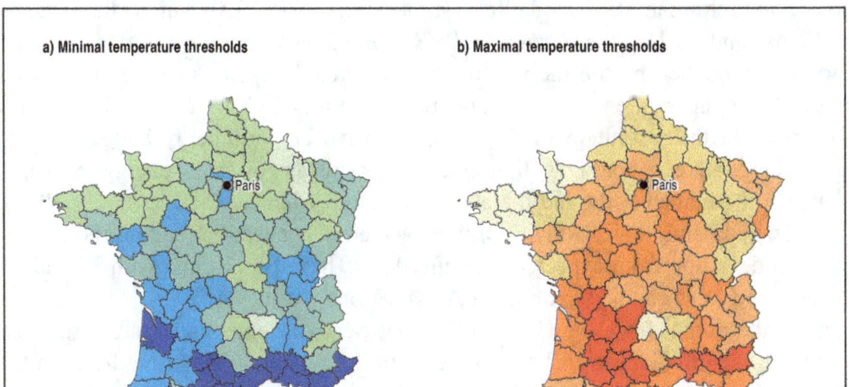

Fig. 13.2 Minimum (**a**) and maximum (**b**) temperature thresholds for French administrative regions ('departments') used in the HHWS for France. (Adapted from: Supplementary material in Pascal et al. (2021). Used with permission under Springer Nature License Number 5704850338102)

Office, evaluates the change in relative risk of mortality in response to an increase in maximum temperature. Beyond a general preparedness level (no alert), there exist three alert levels which are tied to incremental changes in impact as quantified by the relative risk of heat related mortality. The determination of these impact levels stems from an extensive analysis of maximum temperature and mortality data with distinct temperature-related impact thresholds identified for London and several other regions in England. Instead of relying on forecasted temperature values to automatically trigger a heat-health alert, the UKHSA utilises them to initiate a dynamic risk assessment. This assessment weighs up the evolving weather situation outlined by the UK Met Office and considers factors traditionally challenging to quantify such as existing pressures on the health system (e.g., staff availability), the number of previous alerts issued for the summer, the time of year, and any relevant political issues. After completing the dynamic risk assessment, alerts are specifically issued to the health and social care sector, as well as other sectors and agencies involved in responding to heat events (Fig. 13.3). Working in tandem with the UKHSA HHA is the UK Met Office's National Severe Weather Warning Service's (NSWWS) Extreme Heat warnings. The development of NSWWS extreme heat warnings is informed by the heat-health analyses that underpin the UKHSA HHA impact-based system. While the UKHSA issues yellow to red alerts, the NSWWS exclusively provides amber and red alerts for extreme heat. This distinction is made because the health sector, the primary focus of UKHSA's HHA, is likely to experience impacts before other sectors.

Germany's HHWS has been in operation since 2005. Over its lifetime it has undergone various updates (Matzarakis et al., 2020). Innovations include accounting for short-term heat acclimatisation, a heat-health module specific to the elderly and modelling of nocturnal indoor temperatures for urban and non-urban situations to account for urban heat island effects (Matzarakis et al., 2020). Unlike the systems in France and the UK, the German HHWS is based on a measure of human thermal stress as modelled by the thermal index Perceived Temperature (PT). This is an equivalent temperature with units of degrees Celsius (see Chap. 2). Its calculation is grounded in the resolution of the comprehensive human energy budget Klima-Michel model, encompassing all pertinent mechanisms governing energy exchange between the human body and the atmospheric environment (Staiger et al., 2012). Based on epidemiological studies of the association between PT and mortality, significant deviation from expected mortality levels have been found with PT values categorised as strong and extreme heat load. Hourly values of PT are produced as part of the output from the German Weather Service numerical weather forecast model to identify dangerous heat episodes. In the situation that strong heat load is forecast, a warning recommendation is automatically created. However, the warnings are only published once a qualified weather service biometeorologist has undertaken a mandatory confirmation of the heat forecast and done any adjustments related to the evolving heatwave situation. Once confirmed, the warnings are disseminated to a range of specific stakeholders and the public using a variety of means (Matzarakis et al., 2020). Although the heat load thresholds applied in triggering a warning have been established based on PT—mortality associations, the German

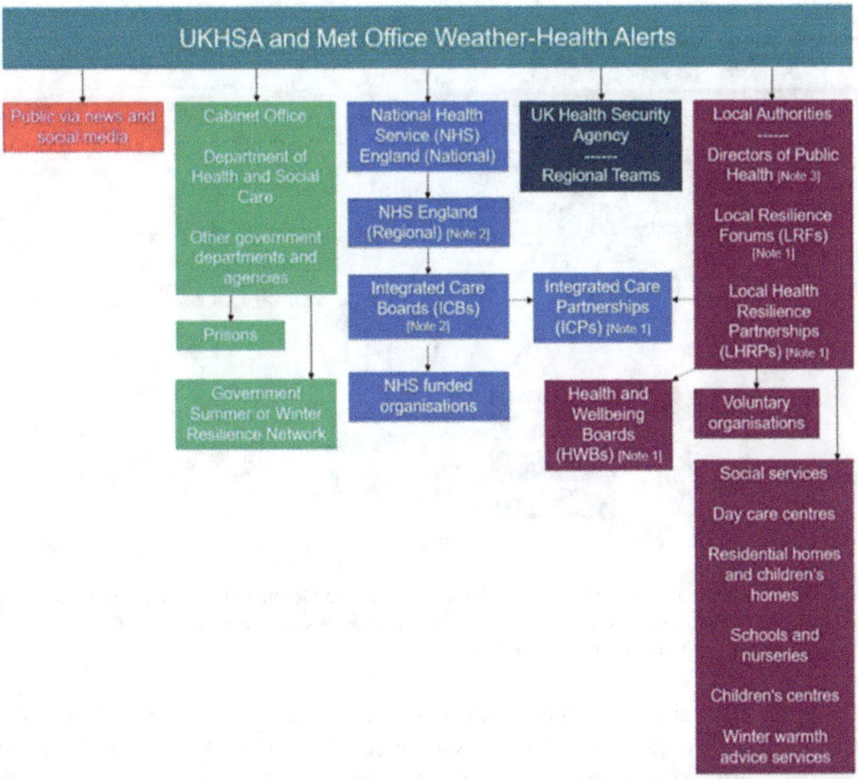

Fig. 13.3 Cascade of health alerts in UK joint UKHSA- Met Office impact-based weather—health alert system. (Source: UKHSA 2023 User guide New Impact-Based Weather-Health Alerting System. Used with permission under the terms of the Open Government Licence v3.0. Available at: https://assets.publishing.service.gov.uk/government/uploads/system/uploads/attachment_data/file/1168272/User-guide-impact-based-weather-and-health-alerting-system-v3.pdf)

HHWS is still essentially a hazard-based system or perhaps alternatively as is the case for the French system an impact-informed hazard-based system. The general flow of activities associated with the issuance of a warning within the German HHWS are illustrated in Fig. 13.4.

Elsewhere game changing heatwave events also prompted action in terms of heatwave response plans. For Australia the devastating January 2009 heatwave event across the states of Victoria and South Australia focused attention on the need for HHWS. For example, the state of Victoria's Department of Health operated a heat health alert system from December 2010 to March 2022. An alert was issued if temperatures were forecast to exceed temperature thresholds based on climate and health relationships with some consideration of acclimatization (Nicholls et al., 2008). However, in the summer of 2022–2023 Victoria aligned its Department of Health's heat health alert system to the Bureau of Meteorology's National Heatwave Service issuing heat health alerts when severe (vulnerable people at risk) or extreme

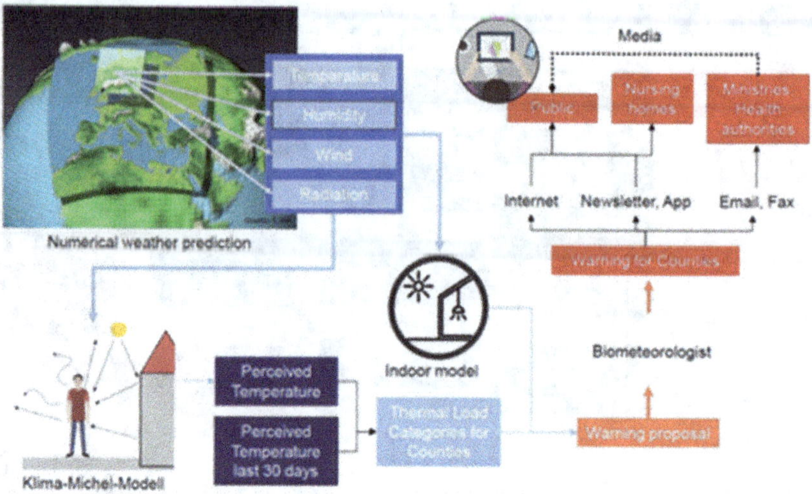

Fig. 13.4 General structure and flow of information in the German HHWS leading to warning issuance. (Source: Matzarakis et al. (2020). Used with permission under Creative Commons CC BY Licence https://www.mdpi.com/openaccess)

(everyone is at risk, even the healthy) heatwaves are forecast. Other Australian states have done the same. South Asia's first HEWS/HHWS was developed in response to the severe heatwave event experienced in the city of Ahmedabad in 2010. For this system a 7-day probabilistic weather forecast is used to trigger the heatwave response plan, in advance of dangerous heat waves defined as when Tmax is forecast to exceed 41 °C (Knowlton et al., 2014).

Canada's first HEWS/HHWS was established in the city of Toronto, Ontario in 2001. This took the form of an airmass based system modelled on that for Philadelphia. This became one of several disparate systems with diverse methodologies used by public health units throughout Ontario to activate and issue public heatwave warnings. In some cases, public health units operated without any systematic warning system (Henderson et al., 2020). The federal government also issued heat alerts using a singular national criterion reliant on the Canadian Humidex, without accounting for intra-provincial differences in climate or population responses. This all led to the situation that amid heatwave events, neighbouring public health units in Ontario and the federal government frequently issued heat warnings at different intervals, accompanied by distinct public messages (Henderson et al., 2020). Understandably this precipitated much confusion about who to listen to and what to do during heatwave events. Recognising the existence of similar situations across Canada the federal government moved to institute Heat Action Response Systems (HARS) across the country and encouraged a harmonised approach to heatwave related warning. While most HARS are developed and coordinated at a local level, at the national level Health Canada provides guidance and information to help local and regional health authorities develop evidence-based extreme heat plans and interventions (Berry et al., 2014). Through a Heat Warning Program (HWP), the Meteorological Service of Canada, Environment and Climate Change Canada (ECCC), supports local and regional HARS protocols. Using pre-defined temperature thresholds, the HWP relays heat alerts to public health authorities across Canada. This new way of doing things resulted in the development of Ontario's harmonised Heat Warning and Information System (HWIS) (Clemens et al., 2022; Henderson et al., 2020), a system in stark contrast to the uncoordinated approach that typified the early days of heatwave warnings for this province. Ontario's HWIS is a single-tiered system such that a heat warning issued by ECCC is the only formal trigger needed to initiate heatwave response actions. In acknowledgment of intra- provincial variations in heatwave sensitivity ECCC has worked with regional health authorities in Ontario to update the national heat alert program to integrate regional climatology, health evidence, heatwave duration, and overnight temperatures (Clemens et al., 2022) as is the situation for other provinces (McLean et al., 2018).

For China, a country with a significant population with high exposure to heatwaves, the first experimental HEWS was set up in the city of Shanghai in the early 2000s ((Tan et al., 2004). Based on the airmass approach of Kalkstein et al. (1996), this 'showcase' system demonstrated some potential. However, it was supplanted by a national level system in the mid-2000s. Presently, warnings at the national level are issued if the maximum temperature surpasses 35 °C for three consecutive days (Yellow alert), exceeds 37 °C (Orange alert), or reaches 40 °C (Red alert) within a 24-hour period. The meteorological departments of various provinces and cities have the flexibility to adjust these thresholds according to their local climate characteristics. For Shanghai, the Shanghai Meteorological Service (SMS) utilizes the daily maximum temperature as the sole indicator for defining a heatwave, without considering other weather characteristics. In 2007, the Shanghai Municipal People's Government implemented an early warning system across all 16 districts of Shanghai. This system employs "Yellow," "Orange," and "Red" signals to alert residents. While it adopts the temperature thresholds from the China Meteorological Administration's warning system, it does not include duration thresholds. In addition to these local warning systems, China introduced a national standard for heatwave indexes in 2012, termed the "grade of the heatwave". However, this heatwave index has primarily found application in research rather than serving as a practical tool for actual heatwave early warnings (Wu et al., 2020). In Japan, the Japan Meteorological Agency (JMA) introduced the 'Hot Temperature Alert' (HA2011) in 2011 as a form of HEWS/HHWS. Alerts were activated when maximum temperatures reached or exceeded 35 °C. The HA2011 was withdrawn in 2020 and replaced by the 'Heatstroke Alert' system (HA2020), jointly launched by the JMA and the Ministry of the Environment, Japan, in the summer of 2020. Based on research indicating a significant increase in heatstroke deaths and ambulance cases when the Wet Bulb Globe Temperature (WBGT) exceeds 33 °C, a "Heatstroke Alert" is issued when the projected daily maximum WBGT for the current or following day is anticipated to reach 33 °C or beyond (Oka et al., 2023). Korea also issues heatwave warnings based on the exceedance of a WBGT threshold (Yi & Yang, 2020). This is despite a fleeting idea that an airmass based system be adopted for this country. For Africa, while some countries issue extreme heat warnings in the form of a hazard-based HEWS in the style of the 'first mile' approach, HHWS remain underdeveloped. There are however moves to put in place experimental systems using absolute and relative measures of temperature and thermal index values as triggers for activating heatwave responses (Kapwata et al., 2022; Thiaw et al., 2022).

That many countries are still some distance from developing HEWS or HHWS is clear from the results of a WMO survey of National Meteorological and Hydrological Services (NMHS). This survey revealed that only 54% of NMHSs produce extreme heat warnings with the majority monitoring, forecasting and warning for extreme temperatures alone. Only a few NMHS were found to provide thermal human heat budget information (e.g., Germany) while those that monitor, forecast and warn for heatwave intensity (e.g., Australia) are in the clear minority (WMO 2023). The rather rich potpourri of HEWS and HHWS flavours will no doubt lead some international agencies to call for the harmonisation of extreme heat hazard definitions and practices. While an attractive idea, as this would facilitate inter-country comparisons of HEWS and HHWS effectiveness, it may prove politically insensitive given the huge country-to-country discrepancies that exist in human and financial resources which form crippling barriers to achieving heatwave risk reduction in many countries. Of course, initiatives such as EWS4All aided by the availability of global adaptation funds could, with the right level of political will and fair and equitable funding distribution mechanisms, go some of the way to solving HEWS and HHWS inequity.

13.3.2 HEWS and HHWS Monitoring and Evaluation

As with all adaptation measures their worthiness needs to be established. Often this is the elephant in the room as many HEWS or HHWS are set up in the absence of a plan for systematic monitoring and evaluation of efficacy with the result there is a general dearth of systematic evaluations. Monitoring involves regularly collecting relevant indicators to assess the functioning of a HEWS or HHWS. These indicators may pertain to either outcomes or processes. Integral to any EWS, monitoring utilizes routine data whenever possible to minimize costs. Regular surveys of key stakeholders may also serve as a valuable monitoring tool. However the data is obtained, and assuming it is robust and fit for purpose, monitoring data forms the basis for evaluations. Evaluation is a more rigorous and time-consuming process than monitoring. It goes beyond assessing effectiveness and reliability often delving into the efficiency and human dimensions of HEWS/HHWS. While system monitoring usually commences immediately, evaluations are typically undertaken after a system has been operational for at least 2–3 years to allow response systems to become well-established and the requisite monitoring data to reach a point in terms of volume deemed suitable for analysis.

Generally, two types of evaluation can be identified, namely process and outcome (Dwyer et al., 2022). Process evaluations centre on the functioning of a warning system and its individual components. This includes examining the entire process, from generating heatwave and heat-health warning forecasts to disseminating alerts to the pertinent agencies responsible for communicating warnings. Governance arrangements linked to HEWS and HHWS play a crucial role in determining accountability during a process evaluation. An outcome evaluation involves assessing the effectiveness of a warning system by measuring the extent to which the impact of a heatwave event has been averted through timely warnings. In the context of HHWS, this evaluation would typically gauge the reduction in heat-related illnesses or fatalities prevented. Historically, outcome evaluations related to HHWS have primarily focused on mortality, with outcomes often reported as odds ratios, mortality rates, excess deaths, relative risk, increased percentage of mortality per centigrade temperature increase, or as a mortality index (Boeckmann & Rohn, 2014). As records of heat-related morbidity improve, it is anticipated that there will be a rise in outcome evaluations focusing on morbidity-related impacts. For HEWS, an evaluation may entail establishing general population response to any heatwave warning that has been issued using social surveys. If HEWS have been tailored to a specific economic sector such as energy, an evaluation may involve establishing the extent to which brown or black outs have been avoided because of strategies put in place to manage energy distribution systems when a heatwave is forecast.

Typically, HEWS and HHWS evaluations focus on establishing dimensions of effectiveness, reliability, efficiency and acceptance. Additional evaluation criteria may also include sensitivity, and specificity (Kovats & Ebi, 2006). Effectiveness if often assessed in relation to whether a HEWS or HHWS is delivering its intended benefits to broader society, either economically or specifically in terms of the health of a designated population. The reliability dimension is concerned with whether a HEWS or HHWS consistently generates precise and reliable predictions of heatwave characteristics or impact levels, while efficiency assessments usually focus on system cost effectiveness often determined via a cost-benefit type of analysis using a variety of methods for economic appraisal of heat warning systems (Chiabai et al., 2018). Establishing acceptance is all about assessing whether the HEWS or HHWS is embraced by the intended sector or population and is inclusive of all societal groups, with particular attention to underserved or socially isolated populations. It could also involve establishing the level of interaction between agencies, participation of agencies other than that issuing the warning and completeness of response of participating agencies (Kovats & Ebi, 2006). An interest to evaluators is the degree to which a system may be beleaguered with false positives and/or false negatives as a high occurrence of these bears implications for system credibility. Establishing system sensitivity and specificity can assist with this. In the case of a HEWS sensitivity indicates the number of times a heatwave warning is issued, and the forecast conditions occurred (true positive) and conversely how often a warning was not issued but adverse meteorological conditions occurred (false negative). Specificity provides insights into the level of false positive forecasts of a warning system (Table 13.3). Complementing the establishment of overall system sensitivity and specificity a range of verification techniques can be applied to assessing the predictability of key meteorological variables which may compose the foundations of single weather variable or thermal index related thresholds for triggering HEWS or HHWS alerts (Jolliffe & Stephenson, 2012).

As noted by Dwyer et al. (2022), whereas HEWS/HHWS process evaluations have been comprehensively reviewed in some detail, those related to the evaluation of outcomes, such as lives saved because of the implementation of heatwave response actions, are meagre in comparison. Generally, process evaluations of HEWS or HHWS at the country or city level have uncovered varied approaches in assigning and describing stakeholder roles, contrasts in the levels of human and financial resources available for developing and implementing heatwave response

Table 13.3 Example of calculating HEWS sensitivity and specificity

Calculating sensitivity and specificity require records on the number of forecasted events and the number of events that occurred. This information is then assembled according to four categories as follows:
True Positives (TP): system correctly predicts a heatwave, a heatwave occurs
False Positives (FP): system predicts a heatwave, but heatwave does not occur (Type I error)
True Negatives (TN): system correctly predicts no heatwave, no heatwave occurs
False Negatives (FN): system predicts no heatwave, but heatwave does occur (Type II error)
Given the above, sensitivity (true positive rate) and specificity (true negative rate) can be calculated as follows:
Sensitivity = TP / (TP + FN)
Specificity = TN / (TN + FP)
So, taking the results from a hypothetical HEWS which operates with a clear definition of a heatwave event or extreme heat event (e.g., at least 3 days with Tmax >95th percentile) the following may have been recorded over several summers:
True Positives (TP) = 21
False Positives (FP) = 6
True Negatives (TN) = 36
False Negatives (FN) = 3
Therefore
Sensitivity = 21 / (21 + 3) = 0.875
Specificity = 36 / (36 + 6) = 0.857
With the interpretation being
Sensitivity tells us that the system correctly predicted a heatwave/extreme heat event 87.5% of the time when such an event occurred, and
Specificity indicates that the system correctly predicted no heatwave/extreme heat event 85.7% of the time when no heatwave occurred
In the case of the hypothetical HEWS, the sensitivity score indicates good performance in relation to predicting the occurrence of heatwave events while the score for specificity indicates the system only occasionally produces a false positive which lends some credibility to the system

Source: Author

plans, differences in general heatwave preparedness and the range of response actions outlined in plans, and variable practices concerning specific attention to vulnerable groups. Added to these are diverse practices concerning system warning parameters (meteorological variables or thermal index values used for triggering warnings), discrepancies in the number of alert levels and their nomenclature as well as the monitoring of critical meteorological variables and health outcomes such as mortality and/or morbidity. Additionally, variations are observed in the timeliness of warnings, the type and content of warning messages and their dissemination methods (Blättner et al., 2020; Errett et al., 2023; Guardaro, 2023; Kotharkar & Ghosh, 2022; Leite et al., 2020; Martinez et al., 2011; Mayrhuber et al., 2018; Price et al., 2018; Randazza et al., 2023; Vanderplanken et al., 2019; Vu et al., 2019). In relation to process, of interest is the way in which the COVID-19 pandemic interacted with heatwave events and the implication this bore for the operation of HEWS/HHWS and the implementation of intervention strategies (Bose-O'Reilly et al., 2021; Jin & Sanders, 2022; Knox-Hayes et al., 2023; Sousa et al., 2022; Wilhelmi et al., 2021; Bose-O'Reilly et al., 2021; Jin & Sanders, 2022; Know-Hayes et al. 2023; Sousa et al., 2022; Wilhelmi et al., 2021).

Probably the first outcome evaluation of a HHWS was that conducted for Philadelphia's PWWS. This evaluation, based on data from 1995 to 1998, aimed to ascertain the system's efficacy in saving lives while maintaining cost-effectiveness. The findings indicated that, on average, 2.6 lives were saved each day a heat warning was issued, as well as on the 3 days after the warning ceased. For the three-year analysis period, this translated to a total of 117 lives saved. Additionally, the evaluation determined that the operational costs of the system were relatively low when compared to the estimated value of lives saved (Ebi et al., 2004). Evaluations for other cities such as Ahmedabad (Hess et al., 2018), Montreal and New York (Benmarhnia et al., 2016; Benmarhnia et al., 2019) and Adelaide (Nitschke et al., 2016) also indicate the effectiveness of HEWS/HHWS in reducing the health burden associated with heatwaves. For Spain there also appears to be similar effectiveness. Here provinces with more actions implemented in their Heat Health Protection Plans (HHPP) have demonstrated stronger decreases in mortality attributable to extreme heat compared to provinces where HHPP actions are underdeveloped (Martinez-Solanas & Basagana, 2019). While these serve as positive examples of HEWS/HHWS efficacy, several evaluations have revealed variable results. For example, while one assessment of the effectiveness of the French system in saving lives during the 2006 heatwave event concluded that the national heatwave plan played a crucial role in saving 4388 lives (Fouillet et al., 2008), another study conducted for nine French cities for the same heatwave event found the impact on mortality of the implementation of heatwave action plan to be mixed and city-dependent with an overall statistically non-significant reduction in the relative risk of mortality of 3.3% (Pascal et al., 2012b). In an assessment of the effectiveness of NWS heat alerts across 20 US cities for the period 2001–2006, Weinberger et al. (2018) found NWS heat alerts were not associated with lower mortality in most cities studied. Building on this study, further investigations were undertaken of the effectiveness of heat alerts in reducing all-cause mortality and cause-specific hospitalizations among Medicare beneficiaries aged 65 years and older across 2817 US counties for the period 2006–2016. Study outcomes indicate that NWS heat alerts are not linked to a reduced risk of mortality, but surprisingly might be associated with increased rates of hospitalization for fluid and electrolyte disorders and heat stroke, an observation that points to heat alerts possibly prompting more individuals to seek or access medical care (Weinberger et al., 2021). Although it could be debated, this apparent negative finding might in fact be interpreted as a positive effect of the heat alerts as increased hospital admissions could be evidence of increased population awareness of the health effects of heat with people making the connection between extreme heat conditions and feeling unwell with a resultant visit to an emergency department. In an assessment of the impacts of Ontario's harmonised heat warning and information system (HWIS) on health outcomes, Clemens et al. (2022) conclude that while emergency department visits for heat-related illnesses might have decreased in certain subpopulations after the implementation of the HWIS, this change did not reach statistical significance. Variable results are also apparent for the heatwave alert system in Korea where heatwave alerts only reduce mortality for a limited number of causes and subpopulations of a certain age and socio-economic status with no overall impact on all-cause mortality (Heo et al., 2019). For Italy although it seems the country's heatwave response plan has played a role in terms of increasing awareness and adaptation to heat it remains uncertain whether the recorded decrease in heatwave related mortality over the period since its inception can be directly attributed to this (De'Donato et al., 2018). For Switzerland the national heatwave response plan has been found to have little if any impact on the reduction of heatwave related mortality for a period of extreme heat in 2015 (Vicedo-Cabrera et al., 2016). Interestingly for the same period, Ragettli and Roosli (2019) and (Ragettli et al., 2019) found an elevated level of emergency hospital admissions for several health outcomes especially across Switzerland's warmer regions, leading these researchers to conclude that a re-evaluation of public health interventions to prevent heat-related illness was necessary, implying that adaptation/mitigation measures associated with the national heatwave response plan were insufficient. In a comprehensive evaluation of the Heatwave Plan for England, Williams et al. (2019) confirmed an association between hot weather and a rise in both death and hospital admissions. However, despite finding a decline in heat-related mortality and illness rates prior to the implementation of the Heatwave Plan, and a continued decrease afterward, these researchers did not find substantial evidence that the Heatwave Plan itself significantly impacted these trends. Of significance is their discovery that a sizable portion of the increase in deaths and hospital admissions occurred during hot weather "outside" of heatwave alert periods when temperatures did not meet the criteria for issuing a heatwave alert. Consequently, they concluded that while the Heatwave Plan demonstrated effectiveness in safeguarding individuals during heatwave alert periods, its efficacy seemed less pronounced in situations where no alert was issued during elevated temperatures. Arising from this analysis is the powerful point that perhaps evaluations of the effectiveness of heatwave related response plans should cast their net beyond a strict focus on extreme heat events. Furthermore, considering that the evaluation of the England Heatwave Plan by Williams et al. (2019) took place prior to the implementation of the new impact-based Heat Health Alert system, it will be intriguing to see if subsequent assessments of the updated system reveal substantial enhancements compared to its predecessor. This is especially pertinent given the case for developing impact-based heatwave related warning systems will need to be supported by clear evidence they are more effective than traditional hazard-based systems.

As might be expected, the robustness of many of the results from outcome evaluations outlined above are heavily dependent on the methodology applied. As yet, no silver bullet has been identified in terms of a single method best suited for delivering rigorous outcome evaluations. Generally the methods applied in HEWS/HHWS outcome evaluations include gauging public awareness of an extreme heat episode related to a wider heat response plan or specific heatwave warning and any subsequent changes in behaviours, documenting and testing for changes in morbidity and mortality attributed to the implementation of heatwave related public health interventions (often referred to as before and after studies), analysis of specific interventions, time series analyses using distributed lag and autoregressive-moving average techniques and so-called differences-in-differences approaches (Bassil & Cole, 2010; Boeckmann & Rohn, 2014; Dwyer et al., 2022; Lou et al., 2021; Toloo et al., 2013). Usefully, Dwyer et al. (2022) have set out the main methodological challenges associated with undertaking outcome evaluations and some of the solutions that might be effectively applied in resolving these (Table 13.4). Addressing these challenges will certainly pave the way for facilitating robust evaluations of heatwave response plans. Added to this, the imbalance of evaluations between high and low to medium income countries needs to be tackled as to date most have focused on European and North American systems. Further while descriptions of HEWS and HHWS processes remain valuable and provide insights into the flavours, virtues, and shortcomings of system processes, what is desperately needed are systematic inter-system comparisons of effectiveness using common protocols for outcome evaluations. With the results of such outcome evaluations in hand, policy makers and practitioners will be in better place to judge whether HEWS and HHWS constitute credible and worthwhile adaptation options.

13.4 Heatwave Prediction

Implicit in the above discussion of the nature of HEWS and HHWS is the criticality of reliable predictions of heatwaves or specific heatwave related thresholds in the form of values of key weather variables or thermal indices that act as warning system triggers. Given this, a response within the atmospheric science community has been to investigate whether heatwave prediction is possible beyond the typical weather timescale of 1–5 days so the timeframe of preparedness for heatwave events can be improved for a range of lead-times (Table 13.5). That predictions are possible stems from an understanding of the physical causes of heatwaves as touched

Table 13.4 Methodological challenges associated with conducting outcome evaluations to establish effective of HEW and HHWS and heatwave intervention strategies in general

Challenge	Explanation	Possible solutions to the challenge	Remaining issues or difficulties
A. Other factors affecting the heat-health association	Noted decreases in health-related consequences following the execution of a heat plan in contrast to periods preceding its implementation might be attributable to factors other than the introduction of the heat plan. These alternative explanations may involve unaccounted variables that underwent changes between the pre- and post-implementation periods, such as enhanced healthcare, improved housing conditions, increased public awareness of extreme heat, and biological acclimatization. These variables could potentially diminish the likelihood of illness or mortality due to extreme heat, altering the observed association between heat and health over the study period. Conversely, unmeasured factors could exacerbate the heat-health association, thereby obscuring the efficacy of an implemented heat plan	Measure factors when possible and include in models Use a short study period to reduce potential for slowly varying factors to change	Obtaining measured data is difficult for many factors Findings might be susceptible to specific characteristics of a limited number of heat events within the analysis timeframe
B. Differentiating the effects of individual interventions on health outcomes	As heat plans usually involve a variety of interventions, frequently introduced concurrently during plan implementation or update, discerning the individual impact of each intervention on mitigating health impacts becomes challenging	Make use of incremental changes to heat plans to evaluate in isolation	It is crucial for modifications to be temporally distinct from other changes, ensuring an adequate dataset for extreme heat events

(continued)

Table 13.4 (continued)

Challenge	Explanation	Possible solutions to the challenge	Remaining issues or difficulties
C. Establishing the counterfactual scenario	Challenges emerge when attempting to identify an appropriate control study group to establish the counterfactual scenario (what would have occurred if the heat plan had not been implemented), and these challenges align with A and B. Frequently, a control study group comprises the population before the heat plan's implementation, with the experimental group being the population after, leading to challenge A. An alternative approach involves using the population of a nearby city not covered by a heat plan but generally exposed to similar heatwaves as a control. However, this alternative is not consistently reliable due to localized differences in the experienced heat events or inter-city variations in demographics and infrastructure, resulting in diverse responses to heat events. In general, achieving contemporaneous studies with reliable control groups is often impractical due to broad geographical coverage under the same heat plan, or it may be ethically unacceptable (e.g., providing increased access to fluids to some care home residents but not others)	Natural experiment to distinguish alerts and non-alerts for similar conditions Difference-in-differences quasi-experimental study to create control and study groups with similar characteristics (heat events, demography, location, time period)	Results may be valid only for a narrow range of conditions The 'other factors' challenge remains
D. Accounting for mortality displacement	Mortality displacement (sometimes referred to as the harvesting effect) occurs when some deaths of frail individuals are imminent anyway but are brought forward by a heatwave event. This effect may complicate the calculation of the effectiveness of response actions	Distributed lag model, e.g. DLNM within time series regression Implicit in autoregressive moving average models (e.g., ARIMA)	None None
E. Lack of evaluations for morbidity outcomes	In contrast to research focused on mortality, there is a scarcity of comparable studies examining morbidity outcomes. This may stem from the difficulty in obtaining relevant morbidity compared to mortality data	Include morbidity or morbidity proxies in evaluation	Morbidity data often harder to obtain

Table 13.4 (continued)

Challenge	Explanation	Possible solutions to the challenge	Remaining issues or difficulties
F. Cross-study comparisons can be difficult	Differences in study designs and methodologies can pose challenges when attempting to conduct cross-study comparisons, as often confronted in meta-analysis	Greater consistency between studies, e.g., use of standardised thermal or heatwave index	May reduce diversity of approaches.

Adapted from: Dwyer et al. (2022) and used in accordance with Creative Commons CC BY Licence

Table 13.5 Forecast lead-times—length of time between the issuance of a forecast and the occurrence of the predicted phenomena

Forecast lead-time	Time range
Short range	1 h—3 days
Medium range	3–15 days
Extended range	10–30 days
Sub-seasonal	3–6 weeks
Seasonal	1–3 months
Long-range	30 days—12—24 months

Source: Author

upon in earlier chapters and the rapid development of modern weather and climate forecasting techniques.

Before venturing any further, first a clarification related to the meaning of forecasting and prediction. While forecasting and prediction are often used interchangeably, they possess subtle differences in emphasis. Essentially forecasting is a subdiscipline of prediction. In a pure sense forecasting involves the analysis of current weather conditions and historical data and running computer models to provide a comprehensive outlook of the atmospheric state ranging from the short- to the medium-range time (Table 13.5) with the output from forecasts generally in the form of deterministic values of weather variables such as temperature. Beyond 10–15 days, generally considered to be the deterministic predictability limit, forecasts become less accurate due to the inherent complexity of atmospheric processes (Golding, 2022). Given this, deterministic weather forecasts are typically used for decision making over timescales of less than 2 weeks—medium range or less (Table 13.5), Prediction is a broader term. Encompassing forecasting, it extends to a more general anticipation of future atmospheric conditions at the extended range timescale and beyond (Table 13.5). Predictions or outlooks at these longer-range timescales are more speculative with predictions often involving considerations beyond meteorology, such as climatology, to offer insights into broader trends.

Predictions at the extended range to longer-term timescales typically include sub-seasonal, seasonal and inter-annual to decadal climate outlooks with predictions presented in the form of probabilistic information. Predictions have a wider scope, incorporating not only immediate planning but decision-making sometimes extending out to a decade. These differ from projections which generally refer to the output from models that focus on establishing the climate outcomes associated with human-induced climate change for a range of time horizons such as 2050 or the end of the twenty-first century.

Numerical weather prediction systems (models) that produce short to medium range forecasts attempt to represent the processes at play arising from the interaction of the atmosphere and ocean with outputs produced at a high spatial resolution. Typically, the data assimilation processes, resolution and complexity of prediction systems that produce short to medium range forecasts differ from those that make extended to long-range predictions. For short to medium-range forecasting, achieving an accurate portrayal of the initial state of the atmosphere is crucial, necessitating the assimilation of the most recent weather observations as gathered via the Global Observing System. Shorter forecast durations also enable the models to operate at a higher resolution, facilitating a more precise representation of atmospheric dynamics and surface features, while processes associated with atmospheric chemistry and the carbon cycle, generally considered insignificant in weather prediction timescales, are not included in short to medium-range models. Sometimes nested high-resolution models, embedded within the broader numerical weather prediction system are used to produce even more detailed short to medium-range forecasts. Such models often include a higher resolution depiction of surface features, including coastlines and orography and represent some specific atmospheric processes more accurately. For example, the short-range forecast produced by the Met Office for the UK employs a variable resolution model. This features a high-resolution inner domain with outputs produced for 1.5 km grid boxes covering the UK and the Republic of Ireland surrounded by a coarser grid (4 km).

Models used in extended to long-range prediction systems operate at a lower temporal and spatial resolution compared to day-to-day weather forecast models. For example, whereas the spatial resolution for the ECMWF's medium range predictions is 9 km it is 36 km for its extended and long-range predictions. Unlike short-term weather forecast models, the models used for producing extended to long-range predictions try and represent the interaction between all the components of the climate system. For this reason they are often referred to as 'Earth System' models incorporating processes related to large scale climate patterns (e.g., ENSO, NAO, SAM), oceanic (e.g. SST), and land surface conditions (e.g. soil moisture), the cryosphere (e.g., sea ice extent) and the biosphere (e.g. seasonal variation in evapotranspiration) all of which may influence trends in the atmospheric state beyond a couple of weeks and play a role in creating conditions conducive to heatwave formation.

Commonly an ensemble approach is used to produce short to medium term forecasts and extended to long-range predictions. Often referred to as ensemble forecasting, instead of relying on a single forecast, this method involves generating multiple forecasts, with each forecast beginning with slightly different initial conditions or representation of model physical processes ('model physics'). For example, in producing their seasonal predictions the UK Met Office and the ECMWF use 41 and 51-member ensembles respectively with in each case one of the members used as the control, that is, with no perturbation of initial conditions or model physics. Such variations are introduced to account for uncertainties in our understanding of the atmosphere and its dynamics. The idea behind ensemble forecasting is to capture the range of possible outcomes and provide a more comprehensive assessment of the uncertainty associated with a particular forecast. By running multiple simulations, analysts can observe how different initial conditions or model configurations lead to divergent forecast outcomes. Ensemble forecasts are usually presented as a range of possible future scenarios, for example a spread of values of maximum temperature anomalies often portrayed in the form of an anomaly plume (Fig. 13.5), from which likelihood (probability) information is derived. Ensemble forecasting is valuable for dealing with the inherent uncertainty in prediction. It therefore helps climate or heatwave risk analysts make more informed decisions and communicate

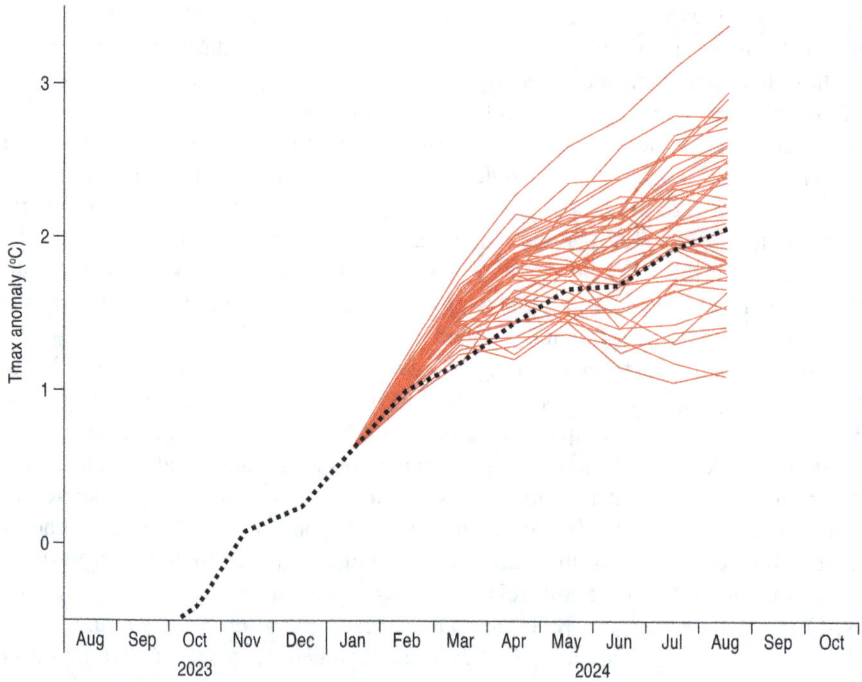

Fig. 13.5 Hypothetical example of a monthly maximum temperature anomaly plume. Each trace represents the evolution of the maximum temperature anomaly, compared to a reference period of 1981–2010, for each ensemble member over a six-month period from model initiation in February ending in August. As for most seasonal predictions the ensemble spread (uncertainty) increases with distance from the time the model was initiated. (Adapted from: Nino Plumes at https://charts.ecmwf.int/)

the range of possible outcomes to stakeholders involved in the heatwave risk management problem.

Many HEWS and HHWS operate in the timeframe of 12–48 h with numerical weather forecast models producing outputs of key heatwave variables on an hourly basis up to 48 hours with these usually transitioning to values at 3–6 hourly time steps at 3–7 days. Because weather forecasting has made spectacular advances in prediction accuracy over the past decades—5-day forecasts are now far more accurate than 1-day forecasts were 50 years ago—the confidence in short-term to medium term forecast accuracy has grown significantly (Stern & Davidson, 2015) which bodes well for heatwave forecasting (Abhik et al., 2023; Naveena et al., 2021; Singh et al., 2017; Xiao et al., 2023) at operational timescales typical of HEWS and HHWS. The European Heat Health Information System (EHHIS) represents a good example of the efforts to address the prevailing gap in existing forecasting systems to support HEWS and HHWS by enhancing predictive capacity for lead times extending beyond a few days. The EHHIS (https://heathealth.eu/index.html) is a prototype climate service that utilizes an extensive dataset comprising daily meteorological and mortality data from 147 European regions across 16 countries. Employing Distributed Lag Nonlinear Models (DLNM), associations between critical weather variables and mortality have been computed, accounting for seasonality, long-term trends, and incorporating various time lags to model delayed effects. The EHHIS is powered by weather forecasts from the European Centre for Medium-Range Weather Forecasts (ECMWF) Atmospheric Model Ensemble 15-day forecast with a focus on lead times of up to 15 days. Experimental health forecasts are generated by amalgamating temperature forecasts, region-specific temperature-mortality associations, and daily mortality data.

Although weather forecasts await advances in science and technology (Brotzge et al., 2023), from a preparedness and early warning perspective knowing the likelihood of heatwave occurrence beyond the typical short to medium term forecasting window could constitute a real leap forward in the ability to manage heatwave and climate related risk in general (Merryfield et al., 2020; White et al., 2017). Although the evolving goal over the last couple of decades has been to achieve seamless weather-to-climate prediction (Brunet et al., 2023; Dirmeyer & Ford, 2020; Ren et al., 2023), the time horizons of sub-seasonal, seasonal and inter-annual to decadal are generally the main focus in relation to heatwave prediction with the ultimate aspiration to develop truly seamless sub-seasonal-to-seasonal (S2S) heatwave forecasts. Commonly, in relation to the prediction of heatwaves beyond the short-range timescale, the thinking is that by working with seasonal forecast information updated monthly, and consulting as time progresses shorter-term forecasts at the extended to medium range, early awareness of potential heatwaves will be facilitated. That seasonal and sub-seasonal climate forecast information may be a game changer in the management of climate related risk in general is evidenced by the World Meteorological Organisations establishment of Global Producing Centres for Long Range Forecasts (GPCs-LRF) in 2006 and Global Producing Centres for Sub-Seasonal Forecasts (GPC-SSF) and a Lead Centre for Sub-Seasonal Forecast multi-model ensemble, as part of the WMO Integrated Processing and Prediction System more recently in 2023. Both of these endeavours will add tangible support to achieve the aspirations of the EW4All programme.

At the extended range to sub-seasonal timeframe, adverse heat conditions and thus evidence for potential heatwaves may become apparent as heatwave indicators (e.g., maximum and minimum temperature anomalies) deviate significantly from their climatological levels. Encouragingly, amongst a range of hydrometeorological hazards, heatwaves are considered the most predictable at extended range to sub-seasonal timescales with evidence of promising forecasts with lead times ranging from 2 to 3 weeks emerging in the heatwave prediction literature (Barriopedro et al., 2023; Domeisen et al., 2023; Vitart & Robertson, 2018). For example, skilful predictions related to heatwave indicators persisting up to a lead time of 3 weeks have been noted for Australia (Marshall et al., 2014), Europe (Kueh & Lin, 2020; Rouges et al., 2023; Wulff & Domeisen, 2019), India (Mandal et al., 2019), China (Xie et al., 2020), Russia (Vitart & Robertson, 2018), South America (Osman & Alvarez, 2018), and North America (Emerton et al., 2022; Tian et al., 2017). Sub-seasonal forecasts also demonstrate notable proficiency in predicting heatwaves in Sahelian and West Africa, with significant skill extending up to the same lead times as elsewhere (Batte et al., 2018; Guigma et al., 2021).That said, it would appear for the extra-tropics and the tropics that successful prediction of heatwave amplitude at the sub-seasonal timescale remains a challenge with, for the meantime, this only possible for a lead time of around 7 days (de Perez et al., 2018; Lavaysse et al., 2019; Pyrina & Domeisen, 2023). To demonstrate the potential utility of extended-range forecasts for advanced heatwave preparedness, Domeisen et al. (2023) present the example of contemporaneous heatwaves across the Northern Hemisphere for June 2021 the most notable of which was the North American Pacific Northwest event (Fig. 13.6). As can be seen from Fig. 13.6 forecasts initiated over the period June 3–10, 2021 suggest temperatures above normal, albeit with an underestimated intensity and the maximum temperatures located in the wrong position. As the lead time decreases, the model progressively captures the accurate location and intensity of both North American and European heatwaves, reaching a more precise representation approximately 1 week before these events begin (Domeisen et al., 2023). Other studies have also noted good predictability of the Pacific Northwest heatwave event at the sub-seasonal timescale (Emerton et al., 2022; Lin et al., 2022; White et al., 2023).

Perhaps not surprisingly, heatwave predictions at the seasonal scale fare less well than their shorter-term counterparts. Lower predictability stems from the overall limited accuracy of models in capturing enduring warm season circulation anomalies, the unclear influence of near and distant oceans and upper-atmosphere factors, and the intricacies of land surface—atmosphere interactions (Domeisen et al., 2023). Notwithstanding these challenges, seasonal forecasts offer insights into whether an upcoming summer may be predisposed to heatwave occurrence often represented by strong positive summer temperature anomalies in ensemble and multi-model predictions. Typically, predictions of temperature or heatwave related indices are averaged over monthly to three-monthly intervals with their probability associated with the tercile categories of below-normal, normal, or above-normal

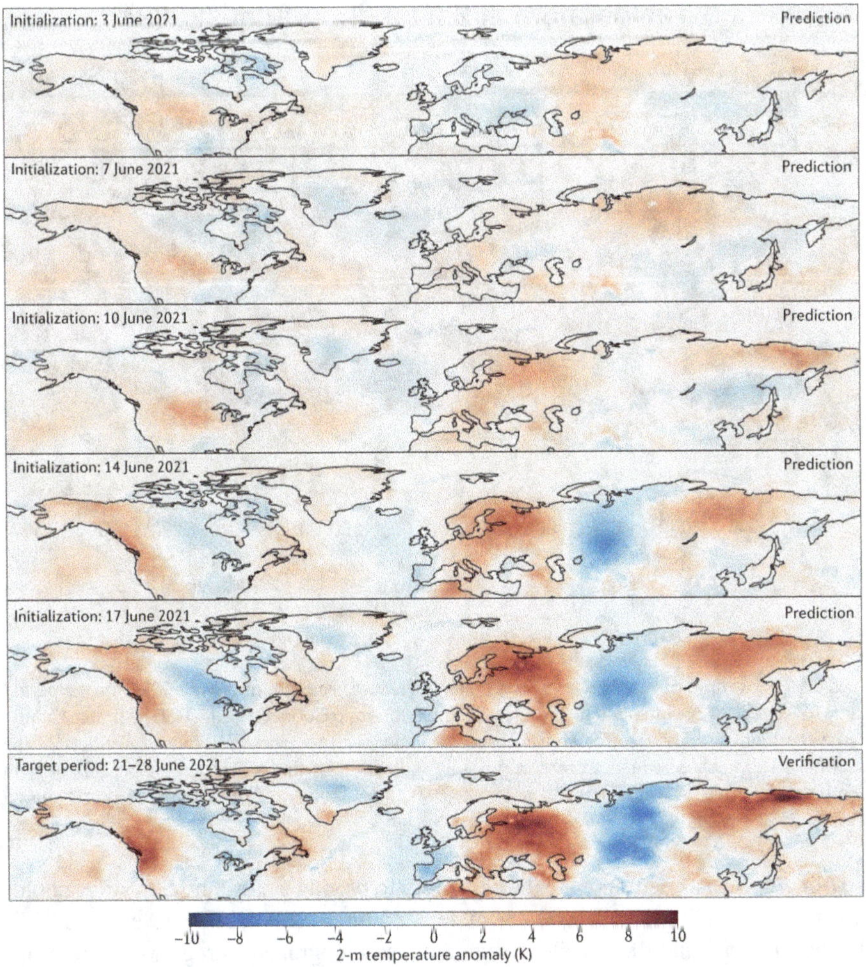

Fig. 13.6 Extended-range forecasts from the European Centre for Medium-Range Weather Forecasts (ECMWF) for weekly mean 2-metre temperature anomalies for the target period of June 21, 2021. Forecasts were initialized on June 3, 7, 10, 14 and 17, 2021. The bottom panel illustrates the validation for the June 21–28, 2021 target period using ERA5 reanalysis. The evolution of the forecasts indicates early signals of warm anomalies 2–3 weeks in advance, with an enhanced prediction of both the intensity and location of the heatwaves across the US Pacific Northwest and Europe emerging at 1–2 weeks' lead time. (Source: Domeisen et al. (2023). Used under permission of Springer Nature Licence 5707640584304)

stated in relation to a climatological reference period (Fig. 13.7). The propensity for heatwave conducive conditions is usually indicated by high probabilities in the above normal category for temperature or high probabilities in the below normal category for soil moisture.

At the global scale, reliable seasonal predictions of daily temperature extremes are achievable over 3-month periods (Hamilton et al., 2012). In West Africa,

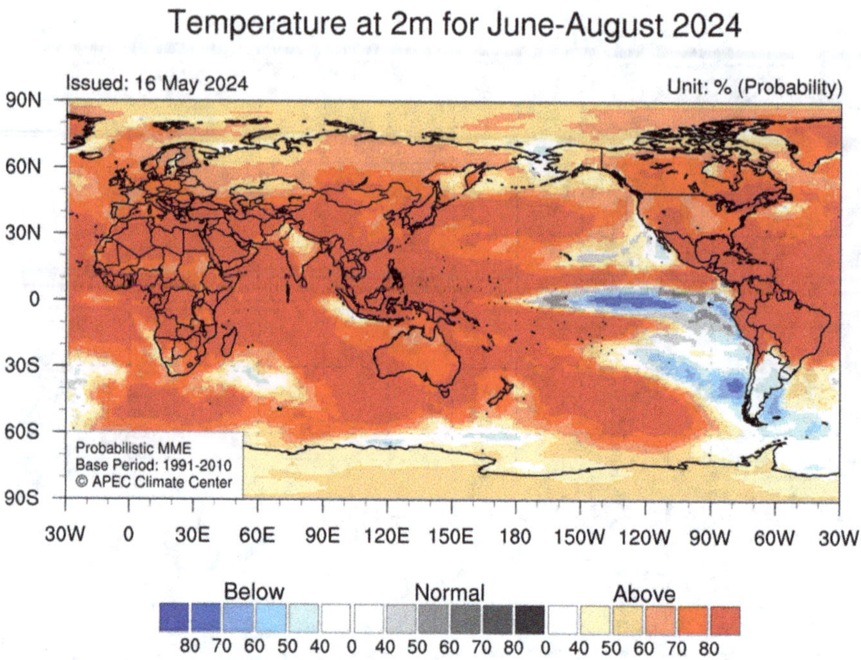

Fig. 13.7 Example of seasonal 2 m temperature forecast issued in mid-May 2024 for the period June to August 2024 generated by the Asia-Pacific Economic Cooperation (APEC) Climate Centre. Note the high probabilities for above normal summer 2 m temperatures over vast tracts of the Northern Hemisphere's land surface, a possible indication of the occurrence of periods of high temperatures that might manifest as heatwaves. (Source: https://www.apcc21.org/?lang=en Accessed 22/05/24)

seasonal forecasts demonstrate the capability to predict a substantial positive anomaly in the duration and frequency of temperature extremes (Batte et al., 2018). Eastern Europe and the Mediterranean are also recognized as regions where proficient forecasts of periods of anomalous heat are possible up to 3 months ahead due the strong influence of land-surface conditions on heatwave occurrence there (Quesada et al., 2012). For the 2021 North American Pacific Northwest heatwave, early signals of unusually high temperatures emerged in forecasts from April and May 2021, indicating a 50–70% probability of 2 m temperatures reaching the top 20% of climatology in June across the Pacific Northwest (Emerton et al., 2022). Similar predictive success was observed for the heatwave impacting the U.S. Central Plains in 2011 (Luo & Zhang, 2012). As well as ENSO, some seasonal predictability of temperature anomalies have been associated with increases in global temperatures (Doblas-Reyes et al., 2006; Livezey & Timofeyeva, 2008; Patterson et al., 2022).

Heatwave predictions, or more accurately predictions of anomalous temperature conditions, are also undertaken for inter-annual to decadal timescales. Although well beyond the timescales typically of concern for immediate or near-term heatwave occurrence and associated response actions, predictions at these timescales may serve long-term strategic planning in relation to infrastructure and a range of mitigation measures designed to manage down chronic levels of heat, especially in urban environments. For both these timescales sources of predictability reside in large scale modes of climate variability such as ENSO and the PDO (see Chap. 5) as well as a range of external forcings (e.g., greenhouse gas concentrations, solar variability, low latitude volcanic eruptions) (Bellucci et al., 2015). At the interannual timescale regions for which ENSO exerts a strong control on climate variability usually demonstrate the best predictability of temperature anomalies (Ferreira et al., 2022; Weisheimer et al., 2022). For some regions reasonably confident decadal predictions of maximum 5-day temperatures and temperature extremes have been demonstrated (Asadi-RahimBeygi et al., 2024; Borchert et al., 2021; De Luca et al., 2023; Delgado-Torres et al., 2023; Eade et al., 2012; Xin et al., 2018).

13.5 Heatwave Naming

As noted earlier, an integral element of EWS is the process of warning dissemination and communication. This process is critical to the effectiveness of EWS because, notwithstanding the sophistication of the other EWS elements, if public awareness of an impending heatwave event is poor the potential for extensive impacts could be considerable. While numerous factors contribute to the effectiveness of heatwave risk communication (Table 13.6), a valid inquiry, possibly influenced by the precedent of naming hurricanes and winter storms, as well as being grounded in exemplification theory (Lachlan & Spence, 2010; Lin et al., 2018; Zillmann, 1999), is whether the practice of naming heatwaves could serve various practical purposes concerning heatwave risk awareness, communication, and the implementation of risk reduction actions. This is a question very much prompted by the naming of a heatwave across southern and eastern Europe in 2017 as Lucifer and more recently the establishment of the world's first pilot programme to name and categorise heatwaves in the city of Seville, Spain in June 2022.

Supported by the US based Arsht-Rockefeller Foundation Resilience Centre (Arsht-Rock), Seville's heatwave naming and categorisation system follows the general model of the original PWWS of Philadelphia, that is, it uses an airmass informed approach to identify dangerous heatwave conditions based on a historical analysis of airmass—mortality associations and then categorises conditions falling into one of four levels depending on the predicted level of excess mortality. For the Seville system three or more consecutive days with high levels of heat are designated as 'heat events' and fall into categories zero and one with predicted excess mortality of 1–10% and 10–19% respectively. Using the same duration criteria, categories two and three are termed 'heatwaves' being typified by extreme heat and levels of excess mortality at 20–29% and greater than 30% respectively. Heatwaves falling within category 3 are given a name.

Table 13.6 Determinants of heatwave risk communication[a]

Determinant	Description
Perceived risk	The way individuals perceive the severity and likelihood of heatwave risk influences their response to risk communication. Levels of receptiveness may be based on their subjective assessment of heatwave risk
Trust	Trust in the communicator or the source of information is critical. People are more likely to accept and act on information if they trust the messenger, whether it's a government agency, scientist, or other authority figure
Credibility of the source	The credibility of the information source is closely tied to trust. People are more likely to believe and act on heatwave information provided by sources they consider credible, such as experts, institutions, well-known organizations or government weather services
Cultural and social factors	Cultural and social contexts significantly influence how people interpret and respond to risk communication. Heatwave messages should be tailored to the specific cultural, linguistic, and social norms of the target audience
Message content and framing:	The way information is framed can impact how it is received. Positive framing (emphasizing benefits of a recommended action) or negative framing (emphasizing risks of inaction) can influence heatwave related decision-making
Audience characteristics	Understanding the demographics, values, knowledge, and attitudes of the target audience is crucial. Tailoring heatwave messages to the characteristics of the audience increases relevance and effectiveness
Media and communication channels	The choice of communication channels, such as traditional media, social media, or community meetings, can affect how information is disseminated and received. Different channels may be more effective for different audiences
Timing and relevance	Communicating information at the right time and making it relevant to the audience's current concerns can enhance the impact of heatwave risk communication
Two-way communication	Encouraging feedback and engagement from the audience promotes a two-way communication flow, allowing for a better understanding of public concerns and facilitating more effective heatwave risk management strategies

[a]Based on: Ettinger et al. (2023), Golding (2022), Grundstein and Williams (2018), Hass et al. (2021), Lee and Kim (2020), McGregor et al. (2015), McLoughlin et al. (2023). VanderMolen et al. (2022), Roberts et al. (2022), Ratwatte et al. (2022), Schoessow et al. (2022), Zottarelli et al. (2022)

In July 2022, not long after the pilot Seville system was launched, a heatwave with temperatures reaching 43 °C over the period July 22–27 and falling within category 3 was bestowed with the name Zoe. In June of the following year Seville named its second heatwave as Yago. Both Zoe and Yago seem to have paved the way for the naming of a period of exceptional heat across southern Europe in July 2023 as heatwave Cerberus, with Cerberus being the Greek mythological multi-headed dog belonging to Hades the god of the underworld and the ruler of the dead. A week after Cerberus another exceptional heatwave occurred across southern Europe with this one named Charon (also appearing as Charonte/Caronte in some media outlets), the ferryman of the dead in Greek mythology who carries souls of the deceased across the river Styx to Hades. Interestingly Cerberus and Charon were named by a private Italian weather company with local and national weather services and learned societies distancing themselves from any naming attribution. While the named heatwaves Zoe, Yago, Cerberus, and Charon garnered media attention, potentially increasing awareness of their severity and extent, the heat generated within the weather and climate and disaster risk reduction communities regarding the wisdom of heatwave naming rivalled, in a metaphorical sense, the intensity of heat associated with the named heatwaves. Such was the intensity of the debate about heatwave naming, following the launch of the Seville pilot naming project, the WMO released a comprehensive statement on this subject (WMO, 2022a). Amongst other matters related to advantages and disadvantages, the WMO expressed caution in adopting or promoting the practice of heatwave naming preferring instead that weather services give attention to fortifying heat-health warning systems, improving technical heat forecasting capabilities, and fostering capacity and partnerships to safeguard vulnerable communities from the impacts of heatwaves (WMO, 2022a).

Historically hurricanes and their equivalents in other regions (e.g., typhoons and tropical cyclones) as well as winter storms have been the only named meteorological phenomena. Clement Wragge, an Australian meteorologist is credited with pioneering the systematic naming of storms and cyclones. In a distinctive approach, he opted for the names of Polynesian females for tropical cyclones. Wragge also assigned names of politicians, who were opposed to his aspirations to establish a meteorological bureau, to severe storms originating from southern latitudes. The practice of naming is also evident during World War II when US Army and Navy meteorologists started to name tropical cyclones in the western Pacific in their forecasts. By 1953, assigning female names to hurricanes in the North Atlantic became a standard practice for American meteorologists with storm-naming expanding to include both male and female names in 1978, beginning in the eastern Pacific Ocean and later extending to the Atlantic Ocean and the Gulf of Mexico (Skilton, 2018). Currently, the World Meteorological Organization oversees the compilation of lists containing names utilized for hurricanes across all five ocean regions worldwide. Generally, names drawn from a pre-defined list are assigned to potential hurricanes when a developing tropical disturbance intensifies into a tropical storm with winds of at least 18 m/s—a category 1 hurricane has sustained windspeeds of at least 33 m/s. Undoubtedly influenced by the practice of hurricane naming, and perhaps the US based The Weather Channel's naming of winter storms from 2021, in 2014 the UK Met Office in association with Met Éireann, Ireland introduced the naming of winter storms. In contrast to naming when a physical threshold such as windspeed is exceeded, the UK and Ireland only named storms if they were deemed to potentially cause a medium to high level of impact in either country. This practice is in some ways impact-based naming. Since 2014 the Netherlands has joined the UK and Ireland forming the western storm naming group, with Spain, Portugal, France, Belgium and Luxembourg making up the south-western storm naming group and

Norway, Sweden and Denmark the northern group. Interestingly, the National Weather Service (NWS), the official weather warning agency in the United States, has chosen not to employ or recognize names for winter storms because of the challenges associated with assessing the risks and impacts of a winter storm, which vary depending on the geographic location (Lin et al., 2018).

While the above overview of hurricane and winter storm naming may be interesting, the reader is probably wondering what relevance it is to heatwave naming. In brief, relevance relates to justification for naming, naming governance and at what point—a physical or impact-based threshold—should a heatwave be named. Borrowing from the general reasons for storm naming, heatwave naming could be justified for several reasons (Table 13.7).

Clearly the content in Table 13.7 has a strong 'normative' flavour about it and serves as an invitation to establish whether the stated normative goals for heatwave naming can be realised. At the time of writing there appears to be an absence of peer reviewed research on the efficacy of heatwave naming. This is apart from a non-peer reviewed report that describes the results of an Arsht-Rock sponsored social survey to establish the impact of naming heatwave Zoe (Table 13.8). The report's overall conclusion was that the naming of this heatwave brought a high level of attention to this event, although it is not stated whether that attention pertains to effectiveness in terms of heatwave risk reduction actions or media response, notwithstanding of course that these may be linked. Further based on its one-off investigation of responses to heatwave Zoe, Arsht-Rock concluded that its pilot study confirmed the effectiveness of its system for heatwave categorisation and naming, demonstrating

Table 13.7 Possible justifications for heatwave naming

Avoiding confusion	In the situation of contemporaneous heatwaves in different parts of the world names could prevent confusion when multiple heatwaves are being discussed, especially in the era of 24-hour news cycles and social media
Clarity and communication	Naming could assist in clearly and concisely communicating about specific heatwaves. Instead of using technical terms, a name may make it easier for the public and media to follow and discuss a heatwave
Enhanced safety messaging	Named heatwaves could facilitate public awareness and safety messaging. People may pay attention to and remember a named heatwave, making it easier for authorities to convey critical information about its trajectory, potential impact, and safety precautions
Historical record	Naming could help with maintaining a historical record. It could allow stakeholders to refer to specific events and track their trajectories, intensities, and impacts over time
International cooperation	Names could be agreed internationally to facilitate communication between meteorological agencies and countries, especially for the case of transboundary heatwaves. This collaborative approach could ensure consistency and avoid confusion if heatwaves cover several jurisdictions
Public engagement	Naming heatwaves could engage the public more effectively. People tend to relate to named events thus fostering a sense of awareness and preparedness within communities exposed to heatwave events

Source: Author

Table 13.8 Outcomes of social survey of 2000 people and a subsample of local media 2 months after heatwave Zoe to establish the possible impact of heatwave naming in Seville, Spain

Impact indicator	Survey finding
Recollection of any heatwaves that received names	Six percent of the total participants and 7% of the regional media market spontaneously remembered the name Zoe. Additionally, a significant number of those who initially couldn't recall the name acknowledged remembering Zoe when prompted. The overall awareness level, including both aided and unaided recall, reached 32%, constituting approximately one-third of the survey's total participants
Recollection of any heatwaves categorized by level of severity	Survey results revealed evidence of potential confusion between the new Arsht-rock heat-health-based early warning system and the well-established ones by the state meteorological Agency of Spain (Agencia Estatal de Meteorología—AEMET). While 80% of respondents affirmed remembering heatwaves categorized by level of severity, a subsequent question revealed an inability of half of the initial 80% to recall whether it was the Seville number-based categorisation system, that of the AEMET (a colour coded as opposed to a number-based system) or one based on the letters A—C. Further the remaining respondents were evenly divided among the three options, two of which were incorrect
Evidence of any significant link between awareness of heatwave Zoe and behaviour change	To establish this, those who spontaneously remembered Zoe were contrasted with those who did not in terms of heatwave risk reduction behaviours. Results revealed that those who recalled the name Zoe were more inclined to stay indoors, increase water intake, and adjust work schedules and locations to minimize exposure to extreme heat as well as participate in pro-social heatwave safety activities, such as alerting others or providing guidance on staying safe during extreme heat conditions. No significant differences related to Zoe recall were observed for other measured heatwave behaviours (e.g., altering leisure plans to avoid heat). Further there was no evidence that those who recalled Zoe were less likely to adopt safer behaviours
Media coverage	Heatwave Zoe triggered the largest burst of media activity during the 2022 pilot. The social listening platform *Infegy,* showed a total of 140 posts mentioning both Zoe and the word heatwave (in English or Spanish), resulting in 37,500 impressions reaching about six million people. Zoe attracted more than 25 regional print, digital, radio, and TV placements, 15 national placements, and 16 international placements. The greatest spike of posts occurred the week of July 24 with 88 posts generating 808 engagements. News about heatwave Zoe also spread across blogs, twitter, Instagram, and the European social network VK. Estimates

Adapted from Arsht-Rock (2023)

its potential to enhance the value of State Meteorological Agency of Spain's (Agencia Estatal de Meteorología—AEMET) existing heatwave warning system.

While some may be tempted to advocate for heatwave naming based on the results of the Arsht-Rock report on heatwave Zoe and a recently published assessment based on the survey data used to develop the Arsht-Rock report (Metzger et al., 2024), in the absence of a substantive volume of peer reviewed literature on heatwave naming, and notwithstanding the fundamental differences between storms and heatwaves as meteorological phenomena, the more cautious may have no other option but to turn to the rather limited literature on the effectiveness of storm naming to establish whether some of the normative goals set out in Table 13.7 might be achievable. Emerging from the few studies on storms is mixed evidence of naming effectiveness. Using an internet-based questionnaire targeted at assessing the impact of storm naming on citizens risk perception and awareness in the eastern Mediterranean, Kotroni et al. (2021) found storm naming had a noteworthy impact, via activation of perceptual and cognitive mechanisms, on the readiness of citizens. For storm Doris, which affected the Republic of Ireland and the UK in February 2017, Charlton-Perez et al. (2019) report an interesting time trend in social media content. In the run up to the arrival of Doris social media activity largely focused on the forecast itself with this rapidly transitioning to postings about impact as the storm crossed affected areas. Similarly for a collection of ten named storms across the Republic of Ireland and UK for the 2017–2018 winter, Spruce et al. (2020) describe how online conversations shift from pre-storm forecasts and warnings to on-the-ground observations and reports as the storms unfold, and later transitions to discussions about damage and disruptions, along with the sharing of news reports in the aftermath of the storms. From a risk reduction perspective both these accounts are interesting because for the storm pre-arrival period most online discussions focus on the forecast with material related to impacts occurring post storm arrival. This raises the question whether the act of naming in the case of the winter storms considered by these two studies induced pre-storm risk reduction activities. Interestingly in the case of hurricanes there appears little literature on the extent to which hurricane naming as opposed to the hurricane forecast itself precipitates preparedness related actions. One intriguing discovery from the Spruce et al. (2020) research is the identification of spikes in Twitter activity related to wind and precipitation, which are not necessarily tied to named storm events. Upon closer examination, these peaks indicate discussions about weather conditions, demonstrating significant public engagement and concern with weather in general, with the level of social media activity sometimes comparable to the levels observed for named storms. This finding holds implications for heatwave naming in terms of whether naming may elevate awareness of key impact-causing heatwave variables beyond that of base level awareness of periods of hot summer weather in general. In contrast to the studies that studied the responses to observed named storms which suggest possible benefits for forecast communication, two experimental studies on storm naming that used fictional storms, suggest otherwise. Lin et al. (2018) find that storm naming may be relatively unimportant, suggesting that demographic and psychographic factors may be far more functional in motivating people to take protective actions. This finding is consistent with that of Rainear et al. (2017) who offer evidence of minimal distinctions in individual perceptions based on storm name use, suggesting that people's perceptions of the severity or susceptibility of a fictional winter storm remain consistent regardless of naming. Setting aside the contrasting views on effectiveness, a common theme across the various studies on storm naming is the methodological challenges associated with discerning robust naming effects. Solving this challenge bears implications for advancing the debate about heatwave naming effectiveness.

While whether to name or not to name remains a fundamental matter, allied issues relate to categorisation as a basis for naming and naming governance. As noted earlier the experimental Seville system is based on a categorisation of heat events and heatwaves according to their estimated health impacts. There is probably nothing fundamentally wrong with an impact-based categorisation. However, such a categorisation and the underlying algorithms used for health impact estimation will be quite location specific because of local acclimatisation effects (Carmona et al., 2017; Díaz et al., 2015) with categorisation intervals possibly varying by climate type and the degree of adaptation to heat extremes. Related to this, and because heatwaves have spatially diffuse or fuzzy boundaries, is the issue of over what geographical area does a named heatwave refer to, or in other words where might a named heatwave start and stop geographically speaking. Drawing a line around the area affected by a named heatwave based on an impact-informed categorisation may prove a challenge in terms of defining where the core and peripheral areas are especially if a mapping like that provided by some national meteorological and hydrological services for storms is imagined. With time there are likely to be non-stationarity effects in relation to heat—health associations (Martinez-Solanas & Basagana, 2019; Pascal et al., 2021) which bears implications for the temporal stability of impact-based categorisation and naming systems and the efficacy of HHWS in general. Moreover, in setting health impact defined category boundaries a judgement call related to what is a tolerable level of risk is inevitable. That is, if a heatwave is only named for example when excess mortality is estimated to exceed 30%, with the idea that the purpose of naming is to increase awareness and thus reduce risk, does this make in a relative sense unnamed heatwaves less impactful in the public's mind's eye. Clearly this is an important decision with the prospect that an unintentional moral hazard may be created—the public prepares less for a period of unusually 'unnamed' high temperatures because it treats the naming of heatwaves as a form of insurance. Given the possible issues associated with naming linked to a specific impact level, an 'unnamed' hazard-based categorisation based purely on exceedance of an observed measure of heat from a climatological reference point, as has been proposed for marine heatwaves (Fig. 2.9), or such a system for which the broad type and scale of impacts is understood for each category may be more appropriate. For example, the Australian heatwave warning service uses the excess heat factor (EHF—see Chap. 4) to categorise heatwaves into one of three classes of intensity (low, severe and extreme) with a general indication of the implications each level of heatwave intensity holds for population health provided (Nairn & Fawcett, 2015; Nairn et al., 2022). This is like the US National Weather Service's extreme heat warnings which are issued based on a categorisation of the Heat Index. By January 2025 the US State of California plans to have a heatwave ranking system in place as part of a state-wide heatwave early warning system. Sanctioned via the State of California Assembly Bill 2238, which was signed into action by the governor in September 2022, it will be interesting to see what form the proposed ranking system takes. For example, if purely hazard-based will it follow something like that applied in the Australian heatwave warning service or if health impact-based will the ranking system be sensitive to intra-state variations in heatwave-health associations and if so at what levels will the health impact informed heatwave category boundaries be set?

Compared to the approach taken to naming hurricanes (tropical cyclones) and winter storms the evolving governance situation for heatwave naming appears rather ad hoc—to date two unrelated authorities in Spain and Italy, both of which are not affiliated with their respective national weather service, have named heatwaves. Added to this, heatwave Charon also appeared in various media outlets using the two alternative spellings of Charonte and Caronte, generating some confusion about whether these named heatwaves were all one in the same. This situation contrasts with that for hurricane naming which is coordinated through the WMO as an authoritative source and guided by a strict set of protocols. Associated with the issue of naming is the importance of the name chosen, especially in the context of the gendering of disaster (Skilton, 2018). For example, there is evidence to suggest that the alternating male-female naming system for hurricanes has undermined weather officials' goal of effectively warning populations of impending disaster with individuals reacting with less commitment to female-named hurricanes than to male-named ones (Jung et al., 2014a). Although this finding has been challenged (Bakkensen & Larson, 2014; Christensen & Christensen, 2014; Maley, 2014; Smith, 2016) and subsequently defended (Jung et al., 2014b, c, d), it points to potential differences in the perception of extreme events based on the gendered attributes assigned to them (Botner et al., 2020; Mecit et al., 2022; Skilton, 2018). Of course, to control for the possible deleterious effects associated with gender-based names and exercise cultural sensitivity in naming practice (Morss et al., 2020), geographical names could be used. For heatwaves this has partly been the case to date with names such as Chicago, European and Russian accorded to major heatwaves in each of these areas in 1995, 2003 and 2010 respectively. This resonates with what has been suggested for marine heatwaves (Hobday et al., 2018) and appears to be the case for wildfires in some parts of the world (Sublett & Sublett, 2023).

13.6 Conclusion

The 1990s marked a pivotal shift in disaster risk management philosophy, emphasizing preparatory action over reactive response, particularly in the context of extreme weather events. The need for this shift was laid bare with the occurrence of the significant Chicago 1995 and European 2003 heatwave events. From the ashes of these events several HEWS and HHWS emerged with HEWS focused on predicting heatwave occurrence while HHWS prioritized forecasting health impacts based on forecasts of heat relevant variables and dose response models describing heat health associations. The theoretical and in some cases increasingly tangible benefits of EWS eventually led to the global call for Early Warning Systems for All (EWS4All). This chapter concludes that despite their potential as an adaptation option, the efficacy of HEWS and HHWS remains uncertain with rigorous process and outcome evaluations required based on fit for purpose data, for example impact data in the case of outcome evaluations. Accurate prediction of heatwaves is crucial for timely action, with advancements in short-range forecasts showing great promise. However, the real potential breakthrough lies in extended-range predictability, though still in experimental stages. A pressing question in heatwave management is whether to categorize and name heatwaves. While this may aid communication, operational challenges abound. The decision to name or not to name hinges on resolving categorization methods and weighing potential benefits against drawbacks. Systematic research is needed to determine the effectiveness of heatwave naming in enhancing risk communication. Ultimately, the adoption of such measures must be grounded in evidence and practicality to ensure their utility in mitigating the impacts of heatwaves.

References

Abhik, S., Lim, E. P., Hope, P., & Jones, D. A. (2023). Multiweek prediction and attribution of the black Saturday heatwave event in Southeast Australia. *Journal of Climate, 36*(19), 6763–6775. https://doi.org/10.1175/jcli-d-22-0833.1

Arsht-Rock. (2023). *Evaluation report Seville heatwave naming and categorisation pilot project*. Arsht-Rockefeller Foundation.

Asadi-RahimBeygi, N., Zarrin, A., Mofidi, A., & Dadashi-Roudbari, A. (2024). Near-term temperature extremes in Iran using the decadal climate prediction project (DCPP). *Stochastic Environmental Research and Risk Assessment, 38*(2), 447–466.

Bakkensen, L. A., & Larson, W. (2014). Population matters when modeling hurricane fatalities. *Proceedings of the National Academy of Sciences of the United States of America, 111*(50), E5331–E5332. https://doi.org/10.1073/pnas.1417030111

Balehegn, M., Balehey, S., Fu, C., & Liang, W. (2019). Indigenous weather and climate forecasting knowledge among Afar pastoralists of north eastern Ethiopia: Role in adaptation to weather and climate variability. *Pastoralism-Research Policy and Practice, 9*. https://doi.org/10.1186/s13570-019-0143-y

Barriopedro, D., García-Herrera, R., Ordóñez, C., Miralles, D. G., & Salcedo-Sanz, S. (2023). Heat waves: Physical understanding and scientific challenges. *Reviews of Geophysics, 61*(2). https://doi.org/10.1029/2022rg000780

Basher, R. (2006). Global early warning systems for natural hazards: Systematic and people-centred. *Philosophical Transactions of the Royal Society a-Mathematical Physical and Engineering Sciences, 364*(1845), 2167–2180. https://doi.org/10.1098/rsta.2006.1819

Bassil, K. L., & Cole, D. C. (2010). Effectiveness of public health interventions in reducing morbidity and mortality during heat episodes: A structured review. *International Journal of Environmental Research and Public Health, 7*(3), 991–1001. https://doi.org/10.3390/ijerph7030991

Batte, L., Ardilouze, C., & Deque, M. (2018). Forecasting West African heat waves at subseasonal and seasonal time scales. *Monthly Weather Review, 146*(3), 889–907. https://doi.org/10.1175/mwr-d-17-0211.1

Baudoin, M. A., Henly-Shepard, S., Fernando, N., Sitati, A., & Zommers, Z. (2016). From top-down to "community-centric" approaches to early warning systems: Exploring pathways to improve disaster riskrReduction through community participation. *International Journal of Disaster Risk Science, 7*(2), 163–174. https://doi.org/10.1007/s13753-016-0085-6

Bellucci, A., Haarsma, R., Bellouin, N., Booth, B., Cagnazzo, C., van den Hurk, B., Keenlyside, N., Koenigk, T., Massonnet, F., Materia, S., & Weiss, M. (2015). Advancements in decadal climate predictability: The role of nonoceanic drivers. *Reviews of Geophysics, 53*(2), 165–202. https://doi.org/10.1002/2014rg000473

Benmarhnia, T., Bailey, Z., Kaiser, D., Auger, N., King, N., & Kaufman, J. S. (2016). A difference-in-differences approach to assess the effect of a heat action plan on heat-related mortality, and differences in effectiveness according to sex, age, and socioeconomic status (Montreal, Quebec). *Environmental Health Perspectives, 124*(11), 1694–1699. https://doi.org/10.1289/ehp203

Benmarhnia, T., Schwarz, L., Nori-Sarma, A., & Bell, M. L. (2019). Quantifying the impact of changing the threshold of New York City heat emergency plan in reducing heat-related illnesses. *Environmental Research Letters, 14*(11). https://doi.org/10.1088/1748-9326/ab402e

Bernard, S. M., & McGeehin, M. A. (2004). Municipal heat wave response plans. *American Journal of Public Health, 94*(9), 1520–1522. https://doi.org/10.2105/ajph.94.9.1520

Berry, P., Yusa, A., Morris-Oswald, T., & Rogaeva, A. (2014). Heat alert and response systems in urban and rural communities in Canada. *Change and Adaptation in Socio-Ecological Systems, 1*(1), 84–97.

Blättner, B., Janson, D., Roth, A., Grewe, H. A., & Mücke, H. G. (2020). Health protection against heat extremes in Germany: What has been done in federal states and municipalities? *Bundesgesundheitsblatt-Gesundheitsforschung-Gesundheitsschutz, 63*(8), 1013–1019. https://doi.org/10.1007/s00103-020-03189-6

Boeckmann, M., & Rohn, I. (2014). Is planned adaptation to heat reducing heat-related mortality and illness? A systematic review. *Bmc Public Health, 14*. https://doi.org/10.1186/1471-2458-14-1112

Borchert, L. F., Koul, V., Menary, M. B., Befort, D. J., Swingedouw, D., Sgubin, G., & Mignot, J. (2021). Skillful decadal prediction of unforced southern European summer temperature variations. *Environmental Research Letters, 16*(10), 104017.

Bose-O'Reilly, S., Daanen, H., Deering, K., Gerrett, N., Huynen, M., Lee, J., Karrasch, S., Matthies-Wiesler, F., Mertes, H., Schoierer, J., Shumake-Guillemot, J., van den Hazel, P., van Loenhout, J. A. F., & Nowak, D. (2021). COVID-19 and heat waves: New challenges for healthcare systems. *Environmental Research, 198*, 111153. https://doi.org/10.1016/j.envres.2021.111153

Botner, K. A., Mishra, A., & Mishra, H. (2020). The influence of the phonetic elements of a name on risk assessment. *Journal of Consumer Research, 47*(1), 128–145. https://doi.org/10.1093/jcr/ucz050

Bower, D., McGregor, G. R., Hannah, D. M., & Sheridan, S. C. (2007). Development of a spatial synoptic classification scheme for western Europe. *International Journal of Climatology, 27*(15), 2017–2040. https://doi.org/10.1002/joc.1501

Brotzge, J. A., Berchoff, D., Carlis, D. L., Carr, F. H., Carr, R. H., Gerth, J. J., Gross, B. D., Hamill, T. M., Haupt, S. E., Jacobs, N., McGovern, A., Stensrud, D. J., Szatkowski, G., Szunyogh, I., & Wang, X. G. (2023). Challenges and opportunities in numerical weather prediction. *Bulletin of the American Meteorological Society, 104*(3), E698–E705. https://doi.org/10.1175/bams-d-22-0172.1

Brunet, G., Parsons, D. B., Ivanov, D., Lee, B. R., Bauer, P., Bernier, N. B., Bouchet, V., Brown, A., Busalacchi, A., Flatter, G. C., Goffer, R., Davies, P., Ebert, B., Gutbrod, K., Hong, S. Y., Kenabatho, P. K., Koppert, H. J., Lesolle, D., Lynch, A. H., Mahfouf, J. F., Ogallo, L., Palmer, T., Petty, K., Schulze, D., Shepherd, T. G., Stocker, T. F., Thorpe, A., & Yu, R. C. (2023). Advancing weather and climate forecasting for our changing world. *Bulletin of the American Meteorological Society, 104*(4), E909–E927. https://doi.org/10.1175/bams-d-21-0262.1

Carmona, R., Linares, C., Ortiz, C., Mirón, I. J., Luna, M. Y., & Díaz, J. (2017). Spatial variability in threshold temperatures of heat wave mortality: Impact assessment on prevention plans. *International Journal of Environmental Health Research, 27*(6), 463–475. https://doi.org/10.1080/09603123.2017.1379056

Casanueva, A., Burgstall, A., Kotlarski, S., Messeri, A., Morabito, M., Flouris, A. D., Nybo, L., Spirig, C., & Schwierz, C. (2019). Overview of existing heat-health warning Systems in Europe. *International Journal of Environmental Research and Public Health, 16*(15). https://doi.org/10.3390/ijerph16152657

Charlton-Perez, A. J., Greetham, D. V., & Hemingway, R. (2019). Storm naming and forecast communication: A case study of storm Doris. *Meteorological Applications, 26*(4), 682–697. https://doi.org/10.1002/met.1794

Chiabai, A., Spadaro, J. V., & Neumann, M. B. (2018). Valuing deaths or years of life lost? Economic benefits of avoided mortality from early heat warning systems. *Mitigation and Adaptation Strategies for Global Change, 23*(7), 1159–1176. https://doi.org/10.1007/s11027-017-9778-4

Christensen, B., & Christensen, S. (2014). Are female hurricanes really deadlier than male hurricanes? *Proceedings of the National Academy of Sciences of the United States of America, 111*(34), E3497–E3498. https://doi.org/10.1073/pnas.1410910111

Clegg, G., Haigh, R., & Amaratunga, D. (2022). Towards an improved understanding of participation in natural hazard early warning systems. *International Journal of Disaster Resilience in the Built Environment, 13*(5), 615–631. https://doi.org/10.1108/ijdrbe-11-2020-0120

Clemens, K. K., Ouédraogo, A. M., Le, B., Voogt, J., MacDonald, M., Stranberg, R., Yan, J. W., Krayenhoff, E. S., Gilliland, J., Forchuk, C., Van Uum, R., & Shariff, S. Z. (2022). Impact of Ontario's harmonized heat warning and information system on emergency department visits for heat-related illness in Ontario, Canada: A population-based time series analysis. *Canadian Journal of Public Health-Revue Canadienne De Sante Publique, 113*(5), 686–697. https://doi.org/10.17269/s41997-022-00665-1

D'Ippoliti, D., Michelozzi, P., Marino, C., De'Donato, F., Menne, B., Katsouyanni, K., Kirchmayer, U., Analitis, A., Medina-Ramón, M., Paldy, A., Atkinson, R., Kovats, S., Bisanti, L., Schneider, A., Lefranc, A., Iñiguez, C., & Perucci, C. A. (2010). The impact of heat waves on mortality in 9 European cities: Results from the EuroHEAT project. *Environmental Health, 9*. https://doi.org/10.1186/1476-069x-9-37

Davis, R. E., & Kalkstein, L. S. (1990). Development of an automated spatial synoptic climatological classification. *International Journal of Climatology, 10*(8), 769–794. https://doi.org/10.1002/joc.3370100802

De Luca, P., Delgado-Torres, C., Mahmood, R., Samso-Cabre, M., & Donat, M. G. (2023). Constraining decadal variability regionally improves near-term projections of hot, cold and dry extremes. *Environmental Research Letters, 18*(9). https://doi.org/10.1088/1748-9326/acf389

de Perez, E. C., van Aalst, M., Bischiniotis, K., Mason, S., Nissan, H., Pappenberger, F., Stephens, E., Zsoter, E., & van den Hurk, B. (2018). Global predictability of temperature extremes. *Environmental Research Letters, 13*(5). https://doi.org/10.1088/1748-9326/aab94a

de Perez, E. C., Harrison, L., Berse, K., Easton-Calabria, E., Marunye, J., Marake, M., Murshed, S. B., & Shampa, Z. F. H. (2022). Adapting to climate change through anticipatory action: The potential use of weather-based early warnings. *Weather and Climate Extremes, 38*, 100508. https://doi.org/10.1016/j.wace.2022.100508

De'Donato, F., Scortichini, M., De Sario, M., de Martino, A., & Michelozzi, P. (2018). Temporal variation in the effect of heat and the role of the Italian heat prevention plan. *Public Health, 161*, 154–162. https://doi.org/10.1016/j.puhe.2018.03.030

Delgado-Torres, C., Donat, M. G., Soret, A., González-Reviriego, N., Bretonnière, P.-A., Ho, A.-C., Pérez-Zanón, N., Cabré, M. S., & Doblas-Reyes, F. J. (2023). Multi-annual predictions of the frequency and intensity of daily temperature and precipitation extremes. *Environmental Research Letters, 18*(3), 034031.

Díaz, J., Carmona, R., Mirón, I. J., Ortiz, C., León, I., & Linares, C. (2015). Geographical variation in relative risks associated with heat: Update of Spain's heat wave prevention plan. *Environment International, 85*, 273–283. https://doi.org/10.1016/j.envint.2015.09.022

Dirmeyer, P. A., & Ford, T. W. (2020). A technique for seamless forecast construction and validation from weather to monthly time scales. *Monthly Weather Review, 148*(9), 3589–3603. https://doi.org/10.1175/mwr-d-19-0076.1

Doblas-Reyes, F., Hagedorn, R., Palmer, T., & Morcrette, J. J. (2006). Impact of increasing greenhouse gas concentrations in seasonal ensemble forecasts. *Geophysical Research Letters, 33*(7).

Domeisen, D. I. V., Eltahir, E. A. B., Fischer, E. M., Knutti, R., Perkins-Kirkpatrick, S. E., Schar, C., Seneviratne, S. I., Weisheimer, A., & Wernli, H. (2023). Prediction and projection of heatwaves. *Nature Reviews Earth & Environment, 4*(1), 36–50. https://doi.org/10.1038/s43017-022-00371-z

Dwyer, I. J., Barry, S. J. E., Megiddo, I., & White, C. J. (2022). Evaluations of heat action plans for reducing the health impacts of extreme heat: Methodological developments (2012-2021) and remaining challenges. *International Journal of Biometeorology, 66*(9), 1915–1927. https://doi.org/10.1007/s00484-022-02326-x

Eade, R., Hamilton, E., Smith, D. M., Graham, R. J., & Scaife, A. A. (2012). Forecasting the number of extreme daily events out to a decade ahead. *Journal of Geophysical Research-Atmospheres, 117*. https://doi.org/10.1029/2012jd018015

Ebi, K. L., Teisberg, T. J., Kalkstein, L. S., Robinson, L., & Weiher, R. F. (2004). Heat watch/warning systems save lives – Estimated costs and benefits for Philadelphia 1995-98. Bulletin of the American Meteorological Society 85 (8):1067–+. https://doi.org/10.1175/bams-85-8-1067

Emerton, R., Brimicombe, C., Magnusson, L., Roberts, C., Di Napoli, C., Cloke, H. L., & Pappenberger, F. (2022). Predicting the unprecedented: Forecasting the June 2021 Pacific northwest heatwave. *Weather, 77*(8), 272–279. https://doi.org/10.1002/wea.4257

Errett, N. A., Hartwell, C., Randazza, J. M., Nori-Sarma, A., Weinberger, K. R., Spangler, K. R., Sun, Y. T., Adams, Q. H., Wellenius, G. A., & Hess, J. J. (2023). Survey of extreme heat public health preparedness plans and response activities in the most populous jurisdictions in the United States. *BMC Public Health, 23*(1), 811. https://doi.org/10.1186/s12889-023-15757-x

Ettinger, J., Walton, P., Painter, J., Flocke, S. A., & Otto, F. E. L. (2023). Extreme weather events as teachable moments: Catalyzing climate change learning and action through conversation. *Environmental Communication – A Journal of Nature and Culture, 17*(7), 828–843. https://doi.org/10.1080/17524032.2023.2259623

Ferreira, G. W., Reboita, M. S., & Drumond, A. (2022). Evaluation of ECMWF-SEAS5 seasonal temperature and precipitation predictions over South America. *Climate, 10*(9), 128.

Filho, W. L., Barbir, J., Gwenzi, J., Ayal, D., Simpson, N. P., Adeleke, L., Tilahun, B., Chirisa, I., Gbedemah, S. F., Nzengya, D. M., Sharifi, A., Theodory, T., & Yaffa, S. (2022). The role of indigenous knowledge in climate change adaptation in Africa. *Environmental Science & Policy, 136*, 250–260. https://doi.org/10.1016/j.envsci.2022.06.004

Fonseca-Rodríguez, O., Sheridan, S. C., Lundevaller, E. H., & Schumann, B. (2020). Hot and cold weather based on the spatial synoptic classification and cause-specific mortality in Sweden: A time-stratified case-crossover study. *International Journal of Biometeorology, 64*(9), 1435–1449. https://doi.org/10.1007/s00484-020-01921-0

Fouillet, A., Rey, G., Wagner, V., Laaidi, K., Empereur-Bissonnet, P., Le Tertre, A., Frayssinet, P., Bessemoulin, P., Laurent, F., De Crouy-Chanel, P., Jougla, E., & Hémon, D. (2008). Has the impact of heat waves on mortality changed in France since the European heat wave of summer 2003?: A study of the 2006 heat wave. *International Journal of Epidemiology, 37*(2), 309–317. https://doi.org/10.1093/ije/dym253

Garcia, C., & Fearnley, C. J. (2016). Evaluating critical links in early warning systems for natural hazards. *Natural Hazards and Disaster Risk Reduction*, 53–67.

Golding, B. (2022). *Towards the "perfect" weather warning : Bridging disciplinary gaps through partnership and communication*. Springer. https://doi.org/10.1007/978-3-030-98989-7

Grundstein, A. J., & Williams, C. A. (2018). Heat exposure and the general public: Health impacts, risk communication, and mitigation measures. In Y. Hosokawa (Ed.), *Human health and physical activity during heat exposure* (SpringerBriefs in medical earth sciences) (pp. 29–43). https://doi.org/10.1007/978-3-319-75889-3_3

Guardaro, M. (2023). Strengthening heat action plans in the United States. *American Journal of Public Health, 113*(5), 465–467. https://doi.org/10.2105/ajph.2023.307260

Guigma, K. H., MacLeod, D., Todd, M., & Wang, Y. (2021). Prediction skill of Sahelian heatwaves out to subseasonal lead times and importance of atmospheric tropical modes of variability. *Climate Dynamics., 57*, 537–556. https://doi.org/10.1007/s00382-021-05726-8

Hadlos, A., Opdyke, A., & Hadigheh, S. A. (2022). Where does local and indigenous knowledge in disaster risk reduction go from here? A systematic literature review. *International Journal of Disaster Risk Reduction, 79*, 103160. https://doi.org/10.1016/j.ijdrr.2022.103160

Hajat, S., Sheridan, S. C., Allen, M. J., Pascal, M., Laaidi, K., Yagouti, A., Bickis, U., Tobias, A., Bourque, D., Armstrong, B. G., & Kosatsky, T. (2010). Heat-health warning systems: A comparison of the predictive capacity of different approaches to identifying dangerously hot days. *American Journal of Public Health, 100*(6), 1137–1144. https://doi.org/10.2105/ajph.2009.169748

Hamilton, E., Eade, R., Graham, R. J., Scaife, A. A., Smith, D. M., Maidens, A., & MacLachlan, C. (2012). Forecasting the number of extreme daily events on seasonal timescales. *Journal of Geophysical Research: Atmospheres, 117*(D3).

Hass, A. L., Runkle, J. D., & Sugg, M. M. (2021). The driving influences of human perception to extreme heat: A scoping review. *Environmental Research, 197*, 111173. https://doi.org/10.1016/j.envres.2021.111173

Hawkins, M. D., Brown, V., & Ferrell, J. (2017). Assessment of NOAA National Weather Service methods to warn for extreme heat events. *Weather Climate and Society, 9*(1), 5–13. https://doi.org/10.1175/wcas-d-15-0037.1

Henderson, D., Aubin, L., Behan, K., Chen, H., Doyle, H., Gower, S., MacDonald, M., Mee, C., Richardson, G. R. A., Rochon, G., Shnabel, M., Storfer, J., Yagouti, A., & Yusa, A. (2020). Developing a harmonized heat warning and information system for Ontario: A case study in collaboration. *Canadian Journal of Public Health-Revue Canadienne De Sante Publique, 111*(3), 426–432. https://doi.org/10.17269/s41997-020-00337-y

Heo, S., Nori-Sarma, A., Lee, K., Benmarhnia, T., Dominici, F., & Bell, M. L. (2019). The use of a quasi-experimental study on the mortality effect of a heat wave warning system in Korea. *International Journal of Environmental Research and Public Health, 16*(12). https://doi.org/10.3390/ijerph16122245

Hess, J. J., Sathish, L. M., Knowlton, K., Saha, S., Dutta, P., Ganguly, P., Tiwari, A., Jaiswal, A., Sheffield, P., Sarkar, J., Bhan, S. C., Begda, A., Shah, T., Solanki, B., & Mavalankar, D. (2018). Building resilience to climate change: Pilot evaluation of the impact of India's first heat action plan on all-cause mortality. *Journal of Environmental and Public Health., 2018*, 1–8. https://doi.org/10.1155/2018/7973519

Hobday, A. J., Oliver, E. C. J., Sen Gupta, A., Benthuysen, J. A., Burrows, M. T., Donat, M. G., Holbrook, N. J., Moore, P. J., Thomsen, M. S., Wernberg, T., & Smale, D. A. (2018). Categorizing and naming marine heatwaves. *Oceanography, 31*(2), 162–173. https://doi.org/10.5670/oceanog.2018.205

Hondula, D. M., Vanos, J. K., & Gosling, S. N. (2014). The SSC: A decade of climate-health research and future directions. *International Journal of Biometeorology, 58*(2), 109–120. https://doi.org/10.1007/s00484-012-0619-6

Hondula, D. M., Meltzer, S., Balling, R. C., & Iñiguez, P. (2022). Spatial analysis of United States National Weather Service excessive heat warnings and heat advisories. *Bulletin of the American Meteorological Society, 103*(9), E2017–E2031. https://doi.org/10.1175/bams-d-21-0069.1

IDNDR. (1994). *International decade for natural disaster reduction Yokohama strategy and plan for action for a safer world*. Yokohama.

Jin, A. S., & Sanders, K. T. (2022). Analyzing changes to US municipal heat response plans during the COVID-19 pandemic. *Environmental Science & Policy, 128*, 347–358. https://doi.org/10.1016/j.envsci.2021.11.022

Jolliffe, I. T., & Stephenson, D. B. (2012). *Forecast verification : A practitioner's guide in atmospheric science* (2nd ed.). Wiley.

Jung, K., Shavitt, S., Viswanathan, M., & Hilbe, J. M. (2014a). Female hurricanes are deadlier than male hurricanes. *Proceedings of the National Academy of Sciences of the United States of America, 111*(24), 8782–8787. https://doi.org/10.1073/pnas.1402786111

Jung, K., Shavitt, S., Viswanathan, M., & Hilbe, J. M. (2014b). Reply to Christensen and Christensen and to Malter: Pitfalls of erroneous analyses of hurricanes names. *Proceedings of the National Academy of Sciences of the United States of America, 111*(34), E3499–E3500. https://doi.org/10.1073/pnas.1411652111

Jung, K. J., Shavitt, S., Viswanathan, M., & Hilbe, J. M. (2014c). Reply to Bakkensen and Larson: Population may matter but does not alter conclusions. *Proceedings of the National Academy of Sciences of the United States of America, 111*(50), E5333–E5333. https://doi.org/10.1073/pnas.1419330111

Jung, K. J., Shavitt, S., Viswanathan, M., & Hilbe, J. M. (2014d). Reply to Maley: Yes, appropriate modeling of fatality counts confirms female hurricanes are deadlier. *Proceedings of the National Academy of Sciences of the United States of America, 111*(37), E3835–E3835. https://doi.org/10.1073/pnas.1414111111

Kalkstein, L. S., Jamason, P. F., Greene, J. S., Libby, J., & Robinson, L. (1996). The Philadelphia hot weather-health watch warning system: Development and application, summer 1995. *Bulletin of the American Meteorological Society, 77*(7), 1519–1528. https://doi.org/10.1175/1520-0477(1996)077<1519:tphwhw>2.0.co;2

Kapwata, T., Gebreslasie, M. T., & Wright, C. Y. (2022). An analysis of past and future heatwaves based on a heat-associated mortality threshold: Towards a heat health warning system. *Environmental Health, 21*(1), 112. https://doi.org/10.1186/s12940-022-00921-4

Kelman, I., & Glantz, M. H. (2014). Early warning systems defined. In A. Singh & Z. Zommers (Eds.), *Reducing disaster: Early warning systems for climate change* (pp. 89–108). Springer. https://doi.org/10.1007/978-94-017-8598-3_5

Khoza, S., van Niekerk, D., & Nemakonde, L. D. (2022). A decade of inaction in the SADC region?-disaster risk data gaps and inconsistencies on the Sendai framework monitor. *Progress in Disaster Science, 16*, 100250. https://doi.org/10.1016/j.pdisas.2022.100250

Kimutis, N., Wall, T., & Darrow, L. (2024). Emergency management short term response to extreme heat in the 25 most populated US cities. *International Journal of Disaster Risk Reduction, 100*, 104097. https://doi.org/10.1016/j.ijdrr.2023.104097

Knowlton, K., Kulkarni, S. P., Azhar, G. S., Mavalankar, D., Jaiswal, A., Connolly, M., Nori-Sarma, A., Rajiva, A., Dutta, P., Deol, B., Sanchez, L., Khosla, R., Webster, P. J., Toma, V. E., Sheffield, P., Hess, J. J., & Ahmedabad Heat Climate Study G. (2014). Development and implementation of South Asia's first heat-health action plan in Ahmedabad (Gujarat, India). *International Journal of Environmental Research and Public Health, 11*(4), 3473–3492. https://doi.org/10.3390/ijerph110403473

Knox-Hayes, J., Osorio, J. C., Stamler, N., Dombrov, M., Winer, R., Smith, M. H., Blake, R. A., & Rosenzweig, C. (2023). The compound risk of heat and COVID-19 in New York City: Riskscapes, physical and social factors, and interventions. *Local Environment, 28*(6), 699–727. https://doi.org/10.1080/13549839.2023.2187362

Kotharkar, R., & Ghosh, A. (2022). Progress in extreme heat management and warning systems: A systematic review of heat-health action plans (1995–2020). *Sustainable Cities and Society, 76*, 103487. https://doi.org/10.1016/j.scs.2021.103487

Kotroni, V., Lagouvardos, K., Bezes, A., Dafis, S., Galanaki, E., Giannaros, C., Giannaros, T., Karagiannidis, A., Koletsis, I., Kopania, T., Papagiannaki, K., Papavasileiou, G., Vafeiadis, V., & Vougioulas, E. (2021). Storm naming in the eastern Mediterranean: Procedures, events review and impact on the citizens risk perception and readiness. *Atmosphere, 12*(11). https://doi.org/10.3390/atmos12111537

Kovats, R. S., & Ebi, K. L. (2006). Heatwaves and public health in Europe. *European Journal of Public Health, 16*(6), 592–599. https://doi.org/10.1093/eurpub/ckl049

Kueh, M. T., & Lin, C. A. Y. (2020). The 2018 summer heatwaves over northwestern Europe and its extended-range prediction. *Scientific Reports, 10*(1), 19283. https://doi.org/10.1038/s41598-020-76181-4

Lachlan, K. A., & Spence, P. R. (2010). Communicating risks: Examining hazard and outrage in multiple contexts. *Risk Analysis, 30*, 1872–1886. https://doi.org/10.1111/j.1539-6924.2010.01511.x

Lass, W., Haas, A., Hinkel, J., & Jaeger, C. (2011). Avoiding the avoidable: Towards a European heat waves risk governance. *International Journal of Disaster Risk Science, 2*(1), 1–14. https://doi.org/10.1007/s13753-011-0001-z

Lavaysse, C., Naumann, G., Alfieri, L., Salamon, P., & Vogt, J. (2019). Predictability of the European heat and cold waves. *Climate Dynamics, 52*(3–4), 2481–2495. https://doi.org/10.1007/s00382-018-4273-5

Lee, K. K., & Kim, I. G. (2020). Social media data analytics to enhance sustainable communications between public users and providers in weather forecast service industry. *Sustainability, 12*(20). https://doi.org/10.3390/su12208528

Lee, D. G., Kim, K. R., Kim, J., Kim, B. J., Cho, C. H., Sheridan, S. C., Kalkstein, L. S., Kim, H., & Yi, S. M. (2018). Effects of heat waves on daily excess mortality in 14 Korean cities during the past 20 years (1991–2010): An application of the spatial synoptic classification approach. *International Journal of Biometeorology, 62*(4), 575–583. https://doi.org/10.1007/s00484-017-1466-2

Leite, A., Santos, A. J., Silva, S., Nunes, B., Mexia, R., & Rodrigues, A. P. (2020). Assessing the use and understanding of the Portuguese heat-health warning system (ICARO). *Journal of Public Health, 42*(2), 395–402. https://doi.org/10.1093/pubmed/fdaa029

Lin, X. L., Rainear, A. M., Spence, P. R., & Lachlan, K. A. (2018). Don't sleep on it: An examination of storm naming and potential heuristic effects on twitter. *Weather Climate and Society, 10*(4), 769–779. https://doi.org/10.1175/wcas-d-18-0008.1

Lin, H., Mo, R. P., & Vitart, F. (2022). The 2021 Western north American heatwave and its subseasonal predictions. *Geophysical Research Letters, 49*(6). https://doi.org/10.1029/2021gl097036

Linares, C., Díaz, J., Negev, M., Martínez, G. S., Debono, R., & Paz, S. (2020). Impacts of climate change on the public health of the Mediterranean Basin population – Current situation, projections, preparedness and adaptation. *Environmental Research, 182*, 109107. https://doi.org/10.1016/j.envres.2019.109107

Livezey, R. E., & Timofeyeva, M. M. (2008). The first decade of long-lead US seasonal forecasts: Insights from a skill analysis. *Bulletin of the American Meteorological Society, 89*(6), 843–854.

Lou, J. N., Ban, J., Zhang, T., Wang, P., Wu, Y. Y., Huang, L., Li, T. T., & Bi, J. (2021). An intervention study of the rural elderly for improving exposure, risk perception and behavioral responses under high temperature. *Environmental Research Letters, 16*(5). https://doi.org/10.1088/1748-9326/abfa4f

Lu, Y. C., & Romps, D. M. (2023). Predicting fatal heat and humidity using the heat index model. *Journal of Applied Physiology, 134*(3), 649–656. https://doi.org/10.1152/japplphysiol.00417.2022

Luo, L., & Zhang, Y. (2012). Did we see the 2011 summer heat wave coming? *Geophysical Research Letters, 39*(9).

Macherera, M., & Chimbari, M. J. (2016). A review of studies on community based early warning systems. *Jamba (Potchefstroom, South Africa), 8*(1), 206–206. https://doi.org/10.4102/jamba.v8i1.206

Maley, S. (2014). Statistics show no evidence of gender bias in the public's hurricane preparedness. *Proceedings of the National Academy of Sciences of the United States of America, 111*(37), E3834–E3834. https://doi.org/10.1073/pnas.1413079111

Mandal, R., Joseph, S., Sahai, A. K., Phani, R., Dey, A., Chattopadhyay, R., & Pattanaik, D. R. (2019). Real time extended range prediction of heat waves over India. *Scientific Reports, 9*, 9008. https://doi.org/10.1038/s41598-019-45430-6

Marchezini, V., Horita, F. E. A., Matsuo, P. M., Trajber, R., Trejo-Rangel, M. A., & Olivato, D. (2018). A review of studies on participatory early warning systems (P-EWS): Pathways to support citizen science initiatives. *Frontiers in Earth Science, 6*. https://doi.org/10.3389/feart.2018.00184

Marshall, A. G., Hudson, D., Wheeler, M. C., Alves, O., Hendon, H. H., Pook, M. J., & Risbey, J. S. (2014). Intra-seasonal drivers of extreme heat over Australia in observations and POAMA-2. *Climate Dynamics, 43*(7–8), 1915–1937. https://doi.org/10.1007/s00382-013-2016-1

Martinez, G. S., Imai, C., & Masumo, K. (2011). Local heat stroke prevention plans in Japan: Characteristics and elements for public health adaptation to climate change. *International Journal of Environmental Research and Public Health, 8*(12), 4563–4581. https://doi.org/10.3390/ijerph8124563

Martinez, G. S., Kendrovski, V., Salazar, M. A., De'donato, F., & Boeckmann, M. (2022). Heat-health action planning in the WHO European region: Status and policy implications. *Environmental Research, 214*, 113709. https://doi.org/10.1016/j.envres.2022.113709

Martinez-Solanas, E., & Basagana, X. (2019). Temporal changes in temperature-related mortality in Spain and effect of the implementation of a heat health prevention plan. *Environmental Research, 169*, 102–113. https://doi.org/10.1016/j.envres.2018.11.006

Matthies, F., & Menne, B. (2009). Prevention and management of health hazards related to heatwaves. *International Journal of Circumpolar Health, 68*(1), 8–22. https://doi.org/10.3402/ijch.v68i1.18293

Matzarakis, A., Laschewski, G., & Muthers, S. (2020). The heat health warning system in Germany-application and warnings for 2005 to 2019. *Atmosphere, 11*(2). https://doi.org/10.3390/atmos11020170

Mayrhuber, E. A. S., Duckers, M. L. A., Wallner, P., Arnberger, A., Allex, B., Wiesbock, L., Wanka, A., Kolland, F., Eder, R., Hutter, H. P., & Kutalek, R. (2018). Vulnerability to heatwaves and implications for public health interventions – A scoping review. *Environmental Research, 166*, 42–54. https://doi.org/10.1016/j.envres.2018.05.021

McGregor, G., Ebi, K., Menne, B., & Bessmoulin, P. (2015). *Heatwaves and health guidance on warning system development*. World Meteorological Organisation.

McKemey, M. B., Rangers, B., Ens, E. J., Hunter, J. T., Ridges, M., Costello, O., & Reid, N. C. H. (2021). Co-producing a fire and seasons calendar to support renewed indigenous cultural fire management. *Austral Ecology, 46*(7), 1011–1029. https://doi.org/10.1111/aec.13034

McLean, K. E., Stranberg, R., MacDonald, M., Richardson, G. R. A., Kosatsky, T., & Henderson, S. B. (2018). Establishing heat alert thresholds for the varied climatic regions of British Columbia, Canada. *International Journal of Environmental Research and Public Health, 15*(9). https://doi.org/10.3390/ijerph15092048

McLoughlin, N., Howarth, C., & Shreedhar, G. (2023). Changing behavioral responses to heat risk in a warming world: How can communication approaches be improved? *Wiley Interdisciplinary Reviews-Climate Change, 14*(2). https://doi.org/10.1002/wcc.819

Mecit, A., Shrum, L. J., & Lowrey, T. M. (2022). COVID-19 is feminine: Grammatical gender influences danger perceptions and precautionary behavioral intentions by activating gender stereotypes. *Journal of Consumer Psychology, 32*(2), 316–325. https://doi.org/10.1002/jcpy.1257

Merryfield, W. J., Baehr, J., Batté, L., Becker, E. J., Butler, A. H., Coelho, C. A. S., Danabasoglu, G., Dirmeyer, P. A., Doblas-Reyes, F. J., Domeisen, D. I. V., Ferranti, L., Ilynia, T., Kumar, A., Müller, W. A., Rixen, M., Robertson, A. W., Smith, D. M., Takaya, Y., Tuma, M., Vitart, F., White, C. J., Alvarez, M. S., Ardilouze, C., Attard, H., Baggett, C., Balmaseda, M. A., Beraki, A. F., Bhattacharjee, P. S., Bilbao, R., de Andrade, F. M., DeFlorio, M. J., Díaz, L. B., Ehsan, M. A., Fragkoulidis, G., Grainger, S., Green, B. W., Hell, M. C., Infanti, J. M., Isensee, K., Kataoka, T., Kirtman, B. P., Klingaman, N. P., Lee, J. Y., Mayer, K., McKay, R., Mecking, J. V., Miller, D. E., Neddermann, N., Ng, C. H. J., Ossó, A., Pankatz, K., Peatman, S., Pegion, K., Perlwitz, J., Recalde-Coronel, G. C., Reintges, A., Renkl, C., Solaraju-Murali, B., Spring, A., Stan, C., Sun, Y. Q., Tozer, C. R., Vigaud, N., Woolnough, S., & Yeager, S. (2020). Current and emerging developments in subseasonal to decadal prediction. *Bulletin of the American Meteorological Society, 101*(6), E869–E896. https://doi.org/10.1175/bams-d-19-0037.1

Metzger, A., Baharav, Y., Nichols, L., Finke, M., Saunders, B., Mitchell, P., Wellenius, G. A., Baughman McLeod, K., & Shickman, K. (2024). Beliefs and behaviors associated with the first named heat wave in Seville Spain 2022. *Scientific Reports, 14*(1), 9055.

Morabito, M., Messeri, A., Noti, P., Casanueva, A., Crisci, A., Kotlarski, S., Orlandini, S., Schwierz, C., Spirig, C., Kingma, B. R. M., Flouris, A. D., & Nybo, L. (2019). An occupational heat-health warning system for Europe: The HEAT-SHIELD platform. *International Journal of Environmental Research and Public Health, 16*(16). https://doi.org/10.3390/ijerph16162890

Morss, R. E., Lazrus, H., Bostrom, A., & Demuth, J. L. (2020). The influence of cultural worldviews on people's responses to hurricane risks and threat information. *Journal of Risk Research, 23*(12), 1620–1649. https://doi.org/10.1080/13669877.2020.1750456

Nairn, J. R., & Fawcett, R. J. B. (2015). The excess heat factor: A metric for heatwave intensity and its use in classifying heatwave severity. *International Journal of Environmental Research and Public Health, 12*(1), 227–253. https://doi.org/10.3390/ijerph120100227

Nairn, J. R., Moise, A., & Ostendorf, B. (2022). The impact of humidity on Australia's operational heatwave services. *Climate Services, 27*, 100315. https://doi.org/10.1016/j.cliser.2022.100315

Naveena, N., Satyanarayana, G. C., Raju, A. D., Umakanth, N., Srinivas, D., Rao, K. S., & Suman, M. (2021). Prediction of heatwave 2013 over Andhra Pradesh and Telangana, India using WRF model. *Asian Journal of Atmospheric Environment, 15*, 3–44. https://doi.org/10.5572/ajae.2020.117

Nicholls, N., Skinner, C., Loughnan, M., & Tapper, N. (2008). A simple heat alert system for Melbourne, Australia. *International Journal of Biometeorology, 52*, 375–384.

Nitschke, M., Tucker, G., Hansen, A., Williams, S., Zhang, Y., & Bi, P. (2016). Evaluation of a heat warning system in Adelaide, South Australia, using case-series analysis. *BMJ Open, 6*(7), e012125. https://doi.org/10.1136/bmjopen-2016-012125

Nogueira, P. J. (2005). Examples of heat health warning systems: Lisbon's ICARO's surveillance system, summer of 2003. *Extreme Weather Events and Public Health Responses:141–159.* https://doi.org/10.1007/3-540-28862-7_14

Oka, K., Honda, Y., & Hijioka, Y. (2023). Launching criteria of 'Heatstroke Alert' in Japan according to regionality and age group. *Environmental Research Communications, 5*(2). https://doi.org/10.1088/2515-7620/acac03

Osman, M., & Alvarez, M. S. (2018). Subseasonal prediction of the heat wave of December 2013 in Southern South America by the POAMA and BCC-CPS models. *Climate Dynamics, 50*(1), 67–81.

Pascal, M., Laaidi, K., Wagner, V., Ung, A. B., Smaili, S., Fouillet, A., Caserio-Schönemann, C., & Beaudeau, P. (2012a). How to use near real-time health indicators to support decision-making during a heat wave: The example of the French heat wave warning system. *PLoS Curr, 4*, e4f83ebf72317d. https://doi.org/10.1371/4f83ebf72317d

Pascal, M., Le Tertre, A., & Saoudi, A. (2012b). Quantification of the heat wave effect on mortality in nine French cities during summer 2006. *PLoS Currents, 4.*

Pascal, M., Wagner, V., Le Tertre, A., Laaidi, K., Honore, C., Benichou, F., & Beaudeau, P. (2013). Definition of temperature thresholds: The example of the French heat wave warning system. *International Journal of Biometeorology, 57*(1), 21–29. https://doi.org/10.1007/s00484-012-0530-1

Pascal, M., Lagarrigue, R., Tabai, A., Bonmarin, I., Camail, S., Laaidi, K., Le Tertre, A., & Denys, S. (2021). Evolving heat waves characteristics challenge heat warning systems and prevention plans. *International Journal of Biometeorology, 65*, 1683–1694.

Patterson, M., Weisheimer, A., Befort, D. J., & O'Reilly, C. H. (2022). The strong role of external forcing in seasonal forecasts of European summer temperature. *Environmental Research Letters, 17*(10), 104033.

Poudel, J. M., Sigdel, M., Chhetri, R. B., & Sudarsan, K. C. (2022). Farmers reading nature's clues to figure out impending weather. *Weather Climate and Society, 14*(3), 801–812. https://doi.org/10.1175/wcas-d-21-0174.1

Price, K., Benmarhnia, T., Gaudet, J., Kaiser, D., Sadoine, M. L., Perron, S., & Smargiassi, A. (2018). The Montreal heat response plan: Evaluation of its implementation towards healthcare professionals and vulnerable populations. *Canadian Journal of Public Health-Revue Canadienne De Sante Publique, 109*(1), 108–116. https://doi.org/10.17269/s41997-018-0020-2

Pyrina, M., & Domeisen, D. I. V. (2023). Subseasonal predictability of onset, duration, and intensity of European heat extremes. *Quarterly Journal of the Royal Meteorological Society, 149*(750), 84–101. https://doi.org/10.1002/qj.4394

Quesada, B., Vautard, R., Yiou, P., Hirschi, M., & Seneviratne, S. I. (2012). Asymmetric European summer heat predictability from wet and dry southern winters and springs. *Nature Climate Change, 2*(10), 736–741. https://doi.org/10.1038/nclimate1536

Ragettli, M. S., & Roosli, M. (2019). Heat-health action plans to prevent heat-related deaths-experiences from Switzerland. *Bundesgesundheitsblatt-Gesundheitsforschung-Gesundheitsschutz, 62*(5), 605–611. https://doi.org/10.1007/s00103-019-02928-8

Ragettli, M. S., Vicedo-Cabrera, A. M., Flückiger, B., & Röösli, M. (2019). Impact of the warm summer 2015 on emergency hospital admissions in Switzerland. *Environmental Health, 18*(1), 66. https://doi.org/10.1186/s12940-019-0507-1

Rainear, A. M., Lachlan, K. A., & Lin, C. A. (2017). What's in a #name? An experimental study examining perceived credibility and impact of winter storm names. *Weather Climate and Society, 9*(4), 815–822. https://doi.org/10.1175/wcas-d-16-0037.1

Randazza, J. M., Hess, J., Bostrom, A., Hartwell, C., Adams, Q. H., Nori-Sarma, A., Spangler, K. R., Sun, Y. T., Weinberger, K. R., Wellenius, G. A., & Errett, N. A. (2023). Planning to reduce the health impacts of extreme heat: A content analysis of heat action plans in local United States jurisdictions. *American Journal of Public Health, 113*(5), 559–567. https://doi.org/10.2105/ajph.2022.307217

Ratwatte, P., Wehling, H., Kovats, S., Landeg, O., & Weston, D. (2022). Factors associated with older adults' perception of health risks of hot and cold weather event exposure: A scoping review. *Frontiers in Public Health, 10*. https://doi.org/10.3389/fpubh.2022.939859

Ren, H. L., Bao, Q., Zhou, C. G., Wu, J., Gao, L., Wang, L., Ma, J. R., Tang, Y., Liu, Y. K., Wang, Y. J., & Zhao, Z. S. (2023). Seamless prediction in China: A review. *Advances in Atmospheric Sciences, 40*(8), 1501–1520. https://doi.org/10.1007/s00376-023-2335-z

Roberts, T., Seymour, V., Brooks, K., Thompson, R., Petrokofsky, C., O'Connell, E., & Landeg, O. (2022). Stakeholder perspectives on extreme hot and cold weather alerts in England and the proposed move towards an impact-based approach. *Environmental Science & Policy, 136*, 467–475. https://doi.org/10.1016/j.envsci.2022.07.012

Romps, D. M., & Lu, Y. C. (2022). Chronically underestimated: A reassessment of US heat waves using the extended heat index. *Environmental Research Letters, 17*(9). https://doi.org/10.1088/1748-9326/ac8945

Rouges, E., Ferranti, L., Kantz, H., & Pappenberger, F. (2023). European heatwaves: Link to large-scale circulation patterns and intraseasonal drivers. *International Journal of Climatology, 43*(7), 3189–3209. https://doi.org/10.1002/joc.8024

Schoessow, F. S., Li, Y. J., Marlon, J. R., Leiserowitz, A., & Howe, P. D. (2022). Sociodemographic factors associated with heatwave risk perception in the United States. *Weather Climate and Society, 14*(4), 1119–1131. https://doi.org/10.1175/wcas-d-21-0104.1

Scolobig, A., Prior, T., Schröter, D., Jörin, J., & Patt, A. (2015). Towards people-centred approaches for effective disaster risk management: Balancing rhetoric with reality. *International Journal of Disaster Risk Reduction, 12*, 202–212. https://doi.org/10.1016/j.ijdrr.2015.01.006

Sheridan, S. C. (2002). The redevelopment of a weather-type classification scheme for North America. *International Journal of Climatology: A Journal of the Royal Meteorological Society, 22*(1), 51–68.

Sheridan, S. C., & Kalkstein, L. S. (2004). Progress in heat watch-warning system technology. *Bulletin of the American Meteorological Society, 85*(12):1931–+. https://doi.org/10.1175/bams-85-12-1931.

Singh, H., Arora, K., Ashrit, R., & Rajagopal, E. N. (2017). Verification of pre-monsoon temperature forecasts over India during 2016 with a focus on heatwave prediction. *Natural Hazards and Earth System Sciences, 17*(9), 1469–1485. https://doi.org/10.5194/nhess-17-1469-2017

Skilton, L. (2018). Gendering natural disaster: The battle over female hurricane names. *Journal of Women's History, 30*(3), 132–156.

Smith, G. (2016). Hurricane names: A bunch of hot air? *Weather and Climate Extremes, 12*, 80–84. https://doi.org/10.1016/j.wace.2015.11.006

Sousa, P. M., Trigo, R. M., Russo, A., Geirinhas, J. L., Rodrigues, A., Silva, S., & Torres, A. (2022). Heat-related mortality amplified during the COVID-19 pandemic. *International Journal of Biometeorology, 66*(3), 457–468. https://doi.org/10.1007/s00484-021-02192-z

Spruce, M., Arthur, R., & Williams, H. T. P. (2020). Using social media to measure impacts of named storm events in the United Kingdom and Ireland. *Meteorological Applications, 27*(1). https://doi.org/10.1002/met.1887

Staiger, H., Laschewski, G., & Gratz, A. (2012). The perceived temperature – A versatile index for the assessment of the human thermal environment. Part A: Scientific basics. *International Journal of Biometeorology, 56 (1):165-176*, 165–176. https://doi.org/10.1007/s00484-011-0409-6

Stern, H., & Davidson, N. E. (2015). Trends in the skill of weather prediction at lead times of 1-14 days. *Quarterly Journal of the Royal Meteorological Society, 141*(692), 2726–2736. https://doi.org/10.1002/qj.2559

Sublett, M. D., & Sublett, J. A. (2023). Creeks and peaks: Wildfire name-giving in the United States. *Names-a Journal of Onomastics, 71*(3), 47–61. https://doi.org/10.5195/names.2023.2484

Sufri, S., Dwirahmadi, F., Phung, D., & Rutherford, S. (2020). A systematic review of community engagement (CE) in disaster Early Warning Systems (EWSs). *Progress in Disaster Science, 5*, 100058. https://doi.org/10.1016/j.pdisas.2019.100058

Tan, J. G., Kalkstein, L. S., Huang, J. X., Lin, S. B., Yin, H. B., & Shao, D. M. (2004). An operational heat/health warning system in shanghai. *International Journal of Biometeorology, 48*(3), 157–162. https://doi.org/10.1007/s00484-003-0193-z

Tan, M. L., Hoffmann, D., Ebert, E., Cui, A., & Johnston, D. (2022). Exploring the potential role of citizen science in the warning value chain for high impact weather. *Frontiers in Communication, 7*. https://doi.org/10.3389/fcomm.2022.949949

Thiaw, W. M., Bekele, E., Diouf, S. N., Dewitt, D. G., Ndiaye, O., Ndiaye, M. K. N., Ndiaye, P. N., Diene, N., Diouf, M., Diaw, A., Diop, S., Badj, F., & Diouf, A. (2022). Toward experimental heat-health early warning in Africa. *Bulletin of the American Meteorological Society, 103*(8), E1843–E1860. https://doi.org/10.1175/bams-d-20-0140.1

Tian, D., Wood, E. F., & Yuan, X. (2017). CFSv2-based sub-seasonal precipitation and temperature forecast skill over the contiguous United States. *Hydrology and Earth System Sciences, 21*(3), 1477–1490. https://doi.org/10.5194/hess-21-1477-2017

Toloo, G., FitzGerald, G., Aitken, P., Verrall, K., & Tong, S. L. (2013). Evaluating the effectiveness of heat warning systems: Systematic review of epidemiological evidence. *International Journal of Public Health, 58*(5), 667–681. https://doi.org/10.1007/s00038-013-0465-2

Tozier de la Poterie, A., & Baudoin, M. A. (2015). From Yokohama to Sendai: Approaches to participation in international disaster risk reduction frameworks. *International Journal of Disaster Risk Science, 6*(2), 128–139. https://doi.org/10.1007/s13753-015-0053-6

UNDRR. (2015). *Sendai framework for disaster risk reduction 2015–2030*. United Nations Office for Disaster Risk Reduction.

UNDRR. (2016). *United Nations Office of disaster risk reduction report of the open-ended intergovernmental expert working group on indicators and terminology relating to disaster risk reduction*. United Nations.

Urban, A., & Kysely, J. (2018). Application of spatial synoptic classification in evaluating links between heat stress and cardiovascular mortality and morbidity in Prague, Czech Republic. *International Journal of Biometeorology, 62*(1), 85–96. https://doi.org/10.1007/s00484-015-1055-1

Vaidyanathan, A., Saha, S., Vicedo-Cabrera, A. M., Gasparrini, A., Abdurehman, N., Jordan, R., Hawkins, M., Hess, J., & Elixhauser, A. (2019). Assessment of extreme heat and hospitalizations to inform early warning systems. *Proceedings of the National Academy of Sciences of the United States of America, 116*(12), 5420–5427. https://doi.org/10.1073/pnas.1806393116

VanderMolen, K., Kimutis, N., & Hatchett, B. J. (2022). Recommendations for increasing the reach and effectiveness of heat risk education and warning messaging. *International Journal of Disaster Risk Reduction, 82*, 103288. https://doi.org/10.1016/j.ijdrr.2022.103288

Vanderplanken, K., van Loenhout, J., & van den Hazel, P. (2019). The health impact of heat waves in Europe: Insights from national plans and key informant interviews. *European Journal of Public Health, 29*, 64–64.

Vanderplanken, K., van den Hazel, P., Marx, M., Shams, A. Z., Guha-Sapir, D., & van Loenhout, J. A. F. (2021). Governing heatwaves in Europe: Comparing health policy and practices to better understand roles, responsibilities and collaboration. *Health Research Policy and Systems, 19*(1), 20. https://doi.org/10.1186/s12961-020-00645-2

Vicedo-Cabrera, A. M., Ragettli, M. S., Schindler, C., & Röösli, M. (2016). Excess mortality during the warm summer of 2015 in Switzerland. *Swiss Medical Weekly, 146*. https://doi.org/10.4414/smw.2016.14379

Vitart, F., & Robertson, A. W. (2018). The sub-seasonal to seasonal prediction project (S2S) and the prediction of extreme events. *npj Climate and Atmospheric Science, 1*. https://doi.org/10.1038/s41612-018-0013-0

Vu, A., Rutherford, S., & Phung. (2019). Heat health prevention measures and adaptation in older populations – A systematic review. *International Journal of Environmental Research and Public Health, 16*(22). https://doi.org/10.3390/ijerph16224370

Weinberger, K. R., Zanobetti, A., Schwartz, J., & Wellenius, G. A. (2018). Effectiveness of National Weather Service heat alerts in preventing mortality in 20 US cities. *Environment International, 116*, 30–38. https://doi.org/10.1016/j.envint.2018.03.028

Weinberger, K. R., Wu, X., Sun, S. Z., Spangler, K. R., Nori-Sarma, A., Schwartz, J., Requia, W., Sabath, B. M., Braun, D., Zanobetti, A., Dominici, F., & Wellenius, G. A. (2021). Heat warnings, mortality, and hospital admissions among older adults in the United States. *Environment International, 157*, 106834. https://doi.org/10.1016/j.envint.2021.106834

Weisheimer, A., Balmaseda, M. A., Stockdale, T. N., Mayer, M., Sharmila, S., Hendon, H., & Alves, O. (2022). Variability of ENSO forecast skill in 2-year global reforecasts over the 20th century. *Geophysical Research Letters, 49*(10), e2022GL097885.

White, C. J., Carlsen, H., Robertson, A. W., Klein, R. J. T., Lazo, J. K., Kumar, A., Vitart, F., de Perez, E. C., Ray, A. J., Murray, V., Bharwani, S., MacLeod, D., James, R., Fleming, L., Morse, A. P., Eggen, B., Graham, R., Kjellström, E., Becker, E., Pegion, K. V., Holbrook, N. J., McEvoy, D., Depledge, M., Perkins-Kirkpatrick, S., Brown, T. J., Street, R., Jones, L., Remenyi, T. A., Hodgson-Johnston, I., Buontempo, C., Lamb, R., Meinke, H., Arheimer, B., & Zebiak, S. E. (2017). Potential applications of subseasonal-to-seasonal (S2S) predictions. *Meteorological Applications, 24*(3), 315–325. https://doi.org/10.1002/met.1654

White, R. H., Anderson, S., Booth, J. F., Braich, G., Draeger, C., Fei, C. Y., Harley, C. D. G., Henderson, S. B., Jakob, M., Lau, C. A., Admasu, L. M., Narinesingh, V., Rodell, C., Roocroft, E., Weinberger, K. R., & West, G. (2023). The unprecedented Pacific northwest heatwave of June 2021. Nature. *Communications, 14*(1). https://doi.org/10.1038/s41467-023-36289-3

WHO. (2004). *Heat waves: Risks and responses*. Health and Global Environmental Change Series.

Wilhelmi, O. V., Howe, P. D., Hayden, M. H., & O'Lenick, C. R. (2021). Compounding hazards and intersecting vulnerabilities: Experiences and responses to extreme heat during COVID-19. *Environmental Research Letters, 16*(8). https://doi.org/10.1088/1748-9326/ac1760

Wilkins, A., Pennaz, A., Dix, M., Smith, A., Vawter, J., Karlson, D., Tokar, S., & Brooks, E. (2021). Challenges and opportunities for Sendai framework disaster loss reporting in the United States. *Progress in Disaster Science, 10*, 100167. https://doi.org/10.1016/j.pdisas.2021.100167

Williams, L., Erens, B., Ettelt, S., Hajat, S., Manacorda, T., & Mays, N. (2019). *Evaluation of the heatwave plan for England final report*. Policy Innovation and Evaluation Research Unit (PIRU).

Wisner, B. (2020). Five years beyond Sendai-can we get beyond frameworks? *International Journal of Disaster Risk Science, 11*(2), 239–249. https://doi.org/10.1007/s13753-020-00263-0

WMO. (2022a). *Considerations regarding the naming of heatwaves – Technical brief.* World Meteorological Organisation.

WMO. (2022b). *Early warnings for all: The UN global early warning initiative for the implementation of climate adaptation executive action plan 2023–2027.* Geneva.

WMO. (2023). *State of climate services: Health* (Vol. WMO-No. 1335). Geneva.

Wright, N., Fagan, L., Lapitan, J. M., Kayano, R., Abrahams, J., Huda, Q., & Murray, V. (2020). Health emergency and disaster risk management: Five years into implementation of the Sendai framework. *International Journal of Disaster Risk Science, 11*(2), 206–217. https://doi.org/10.1007/s13753-020-00274-x

Wu, Y. Q., Wang, X. Y., Wu, J. Y., Wang, R., & Yang, S. N. (2020). Performance of heat-health warning systems in Shanghai evaluated by using local heat-related illness data. *Science of the Total Environment, 715*, 136883. https://doi.org/10.1016/j.scitotenv.2020.136883

Wulff, C. O., & Domeisen, D. I. V. (2019). Higher subseasonal predictability of extreme hot European summer temperatures as compared to average summers. *Geophysical Research Letters, 46*(20), 11520–11529. https://doi.org/10.1029/2019gl084314

Xiao, X., Wu, L. F., Liu, X. G., Zhang, S., Li, S., & Cui, Y. K. (2023). Long-term forecast of heatwave incidents in China based on numerical weather prediction. *Theoretical and Applied Climatology., 155*, 599–619. https://doi.org/10.1007/s00704-023-04654-3

Xie, J. H., Yu, J. H., Chen, H. S., & Hsu, P. C. (2020). Sources of subseasonal prediction skill for heatwaves over the Yangtze River basin revealed from three S2S models. *Advances in Atmospheric Sciences, 37*(12), 1435–1450. https://doi.org/10.1007/s00376-020-0144-1

Xin, X. G., Gao, F., Wei, M., Wu, T. W., Fang, Y. J., & Zhang, J. (2018). Decadal prediction skill of BCC-CSM1.1 climate model in East Asia. *International Journal of Climatology*, 584–592. https://doi.org/10.1002/joc.5195

Yi, C. Y., & Yang, H. J. (2020). Heat exposure information at screen level for an impact-based forecasting and warning service for heat-wave disasters. *Atmosphere, 11*(9). https://doi.org/10.3390/atmos11090920

Zaidi, R. Z., & Fordham, M. (2021). The missing half of the Sendai framework: Gender and women in the implementation of global disaster risk reduction policy. *Progress in Disaster Science, 10*, 100170. https://doi.org/10.1016/j.pdisas.2021.100170

Zia, A., & Wagner, C. H. (2015). Mainstreaming early warning systems in development and planning processes: Multilevel implementation of Sendai framework in Indus and Sahel. *International Journal of Disaster Risk Science, 6*(2), 189–199. https://doi.org/10.1007/s13753-015-0048-3

Zillmann, D. (1999). Exemplification theory: Judging the whole by some of its parts. *Media Psychology, 1*(1), 69–94.

Zottarelli, L. K., Blake, S. A., & Garza, M. T. (2022). Communicating heat-health information to the public: Assessing municipal government extreme heat event website content. *Weather Climate and Society, 14*(1), 311–321. https://doi.org/10.1175/wcas-d-21-0019.1

Chapter 14
Heatwave Responses: Mitigation

Abstract This chapter begins by making a case for a distinction between mitigation (modifying the physical hazard) and adaptation (behaviour change) in relation to heatwave risk reduction in urban environments. Specifically, it asserts that modifying the physical nature of heat, as the fundamental property of heatwaves, via urban surface radiative and energy balance management, sets mitigation apart from adaptation. After briefly exploring the historical context of urban heat mitigation the Chapter moves on to consider modern strategies, categorizing urban heat mitigation into three broad approaches: modifications to urban surfaces, nature-based solutions (NBS), and changes to urban morphology and built form. Cool pavements and cool roofs are discussed in the context of the modification of urban surfaces. Various types of cool pavements are identified. Cool roofs and their impacts on the reduction of heat transfer into buildings and lowering indoor temperatures and energy consumption as well as their efficacy for curtailing air temperature extremes are also examined. The characteristics of a range of NBS such as green roofs, vertical greenery systems, and urban greenspaces and blue spaces, are described with their potential to mitigate urban heat through shading, evapotranspiration and evaporation considered. Their effectiveness in different urban settings and climates is reviewed. The chapter also discusses how alterations to urban morphology and built form might achieve heatwave mitigation. Although the importance of integrating multiple mitigation options to achieve heatwave mitigation, such as combining cool roofs with increased vegetation, is emphasised, the prospect that the level of heatwave mitigation required to achieve meaningful reductions in temperature extremes during heatwaves may not be obtained, even with widespread application of the most effective mitigation options, is raised. The chapter concludes by stressing the need for interdisciplinary research to understand the long-term impacts and scalability of the mitigation strategies discussed. It highlights the potential societal resistance and financial challenges associated with large-scale implementation and calls for collaboration among stakeholders to optimize co-benefits such as improved air quality and energy efficiency. The chapter underscores that while urban heat mitigation is crucial, it must be complemented by global efforts to reduce greenhouse gas emissions to effectively manage global heat in the face of climate change.

G. McGregor, *Heatwaves*, Biometeorology 6,
https://doi.org/10.1007/978-3-031-69906-1_14

Keywords Urban heat mitigation · Mitigation as distinct from adaptation · Urban surface radiative and energy balance management · Modification of urban surfaces · Nature based solutions · Alterations to urban morphology and built form · Combined effects of heatwave mitigation options · Green and blue infrastructure

14.1 Introduction

Mitigation generally means putting in place an action to reduce how harmful, unpleasant, or bad something is or might be. Mitigation can be both ex-ante and ex-post. The concept of mitigation with associated nuances is applied in a wide range of fields including for example sociolinguistics, law, education and insurance. Also used widely in the field of disaster risk reduction, mitigation refers to lessening or minimizing the adverse impacts of hazardous events, in particular natural hazards such as heatwaves. In the context of climate however, the default meaning has generally come to refer to the reduction of the severity of climate change and associated impacts through the prevention or reduction of human related greenhouse gas (GHG) emissions and/or enhancing the capacity of carbon sinks. Primary approaches to achieving this goal include traditional mitigation, negative emissions, and geoengineering. Traditional mitigation technologies prioritize the reduction of carbon dioxide emissions from fossil fuels, while negative emissions technologies strive to capture and store atmospheric carbon, thereby decreasing overall carbon dioxide levels. Geoengineering methods involve actively managing radiative forcing via a variety of means to modify the Earth's energy balance and thus achieve stabilization or a decrease of global temperatures (Fawzy et al., 2020).

Of the three climate change mitigation approaches, it is the broad principles associated with geoengineering, that is modifying the energy balance, which is relevant in the case of heat mitigation in urban environments as focused on here. As a move to clearly distinguish heat and heatwave mitigation actions from those categorised as adaptation, I suggest measures that modify the urban surface radiative and energy balances as well as alter the exchange of mass and momentum between the urban surface and the overlying atmosphere, with the intention of heat reduction, are classed as urban heat mitigation strategies (Fig. 14.1). Set against this assertion, I also contend that some of the adaptation options classed as structural/physical, such as resurfacing roads with cool pavements that alter surface albedo, as well as nature-based solutions, as manifest by green and blue infrastructure, are in fact mitigation as opposed to adaptation options. Such ideas will be expanded on later in this chapter, but first a little historical context.

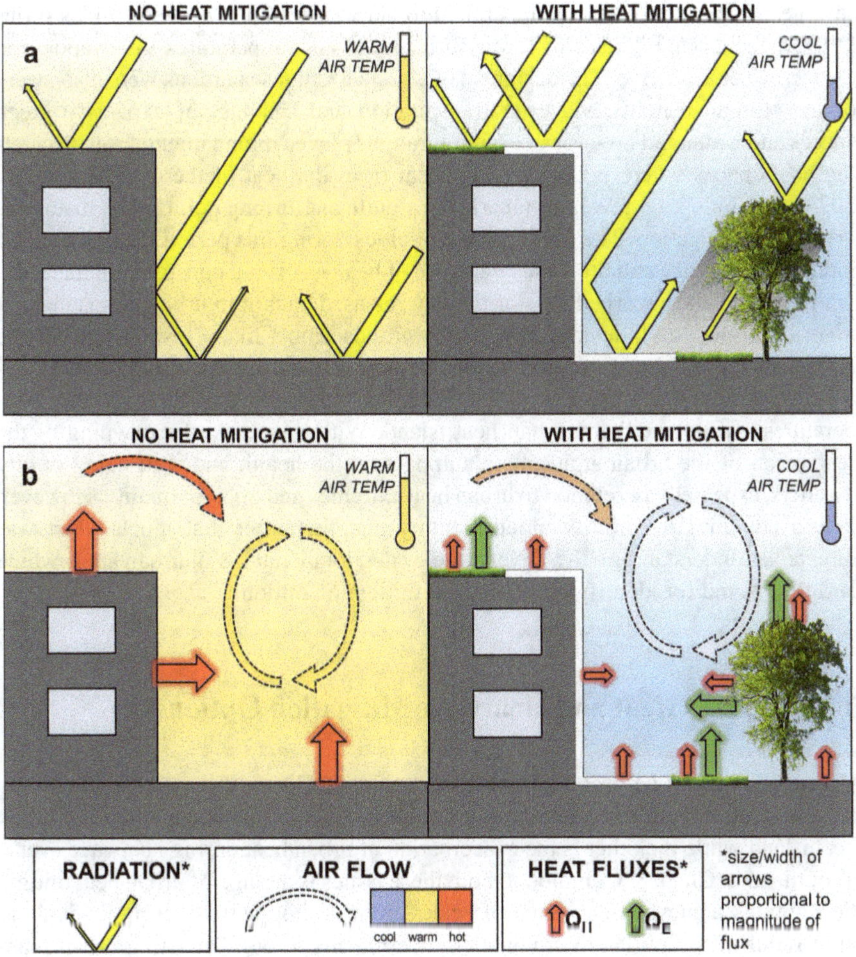

Fig. 14.1 Schematic representation of the way in which the alteration of (**a**) radiative and (**b**) turbulent heat fluxes can promote urban heat mitigation via a combination of cool surfaces and green infrastructure. (Source: Krayenhoff et al. (2021). Used under Creative Commons Attribution 4.0 license. https://creativecommons.org/licenses/by/4.0/)

14.2 Urban Heat Mitigation: The Historical Context

The idea of urban heat mitigation is not new. Although not expressed as such, the roots of urban heat mitigation can be traced back several centuries. For example, the practice of painting houses white in ancient Greece dates to at least the fifth century BCE. During this period the use of white lime wash became popular for coating buildings, not only for its aesthetic appeal but for the practical benefit of increased reflection of sunlight and thus keeping building interiors cooler during the heat of Mediterranean summers. In Japan the tradition of 'uchimizu' stretches back to the seventeenth century. Literally meaning throwing water, this practice involves sprinkling water around houses to lower surface and air temperatures via evaporation (Solcerova et al., 2018). The designers of classical Chinese gardens were also aware of the climate benefits, via evapotranspiration and shading, of expansive green spaces and water and how various configurations played a fundamental role in creating cool spaces where people could retreat from the heat of the day (Huo et al., 2023). The industrial revolution marked a significant turning point in the history of urban heat mitigation. The rapid growth of cities during this period led to increased industrial activities and the widespread use of heat absorbing construction materials with much of the absorbed heat of the day released back into urban atmosphere at night. Many of the industrial processes were also most likely responsible for the release of large amounts of anthropogenic heat. As outlined in Chap. 8, this was around the time that Luke Howard, through his comparisons of rural and city temperatures, 'discovered' the urban heat island. With this came the dawning of the realisation of the urban atmosphere's impact on the health and well-being of city dwellers, especially in relation to urban heat extremes and poor air quality. However, it was not until the mid-twentieth century and thereafter that much effort was directed at understanding the drivers of elevated temperatures in urban areas which laid the ground for identifying urban heat mitigation options.

14.3 Urban Heat and Heatwave Mitigation Options

Urban areas face two broad heat-related problems. One is the increase in chronic heat associated with the interaction between global climate change and the urban heat island while the other is the exacerbation of urban heat during heatwave events as outlined in Chap. 8. Common to both these issues is a range of urban heat mitigation options achieved at a range of scales from ad hoc actions by individuals to purposeful interventions by municipal governments through enacting urban planning and urban design strategies (Aflaki et al., 2017; Akbari & Kolokotsa, 2016; Assaf & Assaad, 2023; Black-Ingersoll et al., 2022; Feng et al., 2023; Han et al., 2023; He et al., 2023; Irfeey et al., 2023; Jamei et al., 2020; Probst et al., 2022; Qi et al., 2020; Rizwan et al., 2008; Wong et al., 2020). Within the urban surface energy balance and the mass and momentum framing introduced above, three broad approaches are visible with regards to urban heat mitigation namely modifications to the facets or elements of urban surfaces, the application of nature-based solutions (NBS) and changes to urban morphology and built form. Of course, while these three analytical categories are presented here to structure thinking related to urban heat and heatwave mitigation it is acknowledged that they are not entirely mutually exclusive with specific options within and across the three categories working as an ensemble to achieve heat mitigation. Because the focus of this section is principally on the outside thermal environment, building design as an indoor heat mitigation option will not be considered.

14.3.1 *Modifications to Urban Surface Facets and Elements*

This category of urban heat mitigation actions generally refers to modifications to the various facets (e.g. building wall, street surface, roof) of the elements (e.g. buildings and roads) that make up the urban surface (Fig. 8.1) the result of which is an alteration of the nature of the radiative and convective and conductive heat exchanges between the facet and the surrounding atmosphere. Generally, such mitigation actions manifest in the form of cool pavements and cool roofs. Although vertical greenery systems could be considered as a modification of walls/vertical surfaces, this mitigation option will be treated under nature-based solutions. The application of 'cool technologies' to hard vertical surfaces, akin to those used for pavements and roofs as described here will not be considered.

14.3.1.1 Cool Pavements

Pavements make up an important part of the urban surface typically comprising 30–40% of the total area in many cities, albeit this percentage can vary depending on factors such as city design, infrastructure development, and land use patterns. Area wise, pavements represent an extensive climatically active surface in cities with a range of factors influencing pavement surface temperatures (Qin et al., 2022). Cool pavements are designed to mitigate urban heat, enhance urban microclimates, and contribute to overall environmental sustainability. A pavement is deemed "cool" when it exhibits higher solar reflectance (albedo- $K\uparrow$ term in the surface radiation budget), facilitates enhanced evaporation (Qe term in the surface energy budget), and/or reduces the release of sensible heat to the urban atmosphere (supresses Qh and enhances Qg in the surface energy budget) compared to a conventional pavement. Variations of cool pavements include reflective pavements, evaporative pavements, and heat-harnessing pavements (Table 14.1). Reflective pavements, characterized by their elevated albedo compared to conventional pavements, contribute to lowering surface temperatures and reducing sensible heat release.

Table 14.1 Categorisation of cool pavements and associated pavement types[a]

Cool pavement category	Type
Reflective Pavements	Reflective asphalt Reflective concrete Thermochromic pavement
Evaporative Pavements	Porous pavers Permeable pavers Pervious pavers Water retaining pavements
Modification of Heat storage Pavements	Phase-change material (PCM) impregnated pavements High conductive pavements Heat harvesting pavements

[a]Based on Qin (2015)

Evaporative pavements promote cooling by retaining water within the pavement surface layers or lower layers, facilitating evaporative processes that help maintain lower temperatures. Heat-harnessing pavements employ innovative techniques to reduce surface temperatures by extracting absorbed heat and converting it into renewable energy for alternative uses (Gong et al., 2023; Kousis & Pisello, 2023; Manteghi & Mostofa, 2020; Santamouris, 2013; Seifeddine et al., 2023; Qin, 2015). In addition to these longer-term cool pavement options is short term watering of pavement surfaces during heatwave periods akin to 'uchimizu' practiced in Japan (Takebayashi et al., 2023).

Cool pavements have received much attention in relation to urban heat mitigation and are often cited in municipal heat risk reduction strategy documents as a key option. Yet little consideration has been devoted to assessing their urban heat mitigation efficacy. In the case of reflective pavements there is no doubt decreasing the absorptivity of pavements through enhancing their albedo achieves a reduction in surface temperatures (Fig. 14.2). Typically, pavement surface temperatures can reach 50–70 °C during the mid-afternoon hours on a summer day. Depending on the time of the day, the sky view factor, weather conditions and the climate setting (Gapski et al., 2023), cool pavements can reduce surface temperatures by 4–25 °C and sensible and latent heat fluxes from the surface to the atmosphere by 100—450 Wm^{-2} (Kousis & Pisello, 2023; Parison et al., 2020; Qin, 2015; Takebayashi & Moriyama, 2012). Impressive as these reductions may be, the

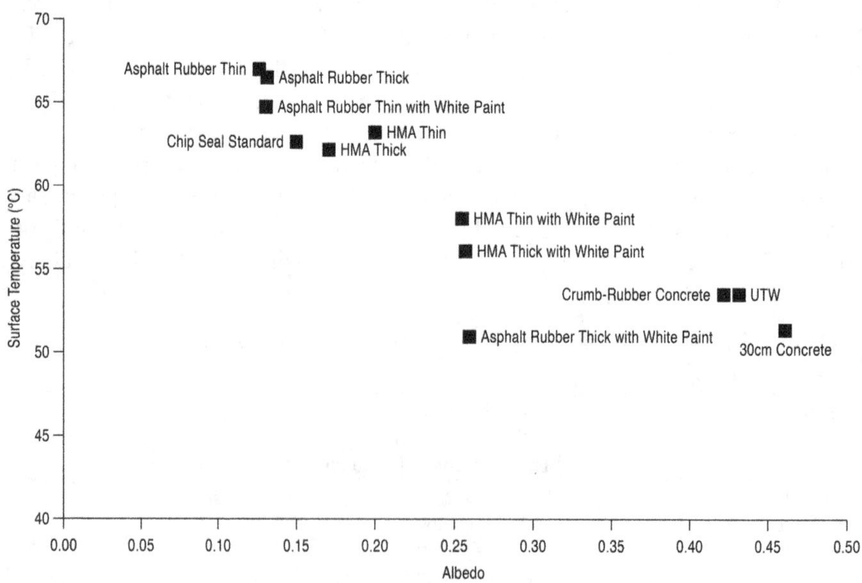

Fig. 14.2 Association between surface albedo for a range of pavement types and surface temperature for Phoenix, Arizona. HMA is hot mixed asphalt, UTW is ultra-thin white-topping. (Redrawn from Dawson et al. (2014) and based on original data from Jay S Golden and Kamil Kaloush, Smart Programme, Arizona State University)

critical question in terms of atmospheric heat mitigation, especially in the context of heatwave events, is do cool pavements have a fundamental impact on air temperature reduction in the urban canopy layer where most human activities take place.

Although laboratory-based investigations of cool pavements mainly focus on changes in surface temperature, some observational/empirical and numerical modelling studies have investigated the impacts of cool pavements on ambient air temperature. Generally, these indicate only very modest impacts of cool pavements on atmospheric heat reduction. Further for tropical environments it would appear other strategies are more effective than cool pavements for urban heat mitigation (Chow, 2024; Jamei et al., 2020). Using a modelling approach, Santamouris et al. (2012) observed for a typical summer day, that cool paving materials led to a reduction in the surface temperature by approximately 12 °C and the peak ambient air temperature by up to 1.9 °C. For the Los Angeles basin, Rosenfeld et al. (1998) found increasing the albedo of pavements from 0.05 to 0.30 could potentially result in a 0.6 °C decrease in local air temperature while Pomerantz (2000) using a simple model showed that changes in pavement area and albedo could contribute to a similar drop in air temperature. For Montreal and Toronto, modelling has demonstrated that cool pavements can contribute to lowering air temperature and dew point temperature by 1–2 °C and 0.2–0.5 °C respectively while slightly boosting near-surface wind speed by 0.01 to 0.4 m/sec. An upshot of this is the transition of approximately 60% of non-benign heat days into benign days with interestingly a possible associated decrease in heat related mortality by 3–7% (Jandaghian & Akbari, 2021). By substituting asphalt pavements with concrete pavements Vásquez-Alvarez et al. (2022) found for the city of Cuenca in Ecuador that the land surface temperature could be lowered by 8 °C resulting in a mean air temperature reduction of up to 0.83 °C. A potentially similar reduction in air temperature has also been found in a model experiment for parking lots in Champaign-Urbana, US where increasing surface albedo from 0.21 typical of parking lots (0.18–0.28) to 0.5 decreased the air temperature over the parking lot by around 1 °C (Sen et al., 2020). For the city of Phoenix in the US cool pavement types have also been found to have a limited cooling effect on air temperature with canyon geometry being a critical determinant of the response of surface temperatures to an alteration of surface properties (Yang et al., 2016). This resonates with the findings of Gapski et al. (2023) for the Brazilian city, Balneário Camboriú and Salvati et al. (2022) for London, UK.

Compared to numerical modelling studies, field-based evaluations of the impact of surface albedo on air temperature are few. Based on a field campaign specifically designed to establish the effect of reflective pavements on a range of critical biometeorological variables such as mean radiant temperature and air temperature, Middel et al. (2020) found the surface temperature of coated asphalt concrete to be 4–6 °C lower than that of regular asphalt concrete. However, coated surfaces exhibited an average increase in reflected shortwave radiation of 118 Wm^{-2}, reaching up to 168 Wm^{-2} or more at noon with the mean radiant temperature over the reflective pavement 4 °C higher during midday and only a 0.5 °C reduction in air temperature. In a field-based analysis of the impact of cool pavement deployment on albedo and temperatures in Los Angeles, Ko et al. (2022) also note possible opposing effects of air temperature reduction and increased solar radiation reflectance with associated biometeorological implications. These results illustrate one of the possible disbenefits of reflective pavements as a cool paving type. Despite achieving lower surface temperatures, cool pavements might only obtain an air temperature reduction which is miniscule and importantly in human heat load terms, the mean radiant temperature increases. This finding corroborates those of several biometeorological modelling studies of the impact of reflective pavements on human thermal comfort with these finding that the air temperature reduction is not adequate to offset heightened radiative heat loads on pedestrians and the like, potentially compromising thermal comfort (Aboelata, 2021; Djekic et al., 2018; Erell et al., 2014; Falasca et al., 2019; Faragallah & Ragheb, 2022; Nasrollahi et al., 2020; Taleghani & Berardi, 2018). Along with thermal discomfort, reflective pavements may also increase glare (Schneider et al., 2023). Similar conclusions have been drawn for the case of increasing the albedo of urban canyon walls (Lee & Mayer, 2018). Of course, countering some of the disbenefits of reflective pavements are a range of benefits allied with other cool pavement types. These along with the costs and barriers associated with cool pavement strategies have been outlined by several researchers who also call for more investigations of the effectiveness of cool pavements as an urban heat mitigation option given this largely remains an unsettled issue (Gong et al., 2023; Kappou et al., 2022; Qin et al., 2022; Schneider et al., 2023; Seifeddine et al., 2023; Wang et al., 2021).

14.3.1.2 Cool Roofs

Cool roofs are defined here in a similar way to cool pavements in that a cool roof is a roofing system designed to reflect more solar radiation and absorb less radiative heat than standard roofs. Cool roofs typically consist of materials with high solar reflectance and thermal emittance properties, The primary goal of a cool roof is to minimize heat transfer into the building below, thereby reducing the need for air conditioning and lowering energy consumption for cooling purposes while an associated goal is to reduce air temperature in the lower parts of the boundary layer and the entire urban canopy later. Cool roofs are distinguished here from green roofs. Although green roofs could be considered a type of cool roof, these function in a different manner by using vegetation to achieve cooling through a combination of evapotranspiration, insulation, and shading as opposed to the application of reflective roof surfaces to reduce heat absorption and minimize heat transfer as in the case of cool roofs.

Like the literature on cool pavements, that for cool roofs has mushroomed over the last couple of decades as many researchers and policy makers have turned their attention to the issue of urban heat and heatwave mitigation (Broadbent et al., 2022). The push for cool roofs as a heat mitigation option has been spurred on by observations that roof temperatures can reach 70–80 °C (Hernández-Pérez, 2021; Vardoulakis et al., 2014) which bears implications for the vertical transfer of heat to the atmosphere and internal building spaces. The extent to which cool roofs might effect changes to vertical heat transfer and thus influence temperatures around and within buildings has been established mainly through numerical modelling with most studies focusing on the building or city scale with few integrating both scales, nor describing the local-scale intra-urban air temperature variability. Further, most studies examine the seasonally averaged impacts of cool roofs and rarely consider what happens during heatwave events (Broadbent et al., 2022; Elnabawi et al., 2022). Notwithstanding the importance of roof form (e.g., flat versus gable), that roofs possessing different levels of reflectance related to the roofing material (e.g., clay tile, asphalt), have a fundamental impact on modelled roof surface temperature is perhaps no surprise, given that albedo is a fundamental determinant of the amount of solar radiation absorbed by a surface (Fig. 14.3). However, of interest in the context of urban heat mitigation is the extent to which roof albedo related reductions in roof surface temperature translate into decreases of atmospheric heat.

For a non-airconditioned building in India Arumugam et al. (2014) have shown that increasing the roof surface reflectance from 0.28 to 0.57 using white paint, the seasonal average indoor air temperature and heat flux decreased by 1.07 °C, and

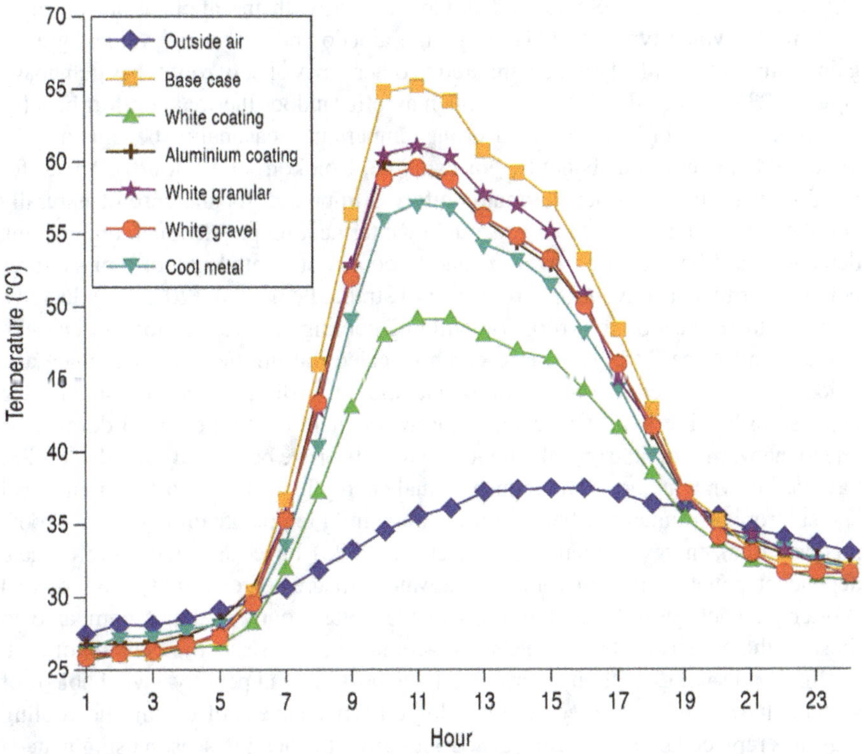

Fig. 14.3 Diurnal variation in air temperature and modelled roof surface temperature for a range of roof surface types. Base case is for conventional dark roof. (Adapted from the New York City Department of Design & Construction Cool & Green Roofing Manual. Available at: https://www.nyc.gov/html/ddc/downloads/pdf/cool_green_roof_man.pdf. Accessed 15/02/2024)

14.4 W m^{-2} respectively with peak reductions of 1.38 °C and 18.3 Wm^{-2} achieved. When compared with the situation for the low albedo case, these reductions translated to an increase of adaptive comfort hours by 8%. Similarly for a garment factory in Dhaka, Bangladesh, cool roof retrofitting achieved indoor air temperatures reductions of 2 °C resulting in the lessening of the number of working days when comfort levels are exceeded by 75 per year (Bach et al., 2023). This parallels the decrease in indoor temperatures accomplished following the application of white paint to the 700 m^2 roof of a non-residential building in Sicily with the reduction of roof surface temperatures by 20 °C (Romeo & Zinzi, 2013) and similarly, for school buildings in Greece where simulation results show indoor air temperature decreases reaching as high as 2.8 °C and an associated decline in the annual cooling load by 25–40% (Stavrakakis et al., 2016; Synnefa et al., 2012). For Biskra southern Algeria, field measurements combined with modelling have revealed that employing a double-skin roof along with cool-reflective paint emerges as the optimal roofing strategy for indoor cooling and energy efficiency. Contrasted with a standard flat roof, significant enhancements were observed, notably a 45% reduction in thermal discomfort hours and a decrease in cooling loads from 1121.91 kWh to 741.09 kWh (Athmani et al., 2023). Some studies have shown with the application of 'super cool' roofs, which typically have very high albedo and emissivity values, greater gains in thermal comfort result compared to other types of cool roof types (Elnabawi et al., 2023). That said, while cool roofs may offer indoor thermal comfort benefits in 'hot' climates, for locations with strong temperature seasonality, benefits may be offset by cool season disbenefits. For example, Kolokotroni et al. (2013) note for London that while summer thermal comfort is improved for the case of naturally ventilated buildings, cool roofs generate a winter heating penalty such that heating demand could be increased by 10% due to cool roof related indoor temperatures being lower than a conventional roof. For Australia, Feng et al. (2022) conducted a study on the effects of cool roofs on building cooling and heating load in Darwin, Sydney, and Alice Springs. The research revealed that in the case of Sydney and Alice Springs, the reduction in cooling load did not sufficiently compensate for the increase in heating load. However, in Darwin, where cooling demand dominates, due to the warm humid tropical climate, cool roofs prove beneficial. Bamdad (2023) has also shown that cool roofs reduce annual energy loads by up to 14% in tropical and subtropical climates, whereas for cool and mild temperate climates, cool roofs increase total energy demand. Chang et al. (2023) have also noted the climate dependent benefits of cool roofs in Taiwan with greater electricity savings and shorter payback periods for buildings in the hotter southern zones compared to those in the cooler northern zones of the country. To preserve the benefits of high cooling load savings without accruing a large heating load penalty, switchable roof systems have been suggested whereby high a reflectance roof during the cooling season is replaced by a low reflectance roof during the heating season using materials that are responsive to changing seasonal conditions (Testa & Krarti, 2017).

In addition to accomplishing gains in indoor thermal comfort and energy efficiency, the extent to which passive cooling via the application of cool roofs can mitigate external atmospheric heat has received mounting attention. Reviews of the effectiveness of cool roofs generally indicate air temperature reductions over the range of 0.1–2.0 °C (Broadbent et al., 2020; Herath et al., 2021; Krayenhoff et al., 2021). While this is of interest from a climatological perspective, more pertinent is the situation for heatwaves given that these represent exceptional conditions for which heat mitigation is essential. For a series of simulated heatwaves in Chicago, Tan et al. (2023) found that the near-surface temperature decreases as the urban fraction covered in cool roofs increases with maximum reduction in peak heat island intensity obtained for areas typified by large low-rise buildings (local climate zone 8—LCZ 8 in Fig. 8.2), a finding that matches that for the large urban agglomeration of the Pearl River Delta in China (Wang et al., 2023b). For the Chinese megacity of Guangzhou, simulations of temperature reductions for the six strongest historical heatwave events in the mid-2000s reveal average urban midday temperature reductions of 1.2 °C for the case of total cool roof coverage of the city (Cao et al., 2015). These parallel those of 1.3 °C found for Al Ain City in Abu Dhabi (Elnabawi et al., 2023). In a different take on what constitutes the conventional cool roof, Shen et al. (2022) present findings for building roofs in a semi-arid city in China covered with photovoltaic panels. For heatwave conditions, temperature reductions of 0.4–0.7 °C are potentially achievable. Building on heatwave simulation studies for the case of super cool roofs with albedo values of 0.95 to 0.96 in the Indian city of Kolkata, which have revealed potential reductions in ambient temperatures and sensible heat flux by up to 5.3 °C and 300 W m^{-2} respectively (Feng et al., 2021; Khan et al., 2022; Khan et al., 2023), Khorat et al. (2024) considered for the same city the impact of the widespread application of cool roofs on an extended range of urban heat related parameters. For this study, existing roofing materials, with a reflectivity of 0.15 and emissivity of 0.85, represented the unmitigated condition with the mitigated case comprising third-generation cool roof materials with a reflectivity of 0.80 and emissivity of 0.85. Notably, during heatwave episodes, substantial reductions in energy flux were found, with respective decreases in sensible heat, latent heat, ground storage, and net inflow radiation of 181.3 Wm^{-2}, 16.6 Wm^{-2}, 56.3 Wm^{-2}, and 251.9 Wm^{-2}. Associated 1400 hr. reductions in ambient temperature, surface temperature, roof surface temperature, and urban canopy temperature were 2.3 °C, 6.1 °C, 21.8 °C and 1.9 °C respectively. Accompanying these alterations were maximum drops in the planetary boundary layer of 1978.5 m at 1400 hrs with an average drop across 24 hours of 870.3 m. The researchers also noted discernible changes in outdoor thermal comfort as measured by the HSI and UTCI (Khorat et al., 2024). In a simulation of the potential impact of cool roofs on the thermal and wind fields for the significant 2018 heatwave in Seoul, South Korea, Baik et al. (2022) show daytime maximum 2 m temperature and 10 m wind speed reductions of 1.0 °C and 0.5 ms^{-1} respectively. Interestingly an implication of these reductions is the convergence of near-surface wind weakens, indicative of the ebbing of the daytime urban breeze, along with a diminished strength of the sea breeze arising from a decrease in the land-sea air temperature difference. On occasions this set of circumstances leads to higher late afternoon near surface temperatures compared to the conventional roof case (Baik et al., 2022). Further evidence for the impact of cool roofs on urban heat reduction during heatwaves comes from a study of the effects of extensive cool roof deployment in the US cities of Atlanta, Detroit, and Phoenix. For the case of extreme five-day heatwave events, maximum city-averaged daytime air temperature cooling of 0.38 °C, 0.42 °C and 0.66 °C is achieved for Atlanta, Detroit and Phoenix respectively (Broadbent et al., 2020). Utilizing a novel metric termed the Albedo Cooling Effectiveness, the researchers highlight that, on average, cool roofs in Phoenix exhibit an 11% greater efficiency in reducing pedestrian-level air temperature compared to those in Atlanta and are 30% more effective than those in Detroit with contrasts in effectiveness related to urban form (Broadbent et al., 2020) a finding similar to that for a heatwave study conducted for Sydney Australia (Jacobs et al., 2018). Beyond considering the urban climate impacts of cool roofs, some studies have investigated the health impacts associated with the application of cool roofs (Bunker et al., 2024). In a noteworthy study of heatwave events during 2003 and 2006 for West Midlands region of the UK, Macintyre and Heaviside (2019) show that cool roofs have the potential to lower city-centre daytime 2 m air temperatures by an average of 0.5 degrees Celsius, with a maximum reduction of up to 3 °C. On average, cool roofs were found to decrease the UHI intensity by approximately 23% and mitigate heat-related mortality associated with the UHI by 25% during heatwave events. Significantly, these workers emphasize that their modelling indicates that cool roof retrofitting of half of all industrial/commercial urban buildings could result in temperature reduction benefits comparable to retrofitting all high-intensity residential buildings in the study region. This finding resonates with other studies that have found that areas with large low-rise buildings possessing cool roofs often achieve maximum air temperature reductions during heatwave events.

While this short review may foster optimism that cool roofs offer some sort of panacea for urban heat mitigation, because reductions in internal and external air temperature have been determined from a range of numerical modelling studies, it is worthwhile noting that many studies have considered reductions in heat from a widespread/city-wide implementation of cool roofs. Notwithstanding potential political barriers, this may prove infeasible not only because of the sheer magnitude of what would be needed for large scale retrofitting to be achieved, but because of the financial resources required if municipal governments are expected to subsidise or totally cover retrofitting costs. The latter challenge will be especially acute in low- and middle-income countries many of which could benefit from such a heat mitigation intervention. Added to this is the building type, roof configuration and climate dependency of the gains afforded by cool roofs especially in relation to indoor thermal comfort and energy savings and the associated trade-off in some climate settings of warm season benefits versus cool season disbenefits (Baniassadi et al., 2018; Tan et al., 2023). Declines in performance attributed to cool roof aging and cleanliness will also bear consideration (Algarni, 2019; Elnabawi et al., 2022; Gaffin et al., 2012; Saber, 2021) as well as colour options for reflective paint (Al-Obaidi et al., 2014). There is also the prospect that highly effective cooling within urban canyons or at the roof level associated with cool roofs may affect the horizontal and vertical exchange of air within urban areas (Baik et al., 2022; Sinsel et al., 2021) and that highly reflective cool roofs may create a glare hazard (Ashtari et al., 2021). In short, much remains to be learned about cool roofs as a heat mitigation option including whether some of the estimated reductions in urban heat at the intra-urban to urban scale are meaningful in terms of the level of heat mitigation required to achieve significant amelioration of heatwave related impacts.

14.3.2 Nature Based Solutions

In the context of urban heat and heatwave mitigation, nature-based solutions (NBS)—variously referred to in the broader context of ecosystem services for climate change adaptation and resilience and ecosystem-based adaptation—can be broadly defined as the utilisation of natural elements and processes within urban areas to reduce heat and cool the environment. NBS approaches to heat mitigation generally fall under the two broad categories of green and blue infrastructure and include strategies such as increasing the amount of greenery which can take on a variety of forms, creating water features and preserving or restoring natural habitats within cities. Within the four broad frames connecting nature and climate proposed by Tozer et al. (2022)—nature for resilience, the integrated benefits of nature, nature first and nature for mitigation—it is the last of these into which heat mitigation options with an overwhelming NBS flavour fit. Although some may disagree, the various actions implemented under the broad heading of NBS in relation to managing heat are considered here to be mitigation, as opposed to adaptation, options. This is because the action (response) of 'installing' or enhancing green and blue infrastructure constitutes an attempt to alter the exchange of heat, mass and momentum between the urban surface and the overlying atmosphere and thus modify not only the nature of background urban heat but heatwaves as a physical event. In brief this is the reduction of heatwave related risk by surface radiative and energy balance management. As a mitigation option NBS in general are considered 'low regret' measures as they potentially offer sustainable and cost-effective ways to mitigate urban heat while also providing additional paybacks via enhancing biodiversity, improving air quality, promoting overall urban resilience to extreme climate events in addition to heatwaves, as well as generating a range of physical and mental health co-benefits (Andersson et al., 2019). Although there is growing understanding of the advantages offered by green and blue infrastructure as evidenced by the rapidly burgeoning literature on NBS and heat mitigation (Aleksandrowicz et al., 2017; Fritz, 2017; Hayes et al., 2022; Sahani et al., 2023; Ramakreshnan & Aghamohammadi, 2024), some studies suggest that nature-based approaches to mitigating the impacts of climate related extreme events in general continue to be inadequately acknowledged and funded in urban planning and development (Matthews et al., 2015; Frantzeskaki, 2019; Frantzeskaki et al., 2019) or at worst are controversial for a range of reasons (Seddon, 2022). Interestingly, Turek-Hankins et al. (2021) found across all regions of the globe, ecosystem-based responses (NBS) account for fewer than 15% of actions taken to mitigate heatwave effects.

Notwithstanding the possible gap between aspiration and reality in terms of the widespread implementation of NBS, the purpose of this section is to provide an overview of various types of green and blue infrastructure approaches to urban heat and heatwave mitigation organised under the headings of vertical greenery systems, green roofs, greenspaces, and blue spaces. Although framed within the idea of water sensitive design Fig. 14.4 provides a useful summary of many of the features touched upon in this chapter in relation to the types of heat mitigation options that fall under the chapeau of NBS.

To gauge the degree to which various NBS moderate urban heat, observation and simulation approaches have been applied. Observational research encompasses a variety of methodologies, such as field measurements, experimental studies, and remote sensing. Traditionally, studies investigating the impact of NBS on urban microclimates have relied heavily on field measurements and remote sensing. More recently however, numerical modelling at the micro- to meso-scale has emerged as an additional tool with its capacity to isolate various influencing factors related to NBS (e.g. size and shape of greenspaces, tree canopy densities, vegetation type) and

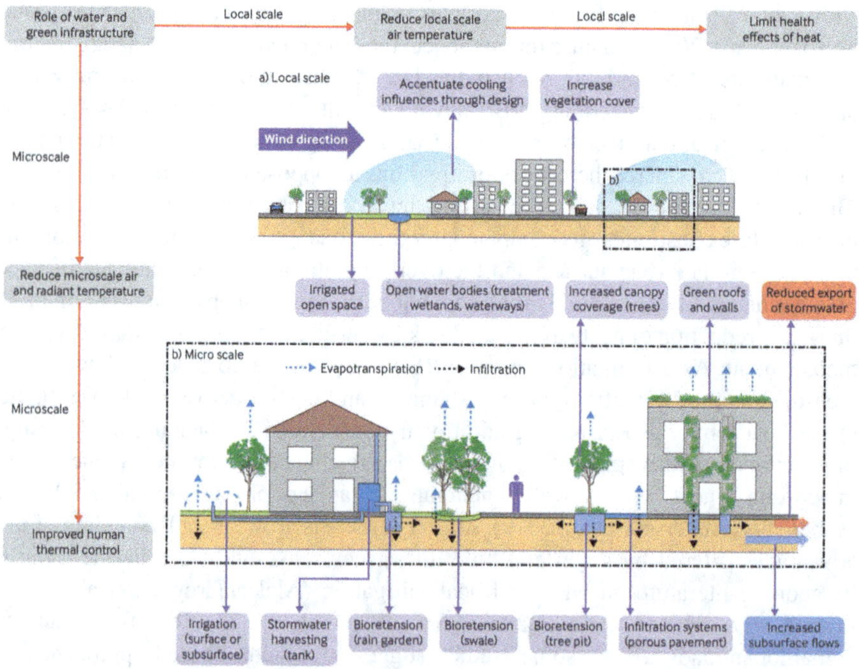

Fig. 14.4 Various green and blue infrastructure elements comprising NBS-based heat mitigation options residing within a water sensitive urban design framework. While some of these are discussed at length in this chapter (e.g. green roofs and walls (VGS) and open water bodies) others are not (e.g., bioretention systems) and represent opportunities for further investigation. (Source: Wong et al. (2020). Used under the Creative Commons CC-BY-NC 4.0 license https://creativecommons.org/licenses/by-nc/4.0/)

evaluate their significance, making this approach valuable in the pursuit of understanding NBS—urban heat and heatwave interactions.

14.3.2.1 Vertical Greenery Systems

Often referred to using a range of terms (Safikhani et al., 2014), vertical greenery systems (VGS) can be categorised into two broad types based on their growing type, namely green façades (soil bound plants) and living wall systems (wall-bound plants) (Köhler, 2008). Although the primary focus of this section is the role of vertical greenery systems in thermal regulation, it is worth mentioning the co-benefits associated with this heat mitigation option. These include air quality improvement (filtering of pollutants), biodiversity support (habitat and food source), noise reduction (absorb and deflect sound) and aesthetic enhancement (Chique (Chiquet et al., 2013; Fonseca et al., 2023; Goel et al., 2022; Roehr & Laurenz, 2008; Shushunova et al., 2022; Treder et al., 2024).

From a heat mitigation point of view, VGS are important because they modify the energy balance through altering the magnitude of the solar and longwave radiative and sensible and latent heat fluxes compared to a non-vegetated vertical surface (Afshari, 2017; Convertino et al., 2020, 2021; Kenai et al., 2020; Lee & Jim, 2019) all of which have a fundamental impact on the interior and exterior thermal environments of buildings through enhanced wall shade and evapotranspiration, and insulation and ventilation. Because of their potential role in the mitigation of background urban heat and the extreme heat associated with heatwaves there has been a rapid growth in research on VGS with studies falling into the three non-mutually exclusive categories of thermal behaviour and energy modelling, urban design and large-scale effects and sustainable buildings management (Blanco & Convertino, 2023; Chafer et al., 2021).

Complementing efforts directed at understanding the energy saving and indoor thermal comfort benefits of VGS is work establishing the extent of external heat reduction. Reviews indicate dependency of the degree of heat reduction on building height, wall area covered by the VGS, vegetation type, leaf area and the type of system installed (Blanco & Convertino, 2023; Koch et al., 2020). One review has found mitigation of the UHI related to VGS of up to 5 °C across a wide range of climate types with greatest reductions of up to 8 °C for highly urbanized areas with narrow and deep urban canyons (Susca et al., 2022). These findings should be considered with some caution, however. This is because it is not clear how the stated UHI reductions were measured (e.g., using air or wall surface temperatures) especially as some empirical studies have found VGS related reductions in ambient air temperature to be nowhere near what might be needed to reduce the UHI by 5–8 °C assuming the reduction is driven by changes in temperature at a reference urban climate site as opposed to a rural one. Added to this, VGS maximum air temperature reductions are usually obtained during the day not at night when the UHI finds it maximum expression. Several small-scale typically short-term experiments that investigate outdoor air temperature reduction due to VGS demonstrate small cooling effects. Using a reduced-scale mock-up of a street canyon, Djedjig et al. (2017) demonstrated air temperature reductions of up to 1.5 °C recorded approximately 0.6 m away from a single 6.5 m^2 green wall, while Widiastuti et al. (2018) by conducting summertime prototype experiments involving a green façade and a green wall in Indonesia have revealed air temperature reductions of 1.2–3 °C. To establish the impact of VGS in street canyons with different height to width ratios over a range of heights, Zheng et al. (2023b) conducted scaled outdoor experiments revealing air temperature reductions of 0.8 °C to 1.6 °C with the greatest reductions associated with the lowest levels in the narrowest canyons effect. Complementing the small-scale experimental studies are a limited number of real world-scale outdoor monitoring studies of air temperature in the vicinity of VGS. For various distances (0, 0.5, 1, and 2 m) from an eastward-facing living wall Daemei et al. (2018) have shown a cooling effect of up to 0.75 °C during summer and an increase of up to 1.26 °C during winter, with these effects diminishing beyond distances of 0.5 m. For Madrid, Spain, de Jesus et al. (2017) measured a cooling effect of up to 2.9 °C during summer when comparing air temperatures adjacent to a green and bare wall of the same area and orientation. While some on-site observations indicate insignificant reductions in outdoor air temperature in front of VGS (Hoelscher et al., 2016) significant reduction in peak ambient temperature have been observed. For example, for a heatwave event in Gush Dan, Israel in 2020, Zuckerman and Lensky (2023) measured a strong cooling effect of up to 0.78 °C at the centre of a street canyon with air temperature reductions of up to 1.58 °C and 1.14 °C adjacent to south and north-facing walls respectively, with the strongest decreases during the evening and early morning. During the hottest part of the day the VGS cooling effect during the heatwave was the least due to a reduction in VGS evapotranspiration rates as plants conserved water (Zuckerman & Lensky, 2023). Cheng et al. (2010) also found that the cooling effect of VGS achieves a maximum on the hottest days. Simulations of the outdoor temperature in response to VGS also demonstrate modest cooling (Su et al., 2022)). In relation to human thermal comfort, the reduction of mean radiant temperature (MRT) at pedestrian level associated with VGS is pertinent. To date the few analyses of this effect, point to MRT decreases of 2–4 °C but in spaces close to the VGS (Alsaad et al., 2022; Kim et al., 2023; Zhang et al., 2019; Zheng et al., 2023b). The limited extension of MRT decreases to distances beyond about 0.5 m from the VGS, as found for air temperature, raises questions about the effectiveness of VGS as a standalone option for obtaining significant heat mitigation. That said and perhaps predictably, more work is required before the jury is totally out regarding VGS as a heat mitigation option. Some of the issues that need to be addressed are set out in Blanco and Convertino et al. (2023), Convertino et al. (2023) and Koch et al. (2020) and include standardisation of investigation and assessment methods, more long-term monitoring of VGS thermal performance in a range of climate settings to complement the large number of simulation studies and the long-term sustainability, cost and benefits and social acceptability of VGS. In the context of heatwaves an important question is: is the thermal performance of VGS during heatwaves sufficient to have a discernible impact on the outside thermal environment in all climate types over the entire diurnal cycle and to what extent during heatwave conditions do VGS need to be actively irrigated to maintain their maximum cooling power?

14.3.2.2 Green Roofs

Green roofs (GR), also known as living roofs or vegetated roofs probably originated in ancient civilisations such as Mesopotamia where they sat atop *ziggurats*, massive ancient religious structures built from circa 4000 to 600 BCE (Osmundson, 1999). Fast forwarding to the near present, modern green roof technology emerged in the twentieth century in Germany when in the 1960s, German landscape architects, popularized the idea of incorporating vegetation into building design to enhance environmental sustainability and aesthetic appeal. During the late twentieth century, particularly in the 1970s and 1980s, green roofs gained traction in Europe, primarily driven by concerns over energy efficiency, urban heat island effects, and environmental degradation with Germany pioneering research and development on green roof technologies (Jim, 2017). In North America, the modern green roof movement gained momentum in the late twentieth century, initially spurred by environmental activists and architects advocating for sustainable urban development. Cities like Chicago and Toronto played pivotal roles in promoting green roofs through policy incentives and pilot projects. Since then, green roof technology has continued to evolve, with advancements in materials, engineering techniques, and vegetation selection (Jim, 2017). Modern green roofs typically consist of various components, including vegetation (referred to as landscape materials), substrate (serving as the growth medium), a filter layer and drainage material (for moisture retention), and insulation, root barrier, and waterproofing membranes, with GR generally categorized into four types: intensive, semi-intensive, single-course extensive, and multi-course extensive. These categories are determined based on factors such as weight, the thickness of the substrate layer, maintenance costs, the plant community supported, and the need for irrigation (Jamei et al., 2021).

The subject area of GR thermal performance is huge and continues to grow at pace given this mitigation option is invariably referred to in discussions related to urban heat and heatwave mitigation. Again, with reference to an energy balance framing, GR potentially influence levels of atmospheric heat for several reasons. Via the thermal insulation GR provide, sensible and conductive heat transfer to the air and building interior respectively is reduced. Evapotranspiration (latent heat flux to the atmosphere) offers cooling of air layers close to GR surface with the magnitude of this process potentially large but dependent on GR moisture status and GR type (e.g., intensive versus extensive) (Silva et al., 2016). Another important heat loss pathway is longwave radiative transfer from the GR surface to the atmosphere with this possibly accounting for around 40% of the heat loss from some GR (Xiao et al., 2014). The shelter provided by GR reduces the shortwave radiative flux received at the roof surface which reduces the surface temperature and has a fundamental impact on upward and downward sensible heat transfer in its turbulent (Qh) or conductive (Qg) forms respectively. Beyond understanding the processes operative on and within the vicinity of GR, of importance is whether the alteration of the radiative, turbulent and conductive heat fluxes induced by GR lead to effective reductions in atmospheric heat, especially in the urban canopy layer.

Many reviews of the effectiveness of GR to mitigate background urban heat, often articulated as the UHI, have appeared in the literature over the last few years (Herath et al., 2021; Irfeey et al., 2023; Jamei et al., 2023; Krayenhoff et al., 2021). However, this is less so for the case of heatwaves with a paucity of studies singling these events out for attention in the context of GR effectiveness. The GR literature tends to be dominated by modelling studies conducted at a range of scales using micro- and meso-scale climate models with a variety of model physics (Assaf & Assaad, 2023). Consequently, the outcomes of modelling-based studies are scale and structure dependent, the result of which is a diversity of conclusions regarding GR effectiveness for atmospheric heat mitigation. For example, Krayenhoff et al. (2021) note for summer afternoon conditions the vegetation cooling efficiency (ratio of decreased surface temperature to the added vegetation fraction) of GR established from mesoscale studies is 1 °C–3 °C, which is several times higher than that found from microscale studies at around 0.5 °C. These values match those found in reviews or location specific studies of the cooling effect of GR (Jamei et al., 2021, 2023; Sharma & Goyal, 2023; Wootton-Beard et al., 2016; Yenneti et al., 2020). In addition to knowing the magnitude of the cooling effect, of interest is the spatial extent of cooling attributed to GR surface temperature reductions. As for the case of VGS the area beyond a GR for which GR have a discernible impact on air temperature is limited. For example, Jin et al. (2018) show how the cooling effect of a GR reduces from 0.6 °C within a 20 m area to 0.3 °C within a 60 m area. At the pedestrian level GR have been found to have effects ranging from negligible to moderate, with such contrasts most likely due to the use of different scale models (Jamei et al., 2021). The literature also indicates a wide range of soil moisture and/or irrigation states applied in modelling studies which adds to the diversity of results (Li et al., 2014). To achieve cooling approaching the upper bound associated with evapotranspiration/latent heat flux, well irrigated GR—in many ways the ideal case—is required (Georgescu, 2015; Heusinger et al., 2018; Krayenhoff et al., 2021). Perhaps unsurprisingly the magnitude of cooling increases with the scale of GR application. For example, for Phoenix Arizona, modelling has shown that UBL temperature reductions of 0.3 °C in the daytime and 0.2 °C at nighttime are achievable per 10% increase in GR coverage (Song & Wang, 2016). The GR cooling effect on air temperature has also been shown to vary by climate type with dry climates demonstrating the highest median cooling effect of 3 °C while that for hot–humid climates is 1 °C (Jamei et al., 2021, 2023).

While this brief review has drawn attention to the candidacy of GR as a heat mitigation option by presenting estimates of the extent to which they may reduce air temperature the reader has no doubt sensed there are many possible dependencies in relation to GR cooling effectiveness. Setting aside factors such as GR (substrate) type, the vegetation possibilities that might comprise GR and level of irrigation, a burning question is at what scale would GR need to be applied to make a difference to heat exposure in cities, especially at the urban canopy level. It seems we are some distance from having the answer to this question for a variety of reasons not least the vast range of modelling approaches adopted which makes developing a consensus view difficult. Added to this, the long-term monitoring of how GR evolve over time, from an eco-climatological perspective, is lacking as is the understanding of how weather patterns, especially those associated with heatwaves, influence GR cooling performance. To make studies of GR heat mitigation effectiveness more relatable there needs to be more of a focus on working out impacts at the pedestrian level—as is the case for VGS—because for many members of the public the UHI is an abstract concept. Other wider matters associated with possible future research agendas include for example GR end of life considerations (Rizzo et al., 2023), willingness to pay ((Cristiano et al., 2023) and full life cycle analyses of GR components. Lastly a critical analysis of the methodologies applied for assessing the possible physical health benefits associated with GR, such as the number of heat-related deaths saved, is overdue as some studies appear to mix scales of climate modelling and health assessment.

14.3.2.3 Greenspaces

Greenspaces are areas within urban environments that are intentionally reserved or designed for the cultivation of vegetation, whether in the form of parks, gardens, urban forests, or other natural landscapes. In contrast to the modification of the urban surfaces by other means as outlined in previous sections, the enhancement of existing or the development of new urban greenspace is generally considered a more environmentally benign option for heat mitigation. Urban vegetation, through its influence on the surface energy balance and its impact on shortwave radiation shading, absorption and reflection and transpiration, actively shapes the physical environment of cities, often creating the so-called 'cool island effect' (Spronken-Smith & Oke, 1998). Trees and other vegetation in greenspaces provide shade, reducing the amount of solar radiation absorbed by the ground surface. This reduces surface temperatures, particularly in densely built urban areas where engineered heat-absorbing surfaces dominate. Vegetation also absorbs a portion of incoming solar radiation for photosynthesis, converting it into chemical energy rather than thermal energy. Vegetation releases water vapour through transpiration and evaporation from soil surfaces. This process of evapotranspiration is an important heat sink and constitutes a significant cooling process effectively reducing near surface temperatures and mitigating heat. Vegetated surfaces can also modify the local albedo often having higher reflectivity compared to built-up areas. For example, light-coloured surfaces such as grass when compared to asphalt can reflect more solar radiation back into the atmosphere, reducing the amount of heat absorbed by the surface.

Research to date has established that the overall cooling effect of greenspace is generally dependent on factors such as greenspace size and type (e.g., entirely tree or grass-covered; mixture of tree and open spaces), the type of vegetation that comprises the greenspace and the overall spatial patterns of greenspace (Demuzere et al., 2014; Wong et al., 2021; Yu et al., 2020; Zhang et al., 2024). While there is general agreement that cooling intensity increases with greenspace size, usually in a non-linear fashion (Bowler et al., 2010), there are mixed views about the importance of shape as a determinant of the level of cooling as there are regarding whether a critical greenspace threshold size can be detected beyond which cooling is clearly discernible (Yu et al., 2020). With regards to a universal threshold value of greenspace size for a clear cooling effect there are no doubt many confounding factors that make its identification difficult including climate setting, the vegetation composition of the greenspace and the nature of the surrounding environment in terms of level of urbanisation and associated morphology and built form (Howe et al., 2017; Makido et al., 2019). To try and quantify the association between greenspace size and cooling effect (temperature difference between the greenspace and the surrounding urban area) the so-called threshold value of efficiency for greenspace has been applied (Yu et al., 2021). For Hanoi, Vietnam and Zhengzhou, China this has been found to be 1 ha and 0.83 ha respectively (Cai et al., 2023; Le et al., 2019) and to be 0.69 ha for the high latitude city of Copenhagen. Interestingly for Beijing the threshold value of efficiency varies according to socioeconomic development levels in the vicinity of the greenspaces (Yang et al., 2022). Beyond individual cities the threshold value of efficiency has been found to range between 0.6 ha to 0.95 ha for seven low latitude Asian cities (Fan et al., 2019), obtain values typically around 0.5 ha for cities in temperate monsoon and Mediterranean climate settings (Yu et al., 2018) with lowest values found for arid climate zones (Wang et al., 2022a). As might be expected the cooling effects of greenspaces are not constant and can vary at diurnal and seasonal timescales ((Probst et al., 2022; Hamada & Ohta, 2010; Santamouris et al., 2018).

Some evidence points to different aspects of greenspace type being important for determining cooling effect in areas adjacent to greenspaces. For example, using a modelling approach and focusing on clear calm nights, Monteiro et al. (2016) while reporting cooling of 3.8 °C for several parks in London has shown that the cooling distance exhibited a linear increase as the area of greenspace, tree canopy, and grass expanded. However, the correlation between these factors and the magnitude of cooling was non-linear. Tree canopy coverage demonstrated the strongest association with cooling distance, whereas the extent of cooling was most closely tied to the coverage of grass. For Leipzig (Germany) greenspaces have been found to provide 3.0 °C of cooling, extending up to 470 m (Jaganmohan et al., 2016), similar to the cooling effect of 3.4 °C extending to 498 m at 1500 hrs local standard time simulated for a park in Antwerp, Belgium (Toparlar et al., 2018). For Prague Kirschner et al. (2023) found that compact urban greenspaces (the interaction of size, shape, and tree density) ranging from 10 to 25 hectares, characterized by dense tree cover, exhibit the most significant cooling effect with this specific type of greenspace linked to an average reduction in land surface temperature of 2.3 °C within a 400-metre radius, compared to the least effective greenspaces characterized by sparse tree cover.

As evapotranspiration is an important cooling process the moisture status of greenspaces can be a critical determinant of cooling effectiveness with irrigated vegetation known to produce greater cooling compared with non-irrigated landscapes (Bonan, 2000). This is particularly so at night-time (Gober et al., 2010). Various studies based on modelling and microclimate observational approaches have explored the impact of greenspace irrigation on cooling during heatwaves. Using meso-scale modelling Grossman-Clarke et al. (2010) demonstrated that urban irrigation can result in a decrease of 0.5 °C to 1.0 °C in daytime maximum and nighttime ambient temperatures during severe heat events in Phoenix, while for Paris simultaneous irrigation of urban green spaces and pavement during heatwave conditions could potentially reduce average nighttime ambient temperatures by 2.6 °C. For a heatwave event in Adelaide, South Australia micro-scale simulation of various hypothetical greenspace irrigation scenarios incorporating different watering rates and timings has shown that irrigation can effectively lower air temperatures with a diurnal average air temperature reduction of up to 2.3 degrees observed. However, the cooling effect of increased irrigation exhibited a non-linear pattern, with minimal additional cooling beyond a certain rate (20 litres m^{-2} per day). Additionally, the extent of cooling was directly related to the proportion of irrigated surfaces, indicating diminished cooling in areas with higher urban development. This study also found that despite the increase in humidity associated with irrigation, the cooling achieved notably enhanced outdoor human thermal comfort during heatwave conditions (Broadbent et al., 2018). Also for Australia, and specifically for the botanic gardens in Melbourne, Lam et al. (2020) showed that under non-heatwave conditions, overnight irrigation of 20–25 mm resulted in a cooling effect ranging from 0.5–1.0 °C, whereas during heatwave conditions, some lawn areas experienced more substantial cooling of 2—4 °C with similar irrigation amounts. Furthermore, the extent and duration of cooling varied across different garden sites, with residual cooling effects persisting for several hours in some locations following nighttime irrigation events. Using an experimental set-up Cheung et al. (2022) also demonstrate the impact of greenspace irrigation on cooling at the scale of sub-urban gardens, an often overlooked greenspace as a heat mitigation option (Cavan et al., 2022). While these few examples point to the possible effectiveness of irrigated greenspaces as a heatwave mitigation option Koc et al. (2018) note the need for more research related to the role of irrigation and associated evapotranspiration for cooling urban areas. This is especially so for some desert cities where the so-called oasis effect is weak or not observed because simply expanding greenspace or introducing plants with a higher leaf area index cities does not achieve heat mitigation with irrigation necessary to promote evapotranspiration related cooling (Gao et al., 2024). That factors other than greenspace physical attributes and the physical processes associated with cooling, such as the type of greenspace infrastructure (Tiwari et al., 2021) and greenspace planning traditions can have a fundamental impact on the degree of cooling achieved by urban greenspaces (Nastran et al., 2019) adds to the complexity of understanding the drivers of greenspace effectiveness as a background heat and heatwave mitigation option.

Further to the interest in greenspaces in general as a heat mitigation option is that related to the role of trees especially in climatically harsh urban street settings—trees are possibly the most explored and recommended NBS related heat mitigation option (Probst et al., 2022). Of course, it is via the same mechanisms described above for vegetation in general (i.e., shading, transpiration etc) that trees situated in urban streets, squares and other open spaces are considered a serious heat and heat-wave mitigation option. By providing canopy shading and transpiration, trees have been shown to reduce outdoor radiant temperature by 7–8 °C compared to open areas and reduce air temperatures on hot summer days by around 3–4 °C (Wang et al., 2015). Air temperature fluctuations can also be significantly reduced with increased tree coverage (Lee et al., 2023) with increases in canopy cover shown to produce incremental increases in cooling of around 0.3 °C per 0.10 canopy cover increase, for afternoon clear-sky summer conditions (Krayenhoff et al., 2021) although the linear decrease in cooling may be slower over transpiring grass surfaces (Rahman et al., 2020). For European cities it has been suggested that to achieve a 1 °C drop in urban temperatures, a tree cover of at least 16% is required (Marando et al., 2022). Trees can also bolster airflow turbulence, modify outdoor wind flow patterns, decrease convective heat and mass transfer coefficients, consequently influencing the heat and moisture equilibrium between trees and their surrounding environment (Zheng et al., 2023a). Trees and their associated microclimate processes also moderate human thermal comfort (Bosch et al., 2021; Duarte et al., 2015; Gillerot et al., 2024; Teshnehdel et al., 2020; Wang et al., 2015; Zölch et al., 2016), although this effect may be highly localized and variable (Coutts et al., 2016) with the possibility that the reduced aerodynamic roughness, increased air temperature and humidification effects arising from dense tree canopies (Meili et al., 2021; Zhang & Dai, 2022) may be detrimental to thermal comfort. Understandably the microclimatic impact of trees is likely to depend on a range of factors such as age, size (diameter at breast height and crown/canopy width), morphology, leaf area, root depth, and leaf reflectance all of which may influence tree cooling efficacy under diverse climate conditions, a matter that needs serious deliberation if trees are considered as a heat mitigation option. For example, a meta-analysis by Rahman et al. (2020) has revealed that transpiration rates of trees planted over grass were found to be ten times higher than those planted in paved cut-out pits, and trees with a diffuse porous wood anatomy and originating from temperate and resource-rich forests exhibit superior cooling potentials. Among leaf traits, those with dark green leaves and thickness less than 0.15 mm show greater transpiration and below canopy surface cooling benefits (Rahman et al., 2020). For the city of Phoenix native trees and palm trees appear to be inferior compared to non-native trees for reducing mean radiant temperature (Middel et al., 2021) while generally alien species adapted to warmer climates survive better in city centres (Géron et al., 2022). Further it seems trees grown in Oceanic and Continental climates demonstrate higher cooling potential compared to those in Mediterranean climates for transpiration cooling, and sub-tropical climates for surface temperature cooling (Rahman et al., 2020). Added to these considerations Lachapelle et al. (2023) note to maximise the cooling benefit of street trees, a balanced approach is required along sun-exposed pedestrian routes because planting a shade tree on a street without existing tree cover provides roughly 1.5–2 times more radiative cooling to pedestrians compared to planting the same tree on a street where most of the sidewalk is already shaded by trees. Consequently, they suggest decisions regarding street tree planting for heat mitigation should be tailored to specific locations. Rendon et al. (2024) have also noted the importance of street tree diversity as a determinant of the degree of heat mitigation.

Overall, the heat mitigation potential offered by greenspaces and particularly trees is clear. However, there is a frustrating lack of research on their efficacy during heatwave episodes. This knowledge gap represents a fruitful avenue for future research with questions such as, for some situations (e.g., different weather situations/heatwave types and climate settings) might a humidifying effect erase the cooling benefit of vegetation and are there threshold air-surface-subsurface conditions beyond which vegetation (e.g., trees) is of little use for moderating temperature extremes and high levels of human thermal discomfort at a range of scales during heatwaves. A mixture of monitoring and modelling is likely to play a key part in answering such questions and unveiling with more preciseness the cooling potential of greenspace related options for mitigating the extreme heat associated with heatwaves. Monitoring of greenspaces, whether they be parks or a collection of street trees, of course poses challenges given the difficulties of setting up and maintaining urban climate measurement sites comprising the requisite instrumentation for harvesting the data essential for confronting micro- to meso-climate models of climate-vegetation interactions. That said, agreements amongst cities comprising networks such as the previously mentioned C40 (see Chap. 12), which span a range of climate settings, to invest in the long-term high resolution (space and time) measurement of climate processes in both non-green and greenspaces, using a common set of protocols to facilitate comparability, would be a step in the right direction. This in combination with the types of modelling enhancements outlined by Zheng et al. (2023a) required to improve understandings of the radiative, heat, mass and momentum exchanges between 'urban vegetation' and the urban atmosphere would also help advance knowledge about the effectiveness of 'installing' greenspace as a heatwave and chronic urban heat mitigation option.

14.3.2.4 Blue Spaces

Urban blue space (UBS) widely referred to as blue infrastructure generally refers to natural and constructed bodies of surface water located in urban areas. Like all terrestrial water bodies, those in urban spaces can be defined as falling into the two broad categories of static (closed) and dynamic (open) distinguished by their flow characteristics with static water bodies further subdivided into ponds, lakes, wetlands and water features such as fountains. In comparison to lakes, urban ponds are typically shallower, smaller and more irregular in shape with smaller surface-area-to-perimeter ratios (Ampatzidis & Kershaw, 2020; Downing, 2010). Dynamic UBS prevalent in urban areas consist mainly of rivers, canals, and streams. UBS are considered a worthy heat mitigation option because they possess the potential to modify the physical properties of the air above and around them to create cooling effects and thus the generation of cool islands in the fashion imagined for greenspaces. Although not discussed here there are multiple co-benefits of blue spaces that straddle the physical and social environment including amongst many those related to water management in water sensitive urban environments (Broadbent et al., 2018; Kalff, 2002; Wong et al., 2020) and well-being (Völker & Kistemann, 2015; White et al., 2020; Yin et al., 2023a).

The cooling effectiveness provided by UBS associates to both the air—water temperature difference and the surface net radiation balance (Q*) which is intimately linked to water surface albedo. The magnitude of Q* influences the amount of energy available at the water surface for warming the UBS surface and for evaporation to occur. The greater the temperature difference between the surface and the air, the greater the flow of sensible heat (Qh) from the water body into the atmosphere. The rate of evaporation from a UBS surface, which is generally dependent on the magnitude of Q*, the UBS surface to atmosphere humidity gradient and the rate of ventilation (degree of atmospheric turbulence), also determines cooling effectiveness. Advection of dry air from nearby urban surfaces and its associated turbulence can affect the rate of evaporation by increasing the surface to atmosphere humidity gradient. As outlined earlier water has a high thermal inertia due to its high specific heat capacity. This means that water can absorb and store a large amount of heat energy without experiencing a significant change in temperature. As a result, water bodies in urban spaces will tend to moderate the temperature of surrounding areas. An upshot of this is UBS typically act as heat sinks during the day but importantly, as a possible counter to their general heat mitigation effects, release heat during the night with evaporation also continuing for several hours after sunset because of thermal inertia effects—the water body remains warm relative to the overlying atmosphere. On a seasonal basis the stratification of static water bodies with large volumes such as lakes may also be important in terms of cooling effectiveness.

Observational, remote sensing and modelling studies as well as meta-analyses have generally found the cooling effect, as expressed by air temperature difference between various UBS and their hinterland, generally spans 1—3 °C albeit higher values have been observed for specific locations and circumstances (Ampatzidis & Kershaw, 2020; Probst et al., 2022; Völker et al., 2013). One of the first observational studies of the impact of UBS on local cooling found that the thermal effect of a 12.7 ha pond in Fukuoka city Japan produced an afternoon temperature difference compared to the surrounding built up area of 3 °C (Ishii et al., 1991). Other field-based studies have also revealed cooling related to urban ponds in the range of 0.5-3 °C with maximum cooling often found downwind of these features (Chen et al., 2009; Cureau et al., 2023; Nishimura et al., 1998; Saaroni & Ziv, 2003; Syafii et al., 2017; Yao et al. 2023a; Xu et al., 2010).

Although some researchers suggest flowing water may have a greater cooling effect than static water bodies, because streams and rivers may take up heat as they flow through urban hotspots and then release the sequestered heat further downstream (Hathway & Sharples, 2012; Kleerekoper et al., 2012), there has been little comparative location-based research on static versus dynamic water body cooling effectiveness (Probst et al., 2022). Rather the focus of the limited number of observational studies on cooling effects of rivers has been on the rivers themselves. Using a statistical method combined with mobile meteorological measurements for built-up areas adjacent to the Yangtze River in Wuhan collected during a hot and humid summer day, Wang et al. (2022d) found a substantial cooling and humidifying effect induced by the river during the daytime, with reductions of up to 3.55 °C in temperature and increases of up to 17% in humidity observed at distances of 1741 meters and 1369 meters from the riverbank, respectively. Their analysis also demonstrated that the river effectively mitigated pedestrian heat stress within a radius of 1200 m from the riverbank during the daytime, with a comparatively weaker influence at night. In another study (Park et al., 2019) detailed field measurements were employed to assess the river cooling intensity and river cooling distance along the Cheonggye River, a restored urban stream, in Seoul, Korea. Findings indicated at 1400 hrs, the river exhibited an average cooling intensity of 0.46 °C and an average cooling distance of 32.7 m; at 2200 hrs these values were 0.37 °C and 37.2 m, respectively. Spatial variations in the cooling intensity were found to be inversely correlated with street width and mean building height in the afternoon underscoring the influence of urban morphology on cooling intensity. Setting aside the possible role of river width, the findings for the Yangtse River at Wuhan and the Cheonggye River in Seoul indicate the potential cooling effects of flowing water (dynamic UBS) and confirm the findings of earlier field-based studies on the cooling effects of rivers such as found for Hiroshima, Japan. Here the Ota River (260 m wide), on clear days during the warm season, reduces the air temperature by 5 °C or more with a cooling effect propagating up to 400 m away from the river aided by a sea breeze (Murakawa et al., 1991). For Melbourne Australia a 270 m wide river has been reported to produce air temperatures about 3–5 °C lower in its vicinity compared to surrounding built-up areas, during the later afternoon hours on warm days (Coutts et al., 2013).

Also considered as observational studies those based on remote sensing techniques point to blue spaces as potentially credible heat mitigation options albeit with a focus on land surface cooling as opposed to air temperature cooling effects. Although conducted at relatively course scales compared to micro-scale observational studies, remote sensing-based studies provide an insight into the cooling effect controls at the larger landscape scale. In a study that considered the nature of the water-cooling effect for 34 mega-cities in China Cai et al. (2022) found both static and dynamic water bodies induce cooling around their littoral zones with the mean cooling distance ranging from 431 to 1350 m and the maximum temperature difference being 2.21 to 7.83 °C with the size of the cooling effect affected by water body spatial distribution, size, location, and background climate setting. Interestingly, these researchers note that as the percentage or water body area within a city increases, mean cooling distance decreases while maximum temperature difference increases. Further, water bodies exhibit the greatest cooling distance and the smallest maximum temperature difference when dispersed throughout the city, whereas the shortest cooling distance and the largest maximum temperature difference are recorded when flowing through the city. A study of the cooling effect of 20 lakes in the city of Hue, Vietnam has also confirmed the importance of landscape factors in setting the general scale of the cooling intensity. Most lakes with a relatively simple shape, high proportion of greenspace and relatively low percentage of impervious surfaces surrounding them were reported to achieve high cooling intensity with a mean intensity of 1.52 °C and a mean cooling distance of 150 m (Le Phuc et al., 2022). Notably these researchers report the negative impacts on cooling intensity of a range of human factors as well as extreme heat conditions associated with a foehn effect that deplete lake water volumes. For the city of Wuhan, the mean cooling intensity of all static water bodies throughout the city has been found to range between 1.1 and 13.3 °C with a mean of 5.5 °C with water bodies possessing a relatively larger water area, more regular boundary and simpler surroundings achieving more efficient cooling of their adjacent built environments (Xie et al., 2023). In another city focused study, using remote sensing techniques (Li et al., 2022), the water-cooling distance and cooling intensity of 34 water bodies in the Chinese city of Hefei with areas ranging from 3.88 to 875.52 ha (average area of 72.62 ha) has been assessed. Findings indicate that the water-cooling distance varies between 140 m and 500 m with an average distance of 343 m with ponds displaying the greatest cooling distance (300 m), followed by rivers (270 m), urban lakes (239 m), and reservoirs (228 m). As for cooling intensity, this was shown to vary between 1.51 and 7.94 °C with an average of 5.21 °C, with the lakes revealing the largest average cooling intensity (5.67 °C), followed by rivers (4.45 °C), reservoirs (3.92 °C), and ponds (3.34 °C). While these results are of interest, relevant in terms of the planning and design of new water bodies is that in the case of Hefei, nine factors appear to significantly impact the variation in water cooling intensity, with road density exerting the greatest influence. Conversely, only a landscape shape index was found to be significantly associated with water cooling distance. Importantly, the interplay among environmental, water body, and socio-economic characteristics was shown to substantially impact the cooling effect of the water bodies, with the interaction between determining factors mutually enhancing their relationships (Li et al., 2022). The cooling effects of wetlands have also been considered using remote sensing techniques. For example, for Chengdu, Southwest China the cooling intensity of wetland areas has been found to rise logarithmically with wetland size with the threshold value of efficiency determined to be 1.47 +/− 0.34 °C per hectare. Moreover, hydrological connectivity was revealed to exert a significant influence, explaining 28.2% of the total cooling effect with wetlands situated in high-density built-up areas exhibiting notably higher cooling levels (Wu et al., 2021).

Complementing observational studies are those based on simulations. These draw mainly on computational fluid dynamics and energy balance modelling techniques (Ampatzidis & Kershaw, 2020). Building on the work of de la Flor and Domínguez (2004), who used the outputs from numerical simulation work to advocate for ponds and other water features as effective natural air conditioning systems in open spaces and acknowledging the paucity of studies that apply numerical modelling techniques, Tominaga et al. (2015) combined computational fluid dynamics and radiative heat transfer analysis to understand the extent to which water surface related evaporative cooling can moderate temperatures at the intra-urban (neighbourhood) scale. They found that within a residential neighbourhood the presence of a pond led to a maximum temperature reduction of approximately 2 °C at the pedestrian level. Furthermore, they observed that this cooling effect extended downwind over an unobstructed distance of 100 m. These findings were established under specific conditions, namely a wind velocity of 3 m/s at a height of 10 m. In their numerical investigation, Yang et al. (2020) determined that water bodies play a significant role in regulating the thermal conditions of urban environments, resulting in a temperature decrease of up to 2 °C while also enhancing wind flow. Based on the simulation of heatwave conditions, Zeeshan and Ali (2023) demonstrated for isolated streets in Karachi, Pakistan that water bodies can improve thermal comfort conditions via reducing ambient air and surface temperatures by around 0.9 °C and 3.5 °C respectively also noting that such changes induce wind velocity increases that reach a maximum at peak times of thermal stress. Also concerned with effects within urban canyons, Ampatzidis et al. (2023) applied a numerical modelling approach to explore how the size and shape of blue spaces with varying temperatures relative to the surrounding air impact wind flow structure. Their findings indicate that inadequately sized warmer water bodies fail to facilitate adequate vertical transport leading to air stagnation and heightened levels of temperature and humidity at street level. Consequently, these researchers conclude that larger water bodies are more effective in the nocturnal removal of accumulated warm air from the urban surface, while smaller water bodies excel in localized evaporative cooling. Their findings also point to a more extensive cooling effect over a broader area for irregularly shaped water bodies (Ampatzidis et al., 2023).

In addition to establishing the explicit effect of water bodies on direct cooling and associated urban canyon physics and dynamics, numerical simulations have been used to shed light on how the arrangement of water bodies in the landscape might play a role in heat mitigation. For example, the numerical modelling results of Theeuwes et al. (2013), notably for a theoretical circular city, reveal that the cooling impact of water bodies is nonlinear and contingent upon factors such as the fractional water coverage, size, and arrangement of individual lakes within the urban landscape relative to wind direction. Larger lakes were found to exhibit a pronounced cooling effect near their perimeters and in downwind regions. Conversely, numerous smaller lakes evenly dispersed throughout an urban area were found to exert a milder temperature influence but extend their impact over a broader area (Theeuwes et al., 2013). For typical heatwave conditions in Wuhan China the work of Cheng et al. (2023) suggests that downtown lake groups, in comparison to similar-sized suburban lake clusters, exhibit higher daytime differences in sensible heat fluxes and lower differences in latent heat fluxes. Consequently, this leads to more pronounced lake cooling intensity effects in downtown areas. Furthermore, lake groups positioned upwind contribute to cooling built-up areas due to the promotion of lake breeze circulations which facilitate the spread of lake cooling over longer distances. However, the evolution of cooling intensity, cooling distance and lake breeze circulations was found to occur asynchronously, with cooling intensities reaching their peak around mid-morning before the onset of lake breezes and consequently cooling distances peaking in the afternoon once the lake breezes became more pronounced.

A possible elephant in the room regarding the heat mitigation effectiveness of UBS is the nocturnal warming and humidifying effect often produced by water bodies and whether this might negate, from a cumulative heat perspective, any daytime urban cooling that water bodies manage to achieve or separately, reduce nocturnal temperature differences between urban and rural locations and thus mitigate the urban heat island effect. As noted earlier, the possible negative nocturnal effects of water bodies on cooling and humidity levels stems from the high thermal inertia of water bodies and the tendency of their waters to remain warm well into the evening thus sustaining heat transfer and evaporation. Based on in-situ measurements of a range of meteorological variables in Nanjing, China, Yao et al. (2023b) show for a 0.775 ha pond, relative to a reference soil-site, that while moderate daytime cooling of 0.6 °C occurred, the pond produced a pronounced nocturnal warming effect of 1.8 °C. This outcome was explained with reference to measured heat fluxes such that during the day 77% of the net radiation (Q*) went into storage ($\triangle$Q) within the pond water column with the remaining Q* (23%) consumed by latent heat flux/evaporation (Qe) with sensible heat (Qh) negligible. However at night, the heat stored in the pond during the day became a heat source, with this heat dissipated via the heat loss pathways of sensible and latent heat flux (5 and 51% respectively) and net radiative flux (44% in the form of L↑). In a similar study for Nanjing and Guangzhou, the same researchers note a significant water body-related nocturnal UHI effect of 1.2–1.3 °C during May–September (warm months), comparable to that achieved for residential sites (1.1–1.3 °C). Further, while lakes were found to provide daytime thermal comfort benefits this was quite the opposite at night because of their intensified warming and humidifying effects (Yao et al., 2023a). Cui et al. (2021) for a medium-sized US city also note that urban lakes result in a deterioration in nocturnal cooling along with an elevation of humidity, while for Rotterdam in the Netherlands Steeneveld et al. (2014) show that water bodies by supressing the diurnal temperature cycle increase rather than decrease the 95th percentile of the daily maximum UHI which bears implications for thermal comfort. Similarly, for the Yangtse River at Wuhan, a negative effect of the river on pedestrian heat stress is apparent at night (Wang et al., 2022d). Field-based measurements carried out in a wetland settlement area in South Sumatra, Indonesia also indicate the propensity of blue space to increase nocturnal temperatures, 1.2–1.6 °C in this case (Triyuly et al., 2021).

Notwithstanding the vast range of methodologies, scales and climate settings for which UBS studies have been conducted and the snapshot views many of them provide of UBS—urban climate interactions, from the material considered here, there appears to be sufficient evidence to support the general idea that UBS constitute a credible heat mitigation option. That said, this brief review has highlighted many of the fundamental factors that affect UBS cooling efficacy which will need to be considered in the design and implementation of UBS. Others have also made this point (Ampatzidis & Kershaw, 2020; Jamei & Tapper, 2019; Probst et al., 2022). Further, as the reader may have detected, estimates of the cooling effect of UBS, especially static water bodies, based on remote sensing techniques use surface temperature as opposed to air temperature as the basis for calculating metrics such as cooling intensity and cooling distance. In terms of heat mitigation such studies only provide an insight into the possible mitigation of surface heat and thus the surface urban heat island effect as opposed to heat in the urban canopy layer and the associated canopy layer urban heat island. Moreover, as pointed out by Ampatzidis and Kershaw (2020) a noticeable gap appears to exist between the cooling potentials identified through remote sensing compared to those derived from field measurements or numerical simulations. They suggest that this difference is probably due to inadequate data collection during nighttime periods, when thermal inertia-induced warming may take place, along with the oversight of latent heat flux considerations (Ampatzidis & Kershaw, 2020). Kumar et al. (2024) in a review of the heat mitigation effectiveness of NBS in general also note the clear discrepancy between levels of cooling reported by remote sensing based studies compared to those described in studies based on monitoring and modelling or a combination of the two approaches. As noted for greenspaces, the number of studies dedicated to understanding of the role of UBS in heatwave mitigation are few and far between with attention to this area desperately needed. Further, as is the case for other NBS, it appears that some of the gains in cooling and thus thermal comfort of around 2-3 °C arising from UBS may fall below those required to make a real difference during heatwave periods when, for example, maximum temperature (afternoon) anomalies may reach 6-8 °C or more. Further that UBS in urban areas, especially in the form of large ponds and small to medium sized lakes may act as a nocturnal heat source and do not ameliorate nocturnal temperatures needs further investigation by combining field-based measurements and modelling. This is especially critical as heatwaves are a 24-hour phenomenon that produce not only extreme heat during the daytime but also nighttime. A characteristic of many of the studies touched upon here is the short duration of either the field- or satellite-based measurements or the periods for which simulation studies are performed. While it is probably impractical to run continuous simulations for sentinel UBS sites, continuous monitoring using micro-meteorological and remote sensing methods would be useful for building a climatological view of the 'cooling climatology' of UBS, especially for static water bodies. Such monitoring would help provide insights into the mean, variability, and extreme event characteristics of UBS and thus their long-term and event-based cooling or indeed warming effects. UBS not considered here include water features such as fountains and constructed wetlands, the reason being, except for occasional cognate studies, there has been limited research conducted on their performance as heat mitigation options (Probst et al., 2022). This also applies to surface- or roof-based bioretention systems such as rain gardens and roof ponds and hydroponic systems (Kasprzyk et al., 2022; Rapisarda et al., 2022; Sharifi & Yamagata, 2015) and vegetated swales, all of which are possible blended green and blue infrastructure solutions to heat mitigation.

In closing this section on NBS it is probably fair to say, there is much enthusiasm in various governmental and non-governmental policy circles about the role that NBS might play in heat mitigation. Worryingly this enthusiasm often arises from a non-critical reading of the literature on the burgeoning subject of NBS for heat mitigation and a lack of distinction between reductions in surface temperature (e.g., wall, roof, ground, or water) and those that are more directly relevant to human health such as urban canopy layer air temperature, mean radiant temperature or some measure of heat stress. For this reason, those involved in heat risk policy development and management should look beyond simple graphical representations of the outcome of meta-analyses of the literature such as that shown in Fig. 14.5 which summarises the analysis of 202 papers that consider the cooling efficiency of 51 types of green and blue infrastructure. To the credit of the authors of this in-depth analysis (Kumar et al., 2024), 95% confidence limits for the average cooling efficiency values for ten key categories of green and blue infrastructure are presented in the body of the published paper—see the caption in Fig. 14.5 for the values associated with the five categories with the greatest cooling efficiency. For example, taking the top cooling efficiency category of botanic gardens, this was found to have a mean cooling efficiency of 5.0 °C with the 95% confidence rage of ±3.5 °C. This value however is only based on five studies with the cooling efficiencies reported ranging between 2.3 °C and 10.0 °C. Further while some of these studies are undertaken in botanic gardens, they do not necessarily refer in a strict sense to the cooling effect attributable to the botanic garden as a contiguous landscape feature. To illustrate this, a cooling efficiency of 10 °C is reported for one of the studies making up the botanic gardens group (Su, 2018). However, this refers to temperature

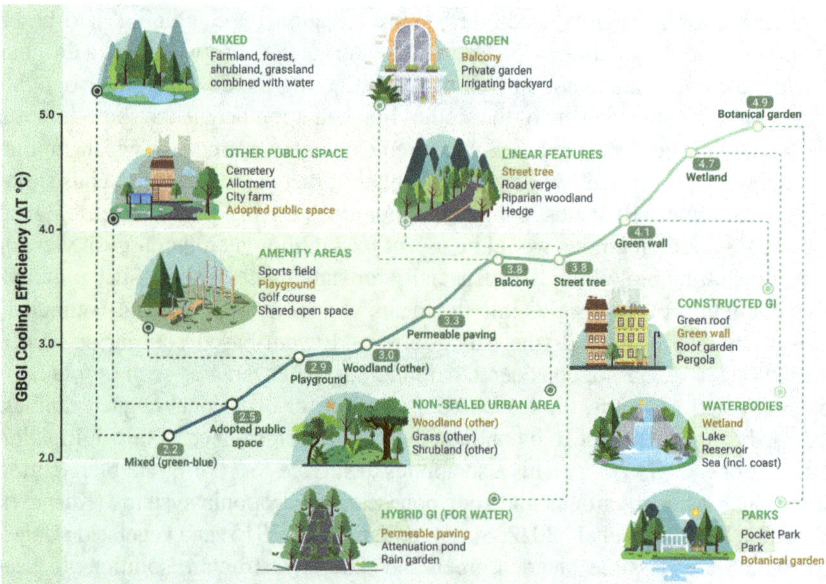

Fig. 14.5 Cooling efficiencies for a range of green and blue infrastructure types. The mean values and associated 95% confidence intervals for the most efficient types are as follows: botanical gardens (5.0 ± 3.5 °C), wetlands (4.9 ± 3.2 °C), green walls (4.1 ± 4.2 °C), street trees (3.8 ± 3.1 °C), and vegetated balconies (3.8 ± 2.7 °C). (Source: Kumar et al. (2024). Used under the terms of the Creative Commons CC-BY licence. https://s100.copyright.com/AppDispatchServlet?publisherName=ELS&contentID=S2666675824000262&orderBeanReset=true)

variability at the intra-botanic garden scale, specifically the temperature difference between a grove of peach trees in a botanic garden and a nearby botanic garden reference climate station, as opposed to the temperature difference between the botanic garden as a whole (establishing this using conventional monitoring is a challenge in itself) and a reference climate station in an adjacent built-up area. Further the cooling effects reported for the other four studies making up the botanic gardens group (Chang & Li, 2016; Irmak et al., 2017; Yilmaz et al., 2018; Yilmaz et al., 2022) are open to interpretation given a close reading of these casts some doubt on the consistency of the cooling values reported with some referring to surface temperature while others report air temperature or physiological equivalent temperature all of which are not directly comparable. This goes to show that caution should be exercised in accepting and indeed acting blindly upon some of the findings that make their way into the literature regarding the efficacy of specific NBS as heat mitigation options.

14.3.3 Urban Morphology and Built Form

As outlined in Chap. 8 urban morphology (physical form, three-dimensional structure, layout, and spatial organization of urban areas) and geometry (shapes, sizes, orientations of buildings, streets, blocks, and other urban features) play a fundamental role in the partitioning of net radiation (Q*) into sensible (Qh and Qg) and latent (Qe) heat and the development of urban climate at the intra-urban to city-wide scale, with the interaction of atmospheric processes with urban form and geometry often leading to the exacerbation of background urban heat during heatwaves. Alterations of the morphology of cities and the geometry of their elements can contribute to urban heat mitigation by enhancing shade provision and ventilation. Generally, the parameters considered in understanding the relationship between urban heat and morphology and geometry include percent of built area, building density, average building height, building height variation, roof shape/profiles, building cover composition and configuration and canyon geometry (street width, aspect ratio, sky view factor, frontal area, facet continuity, percent shading, solar and wind axes) (AlKhaled et al., 2020).

Harnessing prevailing breezes is potentially a crucial element in achieving urban cooling. Accordingly, there has been mounting interest in urban morphology and the role of wind or ventilation corridors as a heat mitigation option in cities, a strategy recognised by the city of Stuttgart some 50 years ago (Ren & McGregor, 2021; Reuter & Kapp, 2021). Ventilation corridors, refer to pathways or zones within urban areas that facilitate the movement of air, promoting natural ventilation and airflow. These corridors are typically designed to allow the free flow of air through the built environment with the co-benefits of urban heat mitigation and air quality improvement. The design of ventilation corridors considers factors such as prevailing wind patterns, building heights, street width, and the layout of streets and open spaces. By strategically placing buildings and incorporating green spaces, such as parks or green belts, planners can create pathways for air movement throughout urban areas. Additionally, features like wide streets and building setbacks can further enhance airflow within these corridors.

There is a growing literature on ventilation corridors some of which focus on analytical approaches to identifying ventilation corridors (Fang et al., 2023; Guo et al., 2023; Nevat & Adelia, 2023; Wang et al., 2022b; Xu et al., 2021) while others apply numerical modelling techniques to the evaluation of the impact of ventilation corridors on urban heat. Interestingly at the time of writing, most of the literature on ventilation corridors appears to stem from China. Useful reviews of the status of this field for China have been provided by Hong and Chien (2018) and Ren et al. (2018) with the latter introducing a technical guide on ventilation corridors entitled 'Specification for climatic feasibility demonstration: Urban ventilation corridors'. Overall, various modelling studies have demonstrated the potential of increased urban ventilation for mitigating urban heat. For example, Guo et al. (2023) in a study of the impact of ventilation corridors on the thermal environment of the mega-cities Beijing, Shanghai, Guangzhou and Shenzhen find maximum land surface temperature reductions of 2.4 °C, 3.2 °C, 2.1 °C and 2.6 °C respectively. Other studies highlight the importance of local climate zone characteristics (see Chap. 8), the magnitude of the building frontal area index and roughness length as well as corridor geometry and building and open space arrangement for influencing ventilation corridor potential and managing the thermal environment (Hsieh & Huang, 2016; Li et al., 2023a; Liu et al., 2022; Ng et al., 2011; Park et al., 2023; Shi et al., 2022; Xu et al., 2022), that rapid changes in the level and location of urbanisation can govern ventilation possibilities (Wicht & Osinska-Skotak, 2016; Wicht et al., 2017) and how small-scale urban renewal projects via the removal of small fragmented buildings can enhance ventilation and mitigate urban heat (Hsieh et al., 2023).

That ventilation corridors are viewed as a worthwhile urban heat mitigation option is evident in the Beijing 2016–2035 Urban Master Plan. This lays out how the establishment of a multi-stage ventilation corridor system will be achieved via the construction and enhancement of central Beijing urban ventilation corridors to bolster air circulation. By 2035, the aim is to create five primary ventilation corridors exceeding 500 m in width, alongside numerous secondary ventilation corridors wider than 80 m. Over the long run, it is envisioned a carefully controlled expansion of the ventilation corridor network, will progressively unlock pivotal nodes to ensure seamless connectivity among the corridors (Qiao et al., 2017). Ventilation corridors are also considered critical for improving the thermal environment of Tokyo. For example, Ashie (2021) describes how urban redevelopment projects around Tokyo railway station have been implemented to enhance ventilation and mitigate urban heat. This involved removal of the station building that blocked the wind flow at street level on the lee side of the station where conditions of thermal discomfort are frequent, with twin towers built on either side of the former station building site to enhance throughflow. Other redevelopments include the removal of an elevated section of highway in the Nihonbashi area which impeded airflow with its relocation underground opening up a much-needed wind path. Elsewhere in Tokyo, urban redevelopment plans specify that new buildings constructed in major wind paths are to be restricted to 50 m in height with existing buildings if redeveloped limited to their current height or less (Ashie, 2021).

Complementing the enhancement of ventilation to achieve urban heat mitigation is the increase of shade through altering urban geometric characteristics mainly in the form of street canyon height-width ratios and thus sky-view (Zhang et al., 2022). Useful reviews of the extent to which changes to urban geometry can mitigate urban heat and improve thermal comfort have been provided by AlKhaled et al. (2020) and Malcoti et al. (2023) with the importance of understanding building height and street width thresholds in determining heating or colling rates as well as the interactive effects of canyon geometry and canyon surface materials emphasised. Equally important these reviews highlight the methodological challenges associated with exploring the degree of heat mitigation induced by alterations to urban geometry. In a comparative analysis of the effectiveness of various street shading strategies Meng et al. (2023) found building shade to exceed the thermal comfort gains offered by trees. For Biskra, Algeria, Ben Ratmia et al. (2024) emphasise the interactive effects of street height-width ratios and building materials for reducing urban heat finding for extreme summer conditions the most effective heat reducing combination to be streets with an east-west orientation, with height-width rations of less than one and canyon walls with high albedo building material such as limestone which contrasts starkly with the situation where brick walls dominate. For heatwave conditions in Hong Kong field-based studies have found a sky view threshold of 0.4 is effective for heat mitigation in urban canyons (Li et al., 2023b), while airflow and air temperature simulations for a range of theoretical urban canyon configurations have revealed maximum urban canyon cooling for 'step-up' building configurations and high building spacing ratios (Aly et al., 2022).

Within the broad field of urban design targeted at urban heat and heatwave mitigation, recent research has begun to unveil the importance of the arrangements of elements such as greenspace and blue space that make up the urban surface (Fu et al., 2022). That the presence and spatial configuration of such potentially 'cooling' spaces is important has been highlighted earlier, especially in relation to the association between the uneven distribution of greenspaces and unequal exposure to heatwaves (Yin et al., 2023b). For Changsha city, China and using land surface temperature as a metric of urban heat, researchers have found that the percentage area of urban greenspace is important for reducing heat in residential, industrial, campus, and public zones. For residential zones the cohesion of urban greenspace patches was also an important factor in determining heat mitigation with the degree of cooling positively related to the degree of greenspace aggregation while for commercial zones, patch density was found to be influential (Wang et al., 2023a). These findings point to the importance of not assuming a one-size fits all approach in designing the layout of greenspace given the possible interaction between greenspace traits and urban functional zones. For one of the most intense heatwaves experienced in the city of Melbourne Australia a microscale climate simulation has revealed a substantial impact of parks' spatial layout on microclimate improvement. However, effects were found to be context-specific with no universally effective design evident even across different sites within the same city with the degree of heat mitigation also dependent on additional factors like surface cover characteristics and building configurations. As noted by the researchers this highlights the necessity of conducting site-specific simulations when implementing green infrastructure to mitigate heat anomalies associated with heatwaves. While specific to Melbourne this finding may have wider relevance (Herath et al., 2024). Recognising the divergent views concerning the effectiveness of urban greenspace arrangement in achieving heat mitigation Li et al. (2023b) based on a review of over 160 studies conclude that aggregated, patchy, larger, and complex-shaped urban greenspaces facilitate better cooling. Importantly they highlight the need for urban planners and designers to move away from applying a binary style approach in planning urban greenspace that does not recognise the temporal and spatial heterogeneity of the urban thermal environment. Likely due to the earlier and more rapid development of literature concerning greenspace arrangement compared to that for blue spaces, the body of research examining the relationship between blue space arrangement and heat mitigation seems comparatively limited. Nonetheless, a few studies have underscored the potential significance of blue space configuration at the landscape scale (Ampatzidis et al., 2023; Ampatzidis & Kershaw, 2020; He et al., 2024; Sheng & Wang, 2024).

14.3.4 Combined Effects of Heat Mitigation Options

From the material presented to this point it is apparent that the assortment of heat mitigation options discussed under the three broad headings of urban surface facets and elements, NBS and urban morphology and built form, can on their own reduce atmospheric heat to varying degrees. As noted at the outset of this chapter, in contrast to the deployment of standalone options an ensemble of options may be more impactful in achieving heat mitigation. For example, simulations of various combinations of urban albedo (roofs and pavements) and urban green space for a planned 10,000m^2 development in Osaka Japan have revealed a combination of urban albedo of 0.3 and greenspace coverage of 20% produce the greatest improvements in urban microclimate as measured by reductions in outdoor air temperature and mean radiant temperature (Yuan et al., 2017). Interestingly these investigators also found for greenspace coverage beyond 40% urban heat increased rather than decreased with this coverage value representing a possible upper bound for greenspace cooling effectiveness. For the same city, but for a centrally located densely built up area, Zheng et al. (2022) found amongst four heat mitigation scenarios (cool pavement, cool roof, greenspace and integrated) that replacing the existing asphalt (albedo = 0.1) with light white concrete pavement with high surface albedo (0.8) and lower heat capacity (emissivity of 0.9)—the cool pavement scenario—produced the greatest cooling compared to a base case (present situation). In investigating the effectiveness of 100% cool and green roof coverage as a heat mitigation option for Berlin, Wang et al. (2022c) found cool roofs are more effective than green roofs in mitigating urban heat, particularly during the nighttime. In addition to temperature reduction, both approaches result in decreased wind speeds, lower mean radiant temperatures, and increased relative humidity. These collective impacts contributed to a city-wide decrease in the Universal Thermal Climate Index (UTCI). Although cool roofs demonstrated a greater reduction in UTCI compared to green roofs, both strategies diminished the duration of intense heat stress from 7 hours per day to 5 hours per day with irrigated green roofs outperforming cooling wise non-irrigated green roofs when combined with high albedo cool roofs. To try and establish the most effective combination of heat mitigation options for heat reduction for one of Melbourne's warmest summers Herath et al. (2021) conducted 52 simulations encompassing changes in vegetation type, built-cover ratio, building height, green roofs, and cool roofs. When compared with a baseline scenario, high albedo roofs demonstrated superior heat mitigation during daytime hours (0.85 albedo; reduction of 1.29 °C), while green roofs exhibited optimal efficacy during nighttime (100% coverage; reduction of 1.15 °C). The researchers also found near-linear negative correlations between vegetation ratio, green roof and cool roof coverage and heat levels. Notably, the distribution of trees within both urban parks and street canyons proved more effective for cooling compared to tree planting exclusively within street canyons. For Mandaue City on Cebu Island, Philippines, Cortes et al. (2022) explored how various heat mitigation options such as increasing vegetation, incorporating open spaces, implementing green roofs, or a combination of these might mitigate heat within a street canyon. Their simulation results indicate that expanding urban greenery and trees could reduce air temperatures by an average of 0.2 °C, while green roofs could lower temperatures by an average range of 0.2 °C–0.4 °C. Combining trees, grass surfaces, and green roofs could further decrease air temperatures by 0.1 °C–0.3 °C. However, despite these interventions, these researchers found that a simulated measure of thermal comfort—the temperature humidity index—remained at a level above the comfortable range (>26 °C), indicating conditions of discomfort would persist no matter which heat mitigation strategy was implemented.

Beyond simulations of climatological conditions representative of summer, of interest is the effectiveness of combinations of mitigation options for reducing extreme levels of heat associated with heatwaves. For a heatwave that occurred in August 2020 across Jeju, Republic of Korea, Lee et al. (2023) tested 10 different configurations of urban green infrastructure, including street trees, grass, green facades, and green roofs to find out what combination was the most effective for heatwave mitigation. When compared to areas devoid of greenery, the combined implementation of NBS-based heat mitigation options resulted in significant reductions in mean air temperature (2.1 °C), mean radiant temperature (25.3 °C), and physiological equivalent temperature (16.1 °C). Among the green infrastructure elements, street trees demonstrated the most substantial impact on both mean radiant temperature and physiological equivalent temperature, both crucial indicators of human thermal comfort. Furthermore, the combined effect of the different NBS options was observed to have a more pronounced influence on mean radiant temperature on south-facing sidewalks—the aspect most exposed to shortwave radiation inputs—compared to those facing north; for south-facing sidewalks, the mean radiation temperature was reduced by 21.2 °C, whereas for north-facing sidewalks, the reduction was 8.0 °C. In a similar fashion Imran et al. (2019) for Melbourne, Australia have assessed the effectiveness of combinations of various urban vegetation patches—namely mixed forest, a combination of mixed forest and grassland, and a combination of mixed shrubland and grassland in mitigating UHI effects during a severe heatwave. Simulations were conducted by incrementally increasing the fractions of vegetated patches by 20, 30, 40, and 50%. Results of this investigation indicate that augmenting vegetated fractions from 20 to 50% leads to reductions in the UHI during the night, with mixed forest reducing the UHI by 0.6–3.4 °C, mixed shrubland and grassland by 0.4–3.0 °C, and mixed forest and grassland by 0.6–3.7 °C. However, no cooling effect was observed for near-surface temperatures during the hottest part of the day. Night-time cooling was attributed to decreased storage heat, as vegetated patches divert more net radiation into latent heat flux via evapotranspiration, while sensible heat flux remains largely unchanged (Imran et al., 2019). The degree of UHI reduction was also found to vary nonlinearly with increasing vegetated fractions, with larger fractions yielding significantly greater reductions, particularly up to 50%, with mixed forest and mixed forest and grassland found to be more effective in reducing the UHI compared to mixed shrubland and grassland (Imran et al., 2019). Also, for heatwave conditions in Melbourne but for various scenarios of cool roof adoption throughout the city, increased vegetation cover city-wide, and combinations of these approaches, Jacobs et al. (2018) conducted 11 simulations to establish heat mitigation effectiveness. Cool roofs were found to exhibit daytime urban temperature reduction, and if widely implemented, their cooling effect extended into the night. Conversely, expanding vegetation coverage lowered nighttime temperatures but offered minimal relief during peak daytime heat under heatwave conditions. The combined implementation of cool roofs and increased vegetation yielded the most substantial temperature reduction during heatwaves, albeit without synergistic effects. This suggests that the cooling achieved through combined strategies may not be simply additive. In the drier, less densely populated western suburbs of Melbourne, increased vegetation led to notable cooling without exacerbating human thermal stress, despite the accompanying rise in humidity. Meanwhile, the leafier, moderately dense eastern suburbs showed greater cooling response to cool roof installations. As noted by these researchers such findings underscore the variability in optimal urban cooling strategies within a single urban area. For a widespread and extremely impactful heatwave event that occurred across Melbourne in January 2009 with average maximum and minimum temperatures exceeded by 15 °C and 11.4 °C respectively, Herath et al. (2023) simulated the relative impact of 17 different combinations of heat mitigation options on the components of the surface energy balance and measures of atmospheric heat. They found that integrating multiple surface parameters in terms of tree coverage, green roofs and cool roofs has the potential to mitigate extreme values of energy budget components (e.g., sensible heat, latent heat, and storage heat fluxes), bringing them closer to the current seasonal averages. Among the scenarios tested, the one featuring the highest ground-level vegetation cover of 20% (5% greater than current), along with 50% coverage of green roofs and cool roofs (0.7 albedo), demonstrated the ability to lower air temperatures by up to 2.4 °C with greater cooling exacted in the CBD compared to elsewhere. Like the study of Jacobs et al. (2018) these simulations indicate a combined strategy incorporating both vegetative cover and high-albedo roofs has the potential to achieve greater cooling effectiveness than standalone options. That said the simulations described here for various heat mitigation combinations are for widespread implementation with the assumption made there is perfect acceptability amongst the population. This of course begs the question whether the universal application of engineered and non-engineered heat mitigations options at the city scale are feasible, achievable or indeed acceptable (Cristiano (Cristiano et al., 2023; Rosenzweig et al., 2009; Seddon, 2022). Moreover, as noted above for some individual heat mitigation options, the heat reductions achieved by a combination approach may not in some circumstances reach the requisite level to significantly diminish heatwave related impacts, especially those associated with peak levels of heat. Of course, it is worth recognising that the outcomes of simulations of the climate impacts of combined or single heat mitigation options are dependent on the modelling strategy applied and the uncertainties associated with the modelling process related to model physics and dynamics. Added to this is the challenge associated with constraining models with real world observations for the base or reference case situation because of a lack of systematic field observations at the intra-urban to urban scales and the associated leap of faith that simulated values of heat (e.g., mean radiant temperature) portray the simulated situation accurately. Undoubtedly with improvements in data assimilation and model efficacy, greater confidence will be lent to simulations of the climate effects of heat mitigation options applied individually or in combination.

14.4 Conclusion

This chapter has underscored the distinction between mitigation and adaptation in urban heat management, focusing on physical interventions altering urban-atmosphere heat, mass and momentum exchanges. It has applied the concept of geoengineering the earth's atmosphere to manage the global energy balance to the case of the urban surface as a basis for making a distinction between adaptation (behavioural change) and mitigation (modifying the physical hazard). Based on a review of three broad categories of mitigation options this chapter concludes that it's essential to recognize that while mitigation solutions may alleviate urban heat, their efficacy may vary especially in relation to the reduction of atmospheric as opposed to surface (roof, road, wall) heat. There is also the potential of societal resistance and financial challenges if large urban scale implementation is imagined by municipal authorities. Interdisciplinary research is crucial to understand the effectiveness, scalability, and long-term impacts of mitigation strategies, alongside their interactions with existing urban infrastructure and their implications for vulnerable communities. Optimizing co-benefits such as improved air quality, energy efficiency and water management is also important. Collaboration among various stakeholders and disciplines will be necessary for integrating these solutions into urban landscapes effectively. Furthermore, assessing policy frameworks and governance structures is vital to support widespread adoption of mitigation options. Importantly this chapter concludes that urban scale mitigation efforts alone may not suffice without addressing the root cause of climate change through greenhouse gas emission reduction. Without global action to curb temperature rise, even extensive mitigation measures may ultimately prove insufficient. Thus, while adaptation remains essential, concerted efforts toward mitigation, both locally and globally, are imperative to manage urban heat effectively in the face of enhanced climate variability and climate change.

References

Aboelata, A. (2021). Reducing outdoor air temperature, improving thermal comfort, and saving buildings' cooling energy demand in arid cities – Cool paving utilization. *Sustainable Cities and Society, 68*, 102762. https://doi.org/10.1016/j.scs.2021.102762

Aflaki, A., Mirnezhad, M., Ghaffarianhoseini, A., Omrany, H., Wang, Z. H., & Akbari, H. (2017). Urban heat Island mitigation strategies: A state-of-the-art review on Kuala Lumpur, Singapore and Hong Kong. *Cities, 62*, 131–145. https://doi.org/10.1016/j.cities.2016.09.003

Afshari, A. (2017). A new model of urban cooling demand and heat Island application to vertical greenery systems (VGS). *Energy and Buildings, 157*, 204–217. https://doi.org/10.1016/j.enbuild.2017.01.008

Akbari, H., & Kolokotsa, D. (2016). Three decades of urban heat islands and mitigation technologies research. *Energy and Buildings, 133*, 834–842. https://doi.org/10.1016/j.enbuild.2016.09.067

Aleksandrowicz, O., Vuckovic, M., Kiesel, K., & Mahdavi, A. (2017). Current trends in urban heat Island mitigation research: Observations based on a comprehensive research repository. *Urban Climate, 21*, 1–26. https://doi.org/10.1016/j.uclim.2017.04.002

Algarni, S. (2019). Potential for cooling load reduction in residential buildings using cool roofs in the harsh climate of Saudi Arabia. *Energy & Environment, 30*(2), 235–253. https://doi.org/10.1177/0958305x18787340

AlKhaled, S., Coseo, P., Brazel, A., Cheng, C. W., & Sailor, D. (2020). Between aspiration and actuality: A systematic review of morphological heat mitigation strategies in hot urban deserts. *Urban Climate, 31*, 100570. https://doi.org/10.1016/j.uclim.2019.100570

Al-Obaidi, K. M., Ismail, M., & Rahman, A. M. A. (2014). Passive cooling techniques through reflective and radiative roofs in tropical houses in Southeast Asia: A literature review. *Frontiers of Architectural Research, 3*(3), 283–297. https://doi.org/10.1016/j.foar.2014.06.002

Alsaad, H., Hartmann, M., Hilbel, R., & Voelker, C. (2022). The potential of facade greening in mitigating the effects of heatwaves in central European cities. *Building and Environment, 216*, 109021. https://doi.org/10.1016/j.buildenv.2022.109021

Aly, A. M., Goodlow, B., & Conerly, J. (2022). Urban heat Island mitigation via geometric configuration. *Theoretical and Applied Climatology, 149*(3–4), 1329–1355. https://doi.org/10.1007/s00704-022-04116-2

Ampatzidis, P., & Kershaw, T. (2020). A review of the impact of blue space on the urban microclimate. *Science of the Total Environment, 730*, 139068. https://doi.org/10.1016/j.scitotenv.2020.139068

Ampatzidis, P., Cintolesi, C., & Kershaw, T. (2023). Impact of blue space geometry on urban heat Island mitigation. *Climate, 11*(2). https://doi.org/10.3390/cli11020028

Andersson, E., Langemeyer, J., Borgström, S., McPhearson, T., Haase, D., Kronenberg, J., Barton, D. N., Davis, M., Naumann, S., Röschel, L., & Baró, F. (2019). Enabling green and blue infrastructure to improve contributions to human well-being and equity in urban systems. *Bioscience, 69*(7), 566–574. https://doi.org/10.1093/biosci/biz058

Arumugam, R., Garg, V., Mathur, J., Reddy, N., Gandhi, J., & Fischer, M. L. (2014). Experimental determination of comfort benefits from cool-roof application to an un-conditioned building in India. *Advances in Building Energy Research, 8*(1), 14–27. https://doi.org/10.1080/17512549.2014.890540

Ashie, Y. (2021). Tokyo heat Island adaptation measures. In C. Ren & G. McGregor (Eds.), *Urban climate science for planning healthy cities* (pp. 119–150). Springer.

Ashtari, B., Yeganeh, M., Bemanian, M., & Fakhr, B. V. (2021). A conceptual review of the potential of cool roofs as an effective passive solar technique: Elaboration of benefits and drawbacks. *Frontiers in Energy Research, 9*. https://doi.org/10.3389/fenrg.2021.738182

Assaf, G., & Assaad, R. H. (2023). Models and methods for quantifying the benefits of engineered heat mitigation initiatives: A critical review. *Urban Climate, 51*, 101654. https://doi.org/10.1016/j.uclim.2023.101654

Athmani, W., Sriti, L., Dabaieh, M., & Younsi, Z. (2023). The potential of using passive cooling roof techniques to improve thermal performance and energy efficiency of residential buildings in hot arid regions. *Buildings, 13*(1). https://doi.org/10.3390/buildings13010021

Bach, A. J. E., Palutikof, J. P., Tonmoy, F. N., Smallcombe, J. W., Rutherford, S., Joarder, A. R., Hossain, M., & Jay, O. (2023). Retrofitting passive cooling strategies to combat heat stress in the face of climate change: A case study of a ready-made garment factory in Dhaka. *Bangladesh. Energy and Buildings, 286*. https://doi.org/10.1016/j.enbuild.2023.112954

Baik, J. J., Lim, H., Han, B. S., & Jin, H. G. (2022). Cool-roof effects on thermal and wind environments during heat waves: A case modeling study in Seoul. *South Korea. Urban Climate, 41*. https://doi.org/10.1016/j.uclim.2021.101044

Bamdad, K. (2023). Cool roofs: A climate change mitigation and adaptation strategy for residential buildings. *Building and Environment, 236*, 110271. https://doi.org/10.1016/j.buildenv.2023.110271

Baniassadi, A., Sailor, D. J., Crank, P. J., & Ban-Weiss, G. A. (2018). Direct and indirect effects of high-albedo roofs on energy consumption and thermal comfort of residential buildings. *Energy and Buildings, 178*, 71–83. https://doi.org/10.1016/j.enbuild.2018.08.048

Ben Ratmia, F. Z., Ahriz, A., Santi, G., Bouzaher, S., Mahar, W. A., Ben Ratmia, M. A. E., & Matallah, M. E. (2024). Street design strategies based on spatial configurations and building external envelopes in relation to outdoor thermal comfort in arid climates. *Sustainability, 16*(1). https://doi.org/10.3390/su16010221

Black-Ingersoll, F., de Lange, J., Heidari, L., Negassa, A., Botana, P., Fabian, M. P., & Scammell, M. K. (2022). A literature review of cooling center, misting station, cool pavement, and cool roof intervention evaluations. *Atmosphere, 13*(7). https://doi.org/10.3390/atmos13071103

Blanco, I., & Convertino, F. (2023). Thermal performance of green facades: Research trends analysis using a science mapping approach. *Sustainability, 15*(13). https://doi.org/10.3390/su15139981

Bonan, G. B. (2000). The microclimates of a suburban Colorado (USA) landscape and implications for planning and design. *Landscape and Urban Planning, 49*(3–4), 97–114. https://doi.org/10.1016/s0169-2046(00)00071-2

Bosch, M., Locatelli, M., Hamel, P., Remme, R. P., Jaligot, R., Chenal, J. M., & Joost, S. (2021). Evaluating urban greening scenarios for urban heat mitigation: A spatially explicit approach. Royal Society open. *Science, 8*(12). https://doi.org/10.1098/rsos.202174

Bowler, D. E., Buyung-Ali, L., Knight, T. M., & Pullin, A. S. (2010). Urban greening to cool towns and cities: A systematic review of the empirical evidence. *Landscape and Urban Planning, 97*(3), 147–155.

Broadbent, A. M., Coutts, A. M., Tapper, N. J., Demuzere, M., & Beringer, J. (2018). The microscale cooling effects of water sensitive urban design and irrigation in a suburban environment. *Theoretical and Applied Climatology, 134*(1–2), 1–23. https://doi.org/10.1007/s00704-017-2241-3

Broadbent, A. M., Krayenhoff, E. S., & Georgescu, M. (2020). Efficacy of cool roofs at reducing pedestrian-level air temperature during projected 21st century heatwaves in Atlanta, Detroit, and Phoenix (USA). *Environmental Research Letters, 15*(8). https://doi.org/10.1088/1748-9326/ab6a23

Broadbent, A. M., Declet-Barreto, J., Krayenhoff, E. S., Harlan, S. L., & Georgescu, M. (2022). Targeted implementation of cool roofs for equitable urban adaptation to extreme heat. *Science of the Total Environment, 811*, 151326. https://doi.org/10.1016/j.scitotenv.2021.151326

Bunker, A., Compoaré, G., Sewe, M. O., Laurent, J. G. C., Zabre, P., Boudo, V., Ouédraogo, W. A., Ouermi, L., Jackson, S. T., Arisco, N., Vijayakumar, G., Yildirim, F. B., Barteit, S., Maggioni, M. A., Woodward, A., Buonocore, J. J., Regassa, M. D., Brueck, T., Sié, A., & Bärnighausen, T. (2024). The effects of cool roofs on health, environmental, and economic outcomes in rural Africa: Study protocol for a community-based cluster randomized controlled trial. *Trials, 25*(1), 59. https://doi.org/10.1186/s13063-023-07804-0

Cai, Z., Guldmann, J. M., Tang, Y., & Han, G. F. (2022). Does city-water layout matter? Comparing the cooling effects of water bodies across 34 Chinese megacities. *Journal of Environmental Management, 324*, 116263. https://doi.org/10.1016/j.jenvman.2022.116263

Cai, X., Yang, J., Zhang, Y., Xiao, X., & Xia, J. C. (2023). Cooling Island effect in urban parks from the perspective of internal park landscape. *Humanities and Social Sciences Communications, 10*(1), 1–12.

Cao, M. C., Rosado, P., Lin, Z. H., Levinson, R., & Millstein, D. (2015). Cool roofs in Guangzhou, China: Outdoor air temperature reductions during heat waves and typical summer conditions. *Environmental Science & Technology, 49*(24), 14672–14679. https://doi.org/10.1021/acs.est.5b04886

Cavan, G., Baker, F., Tzoulas, K., & Smith, C. L. (2022). Manchester: The role of urban domestic gardens in climate adaptation and resilience. In C. Ren & G. McGregor (Eds.), *Urban climate science for planning healthy cities* (pp. 99–118). Springer.

Chafer, M., Cabeza, L. F., Pisello, A. L., Tan, C. L., & Wong, N. H. (2021). Trends and gaps in global research of greenery systems through a bibliometric analysis. *Sustainable Cities and Society, 65*, 102608. https://doi.org/10.1016/j.scs.2020.102608

Chang, N., & Li, D. W. (2016). A study of the temperature-humidity effect and luminous environment design for urban green space. *3rd International Conference on Applied Engineering, 51*, 103–108. https://doi.org/10.3303/cet1651018

Chang, J. H., Huang, K. T., & Wang, J. C. (2023). Energy performance of school roofing materials in hot and humid climates. *Case Studies in Construction Materials, 19*, e02586. https://doi.org/10.1016/j.cscm.2023.e02586

Chen, Z., Zhao, L., Meng, Q., Wang, C., Zhai, Y., & Wang, F. (2009). Field measurements on microclimate in residential community in Guangzhou, China. *Frontiers of Structural and Civil Engineering, 3*(4), 462–468. https://doi.org/10.1007/s11709-009-0066-6

Cheng, C. Y., Cheung, K. K. S., & Chu, L. M. (2010). Thermal performance of a vegetated cladding system on facade walls. *Building and Environment, 45*(8), 1779–1787. https://doi.org/10.1016/j.buildenv.2010.02.005

Cheng, Y. T., Bartesaghi-Koc, C., Tian, Y., Shen, L. D., Teng, M. J., Liu, H. S., Xiao, Z. Y., Zhang, B., & Wu, C. G. (2023). Where and how to cool through blue infrastructure? Large lake groups to ameliorate urban overheating in a typical inland multi-lake megacity. *Sustainable Cities and Society, 98*, 104869. https://doi.org/10.1016/j.scs.2023.104869

Cheung, P. K., Jim, C. Y., Tapper, N., Nice, K. A., & Livesley, S. J. (2022). Daytime irrigation leads to significantly cooler private backyards in summer. *Urban Climate, 46*, 101310. https://doi.org/10.1016/j.uclim.2022.101310

Chiquet, C., Dover, J. W., & Mitchell, P. (2013). Birds and the urban environment: The value of green walls. *Urban Ecosystems, 16*(3), 453–462. https://doi.org/10.1007/s11252-012-0277-9

Chow, W. T. (2024). Reducing urban overheating risks through climate-resilient development in the warming tropics. *One Earth, 7*(1), 6–10.

Convertino, F., Vox, G., & Schettini, E. (2020). Thermal barrier effect of green facades: Long-wave infrared radiative energy transfer modelling. *Building and Environment, 177*, 106875. https://doi.org/10.1016/j.buildenv.2020.106875

Convertino, F., Vox, G., & Schettini, E. (2021). Evaluation of the cooling effect provided by a green facade as nature-based system for buildings. *Building and Environment, 203*, 108099. https://doi.org/10.1016/j.buildenv.2021.108099

Convertino, F., Blanco, I., Schettini, E., & Vox, G. (2023). A nature-based system for improving Mediterranean buildings' performance: Contribution to energy saving by heat transfer reduction and influence of climatic parameters. *Journal of Agricultural Engineering, 54*, 3. https://doi.org/10.4081/jae.2023.1520

Cortes, A., Rejuso, A. J., Santos, J. A., & Blanco, A. (2022). Evaluating mitigation strategies for urban heat Island in Mandaue City using ENVI-met. *Journal of Urban Management, 11*(1), 97–106. https://doi.org/10.1016/j.jum.2022.01.002

Coutts, A. M., Tapper, N. J., Beringer, J., Loughnan, M., & Demuzere, M. (2013). Watering our cities: The capacity for water sensitive urban design to support urban cooling and improve human thermal comfort in the Australian context. *Progress in Physical Geography-Earth and Environment, 37*(1), 2–28. https://doi.org/10.1177/0309133312461032

Coutts, A. M., White, E. C., Tapper, N. J., Beringer, J., & Livesley, S. J. (2016). Temperature and human thermal comfort effects of street trees across three contrasting street canyon environments. *Theoretical and Applied Climatology, 124*(1–2), 55–68. https://doi.org/10.1007/s00704-015-1409-y

Cristiano, E., Deidda, R., & Viola, F. (2023). Awareness and willingness to pay for green roofs in Mediterranean areas. *Journal of Environmental Management, 344*, 118419. https://doi.org/10.1016/j.jenvman.2023.118419

Cui, Y. F., Hu, L. Q., Wang, Z. H., & Li, Q. (2021). Urban nocturnal cooling mediated by bluespace. *Theoretical and Applied Climatology, 146*(1–2), 277–292. https://doi.org/10.1007/s00704-021-03727-5

Cureau, R. J., Pigliautile, I., & Pisello, A. L. (2023). Seasonal and diurnal variability of a water body's effects on the urban microclimate in a coastal city in Italy. *Urban Climate, 49*, 101437. https://doi.org/10.1016/j.uclim.2023.101437

Daemei, A. B., Azmoodeh, M., Zamani, Z., & Khotbehsara, E. M. (2018). Experimental and simulation studies on the thermal behavior of vertical greenery system for temperature mitigation in urban spaces. *Journal of Building Engineering, 20*, 277–284. https://doi.org/10.1016/j.jobe.2018.07.024

Dawson, A., Mallick, R., Hernandez, A. G., & Dehdezi, P. K. (2014). Energy harvesting from pavements. *Climate Change, Energy, Sustainability and Pavements*, 481–517. https://doi.org/10.1007/978-3-662-44719-2_18

de Jesus, M. P., Lourenço, J. M., Arce, R. M., & Macias, M. (2017). Green façades and in situ measurements of outdoor building thermal behaviour. *Building and Environment, 119*, 11–19.

de la Flor, F. S., & Domínguez, S. A. (2004). Modelling microclimate in urban environments and assessing its influence on the performance of surrounding buildings. *Energy and Buildings, 36*(5), 403–413. https://doi.org/10.1016/j.enbuild.2004.01.050

Demuzere, M., Orru, K., Heidrich, O., Olazabal, E., Geneletti, D., Orru, H., Bhave, A. G., Mittal, N., Feliu, E., & Faehnle, M. (2014). Mitigating and adapting to climate change: Multi-functional and multi-scale assessment of green urban infrastructure. *Journal of Environmental Management, 146*, 107–115. https://doi.org/10.1016/j.jenvman.2014.07.025

Djedjig, R., Belarbi, R., & Bozonnet, E. (2017). Green wall impacts inside and outside buildings: Experimental study. Materials & energy I (2015) / materials & energy ii (2016) 139:578-583. https://doi.org/10.1016/j.egypro.2017.11.256

Djekic, J., Djukic, A., Vukmirovic, M., Djekic, P., & Brankovic, M. D. (2018). Thermal comfort of pedestrian spaces and the influence of pavement materials on warming up during summer. *Energy and Buildings, 159*, 474–485. https://doi.org/10.1016/j.enbuild.2017.11.004

Downing, J. A. (2010). Emerging global role of small lakes and ponds: Little things mean a lot. *Limnetica, 29*(1), 9–23.

Duarte, D. H. S., Shinzato, P., Gusson, C. D., & Alves, C. A. (2015). The impact of vegetation on urban microclimate to counterbalance built density in a subtropical changing climate. *Urban Climate, 14*, 224–239. https://doi.org/10.1016/j.uclim.2015.09.006

Elnabawi, M. H., Alhumaidi, A., Osman, B., & Alshehhi, R. (2022). Cool roofs in hot climates: A conceptual review of modelling methods and limitations. *Buildings, 12*(11). https://doi.org/10.3390/buildings12111968

Elnabawi, M. H., Hamza, N., & Raveendran, R. (2023). 'Super cool roofs': Mitigating the UHI effect and enhancing urban thermal comfort with high albedo-coated roofs. *Results in Engineering, 19*, 101269. https://doi.org/10.1016/j.rineng.2023.101269

Erell, E., Pearlmutter, D., Boneh, D., & Bar Kutiel, P. (2014). Effect of high-albedo materials on pedestrian heat stress in urban street canyons. *Urban Climate, 10*, 367–386. https://doi.org/10.1016/j.uclim.2013.10.005

Falasca, S., Ciancio, V., Salata, F., Golasi, I., Rosso, F., & Curci, G. (2019). High albedo materials to counteract heat waves in cities: An assessment of meteorology, buildings energy needs and pedestrian thermal comfort. *Building and Environment, 163*, 106242. https://doi.org/10.1016/j.buildenv.2019.106242

Fan, H. Y., Yu, Z. W., Yang, G. Y., Liu, T. Y., Hung, C. H., & Vejre, H. (2019). How to cool hot-humid (Asian) cities with urban trees? An optimal landscape size perspective. *Agricultural and Forest Meteorology, 265*, 338–348. https://doi.org/10.1016/j.agrformet.2018.11.027

Fang, Y. H., Zhao, L. Y., Dou, B. Y., Li, Y., & Wang, S. X. (2023). Circuit VRC: A circuit theory-based ventilation corridor model for mitigating the urban heat islands. *Building and Environment, 244*, 110786. https://doi.org/10.1016/j.buildenv.2023.110786

Faragallah, R. N., & Ragheb, R. A. (2022). Evaluation of thermal comfort and urban heat Island through cool paving materials using ENVI-met. *Ain Shams Engineering Journal, 13*, 3. https://doi.org/10.1016/j.asej.2021.10.004

Fawzy, S., Osman, A. I., Doran, J., & Rooney, D. W. (2020). Strategies for mitigation of climate change: A review. *Environmental Chemistry Letters, 18*(6), 2069–2094. https://doi.org/10.1007/s10311-020-01059-w

Feng, J., Khan, A., Doan, Q. V., Gao, K., & Santamouris, M. (2021). The heat mitigation potential and climatic impact of super-cool broadband radiative coolers on a city scale. Cell reports physical. *Science, 2*(7), 100485. https://doi.org/10.1016/j.xcrp.2021.100485

Feng, J., Saliari, M., Gao, K., & Santamouris, M. (2022). On the cooling energy conservation potential of super cool roofs. *Energy and Buildings, 264*, 112076. https://doi.org/10.1016/j.enbuild.2022.112076

Feng, J., Gao, K., Khan, H., Ulpiani, G., Vasilakopoulou, K., Yun, G. Y., & Santamouris, M. (2023). Overheating of cities: Magnitude, characteristics, impact, mitigation and adaptation, and future challenges. *Annual Review of Environment and Resources, 48*, 651–679. https://doi.org/10.1146/annurev-environ-112321-093021

Fonseca, F., Paschoalino, M., & Silva, L. (2023). Health and Well-being benefits of outdoor and indoor vertical greening systems: A review. *Sustainability, 15*(5), 4107.

Frantzeskaki, N. (2019). Seven lessons for planning nature-based solutions in cities. *Environmental Science & Policy, 93*, 101–111. https://doi.org/10.1016/j.envsci.2018.12.033

Frantzeskaki, N., McPhearson, T., Collier, M. J., Kendal, D., Bulkeley, H., Dumitru, A., Walsh, C., Noble, K., Van Wyk, E., Ordóñez, C., Oke, C., & Pintér, L. (2019). Nature-based solutions for urban climate change adaptation: Linking science, policy, and practice communities for evidence-based decision-making. *Bioscience, 69*(6), 455–466. https://doi.org/10.1093/biosci/biz042

Fritz M (2017) Impacts of climate change on urban areas and nature-based solutions for adaptation. Nature-Based Solutions to Climate Change Adaptation in Urban Areas: Linkages between Science, Policy and Practice:15–27. doi:https://doi.org/10.1007/978-3-319-56091-5_2.

Fu, J. W., Dupre, K., Tavares, S., King, D., & Banhalmi-Zakar, Z. (2022). Optimized greenery configuration to mitigate urban heat: A decade systematic review. *Frontiers of Architectural Research, 11*(3), 466–491. https://doi.org/10.1016/j.foar.2021.12.005

Gaffin, S. R., Imhoff, M., Rosenzweig, C., Khanbilvardi, R., Pasqualini, A., Kong, A. Y. Y., Grillo, D., Freed, A., Hillel, D., & Hartung, E. (2012). Bright is the new black-multi-year performance of high-albedo roofs in an urban climate. *Environmental Research Letters, 7*(1). https://doi.org/10.1088/1748-9326/7/1/014029

Gao, K., Haddad, S., Paolini, R., Feng, J., Altheeb, M., Al Mogirah, A., Moammar, A. B., & Santamouris, M. (2024). The use of green infrastructure and irrigation in the mitigation of urban heat in a desert city. *Building Simulation., 17*, 679–694. https://doi.org/10.1007/s12273-024-1110-0

Gapski, N. H., Marinoski, D. L., Melo, A. P., & Guths, S. (2023). Impact of urban surfaces' solar reflectance on air temperature and radiation flux. *Sustainable Cities and Society, 96*, 104645. https://doi.org/10.1016/j.scs.2023.104645

Georgescu, M. (2015). Challenges associated with adaptation to future urban expansion. *Journal of Climate, 28*(7), 2544–2563. https://doi.org/10.1175/jcli-d-14-00290.1

Géron, C., Lembrechts, J. J., Nijs, I., & Monty, A. (2022). Woody invaders from contrasted climatic origins distribute differently across the urban-to-rural gradient in oceanic Europe – Is it trait-related? *Urban Forestry & Urban Greening, 75*, 127694. https://doi.org/10.1016/j.ufug.2022.127694

Gillerot, L., Landuyt, D., De Frenne, P., Muys, B., & Verheyen, K. (2024). Urban tree canopies drive human heat stress mitigation. *Urban Forestry & Urban Greening, 92*, 128192. https://doi.org/10.1016/j.ufug.2023.128192

Gober, P., Brazel, A., Quay, R., Myint, S., Grossman-Clarke, S., Miller, A., & Rossi, S. (2010). Using watered landscapes to manipulate urban heat Island effects: How much water will it take to cool Phoenix? *Journal of the American Planning Association, 76*(1), 109–121. https://doi.org/10.1080/01944360903433113

Goel, M., Jha, B., & Khan, S. (2022). Living walls enhancing the urban realm: A review. *Environmental Science and Pollution Research, 29*(26), 38715–38734. https://doi.org/10.1007/s11356-022-19501-7

Gong, X., Liu, Q. T., Lv, Y., Chen, S. C., Wu, S. P., & Ying, H. (2023). A systematic review on the strategies of reducing asphalt pavement temperature. *Case Studies in Construction Materials, 18*, e01852. https://doi.org/10.1016/j.cscm.2023.e01852

Grossman-Clarke, S., Zehnder, J. A., Loridan, T., & Grimmond, C. S. B. (2010). Contribution of land use changes to near-surface air temperatures during recent summer extreme heat events in the Phoenix metropolitan area. *Journal of Applied Meteorology and Climatology, 49*(8), 1649–1664.

Guo, A. D., Yue, W. Z., Yang, J., Li, M. M., Xie, P., He, T. T., Zhang, M. X., & Yu, H. S. (2023). Quantifying the impact of urban ventilation corridors on thermal environment in Chinese megacities. *Ecological Indicators, 156*, 111072. https://doi.org/10.1016/j.ecolind.2023.111072

Hamada, S., & Ohta, T. (2010). Seasonal variations in the cooling effect of urban green areas on surrounding urban areas. *Urban Forestry & Urban Greening, 9*(1), 15–24. https://doi.org/10.1016/j.ufug.2009.10.002

Han, D. L., Zhang, T. T., Qin, Y. X., Tan, Y. F., & Liu, J. (2023). A comparative review on the mitigation strategies of urban heat Island (UHI): A pathway for sustainable urban development. *Climate and Development, 15*(5), 379–403. https://doi.org/10.1080/17565529.2022.2092051

Hathway, E. A., & Sharples, S. (2012). The interaction of rivers and urban form in mitigating the urban Heat Island effect: A UK case study. *Building and Environment, 58*, 14–22. https://doi.org/10.1016/j.buildenv.2012.06.013

Hayes, A. T., Jandaghian, Z., Lacasse, M. A., Gaur, A., Lu, H., Laouadi, A., Ge, H., & Wang, L. (2022). Nature-based solutions (nbss) to mitigate urban heat Island (UHI) effects in Canadian cities. *Buildings, 12*(7), 925.

He, B. J., Wang, W., Sharifi, A., & Liu, X. (2023). Progress, knowledge gap and future directions of urban heat mitigation and adaptation research through a bibliometric review of history and evolution. *Energy and Buildings, 287*, 112976. https://doi.org/10.1016/j.enbuild.2023.112976

He, J., Shi, Y., Xu, L., Lu, Z., & Feng, M. (2024). An investigation on the impact of blue and green spatial pattern alterations on the urban thermal environment: A case study of Shanghai. *Ecological Indicators, 158*, 111244. https://doi.org/10.1016/j.ecolind.2023.111244

Herath, P., Thatcher, M., Jin, H. D., & Bai, X. M. (2021). Effectiveness of urban surface characteristics as mitigation strategies for the excessive summer heat in cities. *Sustainable Cities and Society, 72*, 103072. https://doi.org/10.1016/j.scs.2021.103072

Herath, P., Thatcher, M., Jin, H. D., & Bai, X. M. (2023). Comparing the cooling effectiveness of operationalisable urban surface combination scenarios for summer heat mitigation. *Science of the Total Environment, 874*, 162476. https://doi.org/10.1016/j.scitotenv.2023.162476

Herath, P., Bai, X. M., Jin, H. D., & Thatcher, M. (2024). Does the spatial configuration of urban parks matter in ameliorating extreme heat? *Urban Climate, 53*, 101756. https://doi.org/10.1016/j.uclim.2023.101756

Hernández-Pérez, I. (2021). Influence of traditional and solar reflective coatings on the heat transfer of building roofs in Mexico. *Applied Sciences-Basel, 11*(7). https://doi.org/10.3390/app11073263

Heusinger, J., Sailor, D. J., & Weber, S. (2018). Modeling the reduction of urban excess heat by green roofs with respect to different irrigation scenarios. *Building and Environment, 131*, 174–183. https://doi.org/10.1016/j.buildenv.2018.01.003

Hoelscher, M.-T., Nehls, T., Jaenicke, B., & Wessolek, G. (2016). Quantifying cooling effects of facade greening: Shading, transpiration and insulation. *Energy and Buildings, 114*, 283–290. https://doi.org/10.1016/j.enbuild.2015.06.047

Hong, D.-L., & Chien, S.-S. (2018). Summoning wind for urban cooling: Urban wind corridor projects in China. In A. Cheshmehzangi & C. Butters (Eds.), *Designing cooler cities: Energy, cooling and urban form: The Asian perspective* (pp. 137–150). Springer.

Howe, D. A., Hathaway, J. M., Ellis, K. N., & Mason, L. R. (2017). Spatial and temporal variability of air temperature across urban neighborhoods with varying amounts of tree canopy. *Urban Forestry & Urban Greening, 27*, 109–116. https://doi.org/10.1016/j.ufug.2017.07.001

Hsieh, C. M., & Huang, H. C. (2016). Mitigating urban heat islands: A method to identify potential wind corridor for cooling and ventilation. *Computers Environment and Urban Systems, 57*, 130–143. https://doi.org/10.1016/j.compenvurbsys.2016.02.005

Hsieh, C. M., Yu, C. Y., & Shao, L. Y. (2023). Improving the local wind environment through urban design strategies in an urban renewal process to mitigate urban heat Island effects. *Journal of Urban Planning and Development, 149*(2). https://doi.org/10.1061/jupddm.upeng-3923

Huo, H. Y., Tao, J., Zhang, W. Z., Guo, L., Leng, P., & Li, Z. L. (2023). Urban 3D spatial thermal environment simulation and comparative analysis of Chinese ancient/modern typical garden based on CFD. *International Journal of Remote Sensing.*, 1–27. https://doi.org/10.1080/01431161.2023.2227321

Imran, H. M., Kala, J., Ng, A. W. M., & Muthukumaran, S. (2019). Effectiveness of vegetated patches as green infrastructure in mitigating urban Heat Island effects during a heatwave event in the city of Melbourne. *Weather and Climate Extremes, 25*, 100217. https://doi.org/10.1016/j.wace.2019.100217

Irfeey, A. M. M., Chau, H. W., Sumaiya, M. M. F., Wai, C. Y., Muttil, N., & Jamei, E. (2023). Sustainable mitigation strategies for urban heat Island effects in urban areas. *Sustainability, 15*(14). https://doi.org/10.3390/su151410767

Irmak, M. A., Yilmaz, S., & Dursun, D. (2017). Effect of different pavements on human thermal comfort conditions. *Atmosfera, 30*(4), 355–366. https://doi.org/10.20937/atm.2017.30.04.06

Ishii, A., Iwamoto, S., Katayama, T., Hayashi, T., Shiotsuki, Y., Kitayama, H., Tsutsumi, J. I., & Nishida, M. (1991). A comparison of field surveys on the thermal environment in urban areas surrounding a large pond – When filled and when drained. *Energy and Buildings, 16*(3–4), 965–971. https://doi.org/10.1016/0378-7788(91)90091-g

Jacobs, S. J., Gallant, A. J. E., Tapper, N. J., & Li, D. (2018). Use of cool roofs and vegetation to mitigate urban heat and improve human thermal stress in Melbourne, Australia. *Journal of Applied Meteorology and Climatology, 57*(8), 1747–1764. https://doi.org/10.1175/jamc-d-17-0243.1

Jaganmohan, M., Knapp, S., Buchmann, C. M., & Schwarz, N. (2016). The bigger, the better? The influence of urban green space design on cooling effects for residential areas. *Journal of Environmental Quality, 45*(1), 134–145. https://doi.org/10.2134/jeq2015.01.0062

Jamei, E., & Tapper, N. (2019). WSUD and urban heat Island effect mitigation. In A. Sharma, T. Gardner, & D. Begbie (Eds.), *Approaches to water sensitive Urban Design* (pp. 381–407). Elsevier.

Jamei, E., Ossen, D. R., Seyedmahmoudian, M., Sandanayake, M., Stojcevski, A., & Horan, B. (2020). Urban design parameters for heat mitigation in tropics. *Renewable & Sustainable Energy Reviews, 134*, 110362. https://doi.org/10.1016/j.rser.2020.110362

Jamei, E., Chau, H. W., Seyedmahmoudian, M., & Stojcevski, A. (2021). Review on the cooling potential of green roofs in different climates. *Science of the Total Environment, 791*, 148407. https://doi.org/10.1016/j.scitotenv.2021.148407

Jamei, E., Chau, H. W., Seyedmahmoudian, M., Mekhilef, S., & Hafez, F. S. (2023). Green roof and energy-role of climate and design elements in hot and temperate climates. *Heliyon, 9*(5), e15917. https://doi.org/10.1016/j.heliyon.2023.e15917

Jandaghian, Z., & Akbari, H. (2021). Increasing urban albedo to reduce heat-related mortality in Toronto and Montreal, Canada. *Energy and Buildings, 237*, 110697.

Jim, C. Y. (2017). Green roof evolution through exemplars: Germinal prototypes to modern variants. *Sustainable Cities and Society, 35*, 69–82. https://doi.org/10.1016/j.scs.2017.08.001

Jin, C. Q., Bai, X. L., Luo, T., & Zou, M. (2018). Effects of green roofs' variations on the regional thermal environment using measurements and simulations in Chongqing, China. *Urban Forestry & Urban Greening, 29*, 223–237. https://doi.org/10.1016/j.ufug.2017.12.002

Kalff, J. (2002). *Limnology: Inland water ecosystems*. Prentice Hall.

Kappou, S., Souliotis, M., Papaefthimiou, S., Panaras, G., Paravantis, J. A., Michalena, E., Hills, J. M., Vouros, A. P., Ntymenou, A., & Mihalakakou, G. (2022). Cool pavements: State of the art and new technologies. *Sustainability, 14*(9). https://doi.org/10.3390/su14095159

Kasprzyk, M., Szpakowski, W., Poznanska, E., Boogaard, F. C., Bobkowska, K., & Gajewska, M. (2022). Technical solutions and benefits of introducing rain gardens – Gdansk case study. *Science of the Total Environment, 835*, 155487. https://doi.org/10.1016/j.scitotenv.2022.155487

Kenai, M. A., Libessart, L., Lassue, S., & Defer, D. (2020). Impact of plants obscuration on energy balance: Theoretical and numerical study. *Journal of Building Engineering, 29*, 101112. https://doi.org/10.1016/j.jobe.2019.101112

Khan, A., Carlosena, L., Feng, J., Khorat, S., Khatun, R., Doan, Q. V., & Santamouris, M. (2022). Optically modulated passive broadband daytime radiative cooling materials can cool cities in summer and heat cities in winter. *Sustainability, 14*(3). https://doi.org/10.3390/su14031110

Khan, A., Carlosena, L., Khorat, S., Khatun, R., Das, D., Doan, Q. V., Hamdi, R., Aziz, S. M., Akbari, H., Santamouris, M., & Niyogi, D. (2023). Urban cooling potential and cost comparison of heat mitigation techniques for their impact on the lower atmosphere. Computational urban. *Science, 3*(1). https://doi.org/10.1007/s43762-023-00101-1

Khorat, S., Das, D., Khatun, R., Aziz, S. M., Anand, P., Khan, A., Santamouris, M., & Niyogi, D. (2024). Cool roof strategies for urban thermal resilience to extreme heatwaves in tropical cities. *Energy and Buildings, 302*, 113751. https://doi.org/10.1016/j.enbuild.2023.113751

Kim, E. S., Yun, S. H., Lee, D. K., Kim, N. Y., Piao, Z. G., Kim, S. H., & Park, S. (2023). Quantifying outdoor cooling effects of vertical greening system on mean radiant temperature. *Developments in the Built Environment, 15*, 100211. https://doi.org/10.1016/j.dibe.2023.100211

Kirschner, V., Macku, K., Moravec, D., & Manas, J. (2023). Measuring the relationships between various urban green spaces and local climate zones. *Scientific Reports, 13*(1), 9799. https://doi.org/10.1038/s41598-023-36850-6

Kleerekoper, L., van Esch, M., & Salcedo, T. B. (2012). How to make a city climate-proof, addressing the urban heat Island effect. *Resources Conservation and Recycling, 64*, 30–38. https://doi.org/10.1016/j.resconrec.2011.06.004

Ko, J., Schlaerth, H., Bruce, A., Sanders, K., & Ban-Weiss, G. (2022). Measuring the impacts of a real-world neighborhood-scale cool pavement deployment on albedo and temperatures in Los Angeles. *Environmental Research Letters, 17*(4). https://doi.org/10.1088/1748-9326/ac58a8

Koc, C. B., Osmond, P., & Peters, A. (2018). Evaluating the cooling effects of green infrastructure: A systematic review of methods, indicators and data sources. *Solar Energy, 166*, 486–508. https://doi.org/10.1016/j.solener.2018.03.008

Koch, K., Ysebaert, T., Denys, S., & Samson, R. (2020). Urban heat stress mitigation potential of green walls: A review. *Urban Forestry & Urban Greening, 55*, 126843. https://doi.org/10.1016/j.ufug.2020.126843

Köhler, M. (2008). Green facades—A view back and some visions. *Urban Ecosystems, 11*, 423–436.

Kolokotroni, M., Gowreesunker, B. L., & Giridharan, R. (2013). Cool roof technology in London: An experimental and modelling study. *Energy and Buildings, 67*, 658–667. https://doi.org/10.1016/j.enbuild.2011.07.011

Kousis, I., & Pisello, A. L. (2023). Evaluating the performance of cool pavements for urban heat Island mitigation under realistic conditions: A systematic review and meta-analysis. *Urban Climate, 49*, 101470. https://doi.org/10.1016/j.uclim.2023.101470

Krayenhoff, E. S., Broadbent, A. M., Zhao, L., Georgescu, M., Middel, A., Voogt, J. A., Martilli, A., Sailor, D. J., & Erell, E. (2021). Cooling hot cities: A systematic and critical review of the numerical modelling literature. *Environmental Research Letters, 16*(5). https://doi.org/10.1088/1748-9326/abdcf1

Kumar, P., Debele, S. E., Khalili, S., Halios, C. H., Sahani, J., Aghamohammadi, N., Andrade, M. F., Athanassiadou, M., Bhui, K., Calvillo, N., Cao, S.-J., Coulon, F., Edmondson, J. L., Fletcher, D., Dias de Freitas, E., Guo, H., Hort, M. C., Katti, M., Kjeldsen, T. R., Lehmann, S., Locosselli, G. M., Malham, S. K., Morawska, L., Parajuli, R., Rogers, C. D. F., Yao, R., Wang, F., Wenk, J., & Jones, L. (2024). Urban heat mitigation by green and blue infrastructure: Drivers, effectiveness, and future needs. *Innovation (Cambridge (Mass)), 5*(2), 100588–100588. https://doi.org/10.1016/j.xinn.2024.100588

Lachapelle, J. A., Krayenhoff, E. S., Middel, A., Coseo, P., & Warland, J. (2023). Maximizing the pedestrian radiative cooling benefit per street tree. *Landscape and Urban Planning, 230*, 104608. https://doi.org/10.1016/j.landurbplan.2022.104608

Lam, C. K. C., Gallant, A. J. E., & Tapper, N. J. (2020). Does irrigation cooling effect intensify during heatwaves? A case study in the Melbourne botanic gardens. *Urban Forestry & Urban Greening, 55*, 126815. https://doi.org/10.1016/j.ufug.2020.126815

Le Phuc, C. L., Nguyen, H. S., Dao Dinh, C., Tran, N. B., Pham, Q. B., & Nguyen, X. C. (2022). Cooling Island effect of urban lakes in hot waves under foehn and climate change. *Theoretical and Applied Climatology, 149*(1), 817–830.

Le, M. T., Thi, A. T.C., & Nguyen, A. Q. T. (2019). The role of green space in the urbanization of Hanoi city. In: Tashkent Inst I, Agr Mechanizat E (Eds.), *22nd international scientific conference on construction – The formation of living environment (FORM)*, Tashkent Institute of Irrigation and Agricultural Mechanization Engineers institute, Tashkent, Uzbekistan, Apr 18–21 2019. E3S web of conferences. https://doi.org/10.1051/e3sconf/20199701013.

Lee, L. S. H., & Jim, C. Y. (2019). Energy benefits of green-wall shading based on novel-accurate apportionment of short-wave radiation components. *Applied Energy, 238*, 1506–1518. https://doi.org/10.1016/j.apenergy.2019.01.161

Lee, H., & Mayer, H. (2018). Thermal comfort of pedestrians in an urban street canyon is affected by increasing albedo of building walls. *International Journal of Biometeorology, 62*(7), 1199–1209. https://doi.org/10.1007/s00484-018-1523-5

Lee, H., Lim, H., & Park, S. (2023). Quantitative assessment of green coverage changes under the human-biometeorological perspective: A simulation case study in Jeju, Republic of Korea. *Sustainable Cities and Society, 97*, 104734. https://doi.org/10.1016/j.scs.2023.104734

Li, D., Bou-Zeid, E., & Oppenheimer, M. (2014). The effectiveness of cool and green roofs as urban heat Island mitigation strategies. *Environmental Research Letters, 9*(5). https://doi.org/10.1088/1748-9326/9/5/055002

Li, Y. Y., Xia, M., Ma, Q., Zhou, R., Liu, D., & Huang, L. C. (2022). Identifying the influencing factors of cooling effect of urban blue infrastructure using the Geodetector model. *Remote Sensing, 14*(21). https://doi.org/10.3390/rs14215495

Li, X. S., Lin, K., Shu, Y. L., & Lin, X. D. (2023a). Comparison of the influences of different ventilation corridor forms on the thermal environment in Wuhan City in summer. *Scientific Reports, 13*(1), 13416. https://doi.org/10.1038/s41598-023-40211-8

Li, Y., Ren, C., Ho, J. Y.-E., & Shi, Y. (2023b). Landscape metrics in assessing how the configuration of urban green spaces affects their cooling effect: A systematic review of empirical studies. *Landscape and Urban Planning, 239*, 104842. https://doi.org/10.1016/j.landurbplan.2023.104842

Liu, Y. H., Xuan, C. Y., Xu, Y. M., Fu, N., Xiong, F. L., & Gan, L. (2022). Local climate effects of urban wind corridors in Beijing. *Urban Climate, 43*, 101181. https://doi.org/10.1016/j.uclim.2022.101181

Macintyre, H. L., & Heaviside, C. (2019). Potential benefits of cool roofs in reducing heat-related mortality during heatwaves in a European city. *Environment International, 127*, 430–441. https://doi.org/10.1016/j.envint.2019.02.065

Makido, Y., Hellman, D., & Shandas, V. (2019). Nature-based designs to mitigate urban heat: The efficacy of green infrastructure treatments in Portland, Oregon. *Atmosphere, 10*(5). https://doi.org/10.3390/atmos10050282

Malcoti, M. D., Zia, H., & Kabre, C. (2023). Relevance of urban geometry and outdoor thermal comfort in urban heat Island mitigation – A review. *Journal of Climate Change, 9*(1), 39–52. https://doi.org/10.3233/jcc230005

Manteghi, G., & Mostofa, T. (2020). Evaporative pavements as an urban heat Island (UHI) mitigation strategy: A review. *International Transaction Journal of Engineering Management & Applied Sciences & Technologies, 11*(1). https://doi.org/10.14456/itjemast.2020.17

Marando, F., Heris, M. P., Zulian, G., Udías, A., Mentaschi, L., Chrysoulakis, N., Parastatidis, D., & Maes, J. (2022). Urban heat Island mitigation by green infrastructure in European functional urban areas. *Sustainable Cities and Society, 77*, 103564. https://doi.org/10.1016/j.scs.2021.103564

Matthews, T., Lo, A. Y., & Byrne, J. A. (2015). Reconceptualizing green infrastructure for climate change adaptation: Barriers to adoption and drivers for uptake by spatial planners. *Landscape and Urban Planning, 138*, 155–163. https://doi.org/10.1016/j.landurbplan.2015.02.010

Meili, N., Manoli, G., Burlando, P., Carmeliet, J., Chow, W. T. L., Coutts, A. M., Roth, M., Velasco, E., Vivoni, E. R., & Fatichi, S. (2021). Tree effects on urban microclimate: Diurnal, seasonal, and climatic temperature differences explained by separating radiation, evapotranspiration, and roughness effects. *Urban Forestry & Urban Greening, 58*, 126970. https://doi.org/10.1016/j.ufug.2020.126970

Meng, Y. R., Wang, J. Q., Xi, C., Han, L., Feng, Z. B., & Cao, S. J. (2023). Investigation of heat stress on urban roadways for commuting children and mitigation strategies from the perspective of urban design. *Urban Climate, 49*, 101564. https://doi.org/10.1016/j.uclim.2023.101564

Middel, A., Turner, V. K., Schneider, F. A., Zhang, Y. J., & Stiller, M. (2020). Solar reflective pavements-a policy panacea to heat mitigation? *Environmental Research Letters, 15*(6). https://doi.org/10.1088/1748-9326/ab87d4

Middel, A., Alkhaled, S., Schneider, F. A., Hagen, B., & Coseo, P. (2021). 50 grades of shade. *Bulletin of the American Meteorological Society, 102*(9), E1805–E1820. https://doi.org/10.1175/bams-d-20-0193.2

Monteiro, M. V., Doick, K. J., Handley, P., & Peace, A. (2016). The impact of greenspace size on the extent of local nocturnal air temperature cooling in London. *Urban Forestry & Urban Greening, 16*, 160–169. https://doi.org/10.1016/j.ufug.2016.02.008

Murakawa, S., Sekine, T., Narita, K. I., & Nishina, D. (1991). Study of the effects of a river on the thermal environment in an urban area. *Energy and Buildings, 16*(3–4), 993–1001. https://doi.org/10.1016/0378-7788(91)90094-j

Nasrollahi, N., Ghosouri, A., Khodakarami, J., & Taleghani, M. (2020). Heat-mitigation strategies to improve pedestrian thermal comfort in urban environments: A review. *Sustainability, 12*(23). https://doi.org/10.3390/su122310000

Nastran, M., Kobal, M., & Eler, K. (2019). Urban heat islands in relation to green land use in European cities. *Urban Forestry & Urban Greening, 37*, 33–41. https://doi.org/10.1016/j.ufug.2018.01.008

Nevat, I., & Adelia, A. S. (2023). Urban wind corridors analysis via network theory. *Atmosphere, 14*(3). https://doi.org/10.3390/atmos14030572

Ng, E., Yuan, C., Chen, L., Ren, C., & Fung, J. C. H. (2011). Improving the wind environment in high-density cities by understanding urban morphology and surface roughness: A study in Hong Kong. *Landscape and Urban Planning, 101*(1), 59–74. https://doi.org/10.1016/j.landurbplan.2011.01.004

Nishimura, N., Nomura, T., Iyota, H., & Kimoto, S. (1998). Novel water facilities for creation of comfortable urban micrometeorology. *Solar Energy, 64*(4–6), 197–207.

Osmundson, T. (1999). *Roof gardens : History, design, and construction* (1st ed.). W.W. Norton.

Parison, S., Hendel, M., Grados, A., & Royon, L. (2020). Analysis of the heat budget of standard, cool and watered pavements under lab heat-wave conditions. *Energy and Buildings, 228*, 110455. https://doi.org/10.1016/j.enbuild.2020.110455

Park, C. Y., Lee, D. K., Asawa, T., Murakami, A., Kim, H. G., Lee, M. K., & Lee, H. S. (2019). Influence of urban form on the cooling effect of a small urban river. *Landscape and Urban Planning, 183*, 26–35. https://doi.org/10.1016/j.landurbplan.2018.10.022

Park, C. Y., Choe, H., & Thorne, J. H. (2023). Mapping urban cool air connectivity in a megacity. *Urban Climate, 47*, 101362. https://doi.org/10.1016/j.uclim.2022.101362

Pomerantz, M. (2000). *The effect of pavements' temperatures on air temperatures in large cities*. LBL Publications/Lawrence Berkeley National Laboratory.

Probst, N., Bach, P. M., Cook, L. M., Maurer, M., & Leitao, J. P. (2022). Blue green systems for urban heat mitigation: Mechanisms, effectiveness and research directions. *Blue-Green Systems, 4*(2), 348–376. https://doi.org/10.2166/bgs.2022.028

Qi, J. D., Ding, L., & Lim, S. (2020). Ontology-based knowledge representation of urban heat Island mitigation strategies. *Sustainable Cities and Society, 52*, 101875. https://doi.org/10.1016/j.scs.2019.101875

Qiao, Z., Xu, X. L., Wu, F., Luo, W., Wang, F., Liu, L., & Sun, Z. Y. (2017). Urban ventilation network model: A case study of the core zone of capital function in Beijing metropolitan area. *Journal of Cleaner Production, 168*, 526–535. https://doi.org/10.1016/j.jclepro.2017.09.006

Qin, Y. H. (2015). A review on the development of cool pavements to mitigate urban heat Island effect. *Renewable & Sustainable Energy Reviews, 52*, 445–459. https://doi.org/10.1016/j.rser.2015.07.177

Qin, Y. H., Zhang, X. Y., Tan, K. H., & Wang, J. S. (2022). A review on the influencing factors of pavement surface temperature. *Environmental Science and Pollution Research, 29*(45), 67659–67674. https://doi.org/10.1007/s11356-022-22295-3

Rahman, M. A., Stratopoulos, L. M. F., Moser-Reischl, A., Zölch, T., Häberle, K. H., Rötzer, T., Pretzsch, H., & Pauleit, S. (2020). Traits of trees for cooling urban heat islands: A meta-analysis. *Building and Environment, 170*, 106606. https://doi.org/10.1016/j.buildenv.2019.106606

Ramakreshnan, L., & Aghamohammadi, N. (2024). The application of nature-based solutions for urban heat Island mitigation in Asia: Progress, challenges, and recommendations. *Current Environmental Health Reports., 11*, 4–17. https://doi.org/10.1007/s40572-023-00427-2

Rapisarda, R., Nocera, F., Costanzo, V., Sciuto, G., & Caponetto, R. (2022). Hydroponic green roof systems as an alternative to traditional pond and green roofs: A literature review. *Energies, 15*(6). https://doi.org/10.3390/en15062190

Ren, C., & McGregor, G. (2021). *Urban climate science for planning healthy cities*. Springer.

Ren, C., Yang, R. Z., Cheng, C., Xing, P., Fang, X. Y., Zhang, S., Wang, H. F., Shi, Y., Zhang, X. Y., Kwok, Y. T., & Ng, E. (2018). Creating breathing cities by adopting urban ventilation assessment and wind corridor plan – The implementation in Chinese cities. *Journal of Wind Engineering and Industrial Aerodynamics, 182*, 170–188. https://doi.org/10.1016/j.jweia.2018.09.023

Rendon, P., Love, N., Pawlak, C., Yost, J., Ritter, M., & Doremus, J. (2024). Street tree diversity and urban heat. *Urban Forestry & Urban Greening, 91*, 128180. https://doi.org/10.1016/j.ufug.2023.128180

Reuter, U., & Kapp, R. (2021). Urban climate in urban planning: The experience from Stuttgart. In C. Ren & G. McGregor (Eds.), *Urban climate science for planning healthy cities* (pp. 259–284). Springer.

Rizwan, A. M., Dennis, Y. C. L., & Liu, C. H. (2008). A review on the generation, determination and mitigation of urban Heat Island. *Journal of Environmental Sciences, 20*(1), 120–128. https://doi.org/10.1016/s1001-0742(08)60019-4

Rizzo, G., Cirrincione, L., La Gennusa, M., Peri, G., & Scaccianoce, G. (2023). Green roofs' end of life: A literature review. *Energies, 16*(2). https://doi.org/10.3390/en16020596

Roehr, D., & Laurenz, J. (2008). Living skins: Environmental benefits of green envelopes in the city context. *WIT Transactions on Ecology and the Environment, 113*, 149–158.

Romeo, C., & Zinzi, M. (2013). Impact of a cool roof application on the energy and comfort performance in an existing non-residential building. A Sicilian case study. *Energy and Buildings, 67*, 647–657. https://doi.org/10.1016/j.enbuild.2011.07.023

Rosenfeld, A. H., Akbari, H., Romm, J. J., & Pomerantz, M. (1998). Cool communities: Strategies for heat Island mitigation and smog reduction. *Energy and Buildings, 28*(1), 51–62. https://doi.org/10.1016/s0378-7788(97)00063-7

Rosenzweig, C., Solecki, W. D., Parshall, L., Lynn, B., Cox, J., Goldberg, R., Hodges, S., Gaffin, S., Slosberg, R. B., Savio, P., Dunstan, F., & Watson, M. (2009). Mitigating new York City's heat Island integrating stakeholder perspectives and scientific evaluation. *Bulletin of the American Meteorological Society, 90*(9), 1297–1312. https://doi.org/10.1175/2009bams2308.1

Saaroni, H., & Ziv, B. (2003). The impact of a small lake on heat stress in a Mediterranean urban park: The case of Tel Aviv, Israel. *International Journal of Biometeorology, 47*, 156–165.

Saber, H. H. (2021). Experimental characterization of reflective coating material for cool roofs in hot, humid and dusty climate. *Energy and Buildings, 242*, 110993. https://doi.org/10.1016/j.enbuild.2021.110993

Safikhani, T., Abdullah, A. M., Ossen, D. R., & Baharvand, M. (2014). A review of energy characteristic of vertical greenery systems. *Renewable and Sustainable Energy Reviews, 40*, 450–462.

Sahani, J., Kumar, P., & Debele, S. E. (2023). Efficacy assessment of green-blue nature-based solutions against environmental heat mitigation. *Environment International, 179*, 108187. https://doi.org/10.1016/j.envint.2023.108187

Salvati, A., Kolokotroni, M., Kotopouleas, A., Watkins, R., Giridharan, R., & Nikolopoulou, M. (2022). Impact of reflective materials on urban canyon albedo, outdoor and indoor microclimates. *Building and Environment, 207*, 108459.

Santamouris, M. (2013). Using cool pavements as a mitigation strategy to fight urban heat Island-a review of the actual developments. *Renewable & Sustainable Energy Reviews, 26*, 224–240. https://doi.org/10.1016/j.rser.2013.05.047

Santamouris, M., Gaitani, N., Spanou, A., Saliari, M., Giannopoulou, K., Vasilakopoulou, K., & Kardomateas, T. (2012). Using cool paving materials to improve microclimate of urban areas – Design realization and results of the flisvos project. *Building and Environment, 53*, 128–136. https://doi.org/10.1016/j.buildenv.2012.01.022

Santamouris, M., Ban-Weiss, G., Osmond, P., Paolini, R., Synnefa, A., Cartalis, C., Muscio, A., Zinzi, M., Morakinyo, T. E., & Edward, N. (2018). Progress in urban greenery mitigation science–assessment methodologies advanced technologies and impact on cities. *Journal of Civil Engineering and Management, 24*(8), 638–671.

Schneider, F. A., Ortiz, J. C., Vanos, J. K., Sailor, D. J., & Middel, A. (2023). Evidence-based guidance on reflective pavement for urban heat mitigation in Arizona. *Nature Communications, 14*(1). https://doi.org/10.1038/s41467-023-36972-5

Seddon, N. (2022). Harnessing the potential of nature-based solutions for mitigating and adapting to climate change. Science 376 (6600):1410–+, *376*, 1410–1416. https://doi.org/10.1126/science.abn9668

Seifeddine, K., Amziane, S., Toussaint, E., & Ouldboukhitine, S. E. (2023). Review on thermal behavior of cool pavements. *Urban Climate, 51*, 101667. https://doi.org/10.1016/j.uclim.2023.101667

Sen, S., Fernandèz, J., & Roesler, J. (2020). Reflective parking lots for microscale urban heat Island mitigation. *Transportation Research Record, 2674*(8), 663–671. https://doi.org/10.1177/0361198120919401

Sharifi, A., & Yamagata, Y. (2015). Roof ponds as passive heating and cooling systems: A systematic review. *Applied Energy, 160*, 336–357. https://doi.org/10.1016/j.apenergy.2015.09.061

Sharma, A., & Goyal, H. (2023). A comprehensive review of the empirical and nonempirical literature on green roofs. *Journal of Architectural Engineering, 29*(4). https://doi.org/10.1061/jaeied.aeeng-1633

Shen, L. D., Li, H. D., Guo, L. C., & He, B. J. (2022). Thermal and energy benefits of rooftop photovoltaic panels in a semi-arid city during an extreme heatwave event. *Energy and Buildings, 275*, 112490. https://doi.org/10.1016/j.enbuild.2022.112490

Sheng, S., & Wang, Y. (2024). Configuration characteristics of green-blue spaces for efficient cooling in urban environments. *Sustainable Cities and Society, 100*, 105040. https://doi.org/10.1016/j.scs.2023.105040

Shi, Z. P., Yang, J., Zhang, Y. Q., Xiao, X. M., & Xia, J. C. (2022). Urban ventilation corridors and spatiotemporal divergence patterns of urban heat Island intensity: A local climate zone perspective. *Environmental Science and Pollution Research, 29*(49), 74394–74406. https://doi.org/10.1007/s11356-022-21037-9

Shushunova, N., Korol, E., Luzay, E., Shafieva, D., & Bevilacqua, P. (2022). Ensuring the safety of buildings by reducing the noise impact through the use of green wall systems. *Energies, 15*(21). https://doi.org/10.3390/en15218097

Silva, C. M., Gomes, M. G., & Silva, M. (2016). Green roofs energy performance in Mediterranean climate. *Energy and Buildings, 116*, 318–325. https://doi.org/10.1016/j.enbuild.2016.01.012

Sinsel, T., Simon, H., Broadbent, A. M., Bruse, M., & Heusinger, J. (2021). Modeling impacts of super cool roofs on air temperature at pedestrian level in mesoscale and microscale climate models. *Urban Climate, 40*, 101001. https://doi.org/10.1016/j.uclim.2021.101001

Solcerova, A., van Emmerik, T., Hilgersom, K., van de Ven, F., & van de Giesen, N. (2018). Uchimizu: A cool(ing) tradition to locally decrease air temperature. *Water, 10*(6). https://doi.org/10.3390/w10060741

Song, J. Y., & Wang, Z. H. (2016). Diurnal changes in urban boundary layer environment induced by urban greening. *Environmental Research Letters, 11*(11). https://doi.org/10.1088/1748-9326/11/11/114018

Spronken-Smith, R. A., & Oke, T. R. (1998). The thermal regime of urban parks in two cities with different summer climates. *International Journal of Remote Sensing, 19*(11), 2085–2104. https://doi.org/10.1080/014311698214884

Stavrakakis, G. M., Androutsopoulos, A. V., & Vyörykkä, J. (2016). Experimental and numerical assessment of cool-roof impact on thermal and energy performance of a school building in Greece. *Energy and Buildings, 130*, 64–84. https://doi.org/10.1016/j.enbuild.2016.08.047

Steeneveld, G. J., Koopmans, S., Heusinkveld, B. G., & Theeuwes, N. E. (2014). Refreshing the role of open water surfaces on mitigating the maximum urban heat Island effect. *Landscape and Urban Planning, 121*, 92–96. https://doi.org/10.1016/j.landurbplan.2013.09.001

Su, X. X. (2018). Study on humidification and cooling effect of garden plants. In: *4th International Conference on Education, Management and Information Technology* (Icemit 2018), pp. 1176–1179.

Su, W. X., Zhang, L. K., & Chang, Q. (2022). Nature-based solutions for urban heat mitigation in historical and cultural block: The case of Beijing Old City. *Building and Environment, 225*, 109600. https://doi.org/10.1016/j.buildenv.2022.109600

Susca, T., Zanghirella, F., Colasuonno, L., & Del, F. (2022). Effect of green wall installation on urban heat Island and building energy use: A climate-informed systematic literature review. *Renewable & Sustainable Energy Reviews, 159*, 112100. https://doi.org/10.1016/j.rser.2022.112100

Syafii, N. I., Ichinose, M., Kumakura, E., Jusuf, S. K., Chigusa, K., & Wong, N. H. (2017). Thermal environment assessment around bodies of water in urban canyons: A scale model study. *Sustainable Cities and Society, 34*, 79–89. https://doi.org/10.1016/j.scs.2017.06.012

Synnefa, A., Saliari, M., & Santamouris, M. (2012). Experimental and numerical assessment of the impact of increased roof reflectance on a school building in Athens. *Energy and Buildings, 55*, 7–15. https://doi.org/10.1016/j.enbuild.2012.01.044

Takebayashi, H., & Moriyama, M. (2012). Study on surface heat budget of various pavements for urban heat Island mitigation. *Advances in Materials Science and Engineering, 2012*, 1–11. https://doi.org/10.1155/2012/523051

Takebayashi, H., Mori, H., & Tozawa, U. (2023). Study on an effective roadway watering scheme for mitigating pedestrian thermal comfort according to the street configuration. *Atmosphere, 14*(6), 1014.

Taleghani, M., & Berardi, U. (2018). The effect of pavement characteristics on pedestrians' thermal comfort in Toronto. *Urban Climate, 24*, 449–459. https://doi.org/10.1016/j.uclim.2017.05.007

Tan, H. C., Kotamarthi, R., Wang, J. L., Qian, Y., & Chakraborty, T. C. (2023). Impact of different roofing mitigation strategies on near-surface temperature and energy consumption over the Chicago metropolitan area during a heatwave event. *Science of the Total Environment, 860*, 160508. https://doi.org/10.1016/j.scitotenv.2022.160508

Teshnehdel, S., Akbari, H., Di Giuseppe, E., & Brown, R. D. (2020). Effect of tree cover and tree species on microclimate and pedestrian comfort in a residential district in Iran. *Building and Environment, 178*, 106899. https://doi.org/10.1016/j.buildenv.2020.106899

Testa, J., & Krarti, M. (2017). A review of benefits and limitations of static and switchable cool roof systems. *Renewable & Sustainable Energy Reviews, 77*, 451–460. https://doi.org/10.1016/j.rser.2017.04.030

Theeuwes, N. E., Solcerová, A., & Steeneveld, G. J. (2013). Modeling the influence of open water surfaces on the summertime temperature and thermal comfort in the city. *Journal of Geophysical Research-Atmospheres, 118*(16), 8881–8896. https://doi.org/10.1002/jgrd.50704

Tiwari, A., Kumar, P., Kalaiarasan, G., & Ottosen, T. B. (2021). The impacts of existing and hypothetical green infrastructure scenarios on urban heat Island formation. *Environmental Pollution, 274*, 115898. https://doi.org/10.1016/j.envpol.2020.115898

Tominaga, Y., Sato, Y., & Sadohara, S. (2015). CFD simulations of the effect of evaporative cooling from water bodies in a micro-scale urban environment: Validation and application studies. *Sustainable Cities and Society, 19*, 259–270. https://doi.org/10.1016/j.scs.2015.03.011

Toparlar, Y., Blocken, B., Maiheu, B., & van Heijst, G. J. F. (2018). The effect of an urban park on the microclimate in its vicinity: A case study for Antwerp, Belgium. *International Journal of Climatology, 38*, E303–E322. https://doi.org/10.1002/joc.5371

Tozer, L., Bulkeley, H., & Xie, L. J. (2022). Transnational governance and the urban politics of nature-based solutions for climate change. *Global Environmental Politics, 22*(3), 81–103. https://doi.org/10.1162/glep_a_00658

Treder, M., Joedecke, V., Schweikert, K., Rosenkranz, P., Ruttensperger, U., & Traynor, K. (2024). Vertical greening systems serve as effective means to promote pollinators: Experimental comparison of vertical and horizontal plantings. *Landscape and Urban Planning, 243*, 104951. https://doi.org/10.1016/j.landurbplan.2023.104951

Triyuly, W., Triyadi, S., & Wonorahardjo, S. (2021). Synergising the thermal behaviour of water bodies within thermal environment of wetland settlements. *International Journal of Energy and Environmental Engineering, 12*(1), 55–68. https://doi.org/10.1007/s40095-020-00355-z

Turek-Hankins LL, Coughlan de Perez E, Scarpa G, Ruiz-Diaz R, Nayna Schwerdtle P, Joe ET, Galappaththi EK, French EM, Austin SE, Singh C, Siña M, Siders AR, van Aalst MK, Templeman S, Nunbogu AM, Berrang-Ford L, Agrawal Ta, team TGAMI, Mach KJ (2021) Climate change adaptation to extreme heat: A global systematic review of implemented action. Oxford Open Climate Change 1 (1):13. doi:https://doi.org/https://doi.org/10.1093/oxfclm/kgab005.

Vardoulakis, E., Karamanis, D., & Mihalakakou, G. (2014). Heat Island phenomenon and cool roofs mitigation strategies in a small city of elevated temperatures. *Advances in Building Energy Research, 8*(1), 55–62. https://doi.org/10.1080/17512549.2014.890537

Vásquez-Alvarez, P. E., Flores-Vázquez, C., Cobos-Trres, J. C., & Cobos-Moora, S. L. (2022). Urban heat Island mitigation through planned simulation. *Sustainability, 14*(14). https://doi.org/10.3390/su14148612

Völker, S., & Kistemann, T. (2015). Developing the urban blue: Comparative health responses to blue and green urban open spaces in Germany. *Health & Place, 35*, 196–205.

Völker, S., Baumeister, H., Claßen, T., Hornberg, C., & Kistemann, T. (2013). Evidence for the temperature-mitigating capacity of urban blue space—A health geographic perspective. *Erdkunde, 67*, 355–371.

Wang, Y. F., Bakker, F., de Groot, R., Wortche, H., & Leemans, R. (2015). Effects of urban trees on local outdoor microclimate: Synthesizing field measurements by numerical modelling. *Urban Ecosystems, 18*(4), 1305–1331. https://doi.org/10.1007/s11252-015-0447-7

Wang, C. H., Wang, Z. H., Kaloush, K. E., & Shacat, J. (2021). Cool pavements for urban heat Island mitigation: A synthetic review. *Renewable & Sustainable Energy Reviews, 146*, 111171. https://doi.org/10.1016/j.rser.2021.111171

Wang, C., Ren, Z., Dong, Y., Zhang, P., Guo, Y., Wang, W., & Bao, G. (2022a). Efficient cooling of cities at global scale using urban green space to mitigate urban heat Island effects in different climatic regions. *Urban Forestry & Urban Greening, 74*, 127635.

Wang, W. W., Wang, D., Chen, H., Wang, B. Y., & Chen, X. (2022b). Identifying urban ventilation corridors through quantitative analysis of ventilation potential and wind characteristics. *Building and Environment, 214*, 108943. https://doi.org/10.1016/j.buildenv.2022.108943

Wang, X., Li, H., & Sodoudi, S. (2022c). The effectiveness of cool and green roofs in mitigating urban heat Island and improving human thermal comfort. *Building and Environment, 217*, 109082. https://doi.org/10.1016/j.buildenv.2022.109082

Wang, Y. S., Ouyang, W. L., Zhan, Q. M., & Zhang, L. (2022d). The cooling effect of an urban river and its interaction with the littoral built environment in mitigating heat stress: A mobile measurement study. *Sustainability, 14*(18). https://doi.org/10.3390/su141811700

Wang, Q., Peng, J., Yu, S. Y., Dan, Y. Z., Dong, J. Q., Zhao, X., & Wu, J. S. (2023a). Key attributes of greenspace pattern for heat mitigation vary with urban functional zones. *Landscape Ecology, 38*(11), 2965–2979. https://doi.org/10.1007/s10980-023-01763-2

Wang, X. Y., Liu, G., Zhang, N., Liu, H. N., Tang, X. D., Lyu, M., & Meng, H. D. (2023b). Effects of cooling roofs on mitigating the urban heat Island and human thermal stress in the Pearl River Delta, China. *Building and Environment, 245*, 110880. https://doi.org/10.1016/j.buildenv.2023.110880

White, M. P., Elliott, L. R., Gascon, M., Roberts, B., & Fleming, L. E. (2020). Blue space, health and Well-being: A narrative overview and synthesis of potential benefits. *Environmental Research, 191*, 110169. https://doi.org/10.1016/j.envres.2020.110169

Wicht, M., & Osinska-Skotak, K. (2016). Temporal analysis of urban changes and development in Warsaw's ventilation corridors. *Miscellanea Geographica, 20*(4), 11–21. https://doi.org/10.1515/mgrsd-2016-0021

Wicht, M., Wicht, A., & Osinska-Skotak, K. (2017). Detection of ventilation corridors using a spatio-temporal approach aided by remote sensing data. *European Journal of Remote Sensing, 50*(1), 254–267. https://doi.org/10.1080/22797254.2017.1318672

Widiastuti, R., Bramiana, C. N., Bangun, I. R. H., Prabowo, B. N., & Ramandhika, M. (2018). Vertical greenery system as the passive design strategy for mitigating urban heat Island in tropical area: A comparative field measurement between green facade and green wall. *IOP Conference Series: Earth and Environmental Science, 213*. https://doi.org/10.1088/1755-1315/213/1/012037

Wong, T. H. F., Tapper, N. J., & Luby, S. P. (2020). Planetary health approaches for dry cities: Water quality and heat mitigation. *BMJ-British Medical Journal, 371*. https://doi.org/10.1136/bmj.m4313

Wong, N. H., Tan, C. L., Kolokotsa, D. D., & Takebayashi, H. (2021). Greenery as a mitigation and adaptation strategy to urban heat. *Nature Reviews Earth & Environment, 2*(3), 166–181. https://doi.org/10.1038/s43017-020-00129-5

Wootton-Beard, P. C., Xing, Y., Durai Prabhakaran, R. T., Robson, P., Bosch, M., Thornton, J. M., Ormondroyd, G. A., Jones, P., & Donnison, I. (2016). Improving the impact of plant science on urban planning and design. *Buildings, 6*(4), 48.

Wu, S. J., Yang, H., Luo, P., Luo, C. A., Li, H. L., Liu, M., Ruan, Y., Zhang, S. J., Xiang, P., Jia, H. H., & Cheng, Y. (2021). The effects of the cooling efficiency of urban wetlands in an inland megacity: A case study of Chengdu. *Southwest China. Building and Environment, 204*, 108128. https://doi.org/10.1016/j.buildenv.2021.108128

Xiao, M., Lin, Y. L., Han, J., & Zhang, G. Q. (2014). A review of green roof research and development in China. *Renewable & Sustainable Energy Reviews, 40*, 633–648. https://doi.org/10.1016/j.rser.2014.07.147

Xie, Q. J., Ren, L., & Yang, C. Z. (2023). Regulation of water bodies to urban thermal environment: Evidence from Wuhan, China. *Frontiers in Ecology and Evolution, 11*. https://doi.org/10.3389/fevo.2023.983567

Xu, J. C., Wei, Q. L., Huang, X. F., Zhu, X. Y., & Li, G. M. (2010). Evaluation of human thermal comfort near urban waterbody during summer. *Building and Environment, 45*(4), 1072–1080. https://doi.org/10.1016/j.buildenv.2009.10.025

Xu, Y. P., Wang, W. W., Chen, B. Y., Chang, M., & Wang, X. M. (2021). Identification of ventilation corridors using backward trajectory simulations in Beijing. *Sustainable Cities and Society, 70*, 102889. https://doi.org/10.1016/j.scs.2021.102889

Xu, F. S., Gao, Z., Zhang, J. S., Hu, Y. Y., & Ding, W. W. (2022). Influence of typical street-side public building morphologies on the ventilation performance of streets and squares. *Building and Environment, 221*, 109331. https://doi.org/10.1016/j.buildenv.2022.109331

Yang, J. C., Wang, Z. H., Kaloush, K. E., & Dylla, H. (2016). Effect of pavement thermal properties on mitigating urban heat islands: A multi-scale modeling case study in Phoenix. *Building and Environment, 108*, 110–121. https://doi.org/10.1016/j.buildenv.2016.08.021

Yang, L., Liu, X. D., & Qian, F. (2020). Research on water thermal effect on surrounding environment in summer. *Energy and Buildings, 207*, 109613. https://doi.org/10.1016/j.enbuild.2019.109613

Yang, J. R., Guo, R., Li, D., Wang, X. L., & Li, F. Z. (2022). Interval-thresholding effect of cooling and recreational services of urban parks in metropolises. *Sustainable Cities and Society, 79*, 103684. https://doi.org/10.1016/j.scs.2022.103684

Yao, L. Y., Sailor, D. J., Yang, X. S., Xu, G. Y., & Zhao, L. H. (2023a). Are water bodies effective for urban heat mitigation? Evidence from field studies of urban lakes in two humid subtropical cities. *Building and Environment, 245*, 110860. https://doi.org/10.1016/j.buildenv.2023.110860

Yao, L. Y., Sailor, D. J., Zhang, X., Wang, J. S., Zhao, L. H., & Yang, X. S. (2023b). Diurnal pattern and driving mechanisms of the thermal effects of an urban pond. *Sustainable Cities and Society, 91*, 104407. https://doi.org/10.1016/j.scs.2023.104407

Yenneti, K., Ding, L., Prasad, D., Ulpiani, G., Paolini, R., Haddad, S., & Santamouris, M. (2020). Urban overheating and cooling potential in Australia: An evidence-based review. *Climate, 8*(11). https://doi.org/10.3390/cli8110126

Yilmaz, S., Mutlu, E., & Yilmaz, H. (2018). Alternative scenarios for ecological urbanizations using ENVI-met model. *Environmental Science and Pollution Research, 25*(26), 26307–26321. https://doi.org/10.1007/s11356-018-2590-1

Yilmaz, S., Irmak, M. A., & Qaid, A. (2022). Assessing the effects of different urban landscapes and built environment patterns on thermal comfort and air pollution in Erzurum city. *Turkey. Building and Environment, 219*, 109210. https://doi.org/10.1016/j.buildenv.2022.109210

Yin, J., Ramanpong, J., Chang, J., Wu, C. D., Chao, P. H., & Yu, C. P. (2023a). Effects of blue space exposure in urban and natural environments on psychological and physiological responses: A within-subject experiment. *Urban Forestry & Urban Greening, 87*, 128066. https://doi.org/10.1016/j.ufug.2023.128066

Yin, Y., He, L. Y., Wennberg, P. O., & Frankenberg, C. (2023b). Unequal exposure to heatwaves in Los Angeles: Impact of uneven green spaces. *Science Advances, 9*(17). https://doi.org/10.1126/sciadv.ade8501

Yu, Z. W., Xu, S. B., Zhang, Y. H., Jorgensen, G., & Vejre, H. (2018). Strong contributions of local background climate to the cooling effect of urban green vegetation. *Scientific Reports, 8*, 6798. https://doi.org/10.1038/s41598-018-25296-w

Yu, Z. W., Yang, G. Y., Zuo, S. D., Jorgensen, G., Koga, M., & Vejre, H. (2020). Critical review on the cooling effect of urban blue-green space: A threshold-size perspective. *Urban Forestry & Urban Greening, 49*, 126630. https://doi.org/10.1016/j.ufug.2020.126630

Yu, Z. W., Fryd, O., Sun, R. H., Jorgensen, G., Yang, G. Y., Ozdil, N. C., & Vejre, H. (2021). Where and how to cool? An idealized urban thermal security pattern model. *Landscape Ecology, 36*(7), 2165–2174. https://doi.org/10.1007/s10980-020-00982-1

Yuan, J. H., Emura, K., & Farnham, C. (2017). Is urban albedo or urban green covering more effective for urbanmicro climate improvement?: A simulation for Osaka. *Sustainable Cities and Society, 32*, 78–86. https://doi.org/10.1016/j.scs.2017.03.021

Zeeshan, M., & Ali, Z. (2023). Using a blue landscape to mitigate heat stress during a heatwave event: A simulation study in a hot-humid urban environment. *Journal of Water and Climate Change, 14*(3), 764–777. https://doi.org/10.2166/wcc.2023.363

Zhang, Y., & Dai, M. Q. (2022). Analysis of the cooling and humidification effect of multi-layered vegetation communities in urban parks and its impact. *Atmosphere, 13*(12). https://doi.org/10.3390/atmos13122045

Zhang, L., Deng, Z., Liang, L., Zhang, Y., Meng, Q., Wang, J., & Santamouris, M. (2019). Thermal behavior of a vertical green facade and its impact on the indoor and outdoor thermal environment. *Energy and Buildings, 204*, 109502.

Zhang, J. L., Li, Z., & Hu, D. (2022). Effects of urban morphology on thermal comfort at the micro-scale. *Sustainable Cities and Society, 86*, 104150. https://doi.org/10.1016/j.scs.2022.104150

Zhang, N. H., Zhen, W. A., Shi, D. H., Zhong, C. Y., & Li, Y. (2024). Quantification and mapping of the cooling effect of urban parks on the temperate monsoon climate zone. *Sustainable Cities and Society, 101*, 105111. https://doi.org/10.1016/j.scs.2023.105111

Zheng, T. H., Qu, K., Darkwa, J., & Calautit, J. K. (2022). Evaluating urban heat Island mitigation strategies for a subtropical city Centre (a case study in Osaka, Japan). *Energy, 250*, 123721. https://doi.org/10.1016/j.energy.2022.123721

Zheng, S., He, C. W., Guldmann, J. M., Xu, H. D., & Liu, X. (2023a). Heat mitigation benefits of urban trees: A review of mechanisms, modeling, validation and simulation. *Forests, 14*(12). https://doi.org/10.3390/f14122280

Zheng, X., Hu, W., Luo, S., Zhu, Z., Bai, Y., Wang, W., & Pan, L. (2023b). Effects of vertical greenery systems on the spatiotemporal thermal environment in street canyons with different aspect ratios: A scaled experiment study. *Science of the Total Environment, 859*, 160408. https://doi.org/10.1016/j.scitotenv.2022.160408

Zölch, T., Maderspacher, J., Wamsler, C., & Pauleit, S. (2016). Using green infrastructure for urban climate-proofing: An evaluation of heat mitigation measures at the micro-scale. *Urban Forestry & Urban Greening, 20*, 305–316. https://doi.org/10.1016/j.ufug.2016.09.011

Zuckerman, N., & Lensky, I. M. (2023). Thermal performance of vertical greenery systems (VGS) in a street canyon: A real-scale long-term experiment. *Building and Environment, 244*, 110750. https://doi.org/10.1016/j.buildenv.2023.110750

Chapter 15
Synthesis and Directions for Future Enquiry

Abstract The final chapter of this book synthesizes the material presented in Chaps. 2, 3, 4, 5, 6, 7, 8, 9, 10, 11, 12, 13, and 14 covering the topics comprising Parts I—IV namely, heatwaves as weather and climate phenomenon, the causes and consequences of heatwaves, and responses to them. Heatwaves are distinct from warm spells being restricted to the summer months or heat season. Defined generally as periods of highly anomalous heat lasting several days, they typically develop over 5–10 days at the synoptic time scale. Heatwaves are complex phenomenon, involving multiple interacting factors such as temperature, radiation load, humidity, and wind. To effectively manage heatwave risks, the need for bespoke definitions tailored to specific sectors and regions is emphasised, rather than a universal definition. Material on the causes of heatwaves has explored the atmospheric and oceanic patterns, known as modes of climate variability, that increase the likelihood of heatwaves, as well as the role of synoptic scale processes such as atmospheric blocking, advection, and the foehn effect in heatwave development. Future work on understanding the interactions between different modes of climate variability and their role in heatwave development will enhance extended to long-range prediction of heatwaves, especially in relation to uncovering the physical mechanisms linking 'slow' climate drivers with 'fast' synoptic scale processes. The relative roles of adiabatic warming, diabatic heating, and surface-atmosphere interactions in heatwave formation also requires further investigation. For urban environments, better monitoring of radiative and energy balance processes, including anthropogenic heat flux, is crucial to understand how urban heat is exacerbated by regional scale heatwaves. This book has found heatwaves to have profound consequences for human and natural systems. Regrettably, there remains a dearth of systematic reporting and quality-assured data on heatwave impacts to develop effective risk management strategies. This situation needs to be rectified along with the development of protocols for heatwave impact assessment. Although exposure and vulnerability have been revealed as important moderators of the relationship between heatwave occurrence and their impacts in both human and natural systems, more work on these crucial determinants of heatwave risk and impact is required. This is particularly so for marine and terrestrial ecosystems. Future work directed at understanding the cascading effects of heatwaves on infrastructure and economies and associated interdependencies is also warranted as this will assist with managing the risk associ-

G. McGregor, *Heatwaves*, Biometeorology 6,
https://doi.org/10.1007/978-3-031-69906-1_15

ated with so-called mega-heatwaves. Responses as intentional actions to reduce heatwave-related risks and impacts via adaptation and mitigation have been addressed in this book with a justification to distinguish more clearly between adaptation and mitigation options presented. Associated future lines of enquiry include exploring the effectiveness of a range of adaptation and mitigation strategies. Beyond examining heatwave risk governance and heatwave insurance and other adaptation options, it is essential to continuously evaluate the efficacy of heatwave-related early warning systems, particularly considering the ambitious goal of Early Warning Systems for All (EWS4All). Urgent investigation is also needed concerning the potential benefits and challenges of heatwave naming and categorization to enhance risk communication. There is also a pressing need to address the effectiveness of a range of mitigation options directed at surface radiative and energy balance management to reduce levels of urban heat, especially in relation to the acute levels of heat associated with impactful heatwaves. An allied line of enquiry is investigating the scalability and long-term sustainability of urban heat mitigation strategies, with a consideration of associated enabling and preventative social, political, and economic factors. Importantly, much remains to be learned about what catalyses the development and implementation of planned and autonomous actions related to heatwave adaptation and mitigation. This chapter and hence the book closes by stressing the importance of interdisciplinary research and collaboration among a range of academic disciplines, policymakers, and stakeholders to address the heatwave imperative. It calls for a shift from siloed research to a holistic approach that considers the complex interactions between heatwave causes, consequences, and responses.

Keywords Heatwaves as weather and climate phenomenon · Heatwave causes · Heatwave consequences · Heatwave responses · Future research

15.1 The Heatwave Imperative

This book has made the case for furthering our understanding of heatwaves not only as physical albeit ostensibly invisible objects, but as felt objects in the form of their consequences and associated responses. Cast as the heatwave imperative, the need to address the research and management challenges posed by heatwaves is a compelling one because heatwaves have emerged as a serious threat to society, economy, and the environment particularly since the dawning of the twenty-first century. Indeed, society has reached a pivotal point in terms of its relationship with heat and thus heatwaves as a facet of climate. This is because it has become abundantly clear thanks to heatwave related climate change attribution studies that humans are beginning to exert a discernible influence on the ferocity of heatwaves via anthropogenic climate change. Speaking emotively, the unintentional human interference with the workings of the climate system has unleashed a new generation of heatwaves as manifest by several perilous 'mega-heatwaves' over the last 20–25 years. The spectre that human induced climate change will literally turn up the heat and favour a worsening of the intensity and severity of heatwaves is a chilling one. Further, with much of the world's rapidly growing population concentrated in areas becoming increasing prone to heatwaves, more individuals will become exposed to heatwaves and extreme heat. This poses significant challenges for current and future populations and underscores the urgent need for action to annul the risks associated with heatwaves. Because the basic tenet of this volume has been a better understanding of the present is required if we want to understand the future, the focus has been more on the here and now about heatwaves as opposed to what potential heatwave futures might look like. Accordingly, the compilation of knowledge concerning our current understanding of the physical origins of heatwaves, their resultant consequences, and societal responses in relation to heatwave risk reduction, constitutes the heatwave imperative underlying this work.

15.2 Heatwaves as Weather and Climate Phenomenon

In this book heatwaves have been distinguished from warm spells being limited to the summer months and characterised by extreme levels of heat, levels of heat likely to prove perilous for human and natural systems. In contrast, warm spells are deemed to be periods of unusually warm weather in any season with their 'warmth' not reaching the absolute levels of heat achieved by heatwaves. Broadly understood as periods of highly anomalous heat lasting for several days, heatwaves are both weather (meteorological) and climate (climatological) phenomenon. This is because heatwaves usually develop and decay over the weather timescale of 5–10 days but are also a distinct characteristic of the climate of some regions—they are a climatological feature. Although heatwaves have often suffered from an existential dilemma through being overlooked as serious natural hazards, they have been viewed in this volume as distinct phenomenon worthy of attention. This is because heatwaves can be observed, perceived, or experienced through the senses or by using weather related scientific instrumentation or technology. Further heatwaves are events that occur under certain conditions or circumstances that are generally distinct and recognizable. Heatwaves also exhibit complexity, involving multiple factors, variables, or processes that interact in intricate ways, for example the synergistic effects arising from temperature, humidity and wind. As established in this volume, heatwaves also exert impacts on their surroundings bearing consequences for human and natural systems with impacts being either direct or indirect ranging from subtle changes to significant transformations. As distinct observable and felt events, heatwaves exhibit variability in their behaviour and characteristics over time and space with this variability arising from natural internal variability within the climate system or external forcing as manifest by human induced climate change. This variability contributes to the complexity and dynamism of heatwaves as phenomenon. As a phenomenon, heatwaves have captured the interest of researchers and policy makers attesting to their uniqueness and significance especially in terms of their implications for society and environment, an outcome of which has been the development of specific heatwave-related policy associated with managing the risks arising from heatwaves. To sum up, it is reasonable to assume that in the absence of these traits, heatwaves would not be considered a phenomenon. This is clearly not the case.

Associated with heatwaves as phenomenon is the issue of a succinct definition. This is necessary for clarity, comprehension, and effective communication about heatwaves. As outlined in Chap. 2 a well-defined definition establishes a common understanding among researchers, policy makers and the public concerning heatwaves as a phenomenon, providing a framework for discussion, analysis and action. Further in the case of the heatwave—warm spell continuum, a clear heatwave definition assists in distinguishing between these phenomena and identifying their unique characteristics. As suggested in Chap. 2, two broad frameworks exist for defining heatwaves—the structuralist and the phenomenological. While a phenomenological approach has much to offer, a structuralist approach has been adopted in this work. From the material presented on defining heatwaves it is apparent that quite a 'heatwave definition industry' has developed over the last two to three decades. While a testament to the inventiveness of the heatwave science community, the diversity of heatwave definitions is a frustration to some. From the outset, the goal of this work has not been to offer a 'universal' heatwave definition. Rather, the intention has been to present some basic principles for what constitutes a 'good' physically based definition. Applying the criteria for 'good' to heatwave definitions offered by NMHS has revealed some are less than ideal in relation to exactness. Of course, in not offering a universal heatwave definition some readers may see this as 'an absolution of responsibility'. Assuming "universal" means one definition applied to everything, the answer is clearly no—one size does not fit all. This is particularly true if the intended application of a universal heatwave definition targets a specific sector of society, the economy, or an ecosystem, where the focus is on specific impacts. From this perspective, impact-based definitions should be tailored to fit their specific purposes. In short, bespoke impact-based heatwave definitions are necessary for addressing particular impacts effectively.

Having stated that one size does not fit all, there is probably an argument for a 'good' parsimonious hazard-based definition that would lend itself to a range of objective climatological analyses and thus facilitate comparisons of heatwaves between times and places. A plausible candidate for such a task might be that put forward by Perkins and Anderson (2013). That is, a heatwave is a period of at least three consecutive days when a critical heatwave variable (e.g., Tmax or Tmin) exceeds its local 15-day centred 90th percentile value during the warm, or if identified, heat season. Because heatwave intensity is likely to be an important characteristic for understanding heatwave impacts the intensity-based heatwave definition of Nairn and Fawcett (2015) may well fit the bill. Of course, in putting forward these two heatwave definitions it must be acknowledged they are based on temperature alone and do not incorporate a measure of humidity. That said the principles underlying these definitions appear basically sound. Given this, those interested in understanding humid heatwaves (e.g., in the humid tropics) could well contemplate replacing the pure temperature values with thermal indices that incorporate an absolute measure of atmospheric moisture. In contrast to the atmospheric science community, the physical oceanography community appear to have arrived at a 'universal' definition for marine heatwaves relatively quickly. Whether the broader marine science community will be content with the hazard-based marine heatwave definition described in Chap. 2 or call for complementary bespoke impact-based definitions that acknowledge species specific responses to anomalous ocean temperatures remains to be seen. Notwithstanding the suggestions made here, the need for hazard- or impact-based heatwave definitions will depend on the purpose with the former applied generally to understanding the physical drivers of atmospheric and marine heatwaves and their variability, with the latter relevant if understanding impacts or developing impact-based warning systems is the priority. Overall while the idea of a universal definition for heatwaves may seem beneficial in principle, the practical challenges, and complexities of implementing such a definition, along with the need to account for regional variation and diverse impacts, suggest that a more flexible and context-specific approach may be preferable.

Of interest to heatwave researchers and risk managers are the statistical properties of heatwaves and their constituent variables. As outlined in Chap. 3 statistical analyses using techniques that generally fall within the realm of Extreme Value Theory help in assessing the frequency, duration, intensity, and spatial extent of heatwaves. This information is vital for assessing the associated risks to human and natural systems. To ensure that the description and estimation of heatwave-related statistical properties is as reliable as possible due regard needs to be given to the robustness of data on which estimates are based. As outlined in Chap. 3 this requires an assessment of data completeness, homogeneity, and stationarity. These considerations not only apply to observed climate-related data but also impact data which is often used unquestionably in heatwave impact-oriented studies. As an alternative to climate observations, or perhaps more appropriately in the absence of observational data as is the case for many parts of the world, heatwave analysts can make use of a range of climate reanalysis products. While there is no doubt such products have facilitated an increased understanding of large-scale climate mechanics, there are several issues to consider when using these datasets for heatwave analysis. These relate to possible discrepancies or biases in the assimilated data sources which can introduce errors into the reanalysis outputs, particularly in regions with sparse or inconsistent observational coverage. Uncertainties in model parameterizations may also propagate through the reanalysis process, creating uncertainties in the resulting climate fields. Users of reanalysis products should also be aware of updates and improvements to reanalysis methodologies which could potentially result in discontinuities between different versions of reanalysis products, complicating long-term trend analysis and hindering the seamless integration of data from different sources or time periods. This is especially pertinent as different agencies such as the ECMWF, NOAA and the JMA to name a few, update their reanalysis schemes at different times and perhaps more importantly use different models and assimilation techniques for creating reanalysis products. Given this, ideally researchers should contemplate undertaking reanalysis intercomparisons before advancing any analysis based on a particular reanalysis product. Lastly the spatial resolution of reanalysis products may not be sufficient for the analysis of heatwaves especially at the urban scale requiring the possible use of downscaling techniques to achieve the resolution required which forms another possible source of uncertainty. Despite these challenges and assuming they are understood by heatwave analysts, climate reanalysis products will remain valuable tools for heatwave research, providing insights into temporal and spatial variability characteristics and supporting efforts to contextualise heatwave events and understand their development processes and impacts.

Because heat is the defining physical property of heatwaves, some of the basic concepts associated with heat and the processes related to atmospheric heating were outlined in Chap. 4. As the reader may have detected, considerable attention has been given to the energy balance concept in Chap. 4. This is because the attainment of the temperature extremes in the near surface layers of the atmosphere associated with heatwaves can often be understood through appreciating how the net radiation, the balance between all radiative heat gains and losses from the earth's surface, is partitioned between the turbulent fluxes of sensible and latent heat and the conductive flux subsurface heat. Critical to the partitioning of net radiation between sensible and latent heat fluxes is surface moisture availability with sensible (latent) heat flux, which elevates (dampens) air temperature, dominating over dry (moist) surfaces. Atmospheric heating in the free atmosphere occurs via two processes namely adiabatic and diabatic. Adiabatic processes result in temperature changes caused by the expansion (cooling) or compression (warming) of an air parcel as it rises (descends) in the atmosphere. Under this process temperature changes occur within a volume of air with no exchange of heat with the surrounding air. Warming of the atmospheric column by subsidence warming plays an important role in heatwave development through entrainment of heat into the atmospheric boundary layer complementing the atmospheric heating attributed to sensible heat flux from the surface. Diabatic heating or cooling describes temperature changes because of the gain or loss of heat from a parcel of air. This is opposite to an adiabatic process. There is growing evidence of the importance of diabatic heating in remote upstream regions for the development of heatwaves through its ability to perturb the pattern of the upper westerlies and induce blocking. Local to regional heating can also be enhanced by heat advection and the foehn affect.

While heat in a physical sense is often represented by temperature, the material presented in Chap. 4 emphasises in the case of life forms, especially humans and animals, heat is more than simply temperature with the conflation of temperature and heat discouraged. This is because in biometeorological terms heat is a complex phenomenon resulting from the synergistic effects of air temperature, humidity and ventilation levels, radiation load and metabolic activity. Added to this is clothing in the case of humans and the nature of body cover in animals. All these factors interact to influence the level of heat gain and loss to and from the body as modelled in the case of humans by the human heat balance. A distinction between heat stress and heat strain is made in Chap. 4 with the former generally defined as the negative effect of the thermal energy (heat) environment on individuals while the latter is the physiological response of exposure to heat stress. A plethora of thermal climate indices have been developed to measure and monitor heat stress. Comprising three broad groups—simple, empirical and rational—many of these have been applied in heatwave studies with some forming the basis of heatwave and heat-health warning systems. A burgeoning number of intercomparison studies of thermal climate indices applied in human and animal thermal comfort studies clearly indicate there is no best one size fits all index for understanding the heat stress characteristics and heat-related impacts of heatwaves. Rather the choice of index needs to fit the purpose of the heatwave research problem or application at hand.

15.3 Heatwave Causes

In line with the conceptual framework presented in Chap. 1, certain configurations of atmospheric and ocean climate patterns may set the scene for an increased likelihood of both atmospheric and marine heatwave occurrence. Referred to as modes of climate variability, these are quasi-periodic variations in ocean and atmospheric circulation patterns and processes that oscillate between opposite phases. Although modes of climate variability don't directly cause heatwave events or periods of extreme heat, through shifting atmospheric and ocean patterns away from their 'normal' state they tip the balance in terms of the increased or decreased likelihood of setting the broad atmospheric and land and marine surface conditions for heatwave occurrence. Associated with modes of climate variability are climate teleconnections. Teleconnections, which often manifest as a seesaw like effect in climate conditions between two regions, occur because seemingly 'disconnected' regions are connected by an atmosphere and/or ocean bridge. This bridge 'communicates' changes in one region to the other region and vice-versa. As outlined in Chap. 5 there are a vast number of modes of climate variability. In fact, quite an industry has developed related to the identification of modes of climate variability at a range of spatial and temporal scales, with all modes usually represented by an associated index. Essentially statistical constructs, indices, and the modes of variability they represent, attempt to characterise complex physical processes. This needs to be borne in mind if heatwave analysts are interested in understanding what underpins, in a process sense, a statistical association between heatwave occurrence and a particular teleconnection index. Rather than presenting the plethora of modes of climate variability, a deliberate decision was made to restrict the description of these to the so-called primary modes of variability. This is because these tend to account for a significant percentage of the variance at a range of timescales of those climate variables most relevant to heatwave occurrence. Drawing on the burgeoning literature that addresses the long-term or event-based association between modes of climate variability and heatwave occurrence, it is safe to say that changes in atmospheric and/or ocean conditions can elicit changes in the likelihood of atmospheric and/or marine heatwave occurrence. The modes of variability identified as important for influencing atmospheric and marine heatwave occurrence range across the intra-seasonal to decadal timescales and are likely to interact to influence heatwave occurrence. Unravelling the nature of the heatwave development processes associated with the various modes of climate variability and how compounding between these may enhance the conditions for heatwave development will bode well for endeavours focused on extended range heatwave prediction.

Nestled within the space and time scales associated with modes of climate variability is the synoptic scale which spans a horizontal length scale of 1000–2500 km, a vertical scale of up to 10 km and a time scale of 5–10 days. It is at this scale that heatwaves usually develop and decay. Chapter 6 in providing an overview of the general synoptic scale processes associated with the development of atmospheric heatwaves drew attention to the role of atmospheric blocking, advection and the foehn effect. Of these, blocking which is associated with an area of persistent high pressure, is often cited as a major driver of heatwave generation. As the term suggests blocking prevents the movement of the usual succession of weather patterns over a region. It arises because of a perturbation in the flow of the upper westerly winds or the so called Rossby wave pattern. Some intense blocking events occur when the Rossby waves are highly perturbed and 'break'. Such breaking episodes may be triggered by upstream diabatic heating of the atmosphere due to the release of large amounts of latent heat in the upper atmosphere associated with the deep and rapid ascent of moist air. Such heat sources, because of their magnitude, significantly disrupt the Rossby wave pattern. In the lower latitudes, diabatic heating arising from intense tropical convection can also initiate heatwave development through inducing atmospheric circulation perturbations. Blocking is an important heatwave development process because it often leads to adiabatic warming by subsidence and enhanced radiative heating and the accumulation of heat over a region. Frequently associated with subsidence is a strong inversion which prevents the vertical dissipation of heat, moisture and air pollutants which may exacerbate the levels of discomfort associated with a heatwave. Although blocking is often used as a catch-all phrase, three types of blocking pattern exist namely the omega block, the dipole (Rex) block and the high pressure ridge all of which manifest differently in terms of their visual form on atmospheric pressure maps. Advection of warm and sometimes humid air over long distances is another process often associated with heatwave development with advection frequently occurring along the flanks of omega blocks and high pressure ridges. Similarly, the steering of flows of anomalously warm air along the flanks of mountain ranges from low to mid-latitudes in the form of a low-level jet can enhance heatwave development. Although not often cited as a heatwave development process the downslope flow of winds on the lee side of major orographic barriers and associated compressional warming in the form of a foehn wind can result in strong temperature anomalies and thus heatwave conditions over a region for several days. While the processes of blocking, advection and the foehn effect have been described separately in reality heatwaves are likely to arise from a combination of these processes with, in some cases, one process largely dominating. To bridge the gap between theory and reality a range of heatwave case studies were presented in Chap. 6 to shed some light on the actual processes associated with the development of several important heatwave events across a range of geographical regions. This case study approach has highlighted the geographical variation of the synoptic scale processes that generate heatwaves with some regions demonstrating a strong regional process flavour that reflects not only the climate regime but the geographic setting in terms of oceanic, continental, and orographic influences. Considering this, analysts examining specific heatwaves and the synoptic processes potentially linked to their formation should maintain a flexible perspective, anticipating that heatwaves may stem from variations on a general synoptic model of heatwave development, with these variations reflecting broader environmental factors prevailing at the time, including land and/or marine surface conditions.

Shifting focus away from the large-scale atmospheric processes discussed in the two preceding chapters, Chap. 7 explored the significance of heat, mass, and momentum exchanges between land surfaces, water bodies, and the atmosphere in the formation of atmospheric, marine, lake and stream heatwaves. At the outset the Chapter emphasised the surface refers to a climatically active layer beyond the mere two-dimensional surface. It is this active layer that is involved in the two-way exchange of heat, mass and momentum with the overlying atmosphere and subsurface. In the case of heat and mass exchange, the steepness and direction of the vertical gradients of temperature and moisture respectively is important while the degree of momentum exchange is determined by the vertical gradient of wind speed and surface roughness. Critical to how a surface responds thermally to an energy input from the sun—radiative forcing—depends on the nature of the thermal properties of the materials that make up the surface. These comprise density (ρ), specific heat (c), heat capacity (C), thermal conductivity (k), thermal diffusivity (αh) and thermal admittance (μ). Of these, thermal admittance (μ), a measure of the ability of a surface to accept or release heat, exerts a strong influence on the apportionment of sensible heat flow from the surface to the atmosphere or from the surface to the solid or liquid subsurface. For dry soils, a situation often associated with air temperature related heatwaves, the surface thermal admittance is low, hence large amounts of sensible heat will transfer to the atmosphere raising air temperatures. The surface to atmosphere interactions, important to atmospheric heatwave development, occur in the so-called atmospheric boundary layer (ABL), the layer of the atmosphere below the free atmosphere, While the large-scale weather situation may moderate its depth, the ABL displays clear diurnal to seasonal variations in its vertical extent. A key determinant of heatwave occurrence over land surfaces is how available energy is apportioned between sensible and latent heat in the lower parts of the ABL. This makes surface moisture status critical. With increased drying, as might occur throughout a prolonged dry period, more of the available energy is channelled into surface to atmosphere sensible heat flux with ensuing positive feedback likely to lead to strong temperature anomalies. Exacerbation of near surface temperature anomalies during atmospheric heatwave events may arise due to downward heat transport or heat entrainment from near the top of the ABL or from advection of heat from upwind hot dry areas. Further the type of vegetation cover may also play a role in determining the likelihood of heatwave occurrence via vegetation surface albedo, canopy heating and evapotranspiration interactions.

In the case of marine and lake environments, the nature of energy exchange in a surface mixed layer, which varies in its depth on daily to seasonal time scales, determines water temperature tendency and thus the occurrence of heatwave causing temperature anomalies. Analyses of a range of marine heatwave events points to two main marine heatwave types related to the dominant heatwave generation mechanisms, namely an air-sea heat exchange type and a heat advection type although for many marine heatwaves a mixture of both atmospheric and ocean processes may explain their origin. For lakes it would appear processes akin to those involved in air-sea heat exchange type marine heatwaves predominate. For streams and rivers, Chap. 7 suggests water depth, flow velocity and sky view (shading) are the likely controls determining whether radiative heating of the water column leads to anomalous water temperatures and the occurrence of stream heatwaves.

Chapter 8 opens with the suggestion that nowhere else on the planet are changes in land cover more obvious than in urban settlements with this change leading to the inadvertent alteration of surface—atmosphere exchanges of heat, moisture and momentum with the distinct possibility that during heatwaves these altered exchanges may exacerbate urban heat. For this reason, Chap. 8 presents the urban surface as a special case of the general land surface and suggests that the general modification of the atmosphere over urban areas is the outcome of the synergistic effects of urban surface fabric, structure and land cover. These features underpin the urban surface's morphological complexity and climatic diversity and in concert work to alter fluxes of heat, moisture and momentum giving rise to the unique climates found in cities as manifest by the so-called urban heat island phenomenon. During a heatwave the increase in air temperature in a city, or in other words its thermal response, will depend on the behaviour of all components of the surface urban energy balance. Critical in determining the thermal response of an urban surface is how the available energy (Q*) is partitioned between sensible (Qh) and latent (Qe) heat, as controlled by moisture availability. Because urban environments tend to be dry relative to their rural counterparts, much of the available energy is diverted into sensible heat flux (Qh) into the atmosphere or into storage (Qs) in the urban materials that make up the physical surfaces of urban environments. Consequently, Qe fluxes in urban environments are generally small with the result that cities typically possess high Bowen ratios. A unique feature of the urban energy balance is anthropogenic heat (Qf) which arises from heat emissions from buildings, transport, industrial processes, and human metabolism. This forms an additional energy source to that provided via the net radiative flux.

Amongst other urban climate effects related to the distinct nature of the urban energy balance, including, initiation of local circulation systems, enhanced convection, cloud formation and precipitation, is the much-discussed urban heat island (UHI). Although most studies have focused on the urban canopy urban heat island, it is important to recognise the existence of three further types of urban heat islands namely the surface, urban boundary layer and subsurface heat islands as each of these have a distinct set of processes associated with their formation. Loosely defined as a warm city anomaly in a sea of cool rural air with an island like form, as portrayed by temperature transects from the countryside across cities, the UHI is of special interest in relation to heatwaves. This is because the extra heat contained within cities, arising from the way available energy is partitioned between the various components of the surface energy balance, may be exacerbated during heatwave events leading to an intensification of the urban heat island effect and hence the elevation of urban temperature anomalies above those for surrounding rural areas—the unbearable heat of the city as popularised in books and movies.

From the material presented in Chap. 8 it is clear the debate related to the nature of the synergistic effects of heatwaves and urban heat islands continues. Some studies find no amplification of UHI intensities while others do. This equivocal situation may well relate to the contrasting approaches adopted in addressing this important question. Notwithstanding this, Chap. 8 concludes there is mounting physically plausible evidence that heatwaves enhance urban heat due to alterations in the partitioning of available energy between sensible and latent heat. This arises because of the rapid desiccation of urban surfaces under heatwave conditions leading to sensible heat enrichment of urban air beyond background levels. Added to this is the enhanced flow of heat from storage in the urban fabric, especially at night, because of increased daytime surface heating of urban surfaces under clear sky conditions often associated with heatwaves. That increases in anthropogenic heat flux above background levels also enhances the UHI effect is a distinct possibility given emerging evidence that heat emissions from commercial and residential buildings associated with elevated cooling demand under heatwave conditions have discernible impacts on UCL and UBL temperatures and cognate processes. Although the literature is rife with statements that a warming—air conditioning positive feedback mechanism adds extra heat to that associated with heat emissions from air conditioning systems, this may be more conjecture than fact. This is because the amount of science explicitly directed at distinguishing between the temperature increase simply caused by air conditioner use (i.e., without feedback) and the additional temperature increase caused by feedback is still in its nascent stages. This is especially so in the case of heatwaves.

15.4 Heatwave Consequences

In this book heatwave impacts have been viewed as the realisation of risk, with heatwave related risk a product of the synergistic interaction between four risk determinants namely heatwave attributes, heatwave related vulnerability, heatwave exposure and responses to heatwaves. As outlined in Chap. 9, heatwave impacts can be tangible, intangible, direct or indirect and propagate through systems unfolding over space and time. While understanding the physical processes associated with heatwaves is important, in the absence of knowledge about the nature of heatwave impacts, developing impact management solutions will remain a challenge. This is certainly the situation presently, as the systematic reporting of heatwaves and their impacts either locally, nationally, or globally has not yet been achieved. Added to this is the general lack of a methodology for assessing the impacts arising from heatwaves. These broad issues have served as a motivation for this section on heatwave consequences. To provoke discussion on how to assess the impacts arising from specific heatwave events, some generic steps for undertaking a systematic heatwave impact assessment (HWIA) have been presented in Chap. 9. These include (i) settling on the purpose of the HWIA and establishing context, (ii) undertaking a conceptual mapping of likely impacts and dependencies, developing impact metrics, and identifying stakeholders, (iii) locating evidence of impacts, (iv) assessing the impacts in terms of their nature, evolution, drivers, and modifiers and, (v) developing heatwave risk management strategies. Where possible, practical advice on how to achieve each of the steps to HIWA have been provided.

Acknowledging that exposure and vulnerability are fundamental modifiers of heatwave related risk and thus impact, and that neither exposure nor vulnerability are fixed, the factors that influence these critical determinants of heatwave risk were reviewed in Chap. 10. The Chapter notes that the critical dimensions of exposure are how often, for how long and how much an object is subject to critically damaging heat, or in other words the frequency, duration and intensity of the exposure. The related concept of heat tolerance, which refers to an animate or inanimate object's ability to withstand heat stress without experiencing a heat-related impact was also introduced in this chapter. Importantly the Chapter notes in the absence of impact data, the default has often been the use of a fixed threshold temperature value or value of some other measure of heat as an indicator of the level of heat beyond which impacts manifest, whether they relate to human or natural systems. This of course is rather naïve and ignores the fact that the cumulative effects of exposure below so-called lethal levels may be just as damaging as lethal exposures. In relation to human population related heatwave exposure, the idea of heatwave climates was introduced in Chap. 10 with these empirically defined as those climate regimes where impactful heatwaves have occurred. Vulnerability in this book has been taken to mean the propensity or predisposition to be adversely affected with in the case of heatwaves, vulnerability understood as a modifier of the relationship between the heatwave hazard, heatwave exposure and heatwave related impact. The Chapter notes that while the term vulnerability is equally relevant to animate and inanimate objects, in the case of heatwaves, it is generally discussed in relation to human based activity systems and in particular individuals or specific societal groups who may be exposed to heatwaves. From the material presented in Chap. 10 it is apparent that despite being some distance from a state of nirvana, our comprehension of the factors contributing to human population vulnerability during heatwaves has expanded significantly over the past few decades, encompassing health, socio-economic, institutional, behavioural, and cultural dimensions. However, the situation regarding infrastructure and natural systems is less favourable. Accordingly, and in addition to material on human populations, Chap. 10 presented information on the underlying drivers of heatwave vulnerability within infrastructural systems and marine, terrestrial, and aquatic ecosystems. While the Chapter has unearthed some worthy insights into the factors that influence heatwave vulnerability in infrastructural systems and ecosystems, it notes that significant knowledge gaps exist. This is particularly pertinent with regards to the intricate interplay between the various components of infrastructural systems and the same for ecosystems and how interdependencies in such systems may exacerbate heatwave-related vulnerability. This is especially so in the case of mega-heatwaves, which could precipitate unprecedented impacts through feedback loops and non-linear effects.

Purposefully avoiding speculation about possible heatwave impacts, Chap. 11 focused on observed heatwave impacts. Although the impacts community is some way from establishing a replete picture of the nature of heatwave impacts in human and natural systems it is clear from the limited information available, including that in the non-academic literature, that heatwaves induce substantive impacts, posing serious threats and in some instances precipitating transformative changes in human and natural systems. Setting aside the challenges of when a heatwave is not a heatwave—the definition problem—as well as how heatwave deaths are measured and recorded—the estimation problem—it is highly likely that the impacts of heatwaves on human populations in the form of mortality exceeds at least 170,000 over the period of 1998–2022, according to one disaster data base. Undoubtedly this figure represents a gross underestimate of the actual number of heatwave related deaths, a consequence of the lack of a fully integrated global scale system with a robust protocol for recording heatwave deaths and perhaps a result of heatwaves historically not being considered a serious peril. Heatwaves and their associated heat have serious impacts on human health because when exposed to excessive heat the body's core temperature starts to drift upwards from the normal resting temperature of around 37 °C resulting in a suite of heat-related illnesses that fall into the four broad categories of acute heat illnesses (e.g., heat stroke), chronic disease exacerbations, pregnancy complications, and behavioural and other heat-related emergencies. Unsurprisingly animal populations are also affected by heatwaves. This is nowhere more obvious than in the distress experienced by domesticated animal populations for which high death tolls have been recorded during heatwaves, especially amongst feedlot cattle which are held in large numbers in close quarters, conditions that can create high levels of heat stress during heatwaves, making this a serious animal welfare issue. There is also burgeoning evidence of the impacts of heatwaves on terrestrial wildlife (non-domesticated animals) with concerns that repeated heatwave events in locations for which these hazards have increased in their intensity and duration or have not been a discernible feature of the climate, may threaten the existence of some iconic species.

Beyond human and animal population health effects, heatwaves create significant economic impacts. Although there are only a limited number of estimates of the scale of these impacts, for vulnerable regions such as found across Southern Europe, damages associated with extreme heatwaves may exceed one percent of the national GDP with much off this related to downturns in worker productivity, a burgeoning issue for many economies that are heatwave sensitive. Integral to a smoothly functioning economy is the stability of critical infrastructure such as energy and transport systems. That heatwaves can wreak havoc on infrastructure and precipitate a range of heatwave related impacts is plainly evident from several well documented cases. For example, the precarity of nuclear reactors during heatwaves has been laid bare by the need for some large electricity companies in Europe and elsewhere to shut these down during periods of extreme heat. This is because water used for reactor cooling, often drawn from nearby rivers, becomes too warm for reactor cooling operations to continue, the upshot being a suspension of electricity supply and the occurrence of brownouts or in the worst-case blackouts. Heatwaves not only affect electricity generation in nuclear power plants but also have a wider impact on electricity distribution systems with so-called de-rating a common issue. Physical damage to structural elements of the so-called power grid can also occur during heatwaves. Heatwaves can significantly impact road and rail transport systems, causing disruptions and challenges across various aspects of infrastructure and operations. High temperatures associated with heatwaves can lead to the expansion and heave of road pavement materials and buckling of railway tracks. These effects not only cause damage but pose safety risks for vehicles and rolling stock with the need for speed restrictions on roads and railways to mitigate the risk of infrastructure damage or accidents with associated disruption to transport services a common outcome. Heatwaves also impact the safety and comfort of passengers and workers. Elevated temperatures inside vehicles and trains can sometimes lead to heat-related illnesses or in extreme instances in crowded trains passenger death. The health and productivity of transportation workers are also affected during heatwaves. Presently the observational evidence for the impacts of heatwaves on buildings, airports, and telecommunication and water systems is outweighed by theoretical studies, or those focused on projected impacts related to human induced climate change. That said, as awareness grows of the potential impacts of heatwaves on these often-overlooked aspects of infrastructure, evidence is emerging that such systems are also affected as manifest by the impact of heatwaves on airport runway stability and aircraft lifting power, building indoor thermal comfort, cooling systems in telecommunication centres and water availability.

Heatwaves seriously affect ecosystems and ecosystem services. This is particularly evident in marine and aquatic ecosystems, where unusually warm water in the form of marine, lake and stream heatwaves can disrupt physical and biological processes, harming organisms at various trophic levels. Notable impacts of marine heatwaves encompass substantial changes in the spatial distribution and/or abundance of species and the occurrence of harmful algal blooms that not only affect the nature of the food web but produce knock-on effects for some ecologically and commercially important species. Because of their importance as habitats some attention was given to the impact of heatwaves on seaweed, coral and seagrass colonies. For seaweed colonies there is evidence that invasive species increase in abundance following marine heatwaves. In some cases, intense marine heatwaves have resulted in the extinction of native seaweed species with these rapidly replaced by invasive species. Coral reefs, or the so-called rainforests of the oceans, are particularly vulnerable to acute warming episodes. For intense marine heatwaves, extensive bleaching and mortality have been observed in corals that otherwise would be thermally tolerant during moderate warming events, lending strength to the idea that even heat tolerant coral species can be overwhelmed during intense marine heatwaves. Indirect impacts of coral die-off following heatwaves include changes in the structure of fish communities and microbiomes. Extensive tracts of seagrass mortality have been observed following marine heatwaves with remineralisation of large carbon stocks tied up in seagrass and its substrate and impacts on marine food webs in general noted, with associated impacts on the health of seagrass dependent species. Like seaweed, there is evidence that some native seagrass species are supplanted by non-native ones following intense marine heatwaves. For aquatic ecosystems there is a dearth of observational evidence related to heatwave impacts. However, a steadily growing body of knowledge especially for lakes points to clear impacts of heatwaves including eutrophication and deoxygenation of lake waters, higher levels of cyanobacteria biomass and changes in the volume of chlorophyll-a in lakes. To enhance the empirical evidence base for understanding the effects of marine and lake heatwaves on biota, researchers are increasingly relying on laboratory-based experiments to simulate heatwaves and monitor the responses of specific plant and animal species. The results from such studies generally align with what has been observed to date in the field. For terrestrial ecosystems most of the observational evidence for heatwave impacts relates to trees. While some tree species demonstrate remarkable tolerance to heatwave events because photosynthesis—transpiration decoupling is implemented to avoid lethal leaf temperatures, common heatwave impacts include forest canopy collapse, leaf damage as a result of scorching, reduced photosynthesis, water stress, increased vulnerability to pests and diseases, premature leaf drop and reduced growth, Heatwaves have also been shown to elevate wildfire risk especially if they are embedded within a longer period of drought.

15.5 Heatwave Responses

Responses have been defined in this book as intentional short to long-term and sustained actions and measures to reduce heatwave related risk and thus impact. It is in this context adaptation, as an action in response to heatwave risk and impact, is defined here as adjustments in human systems (e.g., social, economic, cultural, political, technological, institutional) in anticipation of heatwaves and their effects. Adaptation is viewed as a behavioural adjustment influencing levels of heatwave risk over time, with successful adaptation actions reducing general or specific heatwave related impacts, eventually leading to changes in other heatwave risk determinants such as exposure and vulnerability. Of course, there is always the possibility that some well-intentioned adaptation actions may have inadvertent consequences for exposure and vulnerability and exacerbate heatwave related risk giving rise to maladaptation. Allied with adaptation are the concepts of adaptation limits and loss and damage. The former refers to the constraints or boundaries beyond which the ability of a system to adapt to changing conditions is compromised while the latter has been linked to the idea of residual risk, generally referring to the risks and impacts, positioned beyond what adaptation can achieve. For some parts of the world there is the distinct possibility that under changing climate conditions heatwave adaptation limits may be reached.

Amongst the vast range of heatwave adaptation options, heatwave risk governance, heatwave insurance, the deployment of Chief Heat Officers and the development of heatwave risk knowledge networks were chosen for attention as these represent rapidly emerging areas in the broader field of heatwave adaptation. Establishing well-formed heatwave risk governance is considered essential for effective heatwave risk management as this provides the mechanisms to facilitate not only the implementation of available heatwave adaptation options but also those related to mitigation. Heatwave insurance has been introduced as a possible heatwave risk transfer mechanism. Its application potential, in the form of parametric micro-insurance, could be realised in low- and middle-income countries where the heat-related occupational health and safety strategies in place in high-income countries are not available. Complementing insurance applications at the micro-scale is the United Nations Capital Development Fund's aspiration to develop a Climate Insurance Linked Resilient Infrastructure Financing mechanism as an insurance solution designed to encourage cities to allocate resources toward climate resilient infrastructure to de-risk potentially crippling heatwaves as well as other climate-related extreme events. Acknowledging the risk posed by both chronic and acute heat, several cities have appointed Chief Heat Officers. While the appointment of some city Chief Heat Officers, as well as a Chief Global Heat Officer, has been initiated and supported by a not-for-profit organisation, others have been established by city authorities using their own municipal funds with the US State of Arizona being the first to appoint a state-level Chief Heat Officer. The dependencies of being able to garner close collaboration across a diverse group of municipal government departments, agencies, and policy makers as well as civil society interest groups and sustained financing of the positions themselves and related projects, is likely to determine the long-term effectiveness of Chief Heat Officer appointments at the city to global scale.

Reflecting the need to put in place strategies for heatwave risk reduction, and perhaps as a response to the inadequacy of some national and state governments to establish mechanisms to address the risk arising from heatwaves, there has been a rapid growth in heatwave related knowledge networks and portals that focus on extreme heat and the promotion of heat and heatwave related resilience. Examples of such knowledge networks or platforms specifically dedicated to heat and heatwave risk are the Global Heat Health Information Network (GHHIN), the US's National Integrated Heat Health Information System (NIHHIS) and the Extreme Heat Resilience Alliance (EHRA). While possessing wider remits, providing heat and heatwave related information is an integral part of the modus operandi of the Red Cross' Climate Centre, a specialized unit within the International Federation of Red Cross and Red Crescent Societies. Similarly heat and heatwaves as a risk management problem is visible as a priority amongst other international networks such as the C40 Cities Climate Leadership Group (C40) and the Resilient Cities Network (RCN). Opportunities to strengthen heat and heatwave related knowledge as part of their overall vision await structures such as the Local Governments for Sustainability (ICLEI) and the United Cities and Local Government (UCLG) networks.

What precipitates heatwave related responses in the form of active policy change at the city level and beyond has been posed as an open question in this work. Although this topic invites further exploration, evidence suggests that the experience of encountering impactful heatwaves catalyses policy change. The European heatwave of August 2003 serves as a good example of how severe heatwaves might precipitate policy change. Other pertinent examples of so-called focusing events, include heatwaves that occurred in the cities of Chicago, Ahmedabad, Karachi, and across the Australian states of Victoria and South Australia. A range of theories of possible applicability to understanding autonomous decision making related to heatwave risk reduction options were also introduced in Chap. 11. Among these, the health oriented Transtheoretical, Health Belief, and Health Action Process Approach constructs emerge as promising candidates for initiating discussions on the factors influencing autonomous actions in the context of heatwave risk reduction. Additionally, the Positive Deviance model represents an interesting possibility regarding how heatwave adaptation ideas might diffuse throughout communities. Complementing the specific individual and small-group decision making models is the Cultural Theory of Risk. This provides a further lens through which to explore what might determine heatwave responses.

Chapter 13 suggests that the 1990s represent a turning point in the philosophy of disaster risk management in general with an ever-increasing recognition that to improve risk reduction, it is best to get ahead of the game by preparing for extreme events as opposed to responding to and recovering from them once they have occurred. This paradigm shift was undoubtedly hastened by the realisation that the nature of extreme weather and climate events is likely to change for the worst under human induced climate change. In response to mounting evidence of present and future heatwave precarity, many national governments and municipal authorities have developed heatwave and heat-health early warning systems (HEWS and HHWS respectively). Often forming part of a wider multi-hazard early warning system, HEWS and HHWS may also exist as standalone systems. While HEWS tend to be hazard-based and focused on predicting the occurrence of heatwaves alone, HHWS are generally referred to as impact-based because their principal focus is the prediction of heatwave related health impacts. Like most EWS, HEWS and HHWS are typically composed of four inter-related and complementary elements that address risk knowledge, observation, detection and monitoring, warning dissemination and communication, and preparedness and response capability. From the material presented on the general nature of HEWS and HHWS, diversity clearly characterises the nature of systems currently in operation across the globe. Worryingly for many countries where heatwaves have become endemic, HEWS and HHWS are conspicuously absent. Although early warning systems are often hailed as a panacea for reducing heatwave risks, much remains unknown about their efficacy. To address this, process and outcome evaluations of HEWS and HHWS are necessary. While process evaluations typically centre on the functioning of a warning system and its individual operational components, outcome evaluations consider the extent to which the impacts of a heatwave (e.g., number of fatalities or hospital admissions prevented) have been averted because of the issuance of warnings and the resultant implementation of a range of intervention strategies. Notwithstanding the fundamental dependency of the robustness of results on the methodology applied, based on the limited number of comprehensive process and outcome evaluations conducted to date there appears to be mixed evidence regarding the efficacy of HEWS and HHWS.

Central to any HEWS or HHWS is the accurate prediction of periods of extreme heat that may manifest as a heatwave. This work contends that to stay ahead of the curve as much as possible, accurate heatwave predictions well in advance of their arrival, all other things being equal, provide the best chance for moderating the risks associated with heatwaves. For this reason, some attention was given to heatwave prediction in Chap. 13. Encouragingly over the last few decades there has been a vast improvement in short-range weather forecasts which bodes well for heatwave prediction out to 3–5 days. The real game changer however is improvements in predictability at extended-range timescales and beyond. Rather than predicting heatwaves per se at these longer timescales, predictions of temperature anomalies are made. These provide insight into the likelihood of anomalous temperature conditions and thus, the propensity for heatwave occurrence. Promisingly, predictions of the scale of temperature anomalies associated with impactful heatwaves with lead times ranging from 2 to 3 weeks are emerging in the heatwave prediction literature. However, predictions at such timescales remain more experimental than operational. That the WMO in 2023 established Global Producing Centres for Sub-Seasonal Forecasts and a Lead Centre for Sub-Seasonal Forecasts signals there is mounting confidence in the potential for predictions at the extended range timescale to make a real difference in relation to mitigating the effects of extreme weather and climate events. Because heatwaves are the more predictable of a range of hydrometeorological hazards, the hope is that priority will be given to the mainstreaming of extended range heat and heatwave products into operational HEWS and HHWS. Related to the efficacy of heatwave related EWS is effective risk communication. In relation to this, is the unavoidable question of whether heatwaves should be categorised and named with the view to improving heatwave risk communication. As suggested in Chap. 13 although names are given to tropical cyclones (hurricanes) and winter storms this is not justification enough for doing the same for heatwaves as tropical cyclones and severe winter storms are fundamentally different to heatwaves. Further the naming of these extreme events is coordinated at regional to international levels across a range of stakeholder countries. Whether the experimental naming and categorisation of heatwaves initiated by a not-for-profit organisation and rolled out across several cities continues and is supported by or co-developed with their national meteorological agencies, remains to be seen. In giving this topic due consideration, this work has outlined some of the potential positives associated with heatwave naming but equally has highlighted a range of serious operational concerns. Perhaps in the wider context of a precautionary principle approach to heatwave naming, a wise move would be to prioritise solving the challenge of heatwave categorisation and whether this be hazard- or impact-based. This is because categorisation will undoubtedly form an important decision framework for determining whether a heatwave event is named or not.

In addressing adaptation to and the mitigation of heatwave related risks in urban environments this book has made the case for a distinction between mitigation and adaptation. Accordingly at the outset, Chap. 14 contends that a basis for distinguishing between these broad responses is consideration of whether an action results in the modification of the exchange of heat, mass and momentum between the urban surface and the overlying atmosphere, with the intention of heat reduction. In other words, the basis of separating these two responses rests on establishing whether the action is focused on modifying the physical heat hazard at acute or chronic timescales. If it does (does not) then it is considered to fall into the category of a mitigation (adaptation) action or option. To qualify this distinction further and make clear the relationship between mitigation and adaptation, the implementation of the mitigation solutions discussed in this book are regarded as facilitated through the process of adaptative urban planning and design and related policy making—adaptation actions. The hope is that the reader is convinced that actions that modify the physical climate relationship between the urban surface and the adjacent atmosphere, through what is effectively surface radiative and energy balance management, are plainly different to the nature of the options discussed under the broad heading of heatwave adaptation. Evident from the material presented on heat and heatwave mitigation options, organised under the general categories of modifications to the facets or elements of urban surfaces, the application of nature-based solutions (NBS), and changes to urban morphology and built form, is their capacity to alleviate heat in urban environments, either independently or in conjunction. However, this effect may be more pronounced for chronic rather than acute heat and may not be as significant as perceived by those less acquainted with the details of the scientific literature. Further the scale of mitigation option application, for example the combined installation of cool roofs and cool pavements, required to achieve meaningful heatwave risk reduction in terms of dampening urban heat extremes during heatwaves may pose challenges in terms of feasibility, achievability or indeed acceptability. Lastly as emphasised at the end of Chap. 14 and notwithstanding the importance of adaptation, it would be simplistic to assume that the mitigation strategies discussed in this chapter alone will suffice to counteract the escalating levels of chronic urban heat and that associated with heatwaves driven by anthropogenic climate change. Without addressing the root cause of human-induced climate change and without effectively curbing the rise in global temperatures and associated extreme heat events, there may come a point where even the widespread implementation of the type of urban heat mitigation measures outlined in this work proves ineffective.

15.6 Directions for Future Enquiry

Much remains to be done in relation to enhancing our understanding of the nature of heatwaves as weather and climate phenomenon as well as the causes and consequences of heatwaves, and responses to them.

15.6.1 Heatwaves as Weather and Climate Phenomenon

Most likely the debate related to whether a universal heatwave definition should be agreed upon will continue to roll on. While at least one parsimonious definition has been singled out in this book for application in climatological studies of heatwaves as pure physical events, for understanding heatwave related impacts and managing associated risk, heatwave definitions will need to be bespoke. Accordingly, a useful avenue for future location and sector specific research is the identification of workable and efficacious heatwave definitions for the heatwave risk management problem at hand. One can imagine for the same location but for different sectors, bespoke definitions will take different forms (e.g., minimum duration criteria) and be based on different heat metrics (e.g., thermal indices). While not wanting to trivialise the issue of a widely applicable heatwave definition, the acceptance within some quarters that bespoke definitions are a valid way forward and might assist with breaking free from the constraints or limitations associated with the insistence for a universal heatwave definition, something that could be seen as positive and empowering.

A frequently used descriptor for heatwaves at the meteorological (event) or climatological (climate normal/30 year) timescales is heatwave intensity. Frustratingly there are several definitions of this important heatwave attribute in circulation. On top of that, magnitude and intensity are often used interchangeably. This is an area that deserves some attention and in contrast to the above call for bespoke heatwave definitions there is a good case to be made for a standardised definition of intensity. Perhaps an apposite starting point would be to assess whether the intensity metric used in the present national level heatwave warning system in Australia might be suitable for quantifying intensity at both meteorological and climatological scales and indeed applying in heatwave impact studies. Notwithstanding the need to tighten up what is meant by a compound heatwave event—is it a sequence of seemingly independent heatwave events separated only by heatwave definition criteria, or a combination of threshold breaking day and nighttime temperatures—a further area deserving attention is exploration of the physical processes associated with compound heatwave events. Similarly further process-oriented research is required related to heatwave compounding with other hazards. For example, in the case of a heatwave compounding with a tropical cyclone what are the processes associated with heatwave intensification or exacerbation due to its interaction with the tropical cyclone. In relation to the idea introduced in this book that heatwaves occurring over time at a particular location possess multiple forms, with the example of Tokyo used to demonstrate this, further work on heatwave morphology is required not only with the view to developing location or regional heatwave typologies but also understanding the physical processes associated with different heatwave forms. Some exploration of the association between heatwave morphology and the time and space distribution of associated impacts represents an intriguing area of future research.

Critical to understanding heatwaves as meteorological and climatological phenomenon is the availability of the requisite data whether it be for the variable used to describe heatwaves as physical entities (e.g., temperature and humidity) or data related to heatwave occurrence. With regards to the latter, a furtive area for further enquiry is the mining of written accounts of climate contained in historical literature, diaries, and logs. As sources of information about 'other' extreme events these have proved invaluable assisting with the compilation of event histories that benefit the development of extreme event climatologies. As yet, there appears to be little in the literature on the use of historical documents for gleaning information about historical heatwave events and their impacts. For those interested in historical climatology and social and economic history this could prove an interesting area for exploration. The availability of observational data for building an understanding of the nature of heatwaves will be an ongoing issue for many parts of the world because of the absence of climate monitoring. While heatwave analysts have the luxury of turning to reanalysis products as a source of numerically constructed data, caution needs to be exercised in the use of such data. Accordingly, research on the fitness for purpose of reanalysis products for heatwave analysis and the calculation of thermal climate indices will always be required in the form of reanalysis intercomparisons as well as analyses that consider the outputs from different downscaling algorithms in the case that data at finer spatial scales than that offered by reanalysis products are required. Given that some reanalysis products may do better portraying heatwave related variables in some parts of the world and not others, a possible area for exploration might be whether blended reanalyses might serve to improve the quality of data for heatwave research, especially for those regions that have or are likely to become heatwave hotspots. The statistical analysis of heatwaves and their constituent variables and characteristics will benefit from ongoing developments in data homogeneity and Extreme Value Theory (EVT). Continued advancements in statistical modelling and uncertainty estimation techniques, including Bayesian methods, machine learning algorithms, and non-parametric approaches are likely to contribute to improved statistical analyses of heatwaves as extreme events, as will the incorporation of satellite data. Further, the extension of EVT to heatwaves as multivariate extremes and the joint behaviour of heatwave variables such as temperature and humidity as well as wind and radiative load and the incorporation of external covariates will enhance comprehension of heatwaves as both statistical and physical entities. The continued development of non-stationary extreme value models that can capture changes in the distribution of heatwaves as extreme events over time and space will also benefit heatwave analysis as well as advances in computational techniques and software tools for fitting complex extreme value models to large datasets efficiently.

Sustained interest in heat as the defining physical characteristic of heatwaves is likely to be a furtive area of future enquiry. This is especially so in the case of heat stress assessment as attention turns from the large to the micro-scale. To achieve improvements in the measurement of heat stress in a range of environments, such as unventilated houses and factories and outdoor work settings, advancements in wearable technology will play a significant role especially in relation to real-time monitoring of heat stress variables such as radiation load and physiological indicators of heat strain, such as heart rate, body temperature, sweat rate and hydration levels.

This not only applies to humans but domesticated animals as well. Such field data will serve well the advancement of physiological and biophysical modelling approaches that integrate well-established principles from thermal physiology and human and animal biophysics as well as the development of more sophisticated models that capture the complexities of human and animal thermoregulation, including factors like body size, activity levels, clothing/body cover, and physiological and health restrictions that affect heat tolerance.

Research that assesses the efficacy of thermal climate indicators or heat stress indices will no doubt continue at pace. That said it is probably safe to say we have reached a point of realisation as is the case for heatwave definitions, that no single thermal climate index is likely to offer a silver bullet solution to the universal assessment of heat stress. Rather than chasing this elusive holy grail, a more productive use of research resources might be the establishment of which of a range of thermal climate indices best fits the research or application problem with the prospect that even within human and animal populations for any given location the most efficacious index for modelling heat stress may vary depending on a range of heat vulnerability factors such as age and health status. Research directed in this way would yield benefits for the development of bespoke population group or sector specific heatwave related warning systems and a better understanding of the complex interactions between the thermal environment, individual characteristics, and impacts on health, productivity, and overall liveability. In the case of field-based heat stress assessment, attention mainly defaults to humans and secondly domesticated animal populations with natural systems poorly served. In many ways, the material presented in this book is guilty of this trap and highlights the gap in our understanding of heat exposure and heat stress in natural systems. To address this shortcoming, more needs to be done in relation to heat monitoring and modelling in natural systems. While laboratory-based studies will continue to be of value in terms of understanding how specific species may react to imposed heat stress, focused field-based monitoring of heat sensitive habitats is an area which deserves attention. This is especially so if an aspiration is to enhance the understanding of the climatology of heatwaves in, and the associated impact of heatwaves on terrestrial, freshwater, and marine ecosystems and constituent species.

15.6.2 Heatwave Causes

The conceptual framework underlying the material presented in this book recognises that heatwaves occur in response to a set of cascading processes within the climate system that operate at a range of spatial and temporal scales. Notwithstanding the role of external forcings as exemplified by changes in atmospheric gaseous composition because of human activities, the climate context for heatwave development in many ways is set by the evolving interaction of the components the climate system that drive internal climate variability. This interaction often manifests as clearly recognisable alterations of the patterns of atmospheric, land surface and ocean climate fields usually codified as modes of climate variability. As described in this volume the strength of the link between certain modes of climate variability and heatwave occurrence is variable, an indication that for some locations local processes may dominate over remotely driven processes or indeed, unconsidered modes of variability might also be important. Exploring the links between modes of climate variability and the physical drivers of heatwave occurrence will remain a furtive research area which bears implications for heatwave prediction at extended range timescales. Because modes of climate variability do not operate independently, it is likely compounding, with several modes of variability working in concert, will influence the likelihood of heatwave occurrence. Therefore, unravelling the interplay between modes of climate variability at a range of space and time scales and the relative importance of these compared to local processes in the development of heatwaves represents a worthy direction of travel for future research. Understanding the role of climate variability in the co-occurrence of atmospheric/land-based and marine heatwaves is also a research priority. Because most of the long-term associations between modes of climate variability and heatwave occurrence have been discovered through statistical analyses it is imperative that work continues to proceed at pace on uncovering the nature of the physical processes—the communication mechanism—that link where the largest changes in atmospheric, land surface and ocean state occur and the regions where the associated impact in terms of heatwave likelihood manifests. Understanding the physical mechanisms that link modes of climate variability as slow drivers of heatwave likelihood with 'faster' processes at the synoptic scale is also a research priority.

The potential list of heatwave causes at the synoptic scale is long with blocking often emphasised as critical. As highlighted in this volume advection and the foehn effect also play a role in heatwave development at the synoptic scale. Because blocking, advection and the fohn effect are often addressed in completely different studies advancements are needed in understanding their unified and relative effects in the development of heatwaves. A corollary of this of course is an improved understanding of the processes that give rise to these heatwave drivers, particularly in the case of blocking genesis and decay is needed. Understanding the relative roles of adiabatic warming and diabatic heating in heatwave development also deserves further attention. The extent to which upstream/remote diabatic heating of the upper atmosphere via moist processes associated with strong convection and latent heat of condensation release is a particularly furtive area of future enquiry given that such 'moist' atmospheric processes may induce significant perturbations in the upper westerlies which bears implications for the development of downstream blocking episodes and heatwave development in the middle latitudes.

How slow and fast atmospheric processes link with processes at 'the surface' in the frame of surface—atmosphere interaction represents a clear area for continued research given these two spheres are intimately connected through the exchange of heat, mass and momentum and associated feedbacks. Understanding such interactions is relevant to heatwave development over land and within the marine environment. For heatwaves over land knowledge gaps still exist concerning the role of surface and subsurface moisture as a primer for heatwave development. Further investigation is needed to determine whether a scale-dependent tipping point exists, marking the rapid transition from a state dominated by latent heat flux to one dominated by sensible heat flux. This includes examining the initiation of positive feedback mechanisms that promote rapid soil moisture depletion, increased sensible heating of the surface boundary layer, and the development of extreme heatwaves in terms of both intensity and duration. As vegetation plays an important role as a conduit for moisture exchange between the surface/subsurface and the overlying atmosphere the effects of vegetation phenology and dynamics and surface radiative properties requires investigation in the context of broader land surface atmosphere interactions and heatwave development. Complementing understanding the role of surface heat in elevating boundary layer temperatures through a soil moisture—vegetation control mechanism, more remains to be learned about the entrainment of heat from the free atmosphere through the top of the boundary layer with this heat originating via subsidence/compressional warming. For the marine environment, understanding the relative roles of air-sea heat flux and oceanic heat advection process in marine heatwave development warrants further investigation.

Singled out as a specific land surface category, the urban surface has been privileged over other specific land surface types (e.g., forest, grassland or desert) because of its uniqueness and its possible role in exacerbating background levels of urban heat during heatwaves. Since this book emphasizes the significance of radiative and energy balance concepts in understanding surface-atmosphere interactions during heatwave development, the following research directions flow from this thinking. To enhance our understanding of urban climate processes in general and specifically for heatwave events better urban climate monitoring is required, not just in terms of the standard climate variables but the monitoring of radiative and turbulent and conductive heat fluxes, along with attempts to assess at the daily or sub-daily timescale the magnitude of anthropogenic heat emissions as an additional heat source to the urban atmosphere. While easier said than achieved this is critical for making progress with regards to understanding urban surface heatwave interactions. Robust monitoring data from urban and adjacent rural environments will assist with understanding how surface and near surface climate processes evolve throughout heatwaves especially in relation to the apportionment of available net energy between sensible, latent and conductive heat flux (heat storage in the urban fabric) and the associated development of daytime and nocturnal atmospheric heat. Added to this the issue related to the contribution of anthropogenic heat flux to raising levels of heat needs resolution. As outlined earlier this is because despite assertions that heat from air conditioning systems and standalone units raises air temperatures there is very little evidence that actual feedback processes contribute to elevated temperatures. Addressing these questions will help resolve the debate on how heatwaves amplify urban heat levels and the urban heat island effect through changes in surface energy exchanges and increased anthropogenic heat flux and help determine whether the resulting heat exposure for urban dwellers and ecosystems surpasses that experienced in rural areas.

15.6.3 Heatwave Consequences

Future lines of enquiry related to heatwave consequences include the development of protocols for heatwave impact assessment, a better understanding of the role of heatwave related exposure and vulnerability and a vast improvement in the understanding of the nature of single and cascading heatwave impacts in human and natural systems.

The lack of systematically collected and quality assured heatwave impact data for human systems is a standout issue begging attention. Without quality assured heatwave impact data, heatwave risk management solutions will be more guesswork than evidence based. While at the global level some data on heatwave impacts exists in international disaster data bases it is far from comprehensive with some known impactful heatwaves not recorded. Under optimal conditions of collaboration, with the UN initiative of EWS4All providing context, one might hope that the World Meteorological Organisation and the UN Office for Disaster Risk Reduction and its EWS4All partners (IFRC and ITO) might work with the Centre for Research on the Epidemiology of Disasters (CRED), as the developers of the Emergency Events Database (EM-DAT), and the World Health Organisation to improve the chronicling of the human toll of heatwaves. A further way forward would be for some countries or major municipalities to take the lead on chronicling heatwave impacts for a range of sectors, not just health. To achieve this a set of easily attainable heatwave impact metrics would need to be agreed upon which with time could be widely applied to the description and analysis of heatwave impacts. The requirement for heatwave impact data for natural systems is equally important as presently this is piecemeal with most descriptions of heatwave impacts contained in unrelated documents. Amongst a range of natural systems, the oceans are the probably the one for which most evidence exists of the impacts of heatwaves. Because long-term monitoring of ocean health related to past marine disasters has revealed serendipitous benefits for understanding marine heatwave impacts, a worthy corollary could be the establishment of similar monitoring sites in key marine heatwave hotspots. In a perfect scenario, international agencies such as the International Marine Organisation with its 176 member states and three associate members could provide leadership on this matter. High resolution satellite-based observations of vegetation health and land-surface climate may provide the requisite data for establishing the nature of heatwave impacts in terrestrial ecosystems. Information from satellite-based tracking systems for terrestrial wildlife and marine pelagic species may also offer potential for the investigation of possible atmospheric (land-based) and marine heatwave forced migration related to heat avoidance or changing patterns of food availability. For aquatic environments the existence in some cases of long-term daily records of lake and stream temperatures in combination with ecological monitoring offers much potential for building a picture of not only the nature of lake and stream heatwaves but their impacts.

Heatwave exposure and heatwave related vulnerability are important factors for moderating heatwave impacts. The empirical determination of heatwave exposure is a wide-open field rife for development that will be led by some of the developments described above in relation to heat stress assessment. A better understanding of exposure to levels of chronic and acute indoor heat is required. While there is some evidence for research on this topic more needs to be learned about informal housing, a settlement type found extensively across regions where heatwaves have become or will become endemic. The same applies to the shelter provided in refugee camps which have almost become permanent places of settlement for some. Improved information for these settings and indoor environments in general will assist with identifying suitable heatwave adaptation and mitigation options. For human related heatwave vulnerability, needed lines of enquiry include furthering the understanding of the key sensitivity factors known to be important contributors to overall heatwave vulnerability, such as socio-demographic and economic characteristics, 'place-based' factors including lived and work environments, being pregnant, comorbidities, heat risk perception and access to heatwave related information and interventions. Further, work is needed to establish the utility of heat vulnerability indices and mapping and other assessment tools for capturing the dynamic, multifaceted nature of heatwave vulnerability. With the requisite heatwave vulnerability information in hand, a worthy line of enquiry is to what extent might this enhance the prediction of heatwave related health outcomes. Beyond the health sector, lines of enquiry about the determinants of heatwave related vulnerability need to be initiated for other sectors. Prime candidates are the energy and transport sectors on which many sectors of society and economy are dependent. To date most of the work on heatwave vulnerability has been conducted for high-income countries. Therefore, a priority for the lines of enquiry espoused here should be low and middle-income countries, especially those located in evolving heatwave hotspots. Comparatively, little is known about heatwave vulnerability in natural systems. This situation needs to be addressed via a mixture of complementary laboratory- and field-based experimental studies that explore the underlying genetic and physiological mechanisms that contribute to increased biological vulnerability to extreme heat.

15.6.4 Heatwave Responses

Research on heatwave responses as intentional short to long-term and sustained actions and measures designed to reduce heatwave related risk and thus impact is still in its early days. Notwithstanding the need to further understand how natural systems may implement a range of strategies to reduce the risks associated with heatwaves, the lines for future enquiry put forward here refer to human populations and the built environment.

Adaptation to heatwaves is a burgeoning area of research, with its criticality tied not only to the devasting impacts witnessed for recent heatwaves but the expected increase in heatwave frequency, intensity, and duration under anthropogenic climate change. Despite advancements in understanding heatwave risks and developing heatwave risk reduction strategies, several outstanding research questions concerning heatwave adaptation remain. While some efforts have been made to establish inventories of heatwave adaptation actions, work will need to continue in this area, especially with regards to those actions enacted by indigenous communities and first nation people from whom much stands to be learned about behaviour change or the alteration of the physical nature of heat associated with heatwaves. However, to support the development of heatwave adaptation option inventories and intercomparisons for different settings, a succinct definition of heatwave adaptation is required, aside from the general definition provided in this book. While an improved definition will include a taxonomic aspect, which is the prevailing approach today, it must go beyond that. A fully functional definition of heatwave adaptation will need to facilitate measurement of the degree of heatwave adaptation (the nature and extent of behaviour change) and its effectiveness (whether the behaviour change has brought about a significant reduction in heatwave risk and resultant impacts). To fulfil such an aspiration, well-honed common protocols for describing and measuring heatwave adaptation will be required along with associated qualitative and quantitative descriptors of adaptation extent and effectiveness. Achieving this will assist with managing the frustration often faced by policy makers when trying to decipher the outcomes of adaptation studies that employ various analytical methods and examine diverse units and scales. Harmonizing these differences through adopting a widely agreed definition of heatwave adaptation and how to measure it will enhance the assessment of effectiveness and inform future heatwave risk reduction actions.

Allied with investigating the effectiveness of specific adaptation options, is the need for broader level assessments of the efficacy of existing policy frameworks and governance structures for managing heatwaves. For example, analysing the coordination between different government departments and the role of local authorities in developing and implementing public heat action plans, in which HEWS and HHWS sit, is a wide-open area for further investigation. Related avenues for future exploration include what are the enablers and barriers (e.g., finance and knowledge in addition to governance) to implementing heatwave adaptation policies at local, national, and international levels and how can policies be designed to ensure equitable access to heatwave adaptation resources to protect the most vulnerable populations? While heatwave governance continues to develop as a furtive area of research activity, the tendency to date has been for a public health focus with other sectors generally overlooked. Hopefully this situation can be rectified with the conduct of comprehensive heatwave governance studies for other sectors of the economy (e.g., agriculture, manufacturing) and critical infrastructure (e.g., energy and transport) emerging in the coming years. Because there is generally a financial cost associated with putting an adaptation or mitigation option in place a worthy question for investigation is how can financing for heatwave adaptation and mitigation be mobilised and allocated effectively to support the most needed interventions and most warranted communities? In relation to this, a rapidly developing area of interest is heatwave micro-insurance in the form of a range of parametric heatwave insurance products for application in low- and middle-income country settings. Research on this topic awaits in relation to issues such as the adequacy of the historical data and modelling applied in the derivation of risk levels and associated insurance premiums, its overall effectiveness especially in relation to the avoidance of basis risk, and the issue of insurance affordability in the absence of subsidies. A related question is to what extent might heatwave insurance divert attention from the necessity for structural changes in systems that cultivate heatwave exposure and vulnerability? The issues of heatwave adaptation limits (behavioural as opposed to physiological) and maladaptation, also represent extremely important lines of enquiry, especially in relation to the trade-offs that might be necessary to protect human health during extreme heatwave events and under chronic levels of dangerous heat when options other than air conditioning might prove ineffective.

Broadly in line with the 'disaster-reform hypothesis', this work has suggested that some heatwave events, because of their rarity and level of impact, have acted as "focusing events"drawing significant attention from the public and policy makers with policy change following. In this context what might be described loosely as mega-heatwaves could be functional in overcoming policy inertia. This postulate needs testing, along with addressing several related questions. Understanding the variability in policy development responses to heatwaves represents a worthy line of enquiry with a consideration of factors such as heatwave characteristics, the political and social context, and the presence of policy entrepreneurs who can capitalize on an event to push for change, needing to be explored. In relation to heatwave characteristics such as frequency, intensity and duration, research is needed to better understand how these influence the likelihood of policy change. This includes developing a more nuanced typology of heatwaves as focusing events and examining how heatwave characteristics interact with other factors to influence the initiation of policy development.

Because the community and political context in which a heatwave occurs could affect the policy development response, factors such as local governance capacity, pre-existing cognate policies, and the level of public awareness and concern about heatwaves as a threat could be examined to understand their influence on policy change. Undoubtedly the mechanisms through which heatwaves lead to policy change are complex, potentially involving the role of policy entrepreneurs, influential heatwave risk reduction information networks, media coverage, public opinion, interest group mobilisation, specific sectors of the economy or major players in these sectors, and political leadership. Accordingly, research is needed to establish the extent to which these and other possible mechanisms might interact to bring about policy change.

Once heatwave related policy has become established there will be a need for longitudinal studies to track policy changes over time and assess their sustainability and effectiveness. Understanding the distinction between short-term, reactive policy changes and long-term, proactive policy changes is crucial for developing effective heatwave adaptation strategies. That there is already some evidence for this is apparent with the shift from hazard-based to impact based approaches to heatwave related warnings in some countries and the inclusion of specific warnings for highly vulnerable populations in some HHWS. Within the general frame of what makes research exciting, are investigating occasions that have a low likelihood of happening. With regards to this, exploring the conditions under which heatwaves might lead to counter-mobilisation by groups opposed to policy change, resulting in policy stasis or even regression, could form an intriguing line of enquiry. As well as theories related to what might drive policy change at the municipal level and beyond, several ideas concerning what engenders response at the small group and individual levels were introduced in the section on adaptation. These may serve as a starting point for exploring what drives heatwave adaptation and mitigation action related decision making amongst individuals and small groups. Similarly, the Cultural Theory of Risk framework provides an extremely useful and alternative lens through which to explore the cultural dimensions that might influence the decision-making processes associated with heatwave adaptation actions, including the development of policy at the governmental level, or whether individuals or small groups chose to enact options that might be available to them.

HEWS and HHWS are often put forward as an efficacious adaptation solution for heatwave risk reduction, a contention somewhat strengthened by the international call for EWS4All and the implied panacea that EWS might offer for protecting society and economy from the ravages of natural hazards. However, much remains unknown about the ideal structure for HEWS/HHWS and their efficacy. A worthy question in relation to heatwave related EWS is how can the different components of EWS (risk knowledge, observation, detection and monitoring, warning dissemination and communication, and preparedness and response capability) be effectively integrated into a comprehensive end-to-end warning system for heatwaves and heat-health? An allied question is how can diverse and non-traditional data sources, such as social media, community-based observations and indigenous knowledge improve early warning systems? Presently most HEWS and HHWS primarily concentrate on predicting a singular outcome and do not account for cascading effects. This raises the question of how to integrate a deeper comprehension of cascading effects into HEWS and HHWS to anticipate heatwave related secondary and indirect impacts. This will necessarily involve the development of sophisticated algorithms for predicting the downstream effects of heatwaves. To address this and the improvement of HEWS and HHWS in general, the opportunities and challenges of leveraging new technologies (e.g., cloud computing, Internet of Things) requires exploring.

Because there is mixed evidence regarding their efficacy, why or why not specific HEWS and/or HHWS deliver on their aspirations constitutes a fruitful research area. Related to this, systematic process and outcome evaluations will be required to support diagnoses of HEWS/HHWS successes or failures. Presently a range of methodologies exist for process and outcome evaluations yet few intercomparisons between these have been undertaken, something that requires addressing. A much under-researched area is HEWS/HHWS and HAP governance with questions such as what the best governance structures are to achieve warning system and action plan effectiveness for a range of scales and contrasting political structures yet to be answered. To support such an endeavour, information regarding heatwave governance for a variety of jurisdictions is required. Allied with matters such as resilience and adaptation and issues such as equity and social justice are questions such as how

HEWS/HHWS can best facilitate the enhancement of community and sector adaptive capacity and address the needs of marginalized and vulnerable populations. With regards to marginalised and vulnerable populations, heatwave risk communication is an underexplored area especially in relation to warning system efficacy. Establishing how warning communications can best be tailored to such communities and contexts for better understanding and response, including an evaluation of constraints and barriers to communication and message take-up, is an important area for future research activity. Since heatwave naming remains a hotly debated issue, research that explores whether heatwave naming leads potentially affected populations to take actions that reduce their heatwave risk is urgently needed.

To better prepare for heatwaves through acting on early warnings, exploring the potential benefits of heatwave prediction at the intra-seasonal to seasonal timescales constitutes a worthy line of enquiry. Because land-atmosphere interactions, soil moisture feedbacks, and other critical boundary layer processes are potentially important for the initiation, duration, and intensification of heatwaves, research on how improved representation of these processes and feedbacks in models might enhance the prediction of periods of unusual heat on intra-seasonal to seasonal time scales is required. Research directed at achieving advances in model initialization, data assimilation and incorporating remote sources of predictability (e.g., sea surface temperature patterns) could also enhance predictability at intra-seasonal to seasonal scales. Presently, extended to long-range predictions of heatwave related variables are based on a single perturbed ensemble approach in which multiple forecasts are generated by slightly adjusting model initial conditions or representation of model physical processes. A furtive area for future research is the extent to which multi-model ensembles, that is combining predictions from multiple models from different agencies, might improve predictability. Associated lines of enquiry include Identifying optimal weighting of multi-model outputs, calibration, and interpreting multi-model ensemble predictions. Providing reliable uncertainty estimates alongside predictions of heatwave related variables at the intra-seasonal to seasonal timescales will also be crucial for prediction take-up and heatwave risk management decision-making. However, quantifying and communicating uncertainties from various sources (e.g., initial conditions, model structure) remains an ongoing issue for these longer timescales that will require addressing. As the idea of incorporating long-lead heatwave predictions into HEWS and HHWS gains traction, there will be a need to investigate how impact-based predictions that consider sector-specific vulnerabilities, thresholds, and compound risks might benefit heatwave risk management decision making at the sub-seasonal to seasonal timescales.

Although much has been written on heatwave mitigation solutions, more needs to be done to advance our understanding of how to address the urban heat mitigation imperative not only from a physical and engineering science perspective with the application of monitoring and modelling, but also through the lens of the social, political, and urban planning sciences and related systematic research programmes using qualitative and quantitative methods. To make discernible advances to better our understanding of the effectiveness and scalability of different heat mitigation solutions is paramount. This will involve determining how various strategies perform in mitigating urban heat and identifying methods to maximize their efficacy across diverse urban contexts and climates. Although some progress on this issue has been made, the potential elephant in the room is does the public and/or private sector have the appetite for tackling the multiple challenges (e.g., financial, and public acceptability) associated with the widespread/city-wide application of some of the heat mitigation options outlined in this book? Additionally, examining the interactions and synergies between solutions and existing urban infrastructure is crucial. While optimizing these interactions could enhance the overall effectiveness of urban heat mitigation efforts, the associated evidence base needs to be improved. Presently little is understood about how heatwave mitigation solutions might impact vulnerable communities and whether equity, social justice, and community engagement is, or will be, prioritised in their implementation. This issue needs addressing. Investigating the long-term impacts of heatwave mitigation solutions is also essential, especially in the context of their sustainability to meet the challenges associated with enhancing urban resilience to climate change related increases in mean and extreme levels of heat. In short, will their efficacy dwindle over time and if so, what might be the drivers of any fall-off in performance? Understanding the broader services provided by heat mitigation options, such as improved air quality, better water management and biodiversity conservation, and how to optimise these co-benefits is also worthy of exploration. Moreover, identifying optimal design principles and planning strategies for integrating heatwave mitigation solutions into existing urban landscapes is a worthy line of research activity that will necessarily require interdisciplinary collaboration. How this type of collaboration might be fostered also requires examination. Addressing whether current policy frameworks and governance structures and financial mechanisms are adequate to support the widespread adoption of engineered (e.g., cool pavement and roofs) and non-engineered heatwave mitigation solutions (e.g., NBS) is also critical. An intriguing question for future research relates to the likely costs to municipal authorities associated with implementing heatwave mitigation options on a scale large enough to significantly reduce heatwave related temperature extremes. An associated line of enquiry is, might the associated cost be so large that responsibility defaults to the individual with the result that little action is taken? Such questions align with broader inquiries regarding the willingness to pay for actions aimed at reducing risks associated with natural hazards.

15.7 Final Word: Heatwaves Matter

Because of their profound and multifaceted impacts on human and animal health, economies and ecosystems, heatwaves matter. The heatwave imperative has become even more pressing as unabated climate change has already begun to bite and affect heatwave exposed and vulnerable communities, populations, and economic sectors, especially in regions that are fast becoming heatwave hotspots. The same applies to a range of terrestrial and marine and freshwater ecosystems and associated habitats.

Yet the actions related to heatwave risk reduction remains in many cases inadequate and at best fragmented. In the words of the Canadian scientist David Suzuki "We're in a giant car heading towards a brick wall and everyone's arguing over where they're going to sit." Extending this metaphor to heatwaves, the brick wall symbolises the escalating threat of severe heatwaves before us. The seating arrangement issue represents those who variously continue to live in denial about heatwaves as a threat, argue about their importance compared to a range of other natural hazards, or conduct reductionist research that fails to see the larger wicked problem that is heatwaves.

As suggested in the opening Chapter of this book addressing the heatwave imperative requires a new approach. This approach must be non-siloed and interdisciplinary, uniting a wide range of disciplines and stakeholders to achieve a much deeper understanding of the complex associations between heatwave causes, consequences, and responses. Only then will the balance of probabilities shift towards 'likely' or better in relation to efforts to effectively reduce the risks associated with heatwaves and avoid a world in which many people are potentially beleaguered by the 'heatwave quotidian'. The hope of this book has been to provide material relevant to the endeavour of enhancing our knowledge about heatwaves as weather and climate phenomenon that matter.

Index

G. McGregor, *Heatwaves*, Biometeorology 6,
https://doi.org/10.1007/978-3-031-69906-1

T